D0143058

DISCOVERING ASTRONOMY

During the summer of 1994, astronomers had the rare opportunity to view the collision of a comet with a planet in our solar system. All the Earth's great telescopes were focused on Jupiter to observe the actual impact of Comet Shoemaker-Levy 9. The Hubble Space Telescope captured the results; this photo shows the impacts of fragments "D" and "G" from Comet Shoemaker-Levy 9 on July 18, 1994. The "G" impact, which shows concentric rings, has a central spot 1550 miles (2,500 km) across, a thin dark ring 4,660 miles (7,500 km) across, and a dark outer ring about the size of Earth. Analyzing the data collected from this "once in a lifetime" observation will allow astronomers to learn more about the chemical composition and makeup of comets as well as the characteristics of Jupiter.

DISCOVERING ASTRONOMY

THIRD EDITION

R. ROBERT ROBBINS
UNIVERSITY OF TEXAS, AUSTIN

WILLIAM H. JEFFERYS
UNIVERSITY OF TEXAS, AUSTIN

STEPHEN J. SHAWL
UNIVERSITY OF KANSAS

JOHN WILEY & SONS, INC.
NEW YORK   CHICHESTER   BRISBANE   TORONTO   SINGAPORE

Acquisitions Editor: Cliff Mills
Director of Development: Johnna Barto
Marketing Manager: Catherine Faduska
Senior Production Editor: Bonnie Cabot
Interior Design and Electronic Layout: Laura Ierardi
Manufacturing Manager: Susan Stetzer
Photo Researcher: Hilary Newman
Illustrator: Boris Starosta
Illustration Editor: Sigmund Malinowski
Cover Photo: Gary Kaemmer/The Image Bank

This book was set in Meridien Roman by CRWaldman Graphic Communications and printed and bound by Von Hoffmann Press. The cover was printed by Phoenix Color, Corp.

Recognizing the importance of preserving what has been written, it is a policy of John Wiley & Sons, Inc. to have books of enduring value published in the United States printed on acid-free paper, and we exert our best efforts to that end.

The paper in this book was manufactured by a mill whose forest management programs include sustained yield harvesting of its timberlands. Sustained yield harvesting principles ensure that the number of trees cut each year does not exceed the amount of new growth.
John Wiley & Sons, Inc., uses recycled and recycled content papers in the manufacture of our books.

Library of Congress Cataloging in Publication Data:
Robbins, R. Robert (Ralph Robert), 1938-
     Discovering astronomy / R. Robert Robbins, William H. Jefferys,
  Stephen J. Shawl. -- 3rd ed.
        p. cm.
     Includes bibliographical references (p.).
     ISBN 0-471-58437-1
     1. Astronomy   I. Jefferys, William H.   II. Shawl, Stephen J.
III. Title
QB43.R57 1995   94-5244
520--dc20   CIP

Printed in the United States of America

10 9 8 7 6 5 4 3 2 1

ABOUT THE AUTHORS

R. ROBERT ROBBINS studied mathematics at Yale University and received his doctorate in astronomy from the University of California at Berkeley in 1966. He spent a year as a McDonald Observatory Fellow and a year at the University of Houston before joining the faculty at the University of Texas. His research interests have been in the fields of gaseous nebulae, atomic physics, and the transfer of radiation through gaseous media. In recent years his research interests have been turning towards archaeoastronomy and ethnoastronomy—in particular, studying the astronomy we can find in the structures and folklore of the Indians of Middle America and the American Southwest. He has received a variety of awards for excellence in teaching at the University of Texas and has served on a variety of national and international committees concerned with astronomy education. He has taught a variety of Chautauqua courses for college teachers, the most recent ones dealing with Mayan astronomy.

WILLIAM H. JEFFERYS was educated at Wesleyan University and Yale University, receiving the Ph.D. degree from Yale in 1965. He has taught at Wesleyan University and at the University of Texas and is currently the Harlan J. Smith Centennial Professor in Astronomy at the University of Texas at Austin. His main research interests are in the fields of dynamical astronomy (the study of motions of planets, stars, and galaxies), astrometry (the measurement of the positions, orbits, and distances of celestial objects) and statistics. Since 1977 he has been the Astrometry Science Team Leader for the Hubble Space Telescope project. He has taught widely, both at the undergraduate and graduate levels, as well as teaching several National Science Foundation Chautauqua courses on teaching techniques to in-service teachers at both the high school and college levels . He believes strongly in the "hands on" approach to teaching that gives students a direct experience of what it is like to make, analyze, and understand scientific observations.

STEPHEN J. SHAWL was educated at the University of California at Berkeley in the early 1960s and the University of Texas at Austin, where he received his Ph.D. in 1972. He has been teaching all levels of undergraduate astronomy since that time. His research interests have involved the study of the fundamental properties of globular star clusters, the presence of gas and dust in these clusters, and cool variable stars whose light is sometimes polarized. He has attended numerous National Science Foundation Chautauqua short courses for college teachers, the first one having been taught by Robbins and Jefferys. Because he was one of the original reviewers of the manuscript for *Discovering Astronomy*, he is especially pleased now to be a co-author of this text. He has been a member of the Education Advisory Board of the American Astronomical Society (AAS) and is currently editor of the newsletter of the Working Group on Astronomy Education for the AAS.

PREFACE

The beauty of the night sky can be overwhelming. That beauty becomes even more inspiring when you have some understanding of what you are observing. When you see how humans approach an understanding of the universe, you begin to appreciate humanity and nature even more. The overall goal of this text, then, is to provide the reader with the opportunity to understand and appreciate the incredible place in which we exist.

The authors of this book feel that astronomy is not only an exciting and interesting subject, but also a living one in which everyone can directly participate. A considerable amount of astronomy can be experienced personally, and many things can be learned by simple and direct observations. We believe that the introductory survey course in astronomy, for which this book has been written, can be greatly enriched by the addition of observational activities that lead to a direct experience of the reader's universe. Such an approach, pioneered in astronomy by the first edition of *Discovering Astronomy*, reflects current discussions of science education at the national level. The idea of personal involvement is the guiding philosophy behind this book.

THE PROCESS OF SCIENCE

Science can be defined as a process followed by humans involved in a search for an understanding of their surroundings. This understanding is usually expressed in terms of theories whose predictions are verified or falsified by comparison with observation and experiment. The authors believe that an understanding of the scientific enterprise is important for people in a democratic society. For this reason, we do not present only the final results of scientific investigation, but we emphasize the process, which often contains false starts, side roads, and errors along the way to our current understanding, which will also undergo modification and refinement by future generations of scientists. In placing an emphasis on process, students are better positioned to understand discussions on which they will be asked to make informed decisions.

To help the student better understand the process of science, the book takes a discovery approach, which we discuss next.

THE DISCOVERY APPROACH

We have written this book as a comprehensive introductory astronomy text to facilitate unassisted inquiry through active observational experiences for students. Our approach is a discovery- or inquiry-based approach. To translate this philosophy into a unique textbook, we have woven observational and non-observational activities into the text at appropriate locations. There are two kinds of activities — (1) those included in the text and (2) those presented in a separate *Activity Kit* — allowing for a variety of teaching modes. Both are highlighted by colored boxes at pertinent places in the text narrative. Readers not doing the activities (or not possessing the kit) may simply continue with no loss of information or flow. Thus the book can be used effectively with or without the activities and discoveries.

DISCOVERIES

The Discovery Activities are designed to emphasize that the nature of science is one of discovery. Short and simple, these activities include instructions and inquiry questions. An example is the activity on Weightlessness in Chapter 5. They are noted at the appropriate place in the narrative and appear at the end of chapters.

KIT ACTIVITIES

Some observational activities require the student to make simple astronomical measurements using instruments such as a quadrant, cross-staff, spectrometer, or telescope. These die-cut instruments are included in a separate *Activity Kit* along with a manual of instructions for the activities. Also included are lenses, a holographic diffraction grating, and two pieces of polarizing filter. The *Activity Manual* also includes some stereoscopic images to enhance enjoyment and understanding of planetary bodies; two-color stereo viewers are also part of the Kit. An observational kit activity may require a student to measure, for example, planetary positions, average the measurements, and discuss the errors of the measurements. Other activities, such as Classification of Galaxy Types from Sky Survey Photographs, do not require measurement but the use of photographs (or in other cases, data) that are also included in the Kit.

The *Activity Kit*, with instruments and manual, is available from the publisher along with this text and can be ordered by your college bookstore. The kit can also be ordered separately from the text.

The activities and discoveries allow participation in astronomical exploration. They can be carried out by the thoughtful reader with no extra equipment or tutorial assistance.

PEDAGOGICAL STRUCTURE

We have implemented our inquiry-based approach in the pedagogical structure of this book. We have taken every opportunity to emphasize the scientific process and encourage observational activities to add to student understanding and enjoyment.

INQUIRIES

Each chapter contains in-chapter questions, which we call *Inquiries*, to emphasize the inquiry nature of the presentation. Placing these Inquiries within the text asks students to stop and think to themselves, "I just read something important; what was it?" Sometimes, the Inquiry asks the student to use the knowledge already gained and to think about some idea immediately prior to its discussion in the text. In this way, the student is actively participating, not just reading passively. The answers to all Inquiries are provided at the end of each chapter.

CHAPTER SUMMARIES

A new style of Chapter Summary has been developed to emphasize the nature of the scientific method. When appropriate, Chapter Summaries have three subsections: *Observations*, *Theory*, and *Conclusions*. Although many aspects of science belong in more than one category, this structure nevertheless serves to focus student attention onto these fundamental parts of the scientific method.

SUMMARY QUESTIONS

Next are Summary Questions, which ask students straightforward questions about the concepts covered in the chapter. These questions can serve as a review.

APPLYING YOUR KNOWLEDGE

Knowledge should be applied to new and different situations. This section, at the end of each chapter, presents the reader with questions meant to require the application of concepts previously learned. Some, marked by a bullet (■) require the use of simple mathematics.

GLOSSARY AND SCIENTIFIC APPENDICES

The book contains an extensive Glossary of those terms highlighted with bold print in the text. A set of appendices (A-G) presents a review of scientific notation as well as other mathematical concepts deemed unnecessary in the main text but important for complete understanding, as well as further data on planets, satellites, stars, and galaxies.

COLORFUL AND INSTRUCTIVE ILLUSTRATIONS

The art program includes full color photographs throughout to enable students to enjoy and appreciate the beauty of the universe better. Computer generated color diagrams help to clarify complex concepts. In so far as possible within the limits of space, diagrams convey single concepts.

ORGANIZATION

We have organized the book in the way astronomy developed: from observations to an understanding first of the solar system and only later to stars and galaxies. In this standard "inside out" approach, we have divided the book into five parts:

I. Discovering the Science of Astronomy
II. Discovering the Nature and Evolution of the Solar System
III. Discovering the Techniques of Astronomy
IV. Discovering the Nature and Evolution of Stars
V. Discovering the Nature and Evolution of Galaxies and the Universe

PHYSICS PRESENTED ONLY AS NEEDED

Within our organizational structure, the observational basis of astronomy is further emphasized by generally having observation precede theory. Furthermore, we present physical ideas where and when they are needed rather than grouping them together in a single chapter. This approach allows students to see a reason why physics is being presented, and they immediately see applications for the ideas. We have purposely placed the solar system before the chapters on radiation and telescopes for two reasons: (1) to get the student into modern astronomy as soon as possible without their feeling bogged down by physics ("if I had wanted a physics course I would have taken one!"), and (2) because much of a modern survey of our knowledge of the solar system at the introductory level may be presented effectively without prior knowledge of the material on light, spectra, and telescopes. Instructors desiring to discuss electromagnetic radiation first may do so without difficulty.

There are a few sections, noted with an asterisk, that some reviewers felt went beyond what was necessary for basic knowledge but the authors felt added to overall understanding. Such sections can be skipped if the instructor desires.

SCIENCE VERSUS PSEUDOSCIENCE

We believe that the ideas of what distinguishes science from pseudoscience are so important, our Chapter 2 on Science and Pseudoscience comes immediately after the book's introduction to astronomy. In this way, time constraints at the semester's end will not cause it to be skipped.

COMPARATIVE PLANETOLOGY

Part II, on the nature and evolution of the solar system, begins with an overview of observations important to understanding the solar sysem. We then examine both early and modern hypotheses concerning the origin of the solar system. The next logical topic is a study of objects that are mostly unchanged since their formation and that provide astronomers with important information about the conditions at the formation time—comets, asteroids and meteoroids. Instructors preferring to discuss these objects after the planets may do so without loss of continuity.

The discussion of the planets has been completely rewritten and modernized. Chapter 8 on Earth and Moon presents those concepts necessary to understand the other planets and satellites in the solar system. Chapter 9 discusses the terrestrial planets, and Chapter 10 presents the Jovian planets and Pluto. The presentation within the chapters on the Earth-like and Jupiterlike planets differs from those in other books by approaching planets from the view of comparative planetology. Information about the planets is not isolated from comparable information on similar planets but is integrated together. "The Planets One by One," located after each of the terrestrial and Jovian planet chapters, summarize the information on each planet individually.

LOGICAL COVERAGE OF STELLAR EVOLUTION

The Sun is discussed in its role as a typical star at the point where the stellar evolution discussion begins (Part IV). Those instructors preferring to relate the Sun more closely to the solar system than to the stars can easily do so.

Following the discussion of the concepts required for understanding stellar evolution are chapters on star formation and evolution to the main sequence, post mainsequence evolution, and the terminal evolutionary events.

INTERSTELLAR MEDIUM COVERED AS NEEDED

Rather than presenting the interstellar medium in isolation in a separate chapter, we consider it in various places as appropriate. For example, some information on nebulae is given as an example of spectra in Chapter 13. Additional discussion occurs in Chapter 17 on star formation. Finally, the effects of interstellar material on observations of stars and its influence on our knowledge of the structure of the Milky Way comes in Chapter 20.

SUPPLEMENTS

In order to enhance instructor's teaching resources, a comprehensive set of supplements accompanies *Discovering Astronomy*. For more information about these supplements, please contact your local Wiley representative. These supplements include:

- ☐ An *Instructor's Manual*, by Stephen J. Shawl. This manual has been prepared to provide instructors with teaching hints, syllabi, and chapter overviews for effective classroom use of *Discovering Astronomy*.
- ☐ *Student Study Guide*, by R. Robert Robbins, is available through him at Astronomy Department, University of Texas at Austin, Austin TX 78712. Internet address: rrr@astro.as.utexas.edu. This study guide was designed for instructors who are interested in assembling a self-paced, Keller-method course of instruction for their students. It can be used with small classes or with large ones such as that at the University of Texas that serves some 700 students per year.
- ☐ A *Test Bank*, prepared by Stephen J. Shawl, provides instructors with over eight hundred questions of various types. A *Computerized Test Bank* is also available for both Macintosh and PC platforms.
- ☐ 100 *Overhead Transparencies* replicate pedagogically useful line art from Wiley's astronomy archives.
- ☐ 100 *Slides* astronomical photographs enhance instructor's classroom capabilities.
- ☐ A *CD-ROM* containing line art, photographs, Concept Development Images (developed by Dr. Francis Lestingi at SUNY-Buffalo), and selected animations

from Wiley's acclaimed *Cosmic Clips* video permits professors to display a number of different images in the classroom. The CD-ROM is available for both Macintosh and PC platforms.

ACKNOWLEDGMENTS

The size of the supporting cast in making this book, or for that matter any book, possible is surprisingly large.

A large number of astronomers provided input, beginning with detailed reviews of the second edition, various intermediate drafts, and finally the one that went to press. They all helped to remove the many unnecessary errors that somehow seem to creep into a budding manuscript. Their help is warmly and gratefully acknowledged.

Bill Adams, Texas Christian University

Robert Allen, University of Wisconsin—La Crosse

Timothy Beers, Michigan State University

Katherine Bracher, Whitman College

Jeffrey Braun, University of Evansville

Gerald Crawford, Fort Lewis College

Stephen Danford, University of North Carolina—Greensboro

John Davis, Siena College

C. Gregory Deab, University of New Orleans

Alexander Dickison, Seminole Community College

Michael Jura, UCLA

Robert Kennicutt, University of Arizona

Michael Lysak, San Bernadino Valley College

Larry Lebofsky, University of Arizona

Kenneth Mendelson, Marquette University

Mark Miksic, Queens College

Eberhard Mobius, University of New Hampshire

Ronald Olowin, St. Mary's College

John Safko, University of South Carolina—Columbia

Malcolm Savedoff, University of Rochester

Peter Saulson, Syracuse University

Maurice Stewart, Willamette University

Donald Taylor, University of Nebraska

Harold Taylor, Stockton State College

David Trott, Arapahoe Community College

Kay Weiss, Kansas City Kansas Community College

Barbara Williams, University of Delaware

Louis Winkler, The Pennsylvania State University

Nick Woolf, University of Arizona

Others who provided valuable information or comments include: Hans Bethe, David Branch, Humberto Campins, Harold Corwin, J. P. Davidson, Robert Friauf, Martin Gaskell, Ron Gilliland, William K. Hartmann, Norriss Hetherington, Doug McKay, Bob Phillips, Ed Robinson, Kurt Schendzielos, Sumner Starrfield, Dave Tholen, Joe van Zandt, John Wood, Ed Zeller, and Mike Zeilik.

Those at Wiley who guided, cajoled, and helped us produce this new edition are listed on the back of the title page. While everyone's role was important, our specific interactions with some were especially noteworthy. At the top of the list must be Cliff Mills, who has been our steadfast editor providing continuing support. Elisa Adams, who acted as our developmental editor in asking numerous insightful questions while prodding us to clarify, to shorten, and to delete, was a joy to work with. We are very pleased to acknowledge Laura Ierardi, who designed the text. Boris Starosta, who was responsible for the computer-generated illustrations deserves special recognition; he was able to take some extremely rough sketches and produce illustrations that not only teach concepts, but do so beautifully. We also wish to thank the Wiley staff who helped make this revised edition a reality: Johnna Barto, development director; Maddy Lesure, design director; Pam Kennedy, production manager; Bonnie Cabot, production editor; Sigmund Malinowski, illustration supervisor; Hilary Newman, photo editor.

One of us, S. J. Shawl, wishes to thank the numerous students from the 1993–1994 semesters who class-tested photocopied manuscripts at various stages toward completion, and especially those who were kind enough to provide meaningful comments along the way. I am especially thankful to my local Wiley book rep, Gary Diffley, for suggesting me when they were searching for a new co-author. Finally, I want to thank Jeannette, who has put up with more than should have been necessary during this long project.

R. ROBERT ROBBINS
(rrr@astro.as.utexas.edu)

WILLIAM H. JEFFERYS
(Bill@clyde.as.utexas.edu)

STEPHEN J. SHAWL
(shawl@Kuphsx.phsx.ukans.edu)

BRIEF CONTENTS

CONTENTS

PART TWO

DISCOVERING THE NATURE AND EVOLUTION OF THE SOLAR SYSTEM

97

PART THREE

DISCOVERING THE TECHNIQUES OF ASTRONOMY

235

PART FOUR

DISCOVERING THE NATURE AND EVOLUTION OF STARS

295

CHAPTER 19
CATACLYSMIC ASTRONOMY 407

PART FIVE

DISCOVERING THE NATURE AND EVOLUTION OF GALAXIES AND THE UNIVERSE

433

CHAPTER 20
THE MILKY WAY: OUR GALAXY 435

HELPFUL STUDY AIDS

5

THE HISTORICAL QUEST TO MODEL THE SOLAR SYSTEM

The progress of science is generally regarded as a kind of clean, rational advance along a straight line; in fact it has followed a zig-zag course, at times almost more bewildering than the evolution of political thought. The history of cosmic theories, in particular, may without exaggeration be called a history of collective obsessions and controlled schizophrenias; and the manner in which some of the most important individual discoveries were arrived at reminds one more of a sleepwalker's performance than an electronic brain.

ARTHUR KOESTLER, *THE SLEEPWALKERS*, 1959

70

CHAPTER OVERVIEW

A brief paragraph gives an overview of concepts discussed in the chapter. It serves as an organizing device to help you preview chapter coverage.

5.1 GREEK ASTRONOMY

In this chapter we follow the development of our understanding of our own corner of the universe, the solar system. We will not seek to be comprehensive or encyclopedic in our historical coverage. Instead, we will be more interested in how the ideas that developed in one era were a natural outgrowth of the state of astronomical observations and degree of sophistication of the times. In previous chapters we briefly discussed some historical aspects of astronomy; for example, we considered the notion of a round Earth, and how this concept slowly came to be accepted by Greek thinkers through the gradual accumulation of evidence from both terrestrial and astronomical observations. Yet this concept was established relatively rapidly when compared to the slow acceptance of the idea of a Sun-centered solar system.

5.1 GREEK ASTRONOMY

While we saw in the last chapter that a variety of cultures were intimately involved with phenomena of the sky, it was Greek culture that made progress toward the models we have today. We now look more closely at the astronomical contributions of Greek culture.

WHAT THE GREEKS INHERITED

The Babylonians and Egyptians, who for several thousand years kept records that contained much potentially valuable information, bequeathed to the Greeks an extensive body of astronomical knowledge. However, the astronomy of both Egypt and Babylonia was the province of a priestly aristocracy; as a consequence, practical and political considerations often took precedence over theoretical inquiries. Egyptian astronomers, for example, had discovered that when the bright star Sirius could just be seen rising in the east before the Sun, the flooding of the Nile was imminent. Such knowledge gave tremendous power to the priesthood and inevitably involved its members closely with the state. Similarly, in Babylonia, astronomer-priests had acquired a considerable amount of information concerning the motions of the Moon, Sun, and planets and had found certain regularly occurring cycles that enabled them to predict some eclipses, a power that was frequently used for political purposes.

The Egyptians and Babylonians knew the length of the year and the different types of calendars, both solar and lunar. The Egyptians had learned the rudiments of simple mathematics, algebra, and geometry. Sundials had been invented, and systems of timekeeping were in existence. The Babylonians had made systematic observations of the positions of heavenly bodies and had practical methods for predicting the positions of the Moon, Sun, and planets. **Figure 5–1**, for example, shows a table of data for Jupiter. The bottom part of the figure describes the method of calculation. The Babylonian value for the length of the month was not surpassed in accuracy until the end of the nineteenth century. Both cultures had attempted to construct a cosmology that placed Earth and humanity in their proper position relative to the universe and the gods, but they never attempted to construct a truly consistent theoretical framework for their cosmology in the way the Greeks did.

FIGURE 5–1. A clay tablet from ancient Mesopotamia containing astronomical observations of Jupiter in the top part, and a description of the method of calculation in the bottom part.

71

OPENING QUOTE

Each chapter opens with a quotation from popular fiction or a scientific source that helps to set the stage for topics covered in the chapter. These quotes often give a perspective from outside the immediate field of astronomy.

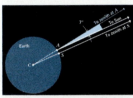

Nearby, small object   Large, distant object

FIGURE 5–4. Two objects with the same angular sizes have diameters that are proportional to their distances.

Inquiry 5–3 Aristarchus was the first to propose that the Earth goes around the Sun, rather than vice versa. Suggest one factor that may have led him to this conclusion.

ERATOSTHENES

Another classic experiment of antiquity was the determination of the Earth's circumference by **Eratosthenes** (c. 200 B.C.). The conclusions of Aristarchus concerning the relative sizes and distances of the Earth, Moon, and Sun were all in terms of the then unknown size of the Earth. Their sizes in customary units (the Greeks used a unit of length called the *stadium*) could not be known until the Earth's size was known.

Eratosthenes had heard that at Syene, near the modern Aswan in Egypt, there was a deep well, and that on a certain day of the year the Sun stood directly overhead so that its reflection could be seen in the bottom of the well. Eratosthenes was also able to observe that on that same day of the year in Alexandria, where he lived, the Sun was not directly overhead but was 7° south of the zenith. He determined this angle with a gnomon, just as

FIGURE 5–5. Eratosthenes' measurement of the diameter of the Earth. The difference in altitude of the Sun is proportional to the distance between the two points, which is known. This allows the length of one degree on the Earth's surface to be determined.

73

you may have done in Kit Activity 4–3. Figure 5–5 shows the geometry of Eratosthenes' experiment. Point *A* represents Alexandria and point *S* is Syene; point *C* is at the center of the Earth. The Sun is so far away that the lines *A*-Sun and *S*-Sun are nearly parallel. The angle at *A* is nearly equal to the angle at *C* at the center of the Earth, and we may write the following proportion:

$$\frac{\text{circumference of Earth}}{\text{distance from } A \text{ to } S} = \frac{360°}{\text{angle at } C}$$

Inquiry 5–4 Assuming that the angle at *C* is 7° and that the distance of Alexandria from Syene is 5000 stadia, what is the diameter of the Earth in stadia? (Stadia is the plural of stadium.) Although the exact length of a stadium is unknown, compute the radius of the Earth assuming it to be about one-tenth of a mile.

Inquiry 5–5 What does Eratosthenes' experiment assume about the shape of the Earth?

Inquiry 5–6 If the Earth were flat, what would be the value of the angle at *C*?

HIPPARCHUS

Perhaps the greatest of all ancient astronomers was **Hipparchus** (c. 150 B.C.). Many of the conclusions he drew were so sophisticated that it takes some knowledge of astronomy to appreciate how great his contributions were. He built an observatory, constructed the best astronomical instruments up to that time, and established a program of careful and systematic observations that resulted in the compilation of a great star catalog, with 850 entries, using a celestial coordinate system similar to our modern one for cataloging the sky. It was Hipparchus who originated a system, which is still in use today in modified form, for estimating the brightness of stars. In addition, he paid much attention to the older Egyptian observations and detected long-term trends in the motions of the celestial sphere that had been previously unsuspected. He deduced Earth's precession, which is so slow that it takes almost 26,000 years for it to complete one cycle. Finally, he greatly developed trigonometry, which was, and still is, a useful tool for astronomy.

OTHER DEDUCTIONS OF THE GREEK ASTRONOMERS: THE DISTANCES OF THE PLANETS

The Greeks estimated the relative distances of the planets from Earth, by means of principles still in use today for determining distances to astronomical objects. They reasoned that the more distant a planet was, the more

INQUIRIES

Throughout the text, Inquiries (questions) are included in appropriate places to ask you to use the information presented in the text. By actively participating and not just passively reading, you gain understanding of the concepts. Answers to all Inquiries are presented at the end of each chapter.

INSTRUCTIVE ILLUSTRATIONS

Computer-generated color diagrams help to clarify complex concepts and encourage an appreciation of the beauty of the universe.

slowly it would move across the sky. The effect is similar to what happens when we compare the apparent motion of a high-flying airplane with that of one that is flying very low. The distant airplane appears to move slowly across the sky, whereas the low-flying one is seen for only a short time and then is gone. In the same way, the Greeks could put most of the naked-eye planets in order of their distance from Earth by assuming that increasing distances corresponded to slower motions. The argument fails with Mercury and Venus, however, because it places Mercury closer to Earth.

Inquiry 5–7 What assumptions are made in employing this argument?

We can obtain another, independent determination of relative planetary distances from their brightnesses. We use an analogy: when you are driving at night and wish to pass the car in front of yours, you pass only if the headlights of the oncoming car are faint. When you do this, you are making an implicit assumption: all car headlights have about the same intrinsic brightness, with their apparent brightness depending on the distance. Similarly, if we assume that all planets have the same intrinsic brightness, then their apparent brightness as seen from Earth would depend on their distances from us. Of course, all the planets do not have the same intrinsic brightness, because their differences in size and distance from the Sun, combined with differences in surface and atmospheric properties, affect the amount of light they reflect in our direction. However, even allowing for these uncertainties, it is still possible to use this principle to rank the planets approximately in order of distance from the Earth.

THE APPARENT MOTIONS OF THE PLANETS RELATIVE TO THE STARS

Three additional observations of planetary motion were important in determining the details of the models the Greeks developed. These observations, which played a prominent role in their models, are:

1. Because the planets are considerably closer to us than the fixed stars, they appear to move against the starry background. Observations of Mars, Jupiter, and Saturn showed them to move generally eastward on the celestial sphere.
2. Occasionally, however, as discussed in Chapter 4, a planet's motion changes from eastward to westward. This retrograde motion would persist for up to several months but would cease as the planet's motion

slowed down and again reversed its direction, resuming its normal easterly motion (see Figure 4–28).

3. Venus and Mercury are never more than 48° and 28°, respectively, from the Sun.

THE GEOCENTRIC MODEL OF THE SOLAR SYSTEM

To describe the observed planetary motions, it was necessary to decide where the center of the system should be. There were really only two obvious candidates—Earth and the Sun. This question was considered carefully by Greek philosophers, and the fact that ultimately they reached an incorrect conclusion provides an interesting example of why the scientific method is not the simple turn-the-crank-and-the-answers-fall-out process that some sources describe it to be. If the Sun is in the center of the solar system, then Earth moves around it in space. Such a hypothesis provides a prediction. As shown in **Figure 5–6**, some of the stars ought to shift their apparent positions in the sky as the Earth moves from one side of its orbit to the other. Such **parallax** effects, as they are called, were looked for by many Greek observers, including Hipparchus, but were never found. The Greeks therefore concluded that the Earth was stationary in space.

Aristarchus, however, apparently espoused the theory that the Earth orbits the Sun, if surviving works of Archimedes and Plutarch are correct. Unfortunately, the work in which he put forth his hypothesis is lost, and apparently no other Greek astronomers held to this opinion. The model of a **geocentric** (Earth-centered)

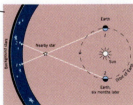

FIGURE 5–6. Stellar parallax. The apparent position of a star with respect to the background stars appears to change as the Earth goes around the Sun. (Not to scale.)

74

(Embedded page 86)

Inquiry 5–19 If the distance between mass M_1 and mass M_2 is made four times *greater*, by what factor is the gravitational force between them changed? Is it increased or decreased?

While the idea that bodies of different masses fall at the same rate was experimentally shown by Galileo to be true, its truth does not make sense to most people. Its validity is shown in Appendix A7.

It is a measure of Newton's genius that he proposed that his laws of motion and gravity applied not only to objects on Earth but to all bodies in the universe. One of the first things he did was to compare the force of gravity exerted by the Earth on a body, allegedly a falling apple, with the gravitational force that would be required to keep the Moon in its orbit around the Earth. As shown in **Figure 5–21**, if there were no gravitational force exerted by the Earth on the Moon, it would travel in a straight line past the Earth in accordance with Newton's first law. To make the Moon travel around the Earth requires a force toward the Earth. The Earth's gravitational force on the Moon causes it to deviate from a straight line and follow a curved path around the Earth. For each kilometer the Moon moves in its orbit, the Moon must fall 0.14 cm towards the Earth's surface in order to stay in its elliptical orbit. Because of the curvature of the Earth's surface, the distance of the Moon from the Earth remains constant. In other words, the Earth's force of gravity causes the Moon to accelerate just enough to maintain its constant distance above the Earth's surface. For this reason, the Moon's orbital motion can be described as resulting from the Moon's falling toward's the Earth's center! From such considerations Newton found that the actual force required to keep the Moon in its orbit around the Earth agreed well with the value he computed theoretically.

FIGURE 5–21. The motion of the Moon around the Earth according to Newton. The Moon "falls" toward the Earth just enough to keep it on a curved elliptical path around the planet.

WEIGHT

Weight is the force a gravitating body exerts on another mass. In particular, your weight is the gravitational force exerted by the Earth's mass on your body. Your weight depends on the mass of your body, the mass of the Earth, and your distance from the Earth's center, which is the Earth's radius. For this reason, your weight would be different should you travel to a different planet having a different mass and size than the Earth; your mass, however, would be the same. Interested readers can find weight expressed mathematically in Appendix A8.

> TO UNDERSTAND WHY AN ASTRONAUT IS WEIGHTLESS EVEN THOUGH THERE IS STILL GRAVITY IN SPACE, YOU SHOULD DO DISCOVERY 5–1 AT THIS TIME.

MOMENTUM

You probably have some intuitive feel for the word **momentum**. A train moving at 20 miles per hour has more momentum than a bicycle moving at the same speed; it would have more impact and effect if it ran into something! For a body moving in a straight line, **linear momentum** is defined as the body's mass times its velocity.

Newton's first law can also be expressed as the principle of the **conservation of linear momentum**. This says that for an isolated system of bodies the sum of the linear momenta of all bodies in a system is always the same. If some bodies in a group slow down, others must speed up by a corresponding amount.

Most bodies move in curved paths, and the concept of **angular momentum** comes into play. The amount of angular momentum possessed by a planet in a *circular* orbit around the Sun is defined by

angular momentum =
    mass × speed × distance from planet to Sun.

Expressed symbolically, if M is the mass of a body and v its speed when at a distance r, the angular momentum is given by

$$\text{angular momentum} = Mvr.$$

Like linear momentum, angular momentum is also conserved for isolated systems. If the mass of an orbiting planet remains constant, the only way for angular momentum to remain constant is for the speed of the planet to increase as the distance decreases, and vice versa. This is exactly what Kepler's second law says. The concept of conservation of angular momentum is important in understanding not only the motions of planets but also such diverse subjects as the formation of the

REFERENCES TO DISCOVERY ACTIVITIES

Although Discoveries appear at the end of the chapter so as not to disrupt the flow of reading, they are referenced at the appropriate places in a blue shaded notation. [It is not necessary to do these activities to understand the concepts.]

(Embedded page 90)

FIGURE 5–25. The relationship between the distance of a star and its parallax. The greater the distance, the smaller the parallax angle.

tant than the nearby stars, the angle *A*-star-*B* of Figure 5–24*a* will be practically equal to the angular shift in position shown in Figure 5–24*b*.

How large a distance to the stars does the small observed parallax imply? Consider the right triangle formed by the Earth, the Sun, and a nearby star, as shown in **Figure 5–25**. The side labeled *A* is the distance from the Earth to the Sun, which is of course 1 AU. The angle *p* is called the **parallax** of the star; it is one-half the total parallactic shift shown in Figure 5–24*a*. Because the triangle is so long and skinny, the distance *D* from the Sun to the star can be computed from the angular size formula used in Chapter 3, namely,

$$D = 57.3° \times \frac{A}{p}$$

if the parallax *p* is expressed in degrees. (We have replaced the variable *R* used in Chapter 3 with the distance *D*.) For example, the nearest star to the Earth, Proxima Centauri, has a measured parallax of 0.76″, which corresponds to 0.00021 degrees, an extremely small angle. Substituting this angle and the length of an astronomical unit, in kilometers, into the formula, we find the distance to be

$$D = 57.3 \times \frac{150,000,000 \text{ km}}{0.00021}$$
$$= 41,000,000,000,000 \text{ km}$$
$$= 4.1 \times 10^{13} \text{ km}.$$

Kilometers or miles are clearly not an appropriate unit for measuring distances to stars. Nor are astronomical units; the distance to Proxima Centauri is over 270,000 AU, for example. Just as we describe distances around town in miles or kilometers rather than inches or centimeters, we describe distances to stars in units that are large enough so that the numbers we use have convenient sizes. This is why we use light-years, because one light-year is about 9.5×10^{12} km or 5.9×10^{12} miles. The distance of Proxima Centauri, then, is 4.3 ly.

> READERS HAVING THE ACTIVITY KIT SHOULD DO KIT ACTIVITY 5–1 (A PARALLAX MEASUREMENT) AT THIS TIME.

5.7
OBSERVATIONAL EVIDENCE OF THE EARTH'S ROTATION

While every schoolchild can tell you the Earth rotates, few college graduates can explain *how* we know this simplest of all astronomical facts. As evidence, most people would cite the observation of the Sun and stars rising in the east and setting in the west. If you were to suggest, just for argument, that the east-west motion of the Sun and stars is produced by the motion of crystalline spheres to which the Sun and stars are attached, great confusion would result.

FOUCAULT PENDULUM

A famous demonstration of the rotation of the Earth is the pendulum experiment first performed by the nineteenth-century French physicist Jean-Bernard-Léon Foucault. In this experiment, a massive pendulum is hung from a long wire and set in motion. Pegs placed in a circle around the pendulum are successively knocked down over time due to the apparently changing plane in which the pendulum swings. To construct the simplest possible explanation of what is observed, imagine that we suspend a pendulum on a long cable from a support that has been placed at the North Pole (Figure 5–26). We make the coupling between the cable

FIGURE 5–26. The Foucault pendulum. (*a*) Setup of the experiment. (*b*) Six hours later, the pendulum still vibrates in the same plane relative to the stars, but the observer has rotated.

REFERENCE TO KIT ACTIVITIES

Activities requiring measurement can be carried out using the Activity Kit available separately. The yellow shaded notation indicates the relevant place in the text for these activities to be performed. [It is not necessary to do these activities to understand the concepts.]

gion during the fall hurricane season. The low pressure in the center is surrounded by regions of high pressure, so winds blow in toward the center. As they do, they are deflected to the right, causing a counterclockwise circulation of wind inside the low pressure region (for Northern Hemisphere hurricanes). If such a disturbance were to move in from the Gulf of Mexico and locate its center of low pressure at the Texas-Louisiana border, central Texas would experience winds from the north, New Orleans would have winds from the south, and Little Rock, Arkansas, would have southeasterly winds.

FIGURE 5-28. Winds will push toward the center of a low-pressure storm region from the surrounding high-pressure regions. As they move, they will be deflected to the right (in the Northern Hemisphere) because of Coriolis effects. The net effect is to make the winds circulate in a counterclockwise direction around the center of the low-pressure region.

DISCOVERY 5–1
WEIGHTLESSNESS

When you have completed this Discovery, you should be able to do the following:

• Describe what is meant by weightlessness.

An astronaut floating in a space shuttle has mass but is weightless. Weightlessness does not occur, however, from a lack of gravity in space.

☐ **Discovery Inquiry 5–1a** Suppose you were standing on a scale in a stopped elevator. What would happen to the scale reading when the elevator suddenly accelerates upward? What would happen to the scale reading when the elevator rapidly descends from rest?

Although the elevator has accelerated, the force of gravity on the elevator occupants has not changed. The scale reading, however, has changed because of the acceleration.

You can easily demonstrate weightlessness for yourself. Take a paper cup or an aluminum pop can and punch two holes on opposite sides near the bottom. With your fingers over the holes, fill the container with water.

DISCOVERY ACTIVITIES

The nature of science is one of discovery. Short and simple Discovery Activities appear at the end of many chapters to reflect this emphasis by getting you actively involved with the material.

☐ **Discovery Inquiry 5–1b** *Before removing your fingers from the holes,* describe what you expect to happen. Then, while taking necessary precautions, remove your fingers from the holes and observe what happens.

Again place your fingers over the holes and fill with water.

☐ **Discovery Inquiry 5–1c** *Before dropping the container* (into a garbage can, the bath tub, or *safely* outside!), describe what you expect to observe. After thinking about the answer, drop the container from as high as you can reach, and describe what you observed.

In this experiment, you should have observed no water flowing from the container while it was dropping. Although gravity was still there, the water was weightless *with respect to the container* because the water and the container accelerated downward at the same rate. Similarly, because the astronaut and the space shuttle both accelerate towards the center of the Earth *at the same rate,* the astronaut is weightless.

CHAPTER SUMMARY

OBSERVATIONS

• **Aristarchus** first suggested that the Earth circles the Sun, and found the relative sizes of the Earth, Moon, and Sun, as well as the relative distances of the Moon and Sun. **Eratosthenes** first found the size of the Earth. **Hipparchus** made a great star catalog, began the magnitude system used to specify the brightness of stars, and discovered precession. **Ptolemy** advanced the theory of planetary epicycles and helped preserve Greek knowledge in the *Almagest.*
• The Greeks observed that the planets moved against the starry background; sometimes changed directions and moved with a **retrograde motion**; and, in the case of Mercury and Venus, were never far from the Sun.
• **Tycho Brahe** built large instruments and made observations over extended periods of time. **Johannes Kepler** found three empirical laws of planetary motion: (1) planets revolve about the Sun in elliptical orbits with the Sun at one focus; (2) planets move more rapidly when close to the Sun than when farther away; (3) if the period of a planet in its orbit is P years and the semi-major axis is A astronomical units, then $P^2 = A^3$.
• **Galileo** made the first telescopic observations and discovered that the Milky Way consists of a large number of stars. He observed lunar surface features, four Jovian satellites, sunspots, and phases of plan-

ets. He showed experimentally that objects of different masses fall to the ground with identical accelerations.
• The **aberration of starlight** is a small apparent shift in the direction to an object caused by the motion of an observer. The observation of this effect illustrates the motion of the Earth.
• **Stellar parallax** is the apparent change in the direction to an object caused by a change in position of an observer.
• The **Coriolis effect** is the apparent rightward drift of a projectile fired from the equator toward the North Pole, or from the North Pole toward the equator. It is caused by the fact that equatorial regions move more rapidly than polar regions.

THEORY

• Greek science and philosophy assumed that planets exhibit uniform motion in circular orbits.
• **Nicolas Copernicus** proposed the heliocentric hypothesis.
• **Mass** is a quantity that measures the amount of **inertia** that a body contains. A body's mass is independent of its location. **Velocity** is a change in location or direction of motion divided by the time over which the change occurs. **Acceleration** is a change in an object's velocity divided by the time over which the change occurs.

DISCOVERY INQUIRIES

To reinforce the key concepts of these Discoveries, questions (Inquiries) are included to guide your thinking.

CHAPTER SUMMARY

The Chapter Summary not only reviews the topics covered but presents the information in a way that reflects the scientific method. The material is organized in subsections of Observations, Theory, and Conclusions to give you practice in thinking in these scientific ways.

CHAPTER 5
THE HISTORICAL QUEST TO MODEL THE SOLAR SYSTEM

- Newton's three laws describe how bodies move both in the absence and presense of forces. They are general laws applicable to a wide variety of situations.
- **The universal law of gravitation**, found by Newton, provided a means to connect the motion of the Moon with the motion of falling bodies on Earth.
- From Newton's laws, a modification of Kepler's third law results that allows astronomers to determine the masses of orbiting bodies:

$$(M_1 + M_2)P^2 = A^3.$$

- The amounts of **linear momentum** and **angular momentum** contained in an isolated system are both conserved.

- The **weight** of an object is determined by the gravitational force exerted on it by the Earth. The weight depends on the object's location.

CONCLUSIONS

- From observations made with the **Foucault pendulum**, astronomers infer the rotation of the Earth.
- From observations of the aberration of starlight, astronomers infer the revolution of the Earth about the Sun.
- Newton's laws provide a model that successfully explains past events and predicts future ones. From them, Kepler's laws of planetary motion can be derived.

SUMMARY QUESTIONS

1. What is the significance of the observations of Aristarchus, Eratosthenes, and Hipparchus? What are the principles used to make their measurements of the sizes of the Earth and Moon?
2. What are two methods by which the order of the planets from the Sun could be determined?
3. What planetary observations must a reasonable model of the solar system incorporate? In your answer use diagrams to show how the Ptolemaic and heliocentric hypotheses explain the observations.
4. What is meant by stellar parallax? Explain its cause. How can parallax be used to distinguish between heliocentric and geocentric hypotheses? Why did the Greeks not observe stellar parallax?
5. What role did the concept of uniform circular motion play in the history of astronomical thought? Give examples from several eras.
6. What were specific astronomical contributions

made by Copernicus and Brahe? What was the importance of these contributions?
7. State and explain Kepler's three laws. Explain Kepler's second law in terms of the conservation of angular momentum.
8. What are the principal astronomical discoveries of Galileo? Explain how these discoveries may have accelerated the acceptance of the heliocentric hypothesis.
9. State Newton's laws of motion and the law of gravity. Explain how, in principle, they can be used to explain the orbiting of one body about another.
10. What are some specific astronomical examples of how Newton's laws have been shown to be valid descriptions of nature?
11. How were the heliocentric hypothesis and the Earth's rotation finally confirmed observationally? How did Newton's laws play a role in demonstrating the Earth's rotation and revolution?

APPLYING YOUR KNOWLEDGE

1. Make up a table that lists, chronologically, the contributions to astronomy made by Aristarchus, Apollonius, Eratosthenes, Hipparchus, Ptolemy, and Pythagoras. Include dates.
2. If you had been a traditional scholar in the mid-1500s, what arguments would you have presented *against* the Copernican system?
3. Of the following people, who in your opinion made the most important contribution to astronomy: Copernicus, Brahe, Kepler, Galileo, Newton? Explain.

4. Use the definition of acceleration to list the accelerators in a standard passenger car.
5. Use Newton's second law to explain why a planet in an elliptical orbit moves more rapidly when near the Sun than when farther away.
6. Explain why your weight would be different on Mars than it is on Earth.
7. Why can a rocket escape from the Moon's surface with a smaller speed than is needed to escape from the Earth's surface?

SUMMARY QUESTIONS

Straightforward questions at the end of the chapter help you to review the material you have read and to check your understanding of it. Answering these questions is good practice for taking tests.

APPLYING YOUR KNOWLEDGE

The best way to test your comprehension of new material is to be able to apply this knowledge to new and different situations. These end-of-chapter questions are designed to require the application of concepts previously learned. Those questions requiring the use of simple mathematics are marked by a square bullet ().

■ 8. What is the distance from the Sun to a planet whose period is 129.14 days?
■ 9. Find the ratio of the gravitational attraction between (a) a man of mass 100 kg and a woman of mass 50 kg who is 10 meters away from him, and (b) the attraction between the woman and the Earth. (Hint: See Appendix A8.)
■ 10. If the Sun is 2×10^9 AU from the center of the Milky Way, and if it has a period of 200 million years, what is the mass of the Milky Way galaxy?
■ 11. If identical galaxies at a distance of 650 million ly and having a mass of 10^{10} solar masses each were observed to have an angular separation of 10 seconds of arc, how long would their orbital period be? Assume the

galaxies are seen at their maximum separation. (Hint: This problem contains a number of steps along the way.)
■ 12. How much would a 100-lb person weigh on the Moon? (Hint: Write an expression for the person's weight on Earth and an equivalent one for the Moon. Then divide one equation by the other; you will find that some quantities drop out of the problem.)
■ 13. How fast would a person of mass 50 kg (about 110 lbs) have to run to have the same linear momentum as a 5000-kg bus (about 5 tons) traveling 60 miles per hour?
■ 14. What would be the parallax of Proxima Centauri if it were observed from a telescope on one of Saturn's satellites?

ANSWERS TO INQUIRIES

5–1. Circular orbits and uniform motion.
5–2. The Earth is four times larger.
5–3. Because he knew the Moon was smaller than the Earth and was in orbit about the Earth, and because he now knew the Sun was farther than the Moon, it was logical that the Earth would go around the Sun.
5–4. 5000 stadia $\times$ 360°/7° = 260,000 stadia.
5–5. That it is perfectly spherical.
5–6. Zero degrees.
5–7. That the planets are moving at about the same speed, so their apparent speeds are due only to their respective distances. While the assumption is incorrect, the result ends up being correct.
5–8. Stars are very far away.
5–9. Only crescent phases would occur.
5–10. There are countless examples. Some are the germ model of disease, economic models used to set government fiscal policy, and educational models used to design instruction.
5–11. On a straight line, with the Earth in the middle.
5–12. Each has some points of simplicity. Ptolemy's has fewer epicycles, but Copernicus's has a neater explanation of retrograde motion, which in addition predicts retrograde motion only when it is actually observed.
5–13. About 1.87 years.

5–14. About 19.2 AU.
5–15. The planets are closer than stars.
5–16. Jupiter was clearly moving and dragging its moons along.
5–17. From $F = ma$, the can with more mass (the one filled with concrete) would have the smaller acceleration if the forces were equal.
5–18. The reaction force on the foot will certainly be felt more in the case of the concrete can.
5–19. The force is 16 times smaller.
5–20. Twice as great.
5–21. $(M_1 + M_2) = 8^3/10^2 = 512/100 = 5.1$ solar masses.
5–22. In the Southern Hemisphere, the pendulum would appear to rotate in the opposite direction from its rotation in the Northern Hemisphere. Therefore, for the direction of oscillation to change as the pendulum is carried from one hemisphere to the other, there must be no rotation at the equator.
5–23. In the Southern Hemisphere, projectiles will always deflect to the left, when fired any direction except directly east or west.
5–24. Yes. Imagining the ball of the pendulum to be a projectile deflecting to the right on each swing gives it the rotation we observe.

ANSWERS TO INQUIRIES

As a check on how you are doing as you progress through the chapter, the answers to the periodic Inquiries are included at the end of the chapter.

PART ONE

DISCOVERING THE SCIENCE OF ASTRONOMY

We begin where astronomy began—with people observing the heavens. Astronomy is first and foremost an observational science; only after observations of phenomena have been made does interpretation based on theory come into play.

The first chapter is a grand tour of the universe. In addition, we take a first look at the range of sizes and distances of objects in the universe.

Astronomy is, in every sense of the word, a *science*. In Chapter 2 we discuss what modern science is and how it allows us to understand the universe. To emphasize what science is, we provide a counterpoint by discussing what is meant by pseudoscience. We examine those qualities that distinguish scientific findings from those of pseudoscience.

We begin learning about astronomical observations in Chapter 3 with a discussion of angles and their measurement on the sky. The relationship between an object's apparent size, true size, and distance is discussed in detail. Because an understanding of random and systematic errors is crucial to understanding the scientific enterprise, we discuss the source and meaning of measurement errors.

Basic observations of the sky made by ancient peoples, and observations you can make, are discussed in Chapter 4. These observations include daily and yearly motions of the Sun, motions and phases of the Moon, and movements of the planets. In addition, we discuss the reasons for the seasons.

To complete Part 1, we look in Chapter 5 at how the observations presented in Chapter 4 were used by various civilizations to produce models leading to an understanding of the universe. This discussion leads naturally to the advances provided by the work of Copernicus, Brahe, Kepler, and Galileo, and finally to Newton's invention of the concept of gravity and his laws of motion. At this point we begin our comparison of observation with theory by looking at how Newton's theoretical ideas have been verified by a variety of observations.

1

BEYOND THE BLUE HORIZON: A GRAND TOUR OF THE UNIVERSE

*My suspicion is that the universe is not only queerer than we supposed,
but queerer than we can suppose.*

J. B. S. HALDANE, *POSSIBLE WORLDS*, 1927

The universe is filled with a bewildering variety of objects. We begin our study of the cosmos with a brief tour to give you a sense of the size and scale of the universe as astronomers now understand it. We will briefly survey the universe beginning at Earth and expanding our horizons outward to ever-increasing sizes and distances. As we progress from figure to figure, our scale will increase until we have covered the presently known universe. This chapter is meant to intrigue you—to raise questions—but not to explain. Explanation and understanding will come from the rest of the book.

As we foreshadow the contents of the book, we will also introduce you to some of the special difficulties that astronomers encounter as they try to understand the universe, and some of the unique problems they face as scientists who are not able to manipulate or experiment with the objects they study.

1.1

OUR VIEW OF THE COSMOS

You probably have grown up with pictures like **Figure 1–1**, taken by the Apollo astronauts. That familiarity may make it difficult for you to appreciate what a milestone in human history such photographs represent. We refer not so much to the effort and expense of landing a person on the Moon—although that was considerable—but rather to the changes in our perspective that have come about in the last few decades.

Most of what we know about the universe has been learned in the twentieth century; indeed, most of our modern astronomical knowledge has been acquired since World War II. As recently as 1920 the best information suggested that the entire universe was only some 10,000 light-years across. A **light-year** (abbreviated **ly**) is the distance that light travels in one year; in more familiar units, it is about 6 million million (or 6 trillion) miles, or about 10 trillion kilometers (abbreviated km). Although this may sound like an enormous distance, by 1930 astronomers had found that the universe was in fact at least a *million* times larger than that. More incredibly, before 1917 the best information we had showed that the Earth's star, the Sun, was at the center of the universe. Our current ideas are about as far from this chauvinistic and provincial notion as possible. They suggest that we are probably just one among billions of planetary systems, located in the peripheral regions of an immense spiral-shaped stellar system containing several hundred billion stars. This stellar system is a **galaxy**, just one of billions of similar galaxies in the universe.

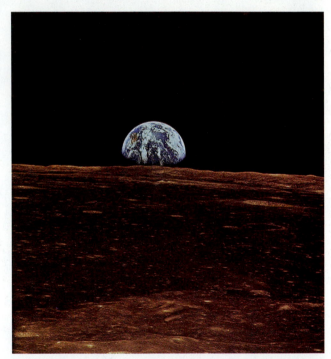

FIGURE 1–1. Earthrise as viewed by a spacecraft in lunar orbit.

These spectacular changes in our understanding of the universe have triggered profound changes in the way we view ourselves. Some argue that the revolution in our perspective is even greater than the much-touted intellectual shake-up that accompanied the Copernican revolution in the sixteenth century. With this new knowledge, we are trying to formulate a cosmic perspective in which we not only consider the possibility that we may not be alone in the universe, but we recognize that our Earth is limited in size and resources and requires care and nurturing if it is to survive as a place for life.

ASTRONOMY USES LARGE AND SMALL NUMBERS

Because astronomy considers both the largest and smallest objects in the universe, astronomy uses both the largest and smallest of numbers. **Table 1–1** shows the equality between ordinary numbers, their powers of ten notation, and the word used for the number. The final column emphasizes that large numbers are combinations of smaller ones.

You can visualize some large numbers using a piece of graph paper 10 centimeters by 10 centimeters (abbreviated cm) with millimeter divisions. There are 100 millimeter squares along each edge, and 10,000 squares in the 10 cm by 10 cm area. Extending this area 10 cm into the third dimension would produce a cube having a million small cubes. If you now increase each side of this 10 × 10 × 10 cm cube 10 times so it is slightly more than one yard on a side, the cube will enclose a billion small cubes! A typical small theater might be 10 times longer on each side and would contain a trillion of the small cubes.

Another method of visualizing large numbers is to count at the rate of one per second; you will reach 1000 in about 17 minutes, a million in 12 days, and a billion in nearly 32 years!

Distances within the solar system are generally measured by using the average Earth-Sun distance as a standard measuring stick. In familiar units, this distance is 93,000,000 miles (or 150,000,000 km[1]), but astronomers generally refer to it simply as one **astronomical unit** (abbreviated **AU**). The average distance of Saturn from the Sun is almost 10 AU. We speak of Saturn, 10 times farther away, as being an **order of magnitude** more distant from the Sun than the Earth. Because the astronomical unit is about 100 times larger than the diameter of the Sun, the astronomical unit is two orders of magnitude larger than the Sun's diameter. The diameter of the Sun is about 100 times, or two orders of magnitude, larger than the diameter of the Earth. These orders of magnitude should help you keep the relative sizes of the Sun, the planets, and the planetary orbits in perspective.

The concept of order of magnitude is closely related to that of scientific, or powers of ten, notation. Using sci-

---

[1]See Appendix A4 for a discussion of conversion from one set of units to another.

TABLE 1–1

Large Numbers

1	10^0	one	
10	10^1	ten	
100	10^2	hundred	
1,000	10^3	thousand	
1,000,000	10^6	million	thousand thousand
1,000,000,000	10^9	billion	thousand million
1,000,000,000,000	10^{12}	trillion	million million

entific notation it is possible to write extremely large or extremely small numbers with ease (Table 1–1). Scientific notation also greatly simplifies the arithmetic of multiplying and dividing large and small numbers together. If you are not familiar with scientific notation, you should read Appendix A1 now, because scientific notation will be used frequently throughout this book.

Astronomy also deals with the smallest objects in the universe. For example, an atomic nucleus is 10^{15} times smaller than a person. In other words, there is a difference of 15 orders of magnitude between the size of a nucleus and a person.

For many purposes it is necessary to know only the approximate size of a number. This is true of most of the numbers that appear in this book. We talked about comparing numbers according to their orders of magnitude, and we used examples of numbers that differed from each other by factors of ten. We compared the relative sizes of such numbers simply by comparing the number of zeros. In the general case, the number of times one needs to multiply a smaller number by 10 to make it of the same order of magnitude as a larger one is the difference in the order of magnitude between the two numbers.

We can now extend this concept to enable us to make rough comparisons between any two numbers. We consider two numbers to be of the same order of magnitude if the larger of them is no more than about 3 times bigger than the smaller (the exact factor chosen depends on the circumstances). On the other hand, if a number is more than about 3 times bigger than another, we would say that it is one order of magnitude bigger. In effect, we are considering "more than 3 times bigger" and "10 times bigger" to mean about the same thing. Often such a statement may be adequate since it is not uncommon for our actual knowledge to be this approximate or even worse. There are many things about the universe that we know only roughly (and, of course, some things we don't yet know at all), and it makes no sense to describe an approximate piece of knowledge by using a number that looks precise.

Inquiry 1-1 For each pair of numbers below state whether they are of the same order of magnitude, or, if not, by how many orders of magnitude they differ. (a) 15,000 and 80,000 (b) 500 and 300 (c) 3,000,000 and 4,000 (d) 10,000 and 30,000

A hydrogen atom is about 10 orders of magnitude smaller than a human being, and its nucleus is several orders of magnitude smaller than that. The nucleus of the atom is approximately one ten-trillionth of a centimeter in extent, or 22 orders of magnitude smaller than the Earth's diameter. The diameter of the observed universe (the distance at which we have seen objects) is approximately 10 billion ly, which is about 19 orders of magnitude larger than the Earth. Thus astronomy deals with sizes that range over more than 41 orders of magnitude. Expressed in decimal form, 41 orders of magnitude is a 1 followed by 41 zeroes. The solar system is approximately in the middle of this range.

Inquiry 1-2 How many orders of magnitude larger than an average person's height is the Earth's diameter?

1.2

THE SOLAR SYSTEM

Our tour beyond the blue horizon of the Earth begins with the nearby planets, which have both similarities and differences when compared with Earth. Then, as our distance increases, we will encounter the planets that are more similar to Jupiter. The Sun, because of its great amount of material, rules over the solar system and must be discussed. Finally, we discuss the outer reaches of our solar system.

THE EARTH-LIKE PLANETS

Earth is an island in the universe, just like every other planet. But it is *our* island. We see that our planet, whose diameter is about 8000 miles (12,000 km), has continents, lots of water, and an atmosphere. The planet is volcanically active. Using a compass, we find it to exhibit magnetism. Further examination shows that the continental bodies and mountain ranges change over time.

Our journey beyond Earth begins with our nearest neighbor, the Moon (**Figure 1–2**), at an average distance of about 240,000 miles (380,000 km). This distance is only 30 times the diameter of the Earth. A small body having roughly a quarter the diameter and con-

FIGURE 1-2. The Moon through an Earth-based telescope.

taining 1% of the material of the Earth, it nevertheless has a strong effect on its parent body. Among other things, its gravitational attraction causes the tides on Earth to ebb and flow. A few questions dealt with in Chapter 8 include: Do the Earth and Moon have identical chemical compositions? Are processes that occur on Earth present on the Moon? Did the two bodies form together or separately?

Mercury, Venus, and Mars show some similarities to Earth. People used to consider the red planet Mars (**Figure 1–3**) a possible location for life. The Viking spacecraft of the 1970s found no life. However, spacecraft have found ancient volcanoes and huge canyons

FIGURE 1-3. Mars. A mosaic of some 100 Viking Orbiter photographs.

spanning the planet. Questions discussed in Chapter 9 include: What processes occur on Mars to produce such large volcanoes? Were the canyons formed from flowing water? What happened to Mars's water?

THE GIANT PLANETS

Figure 1–4 shows the giant planet Jupiter photographed in 1979 by the approaching Voyager 2 spacecraft. Unlike rocky Earth, Jupiter is a huge gas ball; it may not even have a solid core. We will see that solid ground like the surface of the Earth is a rarity in the universe because most of the cosmos is gaseous.

Jupiter contains more matter than all the other planets put together. Its diameter is approximately 10 times larger than Earth's diameter (which means its volume is 10^3 or 1000 times larger). Jupiter's diameter is therefore one order of magnitude larger than that of the Earth; Jupiter's volume is three orders of magnitude larger. The inset of Figure 1–4 shows the Earth to scale.

Jupiter's largest satellites (one of which is seen in Figure 1–4) are perhaps even more fascinating than the giant planet itself. As we will study in Chapter 10, each was found to have its own distinct character. Why is each one unique? Why is the surface of one heavily cratered

while the surface of another is smooth without impact craters? Why does the position of a satellite in its orbit about Jupiter influence the planet's observed radiation?

Smaller than Jupiter, but still a giant by Earthly standards, Saturn (**Figure 1–5**) is famous for its ring system. Several rings are visible from Earth, but 1982 close-up views from the Voyager probes showed an incredibly complex system of tens of thousands of "ringlets" engaged in a complicated gravitational dance. Recent research has shown that the other giant planets (Jupiter, Uranus, and Neptune) also have rings, although none are as prominent as those of Saturn. To what extent are Jupiter and Saturn similar? What are the rings and why do the giant gas planets have them? Why does the ring system of each planet differ from that of the others? These questions are studied in Chapter 10.

While passing Saturn, Voyager 2 received a gravitational boost to propel the space probe on to Uranus and then to Neptune (**Figure 1–6**). In addition to features resembling some of those on Jupiter and Saturn, Neptune

FIGURE 1–5. Saturn and its fabulous ring system.

FIGURE 1–6. Neptune, as seen by Voyager.

FIGURE 1–4. Jupiter, photographed from 17,500,000 miles by one of the Voyager spacecraft. The inset shows Earth in its true size relative to Jupiter.

showed a complete ring system and additional satellites. Its satellite Triton was especially exciting, as you will see in Chapter 10. There, too, you will read about how Neptune differs from Uranus, what surface processes occur on Triton, and in what ways the orbits and motions of some of Neptune's satellites are strange and unexpected.

While some astronomers in the 1970s and 1980s questioned Pluto's stature as a planet, in recent years it has emerged as a body worth studying. In Chapter 10 you will learn what can be so interesting about a body that is so small and distant (6 billion km) that it appears as little more than a point of light in telescopes.

Pictures were transmitted to Earth from Voyager over millions of kilometers as strings of numbers, then were reconstructed into two-dimensional images by computer. The color may or may not be true color because computers can be instructed to manipulate colors to enhance contrast or make certain features more visible. Many of the spectacular images that adorn astronomy books these days are actually enhanced or artificially colored by computer. The colors in the photographs in this chapter are all true colors except as follows: Figures 1–4 and 1–6 are close to real but have been intensified by computer to show features in greater contrast; Figure 1–5 is computer colored.

THE STAR OF THE SOLAR SYSTEM

At the surface of the Sun (**Figure 1–7**) we find violent activity that churns the gases of the solar atmosphere. Sometimes the Sun's activity even reaches out far enough to disturb communications on Earth. The diameter of our star is 1,400,000 km, approximately 10 times that of Jupiter. This means that the Sun's diameter is one order of magnitude larger than Jupiter's and two orders of magnitude larger than the Earth's. Another way of saying the same thing is that the Sun is a **factor** of 100 times larger than the Earth. Judging the order of magnitude simply by counting the number of zeroes in the factor is an easy way to compare the relative sizes of things.

While the diameter of the Sun may be only one order of magnitude larger than Jupiter, the Sun contains 1000 times more material than Jupiter and is 1000 times greater in volume. This enormous amount of material generates extremely high pressures and temperatures in the Sun's central regions, high enough to start nuclear reactions there. The generated energy streams out through the Sun and radiates into space. The tiny fraction intercepted by the Earth's surface generates and sustains life on our planet.

The power of the Sun is difficult to imagine by Earthly standards. The Sun produces energy at the rate of 10 billion nuclear bombs going off every second. If we

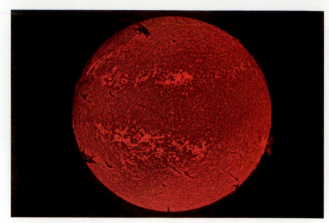

FIGURE 1–7. The Sun, photographed through a filter that passes only the red light emitted by hydrogen atoms. Such a photograph reveals details of the violent surface activity of the Sun.

figured energy costs at approximately seven cents per kilowatt-hour, the Sun would be radiating seven million million million dollars worth of energy into space every *second*. All the bombs exploded during World War II would power the Sun for only a millionth of a second. Yet the Sun is a fairly ordinary star. In Chapter 19 we shall see examples of stars that generate a million times more energy (that is, an amount six orders of magnitude greater). However, there are also large numbers of stars almost a million times less powerful.

After World War II, astronomers began in earnest to develop instruments that would measure other radiation in addition to visible light. Such instruments have provided the data for the rapid growth in astronomical knowledge in the last few decades. **Figure 1–8** shows the Sun as it appears to a detector sensitive to X-rays, a view dramatically different from the view in red light shown in Figure 1–7. Like most other celestial bodies, the Sun radiates energy in the form of gamma rays, X-rays, ultraviolet radiation, infrared radiation, microwaves, and radio energy, as well as visible light. What makes the Sun work? Why is it so hot inside? Will it remain the same forever, or does it change? The discussion of such questions begins in Chapter 16 and continues for a few chapters afterwards.

THE OUTER REACHES OF THE SOLAR SYSTEM

Figure 1–9 is a photograph of Halley's comet taken during its 1986 visit to the inner solar system. Because the previous visit by this comet was 76 years earlier, 1986 was the first opportunity to analyze it with instruments in space, and the comet was closely studied by several

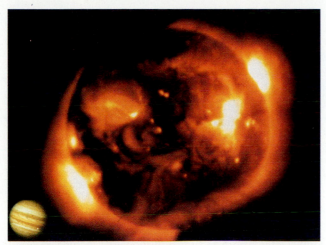

FIGURE 1–8. The X-ray Sun, as photographed by *Skylab*. The temperature of the glowing white regions exceeds 5 million K. Beneath them, areas of intense sunspot activity and magnetic fields are found. The inset shows Jupiter to the same scale.

FIGURE 1–9. Halley's comet.

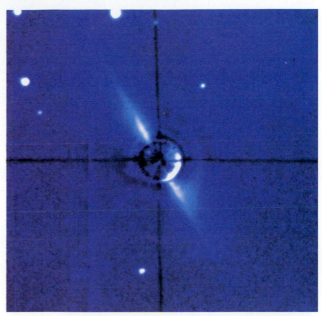

FIGURE 1–10. Beta Pictoris. This recently formed star is still surrounded by a disk of dust. The starlight is blocked to reveal the fainter disk.

frared radiation. The image in **Figure 1–10** was taken by a telescope on the ground and shows what astronomers believe might be a disk of material in the stages preparatory to forming planets. The disk around Beta Pictoris has an outer radius one order of magnitude larger than the radius of Pluto's orbit. Do other planetary systems exist? How can we try to find them? If they do exist, how do they form? Chapter 6 includes a discussion of the possibility of other planetary systems.

1.3

STARS, GALAXIES, AND BEYOND

Stars are the fundamental building blocks for larger structures in the universe, and they are the subject of Chapters 14 and 15. The nearest star beyond the Sun has a distance of about 270,000 AU, or about 25 million times the diameter of the Sun. Clearly, the distances between the stars are great compared to the sizes of the stars themselves. With so much space between them, collisions or even close encounters between stars are rare.

Even for the nearest stars, expressing distances in astronomical units would result in inconveniently large numbers. However, the light-year is ideal, because it is between four and five orders of magnitude larger than the astronomical unit. The distance to the nearest star is about 270,000 AU, so it is easier to talk about 4.3 light-

spacecraft from many nations. Comets merit this kind of intense scrutiny because they consist of dust and frozen gases that may well be remnants of the material from which the solar system formed. Where do comets come from? What are they made of? Did they have any influence on the development of life on Earth? Comets, along with asteroids and meteorites are studied in Chapter 7.

ARE THERE PLANETS AROUND OTHER STARS?

The star Beta Pictoris is a relatively inconspicuous object in the skies of the Southern Hemisphere, and little attention was paid to it until the Infrared Astronomical Satellite (IRAS) identified it as unusually bright in in-years. This distance is also about the average distance between stars in our galaxy. One of the more important questions we can address is how astronomers determine distances to stars. We begin looking at distance questions as early as Chapter 5.

The stars are so far away from us that they appear only as points of light. Even high telescopic magnification does not reveal a surface. (The apparent sizes of star images on photographs, produced by image spreading within the photographic emulsion, are determined by a star's apparent brightness, not its actual size.) But there are some special observing techniques described in Chapter 15 that do allow us to record surface features on a few stars that are not only nearby but also very large. The supergiant Betelgeuse (pronounced beetle juice), in the constellation of Orion, is such a star; indeed, it is one of the largest and brightest stars in our galaxy. Its diameter is about three orders of magnitude greater than the diameter of the Sun. If the Sun were to be replaced by a supergiant star like Betelgeuse, the star would stretch out to nearly the orbit of Saturn. Such a star generates about 25,000 times more energy than the Sun. What does the surface of a star look like? What makes some stars large and others small? These questions and others will be dealt with in Chapters 16, 17, and 18.

STAR FORMATION

Astronomers have concluded that stars are born from large globs of interstellar gas that collapse into stars under the influence of gravity. The dust and luminous gases pictured in **Figure 1–11** form a much-studied region of star formation not too far from the Sun, called the **Orion nebula**. A newly formed cluster of hot stars inside the cloud excites the gases to radiate. These hot stars are in the bright, overexposed central region but are not visible, because the picture was exposed to show best the fainter outer regions of the cloud. A gas cloud like the Orion nebula is approximately a million times (six orders of magnitude) less dense than the best vacuum we can create on Earth. The visible nebulosity seen here is approximately 15 ly across.

STELLAR EVOLUTION

Stars evolve. They typically remain stable and fairly constant in energy output for billions of years. But even stars eventually feel the effects of age. While the Sun has been nearly the same for about five billion years and will be for another five billion years, it will change. Other stars evolve at different rates, depending on the amount of matter they contain.

Betelgeuse is a star that many astronomers feel may be nearing the end of its life. The changing chemical composition inside the star, resulting from nuclear reac-

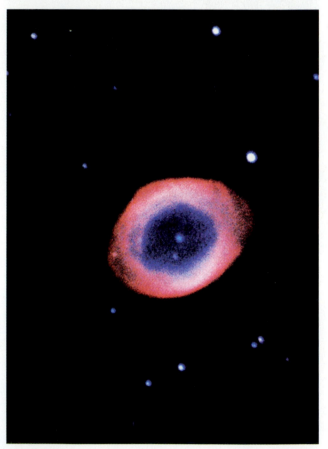

FIGURE 1–12. The Ring Nebula, a cloud of gas ejected by a dying star in the center.

FIGURE 1–11. The Orion Nebula, a region of dust and gas, excited to glow by the ultraviolet light of young stars recently formed inside the gas.

FIGURE 1–13. The Crab Nebula, the remnant of a spectacular stellar explosion witnessed by Chinese astronomers in A.D. 1054.

tions there, eventually causes significant changes in the outer (observable) parts of the star. Betelgeuse is swollen into its supergiant configuration by forces we will examine in Chapter 16, when we begin to discuss stellar evolution in more detail.

Two other objects in advanced stages of evolution are pictured in **Figures 1–12 and 1–13.** The objects in these figures are a few light-years in extent. For differing reasons both these objects are ejecting matter and are close to death. The object in Figure 1–12, called the **Ring Nebula**, is an example of a **planetary nebula**. The star in its center is in the process of sloughing off its outer atmosphere. Perhaps a tenth of a solar mass (an amount of mass equal to 0.1 the mass of the Sun) of gases move outward at a speed of about 30 km per second; this gas then mixes back in with the gases of interstellar space. The central star eventually becomes a **white dwarf**, a star having the amount of material in the Sun compressed into a space with the diameter of the Earth.

A few stars undergo a catastrophic mass ejection, tossing material outward at speeds of tens of thousands of kilometers per second. The result of one such **supernova** explosion is the **Crab Nebula** (Figure 1–13), the remains of an exploding star recorded by Chinese as-

tronomers in the year A.D. 1054. Near the center of the Crab Nebula is an object that has more material than the Sun squeezed into a ball only about 10–15 km across; it is called a **neutron star.** Other supernovae may result in the production of an ultra-dense object from which no light can escape—a **black hole.** If the star Betelgeuse were to become a supernova the Earth would feel significant and perhaps cataclysmic effects from the explosion, because the star is relatively close to the Sun—only about 500 ly away.

CLUSTERS OF STARS

Some gas clouds are too massive to form a single star. Both observations and theory indicate that such a cloud will fragment into condensations and form a cluster of stars. In fact, the best modern studies have shown that *all* stars form in groups of stars, and that most stars not in clusters are part of multiple-star systems. Isolated stars like the Sun are in the minority; they presumably result from the breakup or disintegration of a star cluster.

The large spherical star cluster known as M13 (**Figure 1–14**) is called a **globular cluster**. It is an impressive aggregate of some 100,000 stars bound together by gravity. Such clusters tend to be found in a spherical halo around the center of our galaxy. Globular clusters are among the oldest objects in the galaxy, having formed between 10 and 16 billion years ago.

GALAXIES

Planets, stars, star clusters, and interstellar gas and dust are all combined together into larger units called **galaxies**. The galaxy pictured in **Figure 1–15** is popularly

FIGURE 1–14. The globular cluster M13, a giant aggregate of hundreds of thousands of stars formed early in the history of our galaxy.

FIGURE 1-15. The Andromeda Galaxy, also known as M31, is a spiral galaxy somewhat similar to our own Milky Way galaxy. It is also relatively close, "only" two million light-years away, and under dark skies can be seen by the naked eye as a faint patch of light. It is orbited by two elliptical galaxies.

called the **Andromeda Galaxy** (named for the constellation in which it appears). Equally often referred to as **M31**, it is the thirty-first object in the 1781 catalog of the French astronomer Charles Messier. The Andromeda Galaxy is thought to be similar to our own galaxy, the **Milky Way**. At a distance of roughly two million light-years, the Andromeda Galaxy is the most distant object you can see with the naked eye. The Milky Way and M31 are **spiral galaxies**, thought to be approximately 100,000 ly in their longest dimension. Each galaxy certainly contains a few hundred billion stars, and perhaps trillions. The photograph shows that the Andromeda Galaxy has two smaller satellite galaxies gravitationally

FIGURE 1-16. The Large Magellanic Cloud, a small irregular galaxy orbiting near our Milky Way galaxy.

associated with it. We need to readjust our sense of scale here: the smallest of these companion galaxies is approximately 100 times the size of the globular cluster in Figure 1-14. Like the Andromeda Galaxy, our own Milky Way also has two known companion galaxies orbiting around it.

Galaxies come in forms other than spirals. Two examples of **elliptical galaxies**, which are shaped like giant ellipses, are the companions around M31, shown in Figure 1-15. Others defy easy classification and are known as **irregular galaxies**. The irregular galaxy pictured in Figure 1-16 is a relatively nearby object called the **Large Magellanic Cloud**. Easily visible to the naked eye from Earth's Southern Hemisphere, it is one of the satellite galaxies attached by gravity to our own Milky Way. Explaining the diverse shapes and forms of galaxies is another challenge to modern astronomy. The various colors shown by the galaxies reveal different types of stars in their different parts. Using these observations along with a little detective work, astronomers can document the history of star formation within a galaxy. One important question we can ask is: Why are some galaxies spiral shaped while others are elliptical or irregularly shaped? We discuss galaxy formation in Chapter 21.

The galaxies we have seen are just a few representatives of the vast population of galaxies whose images can be captured with modern telescopes. Using the world's largest telescopes and modern observing techniques, astronomers estimate that they could detect billions, perhaps trillions, of galaxies.

Some galaxies and galactic-sized objects emit little visible light but huge amounts of radio energy. Observations show them to be in violent states of agitation. Astronomers began to detect large numbers of such objects after World War II, when the first generation of radio telescopes was constructed. **Figure 1-17** shows such a **radio galaxy**, known as M82. It contains large amounts of gas and dust surrounding a nucleus undergoing a burst of star formation. Other radio galaxies may be two objects in collision. Radio galaxies might contain black holes at their centers.

Quasars are enigmatic objects. They are all moving away from us at large fractions of the speed of light and are the most distant objects known to astronomers. Furthermore, they emit one to two orders of magnitude more energy than the entire Milky Way. The most important unanswered question is: What produces all the energy emitted by these quasars and radio galaxies? What produces jets of gas observed in many of them? Why are they moving at large fractions of the speed of light? These and other questions will be considered in Chapter 22.

FIGURE 1–17. M82, a galaxy that has clearly been disturbed in some way.

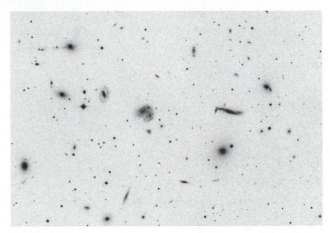

FIGURE 1–18. A cluster of galaxies in the constellation of Hercules.

CLUSTERS OF GALAXIES

As enormous as they are, galaxies are not the largest known entities. In fact, galaxies are generally bound by gravitational forces into **clusters of galaxies** some millions of light-years in diameter. **Figure 1–18** shows a galaxy cluster in the constellation of Hercules. One of the more complex and vexing unanswered questions about galaxy clusters is whether galaxies formed before clustering occurred, or whether matter clustered and then fragmented into galaxies.

While research suggests that all the galaxies in this photograph formed at approximately the same time, we find an amazing diversity of shapes. Because galaxies in clusters, unlike stars in galaxies, are close together relative to their size, we might expect to find examples of galaxies in the act of colliding or otherwise interacting with each other. Such interactions may determine the form of many galaxies.

The Milky Way galaxy is one of about 30 galaxies forming a small cluster of galaxies called the **Local Group**. The Andromeda Galaxy pictured in Figure 1–15 and the Large Magellanic Cloud (Figure 1–16) are members of the Local Group.

In the last 30 years, astronomers have come to appreciate that the clusters of galaxies themselves are collected into even larger clusters of clusters of galaxies, or **superclusters.** These superclusters are the largest objects so far detected in the universe. They are too large to show in a single photograph. Our Local Group of galaxies is affiliated with other relatively nearby galaxy clusters into what we call the **Local Supercluster**.

The most distant objects that have been studied by astronomers are approximately 10 billion ly away, nearly 10 orders of magnitude farther than the nearest star and 19 orders of magnitude larger than the Earth. At a distance of 10 billion ly, a cluster of galaxies is barely resolved into individual objects on a photograph.

From the study of superclusters, astronomers have come to two tremendously important realizations: there are large regions of space called **voids** where fewer galaxies exist, and most of the matter in the universe is invisible. Visible matter is clumped predominantly into stars, which, as we saw, are often organized into star clusters. Star clusters in turn are parts of galaxies, which themselves group into clusters and superclusters. One of the larger goals of modern astronomy is to understand why the universe exhibits this hierarchical structure.

1.4
WHERE DOES ASTRONOMY GO FROM HERE?

The meager amount of light from distant objects that falls on astronomers' telescopes limits our ability to observe the universe. To fully appreciate what the received light tells us, astronomers seek to furnish an explanation for the celestial events and objects based on **astro-** **physics**. This means that astronomers use modern physics to explain how the various forms of radiation received from the cosmos were created. While much of

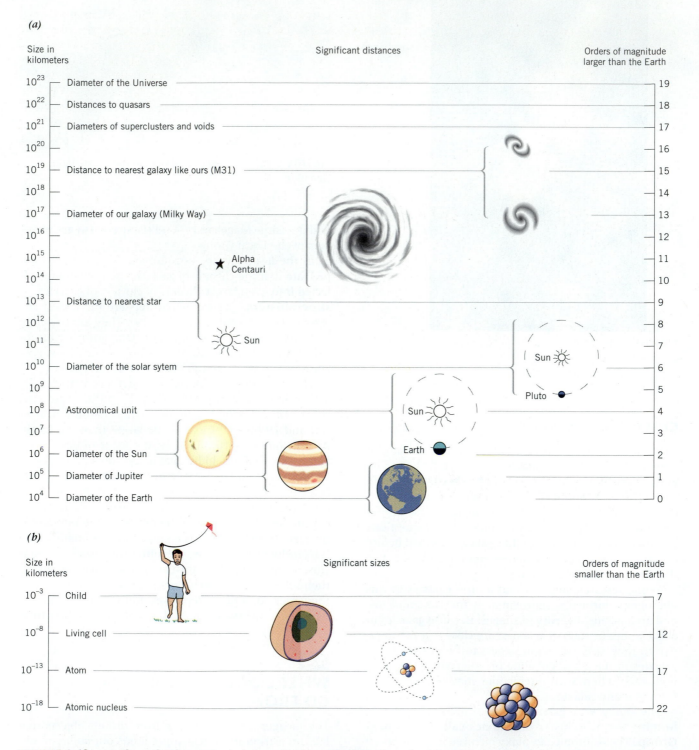

FIGURE 1–19. (*a*) The scale of astronomical sizes. (*b*) The scale of people and microscopic things.

our knowledge of physics comes from Earth-based experiment and observation, all tests done thus far indicate that the laws of physics are valid throughout the universe and throughout all time. Thus an object such as a star can be understood as a collection of an enormous number of atoms bound together by gravity. The radiation given off by the star is actually the total of the radiation emitted by all the individual atoms in the star, so we must descend to the atomic scale to understand the atom's process of giving off radiation.

Smaller than the atomic scale is the nuclear one, where nuclear reactions produce the energy emitted by stars. To understand the first few minutes after the birth of the universe, astronomers also delve into *sub*nuclear physics, the realm of the smallest things in the universe.

Because the material of an atom is concentrated into its nucleus, and most of an atom is relatively empty space, it appears that the hierarchical structure we previously observed continues on down from the largest to the smallest of scales we are able to detect.

All the relative sizes of the objects in the universe that we have been discussing can be put onto a chart that may help in visualizing them. Using orders of magnitude, **Figure 1–19** shows the various dimensions in the universe and how much larger they are than the Earth. Note that each step up the ladder means that the object named is 10 times larger or farther than the previous step; by looking at the chart, you see that the universe is 10,000,000,000,000,000,000 times (19 orders of magnitude) larger than the Earth.

CHAPTER SUMMARY

- The universe contains a variety of objects, including planets, stars, star clusters, galaxies, clusters of galaxies, and clusters of clusters of galaxies (**superclusters**). These objects form a hierarchical structure.
- An **astronomical unit (AU)** is the average distance from the Earth to the Sun. A **light-year (ly)** is the distance light travels in one year.
- An astronomical unit is two orders of magnitude larger than the Sun, which is two orders of magnitude larger than the Earth.
- Stars are born, evolve, and eventually die. In so doing they pass through various stages, beginning

with gas and dust, evolving through middle age like the Sun, and eventually going to their deaths. The final form of a dead star may be a **white dwarf**, **neutron star**, or **black hole**, depending on the amount of matter the star contains.

- An **order of magnitude** is a term used to describe the relative sizes of numbers. Two numbers differ by one order of magnitude for each **factor** of 10 by which one number is larger than the other. The concept is especially useful in astronomy where we deal with a vast range of sizes and distances.

SUMMARY QUESTIONS

1. Briefly describe the appearance of several varieties of galaxies.
2. State the hierarchy of objects in the universe, from the smallest to the largest.

3. Define the astronomical unit (AU) and the light-year (ly) and state their approximate sizes in kilometers. Calculate the size of the light-year in astronomical units.

APPLYING YOUR KNOWLEDGE

Questions marked with ■ require computation.
1. Which photograph in this chapter intrigues or interests you most, and why?
2. From simply looking at the photographs in this chapter, can you say which object is the largest? How do you know?

3. Why do astronomers use astronomical units rather than miles or kilometers?
■ **4.** How many orders of magnitude larger than an atom is a typical person?
■ **5.** How many orders of magnitude are there between each of the following pairs: (a) the diameters of the

Earth and Sun, (b) the diameter of the Earth and an astronomical unit, (c) the length of an astronomical unit and the distance to the nearest star beyond the Sun, (d) the distance to the nearest star beyond the Sun and the diameter of the Milky Way galaxy, (e) the diameter of the Milky Way and the distance to the nearest galaxy like ours, and (f) the size of a nucleus and the diameter of the universe?

■ **6.** Use information in Appendix C to determine the number of orders of magnitude between the diameters of Uranus and Neptune, the Moon and Jupiter, Earth and Jupiter, and Pluto and Saturn.

■ **7.** Write the following numbers in powers of ten notation. (a) 100,000,000 (b) 0.000,000,000,001 (c) 25,000,000,000,000

■ **8.** Use Appendix C at the back of the book as your source of data for placing the planets into groups having diameters of the same order of magnitude. Repeat for the planetary masses.

ANSWERS TO INQUIRIES

1–1. (a) Differ by one order of magnitude
(b) Same order of magnitude
(c) Differ by three orders of magnitude
(d) Same order of magnitude

1–2. Seven orders of magnitude

2

SCIENCE
AND
PSEUDOSCIENCE

Anything's possible, but only a few things actually happen.

RICH ROSEN, *USENET COMPUTER BULLETIN BOARD*, 1985

Some years ago, an astronomer was on a train in Europe. In the dining car, he was seated opposite an elderly gentleman with a long white beard, dressed in a long robe, who looked like one of the patriarchs straight out of the Old Testament.

The elderly gentleman evidently wanted to talk, and so the astronomer just listened, not saying what his profession was. It turned out that the patriarch was the world leader of the Hollow Earth Society, whose adherents assert that Earth is a hollow ball and that we are living on the *inside* of it (**Figure 2–1**). The Hollow Earth movement has many adherents. Our astronomer had happened to meet up with the biggest Hollow Earther of them all!

The patriarch explained to the astronomer why we mistakenly think we are on the outside of the Earth even though we live on the inside; how the Sun, Moon, planets, and stars only *look like* they go around the Earth; and how an optical illusion causes satellites to take pictures of the Earth that make it appear round. Finally, he explained the *direct* evidence that the Earth is hollow. There were only two pieces of evidence of interest to him. The first was a survey made of the Great Lakes around the turn of the century that showed that the surface of the Earth was concave instead of convex. And the second? **Figure 2–2** shows the side view of a shoe. Your own shoe is probably similar. Do you notice how the sole of the shoe curves upward instead of downward? That's because you have been walking around on the *inside* surface of the hollow Earth all these years!

This story may sound made up, but it really happened. The elderly gentleman was a typical example of a pseudoscientist. Although his particular hokum may seem pretty hard to swallow, he had quite a few convinced followers. As psychologist Robert Thouless has pointed out, there is no idea so absurd that you can't find someone who believes it. For this reason, educated people should know what is and what is not science.

Astronomy is a science in every sense of the word. In the first part of this chapter you will learn what science is, using astronomy as an example. In the second part, we further define science by showing what is not science. Finally, we examine various characteristics of what often pretends to be science but isn't.

2.1

AN EXPEDITION TO EARTH (WITH APOLOGIES TO ARTHUR C. CLARKE)

Modern observations of the cosmos started about 100 years ago, when the newly invented photographic emulsion became sensitive enough to be used for celestial observations. Once a photograph has captured a permanent record of an observation, then systematic, repeated studies by more than one scientist can begin. Furthermore, the camera's ability to take a time exposure—something the eye cannot accomplish—enabled astronomers to view much more deeply into space. It can fairly be said, therefore, that we have been seriously observing the universe—whose age is 10 to 20 billion years—for approximately 100 years, only a tiny fraction of its total lifetime. A fairly typical star like the Sun has a life span of approximately 10 billion years, and during

FIGURE 2–1. The Hollow Earth hypothesis. According to this notion, we are living on the inside of a hollow ball.

FIGURE 2–2. The sole of your shoe curves upward. According to its proponents, this is "evidence" for the Hollow Earth theory.

FIGURE 2–3. An interstellar visitor examines data about Earth.

most of this time its characteristics will change little. Even "fast"-evolving stars have lifetimes that run to millions of years. We have recently discovered the existence of explosive, cataclysmic events in stellar evolution that can take place on shorter time scales, but such events are witnessed for only a tiny minority of celestial objects.

To give you a feeling for what science is, and for the difficulties involved in studying objects with extremely long lifetimes, we will discuss an analogous situation—the problems that would confront an alien astronaut who is visiting Earth for only a short period of time, and whose responsibility is to search for life on Earth and try to understand how it has evolved. Suppose we give this visitor a camera and just 15 seconds to take as many photographs as possible (being advanced, the alien's civilization has ultra-high-speed cameras). We choose 15

seconds because it bears approximately the same proportion to the lifetime of a human being as the 100 years we have been observing the universe photographically does to the age of the universe:

$$\frac{15 \text{ seconds}}{\text{human lifetime}} = \frac{100 \text{ years}}{\text{universe's lifetime}}.$$

When the astronaut returns home with the photographs, the scientists at home must try to understand Earth and all its life forms by studying the photos (Figure 2–3). It is almost like studying a still picture, because 15 seconds is not long enough for any serious or important evolutionary changes to show up.

How might the alien scientists determine the dominant form of life on Earth? If size is their main criterion, they might choose whales or elephants for study. If they count sheer numbers, insects might win out. A sophisticated criterion might be to determine the amount of land space controlled by one species, in which case the automobile might be selected as the dominant species, at least in many urbanized areas. Indeed, thinking of ways by which scientists might determine that human beings are the dominant forms of life on our planet is itself a challenging exercise.

Inquiry 2–1 By what criteria might the extraterrestrials decide that humans are living organisms and automobiles are not?

Suppose the alien scientists decide that human beings are worthy of further study. At this point the problem has just begun, because close observation would reveal a considerable diversity of characteristics among

FIGURE 2–4. A visiting alien would find the inhabitants of Earth to show a large variety of characteristics.

TABLE 2–1

Characteristics of People

Size	Color	Sex
Small	Black	Male
Medium	Brown	Female
Large	Yellow	
	White	
	Red	

humans (**Figure 2–4**). The scientists would first have to set up a system for classifying humans on the basis of observable characteristics. (Later on we will see how astronomers have done the same kind of thing with stars and galaxies.) It would be immediately obvious that humans come in a variety of sizes (as do stars). Although the alien scientists would note a fairly continuous distribution of sizes, suppose they decide to classify all humans into the categories small, medium, and large. Suppose they also establish the color categories black, brown, yellow, white, and red. By careful observation and much thought, they might also detect two sexes, call them male and female. But they might initially establish other categories that could prove less fruitful in their analysis of human beings and cause them to waste a great deal of time; categories such as hair length, permanent versus removable teeth, color of clothes, and so on. Each irrelevant category would spawn unproductive hypotheses concerning the evolution of humans, and much thought would have to go into distinguishing important from unimportant characteristics.

Inquiry 2–2 By what criteria might the alien scientists be able to establish that hair length is not relevant, but that sex is, given the fact that in some societies there is a degree of correlation between the two?

Let us consider just the characteristics of size, color, and sex mentioned in the preceding paragraph. The scientists can now begin to construct hypotheses for the evolution of humans, asking questions such as: Do small, brown, female humans evolve into large, red, male humans? Or, instead, do large, black, male humans evolve into medium-sized, yellow, male humans?

Or is there perhaps no evolution at all—the small stay small, and the large stay large? Even in this extremely simple analogy with only three straightforward characteristics to compare (**Table 2–1**), there are $3 \times 5 \times 2 = 30$ different possible combinations, and $30 \times 30 = 900$ possible combinations of starting and ending points for human evolution. Obviously, the universe will present us with many more possible combinations!

Inquiry 2–3 Because the alien astronaut cannot take photographs of every human being, but of only a small number, what fundamental assumption must be made to interpret the data meaningfully?

You can gain some direct personal experience of the dilemma of studying objects with long lifetimes by going outside at night and contemplating the stars for a while. You will undoubtedly be struck by the fact that nothing happens, other than the daily rotation of the sky, which is, of course, just the Earth spinning. Bring back the builder of the pyramids and stargaze with him. He will assure you that all the star patterns still look about the same as they did in his time![1] The naked-eye sky appears eternal and unchanging. For thousands of years, one of humanity's most enduring metaphysical concepts was the idea of the unchanging universe. Only in the twentieth century have we broken through the time barrier and appreciated the evolution of astronomical objects.

To draw one final analogy, astronomers are in the position of being shown just three or four out-of-sequence frames of a movie that runs for about a year and being asked to reproduce the whole plot. Amazingly, to some extent we have been able to do just that.

---

[1]However, he would be surprised to see that the heavens appear to be rotating around a different point in the sky, the star we call Polaris, than the one he remembered. We will return to an explanation of this change in Section 4.6.

ASTRONOMY AS AN OBSERVATIONAL SCIENCE

In this section we will begin to see the ways in which astronomy is a science and how it begins to make advances in understanding the universe. We will also look at how the science of astronomy is fundamentally different from sciences such as physics, chemistry, biology, and geology, to name a few.

THE PROCESS OF DOING ASTRONOMY

Confronted by an almost infinite array of possibilities, an astronomer seems, in many respects, to be the most helpless of scientists. Unlike a physicist or chemist, or even to some extent a biologist, an astronomer cannot perform controlled experiments in which only one thing at a time is allowed to vary, and where results can be verified as often as necessary by repeating the experiment or by doing the same experiment with more ac-

curate equipment when it becomes available. Astronomers cannot even examine their subjects from various angles, as a field worker in archaeology or paleontology can. What an astronomer can and does do is collect the light and other forms of radiation that come from celestial objects and then use all available information, as well as ingenuity, to try to interpret these signals from afar (**Figure 2–5**). An astronomer must become an expert in studying objects from a distance. Indeed, some astronomers feel that once an object has been visited— as the Moon has—it is no longer the subject of astronomical study but should instead be turned over to the geologists and chemists.

A scientist can make observations, which in turn suggest speculations, hypotheses, and perhaps eventually theories. A **hypothesis** is a conjecture that is used as a model for describing the results of observations and experiments. A hypothesis can be wildly speculative, but to be useful, a hypothesis must make predictions about nature that can either be confirmed or refuted by observation (**Figure 2–6a**). For example, I can hypothesize that students completing an introductory astronomy course will know more about the phases of the Moon than students who have not taken such a course. From this hypothesis, I can make predictions that can be confirmed or refuted by giving a test to both students and nonstudents.

In astronomy, predictions coming from a hypothesis usually result in the astronomer's returning to the telescope to determine whether these predictions are supported by further observations. Reasoning from a

FIGURE 2–5. The 2.1-meter telescope at Kitt Peak National Observatory is checked out by astronomers Drs. Catherine Pilachowski and Carol A. Christian in preparation for another night's observing.

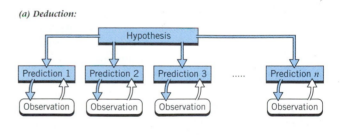

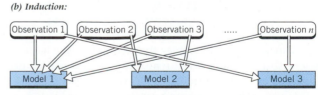

FIGURE 2–6. (*a*) The method of deduction, which begins with a hypothesis and deduces a variety of predictions and observations. (*b*) The method of induction, in which a number of observations produce models that allow explanation of past behavior and the prediction of future results.

hypothesis to a set of particular predictions is called **deductive reasoning**, or simply **deduction**. The new observations may suggest refinements to the hypothesis, which then provides new predictions, and so on.

Another approach, pioneered by Francis Bacon and later by Isaac Newton, begins with a series of observations. These observations are then understood in terms of one or more models. One of the central problems of science is choosing which model is best. As diagramed in Figure 2–6b, model 1 would be considered the best, because it encompasses more of the observations. The fact that observations 2 and 3 do not fit with model 3 serves to falsify model 3. The reasoning involved in this process is **inductive reasoning**, or simply **induction**.

Often the evidence in support of a particular model is indirect, and often the evidence will support more than one model (**Figure 2–6b**). Sometimes a conclusion is obtained by the process of induction, described above. More formally, induction is the process of determining which one of the available hypotheses is most likely to be correct, based on the fact that it does a more satisfactory job of accounting for the available evidence than the competing hypotheses do. Induction is not the same as deduction. As an example, if you go into the kitchen and see the cookie jar broken on the floor, and the cat mewing piteously, you might conclude that the cat pushed it off the shelf. If, on the other hand, there has just been an earthquake, your conclusion might be quite different.

Inquiry 2–4   The following are examples of inductive or deductive reasoning, or situations in which induction or deduction can be used. For each case, which type of reasoning is used? Explain your answer. (a) A detective who finds a blood-stained weapon next to a victim, blood of the same type on the clothes of a suspect, scratches on the suspect's face, and skin under the fingernails of the victim would use what process in concluding who committed the crime? (b) Economist A believes the health of the U. S. economy is dependent only on the money supply allowed by the Federal Reserve. What reasoning process is used in attempting to validate the idea? (c) A social scientist believes that women make better scientists than men. What reasoning process is used to try to validate the idea? (d) A person suggests that UFOs prove the existence of extraterrestrial intelligence. What reasoning process might be used to study this question?

A hypothesis may reach the status of a **theory** once the evidence for its validity becomes strong. Strong evidence is both *repeatable* and *verifiable*. A "good" theory comes from observations and experiments that can be repeated by all scientists in a particular discipline, as well as tested for accuracy. It is not necessarily fixed for all time; new observations may require modification of the theory. Such modification may be made without throwing out the entire theory. The interplay between theory and observation continues indefinitely; in good science, neither can exist without the other. **Figure 2–7** diagrams the process. From a hypothesis or model, one uses deduction to make predictions. Predictions bring about experiments, which produce data. The data, along with inductive reasoning, either verify or falsify the hypothesis. Finally, using creativity, a revised hypothesis or model is made from which the process continues in an endless loop.

But not all theories that agree with the observational data are of equal scientific worth. Some theories are so untestable that they would be in accord with *any* observations that might conceivably be made. An example is the typical newspaper horoscope that is so general it can fit anyone.

Other theories that agree with observations may be unscientific because they are untestable. Consider, for example, a theory that claims that lightning is caused by the god Zeus throwing bolts at the Earth, and that Zeus does this whenever he is angry. There is no evidence that could possibly disprove this idea, even in principle. Supernatural causes are not observable; thus they cannot be tested scientifically. A supernaturally based theory cannot ever be proven wrong, no matter what it claims. Indeed, any theory that appeals to supernatural causes cannot be called scientific, because there is no test we can apply that can rule out supernatural causes for observed facts. Science deals with the natural, observed universe. Experience has shown that speculations on the supernatural do not lead to better scientific understanding of nature. This is not to say that super-

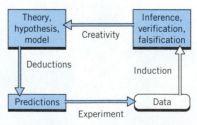

FIGURE 2–7.   Aspects of science fit together in a never-ending cycle of hypothesis, prediction, data gathering, and verification.

naturally based theories are wrong or not of value, only that they are not in the realm of science or scientifically useful. Such theories give us no help in how to proceed further and acquire more understanding.

The scientific use of the word theory is not the same as the everyday, nonscientific use, which is more similar to the word hypothesis—a conjecture. For example, a detective may have a gut feeling (a "theory") about who perpetrated some crime without having any hard supporting evidence in the case. A scientific theory, on the other hand, not only requires detailed supporting evidence, but demands that the theory be able to predict future events.

If two competing theories encompass all observations equally well, which is to be preferred? One answer takes a philosophical view. The philosophy of science tells us that the theory using the fewest arbitrary assumptions is to be preferred. This idea is often called the **Law of Parsimony**, or **Occam's razor**, after the fourteenth-century English philosopher William of Occam. Such a theory is more aesthetically pleasing than one having numerous ad hoc assumptions.

A scientific theory can never be proven to be true. What science can do is to show that observations and experiments are consistent or inconsistent with the theory. Consistency does not prove the theory to be true, but to be reasonable. On the other hand, a lack of consistency can falsify a theory, thereby showing it to be invalid.

THE ASTRONOMER'S CHALLENGE

Astronomers do have some advantages; for example, an absolutely incredible display of phenomena to study, many of which are incapable of being reproduced in any way in Earthly laboratories. We have already pointed out that diffuse gas clouds such as the Orion Nebula (Figure 1–11) consist of gases in such a rarefied (low-density) state that they give off important types of radiation not readily seen on Earth. On the other end of the scale are black holes, objects of such high density that their enormous gravitational fields may prevent light itself from leaving the object.

The range in properties of celestial objects furnishes an almost endless diversity of objects for study. We could not reproduce the energy output of even the most feeble star in a lab on Earth, and the most powerful stars are beacons emitting more energy than a million suns.

The scale of the cosmos is so vast that many things that are unlikely or impossible on Earth become possible. Indeed, in such a vast volume, any behavior that is not specifically *prohibited* by physical law may well be

taking place somewhere, and consequently astronomers can find exotic phenomena simply by searching. The detection of black holes, antimatter, and neutrinos are three examples of this. These objects were predicted by theoretical inquiries, which made people interested enough to go out and look for them.

THE ASTRONOMICAL "TIME MACHINE"

One final advantage astronomers have comes, strangely enough, from the very large distances and time scales that cause them so much difficulty. Because light travels at the finite speed of 300,000 kilometers per second (186,000 miles per second), it takes time for the radiation from celestial objects to reach us. Examination of objects that are a great distance away from us also gives us a certain perspective in time, because the farther an object is from us, the longer it has taken the light from that object to reach us (**Figure 2–8**). Thus we see the Sun now as it was eight minutes ago, the nearest star as it was over four years ago, and relatively nearby galaxies as they were about a million years ago. The most distant galaxies are seen as they were perhaps 10 to 15 billion years ago, an epoch that is not far removed from the earliest moments of the universe itself. It is almost as if we have an astronomical time machine that is capable of transporting our view backward in time to extremely remote epochs and finding out what the universe was like then. For this reason, the analogy with the alien astronaut is not quite perfect, because she has no opportunity to compare different epochs in the situation we put her in.

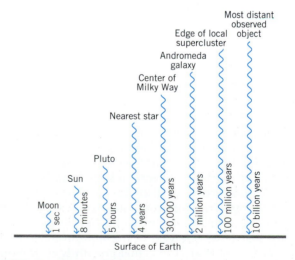

FIGURE 2–8. The length of time that has passed since the light we observe today left the objects shown.

Inquiry 2–5 Making conclusions based on the properties of objects at different distances requires making an important assumption. What might that be?

This ability to look back in time is vital in helping us study galaxies whose time scales for significant change are much longer than the lifetime of an astronomer. To study the evolution of such long-lived objects, astronomers try to use their perspective in time in the following way. If similar galaxies can be observed at different epochs (by observing them at different distances) and still be recognized as being the same type of object, then with patience it may be possible to construct a life history of that particular type of galaxy, because we may be seeing examples of it at different ages and evolutionary stages. Note that this view through time becomes most useful if we push back to the most distant epochs. But it means looking at the most distant objects, which must of necessity also be extremely faint and difficult to observe. So our perspective in astronomical time must be purchased with a great expenditure of human time and effort.

Inquiry 2–6 What would you conclude about galaxies if you were told there was evidence that the most distant galaxies show explosions taking place in their cores, whereas the nearby galaxies do not?

2.3
THE EXPLOSION OF KNOWLEDGE IN THE TWENTIETH CENTURY

Over the centuries, ideas about the nature of the universe have changed dramatically. The universe was once thought to be small and relatively uncomplicated, extending over distances not much larger than a tribe could cover by foot. However, the rate at which astronomical knowledge has been acquired in recent years has increased dramatically, and most of what we now know about the universe has been learned in the past half-century. Such an increase in knowledge is a characteristic of an active, vital science. One possible measure of the growth in astronomical knowledge is the amount of research results published in one year. **Figure 2–9** shows one year of the *Astrophysical Journal* in 1900 and in 1991. The actual growth is even more dramatic than shown because many new journals have come into existence throughout the world in that time span.

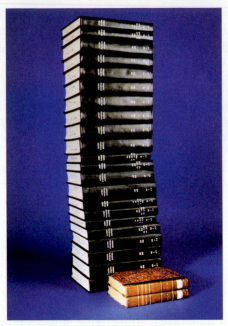

FIGURE 2–9. A complete year of the *Astrophysical Journal* in 1900 and in 1991.

One way to illustrate this explosion of knowledge is to compare the size of the universe as it was estimated in 1920 with today's ideas. In 1920 the universe was believed to be only about 10,000 light-years across, with the Sun located roughly in the center. In many ways, this number is respectably large—it would take an astronaut traveling at 40,000 kilometers per hour (25,000 miles per hour, which is the speed required to escape from the Earth) more than 25,000 years to cover a single light-year, and 200 million years to cross a distance of 10,000 light-years. Yet it turned out to be a gross underestimate of the actual size of the universe. We now believe the universe to be not 10,000 ly across but over 10 billion ly (roughly 100,000,000,000,000,000,000,000 km) containing literally billions of galaxies, each in turn composed of billions of stars.

Inquiry 2–7 Express the estimated size of the universe in kilometers in scientific notation.

These large numbers are intended not to intimidate but merely to illustrate the rapid changes that have taken place in our awareness of the scale of our cosmos. You can better understand these large numbers if you keep track of *relative* sizes and distances using the idea of order of magnitude.

If astronomers were so wrong about the scale of the universe only 70 years ago, what confidence can anyone have that the description of the structure and evolution of the universe as it is given in this book is any more correct? Ultimately, you will have to answer this question for yourself. Our goal is to provide you with enough understanding of basic astronomical ideas and methods so you will be able to appreciate not only the picture of the universe described here but also new discoveries that will be made in the future. In a field expanding as rapidly as astronomy, mere memorization of facts and theories current at one moment is not sufficient, because some of these facts and many theories will soon become obsolete. Instead, you must understand the methods by which astronomers obtain observational data and the underlying physical principles we use to interpret the observations.

It is also necessary to see how observations and theory are combined into a **model** that attempts to describe accurately some aspect of reality. We like to believe that the universe described by current astronomical theory is essentially correct in its broad outlines—for example, that the size and age of the universe are approximately known, and that most of the major types of objects in the universe are at least roughly understood. At the same time, there are some notable exceptions that will be pointed out, and there are doubtless many surprises in store that will significantly alter our understanding of aspects of the universe previously thought to be well understood. We must always keep in mind that the history of science shows that all scientific theories are ultimately refined, and perhaps overthrown in favor of better ones.

Carl Sagan has argued that we are now living in the most intellectually exciting era in the history of the Earth, either past or future, because we are in the first stages of beginning to leave Earth and explore the stars. In a real sense, we are the first people on Earth with the potential for developing a true cosmic consciousness. If not abused, perhaps our heightened awareness will also increase our likelihood of surviving the growing perils of this exciting era in which we live.

THE "GAME" OF SCIENCE

Science may be thought of as an intellectual game, in the sense that it has certain rules that need to be followed. There is nothing sacred about these rules, but they have evolved over time as scientists have learned by trial and error how to develop a procedure that generates new and useful knowledge as efficiently as possible. Those who play by rules different from the rules of science may be engaged in worthwhile activities, perhaps even activities that are more valuable than science,

but they are *not* doing science. The term for those who claim to be doing science when, in fact, they are not playing by the rules of science is *pseudoscientist*, and what pseudoscientists produce is called **pseudoscience**. The prefix "pseudo" literally means "that which deceptively resembles or appears to be something else." The terms pseudoscience and pseudoscientist have something of a pejorative ring, but we know of no neutral term, so in the remainder of this chapter we shall speak of pseudoscience and pseudoscientists despite their negative connotations.

Like all scientists, astronomers are contacted frequently by pseudoscientists who want to explain their latest ideas. Astronomy seems to be one of their favorite subjects. The list of pseudoscientific ideas relevant to astronomy includes astrology, flat and hollow Earthism, geocentrism, creationism, theories about ancient astronauts, and belief in UFOs. Although some of these ideas have small followings (such as flat Earthism), others (like astrology) appear to influence large numbers of people. But the differences between flat Earthism and astrology are differences of degree, not of kind. The problem for most people is in being able to tell whether someone claiming to be a scientist is talking sense or nonsense. After all, both scientists and pseudoscientists use fancy, incomprehensible jargon; how is it possible to distinguish between them?

Despite the fact that most people are not trained as scientists, it is still possible to distinguish between genuine science and pseudoscience. Let us remind ourselves of some characteristics of good science. First, science is a never-ending quest for new knowledge. Second, the results of science must be both repeatable and verifiable. Third, the rules of science dictate that a model or hypothesis should make predictions, and that these predictions are then tested by new observations. The inability of astrology to provide meaningful predictions is illustrated in **Figure 2–10**, which shows the lack of agreement in the predictions made by four astrologers from the same information on human characteristics.

When people begin to find evidence that contradicts established ideas, they may be skeptical at first, particularly if the new evidence conflicts with well-entrenched ideas. Eventually, however, if the evidence is strong enough, and if enough of it is found, the old ideas inevitably give way to the new. It is this tentative but evolving nature that characterizes science—the certainty that at any given moment in history we do not know everything, and the belief that even well-established ideas may be wrong and therefore should be subjected to constant and rigorous verification by experiment and observation. You will see numerous instances of this throughout the book. In contrast, pseu-

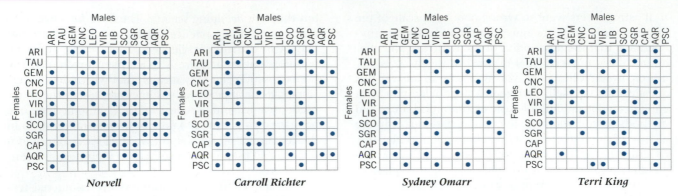

FIGURE 2–10. Sun-sign incompatibilities in marriage as published by four astrologers. If astrology were a scientific model with verifiable and repeatable results, the results of the four astrologers should show a high degree of correlation. However, the total lack of agreement shows astrology's inability to make verifiable predictions. (From Culver and Ianna, *The Gemini Syndrome*, p. 132)

doscientists typically do not volunteer predictions that could endanger their ideas. The predictions they make tend to be ex post facto; that is, made after something is known. Only then can a pseudoscientist "explain" it.

Science is always an unfinished business. There always remain new and undreamed-of discoveries, and there are always opportunies to overturn old and widely believed dogmas. At any moment a certain number of things are unexplained. This does not necessarily mean that current theories are wrong, because there can be many reasons why something is not known (for example, the difficulty of making detailed calculations, or a lack of sufficient data). And, of course, there is always the possibility of inadequate theories. Therefore, although scientists do work within a framework, or **paradigm**, of established theory, they generally strive to maintain a healthily skeptical attitude toward accepted ideas.

A common view among nonscientists (shared by some scientists as well) is that science is engaged in finding the Truth. This is incorrect. The goals of science are really much more limited. To give a simple example, physicists frequently talk about electrons as if they were real physical objects, but in fact no one has ever seen an electron, and our best current theories suggest that we never will. If they exist, electrons are so small that any attempt we make to observe them directly (like trying to "see" them by bouncing light off them) literally knocks them away from wherever they were. The word electron is really just shorthand to describe a set of observations that are related to each other in more or less well-defined ways. These relationships are generally described mathematically, using formulas to predict the outcome of experiments that involve the electron. To the extent that the results of the experiments are pre-

dicted accurately, the physical theory of the electron is a success. We can then say that the atom behaves as if there were things like electrons, but we still do not directly see them. In other words, we create a descriptive model (usually mathematical) and say that something behaves *as if* such-and-such were the case. Do electrons really exist? A good case can be made that they exist only in the minds of scientists. An electron is a concept—a way to model complex observational phenomena.

Many things loosely described as scientific facts are not really facts at all. For example, you might have the impression that this book stated the "fact" that the universe is between 10 and 20 billion years old. But such usage of the word fact is a habit of speech that is seen on close examination to be imprecise. In reality, the age astronomers assign to the universe is an **inference** made from the large amount of observational data that we have, as interpreted with our best understanding of physics and in accordance with those theories that have best stood the test of time and evidence. At present, all the credible evidence now points to an age of 10 to 15 billion years. The creationist's claim that the universe is only 10,000 years old is at present not scientific, *regardless of whether it is true or false*. Most scientific results are inferences.

FROM IDEA TO TEXTBOOK— HOW SCIENCE PROCEEDS

How does science advance from a rough idea to the point where information appears in a textbook such as this? **Figure 2–11** shows the system of filters built into the scientific enterprise that, we hope, prevents the distribution of poor information. Filters are never perfect,

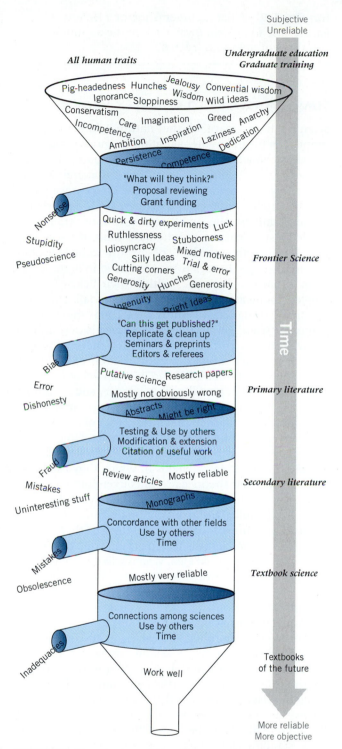

FIGURE 2–11. The knowledge filter, in which deficiencies in science are (ideally) removed as an idea proceeds from formation through publication in textbooks. (Adapted from Henry H. Bauer, *Scientific Literacy and the Myth of the Scientific Method.*)

however, and some results pass further through the system than they should.

Because science is done by people, science contains all the foibles held by those doing the research. The top of Figure 2–11 contains everything that a person contributes to his or her research. In addition to education, we all have human traits that have a bearing on everything we do. Faulty work might be halted when a researcher asks a question like, "What will people think of my idea?" The question could, but probably rarely does, stop an idea from going any further. Because much of modern science requires at least a minimal amount of funds, ideas that are nonsensical might be caught in the process of proposal writing. (A lack of funds, however, often prevents even worthwhile research projects from being done.)

Once the research is done the time to attempt publication arrives. The work is submitted to a scientific journal, which then sends the paper out for review by peer scientists chosen by the editor of the journal. The rate at which papers are rejected depends on both the particular scientific field and the prestige of the journal to which the paper was submitted. Once published, the work becomes part of the primary literature of the field.

The work is now part of the public domain and may be tested by other researchers. If the research can be neither replicated nor validated, or if previously uncaught errors are found, the work will be rejected by the scientific community through the publication of other papers by other scientists. If the work is validated, other researchers may begin to use the results. Furthermore, the results may begin to appear in articles reviewing the status of the field in general. These articles form a field's secondary literature. After getting this far through the system of filters, some of the research may then begin to appear in textbooks, either for the training of future scientists or for the liberal arts education of the majority of our population.

2.4

HOW TO TELL THE SHEEP FROM THE GOATS

How is it possible to tell good science from pseudoscience? While the answer is not always easy, there are a number of characteristics of pseudoscience which, if present in a piece of work, can help you decide.

IS THE HYPOTHESIS AT RISK?

The first test that can help us tell whether we are dealing with pseudoscience or genuine science is to ask: Is the hypothesis at risk? By this we mean, are there practical experiments or observations that might, at least in principle, show that the hypothesis is wrong? If the hypothesis is not at risk of being disproved, then it is not scientific. But by the same token, every time a hypothesis that is at risk is found to be in agreement with new evidence, we gain new confidence in it.

Consider the earlier example of Zeus and the lightning bolts: if there is lightning, Zeus is angry; if not, he is not angry. The theory that lightning bolts are due to an angry Zeus is never at risk, regardless of the circumstances. On the other hand, the early-twentieth-century theory of the observable universe that put the Sun at the center, and which seemed to be so well confirmed by the evidence, fell to pieces when the distribution of globular clusters was studied in 1917. Even the present-day picture of the galaxy we will give you in this book is at risk; our current theories may fall victim at any time to new evidence or better ideas. Pseudosciences, however, do not change with time. They have a fixed body of ideas that rarely changes significantly.

To give an example from an existing pseudoscience, consider the claim of creationists that the universe is only 10,000 or so years old. When astronomers point out that we appear to be receiving light from objects that are millions or billions of light-years away (hence the light has been in transit that long), some creationists reply that all those photons from all the different galaxies were created supernaturally, in motion and at the proper place in space, to arrive at Earth in such a way as to give us that impression. The evidence of Earth's longevity (and of biological evolution) that is contained in the fossil record is dismissed by creationists as having been "put there by the devil to mislead us." We hope it is clear that such responses can never be at risk and are unscientific and unhelpful in acquiring a greater understanding of the world in which we live.

DON'T CONFUSE ME WITH THE FACTS

To pseudoscientists, evidence is interesting only if it tends to support their ideas; indeed, they tend to ignore or even deny the validity of any evidence that does not support their ideas. Consider our friend the Hollow Earther, for example. You could have easily given a dozen or more persuasive pieces of evidence against the Hollow Earth theory, yet the only evidence the patriarch was interested in were an obsolete survey and the shape of your shoes. You can "prove" anything if you ignore all the evidence that disagrees with it.

Inquiry 2–8 What are three arguments that would persuade others that the Hollow Earth theory is wrong?

Inquiry 2–9 What arguments might a Hollow Earther use to explain away each of the arguments you thought of in Inquiry 2–8?

SIMPLE ANSWERS TO COMPLEX PROBLEMS

Scientists have great faith that nature is fundamentally simple. Yet the phenomena that nature presents us with can at first appear to be frustratingly complex. Although everything in the physical world may ultimately be explained by the interactions of a few elementary particles, experience teaches us that the path to knowledge is often a difficult one, and that long, arduous training is needed just to learn to use the basic tools of a scientific field. The growth of the body of scientific knowledge has become so large that scientists are forced to become experts in narrow disciplines within their fields. For example, among astronomers there are specialists in double stars, stellar atmospheres, celestial mechanics, planetary science, interstellar matter, galaxies, and so on. Within each of these specialties there are even smaller specializations. Specialties even exist in the techniques used to observe. No one can become an expert on more than a small part of science, and even then constant study is required just to stay current with the new discoveries made in that little corner of knowledge.

In contrast, pseudoscientific theories tend to be easily understood, even by lay people. Pseudoscientists are usually untrained in the fields in which they claim expertise, and their theories are unconventional in terms of the accepted scientific ideas of the day. In fact, the more they conflict with conventional science, the more the pseudoscientist is convinced of their truth. Not only that, but pseudoscientists usually claim wide applicability for their one, simple idea, which scientists have been too narrow-minded or too stupid to understand. Furthermore, their one idea is supposed to explain a wide diversity of phenomena in a way that anyone can appreciate, once it has been pointed out to them. And, of course, there is no possibility of error. In contrast, although scientists are occasionally arrogant, if they are honest they must admit that even their most important contributions could be overthrown at any time.

Pseudoscientists frequently try to exploit the fact that science has not yet explained everything satisfactorily, but there will always be many things that an evolving understanding will not have explained yet. The problem is that pseudoscientists uncritically give all discordant facts equal weight. They will argue that any discrepancy unexplained by present science is evidence that the whole of modern science is wrong, and that their particular brand of pseudoscience is therefore right. This is simply a fallacy. It is illogical to argue that any evidence against some standard scientific view is evidence in favor of a pseudoscientific view; this ignores the fact that there may be many other much more satisfactory explanations. Science thrives on controversy and disagreement!

PLAYING THE UNDERDOG

It serves a psychological purpose for pseudoscientists to portray themselves publicly as heroic Davids fighting the Goliath of the scientific establishment. They often claim that scientists are prejudiced against their ideas and won't give them a fair hearing. They frequently accuse scientists of being closed-minded because scientists don't unquestioningly accept their claims. The reason scientists often reject pseudoscience is that they understand the issues very well and are in the habit of thinking critically about evidence.

CONSPIRACY THEORIES

A favorite myth among pseudoscientists is that there is a conspiracy of scientists cooperating to lock out new ideas. Perhaps the best example of this is the oft-repeated claim that we have been visited by Little Green Men in Flying Saucers, but that the federal government is keeping them all locked up on a military base so we won't find out! Typically, no rationale is given as to why the government might do such a thing, not to mention how we might be able to lock up someone so technologically superior that they are capable of interstellar travel while we are not. What reward could possibly induce scientists to hide knowledge like this? Any scientist who was the first person to prove that he or she had talked to an extraterrestrial would, for a while at least, become the most famous person in the world. The idea of a conspiracy to hide knowledge neglects the fundamental competitiveness of scientists; they are paid to try and shoot down each other's ideas. If you are a scientist the quickest road to fame and fortune is to rock the established boat, but you have to have the necessary evidence to convince a large body of aggressive and skeptical competitors.

PLAYING ON FEAR AND EMOTION

It is unfortunately true that many people fear and distrust science and scientists, and pseudoscientists typically try to exploit these fears by playing on people's emotions and inducing them to agree with their ideas. Emotional appeals that obscure areas of rational debate provide the reason why scientists usually try to avoid pseudoscientists as much as possible. Yet scientists ought to speak out against pseudoscience. Probably the best way to do this is indirectly, by devoting more time to helping the public understand the power, fascination, and beauty of science. While most people are uncomfortable with uncertainty and yearn to understand themselves and the world around them, science and scientists frequently seem to be incomprehensible, forbidding, and unapproachable. Yet people's desire to understand is so strong that they may turn to the simplistic ideas put forth by various pseudoscientists appearing to have all the answers. People will even ignore the fact that the pseudoscientist profits mightily from their faith in him or her.

DO THEY DO RESEARCH?

Pseudoscientists rarely perform experiments or make observations, as real scientists do. At the 1982 trial of the Arkansas legislation that mandated the teaching of creationism if evolution is taught, leading creationists admitted under oath that they did not do research and did not consider it important. On those rare occasions that they do make observations, pseudoscientists do it primarily for the purpose of finding more support for their ideas, and not with the expectation, much less the hope, that the idea might be shown to be wrong. Often the purpose of the experiments is to garner evidence that can be used for public propaganda. Genuine scientists do not do experiments to affect public opinion; on the contrary, they are usually all too oblivious to public opinion.

Much of the research that pseudoscientists do is library research, trying to dig up more data to support their case. The books they write can be very impressive and full of quotations, although often when the references are checked it is found that the quotations are taken out of context or are otherwise distorted or misinterpreted to make them mean something other than what the original author intended. This is surprisingly easy to do. For example, it is common and fashionable for creationists to quote Darwin out of context to make it seem as if he considered evolution impossible. It can be hard to distinguish such writings from genuine scholarship when the other side of the picture is not presented for comparison.

Inquiry 2–10 Can you find a statement in this chapter that, if quoted out of context, would support a conclusion opposite to that intended by the authors?

Inquiry 2–11 There is only one way to be sure that a quotation has not been made out of context. What is it?

FOR WHOM DO THEY WRITE?

Scientists are professionally interested in advancing scientific knowledge. Although some scientists occasionally write books like this one that are intended for the general reader, scientists generally do not gain professional recognition for their popular writing. Their reputations are made or unmade on the basis of the research that they publish in professional journals especially dedicated to this purpose. Before an article is printed in such a journal, it is subjected to extensive critical reviews from other scientists working in that field. Frequently, the editor of the journal sends the paper to a reviewer who is known to oppose the ideas being put forth, just to increase the hurdles it must overcome.

The articles that scientists submit to such journals are usually highly technical and difficult even for scientists not working in that field to understand, much less a lay person. Indeed, in astronomy, someone working in a narrow field such as interstellar matter, for example, might well have difficulty fully comprehending an article on cosmology. For lay people, the problem is usually the extensive use of mathematics.

Pseudoscientists seldom publish their ideas in the professional journals. Indeed, they seldom even submit their articles to such journals, arguing that prejudiced scientists would reject their ideas anyway. But they hardly ever put this hypothesis to the test, because their real purpose is not the advancement of scientific knowledge but rather gaining converts to their beliefs. Therefore, pseudoscientists normally write books and articles for popular consumption. The publisher is often the author, who is unable to find a reputable publisher of scientific materials to publish his or her ideas. They are very good propagandists, and their books are easy to read and understand. Superficially, their arguments may appear to be convincing, and it may be difficult to sort out truth from fiction. But the best weapon is a skeptical and critical attitude. Bear in mind that authors of pseudoscience profit enormously from public belief, so be careful before you part with your money. Most importantly, always try to read both sides of an issue.

2.5

DO NEW IDEAS THROW OUT THE OLD?

Scientists have good reason to be extremely wary of theories that claim to correct the woes of science, because truly revolutionary ideas in science are few and far between, and even they build on what was known before. Such ideas are revolutionary in enabling us to understand new and unsuspected phenomena, or to understand familiar ones better, but not in contradicting what is already well established.

Newton's theories, for example, completely changed the way people looked at nature, yet Newton himself acknowledged the debt he owed to those who went before him. "If I have seen farther," he wrote, "it is by standing on the shoulders of giants." Newton's theories did not contradict those of Copernicus, Galileo, and Kepler, but perfected them, albeit in a way that they could not have anticipated. After Newton, the planets still moved around the Sun, in orbits that were nearly circles, just as Copernicus and Kepler had said. But observations of bodies such as the satellites of Jupiter, discovered by Galileo, could now be better understood.

In our own century the two great revolutionary ideas in physics have been Einstein's theory of relativity and the theory of quantum mechanics. Yet even though they fundamentally changed the way we look at nature, neither altered by much our understanding of the many areas of knowledge where the earlier concepts of Newton were applicable. To this day astronomers use Newton's equations, with slight modifications at most, to calculate where the planets are. To give another example, although the sun-centered model of the universe was ultimately overturned, the actual *data* were still valid and could be understood in the light of previously known physics once the presence of large amounts of obscuring dust in the galaxy was recognized. In the same way, it is probable that our present-day ideas about the galaxy need revision, but any such revisions are almost certain to fit in with most of our facts and much of our theory, at least to some extent.

CONCLUDING THOUGHTS

We readily admit that science is sometimes wrong, and that some scientists can be both pig-headed and arrogant. This book gives a number of examples from the history of astronomy; many others could be cited. Frequently a mistake by science results when an important part of the scientific process is omitted. For example, it is critically important in science for investigators to publish their studies for all the world to see. This exposes any piece of research to the largest possible critical audience. Secret or classified research done by a small group is much more liable to wander into error, partly because it lacks that exhaustive scrutiny.

An example of the problems that can result from secrecy in research is the recent controversy over cold fusion. As you will learn in later chapters when we study the evolution of stars, there are reasons why scientists believe that fusion will occur only at extraordinarily high temperatures. Were it possible to produce controlled fusion at room temperatures, our energy problems would vanish. The idea of cold fusion is therefore an idea of practical importance, even if fanciful. In March 1989, the chemists B. Stanley Pons and Martin Fleischmann reported in a news conference that they had successfully produced fusion at room temperature. Skeptical scientists the world over attempted to duplicate their work based on the sketchy reports provided. Unfortunately, Pons and Fleischmann did not publish the details of their techniques and their work could not be verified. Researchers were not allowed into their laboratory because they had placed a cloak of secrecy over their work. Pons and Fleischmann had been well-known and respected chemists. However, they failed to play the game of science according to its rules; they tried to go around the knowledge filter shown in Figure 2–11. While the controversy died down, it is not over. From this example we see that even the best of minds will make mistakes, and the best protection against mistakes is to have as many minds as possible thinking about something.

At the end of this chapter we suggest a number of books and articles on both sides of several pseudo-scientific issues related to astronomy. We hope that you will make an opportunity to read some of them.

CHAPTER SUMMARY

- A **hypothesis** is a reasonable supposition made in describing the results of experiments and observations. It attempts to make predictions of future behavior. Once the evidence for its validity is strong, it becomes a **theory**; such validity means that many scientists have repeated and verified (confirmed) the experiments/observations. No theory can be proven to be true. However, data can prove a theory to be false.

- Science progresses through the interaction of observation and experiment with theory. While theory makes predictions, observation and experiment place constraints of reality on theory and thus provide a means of determining the best description of nature.

- In **deductive reasoning** scientists start with a hypothesis from which specific predictions are made.

- **Inductive reasoning** selects as the most likely hypothesis the one that is in best agreement with all of the available data.

- An **inference** is a conclusion reached by incorporating many lines of reasoning, including observation, experiment, and theory.

- A **model** is a description of a phenomenon, based on observation, experiment, and theory. It is not necessarily the truth or reality, but a description that allows prediction of future behavior.

- The characteristics of good science include the continuous search for new knowledge, repeatability and verifiability, and the ability to make predictions and test them against observations.

- Pseudoscience can generally be distinguished from true science by determining whether the ideas espoused can ever be at risk, whether they are open to modification and evolution into new ideas, whether they claim to supply simple answers to complex questions, whether practitioners play the role of an underdog, and whether they do much research and publish it where trained scientists can examine the purported results.

SUMMARY QUESTIONS

1. Distinguish between a theory and a hypothesis.
2. Distinguish between a fact and an inference.
3. Describe what scientists mean by a model and its relationship to "truth."
4. Explain the difference between an observational and an experimental science. List and explain the particular difficulties that face astronomy and describe several advantages that astronomy has over "earthbound" sciences.
5. Describe why the great distances in astronomy are both a disadvantage and an advantage to the study of astronomy.

6. For the hypothetical "expedition to Earth" of this chapter, formulate and evaluate hypotheses that might be made by the alien scientists about life on Earth. Specify the kinds of observations that might be made when attempting to validate the hypotheses.
7. Explain how the position of the alien scientists is similar to that of present-day astronomers (and also how it is different).
8. Compare and contrast the characteristics of pseudoscience presented in the chapter with those of science.
9. Explain how theory and observation interact in the development of scientific ideas.

APPLYING YOUR KNOWLEDGE

1. State and analyze some of the conclusions an alien scientist might make about people on Earth through an analysis of their eating habits.

2. About which of the following sources of information might you tend to be more skeptical on questions of science? (a) *Time* or *Newsweek* magazine (b) *The National Inquirer* (c) *Scientific American* (d) *The New York Times*

3. Which of the following publishing houses might you expect to be the least reliable on questions of science? (a) John Wiley & Sons (b) Bantam Press (c) Creation-Life Publishers (d) Oxford University Press

4. Which of the following true statements are facts, and which are inferred results? (a) The average density of material in the Earth's core is 11 times that of water. (b) Jupiter's average diameter is 11 times that of Earth. (c) All stars like the Sun get their energy from conversion of hydrogen to helium.

5. Critically analyze the following statement: "The first generation of human beings ever to have grown up educated about the true nature of the universe is still alive."

6. Discuss why science is not degraded when talking about it in terms of the "game of science," as done in the chapter.

ANSWERS TO INQUIRIES

2–1. One could note that humans are self-propelled, while cars need a human before they can move; or that human characteristics show a smooth continuum of properties such as size (suggesting growth), while cars are restricted to a small and fixed range of properties. One would have to use such observations to infer that human characteristics could be explained by an evolutionary, biological-type development of growth.

2–2. They would have to examine carefully examples of extreme cases such as women without hair and men with long hair to see that it was a superficial property. If one of the photographs showed the inside of a barber shop, it would probably supply sufficient information.

2–3. It must be assumed that the data gathered fairly represent the entire group and are not biased.

2–4. (a) induction (b) deduction (c) deduction (d) deduction

2–5. We assume that intrinsic properties do not depend on distance.

2–6. The more distant galaxies are seen as they were a long time ago (when they and the universe were younger). One conclusion might be that when galaxies are younger, their cores are more likely to explode than when they are older.

2–7. 10^{23}

2–8. Some arguments might be: (a) if stars are on the inside of a sphere, they cannot be very far away; (b) if you look up, you should see the other side of Earth; (c) we see the Sun rise and set every 24 hours; (d) we can travel all the way around the Earth and measure its size.

2–9. Responses might include: (a) you cannot prove that stars are far away; (b) the great thickness of the atmosphere scatters light and makes it impossible to see the other land on the other side; (c) the rising and setting is caused by a body that revolves about the Sun every 24 hours, thus producing night.

2–10. One example might be "Earth is a hollow ball and we are living on the *inside* of it." The complete statement is in the second paragraph of the chapter opening vignette.

2–11. Go to the original source and see the context in which it was made.

FURTHER READINGS

GENERAL

The Fringes of Reason, Ted Schultz (Ed.). Harmony Books (1989). Short descriptions of many unusual beliefs, and resources to find out more about them.

The Art of Deception, by Nicholas Capaldi. Prometheus Press, 1971. Pseudoscientists frequently make use of dishonest rhetorical tricks to convince their audience. This is an excellent book to alert you to these tactics.

Betrayers of the Truth: Fraud and Deceit in the Halls of Science, by William Broad and Nicholas Wade. Simon and Schuster, 1982. Scientists are humans just like everyone else, and sometimes they are led to do things that are dishonest and contrary to the spirit of science. This book recounts a number of examples and helps to put the scientific enterprise in a more realistic perspective.

Science—Good, Bad and Bogus, by Martin Gardner. Prometheus, 1981. The one to read if you can only read one.

Myths of the Space Age, by Daniel Cohen. Dodd Mead, 1965. Another excellent summary of the arguments against many astronomically related pseudosciences.

Flim-Flam, by James Randi. Lippincott and Crowell, 1980. Has sections on UFOs and ancient astronauts.

Paranormal Borderlands of Science, Kendrick Frazier (Ed.). Prometheus, 1981.

Science and the Paranormal, by G. Abell and B. Singer (Eds.). Scribners, 1981.

Science and Unreason, by D. and M. Radner. Wadsworth, 1982.

ANCIENT ASTRONAUTS

Pro: *Chariots of the Gods?* by Erich von Daniken. Bantam, 1970. This is the book that started it all.

Con: *The Space Gods Revealed*, by Ronald Story. Harper and Row, 1976. Answers the claims made by von Daniken.

Con: *Ancient Astronauts, Cosmic Collisions, and Other Popular Theories about Man's Past*, by William Stiebing. Prometheus, 1984.

VELIKOVSKYISM

Pro: *Worlds in Collision*, by Immanuel Velikovsky. Doubleday, 1950. The classic example of pseudoscience.

Con: *Beyond Velikovsky: The History of a Public Controversy*, by Henry H. Bauer. University of Illinois Press, 1984. The definitive work on Velikovsky. Contains several excellent chapters on pseudoscience in general.

Con: *Scientists Confront Velikovsky*, by Donald Goldsmith (Ed.). Cornell University Press, 1977. A collection of articles, mostly presented at a debate held before the American Association for the Advancement of Science in 1976, between Velikovsky and his critics. The article by Asimov is especially recommended.

ASTROLOGY

Pro: Virtually any bookstore has dozens to hundreds of books on astrology, often in the "occult" section, so specific recommendations seem unnecessary.

Con: *The Gemini Syndrome: Star Wars of the Oldest Kind*, by R. B. Culver and P. A. Ianna. Pachart, 1979. An exellent introduction to why astronomers discount astrology.

Con: *Arachne Rising*, by James Vogt. Granada, London, 1977. An apparent piece of pseudoscience arguing for a 13th constellation of the zodiac that turns out to be a devastating parody of pseudoscience.

Con: "The Scientific Case against Astrology," by Ivan Kelly. *Mercury*, January/February 1981, p. 13, and November/December 1980, p. 135.

CREATIONISM

History: The Creationists by Ronald L. Humbers. Alfred A. Knopf, 1992. History of the modern Creationist movement by a distinguished historian.

Pro: *Scientific Creationism*, by Henry Morris. Creation-Life Publishers, 1974. The definitive statement of what young-Earth Creationists believe.

Pro: *What Is Creation Science?* by Henry Morris and Gary Parker. Creation-Life Publishers, 1982. A more popularized account of Creationism.

Con: "Astronomy and Creationism," by David Morrison. The claims of Creationists are totally at variance with what modern astronomy has discovered. *Mercury*, September/October 1982, p. 144.

Con: "The Science-Textbook Controversies," by Dorothy Nelkin. The "equal time" demands by Creationists are investigated. *Scientific American*, April 1976, p. 33.

Con: *Science on Trial: The Case for Evolution*, by Douglas J. Futuyma. Pantheon Books, 1983. Exposes the fallacies in Creationist arguments. Has an exellent discussion of evolution.

Con: *Scientists Confront Creationism*, by Laurie R. Godfrey (Ed.). Norton, 1983. A collection of articles by scientists in different disciplines that demonstrate the flaws in a number of Creationist arguments. Extensive bibliographies.

Con: *The Meaning of Creation: Genesis and Modern Science*, by Conrad Hyers. John Knox Press, 1984. Written from the perspective of a theologian, this discusses the thesis that Creationism is not only bad science but also flawed theologically.

Con: *Christianity and the Age of the Earth*, by Davis A. Young. Zondervan, 1982. Written by an Evangelical Christian who is also a geologist, this exposes the flaws in Creationist estimates of the age of the Earth.

Con: *Science and Creationism*, by Ashley Montague (Ed.). Oxford University Press, 1984. A valuable collection of essays on the evolution/creation controversy.

No discussion of the evolution/creation controversy would be complete without mentioning Darwin's Origin of Species. There is a continuing series of books by Stephen Jay Gould, including *The Panda's Thumb* and *The Mismeasure of Man*. Like Darwin, Gould is often quoted out of context by Creationists.

UFOs

Again countless books exist at most bookstores to represent the "pro." The most cogent would probably be anything by J. Allen Hynek.

Con: *The Scientific Study of Unidentified Flying Objects*, by Daniel Gillmore (Ed.). Bantam, 1969. The classic Condon Report issued at the termination of the military's extensive review of UFO evidence.

Con: *UFO's Explained* (Random House, 1974) and *UFO's: The Public Deceived* (Prometheus, 1983), both by Philip J. Klass. Klass has for years been the most informed and most aggressive debunker of "flying saucer" theories.

Con: *The World of Flying Saucers*, by Donald Menzel and Lyle Boyd. Doubleday, 1963. A perspective from the point of view of a professional astronomer.

Con: *The UFO Verdict: Examining the Evidence*, by Robert Sheaffer. Prometheus, 1981.

3

ASTRONOMICAL OBSERVATIONS: ANGLES AND ERRORS

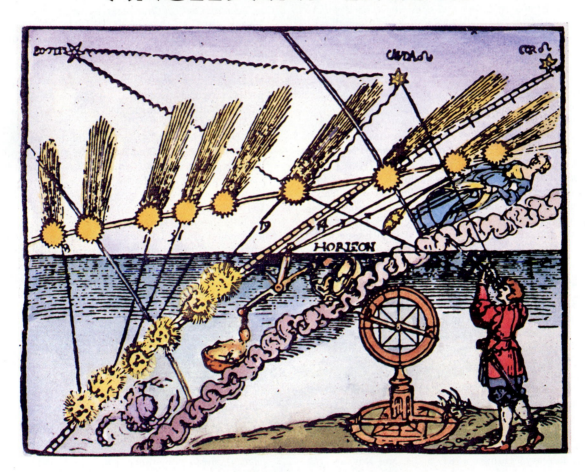

The comet was so horrible and frightful . . . that some . . . [people] died of fear and others fell sick. It appeared as a star of excessive length and the color of blood; at its summit was seen the figure of a bent arm holding a great sword in its hand, as if about to strike. On both sides . . . were seen a great number of axes, knives and spaces colored with blood, among which were a great number of hideous human faces with beards and bristling hair.

AMBROSE PARE, PHYSICIAN, 1528

The chapter opening quotation is a dramatic example of what we would describe today as unscientific thinking. The rapid progress made in the physical sciences over the last few hundred years stems largely from the evolution of a method of investigation in which the systematic and objective measurement of the phenomena of nature is the ultimate arbiter of the validity of our ideas. Speculations in the absence of careful observation can suggest an infinite number of possibilities, with no indication of which one, in fact, has been chosen by nature. On the other hand, data gathered in a random manner without the guiding inspiration of theory may well remain puzzling and may indeed be useless. It is only from the interplay between theory and careful measurement that new knowledge is attained.

Beginning with this chapter, this book contains a variety of Discoveries at the ends of chapters and Kit Activities in a separate kit available at cost from the publisher. Working with these Discoveries and Activities, you can become actively involved in learning astronomy and have the opportunity of making simple but meaningful measurements. Many aspects of naked-eye astronomy—for example, the relative positions of the Sun, Moon, nearby planets, and stars in the sky; and their daily and yearly motions—can be understood by direct observation. Observation also gives a deeper understanding of both the subject matter and the methods of astronomy. It is the philosophy of this book to make observing experiences available wherever it is possible to do so. While some of the activities require the Activity Kit, others are self-contained and allow you the opportunity to do something with available data.

Whenever you observe, do not look directly at the Sun either with the naked eye or with any optical device even if it feels comfortable to your eye.

3.1
ANGLES AND ANGULAR MEASUREMENT

Early astronomical observations measured angles using simple instruments, such as the astrolabe and quadrant shown in the chapter opening photograph and **Figure 3–1**.

The basic unit of angular measure is the **degree**. There are 360 degrees (written 360°) in a full circle. Since there are four right angles in a circle, one right angle is equal to 90°. **Figure 3–2** shows several different angles. Astronomers must frequently deal with angles considerably smaller than one degree, particularly when telescopic observations are considered. Traditionally, one degree has been subdivided into 60 **minutes of arc** (written 60'), and each minute of arc has been further subdivided into 60 **seconds of arc** (written 60").[1] Do not confuse minutes and seconds of *arc* (measures of angles) with minutes and seconds of *time*, which are also based on the sexagesimal system.

Angles can also be expressed as degrees and decimal fractions of a degree, and that is the practice we will follow in this book. For example, instead of 3°12', we will write 3.2°, because 12' is equal to 12/60 or 2/10 of a degree. The use of degrees and decimal fractions of a degree is also more compatible with modern pocket calculators.[2]

---

[1] The traditional method of designating angles in degrees, minutes, and seconds of arc can be traced back to the Mesopotamians, who used the so-called *sexagesimal* notation based on units of 60, rather than our familiar decimal notation based on units of 10.

[2] There is, of course, nothing sacred about decimals, since they probably came about simply from a habit of counting on the fingers. There is a great deal of linguistic evidence for this idea; for example, many primitive tongues use the same words for fingers of the hand and the numbers one through five. The Mayans counted both on fingers and toes and had a number system based on units of 20. Modern computers are most easily designed to work on the binary system, where numbers are expressed in units of 2.

FIGURE 3–3. The northern sky and horizon, showing the Big Dipper, Cassiopeia, and the North Star, or Polaris.

FIGURE 3–1. An old quadrant, used to measure altitudes of stars and planets.

FIGURE 3–2. The sizes of various angles.

3.2
ANGLES ON THE CELESTIAL SPHERE

Models of the universe invented by the earliest civilizations visualized the sky and the bright spots visible on it as something like a large bowl surrounding the Earth,

but at a considerable distance from it, with the bright lights of the heavenly bodies attached to the bowl. This "bowl" is known as the **celestial sphere**. It is not obvious from naked-eye observations of the sky how far away the stars are, but we can certainly map out the relative positions of various stars and constellations by specifying the angles separating them as they appear on the celestial sphere.

For example, most people in the Northern Hemisphere have at some time noticed the grouping of stars called the Big Dipper (in England, the Plough). **Figure 3–3** shows the Big Dipper's stars, with their Arabic names[3] indicated, as they would appear above the northern horizon just after dark in mid-October. Polaris, which is called the North Star or pole star, lies close to the **north celestial pole**, which is the point on the celestial sphere where the Earth's rotation axis intersects it. The stars Merak and Alkaid are separated by an angle of 24°. This means that the angle between the two lines of sight from the eye to the two stars is 24°, as shown in Figure 3–4.

The stars Merak and Dubhe are 5° apart. These two stars are frequently referred to as the Pointer Stars, because a line drawn through them and extended another 30° will pass near the North Star (Figure 3–4).

Objects such as the Sun and Moon, which show a visible disk as opposed to a point of light, can be characterized by their **angular size**, also called **apparent size**. The angular size of an object is the angle between two lines of sight from the eye, one to each side of the object. The angular size of the Moon is 1/2°, as shown in Figure 3–5.

[3]The naming of stars is discussed in Chapter 4.

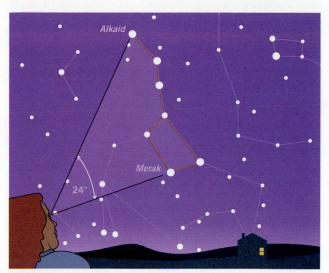

FIGURE 3–4. Angle between Merak and Alkaid, showing a 24° angular separation.

FIGURE 3–5. The angular size of the Moon.

You can measure angles very roughly using the "fist" method; that is, using your hand and fingers held at arm's length as a rough gauge of angles. At arm's length a thumbnail has an angular size of about 1° for most people. The first two knuckles of the fist are about 2° apart, whereas the fist itself spans about 10°. If the hand is spread out as far as it will go, the distance from the tip of the thumb to the end of the little finger is approximately 20° (**Figure 3–6**).

Inquiry 3–1   Assume that a child's hand and arm, being smaller than an adult's, are still in the same proportion to body size. Are the angles given above still correct for the child? Explain.

3.3

THE RELATIONSHIP BETWEEN TRUE AND APPARENT SIZE

Figure 3–7*a* shows an observer measuring the angular size of a person of some height located at some distance. Figure 3–7*b* shows a similar situation with the same person located at a greater distance. You can immediately see that the observed angle decreases as the distance increases. In fact, when the apparent size of the object is expressed in degrees and the true size[4] and distance of the object are expressed in the same units (e.g., meters), then

$$\text{angular size} = 57.3° \times \frac{\text{true size}}{\text{distance}}.$$

This important relationship, which we will refer to as the **angular size formula**, is derived in Appendix B5. The formula gives us the angular size of the object ex-

---

[4]By the term *true size* we mean the actual size measured in, for example, millimeters, centimeters, or kilometers.

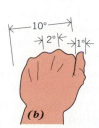

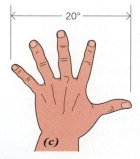

FIGURE 3–6. Using the human hand to measure angles roughly. (*a*) Extend the hand at arm's length and sight along it. (*b*) The thumb subtends about a degree, two knuckles about 2°, and the fist about 10°. (*c*) The hand subtends about 20°.

(a) **(b)**

FIGURE 3–7. The farther the object, the smaller the observed angular size.

pressed in degrees if the true size and distance of the object are expressed in the same units (e.g., meters).

This relationship is *not* valid for large apparent sizes, but for angles less than 20° or so it will give satisfactory results. Small angles occur when the object size is small compared to the distance. Astronomers are mostly concerned with smaller angles, usually a fraction of degree. For this reason, a better unit is the minute or the second of arc. If we express the angle in seconds of arc, the formula becomes angular size = 206265 × true size/distance.

As an example of the use of the angular size formula, let us calculate the angular size of the Moon, given that its true size is 3475 km in diameter and its average distance from Earth is 385,000 km. Substituting into the angular-size formula we find

$$\text{angular size of Moon} = 57.3° \times \frac{3475 \text{ km}}{385,000 \text{ km}}$$
$$= 0.52$$
$$= 31.2 \text{ minutes of arc}$$
$$= 1872 \text{ seconds of arc.}$$

Inquiry 3–2 If the Sun's true size is 1.39 million km in diameter and its distance from Earth is 149.6 million km, what is its angular size (angular diameter) in degrees, minutes of arc, and seconds of arc? How does it compare to the angular diameter of the Moon?

SOLAR ECLIPSES

The equality between the angular sizes of the Sun and Moon means that when the Moon happens to pass directly between the Earth and Sun, a total eclipse of the Sun can take place. At such a time, the Moon moves in front of the bright disk of the Sun (**Figure 3–8**), and the faint, tenuous outer atmosphere of the Sun, called its **corona**, can be seen. The corona is only one-millionth as bright as the Sun's disk and therefore is usually lost in the glare of the Sun. It is best seen at total eclipse time, and for this reason astronomers often go on long and arduous eclipse expeditions to out-of-the-way parts of the globe to study the corona.

FIGURE 3–8. Multiple exposures during a total eclipse of the Sun in July 1991 as observed on Mauna Kea, Hawaii. The solar corona shows itself at mid-totality.

Viewed from space (**Figures 3–9 and 3–10**) the Moon casts a small moving shadow spot onto the surface of the Earth. Only if you are within this shadow do you see the total eclipse. Even then, you do not see it for long, because the shadow spot sweeps rapidly past you. Because the angular sizes of the Sun and Moon are so close, the longest period when the eclipse is total at any one location on Earth is slightly over seven minutes. We will refer to the length of time when the eclipse is total as the period of totality.

FIGURE 3–9. The shadow of the Moon on the Earth during a total eclipse of the Sun (not to scale).

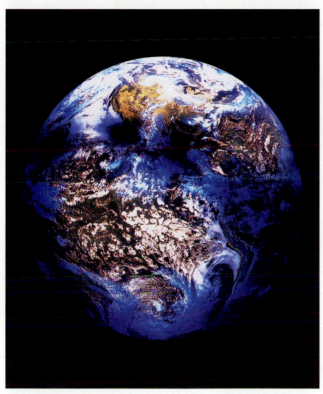

FIGURE 3–10. The Moon's shadow passing over the Earth during the solar eclipse of July 11, 1991. This photograph is a composite of 4 exposures taken by the GEOS-7 satellite.

So you can see whether you might be able to view an eclipse, we list in **Table 3.1** the total solar eclipses for the rest of the twentieth century, where they are visible, and the length of totality. We will return to other aspects of eclipses in Chapter 4.

─────────────── ■ ───────────────

Inquiry 3–3 If the Sun were at the distance of the nearest star—4.3 ly—what would its angular size be? Convert your answer (which the formula gives as a decimal fraction of a degree) into a decimal fraction of a second of arc.

─────────────── ■ ───────────────

TABLE 3–1

Total Solar Eclipses 1995–2000

Date	Location	Length of totality (minutes)
October 24, 1995	South Asia	2.2
March 9, 1997	Siberia, Arctic regions	2.8
February 26, 1998	Central America	4.1
August 11, 1999	Central Europe, Central Asia	2.4

3.4
MEASUREMENT UNCERTAINTY IN SCIENCE

People untrained in the scientific method often tend to imagine that the results of scientific measurements are necessarily exact and without error. Nothing could be further from the truth. Any measurement has associated with it a certain amount of uncertainty or error, no matter how sophisticated or expensive the measuring device. Although it may be true that more elaborate instruments have smaller errors than simple instruments, errors are still present and still need to be considered in detail, especially when the measurement process is being pushed to its limit to extract as much information as possible from the data. Dealing with uncertainties in measurements is one of the most important and critical things scientists do.

The measurement process is fundamental to the scientific method. Without data, there is no way to approach a correct understanding of nature, simply because without observational and experimental data one cannot distinguish among the infinite number of possibilities that exist.

The ancient Greeks made a strong distinction between appearances (or *phenomena*) and reality (what they called *noumena*). They feared that our senses might merely reveal the superficial appearance of the world while missing underlying fundamentals. Hence the school of thought led by the mathematician Pythagoras decided to adopt a purely logical approach without observation, because it would not be subject to such errors. It was not apparent at the time that this approach was subject to other, more serious errors. In the process of evolving the most productive procedures for scientific inquiry, science ended up dropping the distinction between appearance and reality as unproductive. Today, we assume that what we measure is real.

PRECISION AND ACCURACY

When discussing uncertainties in science, scientists make a distinction between **precision** and **accuracy**. A measurement may be precise without being accurate, or accurate without being precise. **Figure 3–11** illustrates the meaning of the words using holes in a target as an example. A tight grouping, away from the center, is precise but not particularly accurate. A wide group centered on the bull's-eye is accurate but not precise.

RANDOM ERRORS

To extract the maximum amount of understanding from measurements, it is necessary to understand the two different types of errors that can occur—**random** and **systematic**. We illustrate these different error types through a specific measurement activity—the measurement with a one-foot ruler of the length of the room in which you sleep.

It is probably evident to you that when you take several readings with a ruler, you rarely if ever get exactly the same answer. You lay the ruler down differently each time, you read it differently, you line it up differently. When you take any measurement repeatedly, there is no way you can do everything in exactly the same way you did it before. These errors are essentially random; that is, some measurements will be larger than the true value and some will be smaller.

But if we get a different number every time we measure something, what should we do to obtain the most accurate value possible? A simple method is to **average** (or take the mean of) all the different measurements. The mathematical process of taking the mean tends to make the values that are too large cancel out the values that are too small.

Inquiry 3–4 Suppose you measured your bedroom five times and got the following values: (*a*) 12 ft 8 in., (*b*) 12 ft 9 in., (*c*) 12 ft 7 in., (*d*) 12 ft 6 in., and (*e*) 13 ft. What would be the best value you could give for the length of your room?

A method by which the random error of a measurement can be found is discussed in Appendix A6. This method is used in the Kit Activities.

SYSTEMATIC ERRORS

It would be nice if we also had a simple method of analyzing systematic errors, but unfortunately such is not the case. A **systematic error** is one that is always in the same direction, as opposed to the random error's tend-

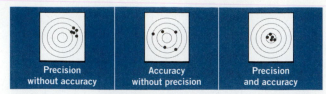

FIGURE 3–11. Accuracy and precision.

ency to be sometimes larger and sometimes smaller than the true value. To illustrate a systematic error in the context of the example discussed above, suppose you measured the length of your room with a meterstick but thought you were using a yardstick (because a meterstick is 39.37 inches in length, they are not much different). You might say the room is four yards across, when in reality it would be four meters, almost a foot longer than you said. Every measurement you made with the meterstick (thinking it was a yardstick) would be systematically too small. Another example of a systematic error would be if the first inch were missing from a yardstick because of a manufacturing error.

Inquiry 3–5 Show that your measurement would be systematically too small by calculating how many inches across the room is if you measure four yards when in fact you had measured four meters.

Inquiry 3–6 What kind of errors, random or systematic, are shown in each part of Figure 3–11? In the diagram showing precision without accuracy, are the random and systematic errors of equal importance?

This example illustrates one reason why systematic errors are dangerous and tricky; they frequently result from mistakes you do not know you are making. Furthermore, there is no standard process by which you can discover with certainty that these mistakes are present. Indeed, a systematic error may sometimes be as subtle as an incorrect assumption, or a failure to consider all the possibilities when a problem is reasoned out. Hidden systematic errors have been responsible for many of the great disasters of the history of science.

> **FOR READERS HAVING THE ACTIVITY KIT:**
> THIS IS THE PROPER PLACE IN THE STUDY SEQUENCE TO CARRY OUT KIT ACTIVITIES 3–1 (CONSTRUCTION OF AN ASTRONOMICAL CROSS-STAFF) AND 3–2 (CONSTRUCTING AND USING AN ASTRONOMICAL QUADRANT). WE STRONGLY URGE YOU TO CARRY OUT THESE MEASUREMENT ACTIVITIES. THEY PRESENT FUNDAMENTAL EXPERIENCES THAT UNDERLIE ALL SCIENTIFIC OBSERVATION AND MEASUREMENT.

FOR READERS NOT HAVING THE ACTIVITY KIT:
SINCE UNDERSTANDING MEASUREMENTS IS CRUCIAL TO UNDERSTANDING THE METHODS OF SCIENCE, WE ENCOURAGE YOU TO DO DISCOVERY 3–1, MEASURING YOUR ROOM, AT THIS TIME.

3.5

SYSTEMATIC ERRORS IN ASTRONOMY: THE KAPTEYN UNIVERSE

In the years before World War I, the principal method astronomers used for mapping the region of space in which we live was to count the number of stars of various brightnesses that could be seen in different directions in space. To make the principle behind this method clear, assume for a moment that all stars have the same intrinsic brightness. Under this assumption, variations in the *apparent* brightness of stars would be due entirely to their differing distances from Earth, with the nearer stars appearing brighter and the more distant ones fainter (**Figure 3–12** illustrates an analogy with approaching cars). In counting fainter and fainter stars, we would be observing stars at greater and greater distances. Once Galileo in 1609 used a small telescope to demonstrate that the visible Milky Way was in fact myriads of distant, pointlike stars, star counts could be used to study the apparent distribution of stars in space.

Star counts were first carried out in a systematic and detailed way by the English astronomer William Herschel in the eighteenth century. Herschel found that beyond a certain brightness the number of stars began to dwindle rapidly, and he interpreted the decline as evidence that there were no stars at greater distances. To put it another way, it was presumed that he had counted out to the edge of our stellar system.

The culmination of this line of research was contained in a paper by the Dutch astronomer Jacobus Kapteyn in 1922. He analyzed decades of laborious research and concluded that the system of stars in which we live was shaped somewhat like a large, fat pancake, as shown in **Figure 3–13**. The long dimension was about 10,000 ly and the short dimension only about one-fifth of that. The Sun was located almost in the center of this system, a "preferential" position that made some astronomers extremely uneasy with this picture. Most astronomers had to admit, however, that Kapteyn had done careful work, and they found it quite convincing. For example, Kapteyn did not make use of the simplifying assumption that all stars had the same energy output. He utilized the extensive body of knowledge that

FIGURE 3–12. When driving at night, we automatically assume that fainter headlights come from more distant cars than for bright headlights. Similarly, if all stars were equally luminous, their brightnesses would depend on their distance from us.

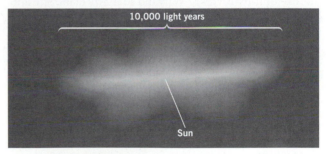

FIGURE 3–13. The Kapteyn Universe, from about 1920, pictured the Sun near the center.

was developing at that time about the differences that exist between stars of different types.

Kapteyn's figures were taken to be not just the dimensions of our own stellar system but those of the entire known universe, because it was not known at that time that there are other systems like our own. So in 1922 the entire universe was considered to be about 10,000 ly in diameter, with the Sun at the center.

The history of astronomy has generally not been kind to chauvinistic concepts such as this one, and sure enough, within five years, developments in astronomical research made astronomers realize that our galaxy was actually 10 times larger in size, and the universe 100,000 times larger in extent, than had been thought. Measured by its volume, the universe was actually a factor of 10^{15} times larger than previously thought. Today,

FIGURE 3–14. Mosaic photograph of the Milky Way.

the most distant object known is thought to be over 12 *billion* light years away. Furthermore, the Sun turned out to be far from the center.

How could astronomers have been so wrong about the size of the universe? Why did so many elaborate and carefully made star counts give such incredibly misleading results? You can understand the answer to this if you have ever had occasion to observe the Milky Way on an extremely clear night, perhaps in the mountains or countryside where you are far from disturbing lights. **Figure 3–14**, a mosaic photograph of the Milky Way, shows what you would observe under those conditions. (Note: This flat photomosaic covers a large section of the sky and would wrap almost completely around a sphere representing the sky.)

Looking at the Milky Way, we see a band of faint light encircling the sky. This light comes from an enormous number of faint stars at great distances, as was first shown by Galileo. But if you examine Figure 3–14 closely, you will see numerous dark regions, called rifts, crossing the Milky Way. These rifts cannot be places where there are simply no stars, because such regions would have to be enormously long, skinny tunnels that cross through our stellar system, all pointing toward the Sun, as illustrated in **Figure 3–15**. If the tunnels were not pointing toward the Sun (direction C), no rift would be seen. Furthermore, even if such tunnels were to exist, they could not persist for long, because the stars are constantly in motion. The explanation of these rifts is that between the stars are clouds of dust particles—small, solid grains of matter floating in space between the stars—that block light from the more distant stars in certain directions just as heavy clouds block the light of the Sun.

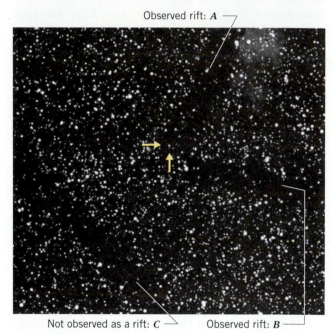

FIGURE 3–15. For the observed rifts to be produced by regions having no stars, the places without stars would have to be long, skinny tunnels (directions A and B) pointed toward the Sun (marked by arrows). For a tunnel in the direction C not pointing at the Sun, a large numbers of stars would still be seen.

Astronomers were aware of the existence of this dust in the galaxy, but they had not discovered a way to tell how much dust there was. The dust will make stars seen through it appear fainter than they really are, but how much fainter? A factor of two? A factor of 10 million?

Kapteyn was aware of the problems that could be caused if absorbing matter were in space, so he did some research to test for the possibility of dust. Along with many other astronomers at that time, he concluded (incorrectly, it turned out) that there were indeed some regions of the sky where heavy dust obscuration was encountered, but that most lines of sight outward from the Earth passed through a space that was almost totally transparent.

We now know that interstellar dust is so pervasive in space that most of our own stellar system is concealed from our optical telescopes. **Figure 3–16** diagrams a galaxy we believe is similar to our galaxy, with the position of the Kapteyn "universe" indicated in it. Because of a serious systematic error—an unsuspected absorption leading to an incorrect interpretation of star counts—for more than a century astronomers were under the impression that what they could easily see in the solar neighborhood was actually the entire universe.

Astronomers today feel more confident about the picture they have developed describing the size and evolution of our physical universe, but the possibility of serious systematic or conceptual errors always remains. To

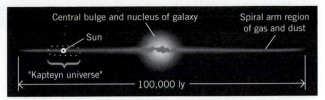

FIGURE 3-16. The modern picture of the galaxy compared to the universe as proposed by Kapteyn.

judge for yourself the confidence you will want to place in our current models, you need an understanding not only of the models themselves, but also of the *process* by which they were constructed. Astronomy is developing so rapidly in the latter half of the twentieth century that any specific set of facts may well be obsolete in a decade; therefore, our goal in this book is not to encourage you to learn dry facts but instead to help you understand how the interaction between theory and observation leads, in a gradual way, to models of the physical universe. Of course, you must learn *some* facts in order to have a context in which to discuss theory and the changes brought about by new observations.

DISCOVERY 3–1
MEASURING YOUR ROOM

After completing this Discovery, you will be able to

- take a set of measurements, find their average, and estimate their random error.

Obtain a ruler; you may use either a meterstick, yardstick, or foot ruler. Make five repeated measurements of the length of your room. Tabulate your data. Find the average and estimate the random error by finding the range in your values (the largest minus the smallest) and dividing by two. Write your results as an average plus or minus your error.

☐ **Discovery Inquiry 3–1** What possible sources of systematic errors might enter into your measurements?

CHAPTER SUMMARY

To emphasize the interrelationships between observation and theory, and the importance of uncertainties, beginning with this chapter we organize summaries into three parts as applicable: (*1*) observations, (*2*) theory, and (*3*) conclusions.

OBSERVATIONS

- Stars appear to be located on the **celestial sphere**. The **north celestial pole** is a point on the celestial sphere where the Earth's rotation axis and the celestial sphere intersect. Any two objects on the celestial sphere are separated from one another by angles measured in **degrees**, **minutes of arc**, or **seconds of arc**.

- Total solar eclipses occur only because the angular size of the Sun and Moon are nearly identical.

THEORY

- The **apparent size** of an object is an angle that is determined by the object's **true size** and distance from the observer.
- All scientific measurements have one or both types of errors: **random error** and **systematic error**. Random errors are handled by taking the **average** of the measurements. Systematic errors are difficult to detect, and scientists must take great care to avoid them. Scientists make a careful distinction between the concepts of **precision** and **accuracy**.

SUMMARY QUESTIONS

1. What do scientists mean by random and systematic errors? Give an example of each.

2. Explain how systematic errors affected Kapteyn's estimates of the size of the universe.

APPLYING YOUR KNOWLEDGE

1. Discuss the errors involved in comparing your car's odometer readings against the one-mile markers on a highway. Assume the comparison continues for 20 miles. (Please do not do this while you are driving!)
2. Suppose you made a number of precise but inaccurate angular measurements with either your fist or a cross-staff. What might you do to improve the accuracy?
3. Think about and discuss the possible errors present the last time you made a measurement of any type. (Such a measurement might include weighing yourself, measuring an ingredient when cooking, filling your car with fuel, etc.)
4. Suppose the dark rifts in the Milky Way were long skinny tunnels without stars. Assume a typical tunnel has a 10 ly diameter, and that a typical star moves with

a speed of 10 km/sec. How long should it take for stars to fill the tunnel? (Remember, the distance something moves is given by its speed times the length of time it travels.)
5. What would be the apparent size of a quarter if it were at a distance of 3000 miles (i.e., across the length of the United States)? (Such an angular size could easily be seen by a modern radio telescope.)
6. What would be the width of the so-called "canals" on Mars if they were to have an angular size of 0.25 seconds of arc when Mars is at its closest approach to the Earth (roughly 50 million miles)?
7. What would be the angular size of the Earth for an observer on the Moon?

ANSWERS TO INQUIRIES

3–1. Yes. All the measurements are ratios, so if the proportion is the same, so is the result.
3–2. 0.53°; 0.53 × 60 = 31.8′; 31.8 × 60 = 1908″
3–3. The angular size would be about 2×10^{-6} degrees, or 7×10^{-3} seconds of arc.
3–4. The average value is 12 ft 8.4 in.

3–5. 4 yards × 36 in/yard = 144 in.
4 m × 39.37 in/m = 157 in.
3–6. (*a*) random and systematic; (*b*) random; (*c*) random. You always have random errors. In this example, random errors are less than systematic errors.

4

BASIC OBSERVATIONS AND INTERPRETATIONS OF THE SKY

The objects which astronomy discloses afford subjects of sublime contemplation, and tend to elevate the soul above vicious passions and grovelling pursuits.

THOMAS DICK, *GEOGRAPHY OF THE HEAVENS*, 1843

The night sky is incredible when viewed from a dark location. Until recent times, most people lived in such locations, and they followed the heavens throughout the year and throughout their lives. In this chapter we will examine some of the simple observations made by ancient peoples (observations you can make if you desire) and the conclusions they drew about the heavens. Those of you doing the Kit Activities will have the opportunity to map out the sky and determine its apparent motions using a cross-staff and quadrant, and to learn to use the seasonal star maps.

READERS DOING THE OBSERVING
ACTIVITIES SHOULD DO KIT
ACTIVITIES 4–1 (MEASURING NORTHERN
STARS WITH QUADRANT AND
CROSS-STAFF) AND 4–2 (MAPPING THE SKY)
AT THIS TIME.

4.1
EARLY OBSERVATIONS OF THE SKY

Three to four thousand years ago, the ancient Chinese, Egyptians, and Mesopotamians were systematically studying the skies and keeping records of their observations. In later centuries, similar activities were carried out on the Indian subcontinent, and also by the great cultures of the New World that flourished in what is now Mexico and most of Central America and in the South American regions we now call Bolivia and Peru. Thanks to their systematic observations, carefully made over a long time, ancient priests were able to detect many regularities in the motions of the heavenly bodies, and to become quite skilled in predicting certain astronomical events, such as eclipses. Even today, some of the ancient eclipse observations are useful in helping astronomers determine the rate at which the Earth's rotation is slowing down (Chapter 8).

The famous Aztec calendar stone shown in **Figure 4–1** is not actually a calendar. The stone is a complex visual representation of a cyclical cosmological model showing the Aztec belief that the Earth and the cosmos undergo alternating periods of creation and destruction. At the time of the Spanish expeditions in the early sixteenth century, Mesoamerican Indians believed themselves to be living in the Fifth Sun, or the fifth such epoch.

Lunar calendars are used to this day by many of the world's major religious groups. The ancient Babylonian calendar, based on the phases of the Moon, began each month at sundown on the day the crescent Moon was first visible in the west. Because the time from one new

FIGURE 4–1. An Aztec calendar stone. The stone is 4 meters in diameter and weighs 24 tons.

Moon to the next is about 29.53 days, the length of the Babylonian months generally alternated between 29 and 30 days. The year of 365.24 days contains an average of 12.37 lunar months, so their years had sometimes 12 and sometimes 13 months. The ancient Hebrews brought the Babylonian calendar back with them after their captivity, and it forms the basis of the present-day calendar of Judaism. Christians use a modified version of this calendar in setting the date of Easter, which was originally associated with Passover. The Islamic calendar is also a lunar calendar, and the holy month of Ramadan is linked to phases of the Moon. The Buddhist New Year, celebrated all over the Far East, is also set by the Moon.

The ancient Egyptian calendar, on the other hand, was a solar calendar. The Egyptian year had 12 months of 30 days each, with a few extra days added at the end of the year. It was important for the Egyptian priests to keep an accurate calendar in order to predict the annual flooding of the Nile, which brought fertility to the land. The Egyptian calendar was a precursor of the conceptually similar Julian calendar, which was ordained by Julius Caesar, and of its successor, the modified Gregorian calendar, which we use today.

At about the same time the pyramids were being built in Egypt, an amazing monument was being constructed in southern England. Stonehenge, as it is called, is one of a number of such constructions that have been discovered in England and in the French province of Brittany (**Figure 4–2**). At Stonehenge, many of the lines of sight from one stone to another point to the places on the horizon where the Sun and Moon rise and set. The principal alignment, shown in Figure 4–2, runs from the center of the monument to the northeast, to a stone known as the Heelstone. On the first day of summer each year, the Sun rises directly over the Heelstone when observed from this central vantage point. The people who built Stonehenge clearly had considerable astronomical knowledge.

One of the most intriguing constructions in North America is the Big Horn Medicine Wheel in Wyoming (**Figure 4–3**), which appears to have numerous celestial alignments. Not only does it seem to indicate the point on the horizon where the Sun rises on the first day of summer, it also shows alignments to the horizon points where the bright stars Aldebaran, Sirius, and Rigel rise just before the Sun at one-month intervals during the summer. Anthropologists have recently discovered that the Indian gods represented by these stars were important participants in the summer solstice rituals of the prehistoric Cheyenne. Like that of Stonehenge, the astronomical interpretation of the Big Horn Medicine Wheel is debated.

Astronomical observations were made by other North American Indian groups as well. A thousand years ago, the ancestors of many of the present-day Pueblo Indian groups, the Anasazi, built elaborate cities of stones and mud mortar that incorporated significant solar alignments in their construction. Some well-known Anasazi sites are Chaco Canyon, Mesa Verde, Hovenweep, and Chimney Rock, all located in the Four Corners area of Utah, Colorado, New Mexico, and Arizona. These constructions were often built with an eye to passive-solar engineering as well, to keep them warm in winter and cool in summer. We also find astronomical alignments in the Cahokia mounds just east of St. Louis, and in the Council Circles in Central Kansas.

Recent studies by one of the authors (R.R.R.) of the art on the burial bowls of the Mimbres Indians, located in southwestern New Mexico, reveals that these Indians had a sophisticated appreciation of many aspects of the Moon's motion, as well as of eclipses. One of the bowls studied (**Figure 4–4**) shows a representation of the su-

FIGURE 4–2. The Sun rising over the Heelstone at Stonehenge on the first day of summer.

FIGURE 4–3. Big Horn Medicine Wheel, Wyoming. This artifact, like Stonehenge, has many alignments with the rising and setting of astronomical objects.

FIGURE 4–4. A burial bowl of the Mimbres Indians, showing an image of the supernova explosion in A.D. 1054 that produced what we now call the Crab Nebula.

pernova explosion of 1054 that produced the object modern astronomers call the Crab Nebula. Astronomers had found many examples of rock art suggesting that the Indians of the Southwest observed this cataclysmic event, but because it is not possible to date rock art without destroying it, these findings had remained speculative. The Mimbres bowl was firmly dated by archaeological tree-ring studies as well, thus confirming its astronomical content.

By far the most advanced of the New World astronomers, however, were the Maya of Southern Mexico, Central America, and, later, the Yucatan. Tragically, nearly all of their records were destroyed by Spanish invaders intent on wiping out the Maya religion. The full extent of their astronomical sophistication may never be known, but it is clear that at the peak of their civilization they were second to none in their understanding of the cosmos. Their complicated cosmology was particularly concerned with cycles of time, and they tracked celestial events on a number of different calendars: a 365–day solar calendar, a 260–day sacred calendar used to schedule rituals (whose origin is not known), and also a sequential count of days elapsed "since the creation of the Universe" that kept time in multiples of 360. This so-called "Long Count" was extended by Mayan astronomers in their calculations forwards and backwards in time for millions of years.

Current research continually increases our awareness of the scope and sophistication of early Asian astronomy. Chinese observations can be traced without difficulty to 1300 B.C. The Chinese, Japanese, and Koreans recorded eclipses, comets, and the appearance of *guest stars*, stars that flared into brilliance and then

slowly faded away. When we look with our telescopes at the place where the guest star of 1054 was seen, we find one of the most fascinating objects studied by modern astronomers—the Crab Nebula (see Figure 1–13), a remnant of the supernova seen in 1054.

The Chinese invented a system for locating the stars in the sky that is similar to the latitude and longitude we use for locating places on the Earth. Introduced into Europe in the Middle Ages, it gained wide acceptance and is in fact the system used by modern astronomers. The Chinese also invented the principle of the clock-driven equatorial telescope mounting, hundreds of years before it was used in the West. This type of mounting makes it possible to follow stars automatically as they move across the sky (**Figure 4–5**).

EARLY GREEK OBSERVATIONS: THE ROUND EARTH

Greek civilization was the first one to organize its observations and then attempt to find a cohesive framework in which to place them. The Greeks made great progress in measuring the positions of stars and determining the relative sizes and distances of the various objects in our solar system. It is fair to say that Greek astronomers were the first to formulate and apply the scientific method successfully to observational sciences such as astronomy. Greek astronomy is discussed in more detail in Chapter 5; here we look at the Greek's earliest attempts to make sense of their observations.

As an example of the way that conceptual models interact with observations to produce further understanding of observed events, let us go back about 2500 years and consider how Babylonian astronomers saw the shape of the Earth. Elementary observations seemed to establish that the Earth was flat. However, no one had ever seen the end of the Earth, and it was not until the sixteenth century that Magellan's ship circumnavigated the globe and directly showed that the Earth had no "end." The Babylonians pictured a finite Earth; they assumed that if one traveled far enough, one would eventually reach the point where Earth and sky meet (**Figure 4–6**). This assumption is not surprising, because the idea of infinity (in this case, an Earth that extended endlessly through space) is a difficult one to grasp and was not invented until about 200 B.C.

The simple model of a flat Earth surrounded by the firmament of the heavens (which comes to the Western world from Babylonia via the biblical account in Genesis) poses some problems. For one thing, if all things fall

FIGURE 4–5. This instrument, preserved at Purple Mountain Observatory, Nanjing, China, operates on the same principle as the modern equatorial telescope mount.

FIGURE 4–6. The Babylonian universe was enclosed in a domelike shell called the Firmament.

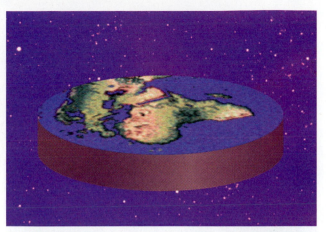

FIGURE 4–7. Anaximander's universe was infinite in extent, with the Earth at the center.

down, as we observe that they do, why doesn't the Earth? To someone who lacks an understanding of gravity, this is a serious question. To avoid a "fall," the Earth would have to extend infinitely downward, but the Babylonians were not yet ready for the concept of infinity. While the Babylonians seem to have ignored the problem, the Greeks did not. Anaximander thought that the Earth was cylindrical in shape (with Greece on the "upper" face) and that it was placed in the exact center of the universe (**Figure 4–7**). Because it was in the center of the universe, it had no particular direction to go and "naturally" remained in the same place. Although Anaximander's concept solved the problem of an infinitely extended Earth, it conflicted with several familiar facts.

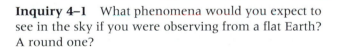

Inquiry 4–1 What phenomena would you expect to see in the sky if you were observing from a flat Earth? A round one?

The Greeks were, of course, seafarers. Sailors had long known that when a ship sailed away from the shore it not only diminished in apparent size but, in addition, appeared to sink into the water. This is easily explained if the Earth is curved. Furthermore, the Greeks knew that if one traveled in a northerly or southerly direction, some stars disappeared from view, whereas other stars became visible. That this is impossible on a flat Earth is easy to see (**Figure 4–8a**), but it can be explained simply if the Earth is round (**Figure 4–8b**).

The Greeks were also well aware that eclipses of the Moon are caused by the Earth's passing between the Sun and Moon, causing its shadow to fall on the Moon (**Figure 4–9a**). Furthermore, they realized that the line

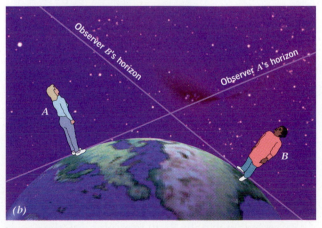

FIGURE 4–8. (*a*) On a flat Earth, all observers see the same stars. (*b*) On a round Earth, observers have different horizons, and they see only those stars above their own horizon.

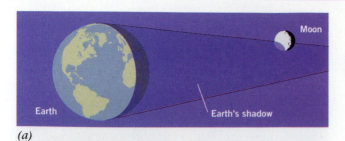

FIGURE 4–9. (a) Eclipse of the Moon (not to scale). The shadow of the Earth on the Moon is always a segment of a circle, proving that the Earth is round. (b) A lunar eclipse showing the Earth's curved shadow in the Moon's face. The eclipsed part of the Moon is illuminated by light refracted through the Earth's atmosphere.

between light and darkness on the Moon's face during the eclipse was observed always to be curved and always to have the same amount of curvature (**Figure 4–9b**). Because the only object that always casts a circular shadow, regardless of its orientation, is a sphere, it followed that the Earth was spherical. The idea of a spherical Earth is attributed to Pythagoras (c. 570–500 B.C.), and the arguments employed here came down to us through the writings of Aristotle some 200 years later.

Inquiry 4–2 Suppose the Earth were shaped like a discus or a Frisbee, which when seen from the Moon is observed from the side. What shape would the Earth's shadow be on the Moon if the eclipse took place at sunset?

By assuming that the Earth was spherical, the Greeks also resolved the question of which way was down: it was toward the center of the Earth. Their model satisfied them completely, because Greek philosophy told them the sphere was the most perfect geometrical form.

The questioning approach that Greek thinkers took toward nature, together with their considerable strength in geometry and mathematics, made it possible for them to progress rapidly in astronomy. The Greeks were able to determine the sizes of the Earth and Moon, and they attempted to determine the size of the Sun (although the method used for this was not accurate). They understood eclipses and determined the proper order of the planets with respect to their distance from the Earth by studying their apparent speeds as they crossed the sky. We will discuss further aspects of Greek astronomy in the next chapter.

Inquiry 4–3 Which of the reasons given above for believing the hypothesis that the Earth is round are primarily supported by observations? Which ones are primarily supported by theory?

4.3
THE OBSERVED MOTIONS OF THE SKY

To understand our observations further we require a better means of describing the sky. We must reconcile into one unified picture of the heavens different apparent motions as seen by observers at different locations on Earth.

THE CELESTIAL SPHERE AND THE POLE STAR

The stars give the appearance of being located on a bowl or sphere with the observer at the center. We call this apparent sphere the **celestial sphere**. The daily apparent east-to-west motion of bodies on the celestial sphere is just a reflection of the Earth's west-to-east daily rotation on its axis. The stars appear to rotate about a point on the celestial sphere called the **north celestial pole** (the **NCP**), as shown in the chapter opening photograph. It is simply the point on the celestial sphere toward which the Earth's rotation axis points. By coincidence, there happens to be a fairly bright star, **Polaris**, relatively close to the NCP. For this reason, Polaris is also known as the pole star. In a similar way, the extension of the Earth's equator to the celestial sphere defines the **celestial equator**.

There is a group of stars that never sets; which stars depends on the observer's latitude. These **circumpolar stars** appear to rotate about the north celestial pole in a circle, always keeping the same angular distance from it. In the chapter opening photograph, they are the stars between the horizon and the center of the star trails.

A star's **altitude** is its angle above the horizon. The altitude of the pole star depends on the observer's location on Earth. For example, **Figure 4–10***a* shows an observer at the North Pole. In this case, Polaris is directly overhead, at the observer's **zenith**. The observer's **horizon** is represented as a plane that is tangent to the Earth where the observer is located. For this observer the altitude of Polaris therefore is 90°. At the North Pole (or South Pole), the stars always stay at the same altitude above the horizon, moving in circles around the pole directly above the observer. Stars near the horizon will appear to move parallel to the horizon (**Figure 4–10***b*). At the poles, therefore, all visible stars are circum-

polar—they never set. Half the stars are below the horizon, and because they can never rise, they are always invisible to an observer at a pole.

Inquiry 4–4 For an observer at the equator, where on the celestial sphere will the north celestial pole be located?

Inquiry 4–5 Do stars move parallel to, perpendicular to, or at some random angle to the celestial equator?

In **Figure 4–11***a*, the observer is at the equator. In this case the observer, sighting on a line parallel to the horizon, sees Polaris just on the northern horizon. The picture is not to scale, and in reality the observer's

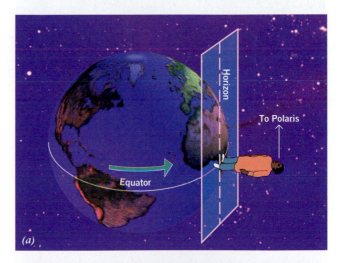

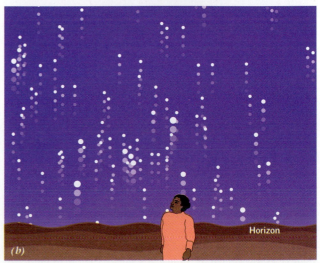

FIGURE 4–10. (*a*) Polaris, as observed from the North Pole, would appear overhead. (*b*) As seen from the pole, stars appear to move on horizontal paths parallel to the horizon.

FIGURE 4–11. (*a*) Polaris, as observed from the equator, would lie just on the horizon. The altitude would be 0°. (*b*) At the equator, stars rise and set vertically.

height is negligible compared with the diameter of the Earth. For this reason, Polaris does not appear above the horizon but, instead, at an altitude of 0°. The celestial equator is now perpendicular to the horizon. Therefore, because a star's motion is parallel to the celestial equator, when a star rises or sets it does so on a path that is perpendicular to the horizon, as shown in **Figure 4–11***b*. Every star in the sky rises and sets for an observer at the equator and is above the horizon half the time and below the horizon the other half. For an equatorial ob-

server, every star in the sky is visible at *some* time of the year, and there are no circumpolar stars.

In **Figure 4–12***a*, the observer is at an intermediate latitude. At such latitudes, the pole star is above the horizon, at an altitude greater than 0° and less than 90°, as in the chapter opening photograph. In the general case, the altitude of the pole equals the latitude of the observer. Some stars are circumpolar and never set, while others rise and set. Over the course of a year, an observer can see more than half the celestial sphere, but there is a significant portion of the sky that never rises and is always hidden from view. Stars rise and set at an angle that depends on the observer's latitude on Earth. **Figure 4–12***b* shows the rising stars for an observer at an intermediate latitude.

A MODEL OF THE SKY

The concepts of the previous section are synthesized in **Figure 4–13**, which shows a way to visualize the celestial sphere. Of course, the Earth is not at the center of the universe, but a model such as this one proves convenient when describing the observed motions of the stars on the celestial sphere. In Figure 4–13, the stars are imagined as fixed to the surface of the celestial sphere, whose radius is so large that the Earth, at the center, is insignificant in comparison. The extension of the

(a)

(b)

FIGURE 4–12. (*a*) Viewed from an intermediate latitude, Polaris would appear at some angle above the north point on the horizon. (*b*) At intermediate latitudes looking east, the stars rise at an angle with respect to the horizon. The angle depends upon the observer's latitude.

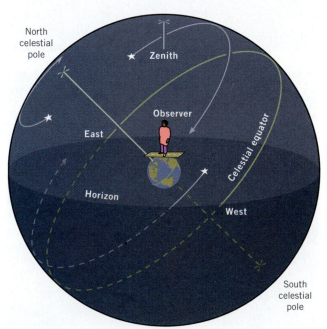

FIGURE 4–13. The celestial sphere, as seen from the "outside" for an observer at intermediate latitudes. The figure is not to scale.

Earth's rotation axis intersects the celestial sphere at the north and south celestial poles, and the extension of the Earth's equator intersects the celestial sphere at the celestial equator, as described before.

4.4
THE MOTION OF THE SUN

Observations of the Sun show it appears to move relative to the observer and relative to the background stars. Such motions determine our methods of timekeeping. In the following sections, we describe these motions.

THE DIURNAL MOTION

The **diurnal** motion of the Sun is its apparent daily motion across the sky resulting from the Earth's rotation. It is the cause of day and night. A **solar day** is the length of time between two successive passages of the Sun through the observer's **meridian**, which is a circle running north-south that divides the celestial sphere into eastern and western halves. By observing the diurnal motion of the Sun during the course of the day, it is possible to obtain a good understanding of the apparent daily motions of other stars as well.

The diurnal motion of the Sun is easy to study simply by using a **gnomon**—a time-honored device that has been used for thousands of years. A gnomon is simply a vertical stick that casts a shadow on the ground. By interpreting the motion of the shadow, one can learn about the apparent motion of the Sun. For example, when the shadow has its shortest length of the day, it

points directly north, the Sun is at its highest point in the sky, and it is crossing the meridian.

> READERS DOING THE OBSERVING ACTIVITIES SHOULD CARRY OUT KIT ACTIVITY 4–3 (THE DIURNAL MOTION OF THE SUN OBSERVED WITH A GNOMON) AT THIS TIME.

THE MOTION OF THE SUN WITH RESPECT TO THE STARS

It is not easy to observe directly the changing position of the Sun with respect to the stars, because the stars are not visible when the Sun is up. But it is not difficult to *infer* the Sun's motion from the observation that different stars are seen in the sky at different times of the year.

We begin with **Figure 4–14a**, which shows the location of the Sun on the celestial sphere at sunset. Just as the Sun is setting in the west, a bright star is rising in the east. One month later, as shown in **Figure 4–14b**, we observe the bright star to be already well above the eastern horizon when the Sun is setting. In other words, the star rose earlier than it did the previous month. The Sun has apparently moved in an *eastward* direction, across the celestial sphere toward the bright star.

The reason for this apparent eastward motion of the Sun is that the Earth moves in an eastward direction around the Sun; the Sun, when viewed from the moving Earth, appears to move eastward against the background of fixed stars. **Figure 4–15** illustrates this. When the Earth is at point *A* in its orbit, the Sun will be seen in the direction of star *A'* when viewed from the Earth.

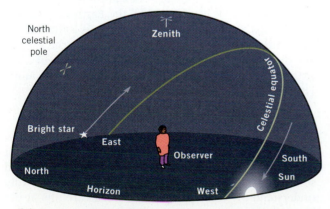

(a) Star rises opposite the sun

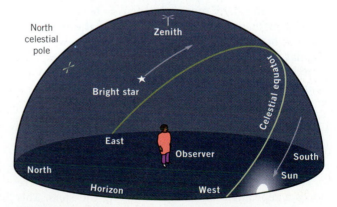

(b) Star rising one month later

FIGURE 4–14. (a) The Sun setting, as a bright star rises on a particular day of the year. (b) When the Sun sets one month later, the same bright star shown in part *a* is already well above the horizon.

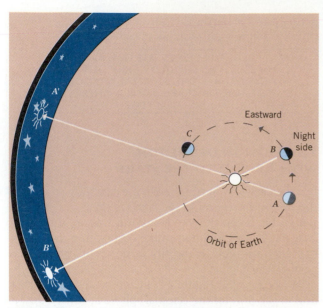

FIGURE 4–15. The apparent motion of the Sun, from point *A′* to point *B′*, against the background of the fixed stars. The view is from above the Earth's North Pole.

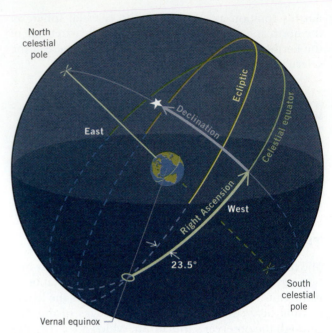

FIGURE 4–16. The celestial sphere, showing the ecliptic tilted 23.5° in relation to the celestial equator, and the vernal equinox. The definitions of the right ascension and declination of a star are illustrated.

The stars on the right side of the figure will be the ones visible at night from the Earth's dark side. One month later, the Earth has moved eastward and is at point *B*; the Sun has shifted its position among the stars and is seen in the direction of star *B′* to the east of *A′*. By the time the Earth moves over to point *C*, the stars on the left side of the figure will be nighttime stars, and the ones on the right side daytime stars.

Because the Earth completes its 360° orbit around the Sun in 365.24 days, the Sun's apparent motion with respect to the background stars is nearly one degree per day. The Sun's daily motion happens to be about twice its angular diameter. The appearance of the night sky changes little from one night to the next because the angular motion of the Sun is small. However, observations made a few weeks apart will show obvious changes in the sky.

> **READERS DOING THE OBSERVING ACTIVITIES SHOULD CARRY OUT KIT ACTIVITY 4–4 (THE MOTION OF THE SUN WITH RESPECT TO THE STARS) AT THIS TIME.**

The movement of the Sun on the celestial sphere traces a repeatable annual path among the stars. This apparent path of the Sun in the sky with respect to the fixed stars is called the **ecliptic**. The entire ecliptic is a circle extending all the way around the celestial sphere.

Inquiry 4–6 The ecliptic has been defined in terms of what the observer on Earth sees. If the observer were located at the Sun and looking at the motion of the planets, how would the ecliptic then be defined?

The relationship between the ecliptic and the celestial equator appears in **Figure 4–16**. The ecliptic plane results from the Earth's *orbital motion* around the Sun. The celestial equator is established by the Earth's *rotation* on its spin axis. Because the Earth's spin axis is tilted 23.5° from the perpendicular to its orbital plane, the ecliptic and the celestial equator are inclined at an angle of 23.5° with respect to each other, intersecting at two points 180 degrees apart.

SOLAR AND SIDEREAL DAYS

If the bright star Sirius rises at 8:00 P.M. one evening, it will rise about 7:56 the next evening, 7:52 the next, and so on—about four minutes earlier each night. To fully understand the reason requires looking carefully at the locations of stars seen by an observer on a rotating, revolving Earth. **Figure 4–17** shows an observer with the Sun directly above, on the meridian; suppose, too, that

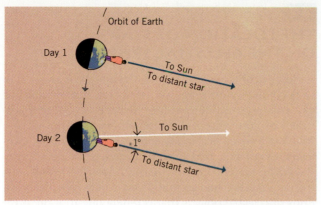

FIGURE 4–17. (*a*) At noon on day 1 the Sun and a distant star are on the observer's meridian. (*b*) One rotation relative to the star later. For the Sun to be on the meridian, the Earth must rotate through about 1°, which requires nearly four minutes. (Note that the angles in the figure are exaggerated.)

a distant star is also on the meridian (but behind the Sun). After a certain interval of time has passed, the same distant star will again be on the meridian. This time interval for a complete rotation of a planet relative to a star is called the **sidereal day.** The Sun, however, will not yet be at the meridian because during the rotation the Earth moved in its orbit to position number *2*. While the Earth moved in its orbit, the Sun appeared to move 1° eastward relative to the background stars. Therefore, for the Sun to be on the meridian again requires the Earth to rotate through an additional 1°. To complete the 1° rotation will require almost four min-

utes. The end result is that the time between successive overhead passages of the Sun is nearly four minutes longer than the sidereal day, which measures the Earth's true rotation period. The cumulative effect of the four minutes per day is that the heavens appear different at different times of the year.

Inquiry 4–7 If a star is on the meridian at midnight on June 1, at approximately what time will it be on the meridian July 1?

4.5
THE REASONS FOR THE SEASONS

In addition to its apparent easterly motion among the stars during the year, the Sun has another apparent motion—one that is responsible for the seasons. Even casual observers in North America find the Sun to be higher in the sky in summer than in winter. These changes in position, which result from an apparent motion of the Sun north and south of the celestial equator, are due to the 23.5° tilt of the Earth's axis of rotation relative to the plane of the Earth's orbit around the Sun.

Figure 4–18 shows how this happens. On the left side of Figure 4–18*a*, the Earth is shown where it would be about December 21, at the beginning of winter in the Northern Hemisphere and summer in the Southern Hemisphere. At this time, the North Pole of the Earth is tilted *away* from the Sun. If you imagine the Earth's equatorial plane projecting out into space, you should be able to convince yourself that the Sun will be south

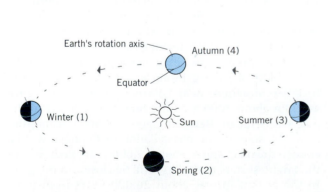

(*a*) Seasons for northern hemisphere

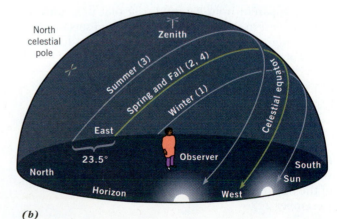

(*b*)

FIGURE 4–18. (*a*) The axis of the Earth's rotation is not perpendicular to the Earth's orbital path around the Sun. This in turn causes the change of the seasons during the course of the year. (*b*) The apparent path of the Sun across the sky during various seasons.

of the celestial equator, and low in the sky at noon. As time progresses, because the Earth's position in its orbit changes, the Sun's position on the celestial sphere moves northward. About March 21 the Sun's location on the ecliptic crosses the celestial equator. At this time day and night are equal, and the Sun is said to be at the **equinox**. This is the first day of spring in the northern Hemisphere; it is called the **vernal** or **spring equinox**. As the Earth continues in its orbit, the Sun's position moves further north of the celestial equator. The Sun is highest in the sky about June 21, when Earth reaches the point in its orbit shown on the right side of Figure 4–18*a*. This point, known as the **summer solstice**, (for the Northern Hemisphere), occurs when the Sun is at its maximum angle north of the celestial equator (23.5°). Hereafter, the Sun's angle north of the celestial equator begins to decrease until the Sun again reaches the celestial equator, at the time of the **autumnal equinox** in September. After another three months, during which the Sun is moving further south on the celestial sphere, the Sun reaches its lowest point again at the **winter solstice**.

A further effect is illustrated in Figure 4–18*b*. In the Northern Hemisphere summer, when the Sun is north of the celestial equator, the Sun is higher in the sky, and its arc across the sky is longer, which means that the Sun is above the horizon a greater fraction of each 24-hour day. Daytime is longer for those in the Northern Hemisphere, which means there is more time for heating of the Earth's surface. More importantly, the Sun's rays are more perpendicular than in the winter, which results in a higher concentration of solar energy per unit area on the Earth's surface. The combination of these two effects produces the heat of summer in the Northern Hemisphere. The Southern Hemisphere, of course, experiences winter at this time.

It is apparent from Figure 4–18*b* that the Sun rises north of east in the northern summer and sets north of west, whereas in the winter it rises south of east and sets south of west.

> READERS DOING THE OBSERVING
> ACTIVITIES SHOULD BEGIN KIT
> ACTIVITY 4-5 (OBSERVING THE SUNSET
> POINT) AT THIS TIME.

Inquiry 4–8 Where in the sky is the Sun at the winter solstice? How would you describe the length of the day (for the Northern Hemisphere) near December 21?

Inquiry 4–9 Do you expect seasonal variations to be large or small at the equator? Explain your answer. How about at the North Pole?

4.6

THE LOCATION OF STARS ON THE CELESTIAL SPHERE

Earlier discussion showed that the apparent location of stars changes with time. Not only that, but the altitude of a given star is different for observers at different latitudes. If you were on the telephone describing to a friend who lived across the country the location of a celestial event you had just witnessed, your description would be invalid for him or her. For this reason, astronomers have had to find a way to describe the positions of celestial objects that is independent of the position of the observer.

A coordinate system that more or less satisfies our needs is the **equatorial coordinate system**. It may be described with Figure 4–16, which is a modification of Figure 4–13 that we used to define the celestial sphere. Just as two coordinates, latitude and longitude, are required to uniquely specify a location on the Earth's surface (Austin, Texas, for example, has a longitude of 97° west and a latitude of 30° north), two coordinates are required to uniquely locate an object on the celestial sphere. The angle of a star north or south of the celestial equator is called **declination**; it is analogous to latitude. Longitude is arbitrarily measured from the Royal Observatory at Greenwich, England. On the celestial sphere, the location where the celestial equator and the ecliptic cross when the Sun moves from the southern into the northern hemisphere, the vernal equinox, is the reference point. This east-west coordinate is called the **right ascension** and is shown in Figure 4–16.

PRECESSION

The pole star, Polaris, is close to but not at the north celestial pole. It is less than 1° away at present. However, observations show that the NCP moves relative to the stars. In fact, the NCP is moving closer to Polaris and, at its closest approach in about the year 2100, it will be within 1/2° of Polaris. In ancient times, however, the NCP was nowhere near Polaris. Referring to **Figure 4–19**, in about 3000 B.C., the bright star closest to the pole was Thuban. Between 2500 and 5000 years from now, the stars in the constellation of Cepheus will be close to the pole, and in about 12,000 years Vega, one of the brightest stars in the sky, will be closest to the pole. In the distant future, about 26,000 years from now, Polaris will again be the pole star.

How can this happen? You may be familiar with the behavior of a spinning top (**Figure 4–20***a*). As the top slows down, its rotation axis rotates as a result of an interaction between the spin of the top and the downward

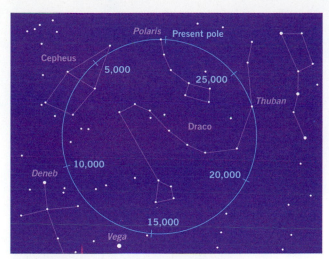

FIGURE 4–19. The path of Earth's rotation axis on the celestial sphere during its 26,000–year precession. Dates indicate the year the pole will be in a given location.

pull of gravity. The axis rotation is called **precession**. The same kind of thing happens to the Earth as it spins in space. The Moon, and to a lesser extent the Sun, exerts a gravitational force on the equatorial bulge of the spinning Earth, which causes the axis of the Earth's rotation to change direction slowly. As a result, the north celestial pole moves slowly around the celestial sphere, taking about 26,000 years to trace a circle among the stars (**Figure 4–20***b*).

The Greek astronomer Hipparchus discovered this motion in about 100 B.C. using Babylonian data. Even in his time, naked-eye observations had accumulated over a long enough time for this very slow motion of the NCP to be detected. This is a dramatic example of the

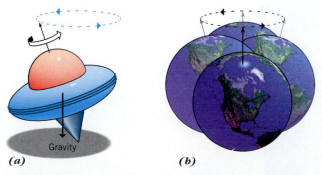

FIGURE 4–20. (*a*) The precession of a spinning top is due to the force of gravity acting on the top. (*b*) The precession of the Earth is due to the force of the Moon and Sun acting on the Earth's equatorial bulge.

value of carefully recorded, continual observations. Casual observations would not have been sufficient to detect precession.

Precession has a practical effect for astronomers: it changes the right ascension and declination of stars in the sky. If astronomers pointing their telescopes to a star did not "precess" the coordinates to the date of observation, the telescope would point to the wrong location.

4.7

THE MOTION AND PHASES OF THE MOON

The Moon is extremely bright compared to the stars, and its changes are easily observed even by the most casual observer. We will now learn the reasons for its changing appearance.

THE MOTION OF THE MOON

The Moon's rapid motion with respect to the fixed stars has given it a unique role in the history of astronomy. From prehistoric times, it has been used as the basis for calendars, and it was the Moon that gave Sir Isaac Newton the crucial information he needed to discover the law of gravity (Chapter 5). Even today, the Moon remains an object of fascination for astronomers and other scientists who study not only its motion, but also the precious rocks brought back from its surface.

The Moon's period of revolution around the Earth relative to the background stars—its sidereal period—is 27.3 days. Its daily eastward motion among the stars amounts to 360°/27.3 days, or about 13° per day. Because this daily eastward angular motion is about 26 times the angular size of the Moon, the motion is easy to see.

Inquiry 4–10 In terms of the Moon's diameter, what is its eastward angular motion with respect to the stars in one hour?

The Moon's movement is eastward relative to the stars because as the Moon's orbital motion carries it eastward in its orbit, an observer on Earth will see it farther to the east with each succeeding hour (**Figure 4–21**).

The Moon's path on the celestial sphere is close to the ecliptic but tilted some 5°. For this reason, sometimes when the Moon crosses the meridian for an observer in the U.S. it is higher in the sky than at other times. This 5° tilt also prevents eclipses from occurring monthly.

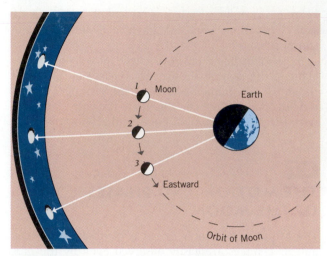

FIGURE 4–21. The revolution of the Moon around the Earth.

We always see the same face of the Moon. While an elementary observation, it leads us to an important conclusion: it proves that the Moon rotates. In particular, it shows that the Moon turns exactly once on its axis each time it revolves around the Earth, just as two people turn while dancing face to face. Because the Moon takes a sidereal month to revolve around the Earth, it takes that long for it to rotate 360° as well.

THE PHASES OF THE MOON

One of the most familiar aspects of the sky is the Moon's changing appearance. The Moon shows **phases**—from **new** to **waxing crescent**, **first quarter**, **waxing gibbous**, and **full**, and then reversing to **waning gibbous**, **last quarter**, **waning crescent**, and new to begin the cycle again (Figure 4–22). It has been known since an-

FIGURE 4–22. The Moon's phases throughout the month.

FIGURE 4–23. The lunar phases. The Moon's illumination as it orbits the Sun is shown. The part of the Moon seen by an observer on Earth is projected onto the plane seen by an observer.

cient times that the phases change as the angle made by lines from the Sun to the Moon to the Earth changes. As this angle changes, the fraction of the Moon's illuminated side visible from Earth changes. **Figure 4–23** illustrates this statement. The figure shows the Moon as it orbits the Sun. The Moon's appearance to an observer on Earth is shown by the Moon's projection onto a plane. The phase changes as the angle between the Sun, Moon, and Earth changes. For example, when the Moon is on the opposite side of the Earth from the Sun, we see all of its illuminated surface and call it "full." New Moon occurs when the Moon is in approximately the same direction as the Sun with its dark side facing us, and quarter Moon is when it is 90° from the Sun. Thus, you can imagine the following situation: the Sun is setting on the western horizon and the first quarter Moon will be on the meridian. Furthermore, if the Sun

were on the meridian (which it is at noon), the first quarter Moon would be rising on the eastern horizon.

Inquiry 4–11 If the Sun sets at 6 P.M., at what approximate time would the full Moon rise?

The time interval from new Moon to the next new Moon is 29.5 days and is called the **synodic period**. It differs from the Moon's 27.3–day sidereal period. The difference between the Moon's sidereal and synodic periods is caused by the Sun's apparent eastward motion in the sky during the month. **Figure 4–24a** shows the situation at new Moon; the Sun and Moon are seen from Earth against the same background stars. In **Figure 4–24b**, one *sidereal* period later, the Moon has returned to the same place among the stars, but the Sun too has

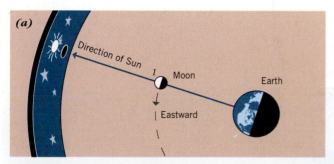

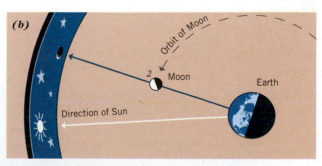

FIGURE 4–24. (a) At new Moon, the Sun and Moon both are seen in the same part of the sky. (b) After the Moon has gone exactly once around the Earth, it is seen in the same location relative to the stars. But the Sun has moved to a different position in the sky, so the Moon is not yet new. (Not to scale.)

moved eastward. Thus the Moon requires about two more days to reach the position of the Sun and establish the new-Moon geometry once again.

> READERS DOING THE OBSERVING ACTIVITIES SHOULD DO KIT ACTIVITY 4–6 (THE MOTION AND PHASES OF THE MOON) AT THIS TIME.

ECLIPSES

Eclipses have awed and frightened observers since the dawn of time. What takes place during eclipses of the Sun and Moon? Every body in the solar system carries a shadow along with it as it moves around the Sun. When one body enters the shadow of another, an eclipse occurs. **Figure 4–25** shows the geometry of the Sun, Moon, and Earth at the time of a **total solar eclipse**. The shadow of the Moon sweeps across the surface of the Earth at a high rate of speed. Observers temporarily located within the dark spot will see the bright disk of the Sun completely covered by the Moon (see Figure 3–8).

A typical shadow consists of a dark inner part, called the **umbra**, and a lighter outer part called the **penumbra** (shown in Figures 4–25 and 4–9 for both lunar and solar eclipses). Should the Moon's umbra pass over an observer, he or she will see a total solar eclipse. However, because the umbra is at most only about 150 miles wide, only those observers along the umbral path will see a total eclipse. Even then the longest period of totality at any one location on Earth is slightly over seven minutes. People not within the umbra but still inside the larger penumbra will see part of the Sun's face blocked, a **partial solar eclipse**. (You can see the partial phases of a solar eclipse prior to totality in Figure 3–8.)

The distance of the Moon from the Earth varies somewhat during the year. Whenever the Moon is relatively far from the Earth, the apparent size of the lunar disk will be smaller than average and may not quite cover the Sun. At such a time, an observer on Earth will see an **annular eclipse**, with a bright ring, or annulus, of sunlight around the dark disk of the Moon. **Figure**

4–26 shows the appearance of the Sun for an annular and a total eclipse.

Eclipses do not occur every month because of the 5° tilt between the Moon's orbit and the ecliptic. Sometimes at new Moon, the Moon is actually above or below the Sun. This means that solar eclipses of all types (partial, total, and annular) occur somewhere on Earth a minimum of two times and a maximum of five times each calendar year. There are from none to three annular or total solar eclipses each calendar year. Should the opportunity arise for you to see a total eclipse of the Sun, take it! It will be one of life's most memorable events.

You should never look directly at the Sun, even during an eclipse. Permanent eye damage can occur rapidly and without your immediate knowledge. If you are fortunate enough to view an eclipse of the Sun, either project the solar image onto a piece of paper, or visit your local library to obtain information on properly protecting yourself from damaging radiation.

Inquiry 4–12 What must the Moon's phase be at the time of a solar eclipse?

When the Moon moves into the shadow of the Earth, a **lunar eclipse** occurs (Figure 4–9a, b). At this time, we can see the curved edge of the Earth's shadow creeping across the illuminated face of the Moon until it is darkened. At the time of a total lunar eclipse, the only light reaching the surface of the Moon is light that has been bent around the Earth by its atmosphere, which behaves like a weak lens. Because of scattering by dust particles in the Earth's atmosphere, the light that reaches the Moon is redder than normal sunlight (this is also what makes the setting Sun red). As a consequence, the Moon's surface has a reddish cast.

Because the Moon can be seen by any observer on the side of the Earth facing the Moon, a lunar eclipse

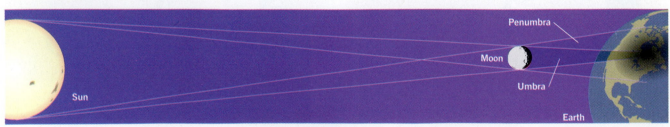

FIGURE 4–25. The geometry of a solar eclipse. (Not to scale)

can be seen by anyone anywhere on that side of the Earth. The conditions for lunar eclipses allow for two to five lunar eclipses of all types per year.

Inquiry 4–13 What must the Moon's phase be at the time of a total lunar eclipse?

(a)

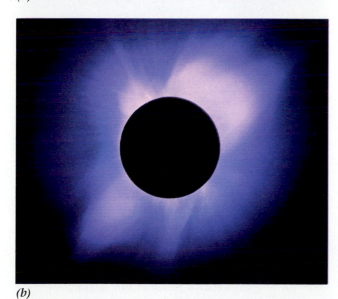

(b)

FIGURE 4–26. (*a*) In an annular eclipse, the Moon's apparent disk is not quite large enough to cover the apparent disk of the Sun. The result is a bright ring around the dark Moon. (*b*) In a total solar eclipse, the Moon completely covers the bright disk of the Sun. Then the sky becomes dark enough for us to see the faint outer part of the solar atmosphere, known as the corona.

4.8
THE MOTIONS OF THE PLANETS

The paths of the planets are more difficult to trace than those of the Sun and Moon, because they generally appear to move more slowly relative to the background stars and are not as bright. In addition, their phases cannot be seen without telescopic assistance. Still, you can learn much by following their motions with the naked eye, just as all astronomers did until the start of the seventeenth century when Galileo turned his telescope to the heavens.

Because the planets appear to move with respect to the fixed stars and vary dramatically in brightness, they readily attracted the attention of sky watchers. In fact, the word **planet** in Greek means *wanderer*. Over a period of time, a planet will appear to move mostly eastward with respect to the background of fixed stars and constellations. **Figure 4–27** shows the positions of Mercury, Venus, and Earth relative to the constellations at the beginning of January and February 1992. In January, while the Sun appeared in Sagittarius, Mercury appeared in Scorpio and Venus between Virgo and Libra.

Inquiry 4–14 Using Figure 4–27, in what constellations were the Sun, Mercury, and Venus during February 1992?

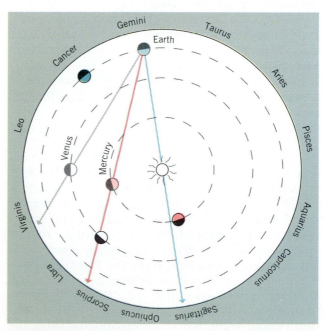

FIGURE 4–27. The solar system as viewed from above on January 1 (light symbols) and February 1 (dark symbols) of 1992. Lines from the Earth to the Sun, Mercury, and Venus show their positions relative to the constellations.

FIGURE 4–28. Retrograde motion of Mars. The dates give an indication of the time interval over which retrograde motion occurs.

If this apparent eastward movement among the stars were all there was to their motion, it is likely that the ancients would not have been so interested in the planets. However, on occasion they appear to slow down, cease their apparent eastward motion, and begin to move westward for a while. Such "backward" motion is called **retrograde motion**. After some weeks or months, the planets begin to move eastward again. They appear to execute a little loop in the sky; **Figure 4–28** shows retrograde loops for Mars. Explaining this looping motion was historically a great challenge, as we will see in Chapter 5.

Two planets, Mercury and Venus, have a rather special motion of their own. They always remain relatively close to the Sun, first moving away from it, then pausing, then moving back toward it. Venus never gets more than 48° from the Sun, whereas Mercury never gets more distant than 28°.

Because of this behavior, Venus and Mercury can be seen in both the morning and evening skies, but they are never seen at midnight (except in polar latitudes, where the Sun can be observed at midnight). For example, if Venus is east of the Sun, it will be an evening object and set after the Sun. If Venus is west of the Sun, then it will be a morning object, rising in the east before the Sun.

The planets do not move all over the sky but are observed only in a rather narrow band near the ecliptic. The ancients noticed this behavior and attributed special significance to this region, which was called the **zodiac**.

Throughout their orbital cycles, planets assume various **configurations** relative to the Earth-Sun line. When a planet farther from the Sun than the Earth—Mars, for example—is closest to the Earth and appears to be opposite the Sun we say the planet is in **opposition** (Figure 4–29). When farthest from the Earth, the planet is said to be in **conjunction** with the Sun. When the planet is 90° (a quarter of a circle) from the Sun, the planet's configuration is eastern or western **quadra-**

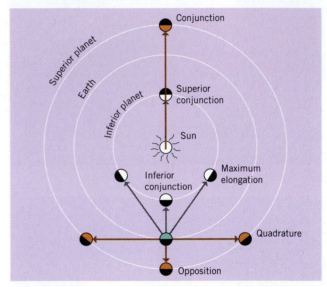

FIGURE 4–29. Planetary configurations.

ture. The two planets closer to the Sun than the Earth—Mercury and Venus—are at **inferior conjunction** when they are closest to the Earth, and at **superior conjunction** when farthest from the Earth. When the angular distance of a planet from the Sun is at its maximum, the planetary configuration is called **maximum (eastern or western) elongation**.

While planets can show phases, not all planets can show all phases. Venus's phase is full at superior conjunction and new at inferior conjunction.

Inquiry 4–15 What phase will Venus have at maximum elongation? Use a drawing to determine your answer. What phase will Jupiter have at quadrature?

The length of time from one opposition to the next is a planet's synodic period. It is what astronomers actually observe. From the observed synodic period, the planet's sidereal period may be derived.

Inquiry 4–16   Why is it easy to observe a planet's synodic period, but difficult to observe its sidereal period?

READERS DOING THE OBSERVING
ACTIVITIES SHOULD BEGIN KIT
ACTIVITY 4–7 (THE MOTION OF THE
PLANETS) AT THIS TIME.

4.9
HOW STARS GET THEIR NAMES

Giving names to stars makes it easier to refer to specific objects in the sky. Stars are named in various ways. Most of the brighter stars have common names dating back to antiquity, such as Castor and Pollux, named for the twins of Greek mythology. A number of stars have Arabic names, such as Betelgeuse, Altair, and Algol, because during the Middle Ages much of the scientific knowledge of the ancient world that was lost in the West survived only because the rising Moslem culture of the Middle East preserved it for posterity. Before and during the Renaissance, astronomical knowledge came to Europe from the Arab world, naturally with many Arabic names.

In 1603, the German lawyer Johannes Bayer published a star map in which a new method for naming the stars was introduced. Bayer named the stars after the constellation in which they appeared, distinguishing one star from another with letters of the Greek alphabet (Table 4–1). In general, the brightest star in a constellation was called alpha (α), the next brightest beta (β), and so on. Thus Sirius, the brightest star in the constellation of Canis Major, the Great Dog, is called α Canis Majoris, using the Latin genitive ending on the name of the constellation. However, Bayer was not perfectly consistent in his schemes. For example, he named the stars in the Big Dipper (part of Ursa Major, the Great Bear) in order from the bowl to the handle, and *not* in order of brightness.

After all 24 letters of the Greek alphabet are used, stars are numbered 1, 2, etc. For example, the star after ω Orionis would be called 1 Orionis. Stars whose brightness varies with time are designated with Latin letters. For instance, a famous variable star in the constellation of Lyra is RR Lyrae. Finally, most stars are so faint that even this notation becomes too cumbersome, and they are known only by their number in a catalog. One such

TABLE 4-1

The Greek Alphabet

α	Alpha	ι	Iota	ρ	Rho
β	Beta	κ	Kappa	σ	Sigma
γ	Gamma	λ	Lambda	τ	Tau
δ	Delta	μ	Mu	υ	Upsilon
ε	Epsilon	ν	Nu	φ	Phi
ζ	Zeta	ξ	Xi	χ	Chi
η	Eta	ο	Omicron	ψ	Psi
θ	Theta	π	Pi	ω	Omega

star that has emerged from obscurity is known as HDE 226868, that is, the 226,868th star in the catalog known as the Henry Draper Extension. This is an object that some astronomers think may actually be two stars, one of which is a black hole.

Thus stars may have many names. For example, the variable star Mira is known as ο Ceti, HD 14386, BD −3° 353, GCRV 1301, BS 681, Boss 2796, IRC +00030, IRAS 02168–0312, SAO 129825, and so on, after various catalogs in which it has been listed. The myriad of names a given object can have often leads to confusion even among professionals! Modern computer data bases have been immensely helpful in keeping names straight.

Finally, it is always possible to refer to an object by its right ascension and declination. Object names such as PSR 0540–69 and 0106+013 are not uncommon.

4.10
THE BRIGHTNESS OF STARS

Astronomers still describe the brightness of stars in terms of a system used over 2000 years ago by the Greek astronomer Hipparchus. He grouped all the brightest stars together, calling them stars of "first magnitude." The stars in the next brightest group were called "second magnitude," and so on down to the faintest stars that he could detect with the naked eye, which he assigned to the class of "sixth-magnitude" stars. Because human eyes are not all the same, magnitudes determined with the eye are highly subjective. Modern astronomers have refined this **magnitude system** greatly and now measure brightness to tenths, hundredths, and even thousandths of a magnitude with precision electronic equipment. In the modern system, a star of magnitude 1.0 is approximately 2.5 times brighter[1] than one of magni-

---

[1]While we will use 2.5, the number is 2.511886, which approximates the fifth root of 100.

TABLE 4-2

The Brightness of Objects in the Magnitude System

Object	Magnitude
Sun	−26.7
Full Moon	−12.7
Venus (at brightest)	−4
Jupiter, Mars (at brightest)	−2
Sirius (brightest star in sky)	−1.5
α Centauri (nearest star)	−0.1
Andromeda galaxy (most distant naked-eye object)	3.5
Naked-eye limit (average)	6.5
Binocular limit	10
6-inch telescope limit	13
Hubble Space Telescope limit	28–29

tude 2.0; a star of magnitude 2.0 is 2.5 times brighter than one of magnitude 3.0; and so on. A star of magnitude 1.0 is 100 times brighter than one of magnitude 6.0, because $2.5 \times 2.5 \times 2.5 \times 2.5 \times 2.5 \approx 100$. In a similar fashion, the magnitude system has been extended to objects fainter than the naked eye can see (magnitude greater than 6), whereas the brightest objects (Sun, Moon, some planets, and a few stars) are so bright that their magnitudes are counted as negative (less than magnitude 0). For example, the brightest star, Sirius, has magnitude −1.5, whereas the Sun's magnitude is −26.7. **Table 4–2** shows the brightness of a variety of objects expressed in the magnitude system.

The word "magnitude" used to describe the brightness of stars is not the same as the factor of 10 "order of magnitude" discussed in Chapter 1.

> READERS WHO WOULD LIKE TO MAKE OBSERVATIONS OF THE NIGHT SKY WILL FIND THE CHAPTER APPENDIX VALUABLE. THOSE HAVING THE ACTIVITY KIT MAY WANT TO MAKE THE TELESCOPE IN KIT ACTIVITY 12-3 AT THIS TIME. THOSE WITHOUT THE KIT MAY USE A PAIR OF BINOCULARS.

APPENDIX
OBSERVING HINTS AND THE USE OF STAR CHARTS

After reading this chapter appendix and observing the sky, you will be able to do the following:

- Use a map as a guide to find the principal constellations and stars in the northern part of the sky at any time of year.

- Use a map of the full sky to locate the principal constellations and stars for your time and season.

BRIGHTNESS OF STARS

The brighter stars are indicated on star maps with larger dots, and the fainter stars with smaller dots. (Note: This is a convention and does not mean that the brighter stars are physically bigger. Brighter stars do appear larger on photographs, but this is because the increased brightness exposes a larger area of the photographic emulsion.)

On the star maps in Appendix G is a key that gives the range of magnitudes corresponding to each symbol used to plot the various stars. The star maps show all the stars down to the third magnitude and some fainter stars, but the faintest (fifth and sixth magnitude) are not shown because most urban observers would not be able to see them. A few other objects of interest, such as bright nebulae and star clusters, are also indicated and can be observed with binoculars or a small telescope.

OBSERVING STARS IN THE NORTHERN PART OF THE SKY

Because the appearance of the sky changes throughout the year, we begin by showing you how to use the star maps to observe the northern sky at one particular time—in September. Afterwards, we look at other times of the year and other parts of the sky.

THE NORTHERN SKY IN SEPTEMBER

The appearance of the northern part of the sky on September 1 at 9 P.M. standard time (10 P.M. daylight saving time) is shown in Figure 3–3. The horizon is drawn for observers at 40° north latitude, but the appearance of the sky would be substantially the same for observers anywhere in the United States, except Alaska and Hawaii.

At this time of year, the two most obvious groupings of stars are the Big Dipper[2] (part of the constellation of Ursa Major, the Great Bear) in the northwest and Cassiopeia (named since ancient times for a queen of Ethiopia) in the northeast. Although most of the stars in the Great Bear are not very prominent, you should have no difficulty locating the Big Dipper. At this time of night and year, the handle of the Dipper points northwest.

[2]The constellation names, and indeed the contellations themselves, vary from one culture to another. In this book, we will use the traditional Greek groupings and Latin names.

The brighter stars of Cassiopeia form a group shaped somewhat like an M or a W depending on the season. They are supposed to resemble the queen's throne, but it is easier to remember the M or W pattern. At this time of year, the W is lying on its left side.

Midway between Cassiopeia and the bowl of the Big Dipper lies the star **Polaris**. This star is located near the north celestial pole in the sky. Notice that two of the stars in the bowl of the Big Dipper (called the Pointer Stars) can be used to locate Polaris, as shown by the arrow in Figure 3–3.

There are stars below Cassiopeia and the Big Dipper, but they are fainter and more likely to be obscured by the haze, dust, and pollution so frequently encountered when looking at the sky close to the horizon. Most city dwellers will not be able to see objects closer than 20° to the horizon, except those that are unusually bright. Objects near the horizon will also tend to twinkle a great deal, varying rapidly in both brightness and color. This is yet another indication of the difficulties that are introduced when observing the sky through an atmosphere.

THE NORTHERN SKY AT OTHER TIMES OF YEAR

If the appearance of the sky were always the same, it would be easy to become familiar with it. But the rotation of the Earth and its orbital motion around the Sun complicate the picture and cause the appearance of the sky to vary. The changes are regular, however, and you can use **Figure 4–30** to help you find the northern stars at any time of year. Facing north, hold Figure 4–30 in front of you so that the name of the month is at the bottom. This will show the orientation of the various stars at 9 P.M. (standard time) on that date. For example, if you put September at the bottom, the configuration will be similar to Figure 3–3.

Inquiry 4–17 What are the locations of the polar star groupings at 9 P.M. on February 1?

Inquiry 4–18 At what time of year is the Big Dipper highest in the sky at 9 P.M.?

Inquiry 4–19 At what time of year would it be hardest to see Cassiopeia because of horizon haze at 9 P.M. standard time, or 10 P.M. daylight saving time?

The Little Dipper, shown in Figure 4–30, is actually part of the constellation of Ursa Minor, the Little Bear. When you observe the stars in this constellation, it will be immediately apparent that they are much fainter than the stars of the Big Dipper and are therefore much harder to see. If either city lights or the Moon make the sky bright, you will probably be able to see only the brightest stars of the Little Dipper—Polaris, at the end of the handle, and the two stars at the end of the bowl. If all the stars in the Little Dipper are visible, then the sky is exceptionally dark and clear, and you have very good observing conditions. On such an evening, you will also be able to see the stars of Draco, the Dragon, snaking between the bowls of the two dippers; and Cepheus, a king of ancient Ethiopia, between the Little Dipper and Cassiopeia. These constellations are shown in Figure 4–30.

OBSERVING THE REST OF THE SKY: USING THE STAR MAPS

The maps in Appendix G will be your guide to the other parts of the sky that can be observed at any time of year. To see how they can be used, select the map that represents the sky on September 5 at 9 P.M. standard time (10 P.M. daylight saving time). Hold the map so that the label NORTH can be read (i.e., turn the page upside down). As shown in **Figure 4–31**, you will see a representation of the stars in the northern part of the sky in the lowest of the four segments into which the map is divided. Compare this with Figure 4–30. Notice that groupings such as the Big Dipper and Cassiopeia are located similarly on the two maps.

Now hold the map so that the label WEST can be read at the bottom. This time, the lower portion of the map shows the appearance of the sky over the western horizon. You will see that the star Arcturus, the brightest star in the constellation of Boötes, the Herdsman, dominates the western sky. In a similar way, the stars of the southern and eastern skies can be found by holding the map with the appropriate direction at the bottom. To observe the stars overhead, hold the map up over your head and look at the central part of it; be sure to match directions on the map with the directions on your horizon.

Like all maps, these try to represent the spherical surface of the sky on a flat piece of paper, and this cannot be done perfectly. You can see how it works by imagining that you have cut out the map along its edges and then taped the edges together. The resulting construction would resemble an inverted bowl, like the sky.

There are twelve maps in Appendix G, each valid for a certain time of year. In addition, each map represents the appearance of the sky for several different combinations of date and time, as shown on the map. Use the map index to choose the map that most closely corresponds to the time and date of your observations. If you are observing at a time not given on the maps, you can easily calculate which map would be the right one using

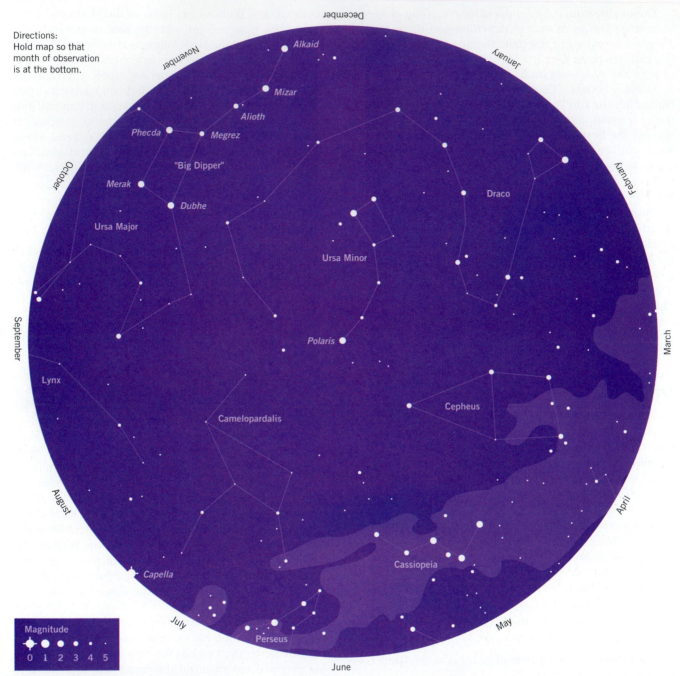

Directions:
Hold map so that
month of observation
is at the bottom.

FIGURE 4–30. Perpetual map of the northern sky. When held with the current month at the bottom, it shows the northern sky at 9:00 P.M. standard time.

the fact that a difference in time of two hours corresponds to a difference in date of one month. For example, the ninth map lists September 5 at 9 P.M. and August 5 at 11 P.M. as the times for its use. It could also be used on July 5 at 1 A.M. or June 5 at 3 A.M.

Inquiry 4–20 In which direction (north, south, east, west, or overhead) would you look to find the Great Square of Pegasus at 9 P.M. on September 5? At 8 P.M. on November 20?

FIGURE 4–31. Using the full-sky star maps.

Inquiry 4–21 At 8 P.M. on March 20, what bright star is closest to the zenith (the point overhead)?

Inquiry 4–22 Where in the sky would the constellation of Orion be found on January 1 at 8 P.M.?

OBSERVING TIPS

You will find it much easier to find your way among the stars if you follow the tips below; these suggestions are the product of much experience.

1. **Observe under a dark sky.** The darker and clearer the sky, the more you will be able to see. It is best to leave the city entirely, but this may not always be convenient. At the very least, you should avoid obvious city lights and find a spot that is as dark as possible. Furthermore, you should try to avoid times of the month when the Moon is nearly full (consult your local newspaper or a calendar for information about when the Moon will be full). It is remarkable how difficult a nearly full Moon makes observations of faint stars.

2. **Adapt to the dark.** The pupil of your eye changes its size to regulate the amount of light it admits. In bright sunshine, the pupil contracts to a small diameter, whereas at night, it opens up wide to admit as much light as possible. It takes time for the eye to become fully adapted to the dark, and you will not be able to see the fainter stars at first. They will begin to appear after about five minutes; full adaption to the dark takes about 15 minutes. Many more stars can be seen when your eyes are fully adapted. (Incidentally, both smoking and the consumption of alcoholic beverages will impair your ability to become fully dark-adapted.)

3. **Use a red flashlight.** A bright flash of light can destroy your dark adaption for many minutes, yet you will need to use a flashlight to read your star map and make notes from time to time. A compromise is to cover a flashlight with red cellophane, both to cut down its brightness and because red light is less likely to harm your dark adaption. When using your flashlight, try to keep it faint and use it as little as possible.

4. **Use averted vision.** Different parts of the retina of the eye have different sensitivities. There is a blind spot where the optic nerve joins the retina, and there is also a region of maximum sensitivity that you can use by looking slightly to one side of the object you are observing rather than directly at it. You can best learn this trick by experimentation: pick a faint object and see how its brightness appears to vary as you look at spots in its immediate vicinity.

5. **Be alert for clouds and haze.** Certain types of clouds, especially a thin layer of cirrus, can be difficult to detect when they appear during an observing session. Even though the brighter stars are easily visible, the fainter ones may disappear completely. Thus you must be on the alert for changing sky conditions. If you are able to see all the stars in the Little Dipper, you know you have a good night to observe.

6. **Take your time.** The most common mistake among beginning observers is hastiness. If you are observing an object or a region of the sky, don't just glance at it and expect to take in all that there is. Not even an experienced astronomer can do that. Instead, concentrate on it and, if what you are observing is large, break it up into smaller regions and observe each smaller region carefully.

7. **Move from the known to the unknown.** It would be difficult to learn all the constellations at once. Start with easily recognized groupings near ones you already know, using a star map as a guide. Look for simple geometrical figures: triangles, squares, rectangles, and so forth. Gauge distances by consulting your map and comparing them with the sizes of familiar groupings.

8. **Make reliable records.** Write down or draw everything you observe. Do *not* entrust anything to memory. Without detailed and accurate written records, you will not be able to observe the various changes in the sky that are discussed in this book. Finally, the process of making a written record will encourage you to observe more carefully.

CHAPTER SUMMARY

OBSERVATIONS

- A variety of ancient civilizations observed celestial phenomena and built structures that allowed them to make astronomical predictions.
- Early observations included the disappearance of ships over the **horizon** and the curved shape of the Earth's shadow on the Moon during a **lunar eclipse**.
- The Sun, Moon, planets, and stars appear to rise in the east and set in the west daily. Night after night the Sun appears to move farther east relative to the background stars. The Moon shows a similar but larger apparent eastward movement. While planets generally have an eastward movement among the stars, they sometimes reverse direction, showing a **retrograde** motion. The Earth's rotation axis shows the effects of **precession**, a 26,000–year apparent motion of the rotation axis on the celestial sphere.
- Observations of the sky can be described in terms of a **celestial sphere** on which celestial objects are located. The **celestial poles** are the projection of the Earth's rotation axis onto the celestial sphere; the **celestial equator** is the projection of the Earth's equator onto the celestial sphere. The path of the Sun on the celestial sphere is called the **ecliptic**. The location of objects on the celestial sphere is specified in terms of **declination** and **right ascension**, which are analogous to latitude and longitude on Earth. The detailed appearance of the sky depends on the observer's latitude.
- Stars that are not circumpolar rise four minutes earlier each day, thus making the stars observable throughout the year change. A **solar day** is about four minutes longer than a **sidereal day**.
- Seasons are produced by the Sun's changing altitude throughout the year. Such changes result from the Earth's rotation axis having a 23.5° tilt with respect to the orbit's perpendicular.
- The Moon exhibits monthly **phases** produced by the changing angle between the Sun, Moon, and observer.
- **Eclipses** occur when the Sun, Moon, and Earth are nearly aligned. **Solar eclipses** occur at **new Moon**, while **lunar eclipses** are at **full Moon**.
- The observed time interval between successive planetary **configurations** such as **opposition** to opposition or **conjunction** to conjunction is a planet's **synodic period**.

THEORY

- The sphere, according to Greek philosophy, is a perfect geometrical form.

CONCLUSIONS

- From a variety of observations (given above), the Greeks concluded that the Earth is round.
- A planet's **sidereal period** is not directly observable but must be derived from the observed synodic period.

SUMMARY QUESTIONS

1. What observations did ancient people use to indicate that the Earth was round? How did these observations interact with the culture's theoretical and philosophical underpinnings?

2. What are the principal motions of the Earth? Explain the effects that each of these motions has on the apparent motions of the stars and other objects in the sky.

3. What effect does an observer's latitude have on what he or she will see in the sky?

4. Why does the Earth rotate faster relative to the stars than it does relative to the Sun?

5. What is precession? What is its effect on the positions of stars?

6. How can you determine your latitude from the observed altitude of Polaris?

7. What do we mean by the terms celestial sphere, celestial equator, ecliptic, meridian, declination, and right ascension?

8. What two factors result in seasonal variations on Earth? How do they operate to create differences between summer and winter in the Northern Hemisphere?

9. How do the different phases of the Moon come about? Your answer should include drawings.

10. What causes solar eclipses? Lunar eclipses? At what phase of the Moon does each occur?

11. What are the observed apparent motions of the planets with respect to the background stars? Describe them.

APPLYING YOUR KNOWLEDGE

1. Is the model of the Earth suggested by Anaximander any better in explaining observations than the Babylonian model? In particular, compare the observations of the sky that could be predicted using the Babylonian model with those that could be predicted using the Greek model.

2. What effects would precession have on the seasons?

3. What would the Earth's phase be for an observer on the Moon if the lunar phase for an observer on Earth is waxing crescent?

4. Where in the sky would you look at midnight to see Jupiter when it is at quadrature?

5. If the star Vega rises tonight at 3:00 A.M., at what time should you be looking for it to rise three months from now? Explain.

6. If the Sun rises at 6:00 A.M., at what approximate time does the waxing gibbous Moon set?

7. Use tracing paper to trace stars from the maps in Appendix G; then try making your own constellations. Are the traditional constellations any better or more useful than yours?

■ **8.** What is the maximum possible lunar altitude for an observer at the North Pole? (Hint: A drawing may be helpful.)

■ **9.** Use a drawing to help you explain which conditions produce the most favorable total solar eclipse: Earth nearest or farthest from the Sun, and Moon nearest or farthest from the Earth.

■ **10.** At what latitude is an observer for whom the stars within 25 degrees of the pole are circumpolar?

■ **11.** Refer to Figure 4–12b and determine the latitude at which the photograph was taken.

■ **12.** How many degrees into the southern celestial hemisphere can an observer at a latitude of 40° north see?

■ **13.** For what southern latitudes will the Sun never be observed at the zenith?

ANSWERS TO INQUIRIES

4–1. *Flat Earth*: same constellations are observed from all locations; height of the Sun above the horizon at noon is constant throughout the year; shadow on Moon during lunar eclipses is a straight line. *Round Earth*: change in constellations as observer moves north or south; height of the Sun above the horizon *might* vary throughout the year; shadow on Moon during lunar eclipses always part of a circle.

4–2. It would produce a straight-line shadow on the Moon.

4–3. *Observational*: Disappearance of boats, different stars seen at different points on Earth, and the shape of the Earth's shadow on the Moon. *Theoretical*: Solves problem of "which way is down" and fits in with notion that the sphere is the most perfect shape.

4–4. On the horizon.

4–5. Stars move parallel to the celestial equator.

4–6. It would be the plane of the Earth's orbit.

4–7. 10:00 P.M.

4–8. The Sun is 23.5° from directly overhead south of the celestial equator, moving low across the southern sky during the day. Most of the diurnal motion of the Sun is below the horizon, giving the shortest number of daylight hours of any day of the year.

4–9. At the equator the altitude of the Sun at noon varies by only 23½° throughout the year; for this reason seasonal variations are small. At a pole, however, the variations are extreme because for six months of the year the Sun never gets above the horizon.

4–10. The Moon moves eastward at about 13° per 24 hours, or about 0.5° per hour. This equals the Moon's diameter every hour.

4–11. 6 P.M.

4–12. New Moon.

4–13. Full Moon.

4–14. The Sun is in Capricornus (Cap); Mercury is between Capricornus (Cap) and Sagittarius (Sgr); Venus is between Ophiuchus (Oph) and Scorpius (Sco).

4–15. Venus would show a quarter phase; Jupiter, gibbous.

4–16. The sidereal period is difficult to observe because the positions of both the planet and Earth change, making it difficult to know when the planet has completed one full cycle. Because the time of opposition is easy to define by an observation, the time interval between two oppositions is easy to get.

4–17. Cassiopeia and the bowl of the Dipper will both be at about 30 degrees altitude, on either side of the pole star.

4–18. May.

4–19. April, May

4–20. East, overhead.

4–21. Castor and Pollux.

4–22. In the southeast, about 50° up from the horizon.

5

THE HISTORICAL QUEST TO MODEL THE SOLAR SYSTEM

The progress of science is generally regarded as a kind of clean, rational advance along a straight line; in fact it has followed a zig-zag course, at times almost more bewildering than the evolution of political thought. The history of cosmic theories, in particular, may without exaggeration be called a history of collective obsessions and controlled schizophrenias; and the manner in which some of the most important individual discoveries were arrived at reminds one more of a sleepwalker's performance than an electronic brain.

ARTHUR KOESTLER, *THE SLEEPWALKERS*, 1959

In this chapter we follow the development of our understanding of our own corner of the universe, the solar system. We will not seek to be comprehensive or encyclopedic in our historical coverage. Instead, we will be more interested in how the ideas that developed in one era were a natural outgrowth of the state of astronomical observations and degree of sophistication of the times. In previous chapters we briefly discussed some historical aspects of astronomy; for example, we considered the notion of a round Earth, and how this concept slowly came to be accepted by Greek thinkers through the gradual accumulation of evidence from both terrestrial and astronomical observations. Yet this concept was established relatively rapidly when compared to the slow acceptance of the idea of a Sun-centered solar system.

5.1 GREEK ASTRONOMY

While we saw in the last chapter that a variety of cultures were intimately involved with phenomena of the sky, it was Greek culture that made progress toward the models we have today. We now look more closely at the astronomical contributions of Greek culture.

WHAT THE GREEKS INHERITED

The Babylonians and Egyptians, who for several thousand years kept records that contained much potentially valuable information, bequeathed to the Greeks an extensive body of astronomical knowledge. However, the astronomy of both Egypt and Babylonia was the province of a priestly aristocracy; as a consequence, practical and political considerations often took precedence over theoretical inquiries. Egyptian astronomers, for example, had discovered that when the bright star Sirius could just be seen rising in the east before the Sun, the flooding of the Nile was imminent. Such knowledge gave tremendous power to the priesthood and inevitably involved its members closely with the state. Similarly, in Babylonia, astronomer-priests had acquired a considerable amount of information concerning the motions of the Moon, Sun, and planets and had found certain regularly occurring cycles that enabled them to predict some eclipses, a power that was frequently used for political purposes.

The Egyptians and Babylonians knew the length of the year and the different types of calendars, both solar and lunar. The Egyptians had learned the rudiments of simple mathematics, algebra, and geometry. Sundials had been invented, and systems of timekeeping were in existence. The Babylonians had made systematic observations of the positions of heavenly bodies and had practical methods for predicting the positions of the Moon, Sun, and planets. **Figure 5–1**, for example, shows a table of data for Jupiter. The bottom part of the figure describes the method of calculation. The Babylonian value for the length of the month was not surpassed in accuracy until the end of the nineteenth century. Both cultures had attempted to construct a cosmology that placed Earth and humanity in their proper position relative to the universe and the gods, but they never attempted to construct a truly consistent theoretical framework for their cosmology in the way the Greeks did.

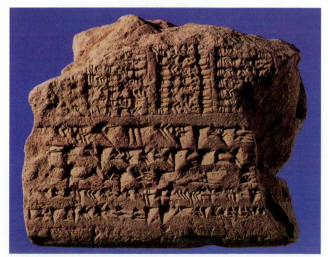

FIGURE 5–1. A clay tablet from ancient Mesopotamia containing astronomical observations of Jupiter in the top part, and a description of the method of calculation in the bottom part.

ARISTARCHUS OF SAMOS

The Greeks enjoyed philosophy, by which they meant a broad attempt to understand all things. When their philosophical thinking was coupled with their highly developed mathematical skills, the Greeks advanced so rapidly in astronomy that their era was without question one of the shining examples of scientific discovery prior to the year A.D. 1500. Progress in understanding our solar system came especially quickly with the emergence of the Alexandrian school of Greek astronomers, particularly the work of **Aristarchus of Samos** (c. 300 B.C.), who combined careful observations and sharp reasoning to draw inferences about the relative sizes and distances of the Earth, Moon, and Sun.

Aristarchus demonstrated that the Sun was many times farther from the Earth than the Moon by using a reasoning process illustrated in **Figure 5–2**. You can see that if the Sun were quite close to the Earth-Moon system, then the time interval from first quarter to third quarter would be *longer* than the time interval from third quarter to first quarter. Aristarchus observed that these two time intervals were in fact nearly equal, implying that the Sun was many times farther away than the Moon. Hence the rays of light from the distant Sun would be arriving in nearly parallel lines for all parts of the Moon's orbit.

Inquiry 5–1 What did Aristarchus assume about the Moon's orbital shape and motion?

With a simple technique, Aristarchus was able to estimate the relative sizes of the Earth and Moon by timing the duration of lunar eclipses. **Figure 5–3** shows the principle of this determination. The length of time it takes the Moon to enter the Earth's shadow is proportional to the diameter of the Moon. The length of time it takes for one part of the Moon—say, its leading edge—to completely cross the Earth's shadow is proportional to the diameter of the Earth. By comparing these two times, we can estimate the ratio between the diameters of the Earth and Moon, as Aristarchus did.[1]

Inquiry 5–2 Suppose it takes an hour for the Moon to enter the Earth's shadow and four hours for the edge of the Moon to cross the shadow. How many times larger than the Moon's diameter is the Earth's diameter?

---

[1]Actually, the diameter of the Earth's shadow at the distance of the Moon is slightly less than the diameter of the Earth; however, this leads to only a small error.

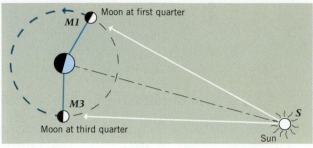

FIGURE 5–2. Aristarchus's method to show that the Sun is farther from the Earth than the Moon. If the Sun were very near the Earth-Moon system, the times between quarter phases would differ measurably. However, if the Sun were much more distant than the Moon, the lines *S-M1* and *S-M3* would be nearly parallel, and the time interval between first and third quarter would be nearly equal to the time interval between third and first quarter.

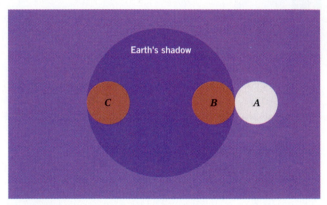

FIGURE 5–3. The principle of Aristarchus's measurement of the Moon's diameter relative to that of Earth. The time it takes the Moon to enter the Earth's shadow (between points *A* and *B*) is proportional to the Moon's diameter. The time the Moon takes to cross the Earth's shadow, between *A* and *C*, is proportional to the Earth's shadow diameter, which is nearly that of the Earth.

Finally, Aristarchus knew that during a total eclipse of the Sun, the Moon was just able to cover the Sun and therefore that the angular sizes of the Moon and Sun in the sky were about the same. He could then reason that the actual sizes of the Moon and Sun were proportional to their distances, and in this way he could estimate the diameter of the Sun (**Figure 5–4**). Unfortunately, although his estimate of the distance to the Moon was quite good, his estimate of the distance to the Sun was about 10 times too small. For this reason, his estimate of the Sun's diameter was also 10 times too small. Nevertheless, he was able to show that the distance to the Sun was considerably greater than that to the Moon, and that the Sun's diameter was much greater than the Earth's.

FIGURE 5–4. Two objects with the same angular sizes have diameters that are proportional to their distances.

Inquiry 5–3 Aristarchus was the first to propose that the Earth goes around the Sun, rather than vice versa. Suggest one factor that may have led him to this conclusion.

Eratosthenes

Another classic experiment of antiquity was the determination of the Earth's circumference by **Eratosthenes** (c. 200 B.C.). The conclusions of Aristarchus concerning the relative sizes and distances of the Earth, Moon, and Sun were all in terms of the then unknown size of the Earth. Their sizes in customary units (the Greeks used a unit of length called the *stadium*) could not be known until the Earth's size was known.

Eratosthenes had heard that at Syene, near the modern Aswan in Egypt, there was a deep well, and that on a certain day of the year the Sun stood directly overhead so that its reflection could be seen in the bottom of the well. Eratosthenes was also able to observe that on that same day of the year in Alexandria, where he lived, the Sun was not directly overhead but was 7° south of the zenith. He determined this angle with a gnomon, just as

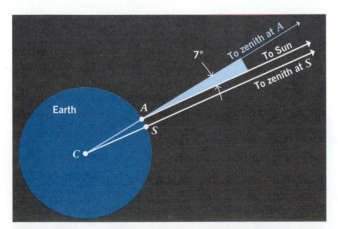

FIGURE 5–5. Eratosthenes' measurement of the diameter of the Earth. The difference in altitude of the Sun is proportional to the distance between the two points, which is known. This allows the length of one degree on the Earth's surface to be determined.

you may have done in Kit Activity 4–3. **Figure 5–5** shows the geometry of Eratosthenes' experiment. Point *A* represents Alexandria and point *S* is Syene; point *C* is at the center of the Earth. The Sun is so far away that the lines *A*-Sun and *S*-Sun are nearly parallel. The angle at *A* is nearly equal to the angle at *C* at the center of the Earth, and we may write the following proportion:

$$\frac{\text{circumference of Earth}}{\text{distance from } A \text{ to } S} = \frac{360°}{\text{angle at } C}.$$

Inquiry 5–4 Assuming that the angle at *C* is 7° and that the distance of Alexandria from Syene is 5000 stadia, what is the diameter of the Earth in stadia? (Stadia is the plural of stadium.) Although the exact length of a stadium is unknown, compute the radius of the Earth assuming it to be about one-tenth of a mile.

Inquiry 5–5 What does Eratosthenes' experiment assume about the shape of the Earth?

Inquiry 5–6 If the Earth were flat, what would be the value of the angle at *C*?

Hipparchus

Perhaps the greatest of all ancient astronomers was **Hipparchus** (c. 150 B.C.). Many of the conclusions he drew were so sophisticated that it takes some knowledge of astronomy to appreciate how great his contributions were. He built an observatory, constructed the best astronomical instruments up to that time, and established a program of careful and systematic observations that resulted in the compilation of a great star catalog, with 850 entries, using a celestial coordinate system similar to our modern one for cataloging the sky. It was Hipparchus who originated a system, which is still in use today in modified form, for estimating the brightness of stars. In addition, he paid much attention to the older Egyptian observations and detected long-term trends in the motions of the celestial sphere that had been previously unsuspected. He deduced Earth's precession, which is so slow that it takes almost 26,000 years for it to complete one cycle. Finally, he greatly developed trigonometry, which was, and still is, a useful tool for astronomy.

Other Deductions of the Greek Astronomers: The Distances of the Planets

The Greeks estimated the relative distances of the planets from Earth, by means of principles still in use today for determining distances to astronomical objects. They reasoned that the more distant a planet was, the more slowly it would move across the sky. The effect is similar to what happens when we compare the apparent motion of a high-flying airplane with that of one that is flying very low. The distant airplane appears to move slowly across the sky, whereas the low-flying one is seen for only a short time and then is gone. In the same way, the Greeks could put most of the naked-eye planets in order of their distance from Earth by assuming that increasing distances corresponded to slower motions. The argument fails with Mercury and Venus, however, because it places Mercury closer to Earth.

Inquiry 5–7 What assumptions are made in employing this argument?

We can obtain another, independent determination of relative planetary distances from their brightnesses. We use an analogy: when you are driving at night and wish to pass the car in front of yours, you pass only if the headlights of the oncoming car are faint. When you do this, you are making an implicit assumption: all car headlights have about the same intrinsic brightness, with their apparent brightness depending on the distance. Similarly, if we assume that all planets have the same intrinsic brightness, then their apparent brightness as seen from Earth would depend on their distances from us. Of course, all the planets do not have the same intrinsic brightness, because their differences in size and distance from the Sun, combined with differences in surface and atmospheric properties, affect the amount of light they reflect in our direction. However, even allowing for these uncertainties, it is still possible to use this principle to rank the planets approximately in order of distance from the Earth.

THE APPARENT MOTIONS OF THE PLANETS RELATIVE TO THE STARS

Three additional observations of planetary motion were important in determining the details of the models the Greeks developed. These observations, which played a prominent role in their models, are:

1. Because the planets are considerably closer to us than the fixed stars, they appear to move against the starry background. Observations of Mars, Jupiter, and Saturn showed them to move generally eastward on the celestial sphere.
2. Occasionally, however, as discussed in Chapter 4, a planet's motion changes from eastward to westward. This retrograde motion would persist for up to several months but would cease as the planet's motion

slowed down and again reversed its direction, resuming its normal easterly motion (see Figure 4–28).
3. Venus and Mercury are never more than 48° and 28°, respectively, from the Sun.

THE GEOCENTRIC MODEL OF THE SOLAR SYSTEM

To describe the observed planetary motions, it was necessary to decide where the center of the system should be. There were really only two obvious candidates—Earth and the Sun. This question was considered carefully by Greek philosophers, and the fact that ultimately they reached an incorrect conclusion provides an interesting example of why the scientific method is not the simple turn-the-crank-and-the-answers-fall-out process that some sources describe it to be. If the Sun is in the center of the solar system, then Earth moves around it in space. Such a hypothesis provides a prediction. As shown in **Figure 5–6**, some of the stars ought to shift their apparent positions in the sky as the Earth moves from one side of its orbit to the other. Such **parallax** effects, as they are called, were looked for by many Greek observers, including Hipparchus, but were never found. The Greeks therefore concluded that the Earth was stationary in space.

Aristarchus, however, apparently espoused the theory that the Earth orbits the Sun, if surviving works of Archimedes and Plutarch are correct. Unfortunately, the work in which he put forth his hypothesis is lost, and apparently no other Greek astronomers held to this opinion. The model of a **geocentric** (Earth-centered)

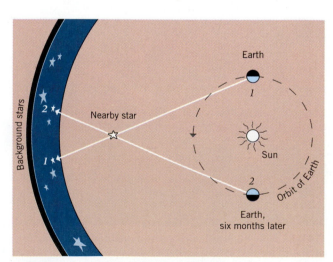

FIGURE 5–6. Stellar parallax. The apparent position of a star with respect to the background stars appears to change as the Earth goes around the Sun. (Not to scale.)

system easily won out over the **heliocentric** (Sun-centered) one.

Inquiry 5–8 What hypothesis might explain why stellar parallaxes were not observed, even though the Earth does in fact orbit the Sun?

THE HYPOTHESIS OF CIRCULAR MOTION

A second feature of the Greek models of the solar system provides another excellent example of the way in which scientific models can go astray—the preconceived notion. Ever since the time of Pythagoras (c. 570–500 B.C.), the circle and the sphere had been considered to be the most perfect geometrical figures. Such perfection was expected in the natural world. (Even today, symmetry and simplicity in scientific thought are important concepts.) Thus the great astronomer Hipparchus continued to subscribe to the then ancient notion that celestial bodies, being perfect, could move only in circular orbits. Furthermore, this prescribed circular motion had to be uniform: a body moved the same distance in its orbit each day. Because of the observations of retrograde motion, however, Hipparchus cleverly extended the idea to include motion that was a combination of circular motions. In this way it was eventually possible to model the retrograde motion of the planets.

The idea of combining circles geometrically is due to Apollonius of Perga (c. 265–190 B.C.), but Hipparchus was the first to apply the idea to actual celestial bodies when he proposed a theory of the motion of the Sun and Moon. However, he did not have enough data to apply it to the planets; this final step was carried out by the astronomer Ptolemy (whom we discuss below).

The explanation of retrograde motion by means of combinations of circular motion is illustrated in **Figure 5–7a**. The planet moves around a small circle called an **epicycle**, and the center of the epicycle moves around on a larger circle called the **deferent**. Because one can adjust the relative sizes of the epicycle and deferent, and the speeds with which motion takes place on each, it is not difficult for the model to produce complicated motions. In particular, if the planet moves around the epicycle faster than the epicycle moves around the deferent, retrograde motion will be observed at certain points along the orbit, as shown in **Figure 5–7b**.

With the invention of the epicycle and deferent, it became possible to explain the special features of the motions of Mercury and Venus by adding only one more feature to the model. If the center of the epicycle of Venus, for example, is attached firmly to the line joining Earth and the Sun, then Venus must always remain near the Sun, as shown in **Figure 5–8**. In a similar fash-

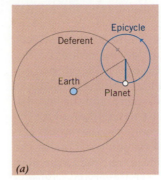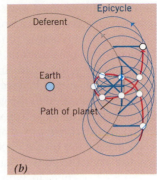

FIGURE 5–7. A geocentric explanation of retrograde motion. (*a*) The planet moves on an epicycle, which moves about the Earth on a deferent. (*b*) The motion resulting from a planet moving about an epicycle as it moves around the deferent.

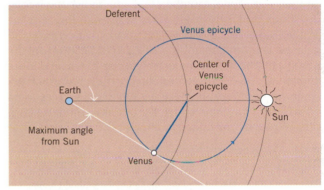

FIGURE 5–8. An explanation of the motion of Venus, using epicycles. According to this model, Venus is always closer to Earth than the Sun is.

ion, Mercury's epicycle is attached to the Earth-Sun line. Now, as the Sun moves along its deferent at a rate of about 1° per day, it carries Mercury and Venus along with it.

Inquiry 5–9 What phases of Venus are possible for the model shown in Figure 5–8?

PTOLEMY

The last of the great ancient astronomers was **Claudius Ptolemy** (c. A.D. 150). In fact, the geocentric system that is discussed above was passed down bearing his name—the **Ptolemaic system**. Ptolemy completed the explanation of planetary motion in terms of combinations of circles and added a number of complex refinements to Hipparchus' system to improve the model's agreement with observations.

Ptolemy's aim was to produce a model that correctly predicted the observations, just as this remains a key goal of science today. In that sense, he succeeded admirably. The Greeks believed that a true knowledge of reality was confined to the gods alone, and the best that humans could do was to produce descriptions of the observed world, the world of phenomena, that would correctly predict the results of observation and experiment. Leaving aside the gods, these ideas are close to the spirit of modern science, in which even everyday concepts such as force and mass, which modern physicists think they understand quite well, have meaning only in reference to models and the measuring processes that are appropriate to them.

Models are necessary not only to interpret observations, but to suggest possible new observations. In fact, an incorrect model can often lead one to ask the wrong questions, and thus it may delay understanding. For example, Kapteyn's model for the galaxy was wrong and delayed our understanding of Earth's position in the galaxy. In astronomy, models necessarily become more and more uncertain as the objects they represent become more distant, because we have less and less data on which to base the model. Despite this, models play a central role in our understanding of the universe.

Inquiry 5–10   What are some models, in the sense just described, that are used in everyday life? (You may want to think about such fields as politics, economics, education, philosophy, psychology, etc.)

5.2
ASTRONOMY DURING THE MIDDLE AGES

The next 13 centuries of astronomical history are notable primarily for their lack of interest. Greek civilization went into decline for various reasons and the Greeks became subservient to the Romans (this had already taken place by the time of Ptolemy). If the Greeks were scientists, the Romans might more properly be called engineers, doing little basic science. Progress in science came to a virtual standstill in the West.

It was a different story in the Middle East. The astronomical knowledge that originated in Greece reached the Arab world, and scientists and scholars there continued to develop a tradition of the pursuit of knowledge. There was a flowering of literature, art, and science throughout the Middle East. One of the most important events was the preservation of much of the

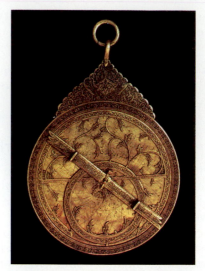

FIGURE 5–9.   A Persian astrolabe. The small leaves mark the positions of bright stars. By suspending the astrolabe by the ring and sighting along the movable sight, we can measure the altitude of a star.

scientific knowledge of the ancient world, at least in translation. Ptolemy's great work, for example, survived as a book called the *Almagest*, Arabic for "the greatest." In this form the Arabs eventually transmitted Ptolemy's opus back to the West.

There was also a considerable amount of original work done in the Islamic world. Much of it was directed toward practical matters, such as navigation. Beautiful and intricate instruments such as the astrolabe (**Figure 5–9**) were brought to a high degree of perfection. Particularly in mathematics, much theoretical work was accomplished, aided by the "Arabic" system of numbers that had come from India, as well as by perfection of the methods of algebra (the word algebra is Arabic).

At the same time in China, an indigenous tradition of astronomy made important progress during long periods of political stability. Astronomy was always closely connected with the state, and changes in dynasty were frequently accompanied by calendar reforms, providing secure employment for court astronomers. Early on, the Chinese developed the view that space was infinite in extent and that the stars floated independently in it. They were assiduous and systematic observers of heavenly events; often our only records of novae and comets are Chinese. They invented advanced instrumentation for observing the sky and measuring time, including elaborate water-driven clocks. In many ways their science was more advanced than what the Greeks had accomplished. Joseph Needham, the eminent scholar of Chinese science and technology, has shown that the influence of Chinese science on the West was much greater than has generally been recognized.

In Europe, on the other hand, intellectual decadence was widespread. The principles of Greek science, received from the Arabs, were adopted as dogma by the increasingly powerful Roman Catholic Church. It was argued that there was only a finite amount of knowledge and that the Greeks had discovered it all since they had come first! Discovery was no longer necessary; humans had only to memorize and copy into manuscripts what the Greeks had found to be true. The Ptolemaic picture became frozen into a rigid image of the universe. In it there was a well-defined place for God and the Devil, for the Seraphim and the planets, for Tygers and humankind.[2] However, there was no place for independent thought and inquiry that challenged this view of the universe.

[2]These references come from the picture of the cosmos detailed in Dante's *Divine Comedy*.

FIGURE 5–10. Nicholas Copernicus (1473–1543) proposed the daring idea that the Sun, not the Earth, was at the center of the universe.

THE HELIOCENTRIC HYPOTHESIS

We will now look at the contribution of three scientists whose work was important in returning Western thought to a Sun-centered view.

COPERNICUS

During the Middle Ages, tinkering with the Ptolemaic model had continued, but in an uncreative way—for example, by adding more epicycles. In fact, this model had become so complicated that King Alfonso of Castile is said to have remarked, when having the contemporary version of the Ptolemaic theory explained to him, that if *he* had been around when the world was created he could have taught the Creator a thing or two.

It is difficult to contain the human mind forever. By the early sixteenth century the Renaissance was shaking the antique castles of Western intellectualism. It was in this changing atmosphere that **Nicolas Copernicus** (1473–1543), a Polish prelate with a strongly mathematical bent as well as a new vision of the heavens (**Figure 5–10**) found himself. As a student in Italy Copernicus read of Aristarchus's heliocentric hypothesis. Wherever he got the idea of placing the Sun at the center of the universe, there is no doubt that he felt this was the proper place for that magnificent body (**Figure 5–11**).

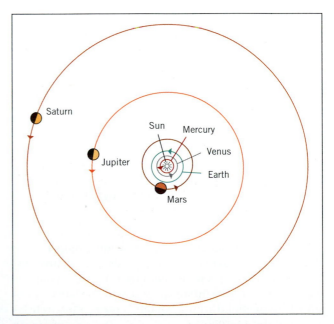

FIGURE 5–11. The Copernican model of the solar system.

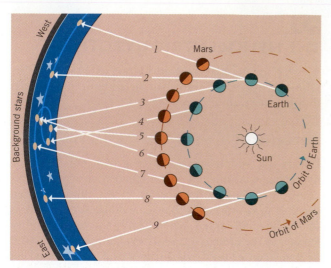

FIGURE 5–12. Retrograde motion is a natural consequence of the heliocentric hypothesis.

Moreover, in a heliocentric model, the apparent retrograde motion of a planet like Mars is a simple consequence of the relative motions of the Earth and Mars. **Figure 5–12** shows how this comes about. At some time Mars will appear in direction *1*. As each planet orbits the Sun, Mars' line-of-sight position relative to the background stars will move toward the east, as indicated by lines *1–4*. However, between times *4* and *6* Mars will appear to move backwards, in a retrograde direction, toward the west. At point *6* Mars will have ceased its retrograde motion and resumed its normal easterly track. As you can see, Mars exhibits retrograde motion whenever the Earth "laps" it in orbit. The greater orbital speed of the Earth results in the apparent backward motion of Mars, which is actually a parallax effect. One sees a similar effect when a fast automobile passes a slower one on the highway. Viewed from the faster car, the slower one appears to move backward when seen against the distant landscape.

Inquiry 5–11 How must the positions of the Earth, Sun, and Mars be related for the retrograde motion to be greatest?

An important benefit of the heliocentric model is that it allowed the sidereal orbital periods and relative distances of the planets to be measured. The resulting systematic increase of orbital period with distance is not only sensible but, through Occam's razor, more satisfying than the alternatives.

Despite its simple explanation of retrograde motion and elegant positioning of the Sun in the center of the solar system, many objections were raised against the Copernican hypothesis. For example, some contended that if the Earth moved it would leave the Moon behind, an argument that was hard to answer because the concept of gravitation had not yet been developed. More seriously, Copernicus's theory in this simple form could not explain the observations as accurately as Ptolemy's theory. It was necessary to introduce a large number of small epicycles to explain numerous small deviations from circular motion. Moreover, Copernicus was opposed to Ptolemy's complicated refinements, feeling that they detracted from the perfection of uniform circular motion and were unworthy of celestial bodies. This conservatism on Copernicus's part meant that he actually needed *more* epicycles than Ptolemy to get equivalent accuracy, and this marred the aesthetic simplicity of his basic heliocentric theory, which used circular orbits.

Inquiry 5–12 Employing the principle of Occam's razor, from the point of view of astronomers of the time, which of the two theories of the universe would have been preferable, Ptolemy's or Copernicus's?

Unquestionably, Copernicus was aware of the difficulties with his model. In addition, he was a rather timid man, and it is possibly for these reasons that publication of his theories was delayed until he was literally on his deathbed. His work did appear, however, in the year 1543, entitled *De Revolutionibus Orbium Cælestium*, which means *On the Revolutions of the Heavenly Spheres*. It contained a preface stating that the book expounded a mathematical model for solely discursive purposes and was not to be construed as a representation of reality. It is certain that Copernicus didn't write the preface and doubtful that he would have sanctioned its inclusion had he known of it.

TYCHO BRAHE

In 1546, a Danish nobleman was born who was to become the first great observational astronomer of the modern era—**Tycho Brahe** (pronounced Tee-ko Brahe) (**Figure 5–13**). Tycho became interested in astronomy as a teenager when he observed a predicted eclipse of the Sun. Later, in 1572, he observed a supernova, an exploding star, that was so bright it could be observed in daytime. Through careful observations, Tycho showed that the daily rotation of the Earth caused no parallactic shifts in the supernova's position—in fact, the star did not move at all during the many months it was visible.

FIGURE 5–13. Tycho Brahe (1546–1601) was the greatest naked-eye observer in the history of astronomy.

Tycho distinguished himself by the great care he brought to his observations. He also repeated observations and averaged them together to cancel random errors. So painstaking was his work that his observations had unprecedented accuracy, not to be improved upon until invention of the telescope. Another valuable aspect of his observations was that he observed continuously and systematically over many years, as Hipparchus had done. As a result, he amassed a large body of data of consistently high accuracy that was suitable for further investigation. It is something of a mystery why an aristocrat such as Tycho could become as obsessed with astronomy as he was, and it is miraculous that his obsession led him to produce such excellent astronomical data. Another fortunate historical accident is that this body of work eventually fell into the hands of Johannes Kepler (1571–1630), who, as it turned out, was uniquely equipped to put it to good use.

FIGURE 5–14. Tycho's great mural quadrant, his principal instrument for measuring star positions.

He therefore concluded that it must be well beyond the Moon, and probably as distant as the stars. This caused instant problems for the older theories of the universe, which had assumed that the heavens were immutable. Five years later, Tycho proved that the comet of 1577 was also more distant than the Moon. Because comets had been believed to be "exhalations of the Earth" this posed additional problems for the older theories.

The fortunate combination of Tycho's noble birth and his astronomical talents soon earned him the finest observatory up until that time in the Western world. It was financed from the coffers of King Frederick II of Denmark and was located on an island off the Danish coast. Since optical instruments had not yet been invented, the instruments were sighting devices similar to the quadrant and cross-staff, but larger in size and constructed with the greatest possible precision. Tycho's principal instrument was the great mural quadrant (**Figure 5–14**). By sighting along the movable pointers in a manner not unlike sighting a gun, he could aim the pointers at two objects and read the angle between them with great accuracy (to about 0.5 minute of arc).

JOHANNES KEPLER

Although Tycho was a great observer, he realized that he was not a strong mathematician, and he longed for a collaborator who could properly interpret his excellent observations. His accumulated observations on the positions of Mars posed the greatest difficulties for theoretical interpretation; Mars deviated from its predicted position more than any other planet. Having moved to Prague toward the end of his life, Tycho acquired a young assistant who by his brilliance in mathematics transcended his lower-class origins. **Johannes Kepler** (**Figure 5–15**) had sufficient genius to overthrow the sterile theories of the geocentric universe and perfect circular motion and to substitute a new and truer description of the solar system.

In his book *The Sleepwalkers*, Arthur Koestler describes the contrast between Tycho and Kepler. Tycho was an aristocrat by birth and arrogant by nature, accustomed to power and privilege. He ran his observatory like a court, arriving for the night in full formal dress and imperiously ordering his assistants about. Always sure of himself, in his college days he had lost his nose in a duel over a point of mathematics, and ever afterward he wore a false noses made of silver or bronze. He had a dwarf servant named Jepp, who followed his master about like a pet and received scraps of food at the supper table. Kepler, on the other hand, had no advantages of birth and reached a position of eminence by

virtue of brilliance and sheer tenacity. It appears that he had little in common with Tycho, other than irascibility and an interest in astronomy. During much of his life, he had to earn his living by casting horoscopes, and in fact he cast his own horoscope every day. He was a compulsive individual who kept meticulous notes on everything he thought and did. A hypochondriac as well, Kepler recorded an hour-by-hour chronicle of his physical maladies. But because his diary also included his scientific efforts, we have been left a detailed account of the paths through which he wandered in making his momentous discoveries.

Kepler took Tycho's observations of Mars (literally took them, because Tycho's heirs had other plans for the data!) and set out to find a geometric curve that would represent its motion accurately. The calculations were extremely tedious, and he made many mistakes. Without calculators or even logarithms, every calculation had to be done by long multiplication or division. Yet something in Kepler's character kept him working persistently at the problem. Time after time he rejected solutions that had taken him months, even years, to work out, because they failed to agree with the observations as accurately as he knew they should. At last, after a total of eight years, he boldly rejected the hallowed notion that planetary motion must take place on circular paths, thus ending two millennia of tradition. He describes in his diaries the fear and trembling he suffered in his mind when he took this step.

FIGURE 5–15. Johannes Kepler (1571–1630), seen here with Käiser Rudolf II, refined Copernicus's heliocentric hypothesis with the laws of elliptical motion.

5.4

KEPLER'S LAWS OF PLANETARY MOTION

Kepler's laws of planetary motion are not "laws" in the sense that we use the word today, because there was at that time no conception of the physical forces that caused these motions. Although he came close to the notion of gravitational force, Kepler didn't quite succeed—he tried to construct a theory in which the force was repulsive rather than attractive. His laws would be better described as empirical descriptions of planetary motions. However, for the first time they gave a description that was as accurate as the best available data allowed.

Kepler found that Mars moved on a mathematical curve called an **ellipse**. Figure 5–16 shows a simple way to draw this curve. The two points through which the pins are pushed are called the **foci** of the ellipse (foci is the plural of **focus**). Ellipses are characterized by the length of their longest dimension (called the **major-axis**—B in the figure) and their degree of noncircularity. If the two foci coincide, we have a circle. The further apart the foci, the greater is the **eccentricity** of the ellipse.

Kepler's first law can be stated in this way:

The orbits of the planets are ellipses, with the Sun at one focus.

Figure 5–17 illustrates Kepler's first law. The flattening of the ellipse is much exaggerated—no planet has an orbit this elliptical. In fact, if one were to observe these orbits from outside the solar system, it would be hard using the naked eye to distinguish most of them from circles. This is the principal reason why it took so long to discover the true shapes of the orbits of the planets. As shown, the second focus is empty. Eccentric ellipses are also found; comets are examples of objects that move in highly eccentric ellipses.

The distance of a planet from the Sun varies as the planet moves. For example, Earth is 3 million miles closer to the Sun in January than in July. Kepler found that as Mars' distance from the Sun varied, its orbital speed also varied, being greatest when the planet was closest to the Sun and least when it was farthest away. He found even more: there is a definite relationship between a planet's distance from the Sun and its speed in its orbit.

Figure 5–18 indicates four positions on an elliptical orbit. During the interval of time between points A and B, a line from the planet to the Sun sweeps out the long, skinny area shaded on the left. Between the *equal* time interval from C to D, the line from the planet to the Sun sweeps out the fat area on the right. Kepler found that

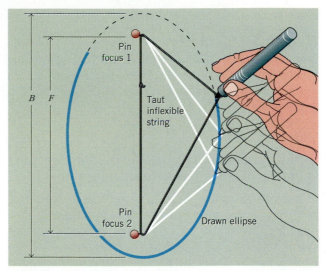

FIGURE 5–16. Drawing an ellipse with two pins and a loop of string. The pins mark the position of the two foci of the ellipse. The eccentricity of an ellipse is defined to be $e = F/B$, where F is the distance between the two foci and B is the longest axis (the major axis) of the ellipse. It characterizes how far the ellipse deviates from being a perfect circle.

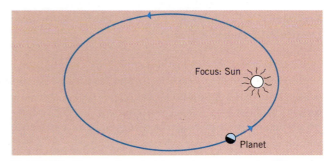

FIGURE 5–17. Kepler's first law: Planets move in ellipses with the Sun at one focus.

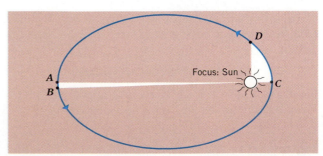

FIGURE 5–18. Kepler's second law: If the planet takes the same amount of time to go from A to B as it does to go from C to D, then the two shaded areas are equal.

this area is equal in size to the area swept out between times A and B. In other words, because the time intervals from A to B and C to D are equal, and because the distance from C to D is greater than that from A to B, the planet must be moving faster from C to D than from A to B. For example, the motion of the Earth is more rapid in January than in July.

Kepler's second law of planetary motion can be formulated as follows:

> **The line joining a planet and the Sun sweeps out equal areas in equal amounts of time.**

Kepler's third law was discovered much later than the first two. It appears, almost as an afterthought, in a rambling, mystical work of Kepler's entitled *Mysterium Cosmographicum*, or *Cosmic Mystery*. The emphasis of this work lay in mystical speculation on the cosmos rather than on the mathematical relationship expressed in the third law.

Kepler's third law is a relationship between the average distance of a planet from the Sun (which equals half the length of the major axis, or **semi-major axis**) and the planet's sidereal period, the length of time it takes for the planet to orbit once around the Sun. The third law is most easily expressed mathematically. It states that if P is the orbital period of a planet measured in years and A its average distance from the Sun measured in astronomical units, then

$$P^2 = A^3.$$

This equation clearly works for the Earth. The Earth's orbital period is 1 year and its average distance from the Sun is 1 AU, so $P = 1$ and $A = 1$, and $1^2 = 1^3$. We can express the equation in equivalent forms that are easier to use for computation:

$$P = A\sqrt{A}, \text{ and } A = \sqrt[3]{P^2}.$$

For example, if a planet's semi-major axis is 9 AU, one finds that its orbital period is

$$P = 9\sqrt{9} = 9 \times 3 = 27 \text{ years.}$$

On the other hand, a planet that has an orbital period of 8 years would have an average distance from the Sun of

$$A = \sqrt[3]{P^2} = \sqrt[3]{8^2} = \sqrt[3]{64} = 4 \text{ AU}$$

because $4 \times 4 \times 4 = 64$. Problems using Kepler's third law are easily solved with a pocket calculator.

Inquiry 5–13 The average distance of Mars from the Sun is 1.52 AU. To the nearest tenth of a year, what is its sidereal orbital period?

Inquiry 5–14 Uranus's sidereal orbital period is 84 years. To the nearest tenth of an AU, what is its average distance from the Sun? (Hint: $20 \times 20 \times 20 = 8000$.)

Kepler's three laws, as described here, are exact only in the idealized case of two isolated bodies. In the real solar system, the large number of bodies that interact with one another causes deviations from the ideal. Nevertheless, the laws are accurate enough to describe the motions of the planets and agree with the best naked-eye observations, which happened to be Tycho's. Had Tycho's observations been somewhat less accurate, or had Kepler been less exacting in his criteria for agreement between theory and observation, Kepler's three laws might not have been discovered until much later in history.

Kepler's three laws overthrew the earlier ideas; no longer did we believe that the planets moved with uniform speeds in circular orbits. These laws are general in that they apply to any two bodies circling each other. They may be applied to satellites orbiting planets, stars orbiting stars, or even galaxies orbiting galaxies. Thus we will return to Kepler's laws at many places throughout our study of astronomy.

5.5
THE SEARCH FOR UNDERLYING LAWS

Kepler's works are purely empirical, meaning they were derived solely on the basis of observation with no theoretical underpinning. The reasons Kepler's laws are valid can be understood because of Galileo's experimental work on the motion of bodies, and Newton's beautiful theoretical development of his laws of motion.

GALILEO

Galileo Galilei (1564–1642), shown in the chapter opening photograph, was a contemporary of Kepler's, was an early convert to the heliocentric hypothesis, and argued vigorously for it. In his later years this brought him into conflict with the Inquisition, and he was forced to recant his beliefs and live under a form of house arrest—a pitiful, partially blind old man, fearful of the instruments of torture that apparently had been shown to him.

Kepler's laws, successful as they were in predicting the positions of the planets in the sky, were no proof that the Earth moves, and skeptics could reasonably claim that the three laws were merely mathematical tricks that happened to give accurate predictions and not in any way a physical theory about the nature of the universe.

Galileo clung to the hypothesis of circular motion for heavenly bodies while conducting experiments in dynamics that eventually overthrew it. Aristotle had claimed that a body sought to be at rest in its natural place, and all natural motions expressed the striving of a body to rest as close as possible to its natural place. Because according to Aristotle the natural place for "earthy" material was on the Earth, such material fell to the ground. Similarly, the natural place for "airy" material was in the atmosphere. He termed all other motions violent motions, and he taught that such motions required continuous contact between the mover and the moving object to perpetuate the motion. Galileo, in a revolutionary move, decided to let Nature, not human reason, be the ultimate arbiter of physical reality. He performed experiments to decide whether a particular hypothesis was true or not. For example, he investigated the motions of objects under various conditions by rolling balls down inclined planes, thus showing that a state of motion was as natural as a state of rest, and that to change either state required an outside influence. In another famous case, it had been claimed by believers in Aristotelian physics that an object that was twice as heavy as a second one would fall twice as fast. Perhaps by actual demonstration, or maybe only with thought experiments, Galileo showed that such objects accelerated at nearly the same rate and hypothesized that only the resistance of the air kept the rates from being identical. The truth of Galileo's position was shown dramatically by the Apollo astronauts, who dropped a feather and a hammer together in the vacuum of the Moon. The two objects reached the lunar surface simultaneously. You, too, can perform Galileo's experiment by simultaneously dropping, say, a paper clip and a heavy shoe.

In performing his experiments, Galileo formulated the concept of **inertia**, which is a body's resistance to a change in its motion. The amount of inertia is measured by what we call **mass**, which is, loosely, a measure of the amount of matter in a body.

Galileo's greatest fame comes from his discoveries in astronomy, even though his contributions to physics were probably more important. Upon learning of the invention of the telescope, he immediately constructed one of his own and turned it to the heavens. His most important astronomical discoveries were the following:

1. He found that the diffuse band of light across the sky, known as the **Milky Way**, actually consists of myriad stars too faint to be seen by the naked eye. This discovery was the first step toward a modern view of the nature of our galaxy.
2. He examined the Moon in detail, and discussed and named many of its surface features—its mountains, *maria* ("seas"), craters, and the like.

3. While observing with a telescope, he discovered that the planets had a visible disk, whereas the stars remained infinitesimal points even at the highest magnification.
4. He discovered the four brightest satellites of Jupiter by following the variations in their positions from night to night and showing that they always remained close to the moving planet.

Inquiry 5–15 What might be the reason for Galileo's third observation?

Inquiry 5–16 One of the arguments against Copernicus's theory was that the Earth could not move because it would leave the Moon behind. How does Galileo's fourth observation help disprove that contention?

---

5. He studied sunspots and, by following them across the visible disk of the Sun, showed that the Sun rotated. Together with his observations of the Moon, this discovery struck a blow to the older theories of astronomy, which had maintained that the heavenly bodies were perfect, without blemishes.
6. He found that the planets could show phases, as the Moon does. This was particularly important in the case of Venus, which exhibits all phases. In the Ptolemaic theory, Venus could only show crescent phases, because it would always be closer to Earth than the Sun (**Figure 5–19a**). Only if Venus were sometimes

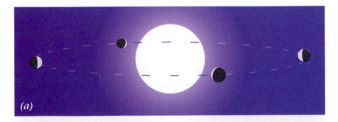

FIGURE 5–19. (*a*) Venus's phases under the Ptolemaic theory, as viewed from Earth. Note that only crescent phases are seen. (*b*) Venus's phases in a heliocentric theory, as viewed from Earth. Because Venus can go behind the Sun, a full set of phases including gibbous phases is possible.

farther from Earth than the Sun could it be more than half-illuminated and exhibit a gibbous phase (**Figure 5–19b**). This discovery showed that, in at least one respect, the Ptolemaic picture of the universe must be incorrect.

NEWTON

As important as the contributions of Galileo and Kepler were in altering long-held ideas about astronomy and physics, they did not create a single comprehensive system that unified the physics of the heavens with the physics of Earth. Such a unification awaited the genius of **Isaac Newton** (1642–1727),[3] who was born in the year of Galileo's death. Newton provided us with a system of physical laws that is so fundamental that even today they are the basis for much scientific activity (**Figure 5–20**). It would be difficult to give an adequate idea of the power of Newton's intellect and the importance of his accomplishments; we can only scratch the surface. As a young man Newton invented the tools of the branch of mathematics known as calculus, did fundamental work on optics and color vision, and developed the reflecting telescope, as well as proposing his famous laws of motion and gravity. He made such an impression on his contemporaries that the poet Alexander Pope was moved to write: "Nature and Nature's laws lay hid in night:/God said *Let Newton be*! and all was light."

To begin to understand Newton's contributions to the study of moving bodies we must define some terms. **Velocity** describes the change in position of a body divided by the time interval over which the change occurs. Velocity describes not only the speed of a body but its direction. For example, while a car traveling 60 miles per hour north moves at the same *speed* as a car traveling 60 miles per hour east, their *velocities* are different because their directions of motion are different.[4] Motion is rarely constant in either amount or direction. Any change in a body's velocity, in either its speed or direction, is called an **acceleration**. When traveling at a constant speed on a Ferris wheel, you are accelerating because the direction is continuously changing.

Newton is most renowned for his three laws of motion and his law of gravity. His famous three laws of motion, first published in his *Philosophiae Naturalis Principia Mathematica* (*The Mathematical Principles of Natural Philos-*

FIGURE 5–20. Sir Isaac Newton (1642–1727) laid a firm theoretical basis in mathematics and physics for planetary motions, which was to reign unchallenged for over 200 years.

ophy), are generalized statements concerning the motions and interactions of bodies. The first law concerns inertia.

> **In the absence of an outside influence, a body at rest will stay at rest, while a body in motion will continue to move at a constant speed in a straight line.**

If this statement (usually called the **law of inertia**) seems obvious to you, a little history might illustrate how revolutionary it was. Since the time of Aristotle, it had been assumed that a body required some continual action on it to remain in motion, unless that motion were a part of the natural motion of heavy or light substances. The philosophers of the Middle Ages had refined Aristotle's view by asserting that an object was imparted a certain quantity of "impetus" by the hand or bow string that launched it, and that it would continue to move until it had used up its impetus. Impetus, however, was an example of what Newton referred to as an "occult quality," for it could not be perceived in any fashion. It was merely supposed to exist in order to preserve Aristotle's theory. Thus all motion seemed to require the invocation of an unknown substance to explain it. Newton's simple statement swept away all that

---

[3]Newton's birthdate was December 25, 1642, according to the Julian calendar used in England at that time. This date is January 4, 1643, according to the Gregorian calendar we use today.

[4]Velocity is an example of a *vector*, a quantity having both size and direction. Speed, on the other hand, is called a *scalar* because it has only a size associated with it.

chaos and mental clutter, replacing it with brilliant simplicity.

What happens if there is an outside influence? In answering this question, Newton invoked the concept of a **force** acting on a body and causing it to accelerate. Thus the second law of motion:

> **The total force on a body is equal to the product of the body's mass times its acceleration.**

This is the famous formula $F = ma$, which forms the basis for studying the motions of particles of all sizes, from submicroscopic molecules of gas to stars in their orbits in the galaxy. With this law, Newton formulated for the first time the concept of force, which turned out to be enormously powerful and gave us an operational definition of how forces act. Namely, forces produce accelerations, and the size of the acceleration depends on the mass. Furthermore, a body can accelerate *only* if there is a force acting on it; remove the force and the body no longer accelerates. Finally, because acceleration has size and direction, so does force.

If a body is accelerated, it means that either its speed or its direction of motion is changing, or both. If a body is accelerated in the direction of motion, speed increases. Acceleration in the direction opposite to the direction of motion decreases the speed. A sideways acceleration changes its direction of motion. Any acceleration (forward, backward, or sideways) requires an external force that points in the direction of the acceleration. Further, the greater an object's mass, the larger the force required to produce a given acceleration.

Inquiry 5–17 Consider a game of kick-the-can played with two cans—one empty and one filled with concrete. Which can has the greater mass? If someone came along and kicked the two cans with exactly the same force, which can would have the greater acceleration? Explain this in terms of what you have learned.

In the third law, Newton shows that forces always come in pairs.

> **For every force that a body exerts on a second body, there is an equal and opposite force exerted by the second body on the first.**

As an example of pairs of forces, when someone touches you, you also touch that person; one touch cannot occur without the other. If you and a friend are in rowboats on a lake and one of you pushes against the other with an oar, *both* boats will move, but in opposite directions. The simple act of sitting in a chair illustrates

this law: you exert a downward force on the chair, and the chair exerts an equal and opposite force on you. While a falling brick exerts a force on the air molecules it encounters, pushing them ahead of it, the air molecules exert an equal and opposite force on the brick, thus slightly slowing its motion. A final example is a rocket. If one force is the rocket engine pushing gases backward, the opposite force is that of the gases pushing on the rocket engine in the forward direction. That is why rockets work in the vacuum of space; rockets *do not* operate because of the exhaust gases pushing on the Earth.

Inquiry 5–18 In the game in Inquiry 5–17, what does this third law predict about the effect on the foot that kicks the two cans?

NEWTON'S LAW OF GRAVITY

Newton's three laws, especially the second, enable us to calculate the acceleration of a body and hence its motion, but we must calculate the forces first. It is evident that heavenly bodies such as the Sun, Moon, and planets cannot exert forces on each other directly, since they do not touch each other. Newton proposed that they exert an *attractive* force on each other at a distance, that is, across empty space. He called this force "gravitation." His ideas were brought together in what is known as the **universal law of gravitation**, which expresses the gravitational force of attraction between *any* two bodies. Consider two bodies, M_1 and M_2. The gravitational attraction between the mass M_1 and the mass M_2 must be the same as the attraction between M_2 and M_1 For this reason, the force must be proportional to the product of the masses, $M_1 M_2$. The variation of the gravitational force with the distance between the centers of the bodies was found using reasoning based on geometry. Newton argued that the force had to vary inversely with the square of the distance between the two gravitating bodies. Combining the variation with mass and distance, Newton suggested that gravitational force could be calculated from the formula

$$F = \frac{G\, M_1 M_2}{d^2}$$

where F is the force exerted by a body of mass M_1 on a body of mass M_2, d is the distance between the centers of the two bodies, and G is a numerical constant called the **universal gravitational constant**. The value of the universal constant depends on the units used to express the other quantities in the equation. From this formula, it is possible to calculate how the force between the bodies varies with the distance between them, and therefore what the acceleration is at each moment.

Inquiry 5–19 If the distance between mass M_1 and mass M_2 is made four times *greater*, by what factor is the gravitational force between them changed? Is it increased or decreased?

While the idea that bodies of different masses fall at the same rate was experimentally shown by Galileo to be true, its truth does not make sense to most people. Its validity is shown in Appendix A7.

It is a measure of Newton's genius that he proposed that his laws of motion and gravity applied not only to objects on Earth but to all bodies in the universe. One of the first things he did was to compare the force of gravity exerted by the Earth on a body, allegedly a falling apple, with the gravitational force that would be required to keep the Moon in its orbit around the Earth. As shown in **Figure 5–21**, if there were no gravitational force exerted by the Earth on the Moon, it would travel in a straight line past the Earth in accordance with Newton's first law. To make the Moon travel around the Earth requires a force toward the Earth. The Earth's gravitational force on the Moon causes it to deviate from a straight line and follow a curved path around the Earth. For each kilometer the Moon moves in its orbit, the Moon must fall 0.14 cm towards the Earth's surface in order to stay in its elliptical orbit. Because of the curvature of the Earth's surface, the distance of the Moon from the Earth remains constant. In other words, the Earth's force of gravity causes the Moon to accelerate just enough to maintain its constant distance above the Earth's surface. For this reason, the Moon's orbital motion can be described as resulting from the Moon's falling toward's the Earth's center! From such considerations Newton found that the actual force required to keep the Moon in its orbit around the Earth agreed well with the value he computed theoretically.

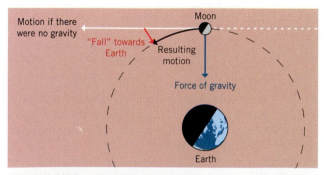

FIGURE 5–21. The motion of the Moon around the Earth according to Newton. The Moon "falls" toward the Earth just enough to keep it on a curved elliptical path around the planet.

WEIGHT

Weight is the force a gravitating body exerts on another mass. In particular, your weight is the gravitational force exerted by the Earth's mass on your body. Your weight depends on the mass of your body, the mass of the Earth, and your distance from the Earth's center, which is the Earth's radius. For this reason, your weight would be different should you travel to a different planet having a different mass and size than the Earth; your mass, however, would be the same. Interested readers can find weight expressed mathematically in Appendix A8.

> TO UNDERSTAND WHY AN ASTRONAUT IS WEIGHTLESS EVEN THOUGH THERE IS STILL GRAVITY IN SPACE, YOU SHOULD DO DISCOVERY 5–1 AT THIS TIME.

MOMENTUM

You probably have some intuitive feel for the word **momentum**. A train moving at 20 miles per hour has more momentum than a bicycle moving at the same speed; it would have more impact and effect if it ran into something! For a body moving in a straight line, **linear momentum** is defined as the body's mass times its velocity.

Newton's first law can also be expressed as the principle of the **conservation of linear momentum**. This says that for an isolated system of bodies the sum of the linear momenta of all bodies in a system is always the same. If some bodies in a group slow down, others must speed up by a corresponding amount.

Most bodies move in curved paths, and the concept of **angular momentum** comes into play. The amount of angular momentum possessed by a planet in a *circular* orbit around the Sun is defined by

angular momentum =
    mass × speed × distance from planet to Sun.

Expressed symbolically, if M is the mass of a body and v its speed when at a distance r, the angular momentum is given by

$$angular\ momentum = Mvr.$$

Like linear momentum, angular momentum is also conserved for isolated systems. If the mass of an orbiting planet remains constant, the only way for angular momentum to remain constant is for the speed of the planet to increase as the distance decreases, and vice versa. This is exactly what Kepler's second law says. The concept of conservation of angular momentum is important in understanding not only the motions of planets but also such diverse subjects as the formation of the

(a)   *(b)*

FIGURE 5–22. An ice skater's spin rate increases due to conservation of angular momentum when the arms are pulled in.

solar system, the motions of stars, and the shapes of galaxies. As applied to the orbits of the planets, the law states that the angular momentum of an orbiting planet remains constant.

A spinning body also has angular momentum. In this case, r represents the extent of the body perpendicular to the rotation axis. As a spinning gas cloud collapses to form a star, conservation of angular momentum requires that it increase in speed, just as an ice skater increases rotation speed as the arms are brought in (**Figure 5–22**).

For simplicity we have defined angular momentum in terms of circular orbits, but the concept can be extended to elliptical orbits.

Inquiry 5–20 Suppose a planet's farthest distance from the Sun is twice as great as its nearest distance. How many times greater is its orbital speed when the planet is nearest the Sun than when it is farthest away?

NEWTON'S FORM OF KEPLER'S THIRD LAW

Newton was able to show, using the calculus that he invented to solve the problem of orbital motion, that Kepler's empirical laws of planetary motion were direct consequences of his own more fundamental laws of motion. For example, he was able to prove that a planet governed by a force of gravity that varies inversely with

the square of the distance from the Sun would follow an elliptical orbit. In addition, using his laws he found that Kepler's third law was even more general than originally proposed. In Newton's formulation, it becomes

$$(M_1 + M_2)P^2 = A^3$$

where again P is measured in years and A in astronomical units, and the additional factor $(M_1 + M_2)$ is the sum of the masses of the two bodies in orbit about each other, expressed in terms of the mass of the Sun. When applied to the planets, $(M_1 + M_2)$ is almost precisely equal to 1, because the mass of the Sun is so much greater than that of any of the planets. (This is why Kepler did not discover it.) However, the form in which Newton wrote Kepler's third law turns out to be much more useful than Kepler's form. For example, if you observe the orbital period and separation of a planet's satellite, you can compute the mass of the planet; or, by observing the size of a double star's orbit and its orbital period, you can deduce the masses of the stars in the binary system. The revised ideas even apply to galaxies in orbit around each other. In fact, much of the observational data we have on the masses of objects in the universe come from applying Newton's form of Kepler's third law in one way or another.

As an example of its use, if two stars identical to the Sun (so the mass of each in solar units is 1) revolve about each other at a distance of 0.1 AU, we can use Newton's form of Kepler's third law as follows:

$$(1 + 1)P^2 = (0.1)^3$$
$$P^2 = \tfrac{1}{2} \times 10^{-3}$$
$$P = 2.2 \times 10^{-2} \text{ years} = 8.2 \text{ days.}$$

If, on the other hand, the binary system had a star identical to the Sun and one six times as massive, the period would be 4.4 days.

Inquiry 5–21 Suppose a double star is observed to have a period of 10 years and an average distance between the two stars of 8 AU. What is the sum of the masses of the two stars in the system?

SUCCESSES OF NEWTON'S LAWS

The publication of Newton's laws of motion set the imagination of the world afire. Educated people of all backgrounds and disciplines set about trying to understand them and then apply them to all kinds of situations to see whether they were indeed as universal as Newton had claimed. Newton himself had difficulty explaining the motion of the Moon with accuracy comparable to observations, because the gravitational force of the Sun, as well as that of the Earth, was important and altered the orbit away from purely elliptical motion. Eventually, however, the mathematical problems of computing the Moon's motion were solved satisfactorily. In a similar way, as telescopes and observational techniques improved, the gravitational effects of one planet on another had to be taken into account to explain observed deviations from purely elliptical motion.

An early triumph of Newton's theory was the discovery by Edmund Halley, a contemporary of Newton, that several comets that had been observed in the past were in fact one and the same body. Halley proceeded to predict that it would reappear in 1758. Although Halley died before the comet returned, its reappearance electrified Europe and led to a great spate of comet hunting. The comet, of course, was named for Halley. It returned again, right on schedule, during late 1985 and early 1986; its next appointment near the Sun will be in the year 2061. Due to gravitational perturbations of the planets, the exact date cannot yet be determined.

In 1781, another discovery was made that at first led to one of Newtonian theory's greatest challenges and later to one of its greatest triumphs. Sir William Herschel, during a routine survey of the sky, discovered a new object that, unlike the stars, showed a clearly visible disk in his telescope. It was soon apparent that the object was moving with respect to the stars, and when the theory of gravity was used to establish its orbit, it was found to be a new planet, farther from the Sun than any previously known. Eventually the new planet was named Uranus, after the Greek god of the heavens.

The discovery of a new planet was not in itself a challenge to Newtonian physics and at first seemed to be a confirmation of it. But after many years of observations, it became evident that something was wrong. No matter what was done, the orbit of Uranus could not be brought into agreement with the predictions of the law of gravity. In late 1845 and early 1846, two young mathematicians simultaneously proposed that there was another, as yet undiscovered planet that was affecting Uranus's orbit and causing it to deviate from the predictions of Newton's laws. The English astronomer John Couch Adams sent his calculations to the Astronomer Royal, who failed to appreciate the importance of the young man's work. The Frenchman U. J. J. Leverrier had better luck. His predictions were sent to the Berlin Observatory, where the new planet was found almost immediately, close to the predicted position. Both Adams and Leverrier now share the honor of Neptune's discovery.[5]

[5]Galileo recorded an object in his notes in the year 1613 that turns out to have been Neptune, according to Charles Kowal of Palomar Observatory. Of course, Galileo did not recognize that he had seen a planet, so the honor of Neptune's discovery rightfully belongs to Adams and Leverrier.

Neptune also showed deviations from Newtonian predictions, and it was tempting to postulate an unknown planet to explain them as well. The deviations were small, so the predicted position of the new planet was much more uncertain. The search was begun in 1906 by Percival Lowell (the brother of the poet Amy Lowell), but the planet was not discovered until after his death by 24-year-old Clyde Tombaugh, working at Lowell Observatory in Arizona. At first, the discovery of Pluto was hailed as another triumph of the method of Adams and Leverrier, but more recently it has become evident that the deviations in Neptune's orbit are really too small to have led to a reliable prediction of its position. It appears that the discovery of Pluto was based more on luck and hard work than on anything else.

5.6
PROOF OF THE HELIOCENTRIC HYPOTHESIS

The great simplicity and many triumphs of the Newtonian theory quickly won over most skeptics to the heliocentric hypothesis. All that is required to prove the heliocentric hypothesis is a clear indication that the Earth moves, and astronomers, delighted with Newton's theories and convinced of their truth, now sought observational evidence for the motion of the Earth. But it was not until the year 1729 that any was discovered. In that year James Bradley, the British Astronomer Royal, who was trying to observe stellar parallax (see Figure 5–6), discovered instead a phenomenon called the **aberration of light** (the name does not imply that there is anything wrong with light). The aberration of light is an apparent shift in a star's position as a result of the motion of the Earth and the finite velocity of light. It is readily understood by means of an analogy. Imagine that you are in a rainstorm, with the raindrops coming straight down. If you hold a pipe vertically, as shown in **Figure 5–23a**, the raindrops will, of course, fall straight down to the bottom of the pipe. On the other hand, if you run along with the pipe, you will have to tilt the pipe in the direction of your motion to keep the drops from hitting the side of the pipe (**Figure 5–23b**).

In the same way, a moving telescope must be tilted very slightly in the direction of its motion if light entering the top of the telescope tube is to pass all the way through and arrive at the eyepiece. The effect is not large: during the course of the year, a star's observed position deviates a maximum of 20 seconds of arc from its true position. However, eighteenth-century telescopes were sufficiently precise to enable this effect to be detected. Because the deviation depends directly on the

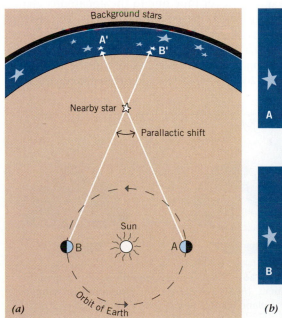

FIGURE 5–23. (*a*) A pipe in the rain. Raindrops fall straight to the bottom. (*b*) Moving a pipe in the rain. It must be tilted to keep the drops from hitting the side.

speed of the Earth relative to the speed of light, a measurement of the deviation gives a numerical value for the speed of the Earth as it moves through space in its

orbit around the Sun. The value is about one ten-thousandth of the speed of light, or 18.6 miles per second. Thus the first actual proof of the moving Earth came in 1729 with the discovery of the aberration of starlight.

The discovery of parallax, predicted by the Greeks and searched for in vain by Tycho, was delayed for yet another century. In the year 1838 the great astronomer Friedrich Wilhelm Bessel finally succeeded in measuring the parallax of the nearby faint star 61 Cygni. Why did parallax take so long to find? The answer is that the stars are so far away that the amount of their parallactic shift is extremely small. In the case of 61 Cygni, for example, the total shift is only about 0.6" (0.00017 degrees), less than a thirtieth of the effect of aberration. This is approximately the angular size of a U.S. penny viewed from four miles away—a very small angle!

Figure 5–24a illustrates the effect of parallax for a star that is located in the plane of the Earth's orbit (the easiest case to visualize). When the Earth is at point *A*, the nearby star will be seen at position *A'* with respect to the more distant background stars. Six months later, when the Earth is at point *B*, the nearby star appears to be located at point *B'*. The **parallactic shift** in the position of the star is the angle *A*-star-*B*. It should be evident that the more distant the star is, the smaller the parallactic shift will be.

If we were to take photographs of the star field when the Earth is at points *A* and *B* in its orbit, they might look like **Figure 5–24b**. The nearby star would appear to shift its position relative to the more distant stars, as shown. Assuming that the background stars are much more dis-

FIGURE 5–24. (*a*) The effect of parallax on a nearby star (not to scale). The apparent direction toward the star shifts slightly as a result of the Earth's motion around the Sun. (*b*) Two drawings of the same star field, six months apart. Note the shifted star.

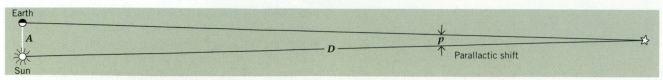

FIGURE 5–25. The relationship between the distance of a star and its parallax. The greater the distance, the smaller the parallax angle.

tant than the nearby stars, the angle A-star-B of Figure 5–24a will be practically equal to the angular shift in position shown in Figure 5–24b.

How large a distance to the stars does the small observed parallax imply? Consider the right triangle formed by the Earth, the Sun, and a nearby star, as shown in **Figure 5–25**. The side labeled A is the distance from the Earth to the Sun, which is of course 1 AU. The angle p is called the **parallax** of the star; it is one-half the total parallactic shift shown in Figure 5–24a. Because the triangle is so long and skinny, the distance D from the Sun to the star can be computed from the angular size formula used in Chapter 3, namely,

$$D = 57.3° \times \frac{A}{p}$$

if the parallax p is expressed in degrees. (We have replaced the variable R used in Chapter 3 with the distance D.) For example, the nearest star to the Earth, Proxima Centauri, has a measured parallax of 0.76″, which corresponds to 0.00021 degrees, an extremely small angle. Substituting this angle and the length of an astronomical unit, in kilometers, into the formula, we find the distance to be

$$D = 57.3 \times \frac{150,000,000 \text{ km}}{0.00021}$$
$$= 41,000,000,000,000 \text{ km}$$
$$= 4.1 \times 10^{13} \text{ km}.$$

Kilometers or miles are clearly not an appropriate unit for measuring distances to stars. Nor are astronomical units; the distance to Proxima Centauri is over 270,000 AU, for example. Just as we describe distances around town in miles or kilometers rather than inches or centimeters, we describe distances to stars in units that are large enough so that the numbers we use have convenient sizes. This is why we use light-years, because one light-year is about 9.5×10^{12} km or 5.9×10^{12} miles. The distance of Proxima Centauri, then, is 4.3 ly.

> READERS HAVING THE ACTIVITY KIT
> SHOULD DO KIT ACTIVITY 5–1 (A PARALLAX
> MEASUREMENT) AT THIS TIME.

5.7

OBSERVATIONAL EVIDENCE OF THE EARTH'S ROTATION

While every schoolchild can tell you the Earth rotates, few college graduates can explain *how* we know this simplest of all astronomical facts. As evidence, most people would cite the observation of the Sun and stars rising in the east and setting in the west. If you were to suggest, just for argument, that the east-west motion of the Sun and stars is produced by the motion of crystalline spheres to which the Sun and stars are attached, great confusion would result.

FOUCAULT PENDULUM

A famous demonstration of the rotation of the Earth is the pendulum experiment first performed by the nineteenth-century French physicist Jean-Bernard-Léon Foucault. In this experiment, a massive pendulum is hung from a long wire and set in motion. Pegs placed in a circle around the pendulum are successively knocked down over time due to the apparently changing plane in which the pendulum swings. To construct the simplest possible explanation of what is observed, imagine that we suspend a pendulum on a long cable from a support that has been placed at the North Pole (**Figure 5–26**). We make the coupling between the cable

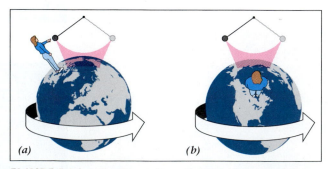

FIGURE 5–26. The Foucault pendulum. (*a*) Setup of the experiment. (*b*) Six hours later, the pendulum still vibrates in the same plane relative to the stars, but the observer has rotated.

and the support as friction-free as possible. If we set the pendulum to swinging back and forth, it will swing along a line that we can mark by driving pegs into the ice. Because of Newton's first law of motion, the plane defined by the swinging cable will be fixed in space relative to the distant stars. Similarly, if the Earth were not turning, the line between the pegs in the ice along which the pendulum would swing would not change. However, if the Earth turns beneath the pendulum, the line will *appear* to rotate (as viewed from the turning Earth) in a direction that is opposite to the Earth's rotation.

At other latitudes, such as the latitude of Paris where the experiment was originally performed, the rotation will be slower but still evident. Only at the equator does the pendulum not rotate at all.

Inquiry 5–22 Why doesn't the plane of oscillation of the pendulum rotate for an observer at the equator? (Hint: Consider how the pendulum would rotate in the Southern Hemisphere.)

Although the motion of Foucault's pendulum is a demonstration of the rotation of the Earth on its axis, the concepts involved in the explanation depend on theories and models from other areas, in particular Newton's first law. Without it, the Foucault pendulum experiment would have made no sense.

CORIOLIS EFFECTS: ADDITIONAL EVIDENCE FOR A SPINNING EARTH

Because each and every part of the Earth rotates full circle in 24 hours, different parts of the surface of the Earth are actually moving through space at different speeds. A point on the equator of the Earth moves about 24,000 miles in 24 hours, or about 1000 miles per hour. On the other hand, a piece of ground 4 feet from the North Pole moves only 24 feet in 24 hours, for a slow speed of 1 foot per hour!

Consider the effects of these speed differences on a projectile (say a cannonball) fired north from the equator of the Earth at 1000 mph (**Figure 5–27**). Not only is the cannonball moving northward at 1000 mph, but it also has an eastward speed of 1000 mph (i.e., the speed of the ground on which the cannon sat). From Newton's first law, the path in space is a straight line. However, relative to the moving Earth's surface, as the projectile moves northward, it will be passing over ground that moves east *less rapidly* than the projectile. As a consequence of the ground's slower motion, *relative to a north-ward facing observer on the ground* the projectile will be deflected toward the right (east).

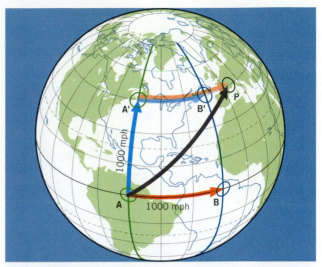

FIGURE 5–27. A cannonball fired northward from the equator (from A toward A') at 1000 mph is also moving eastward at 1000 mph (represented by the arrow AB), because this is the eastward speed of the cannon itself. The land over which the projectile travels moves more slowly than the equator. As a result, the projectile will arrive at P rather than B'. From the point of view of a north-facing gunner at A, the cannonball has been deflected to the right.

This effect was largely a curiosity until World War I, when the Germans built a new generation of long-range cannons and found that they could not hit anything if they aimed directly at it; they had to aim to the left.[6] If you understand this discussion, you should be able to convince yourself that a cannonball fired southward (toward the equator) from some point in the Northern Hemisphere will also be deflected to its right (west). Effects such as these that come about because we live on a rotating frame of reference are referred to as **Coriolis effects.**

Inquiry 5–23 How will projectiles fired north and south in the Southern Hemisphere be deflected?

Inquiry 5–24 Can you explain the motion of the Foucault pendulum as a Coriolis effect?

The general circulation of winds on the Earth and other planets is controlled by Coriolis effects. Most of the time, these effects are too gentle to notice, but during hurricane season we can have dramatic illustrations of their reality. Consider, for example, **Figure 5–28**, which illustrates the wind circulation around a low pressure re-

[6]If the target was exactly east or west of the gun's position, the gun would be aimed directly at it.

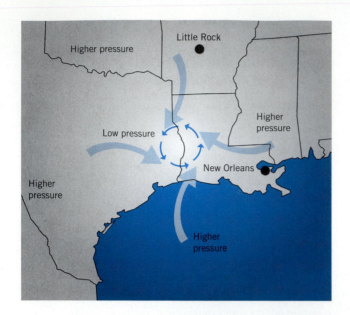

gion during the fall hurricane season. The low pressure in the center is surrounded by regions of high pressure, so winds blow in toward the center. As they do, they are deflected to the right, causing a counterclockwise circulation of wind inside the low pressure region (for Northern Hemisphere hurricanes). If such a disturbance were to move in from the Gulf of Mexico and locate its center of low pressure at the Texas-Louisiana border, central Texas would experience winds from the north, New Orleans would have winds from the south, and Little Rock, Arkansas, would have southeasterly winds.

FIGURE 5–28. Winds will push toward the center of a low-pressure storm region from the surrounding high-pressure regions. As they move, they will be deflected to the right (in the Northern Hemisphere) because of Coriolis effects. The net effect is to make the winds circulate in a counterclockwise direction around the center of the low-pressure region.

DISCOVERY 5–1
WEIGHTLESSNESS

When you have completed this Discovery, you should be able to do the following:

- Describe what is meant by weightlessness.

An astronaut floating in a space shuttle has mass but is weightless. Weightlessness does not occur, however, from a lack of gravity in space.

- **Discovery Inquiry 5–1a** Suppose you were standing on a scale in a stopped elevator. What would happen to the scale reading when the elevator suddenly accelerates upward? What would happen to the scale reading when the elevator rapidly descends from rest?

Although the elevator has accelerated, the force of gravity on the elevator occupants has not changed. The scale reading, however, has changed because of the acceleration.

You can easily demonstrate weightlessness for yourself. Take a paper cup or an aluminum pop can and punch two holes on opposite sides near the bottom. With your fingers over the holes, fill the container with water.

- ☐ **Discovery Inquiry 5–1b** *Before removing your fingers from the holes*, describe what you expect to happen. Then, while taking necessary precautions, remove your fingers from the holes and observe what happens.

  Again place your fingers over the holes and fill with water.

- ☐ **Discovery Inquiry 5–1c** *Before dropping the container* (into a garbage can, the bath tub, or *safely* outside!), describe what you expect to observe. After thinking about the answer, drop the container from as high as you can reach, and describe what you observed.

In this experiment, you should have observed no water flowing from the container while it was dropping. Although gravity was still there, the water was weightless *with respect to the container* because the water and the container accelerated downward at the same rate. Similarly, because the astronaut and the space shuttle both accelerate towards the center of the Earth *at the same rate*, the astronaut is weightless.

CHAPTER SUMMARY

OBSERVATIONS

- **Aristarchus** first suggested that the Earth circles the Sun, and found the relative sizes of the Earth, Moon, and Sun, as well as the relative distances of the Moon and Sun. **Eratosthenes** first found the size of the Earth. **Hipparchus** made a great star catalog, began the magnitude system used to specify the brightness of stars, and discovered precession. **Ptolemy** advanced the theory of planetary epicycles and helped preserve Greek knowledge in the *Almagest*.

- The Greeks observed that the planets moved against the starry background; sometimes changed directions and moved with a **retrograde motion**; and, in the case of Mercury and Venus, were never far from the Sun.

- **Tycho Brahe** built large instruments and made observations over extended periods of time. **Johannes Kepler** found three empirical laws of planetary motion: (1) planets revolve about the Sun in elliptical orbits with the Sun at one focus; (2) planets move more rapidly when close to the Sun than when farther away; (3) if the period of a planet in its orbit is P years and the semi-major axis is A astronomical units, then $P^2 = A^3$.

- **Galileo** made the first telescopic observations and discovered that the Milky Way consists of a large number of stars. He observed lunar surface features, four Jovian satellites, sunspots, and phases of planets. He showed experimentally that objects of different masses fall to the ground with identical accelerations.

- The **aberration of starlight** is a small apparent shift in the direction to an object caused by the motion of an observer. The observation of this effect illustrates the motion of the Earth.

- **Stellar parallax** is the apparent change in the direction to an object caused by a change in position of an observer.

- The **Coriolis effect** is the apparent rightward drift of a projectile fired from the equator toward the North Pole, or from the North Pole toward the equator. It is caused by the fact that equatorial regions move more rapidly than polar regions.

THEORY

- Greek science and philosophy assumed that planets exhibit uniform motion in circular orbits.

- **Nicolas Copernicus** proposed the heliocentric hypothesis.

- **Mass** is a quantity that measures the amount of **inertia** that a body contains. A body's mass is independent of its location. **Velocity** is a change in location or direction of motion divided by the time over which the change occurs. **Acceleration** is a change in an object's velocity divided by the time over which the change occurs.

- Newton's three laws describe how bodies move both in the absence and presence of forces. They are general laws applicable to a wide variety of situations.
- **The universal law of gravitation**, found by Newton, provided a means to connect the motion of the Moon with the motion of falling bodies on Earth.
- From Newton's laws, a modification of Kepler's third law results that allows astronomers to determine the masses of orbiting bodies:

$$(M_1 + M_2)P^2 = A^3.$$

- The amounts of **linear momentum** and **angular momentum** contained in an isolated system are both conserved.

- The **weight** of an object is determined by the gravitational force exerted on it by the Earth. The weight depends on the object's location.

CONCLUSIONS

- From observations made with the **Foucault pendulum**, astronomers infer the rotation of the Earth.
- From observations of the aberration of starlight, astronomers infer the revolution of the Earth about the Sun.
- Newton's laws provide a model that successfully explains past events and predicts future ones. From them, Kepler's laws of planetary motion can be derived.

SUMMARY QUESTIONS

1. What is the significance of the observations of Aristarchus, Eratosthenes, and Hipparchus? What are the principles used to make their measurements of the sizes of the Earth and Moon?

2. What are two methods by which the order of the planets from the Sun could be determined?

3. What planetary observations must a reasonable model of the solar system incorporate? In your answer use diagrams to show how the Ptolemaic and heliocentric hypotheses explain the observations.

4. What is meant by stellar parallax? Explain its cause. How can parallax be used to distinguish between heliocentric and geocentric hypotheses? Why did the Greeks not observe stellar parallax?

5. What role did the concept of uniform circular motion play in the history of astronomical thought? Give examples from several eras.

6. What were specific astronomical contributions made by Copernicus and Brahe? What was the importance of these contributions?

7. State and explain Kepler's three laws. Explain Kepler's second law in terms of the conservation of angular momentum.

8. What are the principal astronomical discoveries of Galileo? Explain how these discoveries may have accelerated the acceptance of the heliocentric hypothesis.

9. State Newton's laws of motion and the law of gravity. Explain how, in principle, they can be used to explain the orbiting of one body about another.

10. What are some specific astronomical examples of how Newton's laws have been shown to be valid descriptions of nature?

11. How were the heliocentric hypothesis and the Earth's rotation finally confirmed observationally? How did Newton's laws play a role in demonstrating the Earth's rotation and revolution?

APPLYING YOUR KNOWLEDGE

1. Make up a table that lists, chronologically, the contributions to astronomy made by Aristarchus, Apollonius, Eratosthenes, Hipparchus, Ptolemy, and Pythagoras. Include dates.

2. If you had been a traditional scholar in the mid-1500s, what arguments would you have presented *against* the Copernican system?

3. Of the following people, who in your opinion made the most important contribution to astronomy: Copernicus, Brahe, Kepler, Galileo, Newton? Explain.

4. Use the definition of acceleration to list the accelerators in a standard passenger car.

5. Use Newton's second law to explain why a planet in an elliptical orbit moves more rapidly when near the Sun than when farther away.

6. Explain why your weight would be different on Mars than it is on Earth.

7. Why can a rocket escape from the Moon's surface with a smaller speed than is needed to escape from the Earth's surface?

■ **8.** What is the distance from the Sun to a planet whose period is 129.14 days?

■ **9.** Find the ratio of the gravitational attraction between (*a*) a man of mass 100 kg and a woman of mass 50 kg who is 10 meters away from him, and (*b*) the attraction between the woman and the Earth. (Hint: See Appendix A8.)

■ **10.** If the Sun is 2×10^9 AU from the center of the Milky Way, and if it has a period of 200 million years, what is the mass of the Milky Way galaxy?

■ **11.** If identical galaxies at a distance of 650 million ly and having a mass of 10^{10} solar masses each were observed to have an angular separation of 10 seconds of arc, how long would their orbital period be? Assume the galaxies are seen at their maximum separation. (Hint: This problem contains a number of steps along the way.)

■ **12.** How much would a 100-lb person weigh on the Moon? (Hint: Write an expression for the person's weight on Earth and an equivalent one for the Moon. Then divide one equation by the other; you will find that some quantities drop out of the problem.)

■ **13.** How fast would a person of mass 50 kg (about 110 lbs) have to run to have the same linear momentum as a 5000-kg bus (about 5 tons) traveling 60 miles per hour?

■ **14.** What would be the parallax of Proxima Centauri if it were observed from a telescope on one of Saturn's satellites?

ANSWERS TO INQUIRIES

5–1. Circular orbits and uniform motion.

5–2. The Earth is four times larger.

5–3. Because he knew the Moon was smaller than the Earth and was in orbit about the Earth, and because he now knew the Sun was farther than the Moon, it was logical that the Earth would go around the Sun.

5–4. 5000 stadia $\times$ 360°/7° = 260,000 stadia.

5–5. That it is perfectly spherical.

5–6. Zero degrees.

5–7. That the planets are moving at about the same speed, so their apparent speeds are due only to their respective distances. While the assumption is incorrect, the result ends up being correct.

5–8. Stars are very far away.

5–9. Only crescent phases would occur.

5–10. There are countless examples. Some are the germ model of disease, economic models used to set government fiscal policy, and educational models used to design instruction.

5–11. On a straight line, with the Earth in the middle.

5–12. Each has some points of simplicity. Ptolemy's has fewer epicycles, but Copernicus's has a neater explanation of retrograde motion, which in addition predicts retrograde motion only when it is actually observed.

5–13. About 1.87 years.

5–14. About 19.2 AU.

5–15. The planets are closer than stars.

5–16. Jupiter was clearly moving and dragging its moons along.

5–17. From $F = ma$, the can with more mass (the one filled with concrete) would have the smaller acceleration if the forces were equal.

5–18. The reaction force on the foot will certainly be felt more in the case of the concrete can.

5–19. The force is 16 times smaller.

5–20. Twice as great.

5–21. $(M_1 + M_2) = 8^3/10^2 = 512/100 = 5.1$ solar masses.

5–22. In the Southern Hemisphere, the pendulum would appear to rotate in the opposite direction from its rotation in the Northern Hemisphere. Therefore, for the direction of oscillation to change as the pendulum is carried from one hemisphere to the other, there must be no rotation at the equator.

5–23. In the Southern Hemisphere, projectiles will always deflect to the left, when fired any direction except directly east or west.

5–24. Yes. Imagining the ball of the pendulum to be a projectile deflecting to the right on each swing gives it the rotation we observe.

PART TWO

DISCOVERING THE NATURE AND EVOLUTION OF THE SOLAR SYSTEM

Because of fantastic images of distant worlds returned in recent years by the Viking, Voyager, and Magellan spacecraft, most people are more at home with the planets than with stars and galaxies. For this reason we begin our study of modern astronomy with the solar system. These spacecraft have provided an unprecedented opportunity for real exploration of the solar system.

We begin in Chapter 6 by looking at regularities observed within the solar system. These regularities lead to questions about how the system as a whole and the bodies within it formed. An important part of our study is a look at various models of the solar system's formation, and the evidence for and against each model. The chapter ends with a brief look at the question of whether there are planetary systems around other stars.

To help our understanding of the solar system's formation, we observe objects whose characteristics preserve both the materials and the physical conditions that were present at the time of their formation. For this reason, in Chapter 7 we examine comets, asteroids, and meteorites. While these objects may constitute only a minor part of the mass of the solar system, they have undergone fewer modifications than the larger planetary bodies and therefore provide crucial

information about the conditions when the solar system formed.

Processes that occur within, on, and around planetary bodies are studied in Chapter 8 on the Earth and Moon. These objects are taken as a "norm" against which the properties of other bodies can be measured. The properties studied include the formation and modification of surface features by volcanic, tectonic, and impact forces; the formation, composition, and properties of the atmosphere; the study of the interior structure with seismic waves; and planetary magnetism.

The Earth-like planets are investigated in Chapter 9 by comparing and contrasting each with the other, and each with the Earth and Moon. While the emphasis is on comparison and the processes that give each body the properties that we observe, we also will examine the unique characteristics of each body.

The final chapter of Part 2, Chapter 10, is similar to Chapter 9 but covers the Jovian planets and Pluto. We also discuss the characteristics of the satellites and the planetary ring systems. Through a comparison of one body with the other, we are better able to make sense of the reams of information sent to Earth by the Voyager spacecraft.

THE STRUCTURE
AND FORMATION
OF THE SOLAR SYSTEM

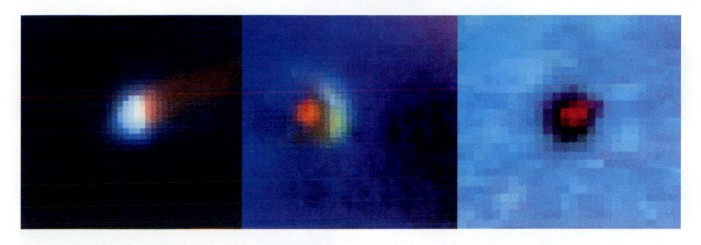

First of all there came Chaos, . . .
From Chaos was born Erebos, the dark, and black night,
And from Night again Aither and Hemera, the day, were begotten . . .

HESIOD, *THEOGONY*, C. 800 B.C.

We begin our survey of the universe close to home with our Sun and its system of planets. In this chapter we will look at the overall structure of the solar system (we defer the details to Chapters 8 to 10). We will find some striking regularities, which can lead us to an understanding about how the solar system formed. We will discuss some of the hypotheses of solar system formation that have been proposed and see how well each fits the observable data. Once we have done this for our own solar system, it will be possible to make some generalizations about possible planetary systems elsewhere in the universe.

6.1

AN OVERVIEW OF THE SOLAR SYSTEM

Let us imagine that the alien astronomer-astronaut introduced in Chapter 2 has been sent to our solar system to investigate the possibility of life in the vicinity of the Sun. Approaching the solar system from its north pole, the alien would have a view somewhat similar to that of the Voyager spacecraft's final image (**Figure 6–1**). Because the planets have a large range in brightness, the first thing she would see after the Sun would be the giant planet Jupiter, about 5 AU from the Sun. After determining Jupiter's period about the Sun, she would then be able to use Kepler's third law to determine the Sun's mass. Translating into metric units, she would find that the mass of the Sun is about 2×10^{33} grams. (A mass of 1 kilogram weighs 2.2 pounds on Earth.)

From a more careful observation, which would show that the Sun is affected slightly by Jupiter's gravity, the alien would discover that the mass of Jupiter is about 1/1000 that of the Sun. However, it would be clear that the Sun is the dominant mass in the solar system and that all other bodies in the system must march to its gravitational tune.

Soon other planets, fainter and less massive than Jupiter, become obvious: first Saturn, with its rings, about 9.5 AU from the Sun; then Uranus, at a distance of almost 20 AU; and finally Neptune, 30 AU distant.

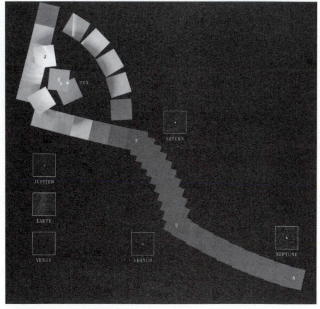

FIGURE 6–1. The solar system as viewed by Voyager 1 at a distance of six billion kilometers. The boxes represent the field view for each of Voyager's images. Each of the planets observed is noted.

Coming even closer, our visitor would notice that these four bodies rotate at different rates and that each of them has satellites and rings. By measuring the satellites' periods and distances from their planets, the alien could also accurately determine the mass of each of the planets. From the planets' diameters, she could determine their volumes, and finally their densities, defined as mass divided by the volume[1] containing the mass. In other words,

$$\text{density} = \frac{\text{mass}}{\text{volume}}.$$

For example, a cubic centimeter of water contains a mass of 1 gram. The same volume of gold contains 19.3 grams; gold is therefore much denser. She would find that these planets have a density about the same as that of water. The density of Saturn, in fact, is less than that of water, and it would float in a bathtub, if one large enough could be found!

Inquiry 6–1 Considering the low density of this group of planets, would you expect them to be composed primarily of low-mass elements such as hydrogen and helium or high-mass elements such as silicon, iron, and nickel?

The astronaut would also notice that these planets have thick atmospheres and heavy cloud cover. Thanks to the thickness of the clouds, she could not determine whether there was any solid surface at all on these planets. Better look elsewhere for a landing site, she would conclude.

By this time, the alien would have detected another group of planets, which she might call "dwarf" planets due to their small size compared to the first group of "giant" planets. She would notice that, with the exception of Pluto, this group of planets was much closer to the Sun than the giant planets. In order of their distance from the Sun, she would find Mercury, Venus, Earth, and Mars.[2] Because Earth and Mars have satellites, she would be able to determine their masses easily; she would find that the largest of the bodies, Earth, has a mass only 1/300 that of Jupiter and a density roughly five times that of water. Consequently, she would conclude that the dwarf planets were composed primarily of heavier elements, such as oxygen, silicon, iron, magnesium, calcium, and the like.

---

[1]If the body is assumed to be a sphere of radius R, the volume is given by $4\pi R^3/3$.

[2]There are a variety of mnemonics for remembering the planets in order of their average distance from the Sun. One, compliments of Earl Phillip, is **M**y **V**ery **E**ducated **M**other **J**ust **S**howed **U**s **N**ine **P**lanets. Another, compliments of Jay Rambo, is **M**y **V**ery **E**asy **M**ethod: **J**ust **S**et **U**p **N**ine **P**lanets.

Compared to the giant planets, the dwarf planets have thin atmospheres, which could be detected in several ways. Surface detail might be blurred a bit, or clouds might be seen, or, when the planet passed in front of a distant star, the star would fade out gradually as its light passed through the planet's atmosphere.

Mercury, the innermost planet, has only an insignificant amount of atmosphere. In attempting to explain this, the astronaut would notice two facts. First, Mercury's mass is the smallest of the planets examined so far. Second, Mercury is so close to the Sun that its atmosphere would be heated to high temperature, giving any molecules of gas in its atmosphere a high velocity.

Inquiry 6–2 Why would the light from a distant star fade gradually as one of the dwarf planets having an atmosphere passed in front of the star? Make a drawing to explain your answer.

Inquiry 6–3 How would Mercury's low gravity and high temperature each tend to cause Mercury's atmosphere to disperse into outer space after a time?

The next planet, Venus, has a much thicker and more extensive atmosphere—so thick that its surface could not be seen directly. Nevertheless, the alien would be able to show that the surface of the planet was not far below the visible cloud layer by bouncing radar signals off the planet. A little probing would show that the principal component of Venus's atmosphere is carbon dioxide (CO_2), and that it is bone dry, with virtually no water vapor. By contrast, the next planet, Earth, which is almost the same size and mass as Venus, has a very different atmosphere—about 78% nitrogen (N_2) and 20% oxygen (O_2), with a readily detectable amount of water vapor. Our visitor would make a mental note to try to understand this difference, because the next planet, Mars, has a very thin atmosphere that is largely carbon dioxide.

Although the alien would find the surfaces of all these planets to be rocky in appearance, the Earth is distinctive in that its surface is largely covered by water, whereas the other planets show little evidence of water. Again, she would wonder about the reason for this.

At this point the alien astronaut might notice an anomaly—the planet Pluto. Like the dwarf planets, Pluto is small. Unlike them it is located a large distance from the Sun, even beyond the giant planets, and has a low density. Moreover, its orbit is unusual and different from the orbits of the other eight planets. Might it be that this planet had an origin different from the rest of the planets?

Summarizing this picture, the astronaut might draw up a table similar to **Table 6–1**. In it are listed the characteristics of the dwarf, or **terrestrial**, planets (after the Latin word for Earth) and the giant, or **Jovian**, planets (after the Latin name for Jupiter). These characteristics must be explained by any successful theory of the origin of the solar system.

Appendix C contains reference tables that summarize some of the fundamental data about each of the planets. The significance of some of the data was suggested in the previous paragraphs and will be discussed further as you learn more about the solar system. To help you better appreciate the relative sizes of the planets, they are illustrated to scale in **Figure 6–2**. Similarly, the relative sizes of the planetary orbits are displayed in **Figure 6–3**. If Jupiter were drawn to scale in Figure 6–3, it would be only about 1/10,000 of an inch in diameter, and the Sun would be about 1/1000 of an inch in diameter. Therefore, to put Figure 6–3 on the same scale as Figure 6–2, we would have to draw it about 10,000 times bigger!

> IN DISCOVERY 6–1 (A SCALE MODEL OF THE SOLAR SYSTEM) YOU WILL MAKE A SOLAR SYSTEM MODEL TO A SCALE THAT YOU CAN WALK.

Inquiry 6–4 On the scale of Figure 6–2, about how far would Earth be from the Sun? How far would Pluto be from the Sun?

TABLE 6–1

Characteristics of the Two Major Groups of Planets

Terrestrial	Jovian
Mercury, Venus Earth, Mars	Jupiter, Saturn Uranus, Neptune
Near the Sun	Far from the Sun
Small diameter	Large diameter
Low mass	High mass
High density	Low density
Primarily composed of heavier elements	Primarily composed of hydrogen and helium (composition similar to the Sun)
Rocky surface	No surface
Thin atmosphere	Thick atmosphere
High temperature	Low temperature

The solar system also contains **minor planets (asteroids)**, **comets**, and other miscellaneous debris. These components give additional insight into the origin of the solar system and are important enough to warrant their own chapter (Chapter 7). However, for the time being we will ignore them as we try to visualize the Big Picture.

FIGURE 6–2. Relative sizes of the planets compared with the Sun.

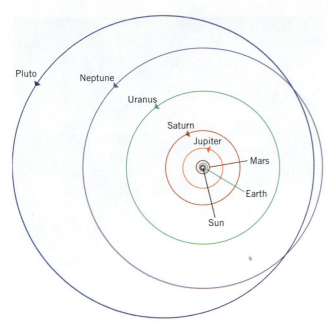

FIGURE 6–3. The solar system, showing the scale of the planetary orbits, and the direction of motion as observed from above Earth's North Pole.

6.2

ORBITAL REGULARITIES IN THE SOLAR SYSTEM

Our space explorer would probably find some of the regularities in the motions of the planets more striking than the various differences between them. For example, she would find that the orbits of the planets all lie nearly in the same plane. (This is why they are always seen near the ecliptic, which defines the fundamental plane of the solar system.) The planets whose orbital planes are farthest from the ecliptic are Mercury and Pluto, as shown in **Figure 6–4**. The other planets have orbital planes closer to the ecliptic than that of Saturn.

Our astronaut would find that, viewed from the north pole of the solar system, all the planets orbit the Sun in the same counterclockwise direction (Figure 6–3). Moreover, with the exceptions of Venus, Uranus, and Pluto, the planets and the Sun rotate on their axes in a counterclockwise direction.

The orbits of the planets are nearly circular. Even in the case of Mercury and Pluto, which have the most highly elliptical orbits, the deviation from a precise circle amounts to only about a 3% difference in the lengths of the long and short axes of the ellipse. For these two planets, the Sun is well offset from the center of the orbits. In the case of Pluto, although its *average* distance is about 40 AU, it is sometimes even closer to the Sun than Neptune.[3] (From Figure 6–3, it might appear that Neptune and Pluto are in danger of colliding; however, the tilt of Pluto's orbit prevents this from happening. Studies have shown that Neptune and Pluto never get closer to each other than 18 AU.)

The rotational (spin) axes of most of the planets are roughly perpendicular to their orbital planes. For example, Earth's axis is at an angle of 66.5° to its orbital plane (**Figure 6–5a**). The inclination for the other planets relative to their orbital planes is shown in **Figure 6–5b**. The axis of Mars is at approximately the same angle as Earth's, and the axes of most of the other planets are relatively close to perpendicular. This is certainly a fact that a good theory of the origin of the solar system must explain. The two striking exceptions to this rule are the planets Uranus and Pluto, whose spin axes are almost in

[3]Pluto will be closer to the Sun than Neptune during the approximate period of A.D. 1979–1999.

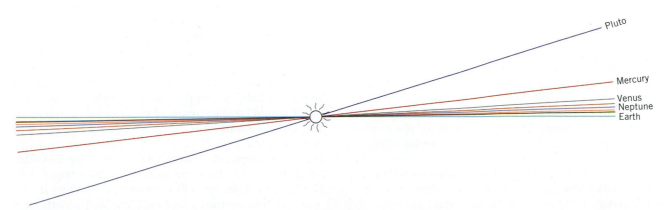

FIGURE 6–4. The solar system viewed from the side, showing the planes of the planets' orbits.

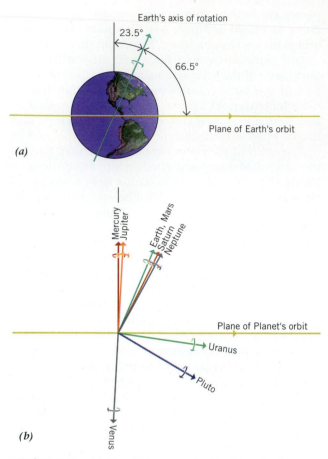

FIGURE 6–5. (*a*) The tilt between the Earth's axis of rotation and its orbital plane. (*b*) Similar to (*a*) for each of the planets. The arrows point in the directions of the planets' north poles. Venus' arrow points down because of the planet's retrograde rotation. The direction of rotation of each planet is indicated.

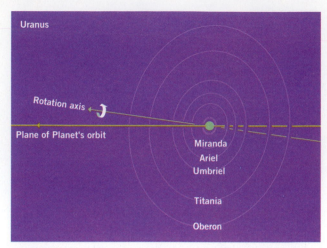

FIGURE 6–6. The anomalous orientation of Uranus and its satellite system.

their orbital planes; **Figure 6–6** illustrates the situation for Uranus and its moons. A really successful theory of the formation of the planets would also be expected to explain this anomaly.

Of the 61 known natural planetary satellites in the solar system, most orbit in planes that roughly coincide with the ecliptic. The moons that are closest to the planets tend to have orbits that lie near the planets' equatorial planes. Importantly, this is true even for the satellites of Uranus and Pluto, whose equatorial planes deviate from the ecliptic plane.

Finally, our astronaut would notice that most of the satellites move around their planets in a counterclockwise direction. Only a few of the outermost satellites of Jupiter, Saturn, and Neptune revolve clockwise.

6.3

THE DISTRIBUTION OF ANGULAR MOMENTUM IN THE SOLAR SYSTEM

Our visitor would increase her understanding of our solar system if she measured the amount of angular momentum (defined in Chapter 5) possessed by each of the bodies in it. A body can possess angular momentum in one of two ways—either by virtue of its orbital motion around another body (*orbital* angular momentum, as discussed in Section 5.5), or by virtue of its rotation about an axis (which is called *spin* angular momentum). The total angular momentum in the solar system is the sum of all the orbital and spin angular momenta of all the planets, their satellites, and the Sun.

One of the most striking properties of the solar system is the observation that 98% of the total angular momentum in the system comes from the orbital angular momentum of the planets, even though the Sun contains 99.8% of the mass. Because the Sun has the lion's share of mass, we might expect it to carry a similar proportion of the angular momentum. But for the Sun to have even half the angular momentum in the solar system it would have to rotate 25 times faster than it actually does, or about once a day. Explaining why the Sun contains such a small percentage of the solar system's angular momentum (2%) has been one of the more difficult tasks facing hypotheses of the solar system's origin; from now on we will refer to this difficulty as the "angular momentum problem." We will return to it later in the chapter.

6.4

HYPOTHESES OF THE ORIGIN OF THE SOLAR SYSTEM

How might our alien astronaut try to put together a hypothesis to explain these broad aspects of the solar system that we have discussed? Her basic task will be to try to explain how a dominant mass, the Sun, might have acquired a system of planets and satellites exhibiting all the regularities mentioned above. In addition, a hypothesis that also explained the anomalies we mentioned would truly recommend itself.

A SIMPLE HYPOTHESIS THAT DOES NOT WORK

Perhaps the simplest hypothesis that might be put forward to explain the existence of the Sun's planetary system is the idea that the Sun, in its motion through the galaxy, has simply acquired its planets and their satellites by encountering these bodies from time to time and capturing them into its gravitational field. How well does this hypothesis fare in explaining the observed regularities in the solar system?

Inquiry 6–5 Would this hypothesis be consistent with the fact that the inclinations of the planes of planetary orbits are nearly the same? With the near circularity of most of the planetary orbits? With the fact that some planets have low mass and some high mass? With the fact that most of the planets and their satellites orbit in the same direction?

Inquiry 6–6 It has been established that the Earth, Moon, and meteorites have the same age. Is this observation consistent with the above hypothesis? Justify your answer.

Inquiry 6–7 What are several objections to this hypothesis? What facts can you think of that are *consistent* with the hypothesis?

In looking carefully at the hypothesis and asking the right questions, we have been able to invalidate it easily. We will now consider some other possibilities.

THE NEBULAR HYPOTHESIS

In a truly original flash of insight, the French philosopher René Descartes suggested in 1644 that space might have been filled with swirling gases and that the collapse of spinning whirlpools of material into denser condensations was responsible for forming celestial bodies.

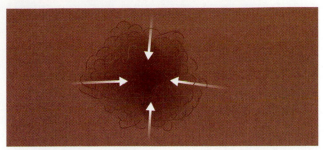

FIGURE 6–7. The collapse of a gas cloud according to the nebular hypothesis.

What is especially surprising is that Descartes made these speculations *before* Newton formulated his fundamental laws of motion and theory of gravity.

A more complete hypothesis was suggested in the eighteenth century by the German philosopher Immanuel Kant; it was later developed further by the French mathematician Pierre-Simon Laplace. Kant's **nebular hypothesis** proposed that a vast cloud of interstellar gas collapsed to form the Sun and planets, as shown in **Figure 6–7**. Once contraction started, the gravitational force of the cloud acting on itself would accelerate its collapse into a relatively small volume.

It is virtually impossible for a cloud to start out with no rotation at all; the motion might be small, but it is not zero. As it collapses, a cloud's rate of spin must increase due to the conservation of angular momentum discussed in the previous chapter. A familiar example of this principle is an ice skater in a spin.

Kant and Laplace pointed out that at a certain stage the rotating mass of gas would have to assume a disklike shape if it had a sufficient amount of angular momentum. The reason is illustrated in **Figure 6–8**, which shows particles at various places on the surface of a rotating sphere. The particle at point A feels a gravitational force pulling it toward the center of the sphere. We can split the force of gravity into two components, one horizontal and the other vertical. The horizontal component of gravity would pull the particle towards the rotation axis were it not for the particle's momentum, which counteracts the pull toward the rotation axis, therefore allowing the particle to remain in the same horizontal location. (Remember, without gravity, Newton's first law tells us that the particle would fly off into space in the direction of its motion). On the other hand, the vertical component of gravity is unbalanced and pulls the particle toward the equatorial plane. Similar situations occur for particles B and C. As shown, the closer the particle is to the rotation axis, the greater the component of gravity toward the equatorial plane. Con-

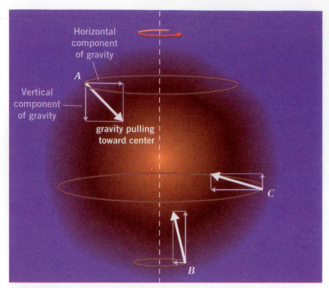

FIGURE 6–8. The forces causing a spinning gas cloud to flatten into a disk. A particle's forward momentum prevents the horizontal component of gravity from pulling the particle toward the rotation axis. The vertical component of gravity is unbalanced, so collapse to a disk occurs.

sidering all particles in a cloud, the result is that the cloud itself collapses into a plane. At this point, Laplace and Kant surmised that the collapsing mass of gas would leave behind a ring of material and that this ring would eventually gather itself into a planet. As the cloud collapsed further, more rings would be left behind. In this way, a planetary system might be formed, with the remaining gas in the middle forming the Sun (**Figure 6–9**).

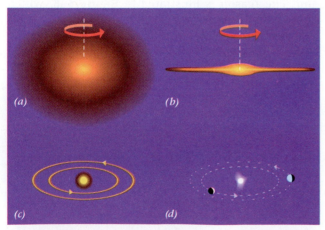

FIGURE 6–9. The original nebular hypothesis of Kant and Laplace. (*a*) Initially. (*b*) Formation of a ring. (*c*) Several rings have formed. (*d*) Condensation into planets.

Inquiry 6–8 Which of the regularities observed in the solar system are consistent with the nebular hypothesis? Which are inconsistent?

Inquiry 6–9 How might you explain the existence of planetary satellites using the nebular hypothesis?

Inquiry 6–10 Would the nebular hypothesis be able to explain the differences in composition between the Jovian and terrestrial planets?

The nebular hypothesis, as *originally* presented, has three fatal flaws. First, it is difficult to think of a way for the gaseous material left behind in the rings to coalesce into planets. It is much more likely that the internal pressure of the hot gas would cause it to disperse into space. Second, the high spin rate required to throw off rings of material would result in a Sun whose spin is so great that it would contain most of the angular momentum of the solar system, instead of the small amount it does have. The final flaw is that the original nebular hypothesis does not explain the differences in composition between the giant and the terrestrial planets. These flaws mean that other hypotheses must be considered.

COLLISION HYPOTHESES

Collision hypotheses, also known as catastrophic hypotheses, of the origin of the solar system have been around for a long time. In 1745, the French naturalist Georges Buffon suggested that the debris left over from a collision between a comet and the Sun was responsible for forming the planets. There was a renewed interest in collisional hypotheses by the late nineteenth century when it became apparent that nebular hypotheses were unable to explain the angular momentum distribution throughout the solar system.

The most popular form of a collision hypothesis is attributed to the geologist Thomas Chamberlin and astronomer Forest Moulton, who worked on the problem at the beginning of the twentieth century. Their hypothesis is illustrated in **Figure 6–10**. Chamberlin and Moulton surmised that when the Sun was relatively young, another star in the galaxy passed close to it and pulled away an enormous streamer of gas (Figure 6–10). The streamer would tend to follow the star, so that when planets formed out of it, they would end up going in the same direction around the Sun. A variation of this hypothesis proposed that at one time the Sun was a binary (i.e., it had a companion star) and that the passing star disrupted the Sun's companion, thereby producing the material that coalesced into the planets.

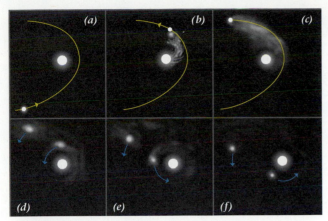

FIGURE 6–10. The Chamberlin-Moulton hypothesis. A passing star pulls matter from the Sun. The matter condenses into planets all revolving in the same direction.

Inquiry 6–11 The distances between stars are enormous compared with their physical dimensions. As a consequence, collisions of the sort envisioned in the Chamberlin-Moulton hypothesis must be extremely rare. What implications would this have for the number of planetary systems in the galaxy? Compare this number to the number that would be expected if the nebular hypothesis were correct.

Inquiry 6–12 Which of the regularities observed in the solar system are consistent with the Chamberlin-Moulton hypothesis? Which ones are inconsistent?

Even though the collision hypotheses were formulated to try to explain the angular momentum distribution in the solar system, they too fell short in this area. Furthermore, they offered no good mechanism to make the material that had been pulled away from the Sun condense into planets. Finally, they offered no explanation for the differences in composition between the Jovian and terrestrial planets. Thus they had basically the same problems as the original nebular hypothesis; neither provided a picture of solar system formation that was consistent with observations.

ADDITIONAL HYPOTHESES

Other hypotheses exist that do not fall directly into either category of nebular or collision hypotheses. One such model supposes the planets formed by accretion[4]

[4]Accretion is growth by gradual addition of material. The verb form is *to accrete*.

directly from interstellar material swept up by the Sun. To keep the material from falling into the Sun, however, requires another nearby star. Yet another model requires the Sun to have passed through *two* interstellar clouds. One was composed of hydrogen, from which the Jovian planets formed, while the other consisted of grains composed of heavy elements that formed the terrestrial planets. A recently suggested model has the Sun encountering a protostar from which material was pulled to form the planets. While such a model appears to successfully account for new data on the abundances of the different elements, it is not accepted by the scientific community at this time.

6.5

MODERN IDEAS

Our modern ideas of the solar system's formation branch out from the nebular hypothesis. Our current understanding, however, is still far from complete. The main questions we will now try to answer involve the role of dust in forming planets, the chemical differences between bodies, and the distribution of angular momentum within the solar system. We begin by discussing the role of energy in the solar system's formation.

ENERGY AND SOLAR SYSTEM FORMATION

A cloud of gas and dust has **potential energy**, which is a body's ability to make something happen based on its position. For example, a rock held above a nail has the potential to push the nail into a board when the rock is dropped onto it. In this case, the source of the potential energy is gravity, and we speak of gravitational potential energy. Once the rock is dropped and it gets closer to the nail, its potential energy decreases while its energy of motion, called **kinetic energy**, increases. The increase in kinetic energy is exactly matched by the decrease in potential energy because the total amount of energy in an isolated system does not change. The last statement is called the law of **conservation of energy**.

A cloud of gas and dust has potential energy because the cloud's gravity gives its mass the ability to collapse. When a cloud collapses, its potential energy decreases. The increase in kinetic energy that must result appears both as an increase in the speeds of the cloud's gas particles and also as an increase in its emitted radiation. The increased motion of the particles is measured as an increase in the cloud's temperature.[5] Thus a collapsing

[5]The temperature of a gas is actually defined in terms of the speeds of the particles in it.

cloud will both heat up and radiate. The loss of energy through radiation causes the cloud to collapse further, thereby increasing the temperature further. These processes continue until others, described later in the chapter, come into play.

THE NEBULAR HYPOTHESIS REVISITED AND IMPROVED

Modern hypotheses of the origin of the solar system recognize the basic soundness of the ideas of Laplace and Kant—that a rotating, contracting mass of gas collapsing under gravity should form a flattened disk with a central condensation. Modern observations (discussed in Chapter 17) indicate that many interstellar gas clouds are in the process of collapsing into stars and star clusters, and the physical arguments that predict the flattening and the increased rate of spin of the cloud are correct. (In fact, recall Figure 1–10, in which we apparently see the remnants of the disk directly.) Where modern hypotheses differ is that, rather than supposing that the disk forms into rings, they suggest other mechanisms that could make the material in the disk coalesce into planets.

In the 1940s, for example, the German physicist C. F. von Weizäcker pointed out that a disk of gas rotating around a massive body should contain a certain amount of *turbulence*—swirling motions that would produce rotating whirlpools within the disk. Both theoretical calculations and laboratory experiments predicted that such whirling eddies would form. This idea is a modern version of Descartes's 1644 suggestion. What are its implications?

Figure 6–11 shows a top view of a hypothetical rotating disk, with its turbulent eddies. Notice that even if all the eddies are rotating in the same direction, in the region *between* two eddies it is possible for two streams of gas to be moving in *opposite* directions. There could be a tendency for material to build up where the two streams collide (e.g., at point *A* in Figure 6–11). If a sufficiently large amount of gas were to collect, it might become self-gravitating, attracting even more material to it and eventually forming into a large body.

The notion of turbulence therefore provides a possible mechanism for forming condensations of matter in the disk. Once several such condensations formed at about the same distance from the Sun, they could collide and form an even bigger body. Eventually one body at that distance would sweep up all the condensations in its vicinity, producing a **protoplanet**. Other bodies would form at other distances from the Sun.

This hypothesis still suffers from some of the drawbacks of the nebular hypothesis in that it is difficult to get enough gas together to form stable, self-gravitating condensations. Except for gravity, almost all imaginable forces would tend to disperse the condensations, and the only way to overcome these disruptive forces is for large amounts of matter to accumulate in the condensations. Unfortunately for the hypothesis, accumulation occurred too slowly in the early solar system for large condensations to build up. We must therefore look closer at the processes available in those early years.

THE ROLE OF DUST GRAINS IN PLANET FORMATION

According to modern hypotheses, dust played a key role in the formation of our planetary system. Dust—fine particles of solid matter on the order of 10^{-5}–10^{-6} centimeters in diameter—is spread throughout interstellar space and is concentrated in regions such as the Orion Nebula pictured in Figure 1–11. Newborn stars, which are formed in clouds of gas and dust, are usually invisible to optical telescopes because dust is opaque to visible radiation. But infrared radiation, like that detected by night-vision scopes, can penetrate the dust and allow us to observe the stellar embryos inside. In the Orion region, whole clusters of collapsing stars not yet visible to the naked eye have been detected by infrared telescopes such as the Infrared Astronomical Satellite (IRAS). **Figure 6–12** shows an area of recent and current star formation in the Monoceros region of the sky. This region is also permeated by both gas and dust.

What was the dust in the embryonic solar system made of? The lowest mass elements, hydrogen and helium, were not a major constituent, for these two elements were too **volatile**, which means they readily vaporized even at low temperature. For this reason, hydrogen would appear only in the form of frozen compounds of materials such as water (H_2O), ammonia (NH_3), and methane (CH_4). The principal constituents of the interstellar dust, then, were the heavier elements such as carbon (C), oxygen (O), silicon (Si), and the like.

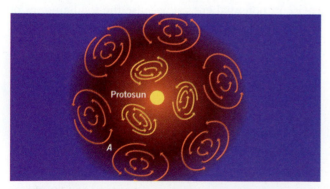

FIGURE 6–11. Turbulence in the disk from which planets form.

FIGURE 6–12. A region of star formation in Monoceros. The cluster of bright stars in the center, NGC 2244, is surrounded by the Rosette Nebula.

Dust was important in the formation of the planets for two main reasons. First, dust particles had a strong tendency to stick together when they collided. Sticking made it possible for the dust particles to accrete into larger and larger aggregates as the material forming the solar system (called the **solar nebula**) collapsed into a **protoplanetary disk**. Second, heavy elements present in the colliding dust particles gave the terrestrial planets their chemical makeup.

In addition to the dust that was already present in the solar nebula before it collapsed, more dust would be produced during the cooling of the nebula. Initially, the collapsing solar nebula heated up as atoms were squeezed into a smaller space and collided with each other. In such collisions, material loses kinetic energy and produces heat. This process of heating by cloud collapse took place most rapidly in the center of the collapsing disk. Here material collapsed most rapidly into the **protosun**, the large aggregate of material that ultimately became the Sun. Eventually, however, the collapsing gas cloud began to cool again, especially in its outer regions as dust radiated energy away. At that time, material that was originally in the form of vapor began to condense into dust grains by itself. This is exactly the same process by which a cooling cloud in the Earth's atmosphere forms ice crystals. In this way, new dust grains formed, and older dust grains (if they survived the heating phases) grew in size through deposits of new dust on their surfaces.

Unlike a cloud forming ice crystals, however, a gas cloud contains a wide variety of chemical elements such as those shown in **Table 6–2** for the Sun. The different

TABLE 6-2

Abundances of Elements in the Sun

Element	Relative Mass of Atoms (percent of total)
H	74.5
He	23.5
C	0.41
N	0.099
O	0.96
Ne	0.17
Na	0.0035
Mg	0.075
Al	0.0065
Si	0.091
S	0.040
Ca	0.0065
Fe	0.16
Ni	0.0081

elements in a gas cloud will condense to solids at different temperatures (**Figure 6–13**). Studies show that all the elements are in gaseous form at 2000 degrees Kelvin (written 2000 K).[6] (A temperature of 2000 K is about 3600°F.) As shown in Figure 6–13*a*, aluminum and cal-

[6]Astronomers measure temperatures in degrees Kelvin (designated by K). A Kelvin temperature is simply centigrade temperature plus 273; on this scale there are no negative temperatures so it is often called *absolute* temperature. See Appendix A3 for the relationships between temperature scales.

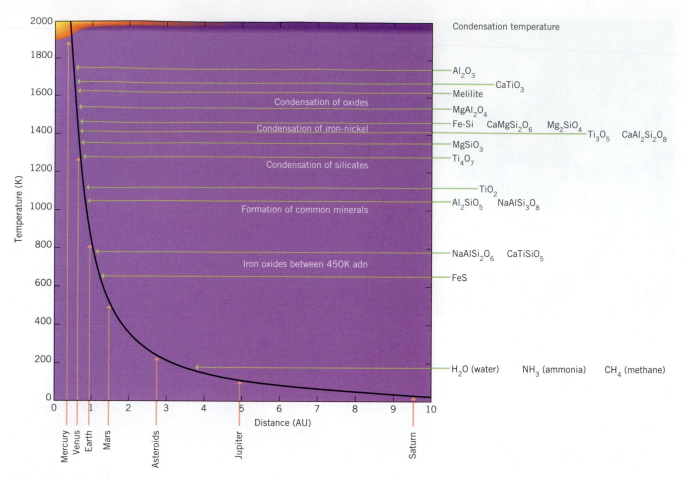

FIGURE 6–13. The condensation temperatures of materials in the solar system depend on temperature. The temperature variation with distance from the Sun in the solar nebula is also shown.

cium begin to form oxide flakes at about 1700 K, whereas iron flakes form at a temperature of about 1450 K and silicate particles at about 1300 K. Because different regions of the collapsing solar nebula would be at different temperatures, different substances would condense out in different parts of the cloud and at different times.

Although this picture is fairly complex, it neatly explains the differences in the densities and compositions of the various planets. The nebular temperatures would decrease with distance from the newly formed Sun (Figure 6–13). In the innermost, hottest parts of the cloud, we would expect only elements that condense at high temperatures, such as iron, nickel, and silicates, to condense out and produce solid materials. Hydrogen-bearing materials, which cannot condense in the hot inner solar system, would be absent. The above discussion is

in agreement with studies of the terrestrial planets that show their principal constituents to be those elements that condense at high temperatures.

On the other hand, in the cooler outer portions of the solar nebula, we would expect to find large amounts of hydrogen-bearing materials. These materials would include ices of water (H_2O) and, even farther out, ammonia (NH_3) and methane (CH_4). Such ices are in fact confirmed by the compositions of the satellites of the outer planets and the chemical compositions of asteroids, meteorites, and comets. Elements that condense at high temperatures did in fact condense in the outer solar system, but at an earlier time. However, because such elements are so much less abundant than hydrogen, hydrogen-bearing materials dominate in the outer solar system. The processes we have just discussed are summarized in **Figure 6–14**.

We can argue that, based on physical and orbital similarities, all the terrestrial planets probably formed by the accumulation of dust grains rather than by the gravitational collapse of gases. In fact, the dust and gas in the solar nebula tend to separate from each other. The motions of gas particles produce a pressure that keeps the gaseous material somewhat extended in a thick rotating disk. The dust grains, which are more massive than the gas particles, are pulled toward both the central regions and the equatorial plane, thus forming a much thinner rotating disk. This packing allows the dust grains to collide more rapidly and to stick together. Studies that model these processes suggest that only a thousand years are required to produce centimeter-sized objects. This is rapid compared with the overall expanse of time for planetary formation.

Finally, clumping of matter in specific parts of the disk was hastened by gravity. The eventual result was the formation of bodies several kilometers in diameter. Such **planetesimals** then collided and merged with each other until only one planet—or a planet with satellites—existed at a given distance from the Sun. Models show that a consequence of the buildup of a planet out of many planetesimals is that it would acquire a rotation in the same direction as the disk. In effect, the direction of rotation of a body was forced to follow the direction of revolution of the disk from which it was formed.

In a similar way, a planet may acquire a disk around itself, consisting of those planetesimals that did not strike the planet's surface. These planetesimals could then coalesce into one or more satellites. Because this disk will orbit the planet in the same (counterclockwise) direction as the planet, so will any resulting satellite (**Figure 6-15**).

The description we have given is not fact per se, but theory. Remember, however, that theory in science is

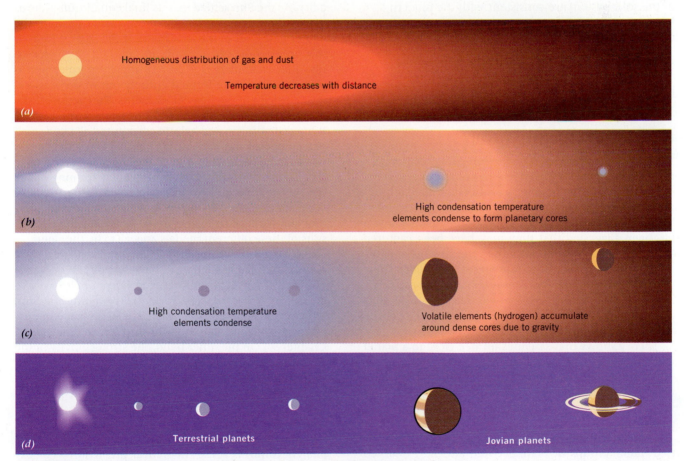

FIGURE 6–14. One possible scenario explaining why the inner planets contain elements that condense at high temperature while the outer planets contain primarily volatile elements. The temperature decreases with distance from the Sun. As time progresses, the temperature decreases further.

FIGURE 6–15. Acquisition of a protosatellite disk by an accreting planet. The disk's direction of rotation is shown.

backed up by observation and experiment. The above ideas are consistent with our knowledge and are likely to be correct in general if not in detail.

Inquiry 6–13 Which of the regularities observed in the solar system are consistent with the modern hypotheses of solar system formation? Which ones are inconsistent?

Inquiry 6–14 How can this theory, as it stands, account for the satellites that travel around the planets in *clockwise* orbits?

REFINEMENTS OF THE THEORY

The basic theory outlined in the foregoing section does not explain all the observed features of the solar system. An example is satellites that revolve in a direction opposite to their planet's rotation. Almost all of these satellites are, as it happens, quite far from their planet, and it has been suggested that they may have been captured by the planet somewhat later than the swarms of particles that formed the inner satellites. Calculations show that far from a planet, satellites revolving in the same direction as the planet's rotation will quickly escape; thus only distant satellites revolving in the direction opposite the planet's rotation will remain.

Theory shows that satellites formed close to a planet will tend to have orbits that lie near the equatorial plane of the planet, whereas those formed far from the planet will tend to be near the fundamental plane of the solar system. Our observations confirm this theoretical expectation.

Some of the differences in composition between the two major groups of planets could also be explained once we realized that the central condensation in the disk would eventually become a star. By studying stars in the process of formation, astronomers have found that just before becoming a stable star, protostars appear

to go through a stage in which they eject a considerable amount of mass. If the Sun went through such a stage, as is likely, it must have started out considerably more massive than it is now. When its extra mass was ejected, gases in the disk that had not been incorporated into planets would have been blown away, along with some or all of the gases in the atmospheres of the protoplanets.

Inquiry 6–15 Which group of planets would be more likely to lose their atmospheres at this time—the low-mass or the high-mass group? What about the group that is nearest the Sun compared with the group that is farthest away? What effect would this have on the chemical composition of the planet?

Because the early chronology of the solar system is far from definitely established, we should mention that the following alternative sequence of events is also a possibility. As the Sun settled into its final equilibrium state as a star, the force of solar radiation started pushing the lighter gases out of the inner part of the solar system. If this occurred before a significant amount of condensation into protoplanets occurred, then the inner protoplanets would have formed entirely out of the ices and dust particles of the heavier elements. Furthermore, the protoplanets would have had negligible atmospheres. Because hydrogen and helium constituted all but a small percentage of the early solar nebula, most of the gas would have been swept out of the inner solar system in this picture, with the inner protoplanets forming out of the tiny residual of heavy elements left behind. **Figure 6–16** summarizes this process, which is different from that summarized in Figure 6–14. Note that the end results of these different models appear to be the same.

No matter which chronology is correct, the terrestrial planets are believed either to have lost all their original **primitive atmospheres** at an early era or else to have formed as atmosphereless rocky balls orbiting the Sun. (Of course, some of the terrestrial planets have atmospheres now, and the reason for this will be discussed in Chapter 9.) On the other hand, the strong gravity of the Jovian planets would have caused them to have retained most of their primitive atmospheres, which would explain why today we see them much as they were when they accumulated their gas from the solar nebula. A comparison of the presently observed compositions of the giant planets with those of the terrestrial planets confirms these theoretical predictions. For example, studies of Jupiter show that its abundances are much like those of the Sun; that is, mostly hydrogen with a little helium and only a tiny fraction of everything else. Earth, on the other hand, has little hydrogen

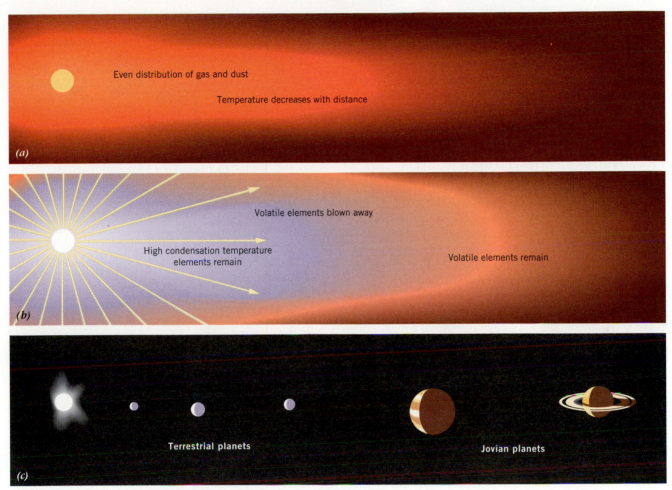

FIGURE 6–16. Another possible scenario explaining why the inner planets contain mostly elements that condense at high temperatures while the outer planets contain primarily volatile elements. Compare with the scheme in Figure 6–14.

and helium (most of its hydrogen is bound up in water, H_2O) and is composed primarily of heavier elements such as silicon, oxygen, and so forth.

LEFTOVER ODDS AND ENDS

There are several irregularities in the solar system that somewhat disturb the orderly picture sketched above. The most vexing, yet intriguing, are the following.

1. Venus's rotation is opposite in direction to that of all the other planets. Its period of rotation is long—243 days.
2. Uranus's spin axis lies nearly in the ecliptic instead of roughly perpendicular to it, as with all the other planets (Figure 6–5b).
3. Pluto's orbital plane is inclined by a rather large angle to the ecliptic (Figure 6–4). In addition, the orbit has

a somewhat large eccentricity (Figure 6–3). Furthermore, Pluto's small mass (only some 0.2% that of the Earth) and diameter implies that its density must be low, near that of the giant planets. Why such an object is beyond the giant planets is not at all obvious.

It has been suggested that the anomalous spin of Venus could be explained if it once had a normal rotation but was slowed down by the tidal effects of the Sun and later by those of the Earth. Collision with another body might also explain it. The reason for its retrograde nature is still unknown. The anomalous rotation of Uranus and its satellite system is more difficult to explain. Possibly many of the blobs of material that formed Uranus collided with the protoplanet at unusual angles, but such speculation is difficult to prove. Similarly, the idea of a single catastrophic collision while the Uranian system was forming might explain why the planet, rings, and satellite system are all tilted, but such ideas are only speculation. Because it is unlikely that the turbulence in the rotating disk of gas will be as regular as Figure 6–11 suggests, it might be possible for such an odd system to form directly out of the gas. Finally, some astronomers have suggested that Pluto is an escaped satellite of Neptune, but recent work has pretty much ruled out this idea. Pluto's formation remains a mystery.

THE ANGULAR MOMENTUM PROBLEM

As it stands, the theory of the origin of the solar system we have described *still* leaves the Sun with too much angular momentum; that is, it predicts that the Sun should be rotating much faster than it actually does. If this theory is basically correct, then some mechanism must have either given the excess angular momentum to the planets or removed it from the system entirely.

Several hypotheses have been advanced to resolve the angular momentum problem. One suggests that the gas in the solar nebula may have contained a magnetic field,[7] as is often the case for the gas in interstellar space. As the cloud collapsed, the effects of the magnetic field would have become stronger.

In addition, the gas in interstellar space is known as a **plasma**, which is a gas that contains many electrons (negatively charged particles) and ions (charged atoms, in this case with positive charge). When such a plasma also contains a magnetic field, the magnetic effects become "frozen into" the gas—the gas and magnetic field are coupled and are forced to move with each other.

If this were the case in the solar nebula, then the rotating protosun would have dragged its magnetic field around with it, and the rotating magnetic field would have dragged the plasma in the disk around with it. This would have transferred angular momentum from the Sun to the disk, and it might have been able to slow the Sun substantially. Such a mechanism is known as **magnetic braking**. While not proven to occur, magnetic braking is the most widely accepted mechanism, although astronomers are investigating other possible processes.

WHICH HYPOTHESIS OF SOLAR SYSTEM FORMATION IS PREFERRED?

The models discussed above can be classified according to their answers to two questions: Were the planets

[7]While the concept of the magnetic field will be presented more completely in Chapter 8, we can describe it for now in terms of the ability of a magnet to attract magnetic materials and charged particles from some distance away. The stronger the attraction, the stronger is the magnetic field.

TABLE 6–3

Classification of Solar System Formation Models

	Planets formed of unaltered interstellar material	Planets formed from processed stellar material
The Sun and the planets formed at the same time	Original nebular hypothesis Modified nebular hypothesis	
The Sun and the planets did not form at the same time	Accretion of interstellar material Encounter with two nebulae Encounter with a protostar whose material formed the planets	Collision theories in which tidal effects of passing star pulls off material from which planets form Sun is part of a binary system; planets formed from a second star

After *The Solar System* by T. Encrenaz, J.-P. Bibring, M. Blanc, Springer-Verlag, page 45, Table 3.2.

formed at the same time as the Sun? Did the solar system form from cooled stellar material, or directly from the gas and dust present among the stars? (By cooled stellar material we mean atoms that have been changed by processes occurring within stars, discussed in Chapters 16 to 19.) A classification scheme is shown in **Table 6–3**. For example, because in the original nebular hypothesis planets were fashioned directly out of interstellar material at the same time as the Sun, the model goes into the upper left box. In the case of the collision hypothesis, planets formed from material pulled from the Sun; that is, material that had been previously processed inside the Sun. Furthermore, in this model, the planets were produced well after the Sun's formation. Therefore, this model belongs in the lower right box. The models involving accretion of planets directly from interstellar material, from an encounter with two nebular gas clouds, and from an encounter with a protostar, all involve unaltered stellar material forming the Sun and planets at different times. These three models therefore go into the lower left box.

Now let's see what the observations tell us, because they have the ability to place severe constraints on our theoretical models. The answer to the question "Did the solar system form from cooled stellar material?" may be answered by looking at the abundance of heavy hydrogen (**deuterium**) relative to the abundance of hydrogen. Deuterium is readily destroyed by the temperatures found in stars. However, because the planets have measurable amounts of deuterium, the material from which the planets formed could not have been previously inside a star. Therefore, all hypotheses on the right side of Table 6–3 may be discarded. The answer to the question "Were the planets formed at the same time as the Sun?" can be examined using radioactive dating techniques (discussed in the next chapter). From such studies, astronomers have determined that the time interval between the protosun's separation from the interstellar medium and planetary formation is less than 10^8 years. This result shows that the Sun and planetary system did form together. The bottom boxes of Table 6–3 are therefore eliminated, strengthening the idea that some type of nebular hypothesis formed the Sun and our planetary system.

MODEL SOLAR SYSTEMS

Is our planetary system the only one that could have resulted from the conditions present when our solar system formed? Astronomers can take the ideas discussed in this chapter—orbital regularities, relative sizes and masses (and their variation with distance from the Sun), the conservation of angular momentum, the elemental condensation sequence—and compute model solar systems. **Figure 6–17**, which pictures the results of several such calculations, shows the distances of the resulting planets from the Sun in astronomical units. The number by each planet is the mass in percentage of the Earth's mass, which is therefore 100. Our solar system is shown for comparison. Because there is a randomness to the accumulation of material into a planetary body, these models show that a variety of solar systems could result, even when the starting conditions are the same.

6.6

DO OTHER STARS HAVE PLANETS?

As we have seen, the more recent hypotheses of the origin of the solar system view the formation of a planetary system as a more or less natural consequence of the formation of a star out of interstellar gas. They suggest that many of the single stars in the galaxy—if not most of them—ought to have planets. On the other hand, hypotheses such as the collision hypothesis, if true, would predict that the formation of planets is an extremely rare event, perhaps unique in the history of the galaxy.

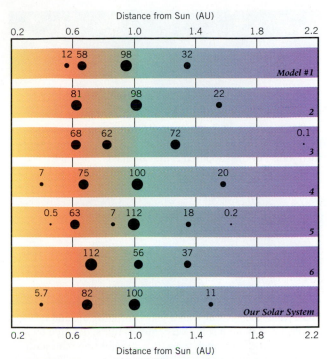

FIGURE 6–17. Model possible solar systems. The number above each dot is the mass of the planet expressed as a percentage of the Earth's mass, which is 100.

However, a hypothesis alone can present us only with possible alternative scenarios. It cannot by itself choose the correct one from among them. For that, it is necessary to appeal to observations.

Unfortunately, astronomers cannot yet directly detect planetary systems around other stars. Such planets, if they exist, would be too faint to be observed, even with the largest of telescopes. Jupiter, for example, appears some 10^9 times fainter than the Sun.

An indirect method involves examination of the apparent motion of a star over a period of many years. Some stars are observed to move across the sky relative to more distant stars. If such a star had an unobservable companion circling it, the observed motion might be similar to that shown in **Figure 6–18** rather than a straight line. The reason for this wiggle motion is shown in **Figure 6–19**, which illustrates two stars revolving about a point in between them called the **center of mass**. With the center of mass (indicated in the diagram by an X) moving in a straight line to the right, each star will appear to oscillate above and below the path traveled by the center of mass, giving the apparent wiggle motion in Figure 6–18.

Such a motion is what astronomer Peter van de Kamp was searching for when he observed a small,

FIGURE 6–18. The expected motion of a star (heavy line) that has a planet orbiting around it. The thin line represents the motion of the supposed center of mass.

FIGURE 6–19. A star with a planet orbiting around it. The objects orbit about a common center of mass, denoted by *X*, which is moving through space in a straight line. The star, which is what astronomers observe, appears to oscillate around the line formed by the moving center of mass. The numbers show the position of the center of mass at different times.

nearby star called Barnard's star. After many years of work in analyzing an effect smaller than a hundredth of a second of arc, he concluded that Barnard's star has a planet that orbits the star with a period of roughly 12 years. Several other astronomers have questioned van de Kamp's results, however. The results are as yet unverified.

Other techniques described later in this book (Sections 14.3 and 15.5) are also being used to search for extrasolar planets. While some tantalizing observations have been made, nothing definitive has yet been observed. Finally, the Hubble Space Telescope (HST) is theoretically capable of discovering extrasolar planets. Stars observed by HST will appear as sharp images, since they will be free of the blurring effects caused by the Earth's atmosphere. The accurate stellar positions that result should be able to resolve the question of the reality of the wiggles that would indicate the presence of planetary systems.

There is some evidence suggesting that planetary systems may be common in the galaxy; namely, the fact that most hot stars seem to rotate very rapidly (in a day or less), whereas cooler stars appear to rotate slowly (taking weeks or months). This may indicate that the cooler stars have transferred their angular momentum to a protoplanetary disk through the process of magnetic braking, and that they might be surrounded by a planetary system. In fact, over 95% of all stars appear to be in the slowly rotating group. This suggests that most of the 100 billion (10^{11}) stars in our galaxy could have planets. But if the difference in rotational speeds has another explanation—for example, if the angular momentum is passed off to the interstellar gases instead—then slow rotation would not necessarily imply a planetary system.

The 1983 launching of the Infrared Astronomy Satellite, IRAS, shed some light on the question of planetary systems about other stars. In the course of doing a survey of the sky, IRAS detected a number of relatively nearby stars that were shining brightly in infrared radiation. Strong infrared radiation suggests the presense of surrounding disks composed of large dust particles. For some nearby objects it was even possible to detect this material in red light with telescopes on the Earth's surface. An example is the star Beta Pictoris shown in Figure 1–10. This picture suggests that we are detecting the "leftover" material of the original cloud from which this stellar system is forming.

These observations are still not definitive proof of other planetary systems, but they are a strong confirmation of the existence of disks of material with associated condensations around many nearby stars. It should be noted that planets themselves would still not be detectable, even by IRAS, because their actual size would be so small.

In 1992 astronomers announced the startling discovery of planets orbiting a pulsar, an exotic type of star thought to result when a massive star explodes. (Chapter 19 discusses pulsars further.) In fact, the pulsar observations indicate the presence of two planets, one having a mass 3.4 times that of the Earth at a distance of 0.36 AU from the pulsar, and another of 2.8 Earth masses at 0.47 AU. The big question now is, how do planets form within or from a supernova explosion—or survive it?

To summarize, some tantalizing pieces of evidence seem to suggest that many stars have planets and that planetary formation is a natural consequence of star formation. The evidence is circumstantial, however. Due to the difficulty of making the necessary observations, it may be some time before conclusive evidence for the existence of other planetary systems is found. The Hubble Space Telescope may provide the evidence we need.

DISCOVERY 6–1
A SCALE MODEL
OF THE SOLAR SYSTEM

When you have completed this Discovery, you should be able to do the following:

- Describe the size of the solar system on a human scale.

A model of the solar system, which can help you understand better the sizes and distances in our planetary system, can be built on a human scale. For this model, let 1 astronomical unit be represented by 25 paces. After completing the questions below, walk your model. Stop at each planet and look back to where you started.

☐ **Discovery Inquiry 6–1a** Make a table of the number of paces required to walk from the Sun to each planet.
☐ **Discovery Inquiry 6–1b** Determine the length of your pace by measuring the length of 10 paces and dividing by 10.
☐ **Discovery Inquiry 6–1c** How large in yards or meters is your model solar system?
☐ **Discovery Inquiry 6–1d** On the scale of your model, how many miles or kilometers are there to the nearest star, which is approximately 200,000 AU away?
☐ **Discovery Inquiry 6–1e** On the scale of your model, determine the sizes in inches or centimeters of the Sun and the various planets.

CHAPTER SUMMARY

OBSERVATIONS

- **Terrestrial planets** are near the Sun, are relatively small in mass and diameter, have high density, are composed of heavy elements, and have a high-temperature, low-density atmosphere. **Jovian planets** are far from the Sun, are relatively large in mass and diameter, have low density, are composed of low-mass elements, and have a low-temperature, high-density atmosphere.
- Planets move in orbits that are nearly circular and lie in the ecliptic plane, which also is in the Sun's equatorial plane.
- With the exception of Venus, Uranus, and Pluto, planets rotate about their axes in a counterclockwise direction. With the exception of Uranus and Pluto, planetary rotation axes are within 30° of the perpendicular to the planet's orbital plane.

- Most planetary satellites revolve around their planet in the same direction the planet rotates. Satellites close to a planet are near the planet's equatorial plane.
- The Sun contains 99.8% of the mass of the solar system but only 2% of the angular momentum.

THEORY

- The **nebular hypothesis** states that a rotating cloud of gas and dust collapsed to form the Sun and left rings of material behind that formed the planets, which are located in a flat plane. The **collision hypothesis**, also known as the catastrophic hypothesis, states that tidal forces from a passing star pulled material from the forming Sun. This material then condensed into the planets.
- Turbulence in a rotating disk may help form planets.

- When dust particles in a collapsing cloud collide, they stick together, thus providing the first step in the eventual growth of planetesimals. Dust particles provided the heavy elements present in the terrestrial planets.
- Satellites forming near a planet should be in the equatorial plane.
- The Sun may have transferred its angular momentum to the planets by means of an interaction of the particles it constantly ejects and its magnetic field (**magnetic braking**).

Conclusions

- The Sun and planets formed at nearly the same time.
- From observations of deuterium, we infer that the planets did not form from previously processed stellar material.
- Chemical differences between the terrestrial and Jovian planets can be understood in terms of condensation of certain elements at high temperatures and volatile elements at low temperatures while the solar nebula was cooling.
- The Sun went through a phase in which it lost much of its original material. This rapid loss caused gas and dust remaining within the inner solar system to be removed from it. This loss of mass may also have contributed to the loss of volatile elements by the terrestrial planets.
- If the Sun and planets formed through some variation of the nebular hypothesis, planets are expected to exist around other stars.

SUMMARY QUESTIONS

1. What are the names of the planets in order of their distance from the Sun?

2. What are the two major groups of planets? List the characteristics that distinguish members of each group.

3. What are the major regularities in the orbits and motions of the planets that need to be explained by any successful theory of the origin of the solar system? What exceptions are there to these regularities?

4. What observational evidence can you use to criticize the early nebular and collisional hypotheses of the origin of the solar system?

5. What are some possible reasons for the dramatic differences in composition between the terrestrial and Jovian planets?

6. Why were dust grains important to the formation of the solar system? What role might turbulence have played in its formation?

7. Where is most of the angular momentum in the solar system located? Why was the distribution of angular momentum a problem for earlier hypotheses of solar system formation? Describe a process thought to solve the angular momentum problem.

8. How might astronomers go about trying to detect planets orbiting other stars?

APPLYING YOUR KNOWLEDGE

1. Hypothesis: The solar system formed when a passing star pulled material from the Sun that then formed the planets. Present evidence both for and against this hypothesis, and state your final conclusion.

2. List all observations about the solar system that are exceptions to the generally observed features.

3. Explain why the following two definitions of volatile materials are the same. (*a*) Volatile materials are those that condense at low temperatures. (*b*) Volatile materials are those that turn to vapor at low temperatures.

4. Use the condensation sequence of Figures 6–13 and 6–14 to explain in your own words the reasons different planets differ in chemical composition.

5. Refer to Figure 6–5*b*, which shows the inclinations of the planets to their orbital planes. For each planet, discuss what you expect for seasonal variations throughout the planet's year.

6. What fraction of planetary orbital angular momentum is held by Jupiter? Saturn? (We are neglecting rotational angular momentum, which is less than that from the orbital motion.)

7. Compute Pluto's density if its mass is 0.0024 the mass of the Earth and its diameter 2302 km. Express the density relative to the density of the Earth. What conclusions about the nature of Pluto might you draw from the observed density?

■ **8.** What would be the angular separation between the Sun and Jupiter if they were observed from the distance of Alpha and Proxima Centauri, the nearest stars, at a distance of 4.3 light years? Compute the distance a dime would have to be for its angular size to match the angle you computed.

ANSWERS TO INQUIRIES

6–1. Hydrogen and helium.

6–2. As the star's light passes through the planet's atmosphere, the light passes through successively denser layers of gas, which causes the light to gradually fade until the planet itself blocks the star's light.

6–3. The low gravity makes it easier for atoms to escape. The high temperature increases the average velocities of the atoms, so that a greater fraction of them is above the speed necessary to escape Mercury's gravity.

6–4. The distance between the Earth and Sun is approximately 10,000 times the Earth's diameter. In Figure 6–2, the Earth's diameter is about 0.4 cm, which means the Earth's distance, on this scale, is 4,000 cm or 40 m. Pluto, which has an average distance 40 times farther, would be about 1.6 km away.

6–5. The hypothesis would not be consistent with the fact that the planes of the planetary orbits are nearly the same, because it would be expected that the planets would be captured into random orbits. To capture a planet in a nearly circular orbit requires special circumstances that would occur only rarely. The hypothesis is consistent with having a large range of planetary masses. The hypothesis is not consistent with the planets and satellites revolving in the same directions.

6–6. The hypothesis would not be consistent with the fact that the Earth, Moon, and meteorites have the same age because the various bodies could be captured at different times. On the other hand, if all these objects had actually been formed at the same time, they would have the same age.

6–7. The hypothesis would not be consistent with the near-circularity of the planetary orbits, because capture would more likely create highly eccentric orbits. It would not be consistent with the fact that the satellites mostly orbit in one direction (the same as the planets),

and it would not place all the giant planets together in the solar system as we see them today. It is consistent with the variable masses and the variable compositions that we see.

6–8. Assuming that planetary masses could actually aggregate in the simple nebular hypothesis, it would explain the basic "traffic pattern" of motions. It would not explain irregularities like the spin of Venus or the odd inclination of Uranus's spin axis.

6–9. Perhaps a process similar to the one proposed for the formation of planets themselves could be used to explain the formation of satellites around a planet.

6–10. The nebular hypothesis as presented would not be able to explain the chemical compositions of the planets (nor the distribution of angular momentum in the solar system).

6–11. The collision hypothesis would lead us to expect that planetary systems are rare in the galaxy, whereas the nebular hypothesis would argue that they are common, a natural companion of star formation. Until we are able to gather some direct observational evidence on the question, these considerations by themselves are not conclusive.

6–12. This hypothesis has the same basic problems as the nebular hypothesis. Also, it is awkward to fit the formation of the satellites of the planets into this picture.

6–13. The modern theory still needs additional hypotheses to explain the anomalous spins of Venus and Uranus. And all hypotheses still have some trouble with Pluto.

6–14. Clockwise rotation is not explained.

6–15. The low-mass group, because the gravity of these planets is lower. Because the low-mass elements present in the disk (hydrogen and helium) would be the ones most readily blown away, the inner planets would tend to be composed of the heavier elements that remain.

7

THE SMALLEST BODIES OF THE SOLAR SYSTEM: COMETS, MINOR PLANETS, AND METEORITES

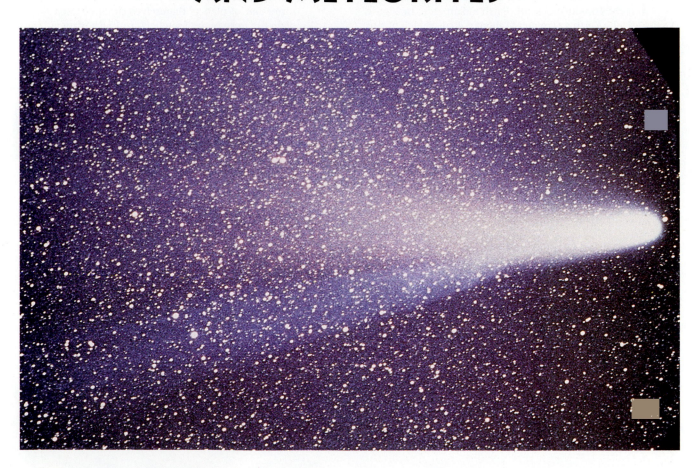

The search for the "primordial" composition [of the solar system] may be likened to the search for the Holy Grail. The analogy is apt insofar as fervor is concerned, though medieval bloodshed has been replaced by mere acrimony.

EDWARD ANDERS, *ANNUAL REVIEW OF ASTRONOMY AND ASTROPHYSICS*, 1971

Comets, asteroids, and meteorites can shed a great deal of light on the origin of the solar system. While the surfaces of the planets have undergone many changes since the solar system was formed, the smaller objects we will discuss in this chapter have changed much less and may be able to tell us much about the early solar system. During this discussion, we will also fill in some of the details of the theoretical scenario proposed in Chapter 6 for the origin of the solar system.

7.1
COMETS

Comets are frequent and sometimes spectacular visitors to our part of the solar system. In former times they caused consternation and even panic among those who viewed them because comets were often interpreted as signs and omens of the future. One such example is the comet of A.D. 1066, which the Saxons saw as an evil omen before their defeat by the Normans at the Battle of Hastings. A portion of the Bayeux Tapestry, which commemorates this event, shows the presence of the comet in the sky (**Figure 7–1**).

Aristotle contended that comets were exhalations from the Earth, and it was Tycho Brahe who showed that they were too distant from the Earth for this to be true. After the discovery of Newton's laws of motion, Edmund Halley (who later became England's Astronomer Royal) calculated the orbit of the comet of

1682; he found that its period was approximately 76 years and that it traveled in an orbit that took it about 35 AU from the Sun (**Figure 7–2**). Halley suggested that the comets of 1607 and 1531 were actually the same object and predicted that it would return in 1758. Although he did not live to see his prediction come true, the comet's reappearance in 1758 caused great excitement and sealed the triumph of Newton's theory by showing that all objects in the solar system are subject to the same physical laws.

The comet was named in honor of Halley; it turned out to be the same comet that was seen in 1066, and sightings of it have been traced in Chinese records as far back as 240 B.C. The Earth actually passed through the extended tail of Halley's comet during its spectacular

FIGURE 7–1. The Bayeux Tapestry. The legend in the tapestry reads: "They marvel at the star."

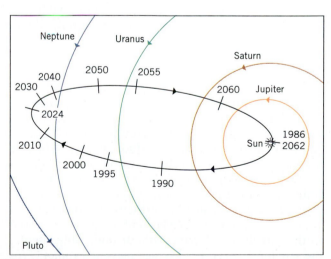

FIGURE 7–2. The orbit of Halley's comet, approaching and receding from the Sun.

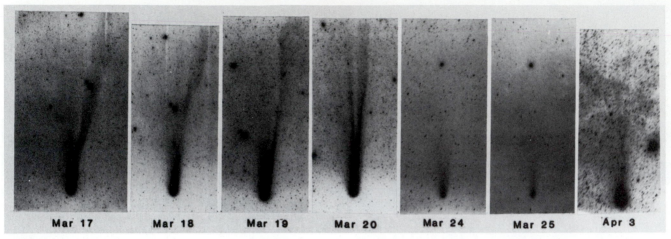

Mar 17   Mar 18   Mar 19   Mar 20   Mar 24   Mar 25   Apr 3

FIGURE 7–3. Halley's comet, as it appeared on various dates during its 1985–1986 appearance.

1910 appearance. Although the comet returned on schedule in 1985–1986 (**Figure 7–3**), that was its least impressive appearance in 2000 years, because Earth and the comet didn't come close to each other, and the comet was low in the sky for Northern Hemisphere viewers. However, during this visit, several spacecraft were sent to rendezvous with it, and we were able to learn more about comets in just a few months than in all previous history.

In any given year an average of about a dozen new comets are discovered and several old ones return. Along with novae—exploding stars that suddenly increase greatly in brightness—comets have always been fair game for discovery by amateurs. Because it is not possible to predict the first appearance of a new comet, a certain amount of luck is required, along with diligence, patience, and enough familiarity with the sky to recognize an intruder.

THE STRUCTURE OF COMETS

Despite their impressive appearance when near the Sun, comets are actually rather small objects. They have been described as ". . . the nearest thing to nothing that anything can be and still be called something." When far from the Sun they do not show a visible disk in the telescope as does a planet, and it is estimated that the cometary body itself, called the **nucleus**, averages only about 10 km in diameter. Furthermore, they are not very massive; comets have been observed to pass close to the satellites of Jupiter without noticeably altering the satellites' orbits. The average mass of a comet is approximately 10^{17} grams, only one ten-billionth that of Earth.

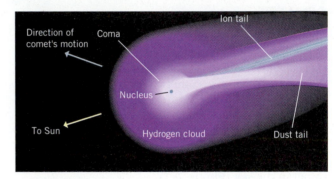

FIGURE 7-4. The structure of a comet near the Sun.

As a comet nears the Sun, it appears to get larger and to develop a fuzzy appearance. The fuzziness comes about because solar energy heats volatile ices located within the nucleus, causing the ices to change directly to a gas. The process is the same as when "dry ice" changes from solid to gas. The central nucleus of the comet appears as a small, bright point surrounded by a nebulous region known as the **coma** (**Figure 7–4**). A coma, when near the Sun, is typically 10^5 km in diameter. As the comet approaches even closer to the Sun, a long **tail** forms, which may extend over 100 million km from the comet's nucleus. The tail can appear highly complex, and there may be more than one tail present (Figure 7–4). The tail (or tails) tends to point *away* from the Sun; the tail thus precedes the comet when it leaves the vicinity of the Sun (**Figure 7–5**). This fact was known to Chinese astronomers at least 900 years before Europeans learned it in the sixteenth century. Finally, observations from satellites have found that comets have a

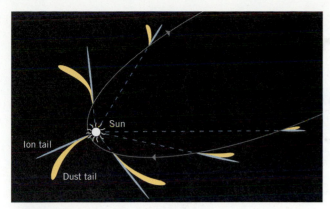

FIGURE 7-5. The tail of a comet always points away from the Sun.

hydrogen halo extending some 1 to 10 million kilometers around them. The hydrogen halo is thought to come from the breakup of H_2O molecules.[1]

What causes these spectacular changes in the appearance of a comet as it rounds the Sun? How can such a small object provide such a fantastic display? To answer these questions, let us look at a model for comets that appears to explain things reasonably well.

THE DIRTY SNOWBALL MODEL

Astronomers now believe that the nucleus of a comet consists of frozen gases and solid materials. The well-known comet expert Fred Whipple has aptly described this body of dust and rocks orbiting the Sun as a sort of "**dirty snowball**" or "dirty iceberg." When the nucleus approaches the Sun, solar energy vaporizes some of the gas, releasing gas and dust from the nucleus into surrounding space. This produces the extended coma.

As the comet approaches even closer to the Sun, two effects that are important in producing the comet's tail increase in significance. First, the Sun's **radiation pressure**, caused by the outward movement of sunlight itself, pushes dust particles away from the Sun. Second, the **solar wind**, which consists of charged particles that are constantly ejected from the Sun, exerts a force on the material released from the coma. Because both the radiation pressure and the solar wind are directed outward, the tail will tend to point away from the Sun.

The two forces that produce tails tend to act in different directions and to affect the gas and dust particles somewhat differently, which is why two tails often are seen. The difference between them can be detected by a

careful study of the light from each tail. For example, because the smooth-appearing tail has light similar to sunlight, this tail must be composed of small dust particles that simply reflect the sunlight. It is, in fact, called the **dust tail**.

The other tail shows a complex and variable structure composed of gases excited into giving off radiation. Within this **gas tail**, we find emissions characteristic of charged molecules such as CH^+, OH^+, N_2^+, and CO_2^+, to name a few. Presumably, the charged molecules are present because the charged particles of the solar wind knock electrons off some of the gas molecules vaporized from the comet's coma. The interaction of the charged particles with magnetism that is spread throughout the solar system produces the complex structure that we observe in comet tails. The smooth dust tail is not charged and is not affected by magnetism.

Inquiry 7–1 It is observed that comets tend to become fainter and fainter with successive returns near the Sun. Can you explain this in terms of the dirty snowball model?

Until recent times the small, faint nucleus could not be analyzed directly, and it was necessary to infer the actual composition of the nucleus's frozen gases from the elements observed in the gas tail—hydrogen, oxygen, carbon, and nitrogen. The molecules we see are "daughter" products formed when the Sun's ultraviolet light and the solar wind break down more complex substances, a fact that was confirmed by the International Comet Explorer (ICE) satellite in a rendezvous with Comet Giacobini-Zinner in late 1985. It is generally thought that the parent gases are frozen methane (CH_4), ammonia (NH_3), and water (H_2O). Other satellite studies of Halley's Comet in the spring of 1986 (see below) enlarged our knowledge of the nucleus greatly.

When a comet is close to the Sun, atomic emissions from metals such as sodium, calcium, silicon, and iron are also observed. These are presumably the result of vaporization of dust particles by the intense solar heat, followed by excitation of the vapor by the Sun's radiation; these emissions are additional evidence of the presence of solid material in the cometary body.

A final piece of evidence in favor of the dirty snowball model is the fact that the motion of some comets is slightly erratic. An example is Comet Encke, which has been well studied because of its short three-year period. It has been observed to slow down or speed up unpredictably, especially when near the Sun. Fred Whipple's model explains this phenomenon: after many passages near the Sun, much of the comet's surface material has evaporated, leaving behind a sort of crust on the surface.

---

[1]A molecule is the smallest particle into which a compound can be divided without changing its physical and chemical properties. It is a group of atoms held together by chemical forces.

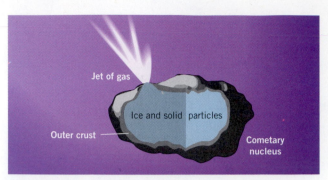

FIGURE 7-6. Whipple's explanation of the erratic motion of Encke's comet. "Jets" of gas squirting out through fissures in the nuclear crust alter the motion of the comet in an unpredictable way.

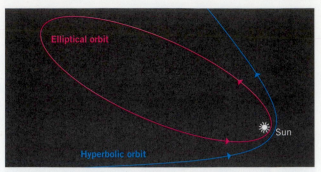

FIGURE 7-7. Cometary orbits can be either elliptical or hyperbolic. The hyperbolic comets escape from the solar system and are never seen again.

Solar heating of pockets of gas near the surface might force the gases through a thin part of the crust as stronger parts held firm (**Figure 7–6**). This released gas acts like a "jet" that accelerates or decelerates the comet, depending on which direction the rotating nucleus happens to be pointed at the time. It is these jets that make cometary behavior difficult to predict reliably.

ORBITS OF COMETS

Cometary orbits tend to be highly elongated ellipses (see Figure 7–2), oriented randomly with respect to the ecliptic plane. In general, it is just as likely that a comet will be found traveling around its orbit in a clockwise direction as in a counterclockwise direction. This is different from the situation for planetary orbits, which are nearly circular, nearly in the ecliptic plane, and always traversed in a counterclockwise direction.

The orbits of comets fall into two basic types. Comets in *elliptical* orbits, such as Halley's comet, are clearly members of the Sun's family, orbiting the Sun indefinitely. In contrast to the closed circuit that a body in an elliptical orbit follows, a body in a *hyperbolic* orbit follows an open path that does not close back on itself (**Figure 7–7**). These comets in hyperbolic orbits have high velocities that cause them to escape from the Sun's gravitational hold and become interstellar outcasts.

Inquiry 7–2 Apply Kepler's law of areas to a comet following a highly elliptical orbit. When is the comet moving most rapidly? Where in its orbit does the comet spend most of its time?

Are comets with hyperbolic orbits interlopers from interstellar space? It appears that this is not the case. Rather, there is evidence that comets with hyperbolic

orbits were once in elliptical orbits but came too close to a planet such as Jupiter, which accelerated them to high speeds and altered the shape of their orbits from elliptical to hyperbolic.

Inquiry 7–3 It is observed that comets come toward the Sun from all directions in space. We also know that the Sun is traveling rapidly toward a point in the constellation of Hercules. If the Sun were picking up comets from interstellar space during its travels toward Hercules, from what direction would the comets tend to come? How does this support the idea that hyperbolic comets were originally part of the Sun's family?

Inquiry 7–4 What might you conclude about the distribution in space of cometary nuclei, given that they are observed to come equally from all directions?

THE OORT AND KUIPER COMET CLOUDS

Because comets come toward the Sun from all directions in space, their distribution around the Sun must be roughly spherical. But because the inner part of the solar system (the part containing the planets) is highly flattened, it follows from their spherical distribution that comets must be associated with the distant outer parts of the solar system. Such reasoning led the Dutch astronomer Jan Oort to hypothesize that there exists a large spherical cloud of comets surrounding the solar system, between about 50,000 AU and 100,000 AU from the Sun. Although well beyond the orbit of Pluto, the Oort cloud is still a part of the solar system. According to Oort, there may be as many as 100 billion objects in this cometary cloud.

All theories about the origin of comets presume that they were formed early in the history of the solar system. For example, they could have condensed out of the solar nebula at an early stage, before it began to flatten. If this idea is correct, then the cloud of gas and dust from which the solar system was formed must have been large. However, the density of material in the outer reaches of the solar system would have been so low that it is difficult to imagine how nuclei some 10 km in diameter would form. For this reason, today we believe the comets formed in a higher-density region, in the general area between the orbits of Saturn and Neptune. Such a model is consistent with the icy composition of the comets. We can then easily imagine that gravitational perturbations from the giant planets (especially Jupiter) could toss them out of the inner solar system, thus forming the Oort cloud. The entire discussion of the Oort cloud is based on inference from observation; it has yet to be actually observed.

Most of the objects in the Oort cloud would have nearly circular orbits, so that they would never get close enough to the Sun to be seen. From time to time, however, a star passing relatively near the Sun might provide the small gravitational nudge required to change these orbits so that comets begin their long trips toward the inner solar system. Astronomers suppose that there are so many comets in the Oort cloud that a great many of them are being perturbed toward the Sun at any given instant.

An additional complexity appears from analysis of mathematical models of the orbits of **short-period comets**, those having periods less than about 200 years. These comets are more confined to the ecliptic than the longer-period ones. The models show that short-period comets could not have come from the spherical Oort cloud. For this reason, astronomers have hypothesized the existence of a nearer, disk-shaped belt of comets some 30–50 AU from the Sun, just outside Pluto's orbit. This belt, called the **Kuiper belt** after the astronomer who first proposed it, is estimated to have about the mass of the Earth, much less than the 100 Earth masses hypothesized for the Oort cloud. Like the Oort cloud, the Kuiper belt has yet to be observed.[2]

Inquiry 7–5 After many passages near the Sun, a comet will eventually "wear out," becoming depleted of gas and unable to produce a coma or a tail. If the period of a typical comet in its orbit around the Sun is 1000 years, and it returns 100 times before using up its material, what is the lifetime of a typical comet? What fraction of the age of the solar system (about 4.5 billion years) is this comet's lifetime?

[2]Two observations were reported in 1992 and 1993 of objects that might be members of the Kuiper belt. No conclusion on their nature is yet available, however.

Inquiry 7–6 Is it likely that the comets we see today have been coming near the Sun since the solar system was created? Explain.

Inquiry 7–7 The short lifetimes of comets, and the fact that we continue to observe them today, presents a dilemma. How might Oort's theory be able to resolve this dilemma?

RECENT SPACECRAFT STUDIES OF COMETS

During 1985 and 1986, several spacecraft missions were sent to rendezvous with comets; ICE made the first close encounter in late 1985. ICE was originally designed to study the Sun, and it had been in orbit for several years doing just that. However, it turned out to be possible to change the orbit of the satellite and send it off to the neighborhood of Comet Giacobini-Zinner, which was passing close to the Earth several months before Halley's comet came through. There was enough spare fuel on board to send the satellite on a trip around the Moon, and from the Moon's gravity it acquired additional acceleration. In fact, it made several passes around the Earth-Moon system before achieving enough of a "slingshot" effect to travel toward the comet. This virtuoso feat of celestial mechanics allowed the satellite to study in detail the interactions between the comet and the solar wind.

After Halley's comet rounded the Sun in the spring of 1986, a veritable flotilla of spacecraft from several different nations (Russia, Japan, and a European consortium) converged on it. Each mission was designed to study the comet from different distances and acquire different kinds of information. Two Russian Vega spacecraft returned the first close-up images of Halley, passing within 5523 miles (almost 8900 km) of the nucleus. Next, the Japanese spacecraft Suisei (the word means "comet") viewed the comet in ultraviolet light. Even though Suisei came no closer to the comet than 94,000 miles (150,000 km), it was still rocked by collisions with two large particles moving through space with the comet.

The most spectacular (and risky) mission was the close pass by the European spacecraft Giotto, only five days after the Vega 2 rendezvous. Named for the Renaissance artist who depicted an earlier passage of Halley's comet in a painting, Giotto passed within 376 miles of the nucleus. Giotto was struck by one collision that aimed its antenna in the wrong direction for a while, but the spacecraft's gyroscopes restored control after 34 minutes of anxiety. No further TV pictures were returned after the collision, but some 2000 images of the comet were collected before Giotto's cameras were literally "sandblasted" away by the dust and debris moving

FIGURE 7–8. Jets of material are seen streaming away from the black, elongated nucleus of Halley's comet in this photograph taken from a distance of 15,950 miles by the European spacecraft Giotto.

through space with the comet. Giotto was designed to resist the destructive forces of the comet as well as its designers could anticipate, but there is still a certain element of luck involved when you are aiming a precision instrument into a sandstorm where particles are moving at thousands of miles per hour!

Giotto confirmed that the nucleus of Halley's comet was about 8 by 16 km (5 by 10 miles) in size and shaped somewhat like a peanut (**Figure 7–8**). It showed jets of evaporating material concentrated at the two ends of the nucleus. Perhaps most surprising was the color of the nucleus. It turned out to be jet black, one of the darkest objects in the solar system, reflecting only 5% of the light incident on it. At least for Halley's comet, the traditional picture of a "dirty snowball" should be modified to something like "a lump of coal with frozen ices inside." Much to everyone's surprise, observations found that the nucleus was hot, with a temperature of 330 K (or 135° F). The surface of the nucleus was definitely not covered with ice or the temperature would have been lower. The dark surface absorbs sunlight efficiently, and the evaporating gases pour out from the interior in jets and streams.

These spacecraft determined that Halley's comet has a low density, only some 0.1 to 0.25 grams/cm^3. Such a low density implies that the nucleus is porous, actually some 90% empty space. Another surprise discovery was the presence of magnetism around the nucleus. The magnetism is produced by the charged molecules in the coma.

Inquiry 7–8 How might studies of the composition of comets help us learn about the primordial composition of the solar system?

On the all-important subject of chemical composition, the instruments on the Giotto and two Vega spacecraft revealed the presence of water, carbon dioxide, and carbon monoxide gases. The gaseous jets, in fact, contained about 80% water vapor. Such a large amount of water, when broken down by energy from the Sun, can explain the extensive hydrogen halo surrounding comets. But observers also noted signatures of heavier compounds containing silicon, carbon, oxygen, sulfur, and iron, to name a few. The amount of carbon is the same as that found in the Sun, confirming the idea that comets formed from raw interstellar material. The dirty snowball model has also been confirmed.

7.2
MINOR PLANETS

Ever since astronomers have had good estimates of the distances of the planets from the Sun, it has been suggested that there is an unusually large gap between Mars and Jupiter and that there ought to be a planet there. In 1801, the Italian astronomer Giuseppe Piazzi discovered a new object in an area of the sky he had been studying. Greatly interested, he followed its motion on successive nights until it neared the Sun and was lost. Enough observations had been made for the German mathematician Karl Friedrich Gauss to determine its orbit, however, and the next year it was found again where Gauss predicted it would be. Piazzi named the object Ceres, in honor of the goddess of Sicily. The orbit Gauss calculated for Ceres lay between Mars and Jupiter and many believed the object to be the long-sought planet. Yet it was an unusual object, because it was much smaller than any of the other planets.

In 1802, a second such "planet," Pallas, was discovered and, in 1804, a third one, Juno. In time, many others were found, and it was slowly realized that the region between Mars and Jupiter was literally swarming with small bodies, all much smaller than the familiar planets. For this reason, they have come to be known as **minor planets**, although the terms **planetoid** and especially **asteroid** are also used. Modern photographic surveys have found thousands of these objects, and orbits have been determined for nearly 2000 of them. The region of space they occupy has come to be known as the **asteroid belt**.

CHARACTERISTICS OF THE MINOR PLANETS

Ceres is the largest of the minor planets, with a diameter of about 950 km (about the size of Texas); next largest is Pallas at 560 km (about the size of Kansas). The sizes must range all the way down to particles as small as dust grains. Because of the large number of asteroids, collisions must take place frequently, slowly grinding them down to smaller and smaller size. When the Pioneer spacecraft passed through the asteroid belt, however, it did not find the region to be particularly dusty; it would appear that other forces are at work sweeping the smaller particles out of the asteroid belt.

Minor planets must be rather irregular in shape, because most of them are not sufficiently massive for their own gravity to force them into spherical shape, as in the case of the planets. They are more like large chunks of rock, and the rigidity of their rocky structure is sufficient for them to maintain whatever shape they originally had. Although asteroids are too small for their shapes to be observed directly from Earth, observations of the asteroid Eros have allowed it to be modeled as an irregularly shaped object about 7 by 19 by 30 km in dimensions. A close-up view of the asteroid Ida (**Figure 7-9**), observed by the Galileo spacecraft from a distance of 10,000 km, shows the 56 km-long asteroid not only to be pitted with craters but to have a 1.5 km diameter satellite only 100 km away.

Inquiry 7-9 It is observed that the brightness of a typical minor planet varies significantly over periods of

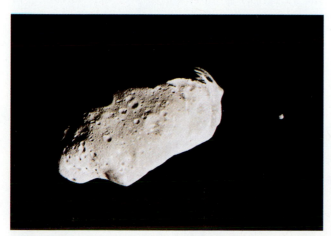

FIGURE 7-9. The asteroid Ida, as imaged by the Galileo spacecraft.

an hour or two. Assuming that these bodies rotate on an axis, as do the planets, what are two possible explanations of these observations?

YOU SHOULD DO DISCOVERY 7-1, ASTEROID BRIGHTNESS VARIATIONS, AT THIS TIME.

ORBITS OF THE MINOR PLANETS

Although the majority of the minor planets have nearly circular orbits situated between Mars and Jupiter and fairly near the ecliptic plane, there are some notable exceptions. One is Hidalgo, whose orbit is shown in **Figure 7-10**. Its orbit is elongated and inclined by over 40° to the ecliptic plane.

Inquiry 7-10 Can you suggest a possible reason why Hidalgo's orbit is so different from those of most minor planets? (Hint: What kind of object does Hidalgo's orbit remind you of?)

An important and unusual group of asteroids are the few dozen known Apollo asteroids, whose orbits come close to the Earth's. These objects could have been sent

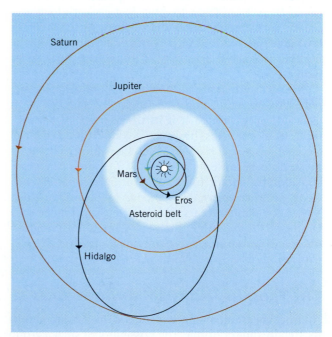

FIGURE 7-10. The orbits of some minor planets. Most minor planets stay between Mars and Jupiter, but there are some exceptions.

into their present orbits by gravitational perturbations from the giant planet Jupiter. There are probably some 1000 such Apollo asteroids larger than one kilometer in diameter. Some objects with near-Earth orbits could in fact be burned-out comets. The risk of collision with one of these objects is not negligible, as we will discuss later.

One of the most important asteroids historically has been Eros, which in 1901 and 1931 came within 26 million km of the Earth. On the second occasion it was possible for astronomers to measure with increased accuracy its distance in kilometers. This in turn made it possible for the length of the astronomical unit, in kilometers, to be determined accurately for the first time, because the distance to the minor planet in astronomical units was well known from Kepler's laws.

Inquiry 7–11   Suppose a beam from a radar dish on Earth is sent to Eros and returns 173.3 seconds later. What would be the distance in kilometers from Earth to Eros?

Inquiry 7–12   If astronomers calculated that Eros's distance from Earth at this time was 0.17 AU, how many kilometers would this indicate that there are in an astronomical unit?

Today, even more accurate methods of measuring the astronomical unit are employed, but the principle is similar. With radar, it has become possible to measure the distance from Earth to other planets with high precision (considerably better than 1 km). Again, Kepler's laws give the same distance in astronomical units, so the number of kilometers in an astronomical unit can be worked out.

Not all asteroids are in the inner half of the solar system. One object, Chiron, was originally classified as an asteroid, even though it is in the vicinity of Saturn's orbit. In recent years, however, a coma characteristic of a comet has been observed around it. Chiron's nature is mysterious because its 200–300 km diameter makes it considerably larger than any ordinary cometary nucleus. Another object, discovered in 1992 and known as 1992AD, takes 93 years to orbit the Sun, longer than Uranus. Its color, redder than any comet or asteroid yet observed, is interpreted as signifying a surface rich in organic materials.

THE TROJANS AND THE KIRKWOOD GAPS

An unusual group of minor planets is the **Trojan group** located in Jupiter's orbit. The possibility of such a group was already known in the eighteenth century, when the mathematician Joseph-Louis Lagrange showed that if a small body were placed in a circular orbit around the

Sun in such a way that it, the Sun, and Jupiter formed an equilateral triangle, then it would stay in that position, moving at the same rate around the Sun as Jupiter (**Figure 7–11**). Such gravitationally stable locations, where the overall gravitational force is zero, are called **Lagrangian points**. The concept will appear again in our discussion of the satellites of the Jovian planets. The Lagrangian points for the Earth-Moon system have been suggested as possible locations for space colonies.

The Trojan asteroids are an example of the important gravitational effects of Jupiter. Another example is found within the asteroid belt itself; there are certain distances from the Sun where few, if any, minor planets are found. These regions are known as **Kirkwood gaps**, after the American astronomer who discovered them. The influence of Jupiter on the formation of these gaps is clear, because they occur at distances from the Sun at which a minor planet (if it were there) would have a period around the Sun that is in a simple ratio to Jupiter's period. For example, as shown in **Figure 7–12**, which shows the variation of the number of asteroids with distance, there is a prominent gap at the place where the period of a minor planet would be just one-half Jupiter's period, and another where its period

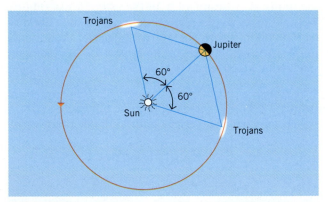

FIGURE 7-11.   The Trojan asteroids, which are located at Lagrangian points in Jupiter's orbit.

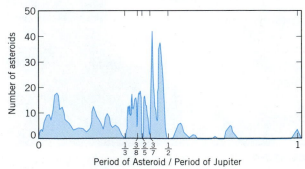

FIGURE 7-12.   The Kirkwood gaps (after Brouwer).

would be just one-third Jupiter's period. An asteroid in these locations would find itself close to Jupiter at regularly spaced intervals in time and would suffer a cumulative perturbing force that would sooner or later send it into an entirely different orbit. The astronomer Jack Wisdom has shown recently that many of the bodies forced out of the Kirkwood gaps by Jupiter would eventually collide with Mars or Earth.

WHY ARE THE MINOR PLANETS SO SMALL?

The bare facts about minor planets raise more questions than they answer. In particular, why are there so many of them and why are they so small? The total mass of all the minor planets put together is considerably less than that of any of the major planets—where did the rest of the mass go, if it was ever present?

There is no question that the presence of Jupiter has been of great importance in answering these questions. Calculations show that if there were a large number of small bodies between Mars and Jupiter early in the history of the solar system, most of the ones fairly near Jupiter would have been so strongly affected by it that they would have either been captured by Jupiter or else thrown entirely out of the solar system. This would have greatly reduced the amount of mass in the region.

We might also speculate that the mere presence of a body as large as Jupiter would have prevented the material between Mars and Jupiter from gathering itself together into a major planet, although it is not clear how this would have happened. Alternatively, several modest-sized bodies may have formed there, despite the presence of Jupiter, and later collided with each other, fragmenting into many smaller bodies. Indeed, calculations show that numerous collisions take place in the asteroid belt at present, so such a scenario is certainly possible.

Inquiry 7–13 Groups of minor planets, called Hirayama families, after the Japanese astronomer who discovered them, have orbits that are remarkably similar to each other, with nearly the same size, shape, and orientation with respect to the plane of the solar system. How might this observation tend to support the theory that the present-day asteroids were formed as a result of the collisional fragmentation of larger bodies?

CHEMICAL COMPOSITION

Chemical information about asteroids can be obtained through detailed studies of their ability to reflect light of different colors. There are four main compositional groups: silicon-rich (S type), nickel-iron-rich (M type), carbon-rich (C type), and the D types, which may be rich in organic compounds. An important finding is that the locations of these groups within the asteroid belt depend on distance from the Sun. The silicon-rich asteroids are reddish, reflect about 15% of the incident light, and are found toward the inner edge of the asteroid belt. The carbon-rich asteroids are dark like a piece of black coal, reflecting only some 3% of the incident light. They are rare in the inner belt, but common in the outer regions, near the orbit of Jupiter. The nickel-iron asteroids, which are only about 5% of the total asteroid population, contain little silicate material and are thought to be the fragmented cores of larger bodies. Such correlations provide important tools to be used in our attempts to understand the formation of these bodies and of the solar system itself.

7.3
METEORS, METEOR SHOWERS, AND METEORITES

Every day in its travels around the Sun, the Earth sweeps up tons of interplanetary debris. Most of this material is in the form of small particles the size of a grain of sand or less, although some of it is substantial in size. Such particles are called **meteoroids**. When a meteoroid traveling at typical speeds of tens of kilometers per second hits the atmosphere, friction with the air heats the meteoroid to a high temperature. The high temperature causes the surrounding atmosphere to glow, creating a bright streak of light called a **meteor**. Meteors are popularly but erroneously known as "shooting stars." The meteoroids usually burn up completely; those few that are massive enough to survive the journey to Earth's surface are known as **meteorites**.

Where do these particles originate? Photographs taken with special automatic meteor cameras indicate that many of them came from the asteroid belt and were perturbed by Jupiter's gravity into an orbit that crosses Earth's.

METEORITE CRATERS

Massive bodies striking Earth at the speeds of tens of kilometers per second are likely to leave visible evidence in the form of craters. A mass moving at a speed of 10 to 30 km/sec or more possesses a formidable amount of kinetic energy. Kinetic energy is energy caused by a body's motion and depends on its mass and speed.[3] The

---

[3]The kinetic energy is given mathematically as $mV^2/2$, where m is the body's mass, and V is the speed.

FIGURE 7–13. Barringer crater (Meteor Crater), in Arizona, is the best known of Earth's impact craters. It is quite young (50,000 years), which explains its excellent state of preservation.

kinetic energy of a meteoroid will be released explosively on impact. For example, the Barringer crater in northern Arizona (**Figure 7–13**), nearly one mile in diameter, was caused by the impact of a body, perhaps the size of a medium-sized apartment house, that may have weighed over 50,000 *tons*. On striking the ground, the enormous kinetic energy of the meteoroid must have vaporized portions of both it and the Earth's surface, causing a tremendous explosion and leaving the crater we see today. The kinetic energy of this event at impact would have been equivalent to about 50 nuclear bombs of the size dropped on Hiroshima.

The Barringer crater is by no means the largest one known; because it has become possible to photograph the surface of the Earth from high-altitude aircraft and satellites, many large craters have been found. Most of these were previously unknown; their great size and the eroded condition due to their age made them difficult to detect from the ground. The largest of these, a couple of hundred kilometers across, was discovered in 1992 off the east coast of Mexico.

The extensive cratering of the Earth is also evidence for the planetesimals we discussed in the previous chapter. After the planets formed, there was a period of intense bombardment from unaffiliated planetesimals until the solar system was swept free of debris. The Moon and other planets and satellites also show the effects of such an epoch of bombardment.

Meteoroids large enough to produce craters more than a hundred kilometers across are extremely rare, and even those as large as your fist are uncommon. The vast majority of meteoroids range downward in size from a pea to a grain of sand to dust particles. The mo-

tion of extremely small particles is rapidly slowed by friction with the atmosphere, and they drift slowly downward to settle on the Earth. These small bodies do not burn up because they are able to radiate away their heat rapidly. They are called **micrometeorites**, and it is difficult visually to distinguish them from ordinary atmospheric dust. Because some micrometeoritic material is iron, and therefore magnetic, it may be gathered using a magnet. The most effective way to study micrometeoroids is with satellites, which can collect them above the Earth's atmosphere on large, specially designed surfaces.

DID AN ASTEROID OR COMET IMPACT SEND THE DINOSAURS INTO EXTINCTION?

An intriguing proposal was put forth by geologist Walter Alvarez and his Nobel-prize-winning physicist father Luis, who suggested that the impact of a large body on Earth was responsible for a massive extinction of lifeforms, including the dinosaurs, about 65 million years ago. It turns out that layers of clay deposited in the Earth's crust at the end of the Cretaceous period contain an anomalously high amount of certain elements (particularly iridium) that are abundant in meteorites but not generally so on Earth. This enrichment in meteoritic material has been found in geological strata from the same era but at many different locations on Earth. Furthermore, researchers have now discovered heavy deposits of soot in those same layers, indicating the occurrence of a global firestorm at this time. The scenario is not unlike the sequence of events that had been proposed to produce a nuclear winter: one or more impacts produce explosions that set off forest fires over extensive regions of the Earth. Enormous clouds of smoke darken the sky, blocking the Sun and bringing on months of deep winter and disruption of the food chain, which in turn could result in massive extinctions of plant and animal life.

Geological strata show evidence of similar worldwide catastrophes at other epochs in the past, and some researchers claim that they have detected a periodicity in these events, coming at intervals of 26 to 34 million years. This alleged periodicity has been strongly challenged by others, who have shown how it could have arisen from faulty data-analysis techniques. However, this hypothesis has triggered speculation as to what type of effect might cause periodic extinctions. A periodic event would seem to argue against asteroids; further, the total energy required for this scenario to work is greater than the typical large asteroid hitting the Earth would carry.

But suppose, for example, that the Sun was gravitationally bound to a massive stellar companion. We can imagine reasons why such a companion star might have escaped detection, so the possibility cannot be ruled out, at least not yet. For example, it could be a star that has lived out its lifetime, used up its sources of fuel, and is now burned out and dark. (In later chapters on stellar evolution we will see how such a thing can happen.) The Sun and its companion would form a binary system, moving about a common center of gravity over a period of many years. It would be difficult to detect this motion, because it would take a long time for one orbit to be completed.

Imagine the effect such a companion would have on the Oort Cloud. The companion star would move through the cloud, and its gravity would perturb nearby comets that would then hurtle off in many different directions. A shower of comets could well be released in the direction of the Sun, resulting in multiple "hits" on Earth. This has prompted some authors to call the hypothetical companion "Nemesis," the Death Star. This kind of overdramatizing of science makes many as-

tronomers uncomfortable. While the hypothesis may be speculative, it is nevertheless deserving of attention. It is hoped that more evidence in the future can resolve this intriguing possibility one way or another.

A spectacular impact that took place in Tunguska, Siberia, on June 30, 1908, might have been due to a cometary remnant. Trees were blown down for more than a thousand square kilometers, animals were killed, and a man 80 km away was supposedly thrown from a chair and knocked unconscious. No meteoritic fragments have been found, leading to the suspicion that the object may have been a fragile one that broke up on its way through the atmosphere. It is hypothesized that the 60 m-diameter body weighed some 100,000 tons and exploded about eight km from the Earth's surface, producing a shock wave equivalent to a 10-megaton nuclear bomb.

The possibility of future devastating impacts on Earth has prompted a search for faint asteroids whose orbits cross Earth's orbit. The pioneer in the field is Tom Gehrels of the University of Arizona, whose Spacewatch camera scans the sky on clear nights for two weeks per month around the time of new moon searching for such objects. In addition to new asteroids, many new comets have also been discovered. Should an object on a collision course with Earth be discovered well in advance of collision, we might have an opportunity to change its orbit.

Inquiry 7–14 Why would a search for faint Earth-crossing asteroids be made at new moon rather than any random lunar phase?

METEOR SHOWERS

In 1845, an unusual event occurred: Comet Biela, with a period around the Sun of about seven years, was observed to split into two fragments. (**Figure 7–14** shows Comet Shoemaker-Levy 1993e, which split into numerous pieces after passing Jupiter at a distance of 0.007 AU

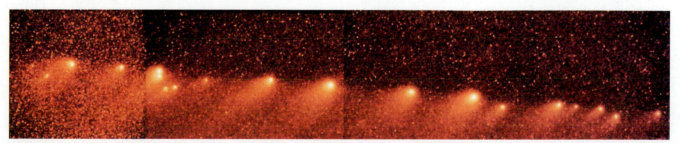

FIGURE 7–14. The nucleus of Comet Shoemaker-Levy 1993e was observed to have broken into numerous pieces, probably from a close passage to Jupiter in 1992.

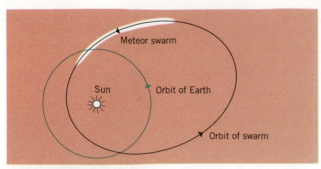

FIGURE 7-15. The path of a meteor stream intersects the orbit of the Earth. Each year, when Earth passes through the region, a meteor shower is seen.

FIGURE 7-16. The radiant of a meteor shower.

in May 1992.)[4] On Comet Biela's next return two comets were observed; on the following return, no comet was seen. Instead, there was a shower of meteors. Nor is this the only example of a known comet being associated with a meteor shower. Spectacular displays in 1833 and 1866 were associated with Comet Tempel, and many other comets, including Halley, have been associated with meteor showers.

How does this happen? Recall that a comet's nucleus is believed to be a rather loosely packed mass of ices and particles, which can lose matter when the ices are vaporized near the Sun. Repeated visits to the Sun's neighborhood will cause considerable shedding of matter that will, of course, continue to orbit the Sun on paths that are similar but not identical to that of the comet. Each particle will have a slightly different speed and eventually the particles will spread themselves all along the comet's orbit (**Figure 7–15**).

What will happen when Earth passes near this stream of meteoroids? If it gets close enough, we should see quite a few meteors, all coming from approximately the same region of the sky. Because the orbit of the meteor stream is fixed in space, one would expect a shower every year on the date that Earth passes near the orbit.

Inquiry 7–15 The particles in a meteor stream should not be distributed uniformly around the orbit but, instead, should be thicker in some parts of the orbit than others. For example, one would expect more particles near the comet's nucleus. These clumps of particles would travel around the orbit in a time roughly equal to the period of the comet. How could this fact help to explain the unusually heavy meteor showers in 1833 and 1866?

[4]This comet collided with Jupiter in July, 1994. See the photograph on the first page of this book.

Individual particles in a stream travel along nearly parallel tracks in space, and as a result meteor trails seen during a meteor shower are not scattered at random all over the sky but appear to emanate from a point in the sky called the **radiant** (Figure 7–16). This is due to an effect of perspective, which can make parallel lines *appear* to radiate outward from a point, much as parallel railroad tracks appear to converge at a distant point. Showers are designated by the position of their radiants. Thus the Leonids appear to radiate from the constellation Leo, the Perseids from Perseus, and so on.

Most meteor showers are modest events, presenting on the average a couple of objects per minute, but some can be spectacular. The 1866 Leonid shower is reputed to have been one of the finest displays of modern times, with over a quarter of a million meteors estimated to have been visible from one observing station, and the 1966 Leonids were almost as abundant. **Table 7–1** lists some prominent meteor showers, giving the location of their radiants and the date in the year that the shower can be expected to be at its maximum.

Meteor showers are best observed on cloudless evenings away from bright lights and when the Moon is not up to brighten the sky. Observing after midnight is best. Just as a car moving into a snowstorm gathers more snow on the front window than the rear one, so too the meteor display is best when Earth is moving into the particles. Such movement occurs after midnight, because at that time the observer is on the side of the Earth that is moving into the meteor stream.

Occasionally, a meteor is observed that is so bright it can be seen even by day. One such **fireball** was sighted by numerous observers in the western part of the United States in July 1972 (**Figure 7–17**). This was an unusual body that, unlike most that the Earth encounters, did not land, but instead skipped back out into space after grazing our atmosphere, much as a stone can be skipped across water. Fireballs sometimes have a

TABLE 7-1

Meteor Showers

Meteor Shower and Radiant Location	Approximate Dates	Associated Comet (if known)
Quantrantids (Bootes)	January 1–4	—
Lyrids (Lyra)	April 19–23	1861 I
May Aquarids (Aquarius)	May 2–6	Halley
July Aquarids (Aquarius)	July 25–31	—
Perseids (Perseus)	August 10–14 (some visible in early August)	1862 III (Swift-Tuttle)
Draconids (Draco)	October 9–19	Giacobini-Zinner
Leonids (Leo)	November 14–19	Tempel[a]
Geminids (Gemini)	December 8–14	—

[a]In 1966 the Leonids again staged a good show after missing 1899 and 1933; they may appear again in 1999.

FIGURE 7–17. A fireball visible in daylight over Jackson Lake at the Grand Tetons.

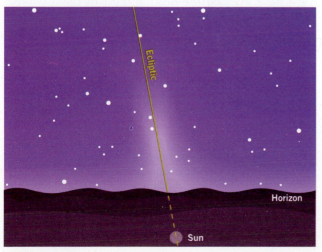

FIGURE 7-18. The zodiacal light is due to light reflected from dust particles in the ecliptic plane.

sound associated them. Such a fireball is called a **bolide** and occurs when the meteoroid explodes within the atmosphere.

7.4

INTERPLANETARY DUST

If you are located in an area blessed by dark, clear night skies, you may be able to detect the presence of dust in the solar system. Approximately 90 minutes before sunrise (or after sunset), look for a faint glow of light extending away from the Sun along the ecliptic (**Figure 7–18**). This zodiacal light is due to dust, concentrated in the plane of the solar system, that reflects the light of the Sun toward our eyes. Reflection of light from dust also produces a faint patch of light in a direction exactly opposite to the Sun, called the gegenschein (from the German *gegen* meaning opposite).

Small dust particles cannot remain in orbit around the Sun forever, because the pressure of solar radiation impinging on them will greatly alter their orbits over a period of time. The smallest of particles will be blown out of the solar system by this radiation pressure. Larger particles, which orbit around the Sun just like the planets, will experience a deceleration from solar radiation and spiral toward the Sun. Eventually they will be vaporized by the intense heat. Because this happens quickly by astronomical standards, somewhere in the solar system there must be a source that is constantly replenishing the dust that has fallen toward the Sun.

Inquiry 7–16 What might be the possible sources of such dust?

7.5

METEORITES AND THE EARLY SOLAR SYSTEM

Until recently meteorites were the only objects from space that we could actually touch. They provided the only direct information on the early physical conditions and early chemistry of the solar system. What can we learn about the early solar system by studying meteorites? To what extent do the meteorites, as random samples of asteroidal material, qualify as primeval material? Such questions are the subject of this section.

CHEMICAL COMPOSITION

There are three basic types of meteorites: **irons**, **stones**, and **stony-irons**. The irons are composed of about 90% iron alloyed with about 10% nickel and traces of other elements. Their composition is similar to what most geologists believe prevails in the Earth's core. The stones are primarily silicate materials similar to the Earth's rocks, whereas the stony-irons are a mixture of the two types.

If a body is observed to pass through the atmosphere to Earth and is successfully tracked and recovered, it is called a meteorite **fall**. On the other hand, if a meteorite is found on the ground but is not known to be freshly fallen, the discovery is referred to as a meteorite **find**. Contrary to popular belief, meteorites are not hot when they hit the ground. In fact, after so many years in cold space the interior is extremely cold, and meteorite falls have been seen with frost on the surface.

Inquiry 7–17 Over two-thirds of meteorite *finds* are irons, but over 90% of *falls* are stones. Why do you think this difference exists?

Inquiry 7–18 Which group is probably most representative of the actual meteorite population in its proportion of stones to irons: the finds or the falls?

Once a meteorite has been found and examined in the laboratory, its composition can be determined with great accuracy. Allowing for the correct proportions of each type of meteoroid, it is found that the proportions of the elements in meteorites are not unlike those of the Earth's crust: perhaps 30 to 40% iron, 30% oxygen, 15% silicon, 12% magnesium, and smaller amounts of sulfur, nickel, calcium, and other elements.

The similarity of this composition to that of the Earth, and the fact that most meteorites originate in the asteroid belt, leads us to suppose that the meteorites we find were once part of larger bodies in the asteroid belt. These bodies were large enough that a certain amount of separation of the elements must have taken place, just as in the Earth the lighter elements formed the crust and the heavier elements the core. If these large bodies broke up (say, by collisions), this would produce some bodies that were primarily iron and others that were primarily of stony material.

On the other hand, if all the mass in the asteroid belt were summed, even the most generous estimates do not put it at more than about one-tenth of an Earth mass. If this is so, there may never have been a body of any substantial mass in the region between Mars and Jupiter, due perhaps to the influence of Jupiter. The stony-irons, in particular, may have come from parent bodies so small that no significant heating and segregation of material took place. However, estimating the total amount of mass that may have been swept out of the asteroid belt over time is difficult, so this argument is not conclusive.

One particular type of meteorite is believed to have remained much as it was when the solar system was formed—the **carbonaceous chondrites**, a relatively rare type of stony meteorite containing an unusually large proportion of hydrocarbons and as much as 20% water. Hydrocarbons and water are both highly volatile materials and tend to vaporize even at low temperatures, so any object that has retained its hydrocarbons must have avoided the heating most of the asteroidal material appears to have undergone. The elemental abundances of carbonaceous chondrites are thought to represent that of the early solar system better than any other bodies we have studied.

It is interesting that organic compounds are found in carbonaceous chondrites. Not only have hydrocarbons been found, but also complex substances such as amino acids, the fundamental building blocks of proteins. Although care must be taken not to identify an amino acid that is actually a contaminant (e.g., a fingerprint of the investigator), it seems clear that some of these compounds are actually of extraterrestrial origin. That the precursors of life can be formed under the conditions prevailing in outer space certainly gives hope to those who would like to believe that life is common in the universe.

Chemical analysis of meteorites has led to a fascinating conclusion: some meteorites that have been picked up in

(a)   **(b)**

FIGURE 7–19. Meteorites found in Antarctica. Chemical analyses indicate that (a) came from the Moon, and (b) from Mars.

FIGURE 7–20. Widmanstätten patterns reveal that iron meteorites once existed in a molten state.

Antarctica definitely came from the Moon, while others may be from Mars (**Figure 7–19**)! Meteorite ALHA 81005 was found to have a chemical composition extremely similar to the rocks collected on the Moon by the Apollo astronauts (Chapter 8) and substantially different from that of other meteorites. Presumably it was a piece of the Moon that was knocked off when a large meteoroid hit the Moon. The conclusion that some meteorites come from Mars is not as definite because Martian surface material has not yet been returned to Earth by spacecraft. However, the relative abundances of magnesium and nitrogen gases trapped inside a few meteorites are similar to those found in the Martian atmosphere.

INTERNAL STRUCTURE

The internal structure of a meteorite can be studied if the object is cut into sections. If an iron meteorite is sliced, polished, and etched with acid, it will reveal a characteristic pattern of lines, called Widmanstätten patterns (**Figure 7–20**). These are actually the boundaries of iron crystals that may have been formed when the material slowly cooled from a state of high temperature and pressure, such as would be expected in the interior of a large body. Recent research suggests that, under some circumstances, bodies as small as 20 km in diameter could form these figures, but their existence still suggests the preexistence of a parent body.

Inquiry 7–19 Why couldn't the Widmanstätten patterns have been formed as a result of the heating of the meteoroid during its passage through our atmosphere followed by subsequent cooling?

The vast majority of stony meteorites, when sliced open, show a beautiful and complex structure (**Figure 7–21**). These structures, called **chondrules**, appear as separate pieces of material that were included in the meteorite at the time it formed. Chondrules exhibit a wide range of chemical composition. Some appear to have formed from rapid cooling of small liquid droplets, while others are highly irregular, indicating slower cooling. The most vexing question in the study of stony meteorites continues to be that of the formation of the chondrules. Because they are probably representatives

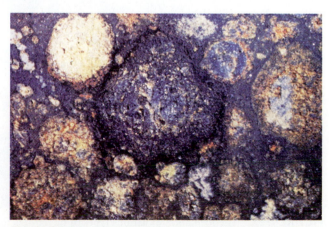

FIGURE 7–21. The inside of a chondritic meteorite. The round inclusions are the chondrules.

of the oldest materials in the solar system, they hold important clues about the history of the bodies in which they are found, and of the solar system itself. They are not giving up that information easily.

Many meteorites show evidence of strong stressing, such as might be expected as a result of collisions occurring at some 20 to 30 km/sec. An example is the presence of small black diamonds in some meteorites. It is not clear whether these structures were formed in collisions within the asteroid belt or on impact with the Earth. Such small diamonds have been found in the layers at the end of the Cretaceous period and provide further evidence for an extraterrestrial origin to mass extinctions at that time.

THE ORIGIN OF METEORITES

Observations that provide hints to the origin of the meteorites include the presence of irons, stones, and stony-iron compositions, the implication that they were originally part of the asteroid belt, the low total mass of all material in the asteroid belt, the properties of the carbonaceous chondrites, the cooling implied by the Widmanstätten patterns, the presence or absence of chondrules, and the stressing that implies the presence of collisions. A schematic drawing of the formation of meteorites is shown in **Figure 7–22**, which suggests a possible scenario for a large asteroid, perhaps 650 km in diameter. Internal heating melts the interior and causes a separation of heavy from light elements. Collisions with other bodies occur, causing the surface to shatter. Further collisions break off large chunks; additional fragmentation produces larger numbers of small chunks of material.

The carbonaceous chondrites are thought to be fragments from the asteroid's surface, because the high abundance of volatiles in carbonaceous chondrites implies they were not subjected to the high temperatures or the high pressures in the asteroid's interior. The broken fragments will have different and complex chemical compositions. Some will show the effects of physical stressing from the collisions, while some will have cooled at different rates from others. Small asteroids, whose interior temperatures never became high enough to produce melting, may have gone through similar shattering processes, but meteoroids from them will have different chemical properties.

While no asteroid collisions have been observed, they no doubt occur. Meteoroids identical to those hypothesized to be produced in such collisions have been observed. From such circumstantial evidence, we can infer that the general picture of meteorite formation is, therefore, correct.

METEORITE DATING

The study of **radioactive elements** and their decay products in meteorites gives us valuable information about when they were formed, and hence about the age of the solar system. For example, uranium-238 (^{238}U)—uranium whose nucleus contains a total of 238 protons and neutrons—decays in a sequence of steps into lead-206 (^{206}Pb). These events can be characterized by what is called the **half-life** of the element. For example, the half-life of ^{238}U is about 4.5 billion years; this is the length of time it takes for half the uranium originally present to decay, no matter what the size of the sample. The greater the age of an object containing ^{238}U, the

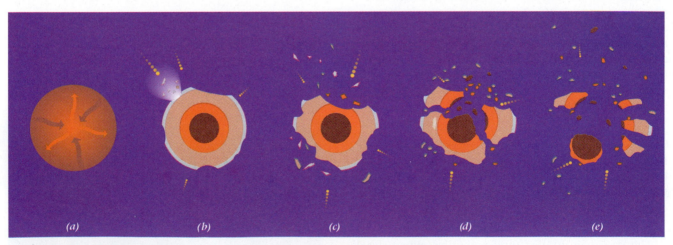

(a)   (b)   (c)   (d)   (e)

FIGURE 7-22. The formation of various types of meteorites from the fragmentation of an asteroid. Different regions of the asteroid have different chemical properties and produce a variety of types of meteorites depending on the details of the internal heating process.

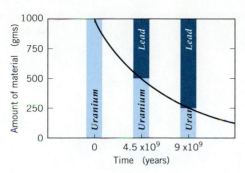

FIGURE 7-23. Radioactive decay of uranium into lead has a half-life of 4.5 billion years.

greater the amount that has decayed to lead and the greater the ratio between the amount of ^{206}Pb and the amount of ^{238}U. By measuring this ratio, we can estimate the age of the object (**Figure 7–23**).

Inquiry 7–20 If the original material of which an object was formed already contained some ^{206}Pb, an incorrect age would be inferred for the object. Would the estimated age be too young or too old in this case? Explain your answer.

By refining the method, we can overcome the uncertainties caused by the presence of primeval ^{206}Pb. For example, one of the products produced during the decay of ^{238}U is helium-4 (^{4}He). If, during the formation of a minor planet, the material was heated to a high temperature, gases such as ^{4}He would be driven off, leaving the object free of gas. After cooling, the ^{238}U would continue to decay, and the ^{4}He produced as a result would become trapped in the solid material. Analyzing the object today, we can be sure that all the ^{4}He we see is due to radioactive decay, and therefore by measuring the ratio of ^{4}He to ^{238}U we can estimate the age of the object more accurately. By such means, the ages of meteorites have been determined to be about 4.5 billion years.

TABLE 7-2

Half-lives of some Common Isotopes

Isotope (Parent → Daughter)	Half-life (years)
Potassium 40 (^{40}K) → Argon 40 (^{40}Ar)	1.3×10^9
Rubidium 87 (^{87}Rb) → Strontium 87 (^{87}Sr)	47×10^9
Uranium 235 (^{235}U) → Lead 207 (^{207}Pb)	0.7×10^9
Uranium 238 (^{238}U) → Lead 206 (^{206}Pb)	4.5×10^9

There are many different radioactive decay processes going on in nature, with half-lives ranging from short to long. **Table 7–2** lists the parent atom of some radioactive elements, the daughter it decays into, and the half-life. Astronomers typically try to construct their chronologies using a variety of radioactive decay cycles of many different chemical elements. By using a variety of decay cycles, uncertainties in the ratio of initial abundances are decreased. The agreement that we find between these various processes gives us confidence that our estimates of primeval abundances are not leading us systematically astray. Among the best results currently available, dating using the decay of ^{87}Rb to ^{87}Sr gives an average age for meteorites of 4.498 ± 0.015 billion years.

Inquiry 7–21 How useful would you expect ^{14}C, which has a half-life of 5568 years, to be in dating astronomical objects?

Yet another piece of evidence can be produced to help us understand the evolution of meteorites. Like all objects in space, meteorites are constantly being bombarded by **cosmic rays**, subatomic particles (generally protons, electrons, and helium nuclei) that move through interstellar space at speeds approaching the speed of light. When a cosmic ray strikes a body in the asteroid belt, atomic nuclei in the body can be fragmented into much lighter nuclei. Among the elements produced are lithium, beryllium, and boron—all rare elements. After being exposed to cosmic rays for many millions of years, the outer skin of a meteoritic particle will develop an excess of these elements, and the time since its surface was first exposed to cosmic rays can be estimated. In this way it is possible to determine how long it has been since the breakup of the parent body from which the asteroid came. Cosmic-ray ages for meteorites range from tens of thousands to millions of years. Such young ages provide further circumstantial evidence for their formation from the collisional breakup of asteroids.

Inquiry 7–22 In some meteorites, the ages determined by the cosmic ray exposure method are different on different sides of the object. What might be the cause of this?

DID A SUPERNOVA EXPLOSION TRIGGER THE FORMATION OF OUR SOLAR SYSTEM?

In the early 1980s, studies of the carbonaceous chondrite meteorite known as the Allende meteorite (named after the Mexican village Pueblito de Allende, where it was found) revealed the presence of some unusual and unexpected chemical isotopes—atoms that resulted from the radioactive decay of unstable elements not normally seen in the solar system. For example, embedded in the meteorite were small chunks of material found to be highly enriched in magnesium-26 (^{26}Mg), a daughter element produced by the radioactive decay of aluminum-26 (^{26}Al).

Interesting and important inferences result from this observation. First, this decay process occurs rapidly (by astronomical standards). Half the atoms in a sample of ^{26}Al will have decayed to ^{26}Mg in only 740,000 years. So if bubbles of ^{26}Mg are found trapped inside a meteorite, it follows that the meteorite must have formed soon after the parent element ^{26}Al was made and before significant amounts of the ^{26}Al had a chance to decay. On the other hand, if the ^{26}Al had decayed while it was still in gaseous form, the resulting ^{26}Mg would have been dispersed in the nebula instead of forming bubbles in the meteorite. Therefore, when we find ^{26}Mg trapped in a solid body, we know that the body must have formed soon after the original ^{26}Al was made. These arguments are shown schematically in **Figure 7–24**.

This is significant because current research into the formation of chemical elements suggests that the most likely way to produce the unusual isotope ^{26}Al is in a supernova explosion, a catastrophically explosive event that occurs toward the end of the lives of massive stars (Chapter 19). During such a cataclysm, extreme temperatures are generated that can create all sorts of exotic chemical elements in brief episodes of nuclear fusion inside the star. The explosion then ejects this material away from the star. At the same time, the detonation wave from the explosion propagates outward from the

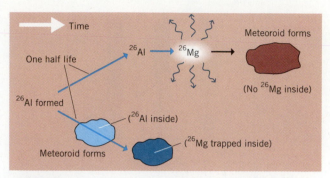

FIGURE 7-24. If ^{26}Al is incorporated into a meteoroid shortly after the ^{26}Al forms, it will decay into ^{26}Mg, which will be found in meteorites. However, meteoroids formed well after ^{26}Al forms will not contain any ^{26}Mg, because it will have dissipated into space.

star through nearby gas clouds in space. Such a blast wave produces a traveling compression called a shock wave, and studies show that it will actually squeeze interstellar gas clouds and cause some of them to collapse into stars (Chapter 17).

Age-dating studies do suggest that the solar nebula collapsed rapidly (in several hundred thousand years), and if the collapse was triggered by a supernova, the primeval gases would indeed be able to capture enough undecayed ^{26}Al to account for the ^{26}Mg found in the Allende meteorite.

At this point in our course of study, this scenario may seem rather exotic, but we will see later on that star formation triggered by stellar explosions is probably a fairly common event in our galaxy and in other galaxies as well.

DISCOVERY 7–1
ASTEROID BRIGHTNESS VARIATIONS

After completing this discovery you should be able to do the following:

- Describe how the brightness of an asteroid changes as it rotates.

To study how an asteroid's brightness changes as it rotates and tumbles, find an object such as a shoe box, a book, or an audio or video cassette box.

☐ **Discovery Inquiry 7–1a**   If one side of your object has an area three times that of another side, what will be the ratio of the brightnesses when one side faces you compared with when the other side faces you?

Hold your object as shown in **Discovery Figure 7–1–1**, with the long axis facing you. Assume the Sun is behind you, so the side of the object facing you receives full illumination. Rotate it about the long axis as shown (the axis coming towards you, labeled **A**), and imagine how the brightness would vary. (If possible, use a bright light to illuminate the object and observe the changes in brightness at it rotates.) Draw a graph of the variation of brightness with time as you rotate the object. Repeat for the other two axes shown in the figure. Think about the relative sizes of the faces as you draw your three graphs. Try to imagine the complications if your object were rotating about all three axes at once!

Curves like those you drew are similar to those that astronomers determine from their brightness observations of asteroids. The name of the game for the astronomer, is to determine the shape of the tumbling asteroid from observations of the brightness changes with time!

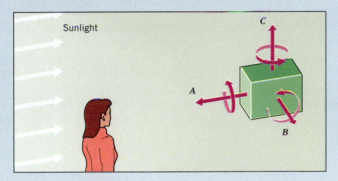

DISCOVERY FIGURE 7-1-1. This figure shows an object that is able to rotate about three perpendicular axes. It is illuminated from behind the observer.

CHAPTER SUMMARY

Observations

- Comets are observed to have a **nucleus**, a **coma**, one or more **tails**, and a **hydrogen halo**. Comet tails are of two types, one consisting of dust particles and the other having charged particles.
- **Long-period comets** follow elliptical orbits that are at random orientations to the ecliptic plane. **Short-period comets** have orbits that are closer to the ecliptic. Some comets may have **hyperbolic orbits**.
- Halley's comet was observed to be dark and to contain a large amount of water.
- The **asteroids (minor planets)** are small rocky bodies found mostly, but not exclusively, between the or-

bits of Mars and Jupiter. Their orbits are generally nearly circular and located in the ecliptic plane.
- The **Trojan asteroids** are two special groups of asteroids located in Jupiter's orbit at distances from Jupiter equal to its distance from the Sun. The Trojan asteroids thus form equilateral triangles with the Sun and Jupiter.
- The asteroid belt is not uniformly filled with asteroids but contains **Kirkwood gaps**, regions devoid of asteroids. These gaps in the asteroid belt occur at locations from the Sun that correspond to simple fractions of Jupiter's period about the Sun.
- There are four chemical groups into which asteroids generally fall: silicon-rich, nickel-iron-rich, carbon-

rich, and those that may be rich in organic compounds.

- Meteorites have composition types that are **iron**, **stony**, or **stony-iron**. The cooling of certain of these materials forms **chondrules** and Widmanstätten patterns inside the rocks.
- Meteor showers, which result when Earth crosses the orbit of a comet, appear to come from a particular direction in space called the **radiant**.
- Interplanetary dust is observed as the zodiacal light and the gegenschein when sunlight is scattered by small dust particles located in the ecliptic plane.

THEORY

- Sunlight acting on dust particles pushes them away from a comet's coma. Particles from the **solar wind** collide with particles in a comet's tail and cause them to become charged. The motions of these charged particles then change in response to magnetism present throughout the solar system.
- Asteroids can sometimes be used to determine the length of the astronomical unit by obtaining their distance in both kilometers and astronomical units.
- The **half-life** of radioactive decay is the time required for half the material to decay from one element into another. From the known half-life of radioactive decay and from the inferred amount of material originally present, ages of meteorites can be found.

CONCLUSIONS

- If a comet has a gravitational interaction with a massive planet, the comet's orbit may change from elliptical to hyperbolic.
- Long-period comets come from the Oort cloud, which is located 50,000 to 100,000 AU from the Sun. Short-period comets come from the Kuiper belt outside the orbit of Pluto.
- Asteroids may be small because Jupiter's strong influence never allowed larger bodies to form, or because numerous collisions fragmented larger bodies.
- The composition types of meteorites are best explained as resulting from the fragmentation of asteroids.
- From observations of excess amounts of iridium in certain rocks, scientists hypothesize that species extinctions may have occurred because of the impact of a large meteoroid with the Earth.
- **Cosmic rays** are energetic subatomic particles that fill the galaxy. Collisions between cosmic rays and asteroids produce chemical changes on the asteroid's surface. From an analysis of the elements present, astronomers can infer the length of time the surface has been exposed to cosmic rays and thus how long ago the meteoroid fragmented from a larger body.
- The dirty snowball model of a comet provides a valid model of a cometary nucleus.
- Certain meteorites have chemical compositions that indicate they came from the Moon and Mars.

SUMMARY QUESTIONS

1. What is the icy conglomerate model of a comet? What evidence is there for it? How can it explain the various phenomena observed in comets?

2. Why does the tail of a comet always point *away* from the Sun? Why may a comet have more than one tail?

3. What evidence do astronomers have in favor of Oort's hypothesis of a comet cloud beyond the orbit of Pluto?

4. What do we mean by the term "minor planet?" What are they? Where are they found? How can they be used to determine the length of the astronomical unit?

5. What do we mean by the term "Trojan asteroid?" What is the relationship between the Trojans and the planet Jupiter? What are the Kirkwood gaps and what was Jupiter's role in forming them?

6. What role did collisions play in the development of asteroid and comet orbits? In the production of meteoroids and the zodiacal light?

7. What is a meteor? Distinguish a meteor from a meteorite. What are the two main sources of meteors?

8. What are the three main types of meteorites? Why is the proportion of stones to irons seen in finds on Earth different from the true proportion in space?

9. How do Widmanstätten figures and chondrules provide information on the conditions under which some meteorites were formed?

10. Why do meteor showers repeat themselves year after year in the same part of the sky? Why do meteors in a shower emanate from a single location in space?

11. How do various radioactive decay processes and cosmic ray exposure times give us clues to the formation and history of the solar system?

APPLYING YOUR KNOWLEDGE

1. Make a drawing to demonstrate that the Earth encounters more meteoroids after midnight than before midnight.

2. Would you expect there to be many asteroids having satellites orbiting them? Explain your reasoning.

3. What arguments might you present to people who believe that comets bring disaster and evil, to convince them their ideas are wrong?

4. From what you know of comets, would you expect their motions to be immediately apparent to the casual observer? Why or why not? How about for a meteor?

5. Meteorites are observed to have an amount of xenon-129 (^{129}Xe) far in excess of that normally found on Earth. ^{129}Xe comes from the radioactive decay of iodine-129 (^{129}I), which has a half-life of 17 million years. Use this information to argue that the formation of the solar system was initiated by the nearby explosion of a supernova.

6. Suppose you came across a rock you thought might be a meteorite. How might you determine whether or not it is a meteorite, given that you had no special equipment? How about if you had all the special equipment you desired?

7. Make a list of easily obtainable household items you might use to make a cometary nucleus.

8. What is the distance of the Trojan asteroids from Jupiter?

■ 9. If the period of Halley's Comet is 76 years, approximately what is its maximum distance from the Sun? Relate this distance to that of the planets.

■ 10. What would be the age of a rock in which you measured the ratio of potassium-40 to argon-40 to be 1 to 7? What assumption are you making? (Hint: a drawing will be helpful.)

■ 11. What would be the age in the previous question if the rock originally contained equal amounts of potassium-40 and argon-40?

■ 12. One of the Kirkwood gaps appears at a location where the asteroid's period is exactly one-half Jupiter's orbital period about the Sun. At what distance from the Sun (in AU) is this gap located?

ANSWERS TO INQUIRIES

7–1. As a comet returns again and again to the neighborhood of the Sun, it loses more and more of its volatile materials. Eventually, they are completely "boiled away" and the object might be difficult to distinguish from a minor planet.

7–2. When the distance from the Sun is small, the comet moves rapidly. Therefore, it spends little time close to the Sun and most of its time far from it.

7–3. One would expect that we would encounter more comets coming from the direction of Hercules, just as when we run in the rain the front of our body gets wetter than the back. The hyperbolic comets must clearly be part of the Sun's family and moving with it towards Hercules.

7–4. They are distributed uniformly around the solar system.

7–5. 100,000 years. Comparison with the age of the solar system: $10^5/(4.5 \times 10^9) \approx 2 \times 10^{-5}$.

7–6. The comets we see today probably started to come near the Sun only recently. If they had always had orbits that brought them near the Sun regularly, they would long since have lost their supply of volatile materials and faded.

7–7. Oort's theory provides a way for new comets to enter our part of the solar system, replacing those that have "worn out."

7–8. If comets are indeed composed of near-primeval material, they could clue us in to the composition of the early solar nebula.

7–9. Irregular shapes and irregular distribution of light and dark regions over the surface.

7–10. Perhaps Hidalgo is a worn-out comet.

7–11. The round-trip distance is 173.3 seconds times 3×10^5 km/sec = 5.2×10^7 km. The distance of the asteroid is one-half this amount, or 2.6×10^7 km.

7–12. 2.6×10^7 km / 0.17 AU = 153 million km/AU.

7–13. If an asteroid is broken up in a collision, one would expect the pieces to continue to follow orbits similar to the original body.

7–14. The sky is much darker at new moon than at full moon, making it easier to find faint objects.

7–15. Presumably, in 1833 and 1866, the Earth passed through a denser region of a stream of particles orbiting the Sun with a period of roughly 33 years. It has been suggested that perturbations of the particles by Jupiter moved the entire orbit in such a way as to substantially miss the Earth in 1899, because there was no spectacular display that year (nor in 1933, but there was in 1966).

7–16. Fragments of asteroids and dust from comets are the two most likely (listed in order of probable importance).

7–17. Meteorite stones are so similar to Earth rocks (especially after undergoing weathering and erosion) that after a time they are difficult to distinguish and so are not noticed. Irons, on the other hand, are distinctive and easily recognized.

7–18. The falls, because we should recover stony and iron meteorites in proportion to the number that actually fell.

7–19. The heating is too brief and only affects the skin of the meteorite.

7–20. Too old. If the original ^{206}Pb is not accounted for, the scientist will conclude that the lead came from radioactive decay over a longer time interval.

7–21. Of no use at all; its half-life is too short.

7–22. Different sides must have been exposed to cosmic rays for different lengths of time. Such an observation is a further indication that meteoroids were produced when colliding bodies broke pieces from larger bodies. Only when fragmentation occurs will a surface that was originally inside a larger body be exposed to cosmic rays.

8

THE EARTH AND THE MOON

Cold hearted orb that rules the night
Removes the colors from our sight
Red is grey and yellow white
But we decide which is right
And which is an illusion

MOODY BLUES, *DAYS OF FUTURE PASSED*, 1967

Earth is a planet, circling the Sun in its own time like all the other planets. Our presence on it makes the Earth of special interest to us. It makes sense to begin our study of planets by learning some of the basic facts about our home body, as well as the processes that have made the Earth the way it is.

The Earth's moon has been an object of study and wonder throughout the ages. During most of history, only naked-eye observations were available; it was not until the seventeenth century that Galileo and other pioneers turned their crude telescopes to the Moon. In the three and a half centuries since then, earthbound observations made important progress in understanding the Moon—progress that was indispensable in the design of the enormously complex and highly successful Apollo missions to the Moon. An understanding of the Moon, and the physical processes that have occurred to make it the way it is, will allow us to understand better what recent spacecraft observations of planets and their satellites have been teaching us.

Our approach to the study of planets and satellites is one of **comparative planetology**. The emphasis is on comparing and contrasting the various bodies with the goal of understanding how the many pieces of the planetary puzzle fit together. For example, the Earth and Moon exhibit both chemical similarities and differences that we must understand if we wish to figure out how each body formed and evolved. Our treatment will not be comprehensive but will consider only those ideas and processes that are necessary for understanding the other bodies of the solar system.

8.1

THE EARTH AS AN ASTRONOMICAL BODY

Earth is an astronomical body. We want to know the same types of information about it as we know about other planets. We therefore first look at how scientists have learned the most fundamental properties of our home planet.

DETERMINATION OF BASIC PROPERTIES

The diameter of the Earth has been known ever since about 200 B.C. when Eratosthenes determined the Earth's circumference (Chapter 5). Modern data show the Earth to be slightly pear-shaped. The diameter measured through the equator is greater than that measured through the poles because rotating bodies are flatter at the poles than nonrotating ones. The average diameter is given in **Table 8–1** along with other basic properties.

The Earth's mass can be determined using a variety of techniques, which are instructive to look at. They all ultimately depend on Newton's laws. For example, Kepler's third law may be applied using the period and distance of the Moon. For the Earth and Moon, whose masses are $M_\oplus$ and M_m, respectively, we have

$$(M_\oplus + M_m)P^2 = A^3.$$

Assuming the mass of the Moon is far less than that of the Earth, we can neglect the Moon's much smaller mass and solve for the Earth's mass:

$$M_\oplus = \frac{A^3}{P^2}.$$

Remembering, too, that the distance must be expressed in AU and the period in years, with an Earth-Moon dis-

TABLE 8-1

Properties of the Earth

Property	Value
Orbital properties	
Average distance from Sun	149,600,000 km (1 AU)
Perihelion distance	0.983 AU
Aphelion distance	1.017 AU
Orbital period	$365^d6^h8^m24^s$ (1.000 years)
Orbital inclination	0° 0' 0"
Physical properties	
Diameter (average)	12,756 km (1.000 $D_\oplus{}^a$)
Mass	5.974×10^{27} gm (1.000 $M_\oplus$)
Density (average)	5.518 gm/cm^3
Surface gravity	980 cm/sec^2 (1.000 Earth gravity)
Escape velocity	11.2 km/sec
Rotation period	$23^h56^m4.091^s$
Tilt of rotation axis from orbit perpendicular	23° 27'
Albedo[b]	0.37
Surface temperature	200 to 300 K (−100 to +117°F)
Satellites	1

[a]The symbol $\oplus$ is the symbol for the Earth. When used as a subscript, such as $D_\oplus$, and $M_\oplus$, it refers to the diameter, and mass of the Earth, respectively.

[b]The albedo, which is defined more thoroughly later in the book, is the fraction of the sunlight falling on a body that is reflected. The number tells us that some 37% of the sunlight falling on the Earth is reflected away.

tance of 380,000 km (2.53×10^{-3} AU), and a period of 27.3 days (7.48×10^{-2} year), we find

$$M_\oplus = (2.53 \times 10^{-3} \text{ AU})^3/(7.48 \times 10^{-2} \text{ year})^2$$

$$= (2.9 \times 10^{-6} \text{ solar masses}) \times (2 \times 10^{33} \text{ grams/solar mass})$$

$$= 5.8 \times 10^{27} \text{ grams.}$$

This simple example gives a result close to the accepted one in Table 8–1.

The density of the Earth is found by dividing its mass by its volume. The Earth's volume is easily calculated from its diameter, and the average density of the Earth is found to be 5.5 grams/cm^3, or 5.5 times that of water. As we saw in Chapter 6, this value is substantially larger than that of a typical Jovian planet, whose density is near that of water. A body's density gives an idea of its overall, average properties, and it allows us to make useful comparisons between different objects.

If you drop a shoe, it accelerates toward the center of the Earth. If you were to drop the same shoe on the Moon, it would accelerate more slowly toward the Moon's center. This acceleration due to gravity is deter-

mined by Newton's second law and his universal law of gravitation. When acceleration is caused by gravity, physicists use the symbol g in Newton's second law instead of a. If the gravitating mass is the Earth ($M_\oplus$) and the distance of the accelerating body is the radius of the Earth ($R_\oplus$), the acceleration caused by gravity is $g = GM_\oplus/R_\oplus{}^2$. Similar equations are used for each planet. Thus, the greater a planet's mass, the faster it will accelerate a falling body. Put another way, the greater a planet's value of g, the more a person would weigh on that body. The acceleration of gravity on a planet is one important characterization of it.

One experimental method of finding the value of g is by dropping an object and observing how its rate of fall changes with time. Using such an experimentally determined value, the known value of the gravitational constant G, and the known value of a body's radius, its mass is readily found using the above expression.

The value determined in this way and the value determined from the use of Kepler's third law are not identical because of observational and experimental uncertainty. Only if measurement could be made without error could they be identical.

We are held to Earth's surface by gravity. To escape Earth's gravity requires a speed in excess of 11.2 km/sec. The speed required to escape is called the *escape velocity* and is proportional to $\sqrt{M/R}$, where M and R are the mass and radius of the planet.[1] The formula shows that high-mass planets have a greater escape velocity than low-mass planets of the same size. Similarly, given two planets of the same mass, the smaller one will have the greater escape velocity because its gravity at its surface will be greater. The escape velocity for each of the planets is given in Appendix C.

8.2

THE EARTH'S INTERIOR

The interior structure of the Earth cannot be studied directly. Although drilling is taking place, the deepest drill hole, which is located in Russia some 250 km north of the Arctic circle, just reached the 12-km mark in 1990, some 20 years after beginning. We can, however, learn about the interior of the Earth from studies of earthquakes.

[1]The complete equation for the escape velocity is $V_{escape} = \sqrt{2GM/R}$.

SEISMIC STUDIES

An earthquake produces waves of energy that travel in all directions, both over the surface and through the interior, to be detected and analyzed by seismograph stations the world over. Earthquake energy compresses and deforms the rocks through which it passes. The rebounding of the compressed rock causes it to transmit the wave throughout the interior. A **compressional wave**, which consists of alternating pulses of compression and expansion (**Figure 8–1a**), is characterized by a wave's motion in the same direction as its vibration. Because the compression causes a change in the pressure on the rock, a compressional wave is also known as a **pressure wave**, or **P-wave**. An example of such a wave in everyday life is a sound wave.

A second type of wave generated by a seismic event deforms rock without compressing it. Because such a stress is called shear stress, the type of wave producing it is often called a **shear wave** or **S-wave**. In contrast to the P-wave, the shear wave's direction of motion is *perpendicular* to its vibration (**Figure 8–1b**).

The properties of S-waves differ from those of P-waves in important ways. P-waves travel with higher speeds than S-waves and thus arrive at a given seismic station first. For this reason, P-waves are also known as **primary waves**, and S-waves as **secondary waves**.

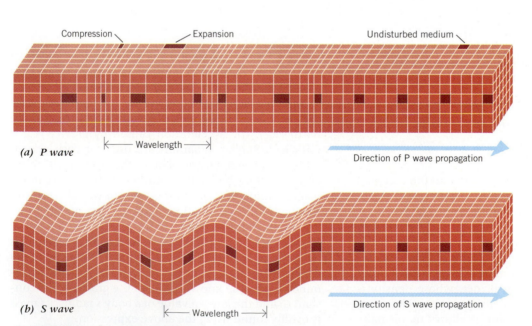

FIGURE 8–1. The two types of seismic waves are illustrated here. (*a*) A P-wave contains regions of high and low compression, which periodically pass through each point. The direction of motion of the vibration is in the direction the wave travels (here, to the right). Although a given square continually becomes a rectangle, it always returns to the original square. (*b*) An S-wave vibrates perpendicularly to the direction in which the wave travels. A given region distorts to a parallelogram, and then back to a square again.

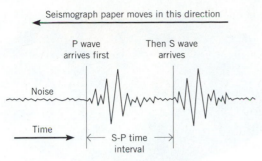

Seismograph paper moves in this direction

P wave arrives first

Then S wave arrives

Noise

Time

S-P time interval

FIGURE 8–2. A seismogram of an earthquake. Note the arrival of the P-waves before the S-waves.

The greater the distance from a seismic event to a recording station, the greater the delay between the P- and S-waves. By observing the delay between the times of arrival of the two waves (**Figure 8–2**), seismol-

ogists can determine the distance to the point where it occurred.

READERS DOING ACTIVITIES SHOULD DO ACTIVITY 8–1, EARTHQUAKES, NOW.

While P-waves easily pass through liquids, S-waves do not. For this reason, a particular seismograph station may only detect P-waves. Knowing this, a comparison of data taken from seismograph stations all over the world has allowed geologists to determine the internal structure of the Earth (**Figure 8–3**).[2] The Earth is found to consist of three general regions: the **core**, the **mantle**, and the **crust**. The core contains 16% of the Earth's volume and is further subdivided into a solid **inner core** and a molten **outer core**. The subdivision occurs because, although the interior temperature is high

[2]The real situation is more complex because waves bend and reflect inside the Earth.

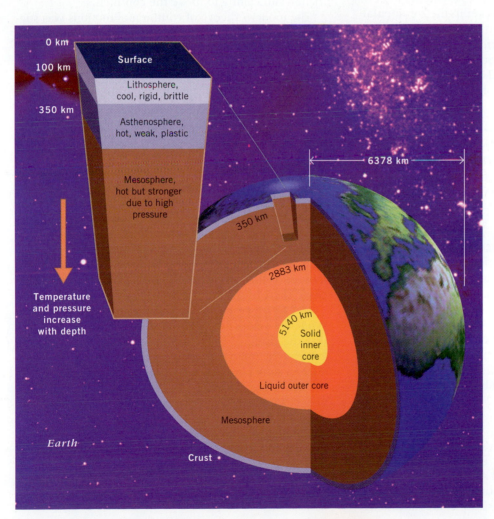

0 km

100 km

350 km

Surface

Lithosphere, cool, rigid, brittle

Asthenosphere, hot, weak, plastic

Mesosphere, hot but stronger due to high pressure

6378 km

350 km

2883 km

5140 km

Solid inner core

Liquid outer core

Temperature and pressure increase with depth

Mesosphere

Earth

Crust

FIGURE 8–3. The interior structure of the Earth.

(thought to be in the vicinity of 3000–5000°C, which is 5400–9000°F), the extreme pressures in the core cause the material in its inner region to solidify.

The mantle is a thick region of dense rock surrounding the core. The lower part of the mantle contains highly compressed rock that has great strength even though the temperature is high. Farther up in the mantle, the pressure decreases, causing the strength of the rock to become weaker. The region of weaker rock is called the **asthenosphere** ("weak sphere"). This region is important because the rock here easily deforms, like soft plastic. Rock on top of the softened asthenosphere will be capable of floating on it. The important consequences of this will be discussed later in the chapter.

The outer 100 km of the Earth is cooler and consists of strong, rigid rock. This "rock sphere" is the **lithosphere**. Although the rock composition of the lithosphere and the asthenosphere are the same, the difference in rock strength distinguishes them.

The crust is the uppermost part of the lithosphere and contains rock of lower density than the material underneath. It is not uniform but is thicker in the continental regions (**continental crust**) than under the oceanic regions (**oceanic crust**). Furthermore, the continental crust is slightly less dense ($2.8 \ g/cm^3$) than the oceanic crust ($3.0 \ g/cm^3$). These observed differences are important for understanding the evolution of the Earth's crust, as we discuss below.

Inquiry 8–1 The average density of the Earth is 5.5 g/cm^3; the density of a typical rock you might pick up is about 3 g/cm^3. What can you conclude from this information about the density of the Earth's interior?

The chemical makeup of the Earth is not uniform throughout. Gravity causes denser materials to sink toward the center, with lighter elements rising toward the top. For this reason the core regions consist of iron and nickel, with the crustal regions containing lighter silicon, magnesium, and aluminum. Bodies whose chemical structure is determined by the effects of gravity have undergone the process of **differentiation**. For example, the inner core has a density of about 13.3 gm/cm^3; the outer core, 11 gm/cm^3; and the mantle, about 4 gm/cm^3. Thus the Earth is highly differentiated.

PLATE TECTONICS

Even a cursory look at the Earth's surface raises an interesting question: Why do the eastern part of South America and the western part of Africa appear to fit together? Furthermore, detailed studies of the flora and fauna of these distant regions show some surprising

FIGURE 8–4. The supercontinent, Pangaea, 200 million years ago.

similarities. For these reasons, some scientists suggested many years ago that these continents had at one time been part of a larger continent that had broken up and moved apart. The suggestion was ridiculed, however, because no forces were known that could overcome friction sufficiently to move the continents.

The modern theory of continental motions was suggested by Alfred Wegener in 1912. Evidence in its favor mounted, until in the 1960s it was accepted as an important theory. Two pieces of evidence helped its acceptance. First, data showed that the oceanic crust moved. Second, scientists discovered the plastic nature of the underlying asthenosphere, which provided a medium on which continents could float. The hypothesis is that some 200 million years ago, the continents existed in the form of a supercontinent we call **Pangaea** (pronounced *Pan-jeé-ah*, meaning "all lands") as shown in **Figure 8–4**. Furthermore, the theory suggests that the continental crust is not a solid sphere but is similar to a loosely fitted jigsaw puzzle whose pieces can be easily moved. These pieces of lithosphere, known as **continental plates**, are diagramed in **Figure 8–5**, along with their velocities in centimeters per year.[3] The study of the motion of the continental plates is what we mean by **plate tectonics**.

But to move the giant continents requires a force. While the details are still not fully understood, we know that the rising of hot, low-density rock along with the subsequent falling of cool, high-density material produces a slow circulatory motion of the rock within the asthenosphere, similar to the movement of boiling water (**Figure 8–6**). This motion, called **convection**, plays a major role in plate tectonics. Because the continental plates contain material of lower density than the mantle underneath, the continents also float, just as

---

[3]This figure provides a good example of the importance of *velocity* compared with *speed* because direction is clearly an important consideration here.

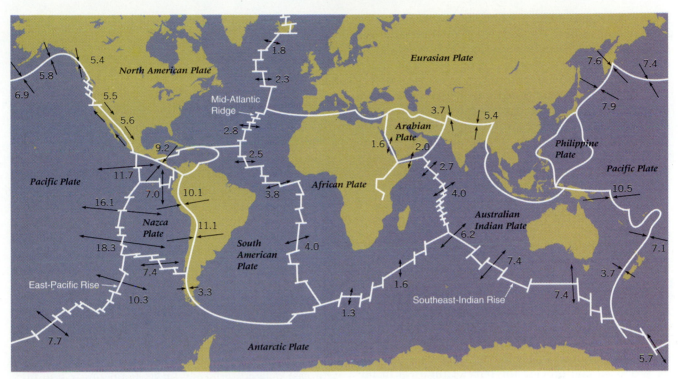

FIGURE 8–5. The motions of the continental plates. The numbers indicate the speed of one plate relative to that of the adjoining plate in centimeters per year. The arrows indicate the direction of relative plate motion.

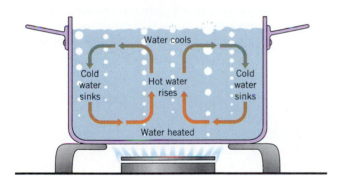

FIGURE 8–6. Convection in a pot of boiling water.

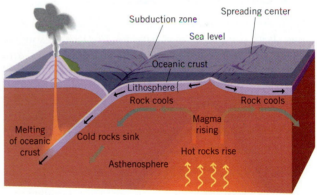

FIGURE 8–7. Schematic diagram showing the separation of two plates at a spreading center and the movement of magma in producing new crust. A subduction zone, where one plate moves under another, is also shown.

lower-density ice floats on top of higher-density water. The combination of continental buoyancy and convective motions within the asthenosphere provides the necessary ingredients for continental motion.

Continental plates interact with one another in various ways. For example, two plates can move apart, leaving a break (called a *spreading center*) in the lithosphere. Asthenospheric material, which has been heated from below, can flow through the break to form new crustal material, as shown at the spreading center in **Figure 8–7.** Another form of interaction occurs when two

plates slide by each other in opposite directions, as along the San Andreas fault in California. Here, the North American Plate on the east side (which contains San Francisco) moves southerly against the Pacific Plate (on which Los Angeles sits), which moves north. The plates grind against one another; sometimes, rock edges catch against each other. The continued motion increases stress, which eventually releases, causing an earthquake.

The other possible type of plate interaction occurs when plates converge on each other. In one instance, the cool lithosphere of one plate can sink underneath another plate, in a process called **subduction**. At the **subduction zone** where this occurs, deep oceanic trenches result (Figure 8–7). In another instance, a piece of low-density continental crust may collide with another piece along a subduction zone and thrust it upward, making spectacular mountain ranges. The Himalaya, Alps, and Appalachians resulted from such collisions.

THE SOURCE OF HEAT IN THE INTERIOR

One of the by-products of the radioactive decay discussed in Chapter 7 is heat. Because radioactive elements in the Earth's core have been decaying ever since the Earth's formation, and because the heat dissipates slowly, temperatures high enough to melt rock are readily produced and provide the bulk of the heat in the Earth's interior.

8.3

THE EARTH'S SURFACE

A planet's surface provides scientists with their first clues about the nature of the planet. For this reason, we will now study a variety of surface features and processes that we can apply as we view the surfaces of other planetary bodies. Our study will include the types of rocks present on Earth, the chemical nature of the surface, processes involving volcanoes and mountains, impact craters, and the effects of tidal forces.

ROCK TYPES, PROCESSES, AND AGES

Geologists classify rocks into three major families that are determined by the formation process. **Igneous rock** (from the Latin for fire) is formed from the cooling and solidification of molten material. Rocks formed from volcanos are a prime example. **Sedimentary rock** forms when loose materials held in water, ice, or air settle onto a surface, stick, and then build up. The final group, **metamorphic rock** (from the Greek *meta* meaning change, and *morphe* meaning form) consists of material whose original form has been modified by high temperature, high pressure, or a combination of both. While 75% of Earth's surface rocks are sedimentary, 95% of the crustal material underneath the surface is igneous or metamorphic rock derived from igneous materials.

Inquiry 8–2 Why does sedimentary rock occur on the surface rather than within the crust?

Rocks undergo a continuous cycling and can undergo change from one type to another. For example, an igneous rock on the surface might have been formed from a volcanic eruption. This rock is subject to erosion from wind, water, ice, and impact with other rocks. The eroded particles form a sediment that can be carried by wind and water, eventually settling, becoming cemented onto the surface, and producing a sedimentary rock. Movement of Earth's crust can carry the sedimentary rock to areas where the temperature and pressure are such that the rock is changed into a metamorphic one. If the rock is brought into yet hotter regions, it may melt and form a new igneous rock. Thus rocks participate in a **rock cycle** involving a slow and continuous interplay of external and internal processes.

Whether a particular material is a solid or a liquid (or a gas) depends critically on its temperature and pressure. An example of the effects of varying temperature and pressure is shown in **Figure 8–8**. Material on the left side (low temperature) is a solid, while material on the right (high temperature) is a liquid. For a given temperature, an increase in pressure can change a material from a liquid to a solid. There is an intermediate region in which solid and melted rock coexist. The concepts il-

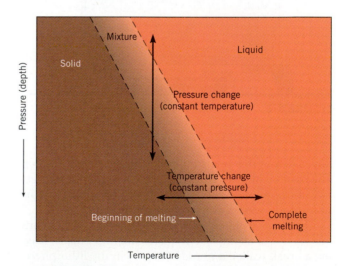

FIGURE 8–8. The change between solid and liquid depends on pressure and temperature. A change in pressure, holding temperature constant (a vertical change in the figure), can change a rock from solid to liquid.

lustrated in Figure 8–8 are factors that determine the detailed internal structure of a planetary body.

Inquiry 8–3 Explain why the inner core of the Earth is solid, while the outer core is liquid.

Rocks on Earth are subject to numerous types of change. For this reason, we should not expect to find any that are as old as the Earth itself. The same radioactive-dating techniques discussed in the context of meteorites in Chapter 7 can be applied to terrestrial rocks. We find that the oldest radiometrically determined age is 4.1 billion years for individual mineral grains in a sedimentary rock from Australia. Ages of 3.9 billion years have been determined for continental crust composed of granite.

CHEMICAL COMPOSITION

The chemical elements most abundant in the Earth's continental crust are shown in **Table 8–2**. These elements combine into some 3000 minerals that comprise the Earth.

Because silicon and oxygen are the most abundant elements in the crust, minerals made from them are the most abundant on Earth. *Silicates* are minerals having silicon, oxygen, and one or more of the other abundant elements. Another major mineral group is the *oxides*, which include quartz (SiO_2) and limonite (Fe_2O_3), also found on Mars. Other elemental combinations produce other mineral groups, including the *carbonates* that we return to later in this chapter.

VOLCANISM

The 1991 eruption of Mount Pinatubo served to remind us that the Earth's surface is highly dynamic. However, volcanic activity need not be so dramatic; sometimes

FIGURE 8–9. A fissure eruption, in which lava flows through a series of parallel fissures, on Mauna Loa, Hawaii, in 1984.

lava erupts nonexplosively from a series of parallel fissures in the crust, as shown in **Figure 8–9** occurring on Mauna Loa in Hawaii. Such eruptions provide the only opportunity to study molten rock, or **magma**, in its natural state. From the study of magma we are able to learn about conditions and processes inside the Earth.

A step toward understanding volcanism comes from examining the locations of volcanos around the Earth. Rather than being random, they form a well defined "ring of fire" around the Pacific basin. This distribution of volcanoes is significant because it corresponds closely to the places where continental plate subduction occurs. As a lithospheric plate sinks, rock is heated by friction and by the hotter asthenosphere. The exact temperature at which the rock melts depends on the pressure and the amount of water present in the rock. The more water, the lower the temperature necessary for melting. Molten rock, under high pressure, then works its way to the surface.

The Hawaiian Islands are volcanic but are located far from plate boundaries. They are thought to have formed as the moving Pacific Plate passed over a midoceanic hot spot that melted the thin oceanic crust and produced successive volcanoes on the moving plate. The process continues today as a new island, still under water, is forming.

MOUNTAINS

Some of the processes that produce mountains on Earth have been mentioned in earlier sections. These include repeated volcanism, and the convergence of continental plates that can fold continental crust and thrust it upwards. Examples of such fold-and-thrust mountain ranges include the Appalachians and Alps mentioned earlier, and the Canadian Rockies.

TABLE 8-2

Elemental Abundances in the Continental Crust by Percentage of the Mass

Element	%	Element	%
Oxygen (O)	45	Potassium (K)	2
Silicon (Si)	27	Titanium (Ti)	0.9
Aluminum (Al)	8	Hydrogen (H)	0.1
Iron (Fe)	6	Manganese (Mn)	0.1
Calcium (Ca)	5	Phosphorus (P)	0.1
Magnesium (Mg)	3	All Others	0.8
Sodium (Na)	2		

What goes up must come down, mountains included! While the downward pull of gravity actually causes the destruction of mountains, the process of degradation may begin by landslides occurring near regions of active volcanism, or simply from the shifting of rocks that occurs over time. Once a boulder begins its downward plunge, it may loosen other material that also will move downward, in a process that helps slowly destroy mountains.

Regions containing water can be eroded in a variety of ways. Flowing water can directly carry away loose material and cause other materials to weaken. Tall mountains, which have continuous freeze-thaw periods, are eroded by pressure from freezing water. Slowly moving glaciers dislodge rocks and grind the mountains through which they move.

IMPACT CRATERS

Any extraterrestrial bodies coming into the Earth's vicinity will be attracted by its gravity. We saw in the previous chapter that the solar system is full of meteoroids, ranging in size from dust grains to boulders kilometers across. Figure 7–13 shows the result of a relatively recent impact in the desert of Arizona some 50,000 years ago. While this crater is the freshest example on Earth, it is certainly not the only one. Since satellite photography of the Earth began, detailed reconnaissance of the Earth's surface has found many ancient craters. Undoubtedly the number of impacts was high in earlier years. The history of these impacts has been erased, however, because the craters have been subjected to the same processes that destroy and build mountains.

OCEAN TIDES

Some 75% of the Earth's surface is covered with water. The large mass of water readily responds to the gravitational pulls of the Sun and Moon, which cause tides. In fact, the "solid" surface also responds to the gravity of the Moon and Sun.

The Moon and Earth are represented in **Figure 8–10**. The gravitational attraction of the Moon on the Earth's water that is on the side toward the Moon is, of course, stronger than the attraction of the Moon on the Earth's center, which is 6500 km more distant (remember Newton's law of gravitation from Chapter 5). Hence, there is a difference in the gravitational force acting on the near side and the force acting on the Earth's center. The result is a net force toward the Moon that we call a **differential gravitational force**, or **tidal force**. The result of the tidal force is that, because water flows so easily, it "piles up" at the point nearly under the Moon (point A in Figure 8–10b).

In a similar manner, the force of the Moon's gravity on the center of the Earth is greater than that on the water located on the side of the Earth away from the Moon. Therefore, the solid Earth experiences a stronger force than the water located away from the Moon, and is "pulled away" from the water. The end result is a bulge of water on the far side of the Earth nearly equal to that on the side facing the Moon.

The Earth is rotating underneath this bulge of water, which always remains pointed nearly toward the Moon. At some point in time, a given spot (point A in Figure 8–10b) will be nearly underneath the Moon and will experience high tide; six hours later, the Earth's rotation will have moved the spot to point B in Figure 8–10c, and it will experience a low tide.

Inquiry 8–4   How many high tides will point A in Figure 8–10 experience in one day? How many low tides?

Inquiry 8–5   Will the time of high tide be the same each day, or will it change? Explain your reasons. If it changes, by how much?

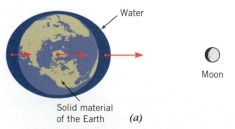
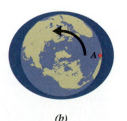
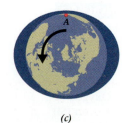

FIGURE 8–10.   The production of tidal bulges on the Earth by the Moon (not to scale). In (a), the straight arrows represent the size of the Moon's gravitational force at each point. In (b), the location A on the Earth is under the bulge and has a high tide. The curved arrow shows the direction of Earth's rotation, such that six hours later (c), the location of point A has rotated relative to the Moon and has a low tide.

The Sun also produces tides. Although the Sun's gravitational attraction on the Earth is greater than that of the Moon, its greater distance means that the difference in the Sun's pull on the two sides of the Earth is less than the difference in the Moon's pull. Therefore, using Newton's second law, astronomers can readily show that the Moon is twice as important as the Sun in producing ocean tides.

8.4

ATMOSPHERE

The Earth's atmospheric volume is composed of 78% nitrogen, 21% oxygen, 1% argon, 0.03% carbon dioxide (CO_2), and a variable amount (less than 2%) of water vapor (H_2O). The nitrogen and oxygen are present as the molecules N_2 and O_2, not as atoms. Ozone, which we have been hearing much about in recent years, is made from three oxygen atoms (O_3). Ozone is present as a trace element, providing only 0.00004% of the volume of the atmosphere. In addition to these gases, the atmosphere also contains varying amounts of solid particles, which come from wind-blown dust, fires, volcanic eruptions, and industrial pollutants.

We measure the amount of atmosphere in terms of its pressure on us: at sea level, a column of atmosphere having a cross-section of one square inch weighs 14.7 pounds. For convenience, we refer to this pressure as 1 atmosphere.

Scientists have divided the atmosphere into layers that are determined by the variation of temperature with height, as shown in **Figure 8–11**. Within the lowest layer, the temperature decreases with height to about 10 km. This layer (the troposphere) is where weather occurs. Within the next layer (the stratosphere) the temperature increases with increasing height, due to absorption of solar ultraviolet energy by the ozone located near its top. Destruction of ozone by the injection of chlorofluorocarbons (CFCs) allows more ultraviolet radiation to reach the surface, where its absorption by living cells, and resulting damage, can occur.

Because there is no ozone above the stratosphere, the temperature again decreases with increasing height. Above about 100 km, the temperature again increases with height because of ultraviolet radiation from the Sun. This ultraviolet energy is capable of ejecting electrons from gas atoms and leaving charged particles behind. Because such charged particles are called ions (the processes involved in producing them are discussed in Chapter 13), this "sphere of ions" is called the **iono-sphere**. Prior to the advent of communications satel-

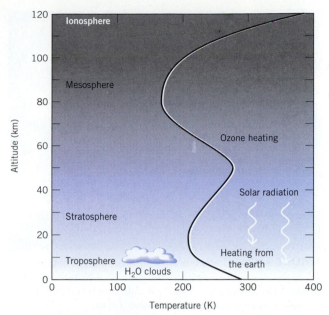

FIGURE 8–11. The temperature structure of the Earth's atmosphere. Note the location of ozone in the region where the temperature increases with increasing altitude.

lites, the only way we could send radio signals beyond the horizon was to reflect them off the ionosphere.

THE GREENHOUSE EFFECT

Although the atmosphere absorbs ultraviolet radiation from the Sun, visual radiation (that which you can see) does pass through and strike the ground, creating heat. This heat energy radiates into the atmosphere, giving it a warm temperature at low altitudes. You have probably seen an effect of reradiation from an asphalt highway in the summertime: the blacktop absorbs visual radiation and then becomes hot. Atmospheric ripples and mirages appear as the reradiation from the road heats the air above it. However, this emerging heat radiation is absorbed by carbon dioxide and water vapor in the atmosphere, which acts like a blanket by not allowing the heat to escape. Because this phenomenon is related to what occurs in a greenhouse,[4] it is called the **greenhouse effect**. A schematic diagram of the process is shown in **Figure 8–12**.

The greenhouse effect occurs naturally. A problem comes about only if the naturally occurring balance be-

[4]The greenhouse effect in a real greenhouse is somewhat simpler than in the atmosphere. In both cases, the infrared radiation is trapped and cannot escape, but in a real greenhouse, the glass cover also prevents the heated air from rising or flowing away.

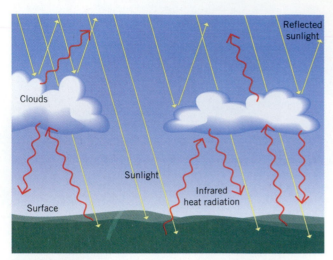

FIGURE 8–12. The greenhouse effect. Sunlight coming through the atmosphere heats the ground, which then emits infrared (heat) radiation. The heat radiation is trapped primarily by carbon dioxide and water vapor.

tween energy input and energy radiated is upset by a dramatic increase in carbon dioxide emissions into the atmosphere.

ATMOSPHERIC CIRCULATION

The atmosphere is far from static. Weather satellites show us the movement of cloud systems, low and high pressure systems, and hurricanes. Atmospheric circulation is driven by heat from the Sun that is reradiated into the atmosphere by the Earth's surface. If this were the only source of energy input, the atmospheric circulation patterns would be simple. The complication is that 75% of the Earth's surface is covered with water, a tremendous storehouse of solar energy. Furthermore, the continents are warmer than the water, thus adding to nonuniform heating of the atmosphere. We now examine the effect of these sources of heat on the atmosphere.

When heat is injected into the atmosphere from water or continents, the atmospheric temperature increases. The heated gas expands and the atmospheric pressure changes. These changes, which also cause the density of air to decrease, force heated air to rise. It is analogous to the low-density continental crust floating on top of the higher-density asthenosphere. Atmospheric convection, therefore, is a process that moves energy through the atmosphere. Because the uneven reservoir of heat on the Earth's surface unevenly heats the atmosphere, the atmosphere contains numerous regions of low and high pressure. For example, air at the

equator increases in temperature and rises. Due to the rising air, the atmospheric pressure at the equator lessens. Atmospheric gases from more northerly and southerly latitudes will move toward the low pressure area at the equator and thus produce surface winds. The gas that rose over the equator will move toward the poles, cool, sink back to the surface, and begin to circulate back to the equatorial regions.

The Earth's rotation complicates the picture dramatically. In Chapter 4, we introduced the concept of the **Coriolis effect**, in which a northward-traveling rocket appears to veer to the east because of the Earth's rotation. In a similar way, the Coriolis effect causes northward-moving surface winds to veer toward the east. Air moving along the surface from the poles toward the equator produces westward-moving winds. The end result is that the atmosphere is broken up into cells, as shown in **Figure 8–13**. Low pressure occurs where the cells rise while high pressure occurs where they fall. The upward and downward movements within the cells at latitudes near 30° and 60° produce high-speed, westward-moving winds (the jet stream we often hear about on TV weather reports). We expect that rapidly rotating

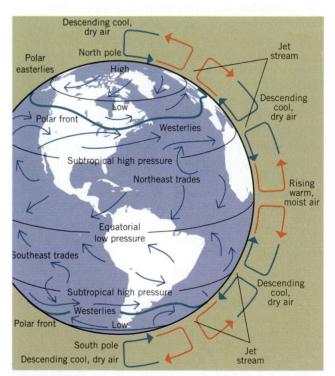

FIGURE 8–13. The complex general circulation patterns of the Earth's atmosphere. The oval shapes with the arrows on the right indicate the cell structure of the atmosphere, in which heated air rises and cool air falls. Jet streams are located where adjacent convective cells meet.

planets would show an even more complex circulation pattern.

ORIGIN AND MAINTENANCE OF THE ATMOSPHERE

The chemical makeup of the Earth's atmosphere is not what we might expect. Most of the universe is made up of hydrogen and helium with all the other elements comprising only some 2% of the total. We might therefore anticipate an atmosphere of hydrogen and helium, with trace amounts of everything else, but this is not the case.

Astronomers believe that the original, primitive atmosphere of the Earth was lost and later replaced by the atmosphere we have today. Until recently, because of the vast amount of hydrogen in the universe, it was thought that the original atmosphere was composed primarily of hydrogen and hydrogen-bearing compounds, such as methane (CH_4), ammonia (NH_3), and water (H_2O). There should also have been an amount of neon approximately equal to the fraction found in the Sun. From the observed lack of neon—which could not have come about as a result of chemical reactions, because neon is chemically inert—astronomers conclude that the Earth's original atmosphere must have been lost.

What happened to the compounds that formed the primitive atmosphere? Gas atoms and molecules are not static. The greater the temperature, the greater their speed. Furthermore, light particles move more rapidly than heavy particles at the same temperature. Therefore, if the temperature of a gas is high enough, the particles may have speeds in excess of the Earth's escape velocity of 11.2 km/sec. At the temperature of the Earth, both hydrogen and helium attain velocities considerably greater than the escape velocity. And because methane and ammonia are readily broken into their constituent molecules, their hydrogen atoms can also escape, leaving behind a planet without an atmosphere.

In the planet formation process, pockets of atmospheric gas of heavier elements can be trapped in the planet's interior. As the temperature increases due to the decay of radioactive elements in rocks, the trapped gases can slowly escape from the interior and form a new atmosphere by a process called **outgassing**. Such gases can also escape through volcanic eruptions. Even today, volcanoes spew large amounts of water, carbon dioxide, sulfur dioxide, hydrogen sulfide, chlorine, and nitrogen into the atmosphere. In fact, a more modern hypothesis on the composition of the primitive atmosphere suggests it was these molecules, and not methane and ammonia, that composed the original atmosphere.

There is little carbon dioxide in the atmosphere because it is tied up in the carbonate rocks described ear-lier in the chapter. You undoubtedly noticed that oxygen has not been mentioned. Little was present until substantial amounts of plant life evolved some 2 to 2.5 billion years ago. Living organisms also enhance one other component of the atmosphere: once death and decay occurs, nitrogen is produced.

8.5
MAGNETISM

A magnetic compass points toward the north *magnetic* pole (which differs from the north *geographic* pole), showing that the Earth has magnetism associated with it. The effects of magnetism are readily seen when small needles of iron are placed over a magnet (**Figure 8–14**). The pattern formed by the iron filings delineates the magnetic force field, and we often speak of the lines formed by the pattern in terms of **magnetic lines of force** or **magnetic field lines**. The closer to the magnet, the stronger the magnetic force, and the closer together the magnetic lines of force.

What produces magnetism? Scientists have known for over 100 years that an electric current, which is produced by electrons moving in a wire, will produce a magnetic field. Whenever electrons move, a magnetic field is produced.

What produces the Earth's magnetic field? While the detailed answer is still not known, we believe it is caused by the movement of iron and other metals in the liquid outer core. Metallic elements such as iron contain a large number of loosely bound electrons; it is these

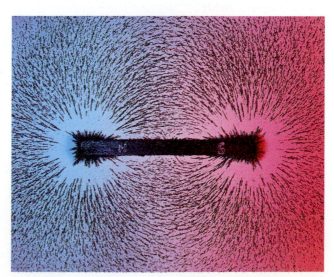

FIGURE 8–14. Magnetic force fields, as shown by the alignment of iron filings around a magnet.

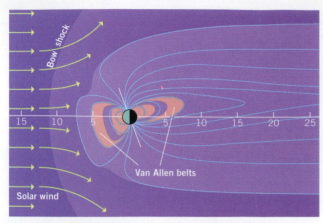

FIGURE 8–15. The Earth's magnetic field as observed from space. The magnetic field lines are distorted by charged particles from the Sun, located in the solar wind. Note the compression of the field on the side toward the Sun, and the stretching on the side away from it. The Van Allen belts are two regions surrounding the Earth in which charged particles are trapped by the magnetic field.

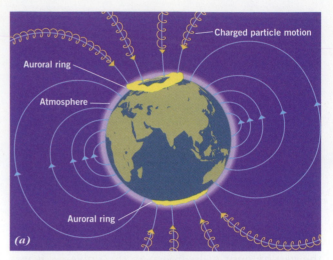

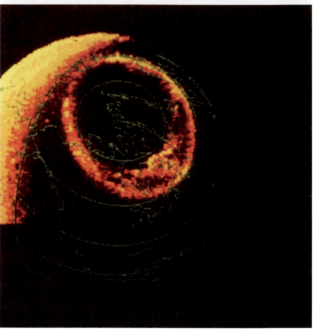

FIGURE 8–16. The aurora borealis. (*a*) Charged particles spiral around magnetic field lines, and the magnetic field intersects the atmosphere in a ring. (*b*) The resulting auroral ring as observed by a satellite about three Earth radii from the North Pole.

electrons that allow metals to conduct electricity and heat so well. The convection of the molten outer core, along with the rapid rotation of the Earth (and by assumption its core), provide a large mass of moving electrons, which we believe produces the magnetic field. This process is known as the **dynamo effect**. While this process explains a number of observations, it does not explain everything. For example, geological measurements show that the north and south magnetic poles switch positions every now and then. The dynamo effect does not easily account for these observed switches in the magnetic field.

The Earth's magnetic field, shown in **Figure 8–15**, is similar to that of the simple bar magnet in Figure 8–14. The magnetic lines of force, which surround the Earth in a "sphere" of magnetism we call the **magnetosphere**, extend out into space. The Earth is bathed in charged particles because, as we will see in Chapter 16, the Sun is constantly ejecting electrons and protons into space in all directions. The place where the particles and the magnetic field collide forms a shock wave, called a *bow shock*. Once these charged particles encounter a strong magnetic field, the particles' path changes.

The charged particles from the Sun modify the Earth's magnetic environment. The solar particles push on the field lines, compressing them in the direction from which the particles came. In addition, those particles that pass near the Earth pull the field lines along with them, thereby stretching them out in the direction away from the Sun. The compression and stretching of the magnetic field distorts the magnetosphere. The pres-

ence of a field, and its distortion, can be detected by spacecraft passing through the magnetosphere.

The path followed by charged particles in a magnetic field is not a straight line but a spiral around the magnetic field lines (**Figure 8–16a**). The particles move toward locations where the field is strongest, such as the magnetic poles. Whenever large concentrations of particles moving along the field lines strike the upper atmosphere, the gas atoms in the atmosphere can be made to glow, forming the **aurora borealis** in the Northern Hemisphere and the **aurora australis** in the Southern Hemisphere. Because the magnetic field is three-dimensional, it intersects the atmosphere in a circle (Figure 8–16a), and an auroral ring is produced surrounding the north magnetic pole (**Figure 8–16b**).

Finally, the Earth is surrounded by two donut-shaped regions of charged particles that are trapped by the magnetic field. These belts of particles were discovered when the first U. S. satellite was launched in 1958. Named the **Van Allen belts** after the scientist who discovered and analyzed them, the two belts are centered on regions about 1.5 and 3.5 Earth radii away and are included in Figure 8–15.

8.6
EARTH-BASED STUDIES OF THE MOON

In the rest of this chapter we explore the Moon. Our discussion naturally divides itself into three parts: (1) what was known about the Moon prior to the inception of the space program, (2) what was learned from unmanned missions, and (3) what has been learned as a result of the Apollo missions.

Inquiry 8–6   Before going any further, imagine that you are an official of the space program planning a mission to the Moon. What would you want to know about the Moon?

BASIC DATA

From a study of eclipses, Aristarchus, in the third century B.C., determined the size of the Moon relative to that of the Earth and found it to be some four times smaller (Chapter 5). Knowing the size and observing the angular diameter to be about 1/2° allows us to find the distance, which is about 384,000 km. The Moon's mass is most easily determined by examining the motion of an artificial satellite in lunar orbit; the Moon's mass is 1/81 that of Earth. From this we find a density of 3.3 g/cm^3. The currently accepted values, along with other data for the Moon, are given in **Table 8–3**.

THE MOON'S ATMOSPHERE

If the Moon had a substantial atmosphere, starlight passing through it would gradually become fainter and fainter until, at last, the body of the Moon would extinguish it completely. However, as the Moon moves in its orbit

---

TABLE 8–3

Properties of the Moon

Property	Value
Orbital properties	
Average distance from Earth	384,401 km (60.4 $R_\oplus$)
Closest approach	363,297 km
Farthest distance	405,505 km
Orbital period	$27^d7^h43^m12^s$
Lunar month (from new moon to new moon)	$29^d12^h44^m3^s$
Orbital inclination	5° 8' 43"
Physical properties	
Diameter	3,476 km (0.273 $D_\oplus$)
Mass	7.35×10^{25} gm (0.0123 $M_\oplus$)
Density (average)	3.34 gm/cm^3
Surface gravity	0.165 Earth gravity
Escape velocity	2.4 km/sec
Rotation period	$29^d12^h44^m3^s$
Tilt of rotation axis from orbit perpendicular	6° 41'
Albedo	0.07
Surface temperature	100 K night to 400 K day (−279 to +260°F)

around the Earth, from time to time passing in front of stars, we observe the starlight to disappear from view extremely rapidly. From such an event, which is called an **occultation**, astronomers have long since concluded that the Moon has virtually no atmosphere. This is no surprise in view of the Moon's low escape velocity and high daytime temperature. Escape of the Moon's atmosphere, if it ever had one, must have been extremely rapid.

Without an insulating layer of atmosphere, the Moon's surface undergoes extreme variations in temperature, from 375 K (which is equivalent to 216°F, just above the temperature of boiling water) during the day down to only 125 K (−234°F) at night.

THE LUNAR SURFACE

The visible features on the Moon's surface seem permanent, indicating that the surface is solid. Is it like a cement slab, a frozen pond, a sand dune, a dust pile, or something else entirely? Galileo, on first observing the dark areas of the Moon's surface (**Figures 8–17** and **8–18**), thought they looked like water and named them **maria** (singular form **mare**, pronounced mar-ay), the Latin word for "sea." However, it has long been clear that there is no water on the Moon and that the dark color of the maria has to have another cause.

A number of things can affect the reflectivity, or **albedo**, of a surface. For example, a smooth surface will directly reflect a large proportion of the light falling on it, whereas a rough surface will reflect the light in various directions, thereby decreasing the surface's overall ability to reflect. Thus surface texture is one possible cause of variations in the Moon's albedo. A second possible cause is that the maria may be covered by a different type of material than that found in other, lighter-colored areas of the Moon. Just as some rocks and soils on Earth are light in color and others are dark, the same could be true of the Moon.

We can infer something about the nature of the lunar surface from observations of how the Moon's temperature changes when incident sunlight is blocked during a lunar eclipse. The lunar surface cools slowly, telling us that it must be composed of a good insulator for it to be able to hold the heat of the lunar day so effectively.

One plausible model that was consistent with the observed temperature variation during an eclipse was that the Moon's surface was covered by a substantial layer of porous dust. There would be no surprise in this, because the Moon is constantly being bombarded by meteoroids of all sizes that would slowly pulverize its surface rocks. Even the extreme temperature changes that the Moon undergoes every month would tend, in time, to weaken

FIGURE 8–17. The full Moon with the principal features indicated.

FIGURE 8–18. The Moon at first quarter. Note how much more visible the craters are than in the previous figure.

and even crumble the rocks by causing them to expand and contract.

The possibility that there might be a thick layer of dust on the Moon raised fears that the first astronauts might simply disappear into the bottom of a dust bowl. It was clear to the planners of the Apollo missions that an unmanned vehicle had to be sent first! Fortunately, it turned out that the dust had a sufficient amount of cohesion to prevent such a disaster from taking place, but this could not have been known for certain before the first lunar lander was flown in 1966.

8.7
EXAMINING THE LUNAR SURFACE

> ALL READERS ARE STRONGLY ENCOURAGED TO DO DISCOVERY 8–1 AT THE END OF THE CHAPTER **BEFORE** CONTINUING TO READ.

> READERS HAVING THE ACTIVITY KIT CAN DISCOVER MORE ABOUT THE LUNAR SURFACE BY DOING ACTIVITY 8–2, SURFACE DETAILS FROM LUNAR ORBITER PHOTOGRAPHS, AT THIS TIME.

The American program of lunar exploration involved numerous phases of unmanned exploration along the way. First was the Lunar Ranger project, in which several probes relayed close-up television pictures to Earth prior to crash landing on the lunar surface. The Lunar Orbiter program provided continuous pictures while orbiting the Moon of the lunar far-side along with high quality images of the near-side. The final unmanned exploration before the manned landings was the Surveyor program, which landed spacecraft to carry out some early experiments.

We will now examine a variety of surface features present on the Moon, paying particular attention to the processes that formed them. We want to find out whether the processes that shaped the Moon are similar to those that shaped the Earth.

CRATERS

Ever since Galileo first discovered lunar craters, scientists have speculated about their origin. Two possible hypotheses immediately present themselves: they may be due to meteoritic impacts, similar to those that formed the Barringer crater in Arizona (Chapter 7); or they might be volcanic.

There is no question that many, if not most, of the lunar craters have their origin in meteoritic impacts. Evidence for impact is plentiful. In **Figure 8–19** the relatively young crater Euler is surrounded by material that has been ejected from the crater; such an **ejecta blan-**

FIGURE 8–19. The impact crater Euler, with a well-defined ejecta blanket and secondary craters. Euler is 27 km in diameter.

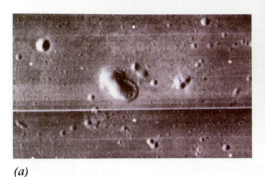

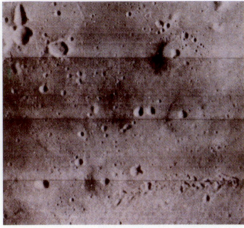

FIGURE 8–20. Lunar volcanic craters. (*a*) This crater is elongated with dimensions of 3 km by 5 km, lacks exterior walls, and is similar to basaltic volcanoes on Earth. (*b*) A volcanic dome with a crater on its summit. The domes are roughly 10 km across.

(*a*)   (*b*)

ket is characteristic of impact. Furthermore, its circular shape is in contrast to the various volcanically produced craters, which are often somewhat elongated. The figure also shows a chain of craters along a line from Euler; such **crater chains** are produced when clots of excavated material from one impact strike the surface, producing **secondary craters**.

Further evidence of the impact nature of craters comes from the light-colored **rays** emanating from the crater Tycho (Figure 8–17), suggesting that material was thrown outward by tremendous forces during impact. The distribution of rays around a crater can be understood in terms of the angle at which the impacting body hit the surface. A perpendicular impact produces a symmetric distribution of rays, while an oblique impact produces asymmetric rays.

Impact can readily explain the presence of central mountain peaks in a significant number of the largest craters (Figure 8–19). These peaks result from rebounding of the compressed surface after the impact. The crater in this figure has terraced walls, caused by a gravitational pull that makes the walls slump downward.

However, not all lunar craters are the results of impact. The craters in **Figure 8–20a** are clearly not impact craters because they lack walls and are not circular. The Moon shows numerous **volcanic domes** similar to those on Earth (**Figure 8–20b**). Such domes result when lava's **viscosity**, its resistance to flow, is high enough to prevent it from readily flowing across the surface. Although the Moon has numerous domes, it does not have any tall volcanoes similar to Mt. Etna or the large Hawaiian volcanoes.

Craters do erode with time. The continuous bombardment by micrometeorites, meteorites, and cosmic rays over billions of years gives old craters a softened appearance.

Much research has been done into how craters form. When a meteoroid, perhaps moving at 20 km/sec, strikes the surface, rapid compression of the surface occurs. Shock waves begin to travel throughout the impacted body and produce molten rock that is squeezed out to the sides at high speed. Some of this molten material moves at speeds greater than the Moon's escape velocity and is ejected into space. Shortly, new outward-moving material is deflected upward by the forming walls. Because velocities are now below the escape velocity, the material falls back to the lunar surface. As the energy of the impacting body is expended, the speed of the ejected material goes to zero, and the cratering event is over. These events are summarized in **Figure 8–21**. Material originally on the surface becomes buried, and material originally under the surface ends up on the surface. For a typical large crater, this process takes about a minute.

MARIA

The maria are easily understood in terms of a gigantic impact that produces a large crater and its subsequent filling-in by flowing lava. The sizes of these maria indicate impacting bodies having diameters from about 30 to 100 km. Their relative flatness comes from fluid flow. For example, in Mare Imbrium (the Sea of Rains), which has a diameter of nearly 800 km, extensive lava flow fronts are visible, coming from the lower left of **Figure 8–22**. Because no volcanic cones are present to account for the flows, the material must have come from long fissures similar to those on Earth, shown in Figure 8–9. Not all lunar volcanoes show flow fronts. This indicates a variation in the lava's viscosity from one place to another. An analysis of rocks returned by the Apollo astronauts shows some rocks, when melted, flow

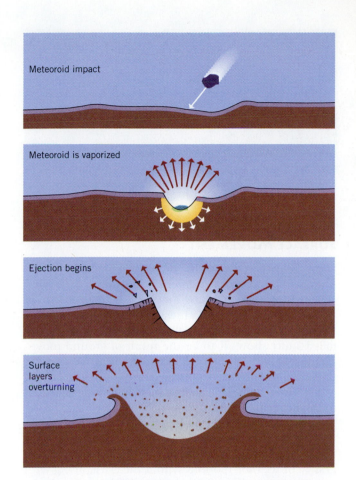

Meteoroid impact

Meteoroid is vaporized

Ejection begins

Surface
layers
overturning

Ejecta blanket

FIGURE 8–21. How a crater is formed. After initial compression, the surface reacts by throwing material into the surrounding area. Note that material originally on the surface ends up buried, while material originally underground is on top.

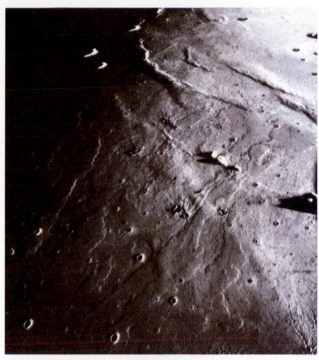

FIGURE 8–22. Lava flows in Mare Imbrium. The lava flows shown here traveled distances up to 600 km. The average flow fronts have heights of 30 m. Since there are no obvious craters from which the lava erupted, it must have come from fissure eruptions, similar to that in Figure 8–9.

FIGURE 8–23. Mare Orientale.

easily compared with lava flows on Earth, and have the viscosity of engine oil. Finally, an analysis of the sizes of crystals in the lunar maria shows that the material cooled relatively rapidly, over a period of a few years.

Probably the most dramatic feature on the lunar surface is Mare Orientale (**Figure 8–23**), located right on the dividing line between the near and far sides. This is an example of a so-called **multi-ringed basin**. It consists of four rings, two of which are easily seen. The inner ring is called the Rook Mountains and has a 620-km diameter, while the outer one, called the Cordillera

Mountains, is 900 km in diameter. The central region has been flooded with lava (as we discuss below).

There are two main hypotheses for the formation of the Orientale Basin. The more widely accepted one is that a gigantic impact produced tidal-like waves that froze in place. The other one looks at the entire feature as a single crater, in which the rings are terraces that resulted from collapse of the walls.

FIGURE 8–25. Mare ridges in the Mare Serenitatis. The region shown is 115 km by 155 km. The ridges are a result of the folding of solidified lava.

SURFACE MOVEMENT

Since the earliest days of lunar observing, nearly straight lines were observed at numerous locations. They were called **rilles**, which means "tiny watercourses." One of the rilles easily observable on ground-based photographs is the Straight Wall (**Figure 8–24**). More detailed study shows it to be the result of ground movement—a fault 115 km long, in which the ground on the one side has dropped 400 m relative to that on the other. Thus, the term rille is a misnomer when applied to this particular feature.

The result of more complex crustal movement can be inferred from ridges visible in some maria (**Figure 8–25**). Note the alignment of the ridges, which is suggestive of tectonic motions within the crust, along with crustal folding and buckling. However, the Moon shows no large-scale plates or plate motion like that on Earth.

LAVA CHANNELS AND TUBES

In addition to the straight rilles discussed above, the lunar surface shows a large number of rilles that appear to wander like a meandering stream across the lunar surface. These are called **sinuous rilles** to distinguish them from the straight ones. The majority of them begin at a rimless crater located higher than the end point. These findings are consistent with their being volcanic lava channels. The rimless crater at which they begin is the source of material, which then flows downhill.

The outer surface of lava flowing in a channel will cool rapidly, forming a skin above the hot lava flowing underneath. This skin thus forms a roof, enclosing the molten rock inside a **lava tube**. The roof insulates the liquid, slowing its cooling and allowing it to flow farther from the source. Once the source of lava runs out, the liquid will continue to flow until the tube is drained. Thereafter, the tube may collapse. Such collapsed tubes are observed over parts of some sinuous rilles and provide additional evidence for lunar volcanism. **Figure 8–26** shows the largest sinuous rille on the Moon, called

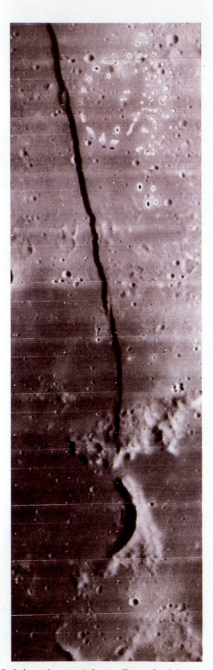

FIGURE 8–24. The Straight Wall resulted from crustal movement.

Schroeter's Valley. It is over 150 km long, 4–6 km wide, and 500 m deep, and contains some collapsed lava tubes. It begins at a crater and eventually empties into a mare.

Further evidence of the volcanic nature of the rilles comes from examples of crater chains inside rilles. It is difficult to see how crater chains could have been formed inside a rille as a result of random meteoritic impacts but easy to understand in terms of volcanism.

HIGHLANDS AND THE FAR SIDE OF THE MOON

Much of our discussion thus far has concentrated on the regions near the dark maria, covering roughly 50% of the observable lunar surface. Because the maria are at lower altitudes than the other areas, the lighter-colored areas are known as the lunar **highlands**. The highlands are higher than the maria because the highlands contain material of lower density than the material underneath. The highlands are saturated with craters.

The near and far sides of the lunar surface are compared in **Figure 8–27**. The most obvious differences are that the far side is more heavily cratered and lacks the

FIGURE 8–26. The sinous rille, called Schroeter's Valley, is an excellent example of the connection between a lava source at a crater, and the channels and tubes that transport it to an outlet at a mare. The valley is over 150 km long, 4–6 km wide, and 500 m deep.

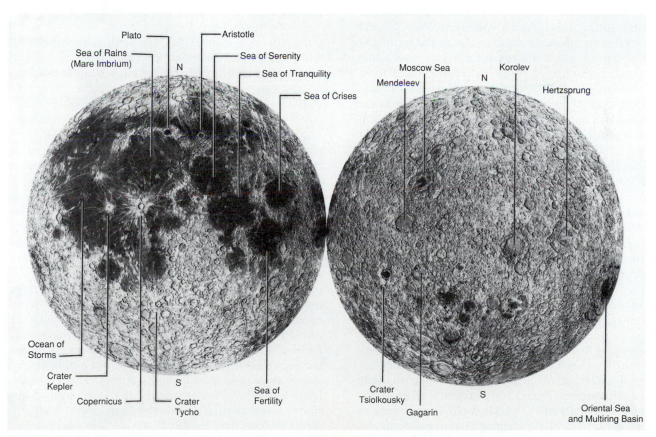

FIGURE 8–27. The far side of the Moon lacks the maria of the near side.

extensive maria of the near side. However, it does contain some large impact basins that were never filled in by lava, such as Hertzsprung, marked in the figure. It appears to have an inner ring.

8.8

MANNED EXPLORATION OF THE LUNAR SURFACE

A primary goal of the manned lunar missions was to return the astronauts safely to Earth. One result of this priority was that the astronauts were sent to "safer" (and hence possibly less interesting) areas of the Moon, where they would be able to land with the least danger. Inevitably, the limited amount of sampling that was done would leave many unanswered questions. Nevertheless, due to the Apollo missions many important new scientific results have been learned that could not have been discovered in any other way.

THE COMPOSITION AND STRUCTURE OF THE LUNAR SURFACE MATERIAL

As the astronaut's footprints in **Figure 8–28** illustrate, the lunar surface is sufficiently cohesive to provide solid support despite its dusty nature. The top layer of the lunar surface is called the **regolith** (rocky layer); it is composed primarily of rocky material ejected from lunar impact craters. The powdery soft nature of some areas is a result of erosion by the constant bombardment of small particles from space. Of course, there are occasional larger impacts too, but, as on the Earth, most of the mass falling upon the Moon today is in the form of particles the size of a grain of sand or smaller. The extreme heating and cooling of lunar rocks is also a source of erosion. Driven by temperature changes, rocks will expand and contract until finally they weaken and crumble.

While some 800 pounds of lunar material was returned to Earth by the various Apollo missions, because of budgetary constraints only a small fraction has been analyzed. Chemical analysis of the lunar rocks shows that they fall into three main categories: **basalts**, **breccias**, and a type that has been called **KREEP**. Maria materials are primarily basalts; that is, igneous rocks that have resulted from the cooling of molten material. Breccias, on the other hand, are formed by the fusing together of rock fragments that may well be older than the breccias themselves. Such fusing occurs because of impacts by external bodies, which increase the pressure and temperature of the region that was impacted. KREEP is a type of basalt that is unusually high in potas-

FIGURE 8–28. The footprints show the surface material to be strong, yet compressible. The retroreflector left on the Moon by the first astronauts is in the foreground.

sium (chemical symbol K), rare earth elements (REE), and phosphorus (P), relative to terrestrial rocks. The breccias and KREEP are abundant in the highlands. **Figure 8–29** shows a lunar basalt and a breccia.

Rocks on Earth almost always include water molecules chemically bonded into the interior structure. Moon rocks, however, are distinguished by their complete lack of water. Any water that was ever present on the Moon must have escaped rapidly during the Moon's early stages.

A detailed comparison of elemental abundances in lunar rocks and Earth rocks provides fundamental information required to understand their origins. When compared with the Earth's rocks, lunar rocks are deficient in volatile elements (those with low melting points). This important finding suggests two possible explanations. First, the Earth and Moon could have formed from materials having slightly different chemical compositions. Second, the materials out of which the Moon was made could have formed at somewhat higher temperatures than was the case for the Earth; a higher temperature would have driven the more volatile elements away from the forming Moon. A higher temperature would have been present if the Moon formed in a hotter location than where the Earth formed, or if it formed earlier than the Earth. At this time, observations do not indicate which idea is preferred.

(a)   *(b)*

FIGURE 8–29. Lunar rocks. (*a*) A basalt from one of the lunar maria. (*b*) A breccia, showing fragments cemented together.

A final comparison with the Earth's materials is a detailed examination of the abundance of the various types of oxygen on the Earth and Moon. While all oxygen atoms have eight protons in the nucleus, some have different numbers of neutrons. Atoms of a given element having different numbers of neutrons are called **isotopes**. A finding of great importance to understanding the evolution of the Moon is that the relative abundances of the various oxygen isotopes are identical on both bodies, although they differ from the abundances in meteorites.

Inquiry 8–7 What might you conclude about the origin of the Earth and the Moon, given the equality of the oxygen isotopes? What would you conclude if the abundances had been different?

THE AGE OF THE LUNAR SURFACE

Relative ages of lunar surface features can be determined using the geological concept of superposition, which says that younger features are on top of older ones. For instance, a crater on top of a mare will have formed after the lava flows in the mare stopped.

Absolute ages can be determined using the same dating techniques that were discussed for meteorites. Lunar rocks became available for radioactive-dating analysis only with the Apollo manned flights to the lunar surface, which began in 1969. The first lunar rocks returned to Earth came from two maria regions. These dark regions of lava flow turned out to be about 3.1–3.9 billion years old (similar in age to the oldest measured Earth rocks). What does this age range mean? The time determined by radioactive dating techniques tells us the time since the rock last solidified. In other words, the radioactive clock is reset to zero when melting occurs. Thus, these ages mean that the maria regions were molten (and then solidified) throughout an 800-million-year period.

Later Apollo landings were more adventurous and also sampled the highland regions. Radioactive ages of highland rocks were found to be about 4.1 billion years—hundreds of millions of years older than the maria. Although the highland region is cratered and does not appear strongly volcanic, the highland material turned out to be composed of solidified magma, just like the maria. The conclusion is inescapable and important: there was an early epoch when the entire surface was molten. The basaltic leftovers that resulted when the surface cooled were then chopped up by heavy cratering to form the breccia material of the highlands.

How can the above observations be interpreted? We can now put together a rough chronology of the Moon's history. The Moon apparently formed about 4.5 billion years ago, about the same time as the planets. Whether it was hot or cold at that time is unknown. Within a few hundred million years, however, the surface must have melted, fusing together the breccias in the lunar highlands. During this time, radioactive decay in the interior regions was producing heat, which after about a billion

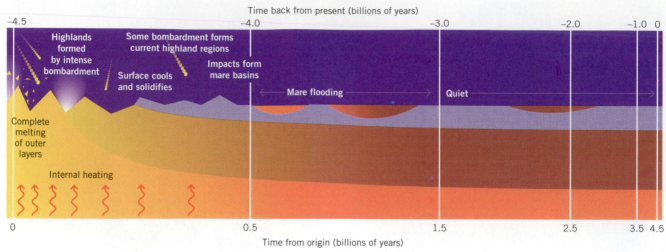

FIGURE 8-30. An evolutionary timeline of lunar history.

years was sufficient to melt the interior. Therefore, the melting of the highlands could have come about either from internal radioactivity whose heating melted rock that then flowed to the surface, or from a short period of intense meteoroid bombardment. (Recall that the solar system was filled with planetesimals while the solar nebula condensed.) After some cooling, during which the crust formed, the bombardment that formed most of the highland craters took place, about 4.1 billion years ago. Some of the impacts at that time would have involved asteroids, and the primitive maria basins would have been formed. Impacts of that large size would have produced deep cracks in the crust. Then, between 3.1 and 3.9 billion years ago, radioactively melted material found its way to the surface through these cracks, filling in the basins and producing the smooth maria plains. For the last 3 billion years, the Moon has remained cool, quiescent, and geologically dead. This history is summarized in the timeline of **Figure 8-30**.

THE MOON'S INTERIOR

Much uncertainty remains in our picture of the Moon, partly because the astronauts were not able to find any truly primeval material in the few regions they sampled. All the rocks they brought back to Earth had undergone some changes during the course of lunar history. Our understanding of the Moon's interior is even less clear.

Some information about the interior can be obtained simply from the Moon's density. The average density of rocks gathered by the astronauts was 2.96 gm/cm^3, only slightly less than the overall average of 3.34 gm/cm^3. From this low density we conclude that the Moon is

lacking in iron relative to the Earth. A further inference is that the interior structure is not nearly as strongly differentiated as the Earth's interior. In fact, the observed properties of the Moon are inconsistent with a percentage iron abundance as high as that in the Sun.

Further information about the interior has been learned from data transmitted back by seismometers left by the astronauts. Although "moonquakes" have been detected, their frequency and strength is far less than that of seismic events on Earth. One important characteristic is that they last much longer than quakes on Earth. The seismic waves bounce around the interior for a long time (up to an hour compared with seconds in the Earth). Lunar seismic events fall into two classes. For one type, the occurrence is correlated with the motion of the Moon around the Earth; roughly half these quakes occur when the Earth and Moon are at their closest, while the other half occur when they are farthest apart. Clearly, the Moon's interior is responding to tidal forces from the Earth, just as the Earth's oceans respond to lunar tidal forces. The second class is unrelated to the Earth, and its events originate deeper in the interior. Some may be triggered by meteoroid impact.

As a result of lunar seismology studies, we now know that the Moon has an outer crust about 60 km deep and a core about 1500 km in diameter. Additional observations show that the crust is thinner on the side facing the Earth than on the lunar far side. The thinness of the crust is consistent with the observed asymmetry of maria on the near side and their lack on the far side. The seismic results show that there *may* be a small region of the core that is at least partially molten; the presence of a molten core has not been unequivocally established (**Figure 8-31**).

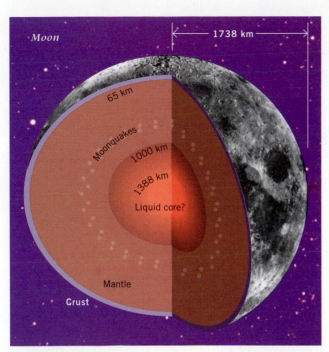

FIGURE 8–31. The interior structure of the Moon.

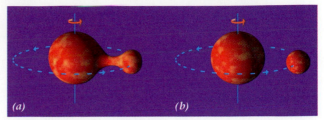

FIGURE 8–32. The fission theory of the Moon's formation. (a) Material is torn out of the Earth and (b) forms into the Moon.

Inquiry 8–8 How might the observed thickness of the crust on the lunar far side explain the lack of lava-filled basins there?

Magnetic field measurements show that the Moon does not have an overall magnetic field as the Earth does. If magnetic fields are formed by the dynamo effect discussed for the Earth, the lack of a lunar field is consistent with the lack of a definite molten interior indicated by the seismic data, and with the Moon's slow rotation. However, there are locations on the Moon's surface where magnetic fields exist that are as strong as 1% of the Earth's field. This surface field might be left over from an earlier time when the Moon did have an overall field—a time when the Moon had a liquid core and rotated more rapidly. On the other hand, perhaps it results from a time in the past when the Moon was closer to the Earth and interacted with its magnetic field, thus magnetizing its own surface.

8.9

THE ORIGIN OF THE MOON

It is difficult to draw firm conclusions about the Moon's origin, because thus far no primeval lunar material has been found. Nevertheless, the question of lunar origin is

an important one and several basic classes of hypotheses have been proposed. Remember, a hypothesis becomes a viable theory when it is consistent with the wealth of data we have.

One group of hypotheses could be called the **fission hypotheses**, because they propose that the material that was to form the Moon was somehow thrown off the rapidly spinning young Earth (**Figure 8–32**). The first fission hypothesis was proposed by the astronomer George Darwin, the son of Charles Darwin, in 1879. Evidence in favor of these theories included the fact that the density of the Moon—about 3.3 grams per cubic centimeter—is close to the density of the outer layers of the Earth. An additional argument was the superficial observation that the Moon might just fit into the space where the Pacific Ocean is now.

There are, however, four good reasons to suspect the fission hypothesis. One is that material thrown from the Earth would probably not have remained in orbit around the Earth but would have fallen back to its surface. The second is that the fission hypothesis would have formed the Moon in the Earth's equatorial plane, not as close to the ecliptic plane as it is today, and calculations show that a transition from one plane to another is not possible. Third, due to continental drift, the present shape and distribution of the Earth's continents is vastly different from what it was at the time of lunar formation. Continental drift clearly indicates that the hole that is now the Pacific Ocean was not created by something being torn away from the Earth.

Inquiry 8–9 There is a fourth reason why the fission hypothesis is weak. How does the anomalous composition of the KREEP material on the Moon help discredit the fission hypothesis?

A second hypothesis explaining the Moon's origin suggests that the Moon was formed somewhere else in the solar system and was later captured by the Earth. Although not impossible, this type of capture is highly improbable. Furthermore, because capture requires the incoming body to lose kinetic energy, the assistance of a third planet-sized body is required to carry away the lost energy.

Inquiry 8–10 Would the existence of KREEP material on the Moon tend to support or discredit the capture hypothesis? What about the abundances of the oxygen isotopes?

The third major class of hypotheses is a group of **accretion hypotheses**. According to these, the Moon was formed at about the same time as the Earth out of material that collected in orbit around the Earth. **Figure 8–33** illustrates this process.

The straightforward accretion hypothesis runs into the same difficulty that the fission hypothesis did—explanation of the KREEP material—because if the Moon and Earth were formed at the same time out of the same material, they should have the same composition. However, it is possible that the Earth formed first. Slightly later the Moon may have formed from small chunks of material that came from elsewhere in the solar system to be captured into orbits around the Earth, where they accreted into the Moon. Such material would have had to have been at a higher temperature than the Earth to explain the Moon's lack of volatile elements. Unlike the improbable capture required by the capture hypothesis, the collection of debris into a ring around the Earth is a likely occurrence. Nevertheless, scientists were discouraged and puzzled for many years when the detailed knowledge of lunar chemical composition determined from rocks returned by the Apollo expeditions did not clear up the question of the origin of the Moon.

In recent years, a new line of thought has been gaining an increasing amount of acceptance among astronomers; originally proposed by William Hartmann of the Planetary Science Institute, it has come to be called the **giant impact theory**. It proposes that the Earth

suffered a major impact during its earliest stages (over 4 billion years ago) when the molten planet was still cooling and developing a surface crust. Recall that the early solar system was probably packed with planetesimals, so that collisions would be more likely at that time. The theory envisions a collision with a body perhaps the size of the planet Mars.

Computer simulations of such a collision (**Figure 8–34**) indicate that the surfaces of the two planets would probably have been vaporized in the collision, and that material would squirt outward in a high-pressure jet with temperatures of thousands of degrees. The simulations further show that some of this material might be able to re-form in Earth orbit, eventually coalescing into the Moon. Thus the Moon would be composed partly of Earth material and partly of material from the colliding body.

This theory has been highly successful in explaining the details of the Moon's chemical composition. For example, a serious problem with all previous origin hypotheses has been the distinct deficiency in lunar iron (when compared to the Earth's chemical composition). This "anemia" is easily explained by the giant impact theory, because the lunar material would have come primarily from the silicon-rich/iron-poor crustal portions of the two colliding objects, whereas the Earth's iron supply is primarily locked up deep in its central regions. Another anomaly of lunar composition that has been hard to explain is a shortage of water, sodium, and other volatile materials (relative to the Earth), but these are precisely the types of substances that would boil away in the vaporization process.

This theory may even explain some long-standing puzzles about the Earth's composition. For example, it has been difficult to explain why the Earth has so much gold and platinum in its surface layers. Typically, such heavy elements should have sunk toward the Earth's core. Now it seems possible that these elements came from the material in the body that collided with the Earth.

Table 8–4 summarizes the various classes of lunar formation theories by assigning a grade to each of the factors that workable theories must explain. Lunar mass is a factor because the Moon is larger relative to its parent planet than any other satellite in the solar system (with the exception of Pluto's moon). Angular momentum is a factor because the Earth-Moon system appears to have more angular momentum than expected from a comparison with other planetary systems. The importance of the depletion of the volatile elements and iron and the equality of the oxygen isotopes were discussed earlier. A viable theory must account for the overall melting of the lunar surface. Lastly, formation mechanisms must be physically plausible.

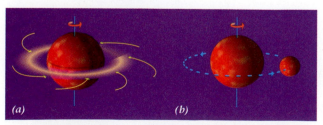

FIGURE 8–33. Stages in the formation of the Moon by accretion. (*a*) Earth gathers a ring of material that (*b*) collects to form the Moon.

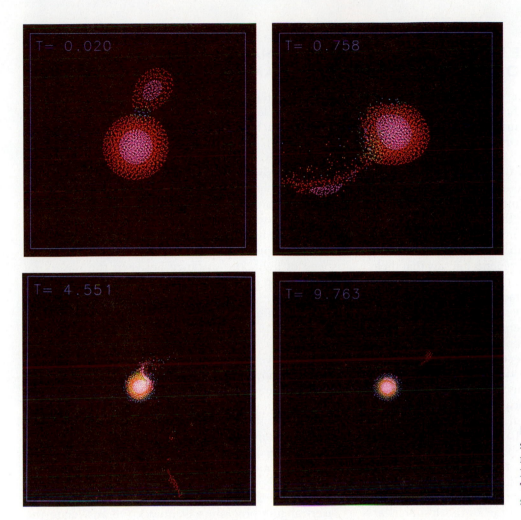

FIGURE 8–34. A computer simulation of the giant impact model. An object one-tenth Earth's mass strikes at 11 km/sec. The time covered by the simulation is 9.763 hours.

TABLE 8-4

A Comparison of Lunar Formation Theories

Factor[a]	Fission	Capture	Accretion	Giant Impact
Lunar mass	D	B	B	B
Earth-moon angular momentum	F	C	F	B
Lack of volatiles	B	C	C	B
Lack of iron	A	F	D	B[b]
Equality of oxygen isotopes	A	B	A	B
Lava covering entire surface	A	D	C	A
Physical plausibility	F	D−	C	Incomplete

After John Wood, in *Origin of the Moon,* edited by W. K. Hartmann, R. J. Phillips, and G. J. Taylor (Houston: Lunar & Planetary Institute). Updated grades in the final column were supplied by John Wood.

[a]The factor in the first column represents the observation that each theory must successfully account for to be considered valid.

[b]This grade depends on the assumption that the Earth's iron core had already formed.

8.10

THE TIDES AND THE FUTURE OF THE EARTH-MOON SYSTEM

As the Earth turns beneath the tidal bulges, it attempts to drag them along with it. This produces a substantial amount of friction, which in turn tends to retard the rotation of the Earth. As the Earth's rotation slows, the day grows longer and longer. The effect is small—the day only lengthens by about 0.0016 of a second each century—but over the course of time this effect accumulates and becomes important.

The first persons to notice the slowing of the Earth were astronomers trying to compare records of ancient solar eclipses with their predictions of those eclipses. They found substantial disagreements but saw that their predictions could easily be explained if they assumed that the Earth rotated more rapidly in the past than it does today. These results have since been confirmed by counting growth rings in ancient corals several hundred million years old. Corals add one growth ring each day, and the width of the ring depends on such things as water temperature, which in turn varies with the season. Ancient corals have been found with over 400 growth rings per year. This means that there must have been over 400 days in a year at one time, implying that the Earth was rotating about 10% faster on its axis than it does today.

If the Earth is rotating on its axis more and more slowly with time, it must be losing angular momentum. But the angular momentum of the Earth-Moon *system* is conserved. The Moon picks up the Earth's lost angular momentum, with the result that its orbit is slowly increasing in size. The reason the Moon picks up the angular momentum lost by the Earth is that just as the Moon exerts a force on the Earth's tidal bulge, the tidal bulge exerts an *equal and opposite* force on the Moon, which tends to accelerate the Moon forward in its orbit.

As a result of the slow increase in the size of the Moon's orbit—only a few centimeters per year—there eventually will come a time when the angular size of the Moon will be too small to completely cover the Sun's face. At this time, Earth will lose the spectacular phenomenon of the total solar eclipse. This is a little sad, for there are few things in nature more beautiful and awe-inspiring than a total eclipse of the Sun.

Eventually the Earth's rotation will slow to the point that it always keeps one face toward the Moon, just as the Moon has been slowed by Earth tides so that it keeps one face toward us.

8.11

THE LUNAR LASER-RANGING PROGRAM

One of the items left by the astronauts on the Moon has made possible an important series of experiments that have far-reaching results. A **retroreflector** (visible in Figure 8–28), is a device similar to a bicycle reflector that has the ability to reflect the light that falls on it directly back to the source from which it came.

Several observatories around the world have been beaming short pulses of laser light at several retroreflectors on the Moon and then detecting the returning pulse of light. The time it takes for the round trip is measured with great precision, and, from the known constant speed of light, the distance between the observing station and the Moon can be calculated. The round-trip time of about 2.5 seconds can be measured to a fraction of a *billionth* of a second, so the error in measurement of the distance to the Moon is small—only a few centimeters.

What have we learned from this program? For one thing, we now know the orbit of the Moon with far greater accuracy than was possible before. This in itself is of great value, but it may also make possible the measurement of some extremely small natural effects that are of great importance. For example, physicists have always assumed that the gravitational constant, *G*, does not change with time, but if it were slowly decreasing as the universe aged, there could be profound effects on the way the universe evolved. It may be that the lunar laser-ranging program will unlock the answer to the question of the variability of gravity with time. Already the program has led to an important confirmation of Einstein's theory of general relativity by showing that a certain motion predicted by some theories of gravity, but not by Einstein's theory, is not present in the motion of the Moon.

Finally, it is now possible to use the Moon as a standard reference point for mapping the Earth. By measuring the distance to the Moon from two different points on the Earth's surface, scientists can measure the distances between the two observing stations to a high degree of accuracy. We then can measure precisely the motions of the Earth's crust by measuring changes over time in the locations of observing stations. There are some important applications of such knowledge: for example, earthquake prediction (from measurements taken along both sides of fault lines) and, most exciting, measurement of continental drift.

DISCOVERY 8–1
LUNAR SURFACE FEATURES, AS SEEN FROM EARTH

After completing this activity, you will be able to do the following:

- Describe different types of lunar surface features.
- Explain why less detail is seen at full Moon than at other phases.
- Determine the sizes of the largest and smallest dark maria and the largest and smallest craters observable from Earth.
- Offer a hypothesis for the formation of lunar rays.

> READERS HAVING A MAGNIFYING GLASS MAY WANT TO USE IT TO
> DISCOVER DETAILS IN THE PRINTS; READERS HAVING THE ACTIVITY
> KIT CAN USE THE LENSES CONTAINED THERE.

Figures 8–17 and 8–18 are photographs of the full Moon and the Moon's first quarter, respectively. Study the photographs carefully, using a magnifier if available, and then do the following Discovery Inquiries.

- **Discovery Inquiry 8–1a** List the different types of features you can see on these photographs—for example, craters, mountains, etc. Don't worry about giving them astronomically correct names; just call them something descriptive of what you see. You may find some kinds of features that don't even have names.

- **Discovery Inquiry 8–1b** Did you find any features in common between the two photographs? How can this be when they show the Moon at different positions in its orbit around the Earth? Draw diagrams of the relative positions of the Earth, Moon, and Sun (a) when the Moon is full and (b) when the Moon is seen at a quarter phase (half illuminated) from the Earth.

- **Discovery Inquiry 8–1c** Examine Figure 8–18 along the **terminator** (the line between the light and dark areas of the Moon). Why do you see more detail in the quarter Moon? Why are craters more easily seen? Use the diagrams you drew in the previous inquiry.

- **Discovery Inquiry 8–1d** Using the photograph in Figure 8–17, how many maria can you count on the side of the Moon facing the Earth? Consider two maria to be separate if each occupies a roughly circular area on the Moon, even if the two regions are connected. (For example, the Mare Serenitatis [Sea of Serenity] and the Mare Tranquilitatis [Sea of Tranquility] count as two distinct maria in Figure 8–17.) If the Moon is 3476 km in diameter, what are the sizes of the largest and smallest of the maria? (Hint: The Moon's diameter is about 3476 km, which corresponds to about 3.6 inches in the picture. Therefore, 1 inch = 695 km and 1 mm = 38 km.)

- **Discovery Inquiry 8–1e** Note the rayed crater Tycho marked in Figure 8–17. The rays extend from it in all directions to great distances from the crater. What could have done this, and why does the material of the rays appear so much lighter than its surroundings?

☐ **Discovery Inquiry 8–1f** What are the diameters of the largest and smallest craters you can see in Figure 8–17 and 8–18? Would you expect there to be smaller craters on the Moon than you can see in these diagrams?

☐ **Discovery Inquiry 8–1g** Do the maria appear to be at the same elevation as the surrounding light-colored material? If not, do they appear to be higher or lower?

CHAPTER SUMMARY

- Table 8–5 presents a general comparison of the properties of the Earth and Moon. Most of the information in this chapter involves observation of well-known phenomena and some inferences from them.

OBSERVATIONS

- The observed average density of the Earth (5.5 gm/cm³), when combined with the density of surface rocks (3.3 gm/cm³) leads to the conclusion that the Earth's core contains dense materials such as iron. The surface chemical composition is primarily silicon and oxygen, with various amounts of other elements grouped together into a variety of minerals. The surface of the Earth is constantly changing due to erosion by water, wind, glaciers, and continental drift.

- The three types of rocks are **igneous**, **sedimentary**, and **metamorphic**. Igneous rocks result from cooling of molten material; sedimentary result from the deposition of material by wind and water; metamorphic rocks result when the effects of temperature and pressure change the characteristics of a pre-existing rock.

- The difference in the strength of the Moon's gravity on different sides of the Earth produces ocean tides. These tides are causing the Earth's rotation to slow and the day to lengthen. Similarly, the Earth produces tides on the Moon, whose distance is slowly increasing due to conservation of angular momentum.

- The Earth's atmosphere contains nitrogen and oxygen with traces of argon, carbon dioxide, and water vapor. The carbon dioxide and water vapor cause the greenhouse effect to be present. The atmosphere comes from outgassing during volcanic emissions. Atmospheric circulation is governed by a combination of heat from the oceans and continents, and movement caused by the Coriolis effect.

- The Earth has a magnetic field that is thought to be formed by rotation and convection in the molten outer core in a process known as the **dynamo effect**. For this reason, the Earth is surrounded by a **magnetosphere**. Charged particles from the Sun, which impinge on the magnetosphere, have their paths changed so that they spiral around the magnetic field into the polar regions, where they produce **aurorae**.

- **Occultations** of stars by the Moon show it to lack an atmosphere.

- Surface features on the Moon include **craters**, **rays**, **maria**, **rilles**, **domes**, and **highlands**. Rays result from material ejected during crater formation from impact. Some rilles result from surface movement, while others, the **sinuous rilles**, result from lava flowing in channels and tubes. Domes result from volcanism.

- The lunar surface is entirely covered with **basalts** in the form of rocks and a powdery **regolith**. In addition, the rocks contain **breccias** and **KREEP**. Lunar rocks are deficient in water molecules and other volatile elements, and also deficient in iron relative to the Earth. However, the relative amounts of various oxygen **isotopes** are the same on the Moon as on the Earth.

THEORY

- **Fission hypotheses** for the Moon's formation are physically implausible; they predict that the Moon would be in the Earth's equatorial plane, and that the Moon should be chemically identical to the Earth. **Capture hypotheses** do not require the Earth and Moon to be chemically identical. While not impossible, capture is unlikely. **Accretion hypotheses** predicts strong similarities between the Earth and Moon. The **giant impact theory** of lunar formation involves the impact of a Mars-sized body on the forming Earth; material ejected into orbit later

TABLE 8-5

The Earth and Moon: A Comparison

	Earth	Moon
Orbital data		
Orbit shape	nearly circular	nearly circular
Orbit inclination to ecliptic (°)	0	5.1
Rotation period	24 hours	27.32 days
Axial tilt (°)	23.5	6.7
Orbital period	$365^d6^h8^m24^s$	$27^d7^h43^m12^s$
Physical data		
Diameter (km; Earth units)	12,756	3,476; 0.273
Mass (gm; Earth units)	5.974×10^{27}	7.35×10^{25}; 0.0123
Average density (gm/cm^3)	5.518	3.34
Surface gravity (cm/sec^2; Earth units)	980	162; 0.165
Escape speed (km/sec)	11.2	2.4
Temperature (surface)	200 to 300 K	100 to 400 K
Surface features		
Craters	few	many
Impact basins	perhaps one	many
Continental-sized areas	yes	no
Mountains	yes	yes
Ancient lava flows	?	yes
Active volcanism	yes	no
Plate tectonics	yes	no
Surface composition	silicates	silicates; iron deficient; deficient in volatile elements
Other characteristics		
Atmosphere	nitrogen, oxygen	none
Magnetic field	yes	no
Differentiated core	yes	no?

coalesced into the Moon. The theory predicts that the Moon would have formed from material that is poor in both volatile elements and iron.

CONCLUSIONS

- From seismic studies, we infer that the Earth has a solid inner core that is surrounded by a liquid outer core. The lower-density mantle exhibits convection in which hot, low-density material rises and cools, and higher-density material descends.
- From a variety of data scientists conclude that Earth's continents were once together in a supercontinent we call **Pangaea**. Thereafter, the land masses separated due to **continental drift**, in which lower-density continental plates float on the higher-density

mantle underneath. In this model, mountains form when plates collide.
- Most lunar craters result from meteoroid impact, although some are volcanic in origin. Maria result from lava flows.
- Younger features are on top of older ones. Absolute ages can be obtained using the methods of radioactive dating, which determine when material last solidified. From such techniques we deduce that the lunar highlands were produced 4.1 billion years ago, and the maria 3.1 to 3.9 billion years ago.
- The Moon's interior contains a core that may be partially melted. The lack of a liquid core along with slow rotation account for the Moon's lack of an overall magnetic field. However, the entire lunar surface was molten at a time early in its history.

SUMMARY QUESTIONS

1. What is the Earth's distance from the Sun? The Earth's diameter? Its average density?

2. What is the Moon's distance from the Earth? The Moon's approximate diameter relative to the Earth? Its average density?

3. What are the meaning and importance of the concepts of the acceleration of gravity and the escape velocity?

4. How do seismic data give us information about the interiors of the Earth and Moon? What were the results of the lunar seismology experiments?

5. What are the various layers of the Earth's interior and their important properties?

6. What is continental drift and what produces it? How is volcanism related to it?

7. What chemical elements make up the Earth's atmosphere? What is the percentage of each?

8. What is the greenhouse effect, and how is it produced?

9. Where did the Earth's current atmosphere come from? Why do scientists believe it is not the same as the original atmosphere?

10. What is the Earth's magnetosphere? Describe its appearance. What effect does the magnetosphere have on charged particles in its vicinity, and what is its role in producing an aurora?

11. How did astronomers know the Moon lacked an atmosphere before NASA sent spacecraft there?

12. What are the characteristics of the various types of lunar surface features? Describe how these features formed and present evidence in favor of the formation mechanism. How do the appearances of the lunar near and far sides differ?

13. What basic types of rock are found on the Moon? What are the implications of these rock types for the origin and history of the Moon?

14. What is the probable scenario for the early history of the Moon's surface as determined from the evidence provided by radioactive dating?

15. What are the four main ideas of the Moon's origin? Describe them and give the evidence for and against each hypothesis. Which one is most likely to be valid? Explain why.

16. Why does the Moon produce two tidal bulges on the Earth? What are the long-term effects of tides on the Earth-Moon system?

APPLYING YOUR KNOWLEDGE

1. Compare and contrast the processes that modify the surfaces of the Earth and the Moon. Consider separately processes that occur internally as well as those involving externally produced events.

2. Apply the idea that theories cannot be proven true, but only falsified, to the theories of lunar formation discussed in the chapter. What is your conclusion?

3. Because the Earth's atmosphere consists of 78% nitrogen (N_2), why does the atmosphere not have substantial amounts of ammonia (NH_3), as does Jupiter?

4. Consider each rock type. If specimens of each type were found on a given planetary body, what would you be able to conclude about conditions and processes on and within that body?

5. How might the Earth have differed if it had been 0.7 AU from the Sun rather than where it is now?

6. The Earth's tidal force acting on the Moon has slowed the Moon's rotation to the point where it is synchronous (the same face always faces the Earth). Given that the Moon also exerts a tidal force on the Earth, why is the Earth not in synchronous rotation?

7. Describe a baseball game being played on the Moon. Include discussions of the distance the ball might be hit, the strides of the base runners, the ease or difficulty of fielding the ball, the likelihood of a rain check or calling the game due to darkness, and how the fans might boo the umpire.

■ 8. Suppose you measured g to be 1000 cm/sec². What, then, is the mass of the Earth?

■ 9. Compute the escape velocities of the Earth and Moon. How much larger is the Earth's escape velocity? (Use data for the Moon from Table 8–3.)

■ 10. Because the uncertainty in the time for a laser beam to reach the Moon and back is about a billionth of a second, compute the uncertainty in the distance by finding how far light travels in a billionth of a second.

■ 11. Using Figure 8–17, determine the diameter of Mare Serenitatis, the Sea of Serenity.

■ 12. What would be the period of a satellite in orbit about the Earth if the satellite were 385,000 km from it?

■ 13. Assuming the South American and African plates have always moved with the same speed they are moving today, compute how long ago the continental land masses were joined. (Make a reasonable estimate of their current separation.)

■ **14.** Explain how you might go about computing the height of the central peak shown in Figure 8–19. What information would you need to have?

ANSWERS TO INQUIRIES

8-1. The interior density must be much greater than 5.5 g/cm³.

8-2. Sediments, which are carried by water, wind, and ice, are located near the surface and will be deposited there, not below the surface. Furthermore, the churning regions below the surface would destroy the soft sediments.

8-3. The pressures are higher in the central regions than they are farther out. According to Figure 8–8, the rock would therefore be solid. However, away from the center, even if the temperature is about as high, the lower pressure would cause it to be a liquid. (We are neglecting any effects of differences in chemical composition.)

8-4. Two high tides and two low tides, because there is a bulge on either side of the Earth and a point on the Earth's surface will pass beneath each one during any 24-hour period.

8-5. It will change due to the Moon's daily eastward movement in its orbit relative to the Sun. As discussed in Chapter 4, the Moon moves eastward some 13° per day;

this corresponds to a delay in the rise time of about 53 minutes each night. The tides will be about an hour later every day.

8-6. Some questions you might think of include: What is the composition of the lunar rocks? Their ages? Are they volcanic in origin? Is there evidence of water either now or in the past? Does the Moon have a magnetic field like the Earth? Does it have seismic activity?

8-7. They must have formed from the same primordial material. Had they been different, the conclusion would be the opposite.

8-8. A thick crust, like that on the lunar far side, is harder to crack than a thin crust, like that on the near side. Once cracked, lava is then free to flow and fill the large basins on the side facing the Earth.

8-9. If the Moon were made of terrestrial material, it would have the same composition as the Earth.

8-10. While the presence of KREEP could support the capture hypothesis, the equality of the various oxygen isotopes would not.

THE EARTH-LIKE PLANETS

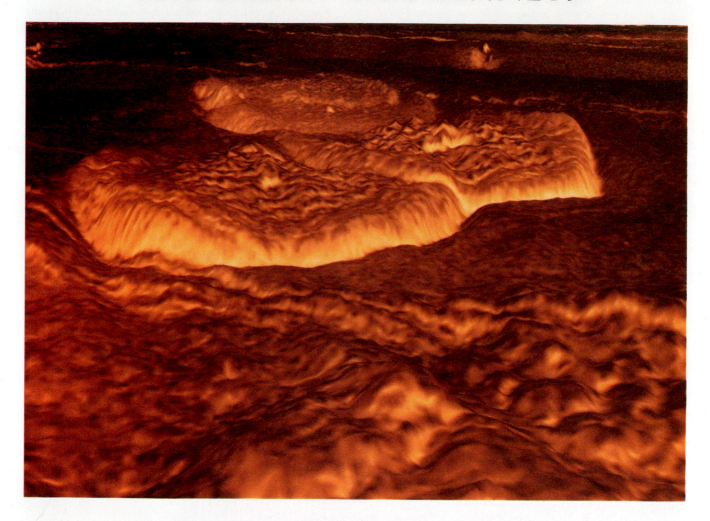

This planet [the Earth] is the cradle of the human mind, but one cannot spend all one's life in a cradle.

KONSTANTIN TSIOLKOVSKY, A RUSSIAN ROCKET AND SPACE PIONEER

A time would come when Men should be able to stretch out their Eyes ... they should see the Planets like our Earth.

CHRISTOPHER WREN, INAUGURATION SPEECH, GRESHAM COLLEGE, 1657

The four planets closest to the Sun exhibit a variety of similarities, including their small size, small mass, high density, and similar chemical composition. Due to these likenesses, astronomers group Mercury, Venus, Earth, and Mars together as the **terrestrial**, or Earth-like, planets. In this chapter, we study the terrestrial planets by comparing and contrasting them with each other.[1] We use Earth as the standard of comparison, simply because we know more about it.

Though our discussion emphasizes comparisons among the terrestrial planets, you should also keep in mind that each planet is a separate and unique world. You will gain great enjoyment through knowing each of these worlds as individuals and understanding why they are the way they are.

9.1

INTRODUCTION TO THE TERRESTRIAL PLANETS

Our study of the terrestrial planets relies on the understanding we have gained of the processes occurring on the Earth and Moon and the observable consequences of these processes. We begin with a general discussion of our knowledge of these bodies gleaned from ground-based observations and the early days of planetary exploration by spacecraft.

EARLY GROUND-BASED STUDIES

The Earth-like planets, Mercury, Venus, and Mars, are shown side-by-side in **Figure 9–1**. Each presents astronomers with its own set of impediments that make Earth-based study difficult. Mercury, for example, is small and never strays far from the Sun. Telescopic observations of it found only indistinct and hazy surface markings. By following their apparent motion, astronomers concluded (incorrectly, as we will see) that Mercury rotated on its axis once every 88 days. Because this rotation period was the same as Mercury's orbital period around the Sun, astronomers inferred that the planet always kept the same face toward the Sun, just as the Moon keeps the same face toward the Earth. The reason in both cases is the same—strong tidal forces. Figure 9–1a is as good an Earth-based photograph of Mercury as we can get.

[1]For detailed information on each of the planets, refer to Table 9–1, and Appendix C.

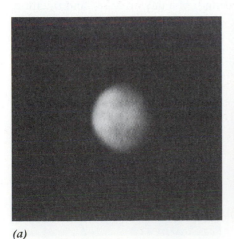  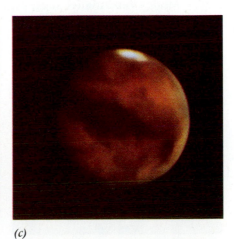

(a)   *(b)*   *(c)*

FIGURE 9–1. Earth-based telescopic views of (*a*) Mercury, (*b*) Venus, and (*c*) Mars.

FIVE PHASES OF VENUS

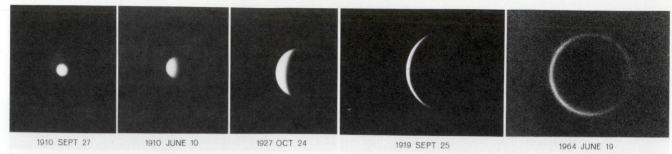

| 1910 SEPT 27 | 1910 JUNE 10 | 1927 OCT 24 | 1919 SEPT 25 | 1964 JUNE 19 |

FIGURE 9–2. Venus, as seen from Earth, shows only a featureless layer of clouds.

While Venus had been observed by Galileo to have phases, no further details could be seen. Like Mercury, it never appeared far from the Sun. However, some indistinct markings observed by numerous observers provided information about an apparent 225-day rotation period. Here too, the rotation was thought (incorrectly) to be the same as the period of orbital revolution. The visual appearance of Venus is indistinct to us because it is perpetually covered by a thick layer of clouds floating in its CO_2 atmosphere (Figure 9–1b). For this reason, the observed rotation period referred to the observed clouds, not the planetary surface itself. Venus's varying phases, which are similar to those of the Moon, are shown in **Figure 9–2**.

While at one time Venus was thought to be a benign place for life to exist, Earth-based studies in the 1960s began to paint a rather unpleasant picture, which included a thick carbon dioxide atmosphere with a surface pressure almost 100 times that of the Earth, a surface hot enough to melt lead, and howling winds. Studies revealed a minute amount of water vapor, suggesting a dry, desertlike terrain—a Death Valley carried to the extreme. More recent studies suggest that the clouds on Venus consist of high concentrations of corrosive compounds such as sulfuric and hydrofluoric acid. To top it off, the winds in the upper atmosphere of Venus can reach 200 miles per hour! A space probe would have a hard time surviving these extreme conditions.

Mars has the liveliest history of study and debate of all the planets. Although Venus is closer to the Earth in mass and distance from the Sun, Mars has provided a greater fascination to the human mind ever since telescopes were pointed at it. When flights of fancy considered the subject of life elsewhere in the universe, the place that usually came to mind was Mars. Telescopic observations of the Martian surface began with Galileo. Observations ever since have provided data from which many speculations have been made. As with all the terrestrial planets, the quality of data about Mars was limited by the planet's small angular size, and turbulence within the Earth's atmosphere. Telescopic observations did show both daily changes due to Mars' rotation and seasonal changes in the colors of certain regions (**Figure 9–3**).

Figure 9–3 shows some of the better photographs of Mars taken from Earth. From such photographs we learned that the length of a day on Mars, and the tilt of

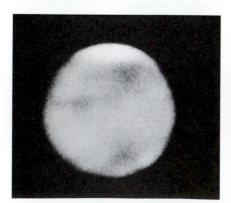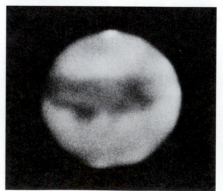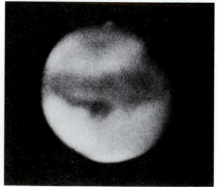

FIGURE 9–3. Several views of Mars as seen from Earth. The photographs are taken during different Martian seasons.

its rotation axis, are close to those of Earth. The axial tilt produces seasons, which last twice as long as earthly seasons, owing to the greater length of the Martian year.

Inquiry 9–1 Can you detect evidence for seasonal changes in the photographs of Figure 9–3 ? What do you find?

One of the earliest explanations of the observed seasonal changes in Mars' coloration from light to dark was that they might be caused by the presence of vegetation. This suggestion was a strong encouragement to the space program to study Mars intensively. It also led to a more speculative school of thought that quickly caught the public's fancy—the notion that Mars harbored intelligent life. This line of thinking had its origin in nineteenth-century observations by the Italian astronomer Giovanni Schiaparelli, who saw the Martian surface as being crisscrossed by an extensive network of straight lines that he called *canali*. Thus began a controversy that lasted until the era of space exploration. The U. S. astronomer Percival Lowell observed similar features on Mars about 1896 and speculated freely about their origin. Although the Italian word *canali* is somewhat noncommittal and can be employed to describe almost any sort of channels or ditches, Lowell interpreted the markings quite literally as *canals*; that is, artificial ditches constructed by intelligent creatures to bring water from the melting polar caps to the presumably populated temperate regions of Mars. But while Lowell continued to see his "canals," other observers did not, and their reality was debated for over half a century.

Martian canals were never photographed, which should be no surprise, because during the time required to make a time exposure Earth's turbulent atmosphere erases fine details. The eye, on the other hand, is a highly sensitive instrument and can see details that are visible only in those rare and fleeting moments when the atmosphere is exceptionally calm.

The hypothesis of intelligent life on Mars received something of a setback when astronomers discovered that its atmosphere was predominantly carbon dioxide, as on Venus. Furthermore, Mars' atmospheric pressure is very low, about one one-hundredth that of the Earth. Observations also revealed only the barest traces of water vapor on the planet, and this fact argued against the existence of plant life on Mars.

THE BEGINNING OF PLANETARY SPACE EXPLORATION

In 1974, the Mariner 10 space probe[2] arrived in the vicinity of Mercury after receiving a gravitational boost on its way as it passed close to Venus (**Figure 9–4**). Employing one planet's gravity to send a probe farther into space than its own rocket could propel it is a commonly used technique. An additional bonus in the case of Mariner 10 was the fact that the spacecraft was given an orbital period of 176 days, just twice Mercury's orbital period. As a consequence, we were able to obtain images[3] when the vehicle passed close to the planet on three separate occasions in 1974 and 1975.

In 1978 the Pioneer-Venus 1 and Pioneer-Venus 2 probes of Venus added considerably to our knowledge of this curious planet. Excellent photographs taken in ultraviolet light (where the view penetrates deeper into the atmosphere) showed more distinct features than Earth-based pictures, although the surface of the planet was still not seen directly. Pioneer featured a "mother ship" and five separate probes that parachuted into the atmosphere of Venus. While floating toward the surface these probes sent back information about temperature, pressure, and chemical composition. Each transmitted as long as its instruments could survive in the planet's extreme heat. The orbiter continuously sent radar mapping information about the surface until it crash-landed in mid-1992.

Hoping to resolve some of the perennial questions about Mars, the United States launched a number of successful spacecraft visits. Controversy about Mars still raged in 1965 when Mariner 4 passed a few thousand

---

[2]Because not all the spacecraft that have been sent to the planets will be discussed, they will not appear in an obvious numerical order. For example, while Mariner 2 was the first probe to Venus and Mariner 4 was the first to Mars, their results have been superseded by more recent probes.

[3]We use the term "image" rather than "photograph" due to the process used in obtaining it. Because the photographic process is not used here, the term "photograph" would be nonsensical.

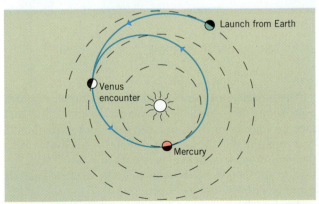

FIGURE 9–4. The Mariner 10 spacecraft used the gravity of Venus to change the spacecraft's orbit and send it on to Mercury.

(a)

(b)

(c)

(d)

FIGURE 9–5. The terrestrial planets as seen by spacecraft. (*a*) Mercury (*b*) Venus (*c*) Earth (*d*) Mars.

miles above the planet and sent back 22 photographs of the surface.

The 1971 Mariner 9 spacecraft took a series of over 7000 magnificent photographs of Mars. Many were of regions that had not been examined before, and Mariner revealed unexpected volcanoes, valleys, craters, and channels that considerably revived interest in the planet, while dramatically increasing our knowledge.

The objective of the 1976 Viking mission was to land a probe on the surface of the planet, examine the surface geology from close up, and carry out experiments that might detect the presence of life-forms in the Martian soil. After some excitement about a possible detection of life processes, biologists upon further analysis concluded that none had been found.

Spacecraft views of each of the terrestrial planets are shown in **Figure 9–5.**

9.2

LARGE-SCALE CHARACTERISTICS

We now look at planetary properties that apply to each body as a whole, including orbital characteristics, rotation, size, mass, density, and seasons.

ORBITS

	Mercury	Venus	Earth	Mars
Orbit eccentricity	0.206	0.007	0.017	0.093
Orbit inclination (°)	7	3	0	2

The elliptical orbits of the terrestrial planets are nearly circular. Mercury's orbit differs from circularity the most, and Venus's orbit is the most nearly circular of all. Earth's orbital plane, by definition, is the ecliptic plane; it defines what we view as the fundamental geometric plane of the solar system. The 7° tilt of Mercury's orbital plane to the ecliptic is larger than that of any planet but Pluto (review Figure 6–4). All these planets orbit the Sun in a counterclockwise direction as observed from above Earth's North Pole.

ROTATION

	Mercury	Venus	Earth	Mars
Rotation period	58.65 days	243 days retrograde	23.9 hours	24.6 hours

In 1965 astronomers used Earth-based telescopes to bounce radar signals off Mercury. In analyzing the structure of the returning pulses, they found to their surprise that the rotation period of Mercury was only 59 days rather than the 88 days that had been thought. Mercury did *not* present the same face toward the Sun at all times as had been thought!

This interesting result provides a good lesson for those who argue that the scientific method leads inexorably to the truth. In actual fact, the path is often quite twisted, with numerous dead ends. Here was a result that had been established by both theoretical and observational evidence and had been accepted for decades, yet it was suddenly found to be completely wrong. There had been clues that something was wrong with the hypothesis of an 88-day rotation period, because measurements of temperature of Mercury's dark side showed it to be unusually high. While suggestions were made to explain the anomalously warm dark side of Mercury, none were particularly successful.

Inquiry 9–2 How would a rotation period of 59 rather than 88 days provide an explanation of the dark side being warm?

How could observers have been so wrong? On Earth, we would see pretty much the same apparent movement of surface features whether Mercury has a 59-day or an 88-day spin. This is so because we see the planet only when it is far from the Sun, at the extremities of its orbit. The ratio of the orbital period to the rotational period, 88/59, which is almost exactly 3/2, would give us the same view of the surface with either rotation period. Because we were unable to observe Mercury continuously, we had a "stroboscopic" effect in our data, analogous to the effect that causes the wagon wheels in the western movies to appear to rotate in a backward direction. The 3/2 ratio of orbital to spin period results in the solar "day" on Mercury (the time for the Sun to go from overhead to overhead) lasting two Mercurian years! This situation is shown in **Figure 9–6**, which shows Mercury starting at perihelion (closest to the Sun) with a feature (indicated by the arrow) pointing toward the Sun at point *1*. After one rotation of the planet relative to the Sun, its orbital position is at point *2*, and so on.

Why does Mercury rotate three times for every two orbits? Just as gravitational tidal effects with the Moon forced its rotation to become equal to its revolution, Mercury has gravitational tidal effects with the Sun that produce the 3:2 rotation to revolution relationship.

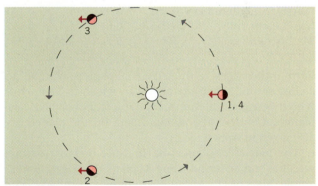

FIGURE 9–6. The rotation and revolution of Mercury. The arrow, which represents a surface feature, points toward the Sun when Mercury is at its closest approach to the Sun, at point 1. One rotation later, Mercury is at point 2 in its orbit, 2/3 of the way around its orbit; one further rotation later, it is point 3 in its orbit. After 1 more rotation, completing three Mercurian rotations relative to the stars, Mercury returns to its original position in its orbit.

Inquiry 9–3 Use Figure 9–6 to determine how many full orbits Mercury completes during three rotations starting from point *1* through point *4*. To which point in its orbit has it returned?

To study Venus's surface rotation, astronomers had to find some way to penetrate the thick clouds. This was first done in the 1960s when scientists started bouncing radar beams from Earth-based telescopes off the planet. From such radar studies, astronomers made two surprising discoveries about Venus: the planet was observed to rotate clockwise, or **retrograde**, relative to the other terrestrial planets; and it rotates at the remarkably slow rate of once every 243 days. This slow rotation means that solar heat is not evenly distributed around Venus's atmosphere, and thus the side of the atmosphere facing the Sun might be expected to be hot. However, unequal heating is expected to cause high winds (as explained in Chapter 8) that would transport heat around the planet and thereby moderate the temperature distribution. The end result is that while the temperature is highest directly under the Sun, it does not vary greatly around the planet.

Observations of the surface of Mars show it to have an Earth-like rotation of just over 24 hours. Thus Mercury and Venus both rotate considerably more slowly than Earth and Mars.

BASIC PROPERTIES: SIZE, MASS, DENSITY

	Mercury	Venus	Earth	Mars
Diameter (Earth units)	0.381	0.951	1	0.531
Mass (Earth units)	0.0558	0.815	1	0.107
Density (gm/cm³)	5.5	5.3	5.518	3.96

Earth is the largest of the terrestrial planets, with Venus a close second. Their diameters are given in **Table 9–1**, which is a general summary table for the terrestrial planets. (These and additional data on the planets are also given in Appendix C).

The masses of planets having one or more natural satellites can be found using Kepler's third law. For example, Mars has two natural satellites, which have been studied to find a mass for Mars only 11% that of Earth. Although neither Mercury nor Venus have natural satellites, their masses have been measured by observing the effects of their gravity on passing or orbiting spacecraft. Mercury's mass was found to be only about 6% that of Earth, while Venus's was 82%.

When we combine planetary masses with their diameters, we can determine average densities. For comparison, remember that the Earth's density is 5.5 gm/cm³, while the Moon's is 3.3 gm/cm³. All the terrestrial planets have nearly the same density. Mercury's is almost exactly the same as the Earth's, with Venus's slightly lower at 5.3 gm/cm³. Mars has the lowest density at just under 4 gm/cm³. As we shall soon see, these basic numbers provide important information about the interior structure of each planet.

Once the mass and radius are known, we can also determine the strength of gravity at the surface, and the escape velocity. For Earth and Venus, this is about 10 km/sec, and only half that much for Mercury and Mars. The values for each of the terrestrial planets is given in Table 9–1.

SEASONS

	Mercury	Venus	Earth	Mars
Seasons	yes	no	yes	yes

Seasons, as we know, are produced by the variation in the concentration of solar heat throughout the year. The primary agent of change in the amount of solar heat is the angle at which the solar energy strikes the planet's surface. For the Earth, this angle comes from the 23.5° tilt of the Earth's axis to the perpendicular of its orbital plane. By contrast, Venus has only a 3° tilt and therefore has essentially no seasonal variations. Because Mars' axial tilt is only slightly greater than the Earth's, seasons similar to Earth's are expected. However, the temperature variations in Mars' southern hemisphere are greater than those in its northern hemisphere or those on Earth, due to Mars' larger orbital eccentricity. As shown in an exaggerated form in **Figure 9–7**, because Mars is closest to the Sun during summer in its southern hemisphere and farthest from the Sun during summer in its northern hemisphere, the summers are substantially warmer in the southern hemisphere than in the

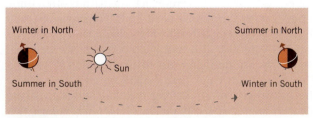

FIGURE 9–7. Seasons on Mars are particularly extreme for winter in the southern hemisphere due to Mars' greater distance from the Sun. (The ellipticity is exaggerated and not to scale.)

TABLE 9–1

The Terrestrial Planets: A Comparison

	Mercury	Venus	Earth	Mars
Orbital data				
Semi-major axis (AU)	0.387	0.723	1.000	1.524
Orbital period (years)	0.241	0.615	1.0	1.881
Orbit shape (eccentricity)	0.206	0.007	0.017	0.093
Orbit inclination (°)	7	3	0	2
Rotation period	59 days	243 days retrograde	23.9 hours	24.6 hours
Axial tilt (°)	28	3	23.5	24
Physical data				
Diameter (Earth units)	0.381	0.951	1	0.531
Mass (Earth units)	0.0558	0.815	1	0.107
Density (gm/cm³)	5.5	5.3	5.518	3.96
Surface gravity (Earth units)	0.38	0.90	1.00	0.38
Surface pressure (Earth units)	———	~90	1.0	~0.01
Escape velocity (km/sec)	4.3	10.3	11.2	5.0
Temperature (surface) (K)	100 to 700	730	200 to 300	130 to 290
Satellites	0	0	1	2
Surface features				
Craters	many	few	few	many
Impact basins	yes	?	perhaps one	yes
Continental-sized areas	no	yes	yes	Tharsis?
Mountains	no	yes	yes	yes
Ancient lava flows	yes	yes	?	yes
Active volcanism	no	perhaps	yes	no
High scarps	yes	no	no	no
Plate tectonics	no	probably not	yes	no
Surface winds	none	slight	yes	strong
Polar ice caps	perhaps (H_2O)	no	yes (H_2O)	yes (H_2O, CO_2)
Surface water	none	none	much	none
Surface composition	silicates?	silicates?	silicates	silicates, iron oxide
Other characteristics				
Atmosphere	very slight	carbon dioxide	nitrogen, oxygen	carbon dioxide
Atmospheric water	none	very little	<2% (variable)	very little
Atmospheric oxygen	none	none	≈20%	trace
Atmospheric carbon dioxide	none	≈95%	trace	≈95%
Atmospheric nitrogen	none	≈3%	≈78%	≈3%
Clouds	none	sulfuric acid	water	water, dust
Magnetic field	yes	no	yes	no
Differentiated core	yes	unknown	yes	no

northern. Likewise, Mars is farthest from the Sun in its southern hemisphere winter, producing colder winters there than in the northern hemisphere.

Inquiry 9–4 What would you expect in the way of seasons on Mercury?

9.3

ATMOSPHERES

The properties of planetary atmospheres that we wish to understand include their chemical compositions, temperatures, and pressures. Other aspects that we examine include the presence of clouds and lightning.

ATMOSPHERIC CHEMICAL COMPOSITION

	Mercury	Venus	Earth	Mars
Nitrogen	none	≈3%	≈78%	≈3%
Oxygen	none	none	≈20%	trace
Carbon dioxide	none	≈95%	trace	≈95%
Water	none	very little	<2% variable	very little

The Mariner 10 encounter with Mercury confirmed ground-based observations that the planet has almost no atmosphere. On the other hand, the spacecraft did find a thin atmosphere of surprising complexity. At its surface, Mercury's atmosphere has a density of only about 100 hydrogen atoms and 6000 helium atoms per cubic centimeter. Ground-based observations have shown that the largest constituent of Mercury's atmosphere is sodium, with a density of 30,000 atoms per cubic centimeter—some 10^{15} times less dense than the Earth's sea-level density. The hydrogen and helium undoubtedly come from atoms ejected by the Sun; the other elements are probably released from rocks during the heat of the Mercurian day.

Venus and Mars have similar atmospheric compositions, which have been known for many years to be about 95% carbon dioxide. Nitrogen contributes only about 3% of the gas in these atmospheres, while 78% in Earth's atmosphere. Oxygen is not present in Venus's atmosphere, and only a small amount is found on Mars. While water vapor is present only in trace amounts in the Venusian and Martian atmospheres, it amounts to about 2% in the Earth's atmosphere. Sulfur dioxide, SO_2, is an important trace element in Venus's atmosphere. These data are summarized in Table 9–1.

ATMOSPHERIC TEMPERATURE

Venus's lack of atmospheric oxygen and ozone causes the temperature in the atmosphere to decrease continuously up to the level of the clouds at 50 km (**Figure 9–8**). A similar situation holds true for Mars. The temperature of the Martian atmosphere rarely rises above the freezing temperature of water. The variations of temperature with height are dramatically different from that of Earth because of differences in the chemical makeup of the atmospheres of the two planets.

ATMOSPHERIC PRESSURE

	Mercury	Venus	Earth	Mars
Pressure relative to Earth	——	~90	1	~0.01

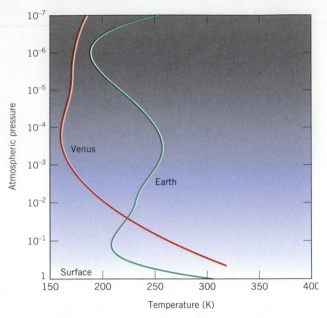

FIGURE 9–8. The variation of Venus's atmospheric pressure with decreasing atmospheric temperature above the surface. The Earth's variation is given for comparison. Pressure is given in units of the Earth's atmospheric pressure at sea level.

Spacecraft observations have found the atmospheric pressure at Venus's surface to be some 90 times greater than the Earth's atmospheric pressure. In other words, whereas the Earth's atmosphere exerts a force of nearly 15 pounds on each square inch, Venus's atmosphere exerts some 1300 pounds per square inch. This is the same as being 3000 feet (nearly 0.6 mile) under the ocean on Earth. Only specially designed research submarines are able to survive to such depths. Mars' average atmospheric pressure is about 100 times less than the pressure on Earth.

LIGHTNING

	Mercury	Venus	Earth	Mars
Presence of lightning	——	yes	yes	none observed

The 1978 Pioneer Venus atmospheric probes confirmed some results that had been suggested by two earlier Soviet Venera missions—that the planet's atmosphere exhibits intense lightning flashes (up to 25 per second) and even occasional crashes of thunder. In this way, it appears similar to Earth. On the other hand, no lightning or thunder has been reported for Mars.

Inquiry 9–5 In visible light, Venus appears to be much cooler than the temperature determined for its surface by radio astronomers. What is the probable reason for this?

CLOUDS

	Mercury	Venus	Earth	Mars
Clouds	none	sulfuric acid	water	water, dust

Those terrestrial planets having substantial atmospheres also have observable clouds. Earth's clouds, of course, consist of water in the form of vapor, droplets, and ices.

Venus's sulfuric-acid clouds are concentrated at three levels about 65, 53, and 48 km above the surface. Below 48 km, the atmosphere is clear of clouds.

Mars shows two different types of clouds, which can be observed from Earth. The white clouds consist of water vapor with some carbon dioxide. Such clouds have been observed by spacecraft to be present in low-lying areas in the early morning, and then to dissipate as the day warms. Extensive cloud systems form over the polar regions as the atmosphere begins to cool in the Martian late summer and early fall. The second type of clouds are yellowish in color; these arise from dust storms that are often so extensive that they cover the entire planetary surface. Such storms arise due to high-speed winds moving over the surface, sometimes well in excess of 100 miles per hour. Without such high-speeds, the thin atmosphere would not be able to lift the dust from the surface.

ATMOSPHERIC CIRCULATION

	Mercury	Venus	Earth	Mars
Winds at high elevation		<200 mph	jet stream 250 mph	
Winds at surface		average 2 mph	0–100 mph	<90 mph

Earth's atmospheric circulation is extremely complicated, as we have seen, because, in addition to energy received directly from the Sun and indirectly from the land masses, large amounts of energy are released by the oceans and absorbed by the atmosphere. Coriolis effects also complicate the atmospheric circulation on Earth.

Although Venus's clouds appear complex, their structure and motions are simple, because atmospheric circulation is determined only by the heating that occurs at the point directly under the Sun. These heated gases rise, then travel around the planet at high speeds (up to 200 miles per hour), until they cool and descend toward the surface in the cooler nighttime regions. Much to our surprise, these great wind speeds fall off to nearly zero at the surface.

Mars' atmospheric pressure may vary by as much as 30% throughout its year because of variations in solar heating. For this reason, high-speed winds of a hundred miles per hour sometimes move across the surface. These winds produce the Martian dust storms discussed previously.

WHY DO THE ATMOSPHERIC COMPOSITIONS DIFFER SO MUCH?

Given that Earth and Venus are nearly the same size and located at nearly the same distance from the Sun, why does Venus have so much carbon dioxide in its atmosphere while Earth has greater amounts of nitrogen, oxygen, and water vapor? How can we understand these substantial differences?

It seems clear that the present-day atmospheres of Venus and Earth are not the same as their original atmospheres (see Chapter 8). Those were blown away by the strong solar wind during the Sun's formative stages. It appears certain that the present atmospheres were formed from volcanic outgassing over geological time. Even today, volcanic outgassing is adding to the atmosphere of the Earth, and probably of Venus.

Among the primary constituents of the volcanic gases would have been carbon dioxide and water vapor, with smaller amounts of nitrogen and no oxygen to speak of. Surely, the original chemical composition of each terrestrial planet was nearly the same.

What processes might have occurred on Earth that would have decreased the amount of carbon dioxide in its atmosphere? As discussed in Chapter 8, over geological history, atmospheric carbon dioxide, catalyzed by liquid water, combined with silicate rocks to form carbonates. This process removed carbon dioxide from the atmosphere and tied it up in rocks. Two additional processes contributed to a decrease of CO_2 in Earth's atmosphere. First when life evolved on Earth, plant photosynthesis chemically removed carbon dioxide. Second, marine animals used CO_2 to create carbonate shells, which sank to the bottom of the sea when the animals died. The accumulated deposits of minerals such as limestone and chalk that are still extensive today contain much of the original carbon dioxide. Thus, once a full accounting is completed, Earth and Venus are found to have the same *total* amount of CO_2. All these processes are still occurring, but an equilibrium was established prior to the industrialization of humanity.

Why was CO_2 not removed from Venus's atmosphere? The fact that Venus is closer to the Sun than the Earth would have given it a somewhat higher surface temperature initially. There might not have been much liquid water to help catalyze the chemical reactions that, on Earth, helped remove carbon dioxide from the atmosphere. Furthermore, Venus's higher temperature would have placed more water vapor into the atmosphere and would have caused a small greenhouse effect. This greenhouse warming would have helped raise the surface temperature and drive carbon dioxide out of the rocks. As more and more carbon dioxide accumulated, the greenhouse effect would have increased even more, thus releasing more CO_2 and increasing the temperature, and so on. Such a *runaway greenhouse effect* would have effectively prevented the removal of carbon dioxide from the atmosphere and eventually led to the situation we see today.

Oxygen on Earth arose as a side benefit of the photosynthetic process that removes CO_2. Only when plants formed on Earth did the slow accumulation of life-giving oxygen begin.

The reason Venus does not have Earth's large amount of nitrogen in the atmosphere is that a substantial fraction of Earth's nitrogen results from the decay of dead organisms.

The difference in water content between the two planets is also explained by Venus's high temperature. While Venus may have formed as a water-rich world, once its temperature became high enough to vaporize any liquid water present, the vapor rose into the atmosphere, where ultraviolet radiation from the Sun could break up the water molecules into hydrogen and oxygen. Venus's gravity was not strong enough to hold onto the hydrogen, which escaped into space.

Mars, on the other hand, had yet a different history. It is colder than Venus or Earth, and much of the water would have been in the form of ice, which does not chemically remove carbon dioxide from the atmosphere. Furthermore, Mars' smaller size limits the volume of outgassed material and results in a smaller escape velocity that allows atmospheric gases to escape easily. Finally, Mars' volcanic activity appears to have ceased early in its history, thus restricting the length of time over which outgassing occurred.

This scenario is our current best understanding of the atmospheres of the terrestrial planets. There are some unexplained parts, however. One example is Venus has a high abundance of argon in comparsion to Earth. In searching for an explanation, some astronomers have revived speculations that some volatile and noble elements (such as argon) came from collisions with comets.

If this is true for Venus, such collisions would undoubtedly have also occurred for Earth, in which case arguments that comets played an important role in bringing biochemical compounds to Earth may be strengthened.

9.4
SURFACES

Studies of planetary surfaces have provided us with detailed information about processes occurring not only on the surface but also within the atmosphere and the interior. In this section, the largest of the chapter, we first look at the surface as seen by orbiting spacecraft and then from landers. Finally, we will take a more detailed look at the variety of surface features visible and make inferences about their causes.

THE VIEW FROM ORBIT

The passage of a spacecraft close to a planet makes it possible to increase greatly the resolution of the planet's surface details. When Mariner 10 made its closest approach to Mercury, scientists were elated as the first pictures arrived on Earth and showed a landscape resembling that of the Moon (Figure 9–5a). But as they began to look in more detail, distinct differences from the Moon appeared, as discussed below.

Our knowledge of Venus is being rewritten as you read this book, thanks to the Magellan spacecraft. The primary goal of this satellite was continuous high-resolution radar imaging of the entire planetary surface. After its arrival at Venus in 1990, Magellan mapped almost all of the surface at least once. Earlier radar studies had shown two-thirds of Venus's surface to be covered by rolling plains, about one-quarter covered by lowland regions, and less than one-tenth by highland areas. The improved resolution of Magellan shows details as small as 100 meters across, about the size of a football field. For the first time we are able to "see" what the surface looks like. **Figure 9–9** shows the best global radar view of Venus yet produced. Radar images do not necessarily look like what your eyes would see. They show the ability of planetary material to reflect radar beams. The reflecting and absorbing properties of a surface depend on a variety of factors, including the composition and sizes of particles. Finely ground materials reflect poorly compared with coarse materials. For these reasons, radar images are difficult to interpret, and great care must be taken to avoid errors.

We can imagine what conditions on the surface of Venus must be like. With its thick atmosphere, the planet is probably quite dark with only about 1% of the sunlight reaching the surface. Red light would be scattered somewhat less than blue, so objects would take on

FIGURE 9–9. A radar view of Venus. Bright regions are those that reflect radar beams well.

a reddish color. Carl Sagan, in his book *The Cosmic Connection*, has described the surface of Venus in this way:

> *...broiling temperatures, crushing pressures, noxious and corrosive gases, sulfurous smells, and a landscape immersed in a ruddy gloom...there is a place astonishingly like this in the superstition, folklore and legends of men. We call it Hell.*

The view of the Martian surface so long awaited by astronomers is shown in Figure 9–5*d*. While about 40% of the planet's surface consists of plains regions, in this view we see giant volcanoes, a long and deep valley, and craters. These features are far larger than anything comparable on Earth. There is no doubt that these mountains are volcanoes, for one can see near the summit the remains of lava flows that took place during past eruptions. The other 60% of the Martian surface is cratered. Most of the plains (and their volcanoes) are confined to the northern hemisphere; the more heavily cratered regions are in the southern hemisphere. The differences in the hemispheres can be seen on the maps of **Figure 9–10**.

But what about the Martian canals suggested by Schiaparelli and Lowell? None of the Mariner photographs support the hypothesis that *canali* exist on the surface of Mars. We must therefore conclude that these features were an optical illusion, perhaps due to the ability of the eye and brain to imagine order where there is none while viewing a scene at the limits of visibility. We might also ask if Lowell and others were not strongly influenced by their *desire* to see canals, because many reputable observers were never able to see them. Sagan has commented wryly that there is no question that intelligence was involved in the problem of the Martian canals, but that the real question was on which end of the telescope the intelligence was located. Clyde Tombaugh, the discoverer of Pluto, has commented that Mars looked "different" to him when he compared the view through Lowell's telescope with the view through other telescopes at the observatory, raising the question whether there may have been some unsuspected instrumental problem at work.

THE VIEW FROM THE SURFACE

The Soviet Venera 13 and Venera 14, which landed in 1982, sent back visible-light photographs of the surface

N

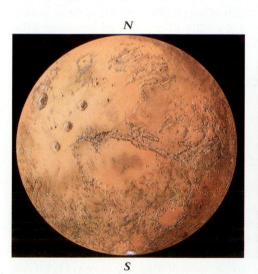

S

N

S

FIGURE 9–10. Maps of the Martian surface. Notice the difference between the northern and southern hemispheres.

FIGURE 9–11. The surface of Venus as photographed by Soviet Venera probes.

of Venus (**Figure 9–11**) showing it to be a barren, rocky plain. From these pictures, taken with a fish-eye lens and therefore covering a large field of view, planetary astronomers can draw some important conclusions. The sharp-edged rocks show a lack of erosional processes, indicating a lack of wind-blown dust on the surface. Instrumented landers had shown previously that the high-velocity winds in the upper atmosphere cease at the lower levels. The presence of shadows shows that some direct sunlight is able to penetrate the multi-level clouds all the way to the surface. A horizon can be seen in the upper left corner, indicating the clarity of the dense atmosphere.

Incredible images from the Martian surface were returned by the Viking spacecraft (**Figure 9–12**). We see a landscape not unlike a desert on Earth, with windblown sand and weathered rocks. Some of the discoveries of the Viking landers are discussed in the coming sections.

FIGURE 9–12. The Martian landscape as seen by Viking 1.

SURFACE TEMPERATURES

	Mercury	Venus	Earth	Mars
Temperature (surface) (K)	100 to 700	730	200 to 300	130 to 290

The surface of Mercury has a daytime temperature at the equator of some 700 K (+800°F), and a nighttime temperature of 100 K (−279°F). This wide range comes about from the lack of a substantial atmosphere to help retain heat.

Venus surprised astronomers, who found unexpectedly large amounts of heat coming from the planet. Analysis of these data showed the average surface temperature to be 730 K (854°F), above the melting point of lead! The fact that Venus is closer to the Sun than Earth cannot by itself account for this remarkably high temperature. It turns out that Venus exhibits an extreme example of the greenhouse effect that was described in Chapter 8. It shows us what can happen if the abundances of greenhouse gases (CO_2, SO_2, and H_2O) get out of hand.

Venus never obtains any relief from its high temperature. The blanketing by the atmosphere and the movement of heat by atmospheric circulation cause Venus's temperature to be nearly the same over the entire surface.

Venus's high surface temperature might have observable effects. Planetary scientists have conjectured that over billions of years at such high temperature, the surface might behave like Silly Putty. In addition to making it difficult for rigid plates to form, the heat would cause material on mountains to flow downward, or slump, due to gravity. The mountain called Maat Mons shown in **Figure 9–13** has such a slumped appearance.

Mars' surface temperature variations range from a rare extreme high of 300 K (80°F) at the equator during summer to 144 K (−200°F) at the poles in winter. (The air temperature, though, rarely rises above freezing.)

SURFACE COMPOSITIONS

	Mercury	Venus	Earth	Mars
Surface composition	silicates?	silicates?	silicates	silicates, iron oxide

From Earth-based observations of Mercury, astronomers concluded many years ago that the nature of the surfaces of Mercury and the Moon must be similar. Spacecraft have shown the similarity in more detail. While no surface samples of Mercury have yet been col-

FIGURE 9–13. A computer-simulated view of Venus, using Magellan spacecraft altimeter data. In this view, the observer is about a mile above the 8-km-high mountain, Maat Mons. The colors are meant to represent the result of sunlight filtered through Venus's sulfuric-acid clouds. Relief is exaggerated.

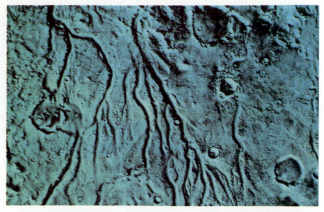

(a)

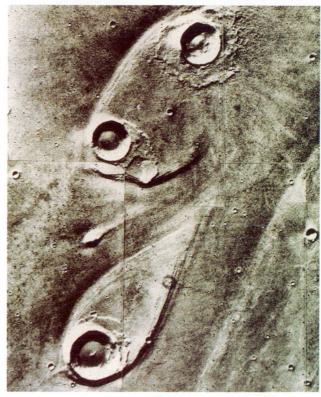

(b)

FIGURE 9–14. Evidence of the past presence of water on Mars. (*a*) Ancient channels are thought to have been produced by a sudden release of water due to melting of permafrost. (*b*) Similar structures are found in rivers on Earth when there is a barrier to water flow.

lected, astronomers infer from the observed planetary density and other data that Mercury's surface materials are most likely basaltic lava.

The only direct measurements we have of Venusian surface materials come from the many Soviet Venera landers. Various detectors measured gamma ray and X-ray emissions from surface rocks. The results show the rocks are basalts and granites, which are igneous rocks formed by volcanism. Venera 13 and Venera 14 revealed that these Venusian basalts contain substantial amounts of calcium, a result of some importance that we will return to later.

The Mars Viking landers finally answered the question of why Mars is red. It has long been suspected that the red color of Mars is due to iron oxides—rust—in its soil. Viking's soil analysis showed the surface to consist mostly of silicate rocks, but with a large fraction (20%) as iron oxide. Another indication of iron is that the soil is magnetic. Such a high surface abundance of iron combined with its overall density means that the planet should not have much of an iron core. In other words, Mars should show little, if any, differentiation.

WATER

	Mercury	Venus	Earth	Mars
Surface water	none	none	much	none

Perhaps the most exciting discovery made by the Mars orbiter Mariner 9 was that the Martian surface has many features that are clearly the result of fluid flows.

Figure 9–14 shows two regions of this type. In Figure 9–14*a* is a sinuous, twisting channel that looks as if it was caused by flows of a mudlike slurry. Such a slurry could be produced if permafrost, which consists of a frozen mixture of soil and ice, were to melt. Such melt-

FIGURE 9–15. A landslide on Mars. Melting of the permafrost allowed the surface to slide. The resulting fluid flowed away toward the left.

ing could be produced by climatic changes or by "geothermal" heat. There is clearly nothing artificial about these Martian regions and, indeed, they strongly resemble the beds of ancient streams and rivers on Earth. Figure 9–14b shows two sand bars around which running water has flowed. Similar formations are seen in rivers all over the Earth.

Inquiry 9–6 What are two mechanisms in addition to those suggested above that could have provided the heating to melt the permafrost?

There is no evidence that large quantities of liquid water exist on Mars now; the low atmospheric pressure prevents it. This raises an obvious question: If there was once water, where has it gone? Sagan and others have suggested that the Martian climate may alternate between ice ages, which are cold and dry with moisture largely locked in the form of permafrost and ice, and temperate ages, which are warmer and wetter. When subsurface permafrost melts, landslides may result (Figure 9–15). The water produced by the melted permafrost flowed away toward the left.

CRATERS

	Mercury	Venus	Earth	Mars
Craters	high density	low density	low density	medium density

Mercury shows extensive cratering, including numerous bright-rayed impact craters and relatively flat regions between craters (Figure 9–5a). Superficially, Mercury's craters resemble those on the Moon, but scientists have noted one significant difference between the two bodies: the entire lunar highlands are extensively cratered and rolling, whereas the analogous regions on Mercury are relatively smooth. This difference is probably explained by the fact that Mercury has a higher surface gravity than the Moon. When a crater formed on Mercury, the ejected material did not travel very far, and the secondary craters remained close to the primary crater. On the Moon, the ejecta could travel farther, and secondary cratering would have taken place over a much wider area.

Venus, too, shows impact craters (**Figure 9–16**), some with central mountain peaks. With a crater density some thousand times less than that of the Moon, Venus is relatively free of impact craters. Such a low crater density implies that planetwide resurfacing has

FIGURE 9–16. Craters on Venus. This is a radar image obtained by the Magellan spacecraft. Note the central peak. The crater is about 35 km in diameter.

been occurring during the past 500 million years. Such resurfacing may have been caused by volcanism.

The Moon, Mercury, and Mars all have craters in a wide range of sizes, from hundreds of kilometers downward. Venus, on the other hand, has no craters less than two kilometers across. Such a lack of small craters is interpreted in terms of a lack of small impacting bodies, which must have been destroyed in passing through Venus's atmosphere. Shock waves that resulted from the passage of meteoroids through the thick atmosphere could have pulverized Venus's surface material into fine dust. This phenomenon might explain the various dark splotches observed to surround some craters on Venus.

IMPACT BASINS

	Mercury	Venus	Earth	Mars
Impact basins	yes	?	perhaps one?	yes

We saw previously that the Moon has basins as large as 1100 km across that resulted from the impact of large bodies. Some of these basins completely filled with lava that oozed from the interior, forming the maria. The Mare Orientale, on the other hand, has not completely filled in but exhibits multiple rings.

No large impact basins are found on Earth, because continental drift obliterated large impact basins long ago. The largest known impact feature appears to be a crater nearly 200 km in diameter discovered in 1992 off the Yucatan Peninsula in the Gulf of Mexico.

In **Figure 9–17**, you see one of Mercury's sizable basins. The **Caloris Basin** is the largest one with a diameter of 1300 km. It is a multi-ringed basin, similar to the Mare Orientale on the Moon (see Figure 8–23).

Mercury's surface on the side opposite the Caloris Basin contains what planetary scientists labeled "weird terrain" because its jumbled appearance had not been observed anywhere else in the solar system. Subsequently, similar but less extensive weird terrain was found on the Moon directly opposite the Mare Orientale. One explanation of weird terrain is that it arose in response to the impacts causing the Caloris Basin and Mare Orientale. Seismic waves traveling through and around the surface caused material opposite the impact to be lifted off the surface, and to fall back in the jumble we now observe (**Figure 9–18**).

More than 20 basins have been identified on Mars. One of them, called Hellas, is a multi-ringed basin. Because the basin floor is some 5 km below the surrounding plains, the greater atmospheric pressure at the bottom allows frost and cloud formation to occur there easily. The high reflectivity of the frost and clouds makes Hellas readily visible from Earth.

FIGURE 9–17. Mercury's giant multi-ringed basin, Caloris Planitia.

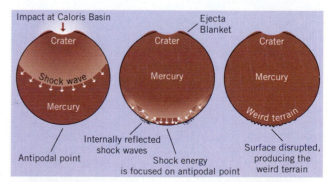

FIGURE 9–18. The formation of "weird terrain" from a gigantic impact. Seismic waves traveling through the planet and over the surface converge on the opposite side.

CONTINENTS, MOUNTAINS, AND VOLCANOES

	Mercury	Venus	Earth	Mars
Continental-sized areas	no	yes	yes	yes
Mountains	no	yes	yes	yes
Ancient lava flows	yes	yes	no	yes
Active volcanism	no	perhaps	yes	no

Relief maps of three of the terrestrial planets are shown in **Figure 9–19**. While about two-thirds of Venus's surface is rolling plains, almost 10% of the surface is concentrated in extreme highland regions. For example, the area known as *Ishtar Terra* is comparable in size to Africa and contains the mountains called Maxwell Montes, which have an elevation of more than 11 kilometers (35,000 feet) above the surrounding plain. This is a mile higher than Mt. Everest! Other continental-sized areas are also present.

Venus shows a range in type and size of volcanically produced features. These include domes, mountains, highland plateaus, and meandering flow channels (**Figure 9–20**). These features can be as large as 500 kilometers across. One channel runs for 6800 km. We do not know what type of lava can flow for such distances

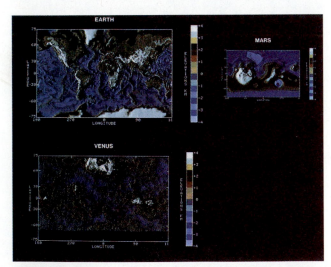

FIGURE 9–19. Relief maps of Venus, Earth, and Mars. The high regions on each planet are the Himalayas on Earth, Ishtar Terra on Venus, and the Tharsis Plateau on Mars.

FIGURE 9–20. (*right*) Volcanism on Venus, as seen by radar on the Magellan Spacecraft. (*a*) Lava flows invading and degrading an impact crater, whose diameter is 65 km. (*b*) Pancake domes, some 25 km across, as observed by the Magellan orbiter. (*c*) A meandering channel that could be produced only by flowing lava. This section is 200 km long. Channels as long as 6800 km have been seen on Venus.

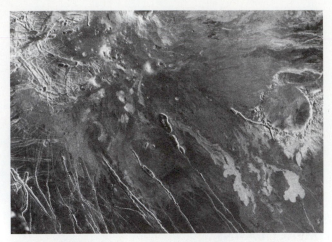

(a)

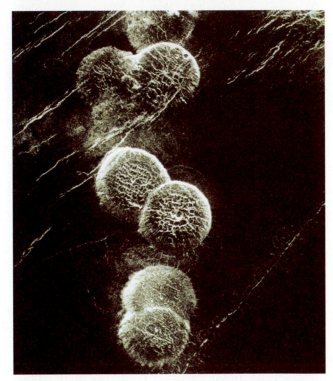

(b)

(c)

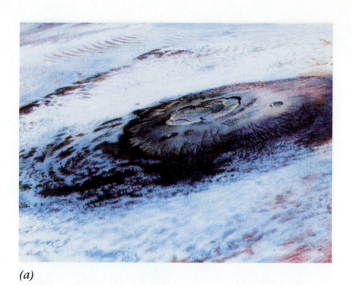

(a)

(b)

FIGURE 9–21.   Olympus Mons, the solar system's highest volcano. (*a*) The summit, at an elevation three times that of Mt. Everest on Earth, is well above the clouds. The large ring is 500 km in diameter. (*b*) Olympus Mons and other Martian volcanoes compared in size with the United States.

without hardening. Figure 9–13 shows a computer-generated false-color view of Maat Mons, the youngest-looking volcano on Venus.

On Mars, volcanoes are located primarily on the Tharsis plateau, an area covering nearly a quarter of the Martian land mass. Tharsis is much higher than the surrounding plain (Figure 9–19*c*). These volcanoes are part of the plains region that covers 40% of the surface. The plains apparently came about from extensive lava flows. In contrast, the southern hemisphere is rough and covered with craters, making it geologically older than the northern hemisphere. Such geologic asymmetries are common on the Earth-like planets.

Mars has the tallest volcano in the solar system. **Olympus Mons** (Mount Olympus) rises almost 27 km (nearly 90,000 feet) above the planet's surface (**Figure 9–21**). The central caldera is 80 km across and resulted from collapse when the lava inside was suddenly withdrawn. The large circular ring surrounding it is a cliff 500 km in diameter. In many places the cliff height is as much as 6 km.

Why are some of the Martian volcanoes so much larger than those on Earth? In the case of the Hawaiian Islands, for example, the volcanoes formed when molten lava from hot spots located on moving continental plates broke through the lithosphere. The plate movement means that new volcanic cones will be continuously formed as the hot spot moves (**Figure 9–22**). On Mars, however, where there is no plate movement, volcanic activity occurs time and again at the same location so that lava piles up to great elevations. The pres-

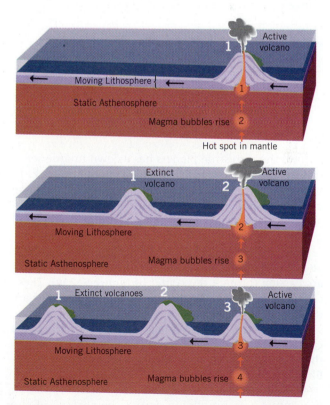

FIGURE 9–22.   Why volcanoes on Mars are larger than those on Earth. Formation on a moving lithosphere limits volcanic sizes on Earth, while the lack of motion on Mars allows them to grow.

ence of such high mountains also tells us the Martian crust is strong enough to hold such a concentrated mass.

Mercury does not show any great volcanoes like those on Venus or Mars. (However, because the Mariner spacecraft observed only one side of the planet, some could exist on the other side.) The resolution of the images Mariner sent to Earth is too low for us to be able to see lunar-type domes and lava flow fronts. Nonetheless, Mercury does have rimless pits similar to those found on the Moon; they are thought to be volcanic vents (see Figure 8–20a). Finally, the plains regions between craters have the general appearance of previously molten material.

Are volcanoes currently active on any terrestrial planets other than Earth? While the answer is a definite no for Mars and Mercury (subject to the caveat that we have seen only half of Mercury's surface), the answer is not so clear for Venus. The Pioneer Venus spacecraft detected 50 times the amount of sulfur dioxide (SO_2) usually present in Venus's atmosphere. Because calcium present in the surface material readily combines with sulfur dioxide, thus removing it from the atmosphere, there must be a source of fresh sulfur dioxide available. We therefore infer that active volcanoes, the most likely source of sulfur dioxide, may be present.

FAULTS, TECTONICS, CLIFFS, AND VALLEYS

	Mercury	Venus	Earth	Mars
High scarps	yes	no	no	no
Plate tectonics	no	probably not	yes	no

Each of the terrestrial planets has its own unique set of surface features caused by its own unique history. In this section, we look briefly at features that on Earth are usually explained in terms of plate motion, but that require different explanations on the other terrestrial planets.

Mercury has cliffs that are several kilometers high and often hundreds of kilometers long (see the darkish arc passing through the center of **Figure 9–23a**). These cliffs are unique among the Earth-like planets. They are thought to have formed when pressures from compression produced during cooling and shrinking of the crust caused local folding of the crust (**Figure 9–23b**). As with the Moon, however, there is no ironclad proof of Earth-like plate motion at any time in Mercury's history.

One of Mars' surprises awaiting scientists was an enormous canyon slicing across the surface of the planet, (**Figures 9–5d** and **9–24**). Named the Valles Marineris (Mariner Valley) for the Mariner spacecraft,

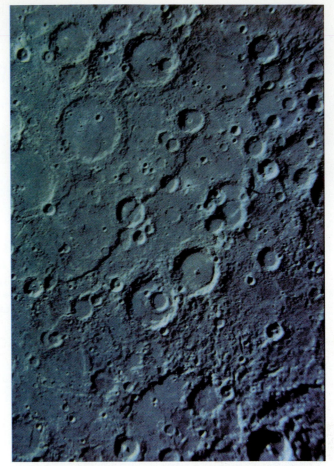

(a)

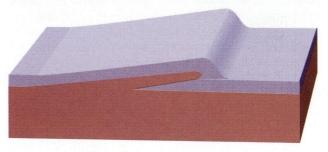

(b)

FIGURE 9–23. (*a*) A close-up of Mercury. The smallest visible details in this figure are about 250 meters across, or the length of three football fields. The dark arc passing through the center is a piece of a high cliff (scarp) named Discovery. (*b*) Mercury's system of cliffs is produced when the surface shrinks, creating compression forces that eventually cause one layer to override another.

(a)

(b)

FIGURE 9–24. (*a*) Valles Marineris is the size of the United States. (*b*) The Candor Chasm region of the Valles Marineris shows erosion and perhaps material deposited on its floor.

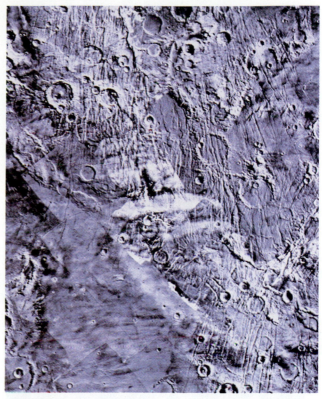

FIGURE 9–25. Martian faults. The area shown here covers 1500 km from top to bottom. One set of faults runs north-south, and another runs northeast-southwest. Most of these large fault systems are radial from the region where the volcanoes formed, indicating a relationship.

this huge feature is about 5000 km long, over 5 km deep, and 80 km wide in places. On Earth, this colossal canyon would stretch from Los Angeles to New York, and Mars is much smaller than Earth! The part of the Valles Marineris called the Candor Chasma appears to have been eroded by both wind and water (Figure 9–24*b*). While ideas about its formation abound, no single explanation is yet accepted by the scientific community.

Mars is covered by a variety of fault regions, one of which is shown in **Figure 9–25**. This is the most intense fault region on the planet and shows two complex sets of faults: one in a north-south direction, the other running northeast-southwest. While the cause of the stresses producing the faults is unknown, they could have resulted from convection within the mantle. Not only is there no direct indication of large-scale crustal motion but, as discussed earlier, the large volcanoes argue against plate motion.

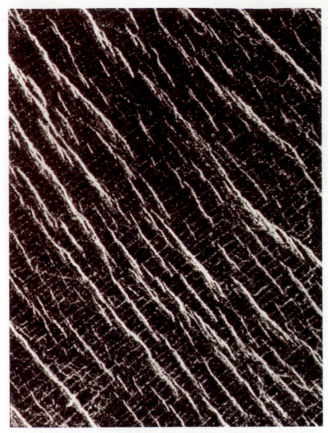

FIGURE 9–26. The "graph paper region" of Venus. The fainter lines, perpendicular to the brighter ones, occur at regular intervals of nearly a kilometer.

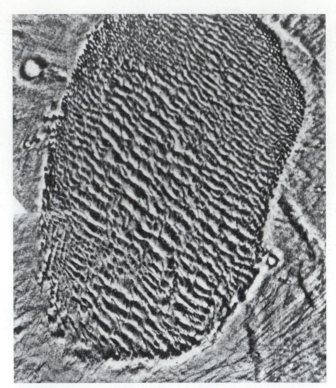

FIGURE 9–27. Wind-blown sand on Mars. A region of Mars covered by sand dunes, evidence of the action of winds in the thin Martian atmosphere.

The Magellan images of Venus confirm that it does not exhibit continental drift. Thus Earth is the only planet having large-scale horizontal plate motions. Nonetheless, Venus does appear to release internal heat by means of cracks in the mantle through which plumes of material flow to form Venusian mountains. Evidence of such cracking is shown in **Figure 9–26**, which shows a region on Venus known as the "graph paper region." This region, unique within the solar system, is thought to be due to an unknown type of subsurface force. Further fault features are visible in Figure 9–20*b*.

wind provides one connection between observable surface features and the atmosphere. Shifting dust, covering some areas and uncovering others, provides the most likely explanation of the changing appearance of Mars' dark markings. An extensive region of wind-deposited sand dunes appears in **Figure 9–27**. Analysis of photographs of such regions shows that winds of up to 90 mi/hr (150 km/hr) exist on the planet.

Inquiry 9–7 What previously discussed observations are there that wind erosion is not important on Venus?

WIND SCULPTING

	Mercury	Venus	Earth	Mars
Wind sculpting	——	little/none	yes	yes

Mars has abundant evidence, both within its atmosphere and over its surface, of wind-blown dust. Such

POLAR REGIONS

	Mercury	Venus	Earth	Mars
Polar ics caps	perhaps	no	yes	yes

Study of planetary polar regions is difficult from Earth because they are viewed from a highly oblique angle. While the Martian polar ice caps had been extensively observed for many years, controversy over whether they were water ice or dry ice (frozen carbon dioxide) continued for a long time. Earth-based observations indicated that Martian ice caps completely melted each summer; would higher-resolution images show this to be the case?

Long-term studies of the temperature of Mars' ice caps by orbiting spacecraft found them to be composed of *both* water and carbon dioxide ices. The more abundant dry ice produces the polar cap viewed from Earth in the views in Figure 9–3; the water ice cap lies underneath it. In the summer, the dry ice in the northern cap completely sublimates and leaves behind a water-ice remnant. Some frozen carbon dioxide remains at the south pole throughout its summer (**Figure 9–28**). These observations are surprising because they are the opposite of what might be expected on the basis of the seasons as illustrated in Figure 9–7. Further photographs showed layering in the polar regions, which provides evidence of periodic sedimentation resulting from long-term, repetitive climatic changes.

Radar observations of Mercury made in 1991 gave astronomers a view of the planet never before seen. At the time these observations were made, the planet's

FIGURE 9–28. Polar regions. Mars' remnant south polar cap is about 400 km across.

north pole was tilted toward the Earth. The observations revealed an oval area, measuring 640 by 300 km around Mercury's north pole, which reflected radar waves strongly. The researchers interpret these data as an ice cap on Mercury's north pole. While surprising, the observations are consistent with recent theoretical calculations indicating Mercury's polar temperatures can be as cold as 125 K (−235°F). Because interpretation of radar images is both difficult and tricky, it is possible the region is bright due to intense cratering or other ragged terrain at the pole. However, most planetary astronomers think such an explanation is unlikely, and that the ice interpretation is correct. Additional observations are required before a definitive statement may be made.

Inquiry 9–8 What might you expect the temperatures of Venus's polar regions to be compared to its equator? Explain.

9.5

INTERIORS

Because a planet's interior cannot be observed directly, the interior structure must be inferred. One of the primary observations used is a planet's magnetic field.

MAGNETIC FIELDS

	Mercury	Venus	Earth	Mars
Magnetic field	yes	no	yes	no

Much to the surprise of planetary scientists, Mariner 10 found Mercury to have a significant magnetic field. With a strength 1% that of Earth's magnetic field, Mercury's field is stronger than that of either Mars or Venus. The magnetic field provides information about the interior structure.

INTERIOR STRUCTURES

In the absence of seismic measurements, a planet's interior structure must be deduced from a variety of observations. These include the planet's shape, density, magnetic field, and effects on orbiting spacecraft.

How can Mercury, which has such a slow rotation speed, show a significant magnetic field? Assuming the dynamo theory for magnetic field production is more or less correct, such a magnetic field can come about only if the planet has a core encompassing a substantially larger fraction of its interior volume than is true for the other terrestrial planets. Models of Mercury's interior show the core to contain 42% of the planet's volume, compared with 16% for the Earth.

Venera 8 found that the level of radioactivity in the surface rocks of Venus was similar to that of terrestrial rocks. This observation supports those who think that Venus is similar to Earth not only in size and mass but also in its stratification into layers: core, mantle, and crust. One difficulty with this model is the lack of a strong magnetic field. However, even if Venus does have a molten core, dynamo action might not have developed due to the planet's slow rotation.

Mars' interior is only partially differentiated. This conclusion results from Mars' large amount of surface iron, which immediately tells us that an extensive inner core of iron does not exist. This observation is consistent with the planet's lack of a magnetic field.

Although each of the Viking spacecraft visiting Mars had a seismometer, only one of them worked, and it showed that Mars does have seismic activity. However, the amount of seismic activity per unit area on Mars is less than that on Earth. The Mars quakes that were observed lasted about a minute, indicating that Mars' internal structure is more similar to that of Earth than it is to the Moon, whose quakes last for hours.

Figure 9–29 presents a scaled figure of the interiors of each of the terrestrial planets and the Moon. Mercury's core is unique in its large fractional volume.

9.6
SATELLITES

	Mercury	Venus	Earth	Mars
Satellites	none	none	1	2

Mars is the only terrestrial planet other than Earth to have natural satellites. Viking succeeded in photographing **Phobos**, whose diameter is some 23 km (**Figure 9–30a**), and **Deimos**, which has a 12-km diameter (**Figure 9–30b**). Their names come from the mythological horses that pulled the war god's chariot and mean Fear and Panic, respectively. The two satellites turned out to be quite irregular in shape and pitted by craters.

An observer on Mars might well be confused by observations of Phobos and Deimos, because Deimos rises in the east and sets in the west, while Phobos does the opposite! The apparent eastward motion of Phobos results from its orbital period of only 7 hours, much less than Mars' rotation period of slightly more than 24 hours.

Phobos has numerous small impact craters and one large one. Due to the satellite's small mass, none of the craters have central peaks or rings. Phobos has strange parallel linear features, many of which are associated with its large crater. Perhaps the parallel lines are indications of a strong impact that might have come close to destroying the satellite, producing deep cracks at the time.

Deimos is much smoother than Phobos and has no large craters. Its surface is dotted with numerous bright patches whose nature is unknown.

The Viking spacecraft passed close enough to the satellites to interact gravitationally and therefore allow masses and densities to be found. While the density of Mars is about 4 grams/cm^3, that of the satellites is only 2 grams/cm^3. They appear to be made of material different from that of Mars. Observations indicate they are similar to large asteroids and the carbonaceous chondrite meteorites. From their chemical makeup, and from their appearance, one might conclude that they are captured asteroids. However, we do not expect randomly captured bodies to be in circular orbits and in the equa-

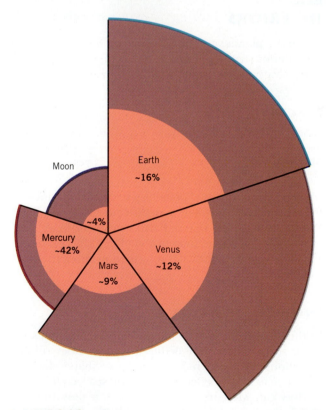

FIGURE 9–29. The interiors of the terrestrial planets and the Moon, drawn to scale. In a relative sense, Mercury has the largest core.

Moon
Earth
~16%
~4%
Mercury
~42%
Venus
~12%
Mars
~9%

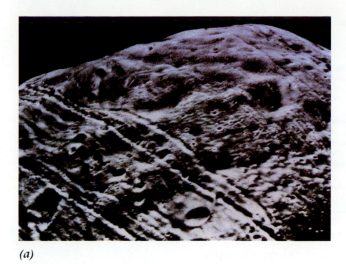

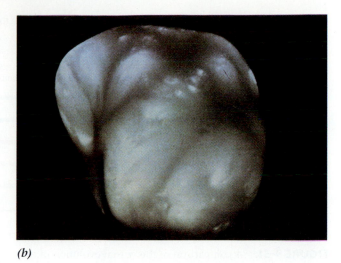

(a)

(b)

FIGURE 9–30. The Martian satellites. (*a*) Phobos (*b*) Deimos.

torial plane, as both Phobos and Deimos are. On the other hand, if they formed by accretion at the same time and place as Mars, why the difference in densities? While various hypotheses have been made, no one answer is clearly better than any other.

9.7
EVOLUTIONARY COMPARISON

The similarities of the terrestrial planets undoubtedly result from their similar evolutionary histories. The differences probably arise from variations in their nearness to Sun, mass, rate of meteoroid impact, and so on.

The evolutionary scheme of Mercury serves as an example of terrestrial-planet evolution. An examination of the Mariner photographs suggests that there was an early period of surface melting. A period of early melting is required because the surface shows the effects of cratering from several billion years, without much "erasure" of features having taken place. The picture of Mercurian history that seems to fit the data is roughly as follows:

1. Creation of the planet out of the primitive solar nebula, as discussed in Chapter 6.
2. Soon after formation, chemical differentiation that led to a heavy iron core and a lighter silicate mantle.
3. Possibly an early "smoothing" of the surface, perhaps by volcanic action, to erase the scars of the formation itself.
4. An episode of meteoritic bombardment similar to

that which took place on the Moon, which created large features such as the Caloris Basin.
5. Thereafter, minor volcanic activity at most, and a long era of relative quiescence with occasional peppering by meteoroids, even to the present day.

We have seen that Mars has a puzzling mixture of surface features. How are we to interpret them? Mars' crustal history has been more active than that of either Mercury or the Moon. Certain regions of Mars are relatively young; for example, there are fewer impact craters near some of its bigger volcanoes than elsewhere, indicating that these volcanoes formed later. There are also other heavily cratered and presumably very old regions, so Mars has a considerably varied history. The different appearance of channels indicate that some of them are a mere 100 million years old, whereas others are as much as 2.5 billion years old.

We are now in a position to sum up our new views of the terrestrial planets. It is clear that after many years of thinking of them as places somewhat similar to Earth, we are forced to conclude that they are unique entities, unlike the Earth or the Moon, and each with its own alien soil, chemistry, history, and geology.

Figure 9–31 emphasizes that the early crustal evolution of the terrestrial planets was the same. In Stage I, they all were formed, heated, and had their cores separate to one extent or another from the remaining material. In Stage II, the igneous crust formed. The lunar highlands, for example, formed during this stage. Heavy bombardment toward the end of this stage produced basins such as Hellas on Mars, Caloris on Mercury, and the large circular basins on the Moon. Stage III involved

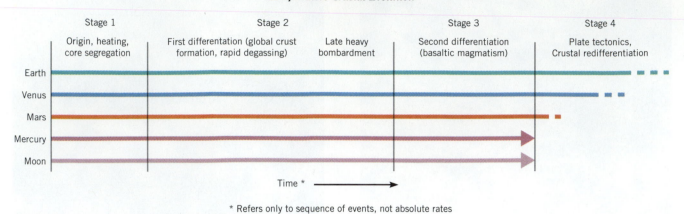

FIGURE 9–31. A comparison of the crustal evolution of the terrestrial planets and the Moon.

the production of relatively smooth, level regions. During this period, the maria basins that were excavated in Stage II were filled with lava, and the smooth plains of Mercury were formed. Crustal evolution of the Moon and Mercury ended here. At Stage IV, crustal evolution becomes unique for each object. We know some crustal evolution continued on Mars because the Valles Marineris is most likely a tectonic depression rather than a canyon formed exclusively from the flow of liquids, such as the Grand Canyon on Earth. Data from Magellan show that crustal development on Venus continued and may still be in progress. On Earth, erosion and extensive tectonic activity (continental drift and volcanism) show that crustal evolution is continuing even today.

Inquiry 9–9   Considering the successes and failures of the Viking missions, what suggestions can you make for future exploration of Mars?

Inquiry 9–10   How does Viking illustrate the differences between space missions carrying astronauts and those carried out by robot probes?

9.8
FUTURE STUDIES

Currently funded future exploration of the terrestrial planets is limited to Mars. Because of the failure of the Mars Observer spacecraft, NASA is beginning a program of Mars exploration called the Mars Surveyor, which will launch small, low-cost probes roughly every two years beginning November 1996. The Russians will launch Mars '94 in 1994 and Mars '96 two years later. These missions had been planned well before the breakup of the Soviet Union, and before the current review of the U.S. budget deficit. Japan will launch a probe called Planet B in 1996.

9.9
MARS LIFE-DETECTION EXPERIMENTS

On its eighth day on the Martian surface, Viking extended its mechanical arm and scooped up a sample of soil that was placed into four life-detection experiments. Three of the experiments tried to detect the presence of microorganisms in the soil by looking for by-products of metabolic activity, such as gases being given off. These experiments all gave positive results; however, in each case, other processes not involving life could have produced the same results. A fourth experiment attempted to detect organic materials in the soil, and it found none. Thus Viking's attempts to find life in Martian soil were inconclusive, and additional missions will be needed to resolve conclusively the question of life on Mars. Future experiments might involve drilling into the permafrost to look for microorganisms there.

CHAPTER SUMMARY

OBSERVATIONS

- Table 9–1, at the end of the chapter, summarizes our factual knowledge of the terrestrial planets.
- Venus's surface temperature is hot enough to melt lead due to the presence of a runaway greenhouse effect. Mars' temperature is generally well below the freezing temperature of water. Mercury is hot because of its closeness to the Sun.
- Mars' surface is red because of large amounts of iron oxide.
- While no liquid water is present on Mars today, there are indications that water exists as **permafrost** below the surface.
- All terrestrial planets exhibit impact craters (although those on Earth have been greatly eroded). Some impacts produced large basins (a number with multiple rings around them) on Mercury and Mars.
- Venus and Mars both have ancient, gigantic volcanoes, and continent-sized raised regions. Neither shows evidence of continental drift. While there is no indication of current volcanism on Mars, there is such evidence for Venus.
- All terrestrial planets show faults that indicate various degrees of surface activity. Mercury has extensive cliffs caused by crustal shrinkage. Venus has a region of crisscrossed faults who origin is not understood. Mars has a variety of fault structures, including an extensive canyon system whose origin is still unknown.
- Mars has high-speed winds traveling over the surface and producing dust storms and erosion. Venus has high-speed winds at the top of its atmosphere, but not at the surface.
- Mars' polar region contains ice caps made of both frozen carbon dioxide and water. Mercury appears to have an ice cap.

CONCLUSIONS

- While the total carbon dioxide content of Venus and Earth are similar, Earth's CO_2 is tied up in carbonate rocks. Venus's water evaporated, at which time the water molecules were destroyed; the rapidly moving hydrogen atoms then escaped the gravitational pull of the planet. Oxygen on Earth came about mostly from plant life. Earth's nitrogen came from volcanic outgassing and also from the decay of dead organisms.
- Mercury's surface composition is thought to be basaltic, like the Moon's surface. Venus's surface composition is basically similar to that of Mercury and the Moon but probably with some as yet unknown chemical differences.
- A variety of observations of Mars point to an extensive amount of liquid water existing for a short period of time in the distant past.
- The planetary interiors are similar in that they have cores surrounded by mantles and crusts. Mercury's magnetic field indicates that the planet has a substantially larger core than the other terrestrial planets.

SUMMARY QUESTIONS

1. What are the major facts known about each of the terrestrial planets? Include distances, general orbital characteristics, size, mass, and density relative to the Earth; and temperature, atmospheric composition, etc.

2. Compare and contrast all the terrestrial planets with one another. In the discussion, include any distinctive features of orbital characteristics, appearance, rotation and revolution, general atmospheric features and composition, interior structure, magnetic fields, and satellite systems.

3. What planetary features were Earth-bound astronomers able to see through their telescopes on each of the terrestrial planets? Describe any changes in these features that astronomers could have observed.

4. What do we mean by the runaway greenhouse effect? Why do we think the atmospheres of Earth and Venus developed as they did? What is the critical role of life in the maintenance of Earth's atmosphere?

5. Describe the history of Martian "canals," paying attention not only to the objective evidence prior to Mariner but also to possible "nonobjective" aspects.

APPLYING YOUR KNOWLEDGE

1. In what ways might Earth and Mars have been different than they currently are if each exchanged places with the other; in other words, if a Mars-sized body were 1 AU from the Sun and an Earth-sized body at 1.5 AU?

2. Examine Figure 9–5*a*. How do you know that this is a picture of Mercury and not the Moon (other than because the figure caption tells you so)?

3. Why does the fact that Venus has a dense core lead to our expectation that the surface rocks probably have a low iron content?

4. On Earth a solar day is a little longer than a sidereal day, but on Venus it is shorter. Explain why. (Hint: Use a *carefully* made drawing.)

5. You have just landed on Mars. Your view is that of Figure 9–12. Describe the view vividly and in detail.

6. The photographs of Venus in Figure 9–2 are at the same scale. Assume the first one was taken when Venus was at its greatest distance from the Earth, and the last one when it is nearly at its closest approach. Make a drawing of the orbits of Earth and Venus, and mark and label the positions corresponding to the photographs.

7. If Mercury's angular diameter is observed to be 0.002 degrees when its distance from Earth is 0.92 AU, what is Mercury's diameter? Express the diameter in kilometers, in units of the Earth's diameter, and in terms of the Moon's diameter.

8. Use the planetary orbital periods in Table 9–1 or Appendix C to calculate the distances of the terrestrial planets from the Sun in AU.

9. How much would a 150-pound person weigh on each of the terrestrial planets, as well as on the Moon?

10. Deimos circles Mars at a distance of 23,500 km in 1.26 (Earth) days. Calculate the mass of Mars, and compare your value with the "true" value. What is the percentage difference?

11. Under the best observing conditions on Earth, the smallest angle astronomers can discern is about 0.25 seconds of arc. Assuming Martian "canals" must have at least this angular size, compute their true size in kilometers. Is such a size reasonable for channels produced by intelligent beings?

12. The photographs of Venus in Figure 9–2 are at the same scale. Assuming the first one was taken when Venus was at its greatest distance from Earth, and the last one when it is nearly at its closest approach, determine the ratio of greatest to least distance from Earth. Does your answer make sense in view of what you know about Venus's orbit?

13. Compute the size in kilometers of the smallest Martian feature on the photographs in Figure 9–3.

ANSWERS TO INQUIRIES

9–1. Two major effects are color changes and variations in the size of the polar caps.

9–2. With a 59-day period, every side of Mercury eventually receives sunlight.

9–3. 2/3 orbit for 1 rotation, 1 1/3 orbits for 2 rotations, and 2 orbits for 3 rotations.

9–4. Mercury's axial tilt of 28° is substantial and would produce seasons similar to those on Earth. While Mercury is hot all the time (except at the poles), the high eccentricity of the orbit will make Mercury's seasonal variations more complex than those on Earth, however.

9–5. Because we see primarily the cloud tops in visible light, we are looking at a cooler layer.

9–6. Volcanism and meteoritic impact.

9–7. Rocks on the surface have sharp edges.

9–8. Because the clouds retain heat and the circulation moves it around, you would expect high temperatures at the poles, too.

9–9. This is an open-ended question. One might think of: better-designed experiments that would unambiguously detect life or its lack; an "intelligent" robot that could move about the surface and sample a wider area of Mars; a probe designed so that it can land safely in the Martian highlands.

9–10. Many questions were left unresolved, because the range of experiments that could be carried out was so limited.

THE PLANETS ONE-BY-ONE

MERCURY[1]

THE VIEW FROM EARTH

Mercury is difficult to observe from Earth because it is the nearest planet to the Sun and is never more than about 27° away from it. For this reason, it is always observed near the horizon and can never be viewed in dark skies far from the Sun. Furthermore, its angular size is always small. For these reasons little information, beyond knowing that it is one-third the Earth's size and 5% the Earth's mass, has been obtained about the planet from ground-based optical observations.

Mercury's orbit deviates considerably from a circle; it has the second most eccentric planetary orbit in the solar system. Likewise, its orbit has the second greatest tilt to the ecliptic of any planet.

Analysis of radar waves reflected from the surface show the planet to rotate three times on its axis for every two revolutions about the Sun. Mercury has no known natural satellites.

THE VIEW FROM SPACECRAFT

Spacecraft images of Mercury show its surface to have numerous craters; it appears similar to the Moon. Because Mercury has a low mass and high temperatures, the planet cannot maintain a substantial atmosphere. The only atmosphere it has comes from the Sun and from gases emitted by its heated surface material.

UNIQUE FEATURES

Mercury has a surprisingly strong magnetic field. Assuming this magnetic field is caused by a dynamo effect similar to that thought to cause the Earth's magnetic field, Mercury must have a large metallic core. Mercury's surface contains a large number of steep cliffs that may have been caused by stresses in the crust while it was cooling. The planet has a large multi-ringed basin called Caloris Planitia, most likely caused by an asteroid impact. The surface area between craters appears to be covered by ancient lava flows.

ADDITIONAL IMPORTANT FEATURES

Mercury lacks mountain ranges similar to those on the Moon. Recent observations indicate a possible ice cap at its north pole.

[1]For detailed information on each of the planets, refer to Table 9–1 on p.183.

VENUS

THE VIEW FROM EARTH

While Venus comes closer to Earth than any other planet, and is nearly the same size and mass as Earth, few details have been observed visually because the planet is covered with dense clouds. Furthermore, Venus is never more than about 48° from the Sun. Radar waves from ground-based telescopes penetrate the clouds to the surface. Such data have shown that Venus rotates in a retrograde manner, in the direction opposite to its revolution about the Sun. Because Venus's rotation axis is nearly perpendicular to its orbital plane, the planet has almost no seasonal changes.

Venus's atmosphere is mostly carbon dioxide. The CO_2 atmosphere produces a greenhouse effect that holds the Sun's heat and results in an average surface temperature of 730 K, high enough to melt lead. There is no water on the surface and only a trace of water vapor in the upper cloud layers. No planet-wide magnetic field has been observed. Venus does not have a natural satellite.

THE VIEW FROM SPACECRAFT

Spacecraft observations show Venus to be encircled by clouds whose circulation is determined by solar heating. The clouds move at high speeds.

The high surface temperature is accompanied by a surface pressure that is about 90 times sea-level pressure on Earth. The clouds contain concentrated sulfuric acid droplets. The Magellan spacecraft has observed ancient volcanoes, long winding channels, and regions of extensive faulting on the surface. No definite indication of continental motions has yet been observed.

UNIQUE FEATURES

Venus stands out from the other terrestrial planets because of its slow retrograde rotation; its high density , carbon dioxide atmosphere; and its runaway greenhouse effect. Its high surface temperature and pressure are also unique features. There is circumstantial evidence for current volcanism, but further data are needed.

ADDITIONAL IMPORTANT FEATURES

Soviet Venus landers found the surface to contain sharp-edged rocks, indicating a lack of surface erosional processes. Lightning occurs often in Venus's atmosphere.

THE PLANETS ONE-BY-ONE

EARTH

THE VIEW FROM EARTH

The third planet from the Sun has a surface showing a wide diversity of features caused by a variety of processes. These include moving continents that float on underlying, higher-density material. When one continental mass moves beneath another, the crust can be uplifted and mountain ranges produced. Seismic events thus occur along such plate boundaries. These seismic events are used to study the interior structure, which is found to consist of a solid inner core, a liquid outer core, and a surrounding mantle. Material that sinks into the mantle can be heated to produce volcanoes along the borders between the continental plates. Earth contains some impact craters, but wind, water, freeze/thaw cycles, and continental motions erode them over geologically short periods of time.

The atmosphere contains 78% nitrogen and 21% oxygen with trace amounts of argon, carbon dioxide, and water. The oxygen on Earth results from plant photosynthesis. The atmosphere's weight yields a surface pressure of 14.7 lb/in.^{2} at sea level. The temperature variation with height above the surface depends on various factors including chemical composition. Heat from the Sun, along with heat emitted from the oceans and continents, produces global wind patterns.

A compass, which in the Northern Hemisphere points to the north magnetic pole, shows Earth to have a magnetic field. The magnetic field traps charged particles from the Sun, forming the Van Allen radiation belts and the aurorae.

THE VIEW FROM SPACECRAFT

Viewed from space, Earth shows changeable clouds and large land masses with surrounding oceans. Spacecraft also detect Earth's magnetic field and Van Allen belts.

UNIQUE FEATURES

Earth is the only planet known to have moving continents, and the only planet known to nurture life.

ADDITIONAL IMPORTANT FEATURES

Earth's surface rocks consist of three types: igneous, sedimentary, and metamorphic. These participate in a continuous rock cycle because of continental motions. Earth's oceans are subjected to gravitational tidal effects caused by the Moon and Sun's gravitational pull being different on opposite sides of the planet.

MARS

THE VIEW FROM EARTH

The fourth planet from the Sun is half the size and one-tenth the mass of Earth. It has a characteristic reddish color that makes it easy to spot during its apparent trek among the background stars. Mars' light and dark areas change appearance throughout its year as winds cover and uncover various surface areas. Polar caps are observed to fade in the Martian summer and grow as winter progresses. Clouds are seen in the Martian atmosphere. Mars has two satellites, Phobos and Deimos.

THE VIEW FROM SPACECRAFT

The northern hemisphere is mostly lava-covered plains, which contain volcanoes as large as many states in the United States. These large volcanoes are considered evidence against continental drift. The southern hemisphere is strongly cratered. A gigantic valley, the Valles Marineris, extends some 5000 km across the Martian surface.

Mars' surface is red because its surface rocks contain relatively large amounts of iron. Thus the planet cannot be as strongly differentiated as other planets, meaning its core is not as extensive. This conclusion is consistent with Mars' lack of a magnetic field.

Evidence for the presence of liquid water in Mars' distant past abounds. Examples include ancient riverbeds showing debris that was carried along with the water, and sandbars shaped by flowing water. Evidence for water in the form of permafrost comes from layering of surface deposits of sand and ice, as well as from landslides. The polar ice caps consist of two parts: the large caps seen from Earth are made of frozen carbon dioxide, while smaller caps that remain for a while after the CO_2 sublimates contain water ice.

UNIQUE FEATURES

Mars' unique features include the volcanoes, valleys, and ancient river beds mentioned previously.

ADDITIONAL IMPORTANT FEATURES

The differences between Martian seasons in the northern and southern hemisphere are greater than those on Earth. This difference occurs because the planet has a more eccentric orbit, which produces greater changes in its distance from the Sun throughout its year. These seasonal changes produce winds that sculpt the surface.

10

THE JOVIAN PLANETS AND PLUTO

Do there exist many worlds, or is there but a single world? This is one of the most noble and exalted questions in the study of Nature.

ALBERTUS MAGNUS, THIRTEENTH CENTURY

The Jovian planets have gross similarities, yet no two are alike. While they all have satellite and ring systems, even these differ dramatically from each other. In this chapter you will learn about some of the exciting results of recent explorations, and you will see that processes studied in earlier chapters are applicable to each of the Jovian planets and their satellites.

Although Pluto is not a Jovian planet (nor a terrestrial planet, either) it fits into this chapter, thanks to its great distance from the Sun, and its similarity to the Jovian satellites.

10.1

INTRODUCTION

We will see what information we can obtain about the Jovian planets with observations from the Earth's surface before we look at the data obtained from spacecraft.

AN OVERVIEW OF GROUND-BASED STUDIES

The four Jovian planets, as observed through Earth-based telescopes, are shown in **Figure 10–1**. Even a quick glance reveals striking similarities and differences. More detail is observable on Jupiter because it is much closer to Earth than any of the others (about 4 AU at its nearest). Neptune, on the other hand, is about 30 AU from Earth at its closest. Jupiter shows distinct, variable, multicolored bands across its face. Markings include a large reddish area in the southern hemisphere, which has become known as the **Great Red Spot**. Its continual presence has been observed ever since first sighted over 300 years ago. Saturn has bands that are similar but less distinct and less colorful. Photographs of Uranus and Neptune show a hint of bands, but they are not as obvious. The relative motions of the bands and the different rotation rates at various latitudes show that we are observing complex atmospheres.

Satellite systems are readily observed around Jupiter and Saturn with even the smallest telescopes. Observations over only a few hours will reveal changes in the locations of the satellites in their orbits. Moderate-sized telescopes are required to observe the satellites of Uranus and Neptune.

Perhaps the most distinctive property of any of the planets is the beautiful ring surrounding Saturn. Although Galileo observed it as fuzzy protrusions, the ring shape was inferred some 50 years later by the astronomer and physicist Christian Huygens. Because Saturn was the only planet known to have a ring system, theoreticians scrambled to explain why rings around planets were rare.

As Saturn and Earth orbit the Sun, we are able to view the ring system from different orientations, as shown in **Figure 10–2**. The rings are so thin that when we see them edge-on, they practically disappear from view. The rings are only a few tens to hundreds of meters thick and are composed of many small bodies—ice, dust, rocks, and boulders up to a few meters in diameter. We know that the rings are not solid, because bright stars can sometimes be seen through them. Furthermore, the inner parts of the ring system rotate more rapidly than the outer parts, in accordance with Kepler's third law. It is estimated that the total mass of Saturn's rings is perhaps one-fourth that of our own Moon.

Inquiry 10–1 If Saturn's rings were solid, how would you expect them to rotate? Would the inner or the outer parts of such a ring take longer to go around?

The visible light we receive from each Jovian planet is reflected sunlight. However, because each planetary atmosphere affects the light in ways that depend on its chemical makeup (which will be described in Chapter 13), we have been able to discover, from Earth-based observations, the presence of methane and ammonia in each planet's atmosphere.

Astronomers were surprised in the middle 1950s when they recorded intense bursts of radio energy coming from Jupiter. These bursts of energy were found to depend on the orbital location of the satellite Io. Energetic discharges were interpreted as being caused by charged particles trapped in an intense Jovian magnetic field, forming Jovian "Van Allen belts," as described for Earth in Chapter 8. The planetary magnetic fields are discussed further in Section 10.4.

While Jupiter and Saturn were known to the ancients, Uranus and Neptune were not. Uranus was dis-

FIGURE 10-1. Ground-based photographs of the four giant planets: (*a*) Jupiter, (*b*) Saturn, (*c*) Uranus, and (*d*) Neptune. (Not to the same scale.)

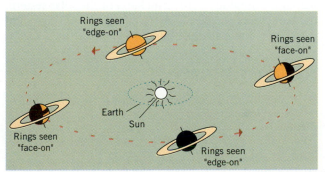

FIGURE 10-2. The apparent orientation of Saturn's rings as seen from Earth varies with its position in its orbit. The small circle around the Sun represents the orbit of the Earth.

covered by the English astronomer William Herschel in 1781. It was immediately apparent to him that it was not a star, because it moved from night to night with respect to the stars. Furthermore, it was blue-green. The story of the discovery of Neptune in 1845 and 1846, and its importance as a verification of Newton's ideas, was discussed in Chapter 5.

SPACE EXPLORATION OF JOVIAN PLANETS

The first spacecraft launched to the outer planets were Pioneer 10 and Pioneer 11 in 1973 and 1974. These were small spacecraft carrying instruments to study the charged particle and magnetic environments of Jupiter and Saturn. In addition, they were to produce images of higher resolution than were possible using telescopes on the Earth's surface. Even today these spacecraft continue to send data back to Earth from regions of the solar system never before studied. A number of important discoveries were made that paved the way for the more sophisticated and better equipped Voyager 1 and Voyager 2 probes launched in 1977.

The two Voyager missions were incredible testimony to both scientific curiosity and knowledge, and to our willingness to explore the unknown. Originally designed to last only long enough to study Jupiter and Saturn, the spacecraft proved to be hardy and long-lived. Voyager 1 flew by Jupiter in March 1979 and Saturn in late 1980. Voyager 2 passed Jupiter in July 1979, Saturn in August 1981, Uranus in January 1986, and Neptune in August 1989. Both are traveling through the outer solar system; we hope to continue to receive data for the next two to three decades as they exit the solar system and enter interstellar space.

Voyager 2 passed within 80,000 km of Uranus (and within 16 km of its designated target point, a feat that has been compared to bowling from Los Angeles to New York and making a strike) and was able to tell us more about Uranus in a few minutes than we had learned since 1781. Voyager 2 then passed 5000 km above Neptune's clouds. Most of the rest of this chapter is a testament to the successes of the two Voyager missions.

10.2
LARGE-SCALE CHARACTERISTICS

In this section we investigate such basic macroscopic properties as orbital and rotational characteristics, mass, size, density, seasonal changes, and energy emission.

ORBITS AND ROTATION

	Jupiter	Saturn	Uranus	Neptune	Pluto
Orbital eccentricity	0.048	0.056	0.047	0.009	0.249
Inclination (degrees)	1.3	2.5	0.8	1.8	17
Rotation period (hours)	9.92	10.66	17.24	16.11	153.3

The orbital characteristics of all the Jovian planets are similar and distinctly different from those of Pluto, whose orbit is both eccentric and inclined to the ecliptic.

Continuous observations of spots moving across a planet's disc allow astronomers to determine the rota-

tion period. Spots at Jupiter's equator complete their journey around the planet in only 9^h55^m. Regions closer to the pole take longer to complete their cycle. Saturn, too, completes its rotation in a short time, 10^h39^m. The small angular size of Uranus and Neptune and indistinctness of their features make their periods difficult to determine accurately. Prior to the Voyager missions, astronomers thought their periods were about 18 hours.

> READERS HAVING THE ACTIVITY KIT
> SHOULD DO ACTIVITY 10–1, THE
> ROTATION OF JUPITER, AT THIS TIME.

A rotation period of roughly 10 hours for Jupiter and Saturn means that the equatorial regions are moving at high speeds. In the case of Jupiter, this means nearly 45,000 km/hour. Such high rotation speeds easily explain the observed flattening in the appearance of the planet (Figure 10–1).

Inquiry 10–2 Why does rapid rotation cause a planet to flatten?

Uranus's rotation has been known for many years to differ from the general pattern in the solar system, in which rotation axes are nearly perpendicular to the orbital plane. For Uranus, the rotation axis is nearly *in* the orbital plane, with an inclination angle of 98°. If current theories are correct, the solar system was formed from turbulent gas, and it is entirely possible that local irregularities in the motion of this material caused the unusual rotation of Uranus. We have seen that, once established, motions persist until other forces enter the picture. Although this allows us to understand and predict the behavior of bodies moving under the influence of gravity, it does not tell us how the bodies were first set into motion. By analogy, suppose we see a baseball come crashing through the living room window. Looking outside, we can only conjecture—albeit in this case reasonably—that the ball was set into motion by the scared kid holding the baseball bat. For the solar system, the game was played nearly five billion years ago, and the players have all gone home.

SIZE, MASS, DENSITY

	Jupiter	Saturn	Uranus	Neptune	Pluto
Diameter (Earth units)	11.19	9.41	4.01	3.89	0.18
Mass (Earth units)	317.9	95.1	14.6	17.2	0.0024
Density (gm/cm³)	1.33	0.70	1.24	1.61	2.10

In Chapter 6 we found that the first object a visitor from space would notice in our solar system, after the Sun, would be Jupiter. Jupiter is brighter than the other planets because it is considerably larger. At 11 times the diameter of the Earth, Jupiter could contain some 1300 Earths inside it. Saturn, Uranus, and Neptune have diameters of 9, 4, and 4 times the Earth's diameter, respectively.

The masses of all the Jovian planets are easy to obtain thanks to the presence of their many satellites. Jupiter, with a mass more than 300 times Earth's mass, contains more material than all the other planets put together, about 70% of the mass of the solar system that is not bound up in the Sun. Jupiter is so massive that it is perhaps more appropriate to think of it as a binary companion to the Sun rather than just the largest of the planets, even though it is not massive enough to be a star in its own right. The other Jovian planets, while smaller, still contain substantial amounts of material.

> READERS HAVING THE ACTIVITY KIT
> SHOULD DO KIT ACTIVITY 10–2, THE
> GALILEAN SATELLITES AND THE MASS OF
> JUPITER, AT THIS TIME.

The density of each of the Jovian planets is low. In the case of Saturn, which has a mass of 95 Earth masses and a size of 9 Earth radii, the density is only 0.7 g/cm³. This value, which is less than that of water, means that Saturn could float if placed in an appropriately sized tub! The densities of the other Jovian planets are between 1.3 and 1.6 g/cm³, considerably less than the 5.5 g/cm³ for Earth.

Inquiry 10–3 Considering their low densities, what two elements would you expect to make up the bulk of the mass of the Jovian planets?

SEASONS

	Jupiter	Saturn	Uranus	Neptune	Pluto
Seasons	no	yes	extreme	yes	extreme

The existence of seasons, of course, is determined by the tilt of the rotation axis to the perpendicular of the orbital plane. The tilts were discussed and shown in Figure 6–5b. While Jupiter's angle of only 3° produces no seasons, Saturn and Neptune have axis tilts between 25 and 30°. These tilts, therefore, produce seasonal changes in the amount of energy received by every part of the planets' atmospheres.

Uranus is unique among the Jovian planets with its 98° tilt. There is a time in its orbit when the north pole points almost directly at the Sun; half of its 84-year orbital period later, the south pole points at the Sun. Midway between these times, the Sun shines directly onto the equatorial regions.

Inquiry 10–4 For approximately how long would the Sun be above the horizon for an observer at Uranus's north pole? Would you describe Uranus's seasonal changes as mild or harsh? Why?

EXCESS ENERGY

	Jupiter	Saturn	Uranus	Neptune	Pluto
Excess energy	yes	yes	small	yes	no

Ground-based observations of Jupiter, later confirmed by the two Pioneer spacecraft, had indicated that Jupiter emits about twice as much energy as it receives from the Sun. Jupiter is not, therefore, just a passive reflector, but has an internal energy source of its own. For this reason, some astronomers have characterized Jupiter as "almost a star," although its mass is more than 10 times too small for it to generate energy by means of nuclear fusion, as real stars do.

The source of Jupiter's excess energy is heat left over from the time of formation, produced when the gravitational potential energy in the protoplanetary cloud was converted into trapped heat during the collapse. The slow rate at which heat escapes the planet indicates that Jupiter's interior is probably in a turbulent, boiling state, with much of the internal heat being carried outward by convection.

Saturn, too, radiates about twice as much energy as it receives from the Sun. In fact, relative to its mass, it emits more excess energy than Jupiter. The amount of Saturn's excess energy requires an additional energy production process, which seems to be the ongoing gravitational settling of helium toward the interior. This settling, or differentiation, releases gravitational potential energy. An observation in favor of this idea is the smaller abundance of helium in Saturn's atmosphere than in Jupiter's.

Of the outer two Jovian planets, Uranus emits only a small amount of excess energy. Neptune radiates more than 2.7 times as much energy as it receives from the Sun. But if these planets are as similar as we have always thought, why does Neptune emit more energy? Part of the reason may be that Uranus's south pole is currently pointing at the Sun. In addition, Uranus and Neptune are not quite as similar as we had thought, as we will see in the next section.

10.3

ATMOSPHERES

Each spacecraft returned dramatic new images. Worlds that had never before been seen as anything more than a point of light were turned into unique "places." This section begins our more detailed examination of results from the Voyager missions. The discussion of the satellites comes in Section 10.5.

APPEARANCE AND CIRCULATION

Full-disc images of each of the Jovian planets, obtained by the Voyager spacecraft, are shown in **Figure 10–3**. Jupiter revealed a varied and continually changing "weather" pattern of great complexity. The banded structure within Jupiter's atmosphere was once thought to be relatively stationary, but now material has been observed moving between bands. In addition, narrow bands have been seen to consolidate and wide bands to come apart.

The spotted regions are thought to be Jovian storm systems, not unlike those on Earth, although larger in scale and more persistent. The Great Red Spot, for example, is similar to a terrestrial cyclone (actually an anticyclone) that contains rising gases. The upward motion of the gas is accentuated by the excess energy flowing from the interior. This motion causes the Great Red Spot to rise to about 8 km above the surrounding gases. Forces generated by the rapid rotation of the planet cause the inflowing gas to swirl around like a whirlpool (the Coriolis effect of Chapters 5 and 8). **Figure 10–4** shows a close-up of the Great Red Spot. About 30,000 km in length, it would easily swallow the Earth!

The Great Red Spot rotates, and small spots approaching the storm have been observed to accelerate and be "gobbled up" by it. This is not a temporary phenomenon; the spot has been observed for about 300 years, although lately it has diminished somewhat. The long duration of this giant storm turns out to be less surprising than it might seem at first, because one consequence of Jupiter's large size and dense atmosphere is that its weather takes place on time scales much longer than those on Earth. Calculations indicate that 300 years is roughly the length of time one would expect a major storm on Jupiter to last. Some theoreticians believe that the Great Red Spot may be a permanent feature of Jupiter's atmosphere.

Regions of the Jovian atmosphere outside the Great Red Spot (Figure 10–4) exhibit fantastic sawtooth-shaped patterns of gas that are remarkably similar to the turbulence created in air by the passage of an airplane. The diversity of color shown in the convoluted flow patterns, from reds and oranges to yellows and browns and even blues, provides a challenge to students of chemistry and meteorology. At present, at least three major cloud levels have been distinguished on Jupiter: a high layer of frozen ammonia crystals, a middle layer that appears brownish, and regions where our line of sight apparently penetrates deep down into the planet's atmosphere, where there is a blue appearance.

Saturn shows many features similar to those on Jupiter. However, Saturn's features have a muted appearance because the color contrast with the surrounding regions is not as strong (**Figure 10–5**). Voyager detected massive electrical discharges coming from a region near the equator, and it is suspected that these are lightning discharges from a giant storm.

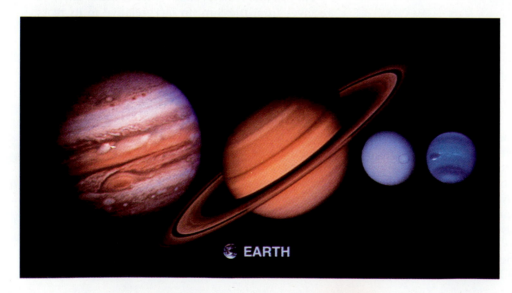

FIGURE 10–3. Voyager images of the Jovian planets: Jupiter, Saturn, Uranus, and Neptune. The Earth is shown for a size comparison.

FIGURE 10–4. Jupiter's Great Red Spot, showing the turbulent gases produced in its wake.

FIGURE 10–6. Neptune as observed by Voyager. The full disc of the planet shows high-altitude white clouds in addition to the Great Dark Spot, and a bright, rapidly moving bright spot called Scooter.

FIGURE 10–5. Saturn's atmospheric features appear less distinct than those on Jupiter.

Voyager's view of Uranus (Figure 10–3) was supposed to give us the same dazzling views of atmospheric features that we saw on Jupiter and Saturn, but such was not to be. Uranus turned out to be enveloped in a featureless haze. There are no stripes, bands, or spots of the sort that we might have expected by analogy with Jupiter, even though Uranus is also rotating rapidly. In spite of careful analysis, only four small clouds were seen. We are apparently witnessing the effects of hydrocarbon compounds in the upper atmosphere, molecules that we would call "smog" if they appeared over Los Angeles! These molecules are highly effective at scattering light; just as in a fog on Earth, detail is lost. The blue regions are indications of methane in the atmosphere, a compound that scatters blue light effectively while absorbing red light.

The absence of details in Uranus's cloud structure may come from the fact that solar radiation is currently incident directly on the planet's north pole (due to the extreme tilt of the planet's spin axis). Perhaps this unusual situation works against the formation of jet streams and other weather patterns familiar from Earth. The lack of substantial excess energy may contribute to the lack of detail in the atmosphere.

Neptune thrilled planetary scientists with its appearance (Figures 10–3 and 10–6). Neptune's blue color, which comes from absorption of red light by methane, is richer because we see to deeper layers of its atmosphere. Neptune is more dynamic than Uranus, with wispy clouds floating at high levels in the atmosphere (**Figure 10–6**) and spots similar to Jupiter's Great Red Spot. The largest, dubbed the **Great Dark Spot** (Figure

10–6), rotates and is undoubtedly a long-lasting storm system. Other spots, such as the white spot called Scooter, move rapidly; winds in excess of 2000 km/hour have been measured.

CHEMICAL COMPOSITION

The low overall density of the Jovian planets points toward a high abundance of the lighter elements, such as hydrogen and helium. We also expect nitrogen and carbon atoms to be present because they are present in the Sun. However, at the cool temperatures at the higher levels of the Jovian planetary atmospheres, hydrogen will combine with carbon and nitrogen to form the volatile compounds methane (CH_4) and ammonia (NH_3). Ground-based observations showed these compounds to be present in the Jovian planets' atmospheres. While observations indicate less ammonia in Saturn, Uranus, and Neptune than in Jupiter, the lack is readily explained: at the lower temperatures of the outermost planets, ammonia condenses into ices that are not readily observed.

While hydrogen (in the molecular form) had been observed in Jupiter and Saturn in the early 1960s, helium could not be detected from the Earth's surface. Helium has been observed in the expected amounts, however, by the planetary probes.

Voyager observations, along with information about temperature and pressure at various depths within the atmospheres, allow astronomers to compute models of the planetary atmospheres. The atmospheric structure of each of the Jovian planets is shown in **Figure 10–7**. The models show that Saturn's cloud structure is different from that of Jupiter. These differences can explain the muted appearance of Saturn's features.

10.4
PLANETARY INTERIORS

The interior structure of a planet is not directly observable but must be inferred from models. The model computations are based on the laws of nature thought to be important in the interior of a gaseous planet and the constraints imposed by a variety of observations. Some of these observations include those previously discussed—density, flattening due to rotation, excess energy, temperature—and the presence of strong magnetic fields.

MAGNETIC FIELDS

	Jupiter	Saturn	Uranus	Neptune	Pluto
Magnetic field	strong	strong	yes	yes	no?

Unexpected radio emissions were observed in the middle 1950s emanating from Jupiter. The observations, which are unrelated to the excess energy previously de-

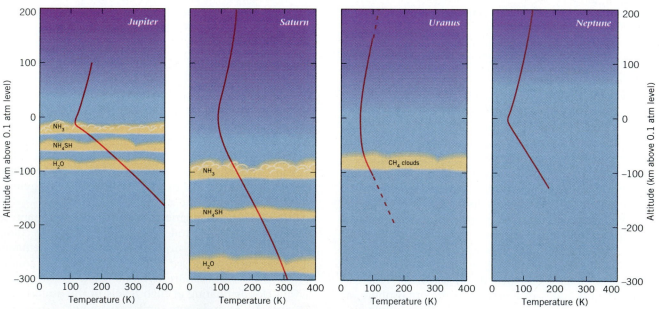

FIGURE 10–7. A comparison of the cloud structures of the Jovian planets. The thicker clouds on Saturn help mute the appearance of atmospheric structure.

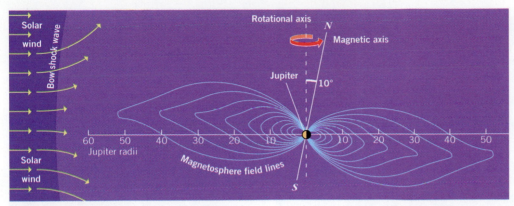

FIGURE 10-8. Jupiter's magnetosphere.

scribed, were inconsistent with energy coming from either a hot or a cool body, but consistent with energy coming from an object having a strong magnetic field in which charged particles were moving at speeds near that of light. Such radiation is called **synchrotron radiation**. In other words, the observation of synchrotron radiation told astronomers that Jupiter has an intense magnetic field. (We will see later that synchrotron radiation also appears elsewhere in the universe.)

A detailed study of the charged particle and magnetic environment around each of the Jovian planets was one of the major goals of the Pioneer and Voyager programs. Jupiter's magnetic field, for instance, is shown in **Figure 10-8** and is found to be extensive. It is so large, in fact, that if you were able to see it visually, its angular extent would be larger than the full moon! The magnetic field is flattened by Jupiter's rapid rotation. Charged particles trapped by the magnetic field form radiation belts, analogous to Earth's Van Allen belts. Some of the particles spiral into the magnetic poles, producing aurorae. **Figure 10-9** shows an auroral ring like that for Earth as seen in Figure 8–16b. The Jovian atmosphere also has an electrical environment that produces extensive lightning.

The other Jovian planets also have magnetospheres, although not as strong or as extensive as Jupiter's. For example, Jupiter's magnetosphere extends a distance of 65 times the planet's radius, but the Earth's magnetosphere extends only 10 times its radius. Also, the magnetic field at Jupiter's equator is nearly 14 times as strong as it is at the Earth's equator. **Figure 10-10** compares the magnetic field information for each of the Jovian planets. Just as Earth's magnetic axis is tilted with respect to its rotation axis (by 11°), the magnetic fields of the Jovian planets generally are not aligned with their rotation axes. Saturn is the main exception; its rotation axis and magnetic field axis are nearly identical. Uranus and Neptune provide a real surprise: in both cases, the magnetic axis is tilted 47° to 60° from the rotation axis.

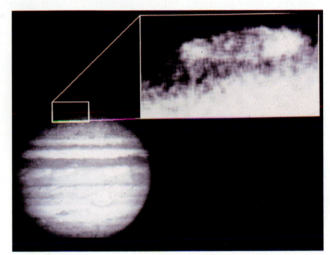

FIGURE 10-9. Jupiter's auroral oval, as observed in ultraviolet light by the Hubble Space Telescope.

Furthermore, none of the magnetic field centers are at the planets' geometric center. The center of Neptune's field is offset farther from the geometric center than for any other planet. The cause of the field is discussed in the next subsection.

Inquiry 10-5 For review, what two conditions are required to produce a planetary magnetic field?

The magnetic fields of the planets rotate at rates that differ from what is observed of the planetary "surface." Because the rotation period of the observable clouds also depends on their latitude, such periods are not necessarily fundamental to the planetary body itself. The magnetic field rotation period is, however, related to the planetary body, and provides the most accurate mea-

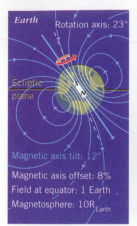
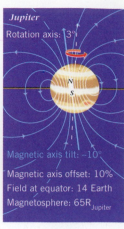

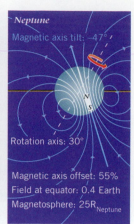

FIGURE 10–10. Summary and comparison of the magnetic environments of each of the Jovian planets. Note the relative strengths of the magnetic field at each planet's equator, the tilt of the magnetic field relative to the rotation axis, and the offset of the field from each planet's geometric center.

sure of rotation. The approximate rotation periods of Jupiter, Saturn, Uranus, and Neptune are 9^h55^m, 10^h40^m, 17^h14^m, and 16^h7^m, respectively.

INTERIOR STRUCTURE

	Jupiter	Saturn	Uranus	Neptune	Pluto
Liquid metallic core	yes	yes	no	no	no
Differentiated	yes	yes	yes	yes	?

Magnetic field observations are telling planetary scientists a great deal about the interiors. Their job now is to understand what they are being told.

Magnetic fields are produced by moving materials that are good conductors of electricity, such as a metal in the case of the Earth. The Jovian planets, however, are mostly hydrogen and helium, which are not metals. A possible answer to the apparent dilemma appears when models of the interior structures are computed, as shown in **Figure 10–11**. Both temperature and pressure increase strongly with increasing depth. Underneath the

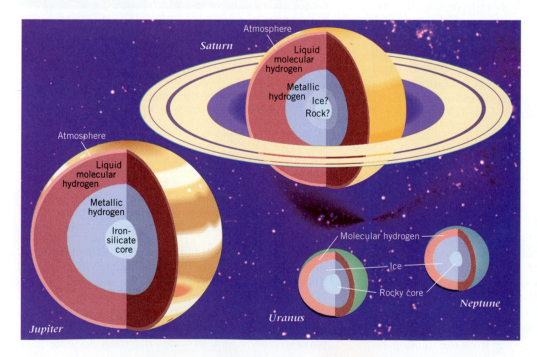

FIGURE 10–11. The interior structure of each of the Jovian planets. Note in particular the similarities of Jupiter and Saturn, and of Uranus and Neptune, and the differences between these two groupings. (Not to the same scale.)

visible clouds of all four Jovian planets, hydrogen is compressed into liquid molecular hydrogen.

Interior models of Jupiter show the pressure to be greater than a million times the sea-level pressure on Earth—so high that hydrogen is compressed into an exotic state not yet duplicated on Earth known as **liquid metallic hydrogen**. In this state, hydrogen is a liquid but behaves like a metal, so it becomes an excellent conductor. Its conducting properties, along with the interior's rapid rotation, are thought to produce a magnetic dynamo such as we discussed for the Earth, and thus to generate the observed magnetic field. Jupiter's central region is thought to consist of a hot (30,000 K), high-pressure solid core consisting of the heavier atoms that sank to the center because of gravity. This 10-to-20-Earth-mass core might also contribute to the observed magnetic field.

A similar interior model results for Saturn. However, the lesser pressures decrease the size of the region of liquid metallic hydrogen.

The interior models for Uranus and Neptune do not include liquid metallic hydrogen; the interior pressures are not great enough. Therefore, the magnetic field must be caused either by the core or by as yet unknown conditions in the interior. In any case, the tilts between the magnetic and rotation axes and the offset from the geometric centers remain mysteries.

10.5
SATELLITES

	Jupiter	Saturn	Uranus	Neptune	Pluto
Satellites	16	18	15	8	1

The views of the outer planets as observed by the Voyager spacecraft were remarkable. However, they were almost upstaged by the magnificent images of their satellites. The satellites generally fall into two groups: those having nearly circular orbits in the planet's equatorial plane, and those having orbits that are peculiar in one way or another (such as orbital shape, inclination, or direction of motion). Most of them have densities near that of water, although some are higher. The densities indicate a mix between water ice and rock.

Perhaps the most surprising fact discovered about these satellites is their lack of similarity in outward appearance. Each seems to have its own dramatic and individual personality. In this section, we will examine each satellite and satellite system separately, comparing individual bodies when appropriate.

THE SATELLITES OF JUPITER

While Jupiter has a complement of 16 known satellites, the 4 largest were first seen by Galileo and are known in his honor as the **Galilean satellites**. Io, Europa, Ganymede, and Callisto have densities that are higher than almost all other Jovian planetary satellites, indicating surfaces covered with ices of methane and ammonia.

Io, the innermost, is red and yellow, looking very much like a pizza (**Figure 10–12a**)! Its surface appears to

(a)

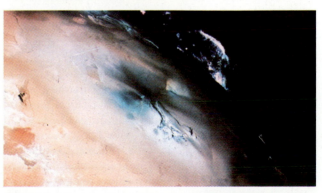

(b)

FIGURE 10–12. (a) Io. (b) A volcano during eruption, along with a volcanic vent in the middle of the figure.

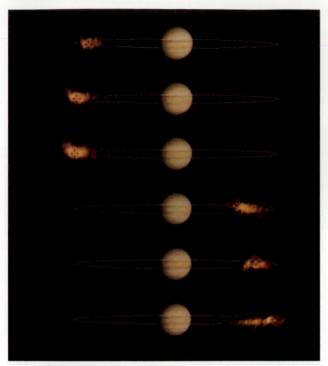

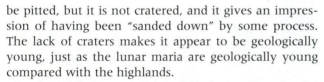

FIGURE 10–13. Ground-based photographs showing sodium vapors surrounding Io. As the satellite circles Jupiter, sodium forms a doughnut-shaped cloud known as the Io torus around the planet. The figure is to scale, in which Io is the dot inside the cross-hair over the cloud.

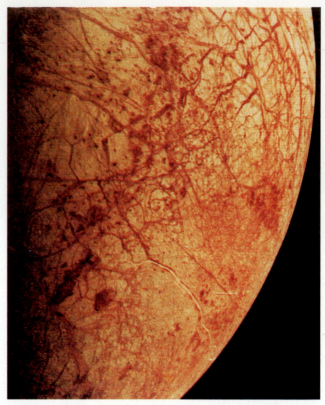

FIGURE 10–14. Europa with a resolution of 10 km.

be pitted, but it is not cratered, and it gives an impression of having been "sanded down" by some process. The lack of craters makes it appear to be geologically young, just as the lunar maria are geologically young compared with the highlands.

Large tidal stresses are raised in the interior of Io by the powerful Jovian gravitational field. The interior is therefore heated to the point at which material is melted and vaporizes. One of Voyager's most fantastic results confirms the nature of Io: images show evidence of many simultaneously active volcanos on its surface (see **Figure 10–12***b* for an example). This discovery makes Io the most geologically active body in the solar system.

Io's volcanoes emit sulfur and oxygen atoms, which become charged. These particles result in a glowing cloud of sodium atoms and a doughnut-shaped region of glowing charged sulfur particles known as the **Io torus** that encompasses its orbit (**Figure 10–13**). The charged particles on its surface interact with Jupiter's intense magnetic field and induce a flow of electricity between Jupiter and Io's surface. The charged particles then interact with the Jovian magnetic field to produce the aurorae discussed previously.

Europa, the smallest of the four satellites, has an orangish, off-white color and a high reflectivity that suggests the possibility of an icy surface (**Figure 10–14**). Its most distinguishing and intriguing features are lines thousands of kilometers long and a hundred kilometers wide running across the surface. Their depth is unknown. They may be extensive surface cracks in the 100-km-thick surface ice. The satellite has few craters and no mountains, indicating a surface that is not strong enough to hold a mountain's weight without sinking. Water ice has been observed on its surface.

The third Galilean satellite, **Ganymede**, is bigger than the planet Mercury and is, in fact, the largest satellite in the solar system (**Figure 10–15**). Whereas the inner two satellites have densities similar to those of the terrestrial planets, the outer two have densities only about twice that of water, suggesting that a significant portion of these satellites may in fact be water. Ganymede shows evidence of an icelike surface along with many features somewhat similar to those on our own Moon, such as maria, highlands, and craters. The icy surface may have resulted when ice in the interior melted, worked its way to the surface, and refroze.

(a)

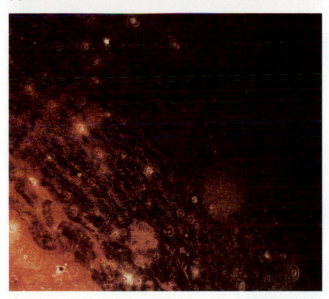

(b)

FIGURE 10–15. Ganymede as seen face-on (*a*) shows craters and rays. (*b*) The surface also shows younger grooves tens of kilometers wide and extending over thousands of kilometers. Abrupt discontinuous lines often occur and indicate surface motion.

While Ganymede has more craters than the lunar maria, there are no big craters. Perhaps of most interest to geologists is the existence of discontinuously broken

FIGURE 10–16. Callisto shows the most heavily cratered landscape of all the Galilean satellites, indicating an old surface. Callisto also shows a multi-ringed basin called Valhalla.

lines; these are extensive transverse fault regions, indicating the probable presence of internal upheavals.

The outermost Galilean satellite, **Callisto**, is heavily cratered (**Figure 10–16**). This appearance could result from the fact that Callisto is the coolest of Jupiter's satellites and hence the one whose surface solidified first. However, its smooth-appearing edge and lack of craters larger than 50 km in diameter suggests that it has a rather plastic surface that may not be able to support features as large as mountains. Callisto has one major impact basin, Valhalla, (Figure 10–16), which is a multi-ringed basin like Mare Orientale on the Moon and the Caloris Basin of Mercury. Those structures, however, were formed on rocky surfaces.

Not all Jupiter's moons are as large as the Galilean ones. Amalthea, one of Jupiter's inner satellites, is both small (only 130 by 80 km) and irregular.

Inquiry 10–6 From the size and shape of Amalthea, what might you conclude about its origin as a Jovian satellite?

THE SATELLITES OF SATURN

The Voyager mission also spent considerable time examining the satellites of Saturn, which turned out to be every bit as diverse as those of Jupiter.[1] They are shown

[1]The names of the nine "classical" moons, in order of distance from Saturn, may be remembered by the mnemonic *MET DR THIP*, for **M**imas, **E**nceladus, **T**ethys, **D**ione, **R**hea, **T**itan, **H**yperion, **I**apetus, and **P**hoebe.

in Figure 10–17. Most of these satellites have densities between 1.0 and 1.4 gm/cm^3, indicating interiors of water ice. Other data indicate the surfaces are water ice and not the heavier methane and ammonia of the Galilean moons. Each has one side that is more heavily cratered than the other side. This asymmetry is easy to understand: all of them have **synchronous rotation**, so that the same side always faces Saturn. Even more important is that the same side always faces toward the

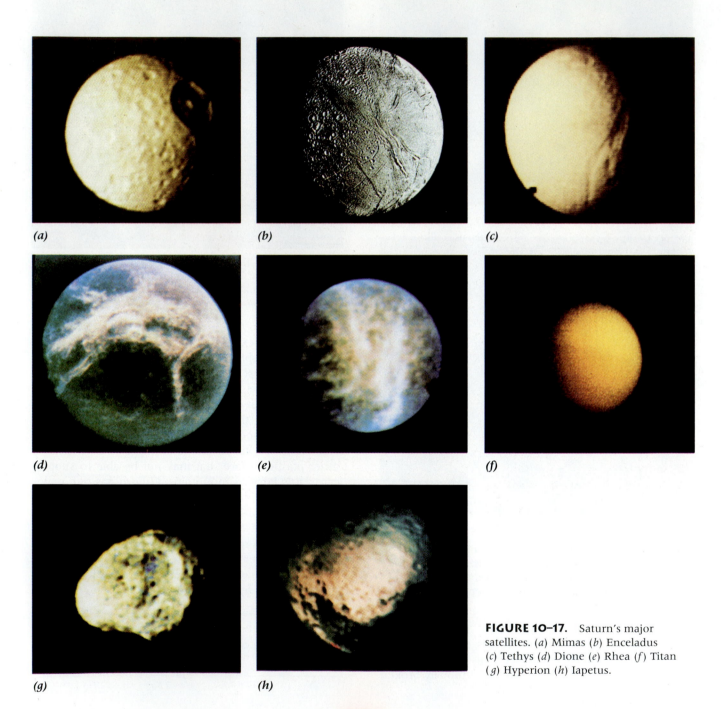

(a)

(b)

(c)

(d)

(e)

(f)

(g)

(h)

FIGURE 10–17. Saturn's major satellites. (*a*) Mimas (*b*) Enceladus (*c*) Tethys (*d*) Dione (*e*) Rhea (*f*) Titan (*g*) Hyperion (*h*) Iapetus.

direction of orbital motion, thus receiving more impacts than the opposite side.

Mimas, the smallest satellite at only 390 km in diameter, looks like a Cyclops with a distinguishing giant (100-km) crater on its surface. The crater, whose diameter is one-third that of the satellite itself and 10 km deep, has a central peak 6 km high. The impacting body is estimated to have been 10 km across. Furthermore, Mimas is scarred by troughs 100 km long, 10 km wide, and 1 to 2 km deep—a possible residue of fracturing from past meteoroid impacts. Its density, 1.2 gm/cm^3, implies it contains primarily water ice.

Enceladus is one of the more complex satellites in the solar system. Its surface is geologically young and diverse. Reflecting more than 90% of incident sunlight, it is brighter than newly fallen snow! This high reflectivity, along with the body's low density, points toward water ice. While it has some impact craters 35 km in diameter, it is not densely cratered, implying that continual resurfacing occurs. However, there is no active volcanism, and no landforms indicative of past volcanic activity. The variety of surface features is taken to imply some unknown source of internal heating.

Tethys is distinguished by a 2000-km-long gorge covering three-fourths of its circumference. Planetary scientists are unsure whether the gorge was caused by an impacting body or came from internal activity. Furthermore, Tethys has a 400-km-diameter crater, which is larger than Mimas's crater (**Figure 10–18a**). Tethys's more densely cratered landscape indicates an old surface.

Dione shows bright wisps across its surface, which are probably (geologically) fresh ice. Its higher-than-average density (1.4 gm/cm^3) means it contains more rock than the other satellites. Its winding valleys are suggestive of internal processes; its plains regions suggest some type of unknown surface renewal. Cratering was more intense on the side facing its direction of orbital motion, while the other side is darker and contains the wispy streaks.

Rhea, the largest of the innermost satellites at 1530 km in diameter, is nearly a twin of Dione. Its forward-facing hemisphere, which is covered with almost pure water ice, has two sections: one consists of large craters without small ones, while the other has small craters without any large ones. This side also contains the remains of three multi-ringed basins. Rhea's trailing hemisphere also has wisps. Planetary scientists interpret the observations of Rhea in terms of a period of extensive volcanism that forced hot liquid water to the surface, where it froze solid to provide the surface we see today.

Titan received its name when it was thought to be the largest satellite in the solar system. It is larger than

(a)

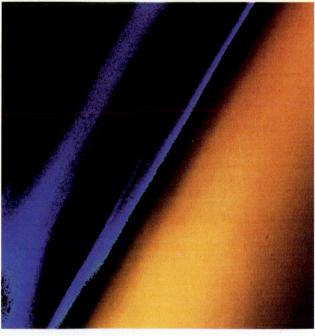

(b)

FIGURE 10–18. Additional views of Saturn's satellites. (*a*) Tethys has a crater whose diameter is 400 km, in addition to a long valley. (*b*) Titan shows the presence of separate haze layers within its atmosphere.

Mercury but slightly smaller than Ganymede. Titan was studied in some detail because Earth-based telescopic data had shown it to have its own atmosphere. **Figure 10–18b**, for example, shows separate haze layers within the atmosphere.

As the photo also shows, there was little detail to be seen in Titan's atmosphere because the atmospheric haze was far thicker than anticipated. The atmosphere is 99% nitrogen, and is so extensive that the pressure on the surface is 50% greater than on Earth. Temperatures are not quite low enough for liquid nitrogen to exist on Titan, but it is possible that puddles and small lakes of liquid hydrocarbons exist on the surface. As a result of these atmospheric conditions the chemistry on Titan can be expected to be highly complex. Indeed, with a composition rich in methane and other organic molecules, the chemistry may not be too dissimilar from that which prevailed on the primitive Earth before life arose.

Titan's density of 1.90 gm/cm³ implies it has equal amounts of water ice and rock. Thanks to its atmosphere, the surface remains unseen and mysterious.

Inquiry 10–7 How would you interpret Titan's lack of an observed magnetic field?

Hyperion is a small, irregularly shaped chunk of material of unknown density (Figure 10–17*g*). It probably resulted from a collision and was later captured by Saturn.

Iapetus was known from Earth-based studies to have two dramatically different sides—one light and the other dark. This interpretation of the observations was confirmed by Voyager; in Figure 10–17*h*, the dark-appearing side is in full sunlight. This side faces the direction of orbital motion and contains material that is as dark as black tar. It either is thick or is replenished often, because no bright spots or craters are seen on top. Whether its source is internal or external is uncertain. One other possible explanation is that some materials darken when exposed to cosmic rays.

Phoebe is small and dark. Its darkness is reminiscent of the dark side of Iapetus. Given that the nucleus of Halley's comet was found to be dark and similar to the carbonaceous chondrite meteorites, we can wonder whether there is a relationship between these bodies.

Pioneer and Voyager discovered a number of new satellites. One of these is called **Dione B**, because it occupies Dione's orbit but leads Dione by a 60° angle. Tethys has two satellites in its orbit—one 60° in front and one 60° behind. These three so-called **co-orbital satellites** are all small and irregular and are probably fragments from collisions. The existence of the co-orbital satellites is readily understood in terms of the La-

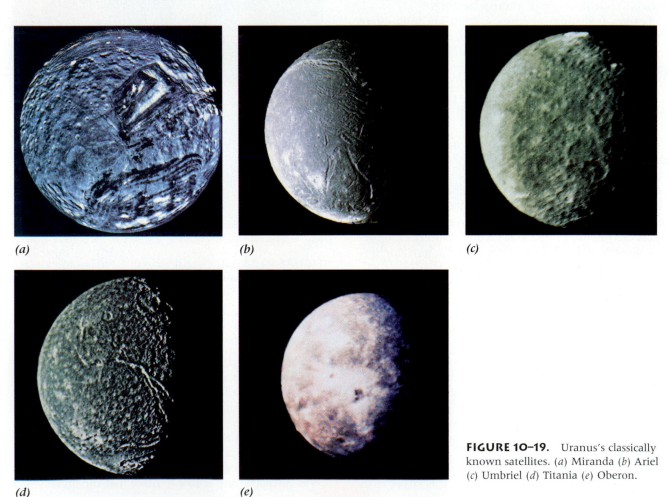

(a) (b) (c)

(d) (e)

FIGURE 10–19. Uranus's classically known satellites. (*a*) Miranda (*b*) Ariel (*c*) Umbriel (*d*) Titania (*e*) Oberon.

grangian points described in Chapter 7 for the Trojan asteroids. In this case, however, the triangle is formed by Saturn, Dione, and Dione B. Locations 60° ahead and behind are where the net gravitational forces of the planet and the Sun cancel, allowing bodies there to remain in stable positions.

THE SATELLITES OF URANUS

Prior to Voyager, Uranus was known to have 5 moons; now 15 satellites are known.[2] Images of some of these satellites (**Figure 10–19**) show the same pattern of heavy bombardment exhibited by Saturn's Mimas. They all have densities slightly higher than that of Uranus, 1.5 to 1.7 gm/cm^3, and probably consist of methane and ammonia ices. But their ability to reflect light is lower than that of Saturn's satellites, perhaps hinting at the presence of small dark particles in the outer solar system.

The satellites' orbital planes are in the strongly inclined equatorial plane of the planet, not in the ecliptic plane. This finding is significant when it comes to understanding the formation of the Uranian system, as described below.

Because **Miranda** is less than 500 km across, astronomers thought it was too small to be particularly interesting. How wrong they were! Its appearance is strikingly odd (**Figures 10–19a, 10–20**). Miranda's surface contains 5-km-high cliffs, deep gorges, faults, craters, and valleys. Abrupt changes in the surface occur from one area to another. The light colored L-shaped feature near the center, called the **chevron**, is located within a well-defined trapezoidal block having 200-km sides. In another area is an elliptical region called the **Circus Maximus** (because it is reminiscent of a Roman racetrack). The 5-km-high cliff is shown in more detail in Figure 10–20. Planetary scientists have been driven to extreme lengths of speculation to understand Miranda's surface. One hypothesis suggests that sometime in its past Miranda was actually broken (perhaps by a collision) into a number of pieces, some of rock and some of ice. The pieces re-formed into a haphazard assemblage, and there was a tendency for the lower-density ice to rise and the higher-density rock to sink. This process might generate enough heat to melt portions of the crust and also create surface stress fractures. Another hypothesis is that tidal interactions with Uranus heated the interior. Partially molten material in the core underwent convection and rose to the surface, broke it open, and then froze in place.

Ariel contains a complex system of faults, along with large flat regions that resemble mudflows. **Umbriel** has

[2]The five classical moons can be remembered with the mnemonic *MAUTO*, or **M**iranda, **A**riel, **U**mbriel, **T**itania, and **O**beron.

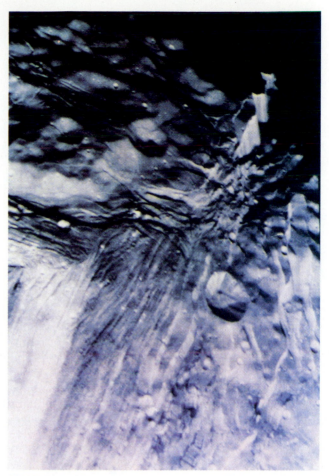

FIGURE 10–20. Miranda. The satellite's distinct surface shows ridges, valleys, and cliffs indicating a unique history.

a surface with few craters. It has a large bright ring of unknown origin. **Titania**, too, has faults that must indicate past internal activity. Some of its craters show rays, while others appear to contain a black substance that oozed from the crater's bottom. The outermost moon, **Oberon**, shows impact craters with bright rings of associated ejecta. Some of the craters have dark centers, perhaps dirty water erupting from the icy interior of the moon. Oberon is distinctive in showing a 6-km-high mountain. Its surface structure must be strong to prevent such a mass from sinking.

The mysteries of the Uranian moons are ultimately tied up with the mystery surrounding the origin of the 98° tilt of the spin axis of Uranus. If the planet's tilt was caused by a collision (possibly with a huge asteroid), and the moons were already in existence around the (previous) equatorial plane of the planet, their orbits would have been severely perturbed; they could have repeatedly collided with each other, ultimately re-forming into the bodies we see today orbiting around the new, tilted equatorial plane. But this process would be a slow one, and at least one satellite of Uranus (Umbriel) shows very old features that have remained undisturbed for a long period. On the other hand, if Uranus spun off gas to form the satellites after its tilt had been changed, then it is difficult to explain where the rocky parts of the satellites came from.

THE SATELLITES OF NEPTUNE

Neptune's number of known satellites jumped from two to eight after Voyager's encounter. The recently discovered ones are all small; one is 400 km in diameter, but the others are less than 200 km in diameter.

Triton, whose radius is nearly 1450 km, has been known since the 1970s to have an atmosphere composed of nitrogen and methane. Its surface pressure is low, only 0.001% the sea-level pressure on Earth. It seems to contain a haze, perhaps of frozen hydrocarbons.

The thinness of the atmosphere allowed observations of the surface, which is composed mostly of frozen nitrogen. Triton's surface rivals those of Enceladus and Miranda in complexity and wonder (**Figure 10–21***a*). While one part is smooth and dimpled like a cantaloupe rind, the other part is heavily cratered and contains grooves and ridges traversing the surface, indicating tectonic processes. Its few craters indicate a relatively young surface. The images show traces of flows; a solidified liquid appears to fill craters (**Figure 10–21***b*). Perhaps the most dramatic discovery is the presence of some small dark streaks, which astronomers have interpreted as carbon-rich geysers shooting material 8 km above the surface. The energy to produce the geysers is thought to come from heat generated during changes in the frozen nitrogen under the surface.

Triton shows aurorae in its upper atmosphere. These are caused by an interaction between charged particles in Neptune's radiation belts and Triton's atmosphere.

Voyager provided data for the first accurate determination of Triton's mass, from which a density of 2 gm/cm^3 was found. It is thus similar to Titan in being about half water ice and half rock.

Triton's orbital motion is retrograde. Why should a large satellite be in such an orbit? Was it formed in a retrograde orbit or did it undergo a collision with one or more bodies? The astronomical community does not yet agree on the explanation. **Nereid**, Neptune's other satellite known prior to the Voyager mission, is small, with a diameter of only 340 km. It moves in a direct (nonretrograde) orbit that is highly eccentric and inclined at a large angle (28°) to the equatorial plane. The orbital characteristics of the satellites raise difficult questions about their origin.

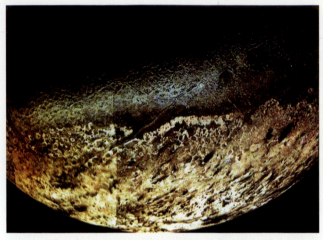

(a)

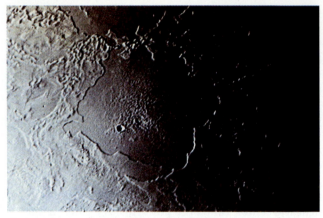

(b)

FIGURE 10–21. (*a*) Neptune's satellite Triton. The south polar region of Triton shows a complex and rough terrain with dark regions thought to be ice geysers. There is also an extensive dimpled region containing long cracks in the icy surface. (*b*) A lake of frozen water, methane, and ammonia.

Inquiry 10–8 What orbital characteristics would you expect if these satellites were formed from natural processes at the time of Neptune's formation? How might you explain the observed orbital characteristics?

10.6
PLANETARY RINGS

	Jupiter	Saturn	Uranus	Neptune	Pluto
Rings	yes	yes	yes	yes	no?

The sight of the rings of Saturn through a telescope on a clear night is a thrill not to be missed. While telescopic pictures and images from spacecraft may show detail, seeing the planet itself with your own eyes is a rewarding experience.

The telescopic view of Saturn's rings (refer back to Figure 10–1) generally shows two rings. (There is a faint inner ring that is sometimes visible.) Between the outer two rings is a region named the **Cassini division**, which appears to lack reflecting particles. The cause of this division was thought to be similar to the process that produced the Kirkwood gaps in the asteroid belt. The Cassini division is located at a distance from Saturn corresponding to a particle moving with one-half the period of the satellite Mimas. In other words, a particle in the Cassini division would orbit Saturn twice while Mimas orbits once. Because such a particle would be directly aligned every two orbits, gravity would provide an extra tug, and the particle would be removed from that orbit, leaving the gap we observe.

Inquiry 10–9 How might you determine that the ring system is not composed of solid rings? (What would be observed if solid rings passed in front of a star?)

Do rings occur around other planets? Rings around Uranus were unexpectedly discovered in 1977; the story of the discovery is an interesting tale of how scientific progress often occurs. In the 1970s, attempts were made to observe occultations of stars by the outer planets on a regular basis. An occultation takes place when a planet passes in front of a star, hiding it from our view. By timing the duration of the occultation, astronomers can obtain both an accurate measure of a planet's diameter and information about its atmosphere's structure. In early 1977, while awaiting an occultation of a star by Uranus, astronomers flying in the Kuiper Airborne Observatory were surprised to notice that the light from the star dimmed and brightened several times before the planet occulted the star. At first they suspected clouds, but they quickly concluded that the events were caused by five thin rings surrounding the planet. Several other observatories around the world independently recorded these events, thus verifying their reality.

With the discovery of Uranus's rings, some astronomers suggested that *all* the Jovian planets would have rings. For this reason, the Voyager spacecraft made a long-exposure image slightly offset from Jupiter itself, specifically looking for a faint ring. One was found and is shown in **Figure 10–22a**. (Evidence for the existence of a faint ring around Jupiter was published as early as 1960 by the Soviet astronomer S. Vsekhsvyatskij, but it had never been confirmed.) Jupiter's ring is at most a

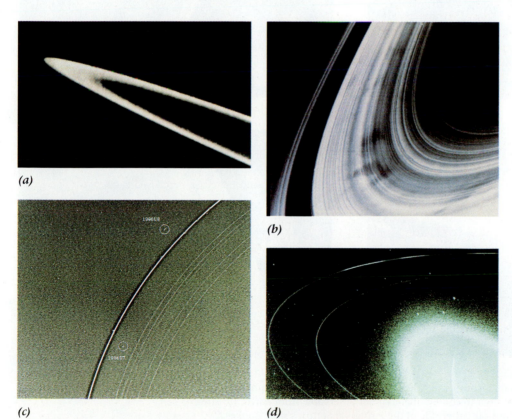

(a)

(b)

(c)

(d)

FIGURE 10–22. Comparison of ring systems for Jupiter (*a*), Saturn (*b*), Uranus (*c*), and Neptune (*d*). Note the shepherd satellites for Uranus.

few tens of kilometers thick, with little structure. The particles are small, most less than 0.01 mm in size. Such particles are subject to a variety of forces in the Jovian environment and will not last for long. For this reason, there must be a continual replenishment of particles, perhaps from meteoritic bombardment of nearby satellites, from micrometeorites trapped in Jupiter's magnetic field, or from material ejected by Io's volcanoes.

During late 1980, the Voyager I spacecraft passed near Saturn, and its sophisticated cameras revealed a wealth of new and surprising information about the rings. The greater resolution of the cameras revealed that the ring system actually consists of myriads of thin ringlets, as the beautiful photo shown in **Figure 10–23a** indicates (see also Figure 1–5). Additional rings were found both inside and outside those visible from Earth. Voyager confirmed the overall flatness of the rings; the thickness is about 1 km over the 300,000-km radius. There is a range in particle size, from 0.05 mm up to meter-sized boulders.

Saturn's F ring surprised project scientists by its twisting, winding structure and its nonuniform brightness (**Figure 10–23b**). At first it was thought that gravity alone could not explain this behavior. But it turns out that there are two previously undetected satellites that straddle the braided F ring (**Figure 10–23c**). They are called **shepherd satellites** because their combined gravity keeps this ring confined to a narrow band. They may also cause its wavy shape. A shepherd satellite found just outside the outermost ring visible from Earth is thought to

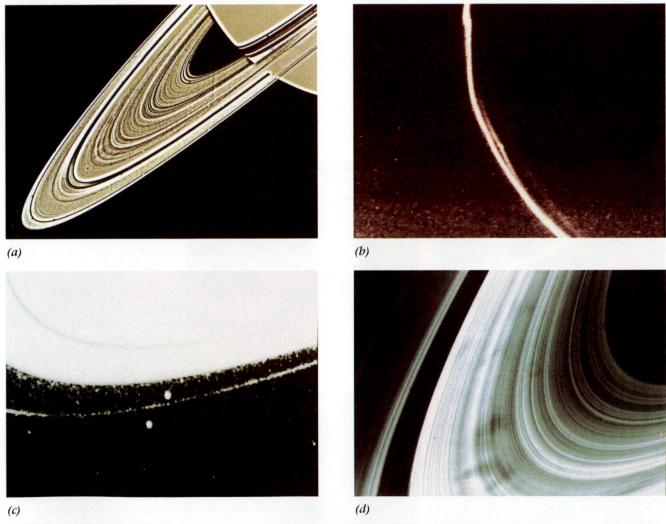

(a)

(b)

(c)

(d)

FIGURE 10–23. Characteristics of Saturn's ring system. (*a*) The rings consist of hundreds of "ringlets." The Cassini division is shown to contain material not observable from Earth. (*b*) The braided F ring. (*c*) Shepherd satellites on either side of the F ring. (*d*) Saturn's ring "spokes."

maintain that ring's sharp outer boundary.

However, most of the rings do not appear to have such companions, so there must be additional, unknown mechanisms to explain the narrowness and sharpness of the rings' edges.

Voyager also observed dark ring structures pointing radially away from Saturn, like spokes of a wheel (**Figure 10–23***d*). The spokes rotate with the rings, appearing within a few minutes and remaining for some tens of hours at the most. The material forming the structures appears to be above the ring plane and to be partly guided by electromagnetic forces. The spokes consist of small particles that have acquired a net positive charge by losing electrons through collisions. This charge makes the particles repel one another and lifts them above and below the ring plane.

Inquiry 10–10 The ring "spokes" were a surprise because according to Kepler's third law they should not last as long as they do. Why does Kepler's third law indicate the spokes should rapidly dissipate?

After the spectacular view of Saturn's rings, astronomers were extremely excited by the prospect of actually seeing Uranus's rings. Voyager confirmed the presence of nine major rings, and found them to be composed of dark particles that reflect only 3% of the incident light. Some planetary scientists have suggested that they are coated with dark carbon compounds. We wonder whether their darkness is in any way related to the dark material of Saturn's satellites Iapetus and Phoebe, or the dark material in the bottoms of craters on Titania and Oberon.

Planetary rings will appear differently when a spacecraft approaches compared with when it leaves because of the complex way small particles scatter light. From a comparison of different views of the rings, it is possible to draw conclusions about the sizes of the particles in them. Surprisingly, the rings of Uranus showed little in the way of tiny dust grains; most of the "particles" in the ring system appear to be about one meter across or larger. Even particles a few centimeters in size are rare.

While narrow, the rings of Uranus do have breadth, and the breadth of the outermost ring is variable. The narrowness of the outermost ring was readily explained by Voyager when it discovered two shepherd satellites on either side (see Figure 10–22*c*). As with Saturn, the lack of shepherd satellites near the other ring means other mechanisms must also be occurring.

By this time (1986), astronomers were in a frenzy trying to observe rings around Neptune! Conflicting reports appeared. Some researchers inferred the presence of ring *arcs*, which are incomplete pieces of rings. These scientists were certainly pleased when the images shown in Figure 10–22*d* were returned by Voyager in 1989. While the images show complete rings, there are also bright concentrations that could easily be interpreted as ring arcs. These arcs are now understood in terms of complex gravitational interactions between the ring and the 150-km-diameter satellite Galatea. Located 1000 km inside the outer ring, Galatea acts as a single shepherd of the fine dust particles composing the ring.

Inquiry 10–11 Ring arcs are unexpected because a clump of particles only 15 km long (in a direction away from a planet) will spread completely around a planet in only three years. What concepts discussed in this book explain why clumps of particles will not exist for long?

Inquiry 10–12 What characteristic distinguishes all the planets that so far have been discovered to have rings?

What is the chemical composition of ring particles? The Jovian ring particles appear to be relatively high-density mineral grains while those of the other planets are low-density ices. This difference is understandable if the more massive Jupiter had a higher temperature than the other planets.

Inquiry 10–13 Why would a higher temperature at Jupiter produce rings composed of minerals rather than ice?

Why do these planets have rings? It is simple to demonstrate from the law of gravity that if a body more than a few tens of kilometers in diameter were located at the distance of the rings, a planet's large mass would produce tidal forces that would eventually tear the smaller body apart. Outside a critical distance, called the **Roche limit**, this would not occur. This does not necessarily imply that the rings are the residue of a broken-up satellite, however. Instead, it could mean that while the protoplanet was forming into a planet, material close to it was prevented from becoming a single, large satellite. The major rings of all the planets are within the Roche limit for each planet.

Current thinking is that ring systems do not persist forever but are a transient phenomenon. Rings are subject to bombardment by meteoroids. Some of the resulting residue will remain in the ring system, but some will be ejected or fall into the planet. Interactions between rings, and between rings and shepherd satellites, appear to set up waves, which influence the rings' long-term stability. For these and other reasons, astronomers no longer believe that ring systems formed at the time the parent planet formed. We now believe ring systems form from the catastrophic break-up of a satellite and then dissipate over time intervals of perhaps 100 million years—a short lifetime in the history of the solar system.

10.7
PLUTO

Pluto is so small and distant that it is difficult to obtain even the most basic information about the planet. It is clearly not one of the gas giants, but neither is it a terrestrial planet. The search for it began because of predictions based on perturbations observed in the motion of Neptune. A new planet was discovered during an exhaustive search in 1930 by the 24-year-old **Clyde Tombaugh** (Figure 10–24), whose systematic search method included viewing two photographs at a time with an instrument called a **blink comparator**. This machine uses a mirror that flips between two aligned photos; an object that has moved between the time the photographs were taken appears to jump back and forth, or "blink." The photographs on which Tombaugh made his discovery are shown in **Figure 10–25**.

Until recent years, all that was known about Pluto was that its orbit was highly elliptical and inclined to the ecliptic. The ellipticity is such that between about 1979 and 1999 the planet is closer to the Sun than Neptune. There is no chance of Pluto and Neptune colliding, however, because the orbits are inclined to one another. Observations of Pluto's brightness showed it to vary regularly, with a period of 6.39 days.

Inquiry 10–14 What are some possible explanations for the regular variations in Pluto's brightness?

January 23, 1930

January 29, 1930

FIGURE 10–25. Pluto's discovery plates obtained by Clyde Tombaugh.

FIGURE 10–24. Clyde W. Tombaugh, the discoverer of Pluto, in a photograph taken when the University of Kansas Observatory was dedicated in his honor in 1980 (the 50th anniversary of his discovery).

Only in late 1992 did astronomers detect frozen ices on Pluto. We now know the surface is composed of 97% nitrogen, 1–2% carbon monoxide, and 1–2% methane. Chemically, then, Pluto's surface is similar to Neptune's satellite Triton. Because Pluto's surface has remained virtually unmodified since its formation, an understanding of its surface chemistry can provide a look at the chemistry of the early solar system. And if, as some scientists believe, some of Earth's atmosphere resulted from collisions with comets formed from the same material as Pluto, we will be better able to understand Earth's primitive atmosphere.

Pluto has a weak atmosphere containing mostly nitrogen with some methane. It may have been produced only in recent years, because Pluto's elliptical orbit has brought it closer to the Sun now than it normally is.

Inquiry 10–15 Hypothesize that Pluto is covered by an extensive atmosphere. In this case, one would expect its disk to be uniform in brightness. Would it be possible for the planet to exhibit the regular variations in brightness that are observed if this were the case?

To everyone's great surprise, tiny Pluto turned out to have a moon. Discovered in 1978, it has been named **Charon**, after the mythical boatman who ferried the dead to the underworld. It is small and difficult to observe; nevertheless, it has proven possible to estimate the mass of Pluto with moderate accuracy from the satellite's motion. Robert Harrington of the U. S. Naval Observatory estimated the mass of Pluto to be a low 0.0024 of the Earth's mass. **Figure 10–26** shows the best ground-based image of Pluto and Charon, as well as a recent image taken with the Hubble Space Telescope. (The rings are not real!)

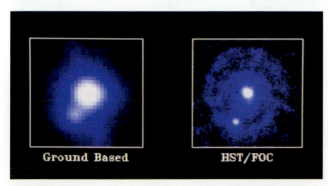

FIGURE 10–26. Pluto and Charon. (*a*) The best ground-based image of Pluto and Charon. (*b*) A similar image from the Hubble Space Telescope. (The rings are *not* real planetary features.)

By pure luck for modern-day astronomers, in recent years Charon and Pluto were involved in a series of mutual eclipses with each other. These events allowed a reliable diameter to be determined for each body for the first time. The presently accepted value for Pluto is about 2300 km. With this value, we can finally determine its density, which is 2.13 gm/cm³. This means that Pluto contains somewhat more rock than ice. Nonetheless, Pluto's surface certainly consists of frozen gases, such as methane and water. Charon's diameter of 1200 km means its density is only about 1.3 gm/cm³. Because Pluto and Charon are of similar size, they can be thought of as a double planet, as is often done for the Earth and Moon.

The mutual eclipses have provided astronomers with data to model the appearance of Pluto's surface. This model shows the presence of a bright south polar cap that reflects 95% of the incident radiation. The darkest regions reflect less than 15%.

Pluto's orbit is the most eccentric and the most inclined to the ecliptic of any planet in the solar system. The existence of a small planet at the outer edge of the solar system has led some astronomers in the past to wonder whether Pluto might not be an escaped satellite of Neptune. Its properties are, as we have seen, reminiscent of many of the satellites of the outer planets. In fact models have been computed considering the possibility that Pluto escaped from Neptune, leaving Triton in its retrograde orbit and Nereid in its high-eccentricity orbit. Escape models also show that Pluto would consist mostly of methane and have a low density. But the recently determined domination of nitrogen and the observed density of 2.13 gm/cm³ indicate that Pluto formed directly out of the solar nebula, just like all the other planets. The fact that Pluto has a moon of its own also strongly argues against the escaped-satellite hypothesis.

10.8
FUTURE STUDIES OF THE OUTER SOLAR SYSTEM

Future studies of the solar system beyond the terrestrial planets include space probes that will increase our knowledge of the objects we already know about. There is, too, the question of whether there are planets beyond Pluto.

FUTURE SPACE PROBES

Our knowledge of Jupiter is due for another boost in December 1995 when the Galileo spacecraft reaches the planet. Galileo's scientific objectives are to improve our knowledge of Jupiter's (1) chemical composition; (2) variation of temperature, pressure, and density with depth; (3) energy balance; (4) electrical discharges; (5) satellite system; and (6) details of the clouds. These goals will be accomplished through a two-pronged attack. An entry probe will be deployed to float into the atmosphere and obtain data during its descent to layers not visible from above. During the descent, the heat shield's temperature will rise to some 25,000 K while the spacecraft slows its speed from over 100,000 miles per hour to 2000 miles per hour in only two minutes! While the probe descends, an orbiter will collect data for later relay to Earth. Thereafter, the orbiter will continue to move around the planet in highly elliptical paths. During this time, it will make close encounters with numerous satellites. With each satellite encounter, the spacecraft will receive a gravitational boost to move it on to the next satellite. Successive satellite encounters and orbital changes will occur until the four Galilean satellites have been visited.

While these events take place, one problem may strongly diminish the overall impact of the mission. The high-gain antenna, which relays data back to Earth, has not completely opened. NASA has tried a number of ways to open it, but to no avail. Painful compromises may have to be made in mission objectives so that at least some new results are obtained.

The final U.S. launch of this century is the 1997 **Cassini/Huygens** mission to Saturn and its satellites, due to arrive at Saturn in November 2004 and return data for four years. The Huygens probe will go to Titan's surface in June 2005.

TABLE 10–1

The Jovian Planets and Pluto: A Comparison

	Jupiter	Saturn	Uranus	Neptune	Pluto
Orbital data					
Semi-major axis (AU)	5.203	9.54	19.18	30.06	34.44
Orbital period (years)	11.86	29.46	84.01	164.79	248.68
Orbit shape (eccentricity)	0.048	0.056	0.047	0.009	0.249
Orbit inclination (°)	1.3	2.5	0.8	1.8	17
Rotation period (hours)	9.92	10.66	17.24	16.11	153.3
Axial tilt (°)	3.1	26.7	97.9	28.8	96
Physical data					
Diameter (Earth units)	10.85	8.99	3.96	3.85	0.18
Mass (Earth units)	317.89	95.18	14.54	17.15	0.0024
Density (gm/cm^3)	1.33	0.69	1.27	1.64	2.13
Surface gravity (Earth units)	2.64	1.13	0.89	1.13	0.06
Escape velocity (km/sec)	60.0	36.0	21.2	23.5	1.2
Temperature (surface) (K)	165	134	76	74	40
Satellites	16	18	15	8	1
Visible features					
Banded appearance	strong	muted	none	slight	no
Storm systems	strong	yes	no	yes	no
Ring system	yes	yes	yes	yes	no
Other characteristics					
Clouds	thick	thick	thick	thick	?
Magnetic field	strong	strong	yes	yes	unknown
Differentiated core	yes	yes	yes	yes	no
Excess energy	yes	yes	small amount	yes	no
Atmospheric composition	H, He, NH$_3$, CH$_4$	H, He, NH$_3$, CH$_4$	H, He, NH$_3$, CH$_4$	H, He, NH$_3$, CH$_4$	N, CH$_4$

ARE THERE PLANETS BEYOND PLUTO?

Pluto's discovery took so much time and effort that few astronomers are eager to search for planets beyond Pluto. In fact, after discovering Pluto, Tombaugh spent many years searching for other planets. More recently, the catalog of objects found by the Infrared Astronomy Satellite has been examined unsuccessfully for possible planets. This is not surprising, because there are theoretical reasons not to expect major planets beyond Neptune. Finally, recent reanalysis of planetary orbits, in which a few discrepant data points are neglected in the orbital calculations, shows no reason to expect the presence of more distant planets. The old argument for there being a "Planet X" beyond Pluto is no longer considered valid.

CHAPTER SUMMARY

OBSERVATIONS

- The characteristics of the Jovian planets and Pluto are summarized in **Table 10–1** at the end of the chapter.
- The Jovian planets—Jupiter, Saturn, Uranus, and Neptune—are all large gas balls composed of hydrogen, helium, methane, and ammonia. They all have low densities, satellite and ring systems, magnetic fields, and rapid rotations. Their nearly circular orbits have low inclinations to the ecliptic.
- The atmospheres of Jupiter, Saturn, and Neptune contain numerous large and small spots, which are storm systems of various sizes. Uranus's atmosphere is featureless.
- The magnetic fields of Jovian planets are often tilted relative to the rotation axis and offset relative to the planet's center.
- The satellites of the Jovian planets have icy surfaces composed of either methane and ammonia, or water ices. Some satellites show volcanoes and geysers; others exhibit craters, valleys, and cliffs that suggest all sorts of complex histories. Some satellites have retrograde orbits; some have orbits that are highly elliptical.
- Certain satellites interact with the ring systems as **shepherd satellites**. Other satellites form **co-orbital** systems in which groups of satellites occupy a single orbit.
- Pluto is in an elliptical orbit that has the greatest inclination of any planetary orbit to the ecliptic. It has a surface of frozen nitrogen and an atmosphere of nitrogen and some methane.

THEORY

- Planetary magnetic fields cause charged particles to become trapped. When moving near the speed of light, they emit **synchrotron radiation**, which is recorded as intense bursts of radio energy. Observation of such radiation shows the presence of magnetic fields.

CONCLUSIONS

- Planetary rings are not solid bodies but are composed of particles ranging in size from small dust grains to meters across. Each individual particle orbits its planet governed by Kepler's laws.
- Complex planetary storms are thought to be caused by an interaction between heating from the Sun, the **Coriolis forces** that result from rapid rotation, and the **excess energy** radiated from the interiors of Jupiter and Saturn.
- The interiors are **differentiated**. High interior pressures in Jupiter and Saturn produce a region in which **liquid metallic hydrogen** exists. These regions, and especially the central cores, are thought to produce the extensive magnetospheres around each of the Jovian planets.
- Some outer-planet satellites have surfaces as bright as newly fallen snow, while others are as dark as tar.
- Rings are thought to form and re-form during periods of 100 million years in which satellites break up, either by collisions or by tidal forces when a body comes inside the planet's **Roche limit**.
- Pluto's satellite, **Charon**, has allowed us to determine Pluto's mass, size, and density. These results indicate that Pluto formed directly from the solar nebula, and that it is not a satellite ejected by Neptune.

SUMMARY QUESTIONS

1. In what ways are each of the Jovian planets similar to and different from one another? In your discussion, include the orbital characteristics, appearance, rotation and revolution, general atmospheric features, and chemical composition.

2. In what ways are the observed characteristics of the major satellites of the Jovian planets similar to and different from one another?

3. In what ways are the interior structures of the Jovian planets similar to and different from one another? Why do the interior structures differ?

4. What are the magnetic field properties of each of the Jovian planets? Where do we think the magnetic fields come from?

5. How do the appearances of the ring systems of the Jovian planets differ from one another?

6. What is meant by shepherd and co-orbital satellites?

7. Explain how astronomers are able to determine the interior structure of a Jovian planet.

8. Describe the methods used to discover Uranus, Neptune, and Pluto. In what years were each discovered?

9. Describe the magnetic fields of the Jovian planets in comparison with that of Earth.

10. Make a table of the characteristics of the major satellites of the Jovian planets as follows: Along the left edge place the satellite names; along the top, write "craters," "faults or cracks," "mountains," "valleys," "scarps," "atmosphere," "volcanic activity," and "density." Fill in the table with an "X" for each characteristic shown by each satellite. If the density is known, place it in the "density" column. What general conclusions about groups of satellites and satellite systems can you make?

11. In what ways is Pluto similar to and different from both the Jovian and the terrestrial planets?

APPLYING YOUR KNOWLEDGE

1. The Jovian planets can be described as objects that have changed little since their formation, while the terrestrial planets have changed a great deal. Explain the meaning of this statement, and give examples of its validity.

2. What is the distance in kilometers from Dione to Dione B? From Tethys to its co-orbital satellites? (Note: You can answer this question without mathematics!)

3. What would you conclude about the formation of the solar system if Jupiter had the same composition as Earth?

4. Speculate on what the Voyager spacecraft might have found if it had gone on to Pluto.

■ 5. Compute Saturn's density relative to Earth's, given its diameter of 9 Earth radii and 95 Earth masses. What is its density relative to that of water (1 gm/cm^3)?

■ 6. Saturn's Roche limit is located at 2.4 times the planet's radius. Measure the inner and outer radii of Saturn's rings from Figure 10–1, and express them in terms of Saturn's radius. Are these distances inside or outside the Roche limit?

■ 7. Saturn's innermost ring begins at about $1.11R_S$ (Saturn radii), and the outermost one extends to about $6R_S$. How much longer does it take an outer-ring particle to revolve around the planet relative to a particle at the inside of the innermost ring?

■ 8. Why would you weigh nearly the same on Saturn as you do on Earth? Make the computation.

■ 9. How far above Io's surface is the volcano in Figure 10–12b? What implicit assumption are you making in solving the problem?

■ 10. Determine the size of the smallest feature you can see on the Earth-based photograph of Jupiter in Figure 10–1.

■ 11. Compute the ratio *satellite radius/planet radius* for the largest satellite in each planetary system throughout the solar system. Which, if any, stand out from the rest?

ANSWERS TO INQUIRIES

10–1. If they were solid, the inner and outer edges of a ring should go around the planet in the *same time*, just as the hub and end of a clock hand go around a clock in the same time.

10–2. The reasons were discussed in detail in Chapter 6 in the discussion of flattening of the solar system during formation.

10–3. The two lightest elements, hydrogen and helium.

10–4. Half of Uranus's period of revolution around the Sun, or 42 years. Because a planet whose rotation axis is perpendicular to its orbit will not have seasons, while a planet whose angle is 90° will have the largest possible seasonal changes, Uranus's seasons will be nearly the maximum possible.

10–5. Electrically conducting region in the interior, and rotation.

10–6. It might be a captured asteroid.

10–7. It lacks a metallic core and/or its rotation is too slow.

10–8. The orbits would be in the planet's equatorial plane and would be nearly circular. Collisions of some kind are probably required to explain the observations.

10–9. In addition to the answer given in Inquiry 10–1, stars can in fact be seen through the rings.

10–10. From Kepler's third law, the innermost particles move faster than the outer ones, just as on a track the inside runners advance on the outside ones. Therefore, the particles should be spread evenly around the rings.

10–11. The answer is the same as for Inquiry 10–10.

10–12. They are all members of the Jovian group of planets.

10–13. Ices are volatile. Higher temperature would cause the ice to vaporize and the volatile elements to escape more readily.

10–14. Pluto is probably rotating once in 6.39 days, and the brightness varies because the surface is not uniformly bright but has bright and dark areas. A satellite with a 6.39-day period could also produce such a variation.

10–15. No, *unless* there is a satellite or ring system successively adding or blocking light.

THE PLANETS ONE-BY-ONE

JUPITER

THE VIEW FROM EARTH

Jupiter is a giant planet by any means of measurement. Located fifth from the Sun, its 318 Earth masses and 11 Earth diameters give it a density 1.3 times that of water. Its appearance through ground-based telescopes is that of a gas giant flattened from its rapid rotation, with reddish and whitish belts and zones. Located within one of the zones in the southern hemisphere is the Great Red Spot, a feature about three times the size of Earth. Ground-based observations show the planet to be escorted by a retinue of satellites. The largest, the Galilean satellites, are large enough and bright enough to have been seen by Galileo; hence their name. Their changing positions are readily observed within a period of hours.

Ground-based observations with radio telescopes found synchrotron radiation, which occurs only when charged particles accelerate near the speed of light in a magnetic field. Such radiation points to the presence of a strong magnetic field.

Other observations show that Jupiter emits nearly twice as much energy as it captures from the Sun. This excess energy comes from internal heat left over from the time of Jupiter's formation.

THE VIEW FROM SPACECRAFT

From the Pioneer and Voyager probes, features become more highly resolved and show greater detail within the turbulent and stormy atmosphere. Lightning and aurorae have been observed. These spacecraft confirmed the expected presence of hydrogen and helium as the main constituents of the planet. Voyager discovered a ring consisting of small particles.

The magnetosphere was found to be extensive and to change with time. The magnetic axis is tilted 10° to the rotation axis.

Models of the interior structure show the planet to contain a region composed of liquid metallic hydrogen. Because this material behaves like a metal, its rotational and turbulent motions are thought to produce the planet's magnetic field.

The Galilean satellites are each unique and show a variety of geological features. Io has volcanoes. Some of the gases form a cloud around the satellite and its orbit. Europa exhibits straight-line features within its icy methane and ammonia surface. Ganymede, the largest satellite in the solar system, is cratered and contains faultlike features indicative of some type of internal motions that move the crust. Callisto is saturated with craters and contains a large multi-ringed basin.

SATURN

THE VIEW FROM EARTH

Saturn's most distinctive feature when observed with ground-based telescopes is its glorious ring system. Although farther from the Sun and smaller in mass and diameter than Jupiter (95 Earth masses and 9 Earth radii, respectively), the planet is observed to be flattened from its rapid rotation and to contain faint bands parallel to its equator. Saturn's axial tilt implies that it has seasons. Satellites revolving around the planet can be seen even with small telescopes.

THE VIEW FROM SPACECRAFT

Saturn's appearance is muted compared to Jupiter's. Its storm systems are smaller. The planet emits energy in excess of that received from the Sun. In fact, its excess energy is greater than that from Jupiter.

Saturn's ring system is complex. Composed of hundreds of narrow ringlets, the particles range in size from dust grains to boulders. They show a variety of features. Not all rings are perfectly symmetrical. They exhibit spokes, which are caused by the charging of dust particles exposed to solar emissions. The F ring shows both an inhomogeneous distribution of material and a transient braiding or twisting. The sharpness of the ring system is caused by a complex interaction with some of Saturn's many satellites. The ring system may be a relatively young feature whose existence comes and goes.

Saturn's satellites consist of a variety of unique and interesting bodies that rotate synchronously. Their surfaces consist of water ice. Mimas's heavily cratered surface has one crater nearly one-third the planet's diameter. Enceladus's surface is as bright as newly fallen snow and has a region that is geologically young, in that resurfacing appears to be a continuous process. Tethys has a gouge covering a large part of its total circumference; it also has a gigantic crater covering a large part of its surface. Dione and Rhea both contain wisps of relatively freshly produced ice. Titan has a dense nitrogen atmosphere that is divided into observable layers. Iapetus has one side that is highly reflective and another side that is black. Phoebe is as dark as tar.

Two of Saturn's satellites, Dione and Tethys, have companion satellites orbiting Saturn. These are co-orbital satellites that occupy Lagrangian points in Dione's and Tethys' orbits around Saturn. Some of the satellites also act as shepherds to the ring by using their gravity to prevent ring particles from escaping.

Saturn has a magnetic field axis aligned with its rotation axis. Models of the interior predict a zone of liquid metallic hydrogen.

THE PLANETS ONE-BY-ONE

URANUS

THE VIEW FROM EARTH

Discovered in 1781, Uranus is the least massive of the Jovian planets, at nearly 15 times the Earth's mass. With a radius of somewhat less than 4 times the diameter of the Earth, its density is lower than that of Jupiter or Neptune. Its great distance gives it a small angular diameter so that few markings had been seen using Earth-based telescopes.

The most significant single piece of information determined from ground-based data was that its rotation axis lies nearly in the ecliptic.

Five satellites were discovered using Earth-based telescopes.

Ground-based observations of Uranus's occultation of a star found the planet to have a system of nine thin, dark rings. It was the first planet other than Saturn known to have a ring system.

THE VIEW FROM SPACECRAFT

The Voyager 2 mission, which arrived at Uranus in 1986, found a planet devoid of surface markings. The planet emits little excess energy.

Voyager found Uranus to have a magnetic field whose axis is tilted 59° to the rotation axis. Furthermore, the center of the magnetic field is offset from the center of the planet by one-third of the planet's radius. The planet's mass is too low to allow liquid metallic hydrogen to exist in the interior, and the source of the magnetic field is therefore unknown.

Uranus's ring structure is at least partly determined by shepherd satellites on either side of its outermost ring. At least one of the rings varies in its width as it circles the planet.

Each of the satellites is small, none being larger than 0.08 the size of Earth's Moon. Nevertheless, the larger ones are as interesting as the satellites around Jupiter and Saturn. Miranda shows a surface that appears to have been put together from random pieces of a puzzle. It contains a 5-km-high cliff in addition to faults, craters, and valleys. Ariel and Titania have faults. Titania's craters sometimes contain a dark-appearing material in their bottoms. Umbriel has a light-colored ring on the surface that may be (relatively) freshly deposited ice. Oberon has a 6-km-high mountain. Voyager discovered 10 new satellites.

UNIQUE FEATURES

The placement of the rotation axis near the ecliptic is unique. This angle produces extreme seasons on the planet and its satellites.

NEPTUNE

THE VIEW FROM EARTH

Discovered in 1845/1846, Neptune is the smallest of the Jovian planets. With a mass of 17 times the Earth's mass, the density of 1.64 is the largest of all the Jovian planets. Its small angular size allows little to be seen from Earth. The two satellites known from Earth-based observations have strongly differing characteristics. Triton, which is nearly as large as Earth's Moon, has a mass of 0.3 times the mass of the Moon and a retrograde orbit. Nereid is small and revolves in a highly elliptical, highly inclined orbit.

Ground-based observations of occultations of stars hinted at a possible ring system consisting of pieces of rings rather than complete entities.

THE VIEW FROM SPACECRAFT

Voyager 2 revealed clouds and storm systems of various sizes. High-level clouds were observed to move and layering of the cloud structure was apparent. Neptune emits more excess energy than any of the other Jovian planets.

Neptune's magnetic field, which is tilted 47° relative to its rotation axis and offset from the planet's center by half the radius, is similar to that of Uranus. The planet's low mass means that it lacks the liquid metallic hydrogen thought to produce the magnetic fields in Jupiter and Saturn.

The rings are both thin and dark, and thus somewhat similar to Uranus's rings. The ring material is not uniformly distributed; bunched material gives the clear impression of ring arcs. These arcs may be caused by a single satellite that acts as a shepherd.

In addition to the discovery of six new satellites, Voyager obtained detailed images of Triton. With only a thin nitrogen atmosphere covering it, the surface of frozen nitrogen revealed grand complexity. One large area is relatively smooth; an adjoining area around a pole is heavily cratered and contains dark, streaky areas that appear to be geysers of ice emanating from the surface. Its density of 2.0 gm/cm³ shows Triton to be about half water ice and half rock.

UNIQUE FEATURES

The ice geysers have been seen nowhere else in the solar system. Triton is one of only two satellites in the solar system to have an atmosphere. The orbital characteristics of Triton and Nereid are unique in the solar system.

THE PLANETS ONE-BY-ONE

PLUTO

THE VIEW FROM EARTH

Pluto was found in 1930 by Clyde Tombaugh after a systematic, exhaustive search. From Earth, Pluto appears as little more than a point of light. The planet's brightness was observed to vary with a 6.39-day period, which was assumed to be caused by variations of brightness across the surface. Photographs of the planet made in 1978 showed the presence of a satellite, later named Charon. Its period, of 6.39 days partially explains the brightness variations observed for Pluto. Furthermore, both Pluto and Charon each rotate with the same 6.39-day periods.

Only in 1980 was a thin atmosphere detected. The atmosphere may have been found only recently because Pluto is now closer to the Sun than it usually is in its highly elliptical, highly inclined orbit.

In recent years, Pluto and Charon underwent a series of mutual occultations. Analysis of the brightness changes allowed astronomers to determine an accurate diameter: 0.18 the size of the Earth. Pluto's mass was also determined from an analysis of Charon's motion; Pluto's mass turned out to be 0.0024 Earth masses. The resulting density of 2.3 gm/cm^3 indicates that Pluto formed directly out of the solar nebula rather than as an escaped satellite from Neptune.

THE VIEW FROM SPACECRAFT

No spacecraft has yet visited Pluto.

PART THREE

DISCOVERING THE TECHNIQUES OF ASTRONOMY

Because astronomy is an observational science, astronomical data for objects beyond the solar system are limited entirely to the diverse types of radiations received from the distant universe. Only through an understanding of the properties of these forms of radiation can astronomy progress beyond descriptive observations. In Chapter 11 we discuss the observed properties of light, and how these properties lead scientists to models that allow us to describe the behavior of light.

The radiation from planets, stars, and galaxies is collected by means of telescopes and analyzed by a variety of auxiliary instruments. Great advances are being made in telescopes and the electronic instruments used to detect radiations from beyond the Earth. These topics are the subject of Chapter 12.

In our final chapter of Part 3, Chapter 13, we break starlight into its component parts as we begin to see how astronomers determine the characteristics of radiating bodies.

11

THE NATURE OF LIGHT

It is indeed wrong to think that the poetry of Nature's moods in all their infinite variety is lost on one who observes them scientifically, for the habit of observation refines our sense of beauty and adds a brighter hue to the richly coloured background against which each separate fact is outlined. The connection between events, the relation of cause and effect in different parts of a landscape, unite harmoniously what would otherwise be merely a series of detached scenes.

M. Minnaert, *The Nature of Light and Colour in the Open Air*, 1954

Until recently, our knowledge of the universe was obtained exclusively from a study of the visible light that happened to arrive on Earth. Since the 1930s, it has become possible to study other kinds of radiation and particles—radio waves, X-rays, gamma rays, cosmic rays, as well as exotic items such as neutrinos and gravitational radiation, as we will see later. But even in this era of exploding technology, it is still true that significant information about celestial objects comes from an analysis of their visible light. To understand the methods that astronomers use to extract this information, we will now examine a few of the fundamentals of light and other radiations.

11.1

LIGHT AS A RAY

The ray model of light is most conveniently illustrated by the properties of reflection and refraction.

REFLECTION

A basic property of an isolated light beam is that it travels in a straight line. This is why we cannot see around corners under normal conditions—it would certainly be hard to make sense of the world if it were otherwise! However, light *can* change direction under certain conditions, for example, when it is reflected from a surface.

The concept of **reflection** explains in a simple way why we see objects as we do. We see a chair, for example, because it reflects some of the light that falls on it toward our eyes; it acquires form to our eyes because we can distinguish the light rays that come from various directions. We see color by virtue of the fact that the chair does not reflect all colors equally well; a blue chair is blue because it reflects blue well while tending to absorb other colors.

> YOU SHOULD DO DISCOVERY 11-1, IMAGES IN A MIRROR, AT THIS TIME.

REFRACTION

The direction a ray of light travels also changes when it passes from one transparent medium, such as air, into another, such as glass. This bending of light as it passes from one medium to another is called **refraction**.

> YOU SHOULD DO DISCOVERY 11-2, REFRACTION, AT THIS TIME.

11.2

LIGHT AS A WAVE

The wave model of light is demonstrated not only by reflection and refraction, but also by a variety of other phenomena. In describing light with the wave model, we will examine waves of different lengths, and discuss how fast they travel.

DIFFRACTION

The phenomena of reflection and refraction can be explained by a model that considers light to be rays that travel in a straight line unless they encounter change in physical conditions, such as a mirror, or a change in the medium, as from air to water. However, there are other characteristics of light that cannot be explained with this model.

For example, if light consisted of parallel rays, they could travel through a small pinhole and make a small bright spot on a screen a short distance away (**Figure 11–1a**). When we actually observe the light that passes through such a hole, however, we see the rather unexpected pattern of **Figure 11–1b**. The spot is larger than the pinhole and not uniformly bright. What can be happening? Somehow, the presence of the pinhole has done more than just prevent some light from passing through; it has also affected the light that did get through.

We can observe a similar phenomenon when sound waves or waves of water pass through a narrow opening (**Figure 11–2**). This analogy leads us to conclude that light, too, must have some of the properties of waves; indeed, the pattern we observe when light passes through a pinhole can be explained only by assuming wave properties rather than ray properties for light.

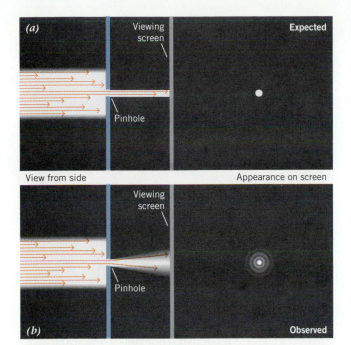

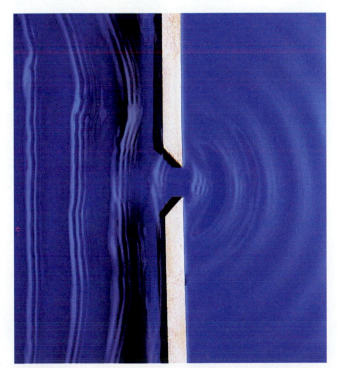

FIGURE 11–1. Light passing through a pinhole. (*a*) If light passed through a pinhole moving in straight lines, the expected pattern would have the same size as the pinhole. (*b*) The observed pattern is larger than the pinhole, and consists of rings of decreasing intensity.

FIGURE 11–2. The bending of water waves passing through a slit.

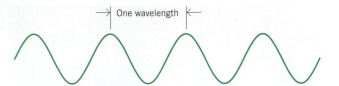

FIGURE 11–3. The definition of wavelength: the distance between two crests of a wave.

The pinhole through which the wave (light or water) passes somehow causes the wave to bend. The bending of a wave as it passes through a hole or around an obstacle is known as **diffraction**. A related phenomenon occurs when identical waves combine with one another, and we have what is called **interference**. Interference was first studied in detail by Thomas Young, a trained surgeon who was also an amateur physicist and archaeologist. On the basis of Young's experiment, scientists produced a new model for light, one known as the **wave theory of light**, which accurately predicts observed diffraction phenomena. The effects are small as far as our everyday experience is concerned. Nevertheless, the effects are important for astronomy; in fact most of what we know about the universe comes about because of the way astronomers use diffraction and interference.

We describe a wave, such as that shown in **Figure 11–3**, in terms of its **wavelength**, the distance between wave crests. We find that the bending effects of diffraction are greatest when the opening is comparable in size to the wavelength.

> READERS HAVING THE ACTIVITY KIT
> SHOULD DO KIT ACTIVITY 11–1,
> DIFFRACTION OF LIGHT PASSING
> THROUGH A SLIT, AT THIS TIME.

If you don't have the kit you can still see diffraction by looking at a distant bright street light through the space between your fingers held directly in front of your eyes. While looking at the lamp, squeeze your fingers together slightly so as to change the opening size; diffraction will cause the light to spread out in a direction perpendicular to your fingers.

ELECTROMAGNETIC RADIATION

Young's work demonstrated that light possesses some of the properties of waves. It now became important to understand just what sort of wave light is. We can understand more about the wave theory of light through the analogy with water waves. If you drop a rock into a pond, a series of waves spreads from the point of impact,

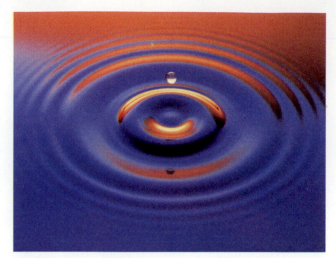

FIGURE 11–4. Water waves spread outward from the point of contact when an object is dropped into a pool.

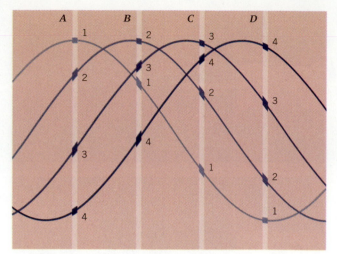

FIGURE 11–5. A water wave seen from the side at four instants of time (1, 2, 3, 4). The lines labeled *A*, *B*, *C*, and *D* are fixed locations in space that we use to measure the position of the wave. A leaf bobbing in the water at position *A* will move continuously downward during the time interval, while one at position *B* will move upward and then downward. A leaf at position *D* will move continuously upward during the time interval shown. The figure also demonstrates that the wave, not a point on it, moves to the right.

as illustrated in **Figure 11–4.** The crests of the waves spread in circles, but the water itself does *not* move outward with them, as you can see if you watch a leaf floating in water; the leaf just bobs up and down as the waves pass by, but it does not move with them (**Figure 11–5**).

Yet it is clear that *something* is moving with the wave. For one thing, the exact point where the wave is highest moves, as illustrated in Figure 11–5, and we observe this as the motion of the wave. Furthermore, as anyone who has been to the beach can attest, water waves can release a great deal of energy when they break on the shoreline, enough to move large amounts of sand and even permanently alter the coastline, given enough time. Therefore, we see that waves transport energy.

To understand the next step in our search for the nature of light, we must make a small digression into the history of science. In the eighteenth and nineteenth centuries, physicists were experimenting with static electricity. Scientists knew that positive and negative electrical charges existed and that like charges repel and dissimilar charges attract. Just as a gravitating object is surrounded by what we call a gravitational field that describes the force at each point around a mass, and a magnet is surrounded by a magnetic field that describes the strength at each point around a magnet (as we studied in Chapter 8), so too an electrical charge is surrounded by an electric field that describes the force at each point around it. When the Danish physicist Hans Oersted (1777–1851) noticed magnets respond to a nearby wire carrying a current, he discovered a connection between electric currents and magnetic fields, a connection to which we will soon return.

These experiments with electrical charges and magnetism were little more than that—experiments. Theoretical understanding of these phenomena was slight, just as prior to Newton there was no understanding of the experiments concerning motion.

The grand synthesis of these ideas was made by the Scottish scientist James Clerk Maxwell (1831–1879), who was the "Newton" of electricity and magnetism. Maxwell's synthesis is in the form of four equations, which, along with the theory of forces between charged particles, describe everything we know of electric and magnetic fields. Among the many findings that result from an analysis of these equations are four of importance to our discussion. These equations describe electric and magnetic fields in terms of waves, show that waves move with a particular speed, require that electric and magnetic waves vibrate perpendicularly to the wave's direction of motion (**Figure 11–6**), and establish that waves carry energy.

Maxwell's equations show that electric and magnetic fields become unified with each other to produce what we call **electromagnetic radiation**. One manifestation of electromagnetic radiation is what our eyes perceive as light. Other manifestations are what we call gamma rays, X-rays, and ultraviolet, infrared, and radio waves. All these types of radiation taken together form

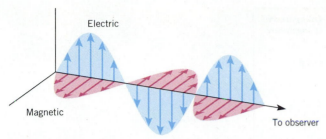

FIGURE 11–6. Electric and magnetic waves vibrate at 90-degree angles to the direction of motion, and to each other.

what we call the **electromagnetic spectrum**. All these types of electromagnetic energy travel at the same speed, the speed of light. The only thing that distinguishes one from another is the wavelength of the particular form of radiation.

The electromagnetic nature of radiation can be demonstrated by observing the motion of a charged particle, such as an electron, as an electromagnetic wave passes by. The radiation's electric field produces a force on the charged particle, which bobs up and down like the leaf on the water. Of course, an individual electron would be too small to observe. But if we put a long wire in the path of the wave, the wave will force the wire's electrons to oscillate as the wave passes. The oscillating electrons in the wire will make an electrical current, which can be amplified and detected by suitably sensitive equipment. We see this connection between waves falling onto wires and the subsequent production of an electrical current every time we turn on a radio, because electrons in a radio antenna vibrate in response to the radio waves falling on it.

Inquiry 11–1 Why are electromagnetic waves relevant to a discussion of light?

What we call "color" is simply a physiological response to light of some wavelength striking the eye. What is deep red to one person may be pale red to someone else. For this reason, specifying a wavelength is considerably more objective than naming a color.

The wavelength of what we call visible light is extremely short. For example, what the average person calls green is produced by radiation having a wavelength around 5×10^{-5} cm. Because such numbers are not particularly convenient to use, a special unit of length, the *angstrom*, is often used to measure it. An angstrom, abbreviated Å, is equal to 10^{-8} centimeters, or a hundred-millionth of a centimeter. The wavelengths of light that appear reddish are roughly 6500 to 7000 Å, whereas radiation having wavelengths in the

range of 4000 to 4500 Å would appear bluish. Many physicists prefer to measure wavelengths in nanometers (i.e., billionths of a meter) because these are the standard SI (Systéme International) units. Angstrom units are used here because most astronomers use them. In these units, red light at about 6500 Å would be 650 nm because 1 nm = 10 Å.

The wavelengths of ultraviolet radiation, X-rays, and gamma rays are shorter than those of visible light, whereas those of radio and television waves, microwaves, and infrared radiation are longer. Yet all are essentially the same in their physical nature—all consist of traveling waves of electromagnetic energy moving at the speed of light. **Figure 11–7** diagrams the entire electromagnetic spectrum and shows the approximate wavelength range occupied by each type of radiation.

Not all parts of the electromagnetic spectrum reach the Earth's surface, though, because atoms and molecules in the Earth's atmosphere absorb certain wavelengths and transmit others. Figure 11–7 also indicates the transparency of the Earth's atmosphere to each form of radiation. Notice what a small portion of the spectrum is occupied by visible light. Regions of the electromagnetic spectrum for which our atmosphere is partially transparent are called **atmospheric windows**. It is no coincidence that the visible and then the radio regions of the spectrum were the first to be exploited by astronomers.

Inquiry 11–2 If you were in charge of the astronomical program for a space station orbiting Earth above the atmosphere, to what portions of the electromagnetic spectrum would you give highest priority for observing? Why?

THE SPEED OF LIGHT

Maxwell concluded that all electromagnetic waves, when traveling through a vacuum, move with the same speed, and that this speed was the same as the speed of light. Usually designated by the letter c, it is about 3×10^5 km/sec. Using centimeters as the unit of length, the speed is 3×10^{10} cm/sec.

Inquiry 11–3 How far does light travel through the vacuum of space in one year? Express your answer in kilometers.

Considerable physical theory and experimental evidence have led to the conclusion that the speed of light in a vacuum cannot be exceeded by any type of radiation or any object in the universe. Certainly, nothing is known that moves faster.

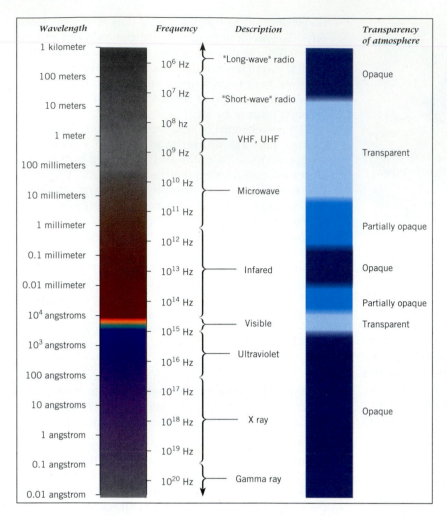

FIGURE 11–7. The electromagnetic spectrum extends from the very long radio waves to the shortest gamma radiation and includes many kinds of radiation in between. Because the Earth's atmosphere is not equally transparent to all wavelengths, not all of them reach the ground.

Inquiry 11–4 How many minutes does light take to reach the Earth from the Sun, which is at a distance of about 150,000,000 km?

Inquiry 11–5 Observations of radar signals reflected from the Moon take about 2.6 seconds to travel from the Earth to the Moon and back. What is the approximate distance to the Moon?

When working with visible light and other short-wavelength electromagnetic radiation, it is normal to speak of the *wavelength* of the radiation as we have been doing. On the other hand, astronomers who observe radio radiation from the universe generally find it more convenient to make reference to the **frequency** of the radio waves they are studying. By definition, the frequency of a wave is the number of complete wave crests that pass the observer per second. For example, if we throw a rock into water, we would probably count a few crests per second passing by. On the other hand, light

waves have a much higher frequency, on the order of 10^{15} waves per second, or 10^{15} *hertz*, to use the unit of measure that has been adopted for frequency. One hertz means one wave passing per second. A radio station broadcasting at 91.5 megahertz is emitting 91.5×10^6 waves per second.

Inquiry 11–6 Consider a wave traveling with a velocity of 100 m/sec. Would its frequency increase or decrease if the wavelength were to decrease?

Inquiry 11–7 Consider a wave having a wavelength of 5000 Å. Would the frequency increase or decrease if the wave's speed were to decrease?

The wavelength, frequency, and speed of any wave are related to each other in a simple but important way. Consider again water waves in a pond. Suppose that the wave crests are 5 cm apart and that we count 3 crests per second. **Figure 11–8** shows that in 1 sec, the wave must

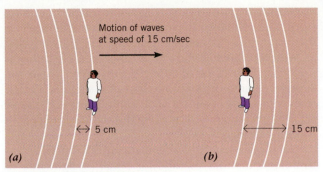

FIGURE 11–8. The relationship that exists between wavelength, frequency, and speed. (a) Initially. (b) One second later.

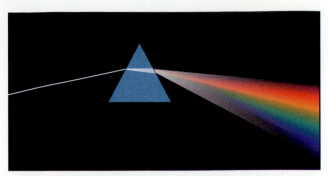

FIGURE 11–9. The dispersion of light into its constituent wavelengths by a prism. Notice that blue light is most strongly refracted.

have moved (3 waves) × (5 cm/wave), or 15 cm in all. The speed of the wave is therefore 15 cm/sec. More generally, *for any type of wave*, it is true that

wavelength × frequency = speed of wave.

Symbolically, if λ (the Greek letter lambda) is the standard symbol to represent the wavelength, f is the frequency, and c is the speed of light, then

$$\lambda f = c .$$

Inquiry 11–8 Using this relationship, what is the speed of a water wave if its frequency is 2 hertz and its wavelength is 10 cm?

Inquiry 11–9 What is the wavelength of radio radiation having a frequency of 15 megahertz (1.5×10^7 hertz)?

Inquiry 11–10 What is the frequency of microwave radiation with a wavelength of 3 cm?

In glass, air, or other transparent materials, electromagnetic waves interact with the charged particles in the matter and travel more slowly than in a vacuum. For example, in glass, visible light travels about two-thirds its speed in a vacuum; in water, light waves travel at only three-fourths of their speed in a vacuum. Because air has low density, the speed of a light wave through air is only slightly less than its speed in a vacuum. While not at all obvious, this slowing of the wave causes it to refract. In other words, the change in speed that occurs when light passes from one medium to another causes refraction. Furthermore, because the speed of an electromagnetic wave in a medium changes with wavelength, the amount of refraction depends on wavelength and is called *dispersion*. Dispersion is what makes a prism work; the dispersion of light passing through a prism is illustrated in **Figure 11–9.**

INTERFERENCE

What happens if two waves run into each other? Waves can interact and combine with each other, resulting in a composite form to which both have contributed. Most people have observed the interactions of water waves, for example, when two waves coming from different directions meet each other. The combination of the two waves can be quite complex. Places where the tops and bottoms of two waves meet will combine into a stronger wave; places where the top of one wave meets the bottom of the other will diminish in amplitude, perhaps even canceling one another.

The same things can happen with electromagnetic waves. In **Figure 11–10a**, two waves are shown that are "in phase" with each other—that is, their crests (the high points of the waves) line up as do their troughs (the low points). In this case, the effects of each wave add to

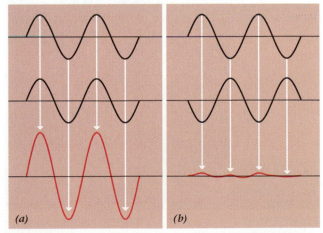

FIGURE 11–10. (a) Constructive interference. The two in-phase waves add up, reinforcing the effect. (b) Destructive interference. The two out-of-phase waves cancel each other out.

produce a wave that is twice as high as the individual waves. If the two waves combine in such a way as to reinforce each other, the result is called **constructive interference**. On the other hand, Figure 11–10*b* shows the case where the two waves are "out of phase" so that the crests of one line up with the troughs of the other. The two waves cancel each other out, giving **destructive interference**.

When light passes through a slit, the waves that come from one part of the slit will interfere with the waves from another part to produce light and dark bands, called diffraction fringes, seen in **Figure 11–11***a*. The explanation can be seen in **Figure 11–11***b*, which shows the waves from the top and bottom parts of the slit. Notice that in certain directions (marked with solid lines), the crests of one wave align with the crests of the other wave. An observer looking in these directions would see that the light was especially bright, because constructive interference is taking place. On the other hand, in other directions (marked with dashed lines), the crests of one wave line up with the troughs of the other, producing destructive interference. An observer looking in these directions would see nothing. Because there are several

directions in which constructive interference takes place, separated by others with destructive interference, the result is the "banded" appearance in Figure 11–11*a*.

With an understanding of interference we can now understand the cause of diffraction through a single slit. A wave passing through one part of a single slit interferes with a wave coming from another part of the same slit. The resulting pattern of bright and dark regions is the diffraction pattern.

The properties of diffraction and interference give astronomers another method for breaking light into its spectrum, by using a **diffraction grating**. This is a surface on which thousands of fine, parallel lines have been drawn by a precision machine, producing many very narrow slits, each capable of diffracting light. The amount of diffraction depends on the wavelength of the light, and the many narrow slits all acting together produce a very good separation of the wavelengths that make up the incoming light. A complete instrument that uses a diffraction grating or a prism as a component to produce a spectrum is called a **spectrograph**. Astronomers attach a spectrograph to a telescope to obtain the spectra of planets, stars, galaxies, and so forth.

> READERS HAVING THE ACTIVITY KIT
> SHOULD DO KIT ACTIVITY 11–2, THE
> DIFFRACTION GRATING, AT THIS TIME.

A final example of interference is provided by sound waves. Stereo systems emit sound waves that can be in phase with each other and produce areas of constructive and destructive interference throughout the room. Without careful placement of speakers, you could place yourself in a location where destructive interference occurs and the quality of the sound is diminished.

POLARIZATION

If you were able to examine the form of the electric part of an electromagnetic wave,[1] you would notice that the wave was vibrating in a single plane (see the electric wave in Figure 11–6). A photograph of a different wave taken 10^{-8} seconds later would find this wave vibrating in a different plane. A series of such photos would show successive waves vibrating in different (random) directions. Situations often arise in nature that cause the wave to vibrate in one plane; when this happens, we have **polarized light**.

Polarization of light demonstrates two important properties. First, the wave model of light is valid, because the wave model accurately predicts polarization phenomena. Second, light waves are similar in type to

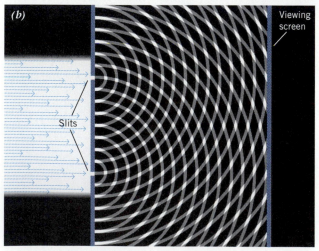

FIGURE 11–11. (*a*) Observation of diffraction fringes when light passes through a narrow slit. (*b*) Diffraction explained. The wave crests from the two sides of the slit interfere constructively where the wave crests meet and destructively in between.

---

[1]The same is true of the magnetic part. We consider the electric part here only for convenience.

FIGURE 11–12. The Crab Nebula photographed in polarized light. The arrows in the lower right corners show the orientations of the polaroid filters.

the seismic S waves of Chapter 8, not the compressional P waves, which cannot be polarized.

> **READERS HAVING THE ACTIVITY KIT SHOULD DO KIT ACTIVITY 11–3, POLARIZED LIGHT, AT THIS TIME.**

The study of the polarization of light is important to astronomy because many processes in the universe serve to polarize light. For example, when molecules in Earth's atmosphere scatter light, polarization results. Similar effects occur for light scattered from the atmospheres of the planets. Light reflected from the surface of a planet can be polarized in ways that depend on the surface characteristics. Processes that emit polarized radiation occur in a variety of astronomical objects, such as the Crab Nebula shown in polarized light in **Figure 11–12.** The differences in the appearance of the two photographs provides astronomers with information about the processes taking place within the object. Further examples of objects emitting polarized light are discussed later in this book.

11.3

LIGHT AS A STREAM OF PARTICLES

In the previous sections we examined two models describing the behavior of light: rays and waves. We have described experiments involving reflection, refraction, diffraction, interference, and polarization, for which these models provide predictions of past and future behavior. There are other experiments, however, for which neither the ray model nor the wave model are satisfactory.

At the beginning of the twentieth century, physics had reached some apparently insurmountable obstacles to further progress. While experiments showed the intensity of electromagnetic energy from hot bodies to rise to a peak and then fall off toward shorter wavelengths, physical theory of the day predicted that such a body ought to radiate more and more energy at shorter and shorter wavelengths (**Figure 11–13**). This dilemma obviously meant that there was something wrong with the theory, but what?

The answer was provided in the last week of 1899 by Max Planck, who ushered in a new era of physics by showing that agreement with theory could be obtained by assuming a model in which electromagnetic radia-

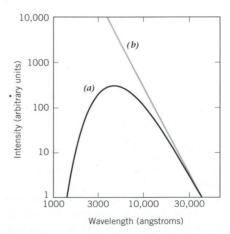

FIGURE 11–13. Radiation from a hot body. (*a*) As observed. (*b*) According to early theory.

tion was always emitted in specific amounts of energy, called **photons**. Photons are bundles of energy and can loosely be thought of as particles. In addition, Planck's photon model suggested that the amount of energy contained by each photon is proportional to the frequency of the radiation. Expressed mathematically, this sentence says that

$$E = hf,$$

where E is the energy of a photon corresponding to a wave of frequency f, and h is a quantity called *Planck's constant* (given in Appendix A2). This equation links the wave and photon models of light by linking the wave property of frequency with the photon property of energy. The constant h, relating the photon energy to the wave property of frequency, is one of the more important fundamental constants of physics. Owing to the relation between frequency and wavelength ($\lambda f = c$), we can also express this equation as

$$E = hc/\lambda.$$

This concept tells us that the higher the frequency of the radiation (i.e., the shorter the wavelength) the more energetic are the photons composing the beam of radiation. Planck's proposal was the first step toward developing the twentieth-century theory called "quantum mechanics," which assumes that most aspects of the natural world resolve themselves into discrete bits and pieces when viewed on the smallest scales. Once the photon nature of radiation was understood, theory and the observations in Figure 11–13 matched perfectly.

Inquiry 11–11   What do you conclude about the nature of light now that observations and theory described in Figure 11–13 are in agreement?

THE PHOTOELECTRIC EFFECT

In 1905, the young Albert Einstein applied this daring idea to explain another phenomenon that had been puzzling physicists for years: the photoelectric effect. In the photoelectric effect, light falling on certain substances—for example, cesium or selenium—forces electrons from the surface, thus making an electric current flow. This phenomenon is the basis of a number of devices in common use (e.g., the exposure meter in modern cameras, and the automatic door openers in many business establishments).

Prior to Einstein's work, there was a discrepancy between the theory accepted at that time and the observations of the photoelectric effect. Theory predicted that an electric current ought to be given off by the photoelectrically active substance no matter what the wave-

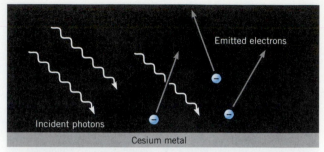

FIGURE 11–14.   The photoelectric effect. When light falls on certain metals, electrons are dislodged from the metal, causing an electrical current to flow.

length of the light, provided that the light's intensity was large enough (**Figure 11–14**). (By *intensity* we mean the number of photons per second striking the surface.) Instead, experiments found that the electrons were not knocked loose unless the wavelength of the light was less than some critical value, *no matter how intense the light.* In addition, light that had a sufficiently short wavelength caused an electric current to flow, *no matter how feeble the intensity.* The energy of the electrons knocked loose therefore depended only on the wavelength and not on the intensity of the light.

Einstein reasoned as follows. If the energy of each photon depends on its wavelength, then only those photons that have a sufficiently great energy will be capable of knocking an electron loose from the surface. If the electron acquires only a small amount of energy, it will not have enough energy to leave the surface. Because shorter-wavelength photons have the larger amount of energy, only they will be effective in generating an electric current. While just one photon could knock an electron loose if it had sufficient energy, photons with too little energy (that is, an amount of energy below some critical amount for each substance) could *never* knock the electron loose, regardless of how many there were. The photon nature of light, therefore, explains all observations concerning the photoelectric effect.

Inquiry 11–12   As the wavelength of the incident photon gets shorter, what should happen to the kinetic energy of the ejected electron?

Inquiry 11–13   For the following types of radiation, what is the order of increasing wavelength? Of increasing photon energy? Gamma rays, infrared radiation, visible light, X-rays, radio radiation, ultraviolet radiation.

MODELS OF LIGHT

At this point, you may be asking, "What *is* light, really?" We have talked about it as a ray, as a wave, and now as a stream of particles. It seems contradictory that light could behave both as a wave, which cannot be localized to a single point, and as a particle, which by definition is localized. It turns out that light is a very complex phenomenon. When we describe light as a wave or as a particle, we are saying that under certain circumstances it can behave as a wave and under other circumstances as a particle. The *models* we call "waves" or "particles" have a certain intuitive meaning in our everyday experience, but we are applying them to describe a submicroscopic phenomenon that is far removed from that experience. In fact, *nothing* in our macroscopic world is, or even can be, a good analogy to explain completely the nature of light. The best models explaining the phenomenon of light are abstract mathematical models that enable phenomena to be predicted but have no easily visualized analogies.

The **wave-particle duality** is not confined to light, but is a property of all matter. Material particles, such as electrons and neutrons, also exhibit wavelike behavior. For example, if a beam of electrons passes through the closely spaced atoms of a crystal and then falls on a photographic plate, the images of the electrons will

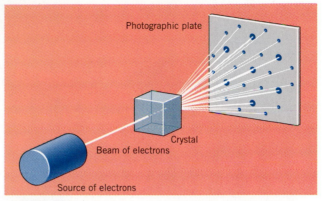

FIGURE 11–15. Electrons passing through a crystal are diffracted, demonstrating that electrons also exhibit some wave properties.

not form a single spot in the center of the plate but will be diffracted and will form a pattern of spots (**Figure 11–15**), showing that electrons have wavelike characteristics, too.

We will apply the photon nature of light to an understanding of spectra in Chapters 13 and 14. First, however, in Chapter 12 we will look at the telescopes astronomers use to collect the electromagnetic radiation we studied in this chapter.

DISCOVERY 11–1
IMAGES IN A MIRROR

After completing this activity, you should be able to do the following:

- State the relationship between incident and reflected rays and use it to trace rays reflected from a mirror.
- Define "virtual image" and identify situations in which a virtual image occurs.
- Diagram and explain the formation of a virtual image by a mirror.

Take a mirror and observe the reflection of a printed page in it. What do you see? The familiar fact is that the mirror reverses the page and the writing. In the "looking-glass world," everything is reversed: left-handed shoes become right-handed, people hold out their left hand, and bottle tops screw on and off in the wrong direction.

To study this phenomenon, shine a flashlight along a piece of paper so that the beam from the flashlight makes a long streak on the paper (**Discovery Figure**

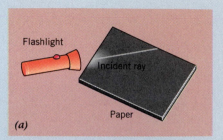
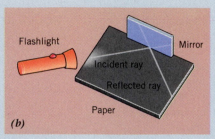
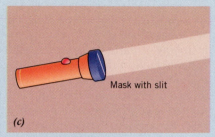

DISCOVERY FIGURE 11–1–1. Making a ray of light reflect off a piece of paper (a) without and (b) with a mirror in place to reflect the beam. Placing a slit over the end of the flashlight (c) helps to define the light beam better.

11–1–1). This will work better if you make a mask for the flashlight so that you allow only a thin slit (rectangle) of light to pass, as shown in part *c*.

☐ **Discovery Inquiry 11–1a** What property of light is illustrated by the fact that the beam of light from the flashlight makes a long streak on the paper?

Now place a small mirror at the edge of the paper so that it intercepts the beam and sends it off in another direction. Tilt the mirror so that the outgoing beam can also be seen on the paper, as shown in Discovery Figure 11–1–1b. By varying the angle A between the incoming beam and the line perpendicular to the mirror, you can vary the direction in which the reflected beam will go. For various incoming angles, A, roughly estimate the size of angle B between the reflected beam and the line perpendicular to the mirror. (In the language of physics, angle A is called the **angle of incidence** and angle B the **angle of reflection**).

You can also do this activity with a flashlight and a full-sized mirror. Place a removable mark or a piece of tape on the mirror. Move straight toward the wall (at an angle to the spot as shown in **Discovery Figure 11–1–2**) as you continue to shine your flashlight on the marked spot. As you approach the wall, you are constantly increasing the angle of incidence. Watch the reflected beam.

☐ **Discovery Inquiry 11–1b** As nearly as you can tell, what is the relationship between the angle of incidence and the angle of reflection?

The rays of light behave as if they consisted of a stream of particles all traveling in the same direction. When hitting a surface, they bounce away in another direction

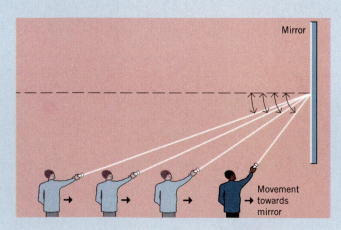

DISCOVERY FIGURE 11–1–2. Studying the law of reflection. The observer shines a flashlight at a point on a mirror while walking toward the wall.

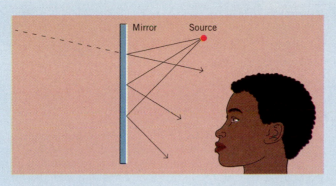

DISCOVERY FIGURE 11–1–3. Reflection of a source of light in a mirror. (Note: Completing the missing rays is part of the activity.)

determined by the direction from which they came. Billiards players will recognize that the answer to the question above is used to estimate the angle at which a ball will rebound from the cushion. Thus you have found the law of reflection: the angle of reflection always equals the angle of incidence.

We are now in a position to understand the formation of an image in a mirror. **Discovery Figure 11–1–3** shows a source of light in front of a mirror. Several rays from the source are shown, and their paths are sketched, using the relationship discussed above. One of these rays, denoted by a dashed line, is extended backward from the mirror surface. *Trace or photocopy the figure;* on the copy, take a straightedge and extend the other two lines straight back to where all three lines intersect. This point of intersection is the point from which the rays *appear* to come—from behind the mirror at a distance equal to the distance of the actual object in front. An observer in front of the mirror has no way of knowing what has happened to the rays along the way and can only sense the *direction* from which the rays *appear* to come.

☐ **Discovery Inquiry 11–1c** Describe the location behind the mirror from which the light appears to come.
☐ **Discovery Inquiry 11–1d** Explain why the eye is fooled into thinking that the source of light is *behind* the mirror.

If you were to place a camera at the point where the image appears to be located and then tried to photograph it, no image would result because there are no actual light beams there. For this reason, the kind of image we are discussing is called a **virtual image**. This means that the rays of light that enter our eyes project from an apparent source, as if they have traveled in straight lines with no deflections. It is an apparent image only, which is what is so puzzling to a kitten trying to "catch" its own reflection in a mirror. Virtual images are to be distinguished from the so-called **real images** we will study later.

DISCOVERY 11–2
REFRACTION OF LIGHT

After completing this activity, you should be able to do the following:

- Diagram and explain the formation of a virtual image by refraction at the surface of a glass of water.
- Diagram and describe the effect of refraction on a ray of light as it passes from one medium to another.

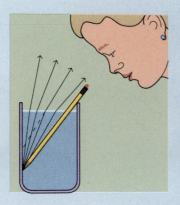

DISCOVERY FIGURE 11–2–1. The pencil experiment, with rays partly drawn in. The dashed line shows the apparent origin of one of the rays. The drawing does not show what you will see if you view the glass from the side.

Fill a clear glass with water. Put a pencil into the water at an angle and observe the appearance when you look directly along the part of the pencil that is out of the water. Draw the appearance of the pencil. (Important: Look directly *along* the part of the pencil that sticks out of the water, not from the side.)

☐ **Discovery Inquiry 11–2a** What appears to have happened to the pencil?
☐ **Discovery Inquiry 11–2b** Where does the tip of the pencil appear to be relative to its actual position?

When light rays pass from one medium to another, as from air to water or from water to air, they bend. The rays of light bend *toward* the line perpendicular to the surface when they enter a denser medium (water) from a less-dense medium (air). They bend *away* from this line when they leave the denser medium, as shown in **Discovery Figure 11–2–1.**

Discovery Figure 11–2–1 shows the rays of light leaving the tip of a pencil and entering the air from the water. Trace or copy the figure and extend the rays backward along straight lines to the point from which they appear to come, as illustrated by the dashed line.

☐ **Discovery Inquiry 11–2c** Is the apparent source of the rays of light from the tip of the pencil in agreement with your answer to the previous Inquiry? Where is the apparent source relative to the actual position of the pencil tip?
☐ **Discovery Inquiry 11–2d** What kind of image (real or virtual) are you observing when you look at the pencil tip in the water?

Now imagine a ray of light that enters a piece of glass with parallel faces, as illustrated in **Discovery Figure 11–2–2.** The path of the light ray inside the glass is determined using the rule stated above that light entering a denser medium is bent toward the perpendicular. Using a pencil, draw the path of the ray as it leaves the glass in the figure.

☐ **Discovery Inquiry 11–2e** After the ray leaves the glass, is it traveling in the same direction it was originally?

The answer to Discovery Inquiry 11–2e depends on the fact that the two faces of the piece of glass are parallel. For example, in a prism (**Discovery Figure 11–2–3**)

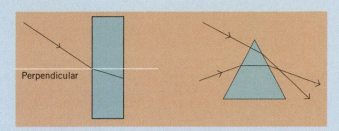

DISCOVERY FIGURE 11–2–2. *(left)* Light entering a piece of glass with parallel faces.

DISCOVERY FIGURE 11–2–3. *(right)* Light entering and leaving a prism.

the two faces are not parallel. A ray of light that enters the first face perpendicularly will not be bent at that face but will be bent at the other one. Even if it enters the first face at an angle, there will be a net change in the direction of travel of the ray.

CHAPTER SUMMARY

OBSERVATION

- Light exhibits properties of rays through **reflection** and **refraction**, and properties of waves through **diffraction, interference**, and **polarization** (in addition to reflection and refraction).
- Light exhibits the properties of particles (**photons**), while particles show the properties of waves, thus giving a "dual nature" to light and particles.
- All types of electromagnetic waves in a vacuum move with the **speed of light** and are characterized by a **wavelength** and **frequency**. These three quantities are related to one another by $\lambda f = c$, where λ is the wavelength, f is the frequency, and c is the speed of light.
- Light, gamma rays, X-rays, ultraviolet and infrared radiation, and radio waves all travel at the speed of light and differ only in their wavelengths.
- Not all types of electromagnetic radiation pass equally well through the Earth's atmosphere.
- The photoelectric effect is a phenomenon in which electrons are ejected from the surface of a material when photons having high-enough energy fall upon the material.

THEORY

- We describe light by means of various models, none of which say what light "is."
- Maxwell's theory of electromagnetism predicts the behavior of electromagnetic waves. It predicts that these waves vibrate perpendicularly to their direction of motion (and can therefore be polarized), that they move with the speed of light, and that they show the properties of reflection, refraction, diffraction, and interference. It also predicts that waves carry energy.
- Nothing can move faster than the speed of light in a vacuum.
- Planck showed theoretically that the energy possessed by a photon is directly related to the frequency associated with it, and inversely related to the wavelength: $E = hf = hc/\lambda$, where E is the photon energy, and h is Planck's constant.

CONCLUSIONS

- Both theory and observation indicate the wave-particle duality of light and matter.

SUMMARY QUESTIONS

1. What is meant by *reflection* and *refraction*?
2. What is meant by *diffraction*? In what situations does it occur? How does diffraction vary with the wavelength of the light and the width of the aperture through which it travels?
3. What is the meaning of *wavelength* and *frequency*?

What is the relationship among wavelength, frequency, and speed of a wave?

4. What does the word *spectrum* mean?

5. What is the importance of the concept of atmospheric "windows" for astronomy?

6. What is the meaning of *interference*? How does the concept explain the slit experiment?

7. What is meant by the polarization of light? Why is it important to astronomy?

8. What is meant by the wave-particle duality of both light and matter?

9. In what way does the energy of a photon depend on its wavelength? Arrange radiations of various types in order of increasing photon energy.

APPLYING YOUR KNOWLEDGE

1. Summarize the models we use to understand light. What evidence is used to validate each model?

2. Explain the differences between the bending of light by refraction, and the bending by diffraction.

3. What is the importance of the photoelectric effect in providing evidence in favor of the photon description of light?

■ 4. Suppose your favorite radio station's frequency is 91.5 megahertz (millions of cycles per second). What is the wavelength of the signal?

■ 5. How much more energy does a 1000 Å photon have than one having a wavelength of 10,000 Å?

■ 6. What is the frequency and the energy of a photon having a wavelength of 21 cm?

ANSWERS TO INQUIRIES

11–1. This inquiry is to get you to think about, and summarize in your own words, the previous paragraphs.

11–2. Very high priority should be given to observing wavelengths that are absorbed by the atmosphere and hence inaccessible at the Earth's surface: gamma-ray, X-ray, ultraviolet, infrared. Observations that are possible from the Earth's surface should be done from there, because space operations are extremely expensive.

11–3. 3×10^5 km/sec $\times$ 1 year $\times$ 365 days/year $\times$ 24 hours/day $\times$ 60 minutes/hour $\times$ 60 seconds/minute = 9.5×10^{12} km.

11–4. 1.5×10^8 km / 3×10^5 km/sec = 500 seconds = 8.3 minutes.

11–5. 3×10^5 km/sec $\times$ 2.6 seconds / 2 (round trip) = 3.9×10^5 km.

11–6. Decreasing wavelength means increasing frequency.

11–7. Decreasing speed means decreasing frequency.

11–8. Speed = 2 hertz $\times$ 10 cm = 20 cm/sec

11–9. $\lambda f = c$; $\lambda = 3 \times 10^{10} / 1.5 \times 10^7 = 2 \times 10^3$ cm, or 20 m (0.02 km).

11–10. $\lambda f = c$; $f = 3 \times 10^{10}$ cm/sec / 3 cm = 10^{10} hertz or 10^4 megacycles/sec or 10 gigacycles/sec.

11–11. Because the theory depended on the assumption of discrete photons, you should conclude that the description of light in terms of photons has some degree of validity.

11–12. The kinetic energy of the ejected electron should increase as the wavelength of the incident photon decreases.

11–13. Increasing wavelength: Gamma rays, X-rays, ultraviolet radiation, visible light, infrared radiation, radio radiation. Increasing energy is the reverse order.

Discovery Inquiry 11-1b. The angle of incidence equals the angle of reflection.

Discovery Inquiry 11-1d. To the eye, it appears as if the *rays* originate at a point behind the mirror.

Discovery Inquiry 11-2b. It appears to be *above* its true position.

Discovery Inquiry 11-2d. Virtual image.

Discovery Inquiry 11-2e. Yes, but it is displaced slightly, having first been bent toward the perpendicular upon entering the glass and away from it upon leaving.

12

TELESCOPES:
OUR EYES OF DISCOVERY

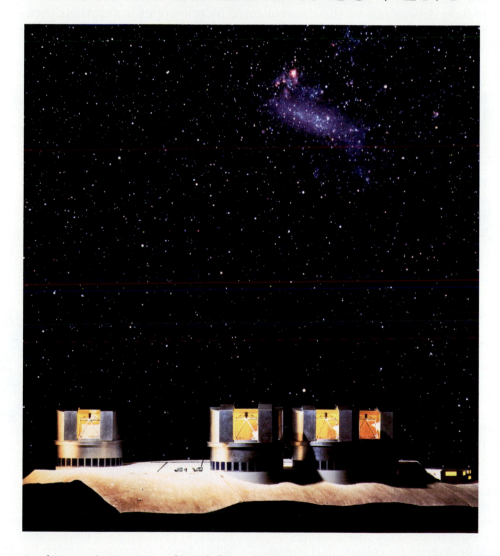

*O telescope, instrument of much knowledge, more precious than any sceptre!
Is not he who holds thee in his hand made king and lord of the
works of God?*

GALILEO GALILEI, 1610

Telescopes provide humanity with a view of the universe not imagined by our ancestors. The telescope is the most basic and fundamental instrument that astronomers use. It has existed as a research tool since the time of Galileo, approximately 350 years. A well-constructed telescope is a versatile instrument with perhaps the longest useful lifetime of any major piece of scientific equipment. The detectors and analyzers of radiation mounted on telescopes are constantly being improved by the rapid progress of modern technology, but a soundly constructed telescope will continue to perform its basic functions for many decades.

There are, of course, many astronomers who never use telescopes. Theoreticians regard large, powerful computers as their principal research tools, but they, along with observational astronomers, are ultimately dependent on the data collected by telescopes to test their theories and hypotheses.

The main function of a telescope is to gather light from a celestial object and bring it to a focus for further study. Secondary functions are to resolve the detail in the image, and to magnify the angular scale of celestial objects. The telescope is placed on a mounting that can be moved, thereby enabling it to follow the apparent paths of objects across the sky.

Many different kinds of telescopes have been used by astronomers since Galileo first turned his small instrument skyward in 1609. Modern astronomers use telescopes of all types and sizes. Some are sent above the atmosphere for a better view, and specialized instruments are designed for certain observing projects. In this chapter, you will study how lenses and mirrors form images, and how they can be combined to form telescopes.

12.1

THE FORMATION OF IMAGES

A simple device that illustrates the basic process of image formation is the pinhole camera. **Figure 12–1** shows the principle of its operation. It forms an image on a screen that can be observed directly with the eye. Light travels from a particular part of the scene being viewed through the pinhole and then (except for diffraction effects) travels onto the screen, where it is observed. The image is upside down, which is easily understood by a glance at the figure.

If a piece of photographic film is used instead of a screen, we can take a picture of the scene. Pinhole cameras are still used today to take very wide-angle photographs that would be difficult for conventional lenses. But the pinhole camera suffers from some obvious drawbacks. The image is faint and hard to see, because not much light gets through the pinhole, and enlarging the pinhole to let in more light would blur the image.

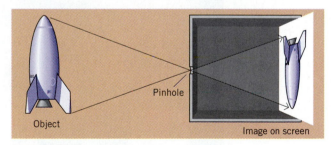

FIGURE 12–1. The principle of the pinhole camera. An image is formed because light travels in straight lines through the pinhole.

Therefore, small holes and long time exposures are necessary to produce good photographs. To photograph faint celestial objects, astronomers need something that has the ability to collect a great deal of light but that can still concentrate the light to a focus, so that a sharp photograph can be made. These goals can be accomplished using either lenses or curved mirrors.

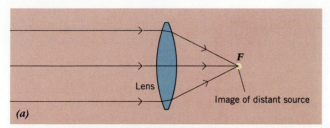

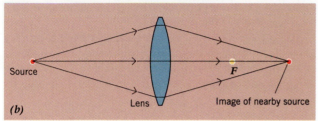

FIGURE 12–2. Image formation of a point source of light. (*a*) A distant source and (*b*) a nearby source.

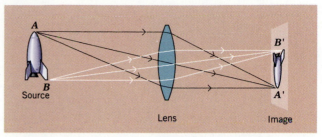

FIGURE 12–3. Formation of an image of an extended source by a lens.

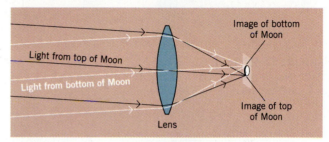

FIGURE 12–4. Formation of an image of the Moon by a lens.

> READERS HAVING THE ACTIVITY KIT
> SHOULD DO ACTIVITY 12–1, HOW LENSES
> FORM IMAGES, NOW.

It is not difficult to see how a lens works. In our discussion, we will consider a *point source* of radiation, which is an object whose distance is so much greater than its size that it looks like a point of light. A star is an example of such a source. **Figure 12–2a** shows parallel light rays from a distant point source brought to a focus by a lens. The distance from the lens to the focus is the most important quantity used to describe a lens and is called the **focal length**. The symmetrical lens shown here has foci placed equal distances from the center of the lens. The lens brings the light to a focus because each ray encounters the glass at a different angle, and each is bent by a different amount. **Figure 12–2b** is similar to Figure 12–2a except that it shows several rays of light from a *nearby* point source entering the same lens; in this case, the rays are brought to a focal point farther from the lens than that defined by the focal length. In both cases, the net result is that all the rays are focused to a single point. Such an image is called a **real image**, because the light rays actually go through this point; you can record it by placing film there. It is different from the virtual image discussed for mirrors in Discovery 11–1.[1]

---

[1]For those of you who did not do Discovery 11–1, an example of what we call a virtual image is the image you see when you look at yourself in a mirror. It is called a virtual image because if you were to place a piece of film at the location where your image appears to be, no photographic image would form because there are no light rays passing through that point.

Any object that is not a point source is called an *extended source* (**Figure 12–3**). Light from various parts of an extended object are focused onto different parts of the observing screen. Light from the subject's head at point *A* is imaged onto the screen at point *A'*, whereas light from the subject's feet at *B* is imaged at *B'*. The image is inverted.

When an object is at a great distance, light rays from any part of an object will arrive at the lens in parallel rays. To visualize why this is true, imagine the source of light in Figure 12–2b moving away from the lens until it was many miles away. The rays of light would clearly be almost perfectly parallel and images would form exactly one focal length behind the lens. If we were to put a piece of film there, we could photograph the image. **Figure 12–4** shows what happens when light from a distant extended source is focused by a lens. In the case of the Moon, one set of parallel rays comes from one side of the Moon and another set from the other side. Each set of parallel rays will be focused to a particular point by the lens. The image is, of course, inverted.

The image sizes of the rocket in Figure 12–3 and the Moon in Figure 12–4 are directly proportional to the focal length of the lens. **Figure 12–5** shows why. Both parts show images being formed by lenses having different focal lengths, with the result that the longer-focal-length lens produces the larger image.

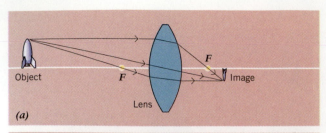

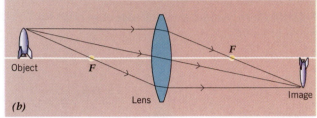

FIGURE 12–5. Image size depends upon the focal length. Each lens shown here has a different focal length and therefore produces images of different size.

Inquiry 12–1 If a lens having a 50-mm focal length produces a 1-cm-high image of a person, what will be the image size when a 150-mm focal length lens is used?

> READERS HAVING THE ACTIVITY KIT
> SHOULD DO ACTIVITY 12–2, OTHER
> PROPERTIES OF LENSES, NOW.

12.2
TELESCOPES

Lenses have been in use for a long time to correct vision. During the Middle Ages, spectacle makers ground lenses of various types to correct human eyes that were either too weak or too strong. Although probably not the originator of the telescope, Hans Lippershey, a Dutch spectacle maker, is usually given credit for making the principle of the telescope widely known in the early 1600s.

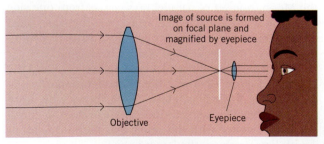

Image of source is formed on focal plane and magnified by eyepiece

Objective

Eyepiece

FIGURE 12–6. How a telescope works.

The basic telescope has two parts, as illustrated in **Figure 12–6**. First, there is an objective lens or objective mirror (often referred to generically as the **objective**), which is a lens or mirror of longer focal length and (usually) larger diameter than the second lens. Telescopes with objective lenses are called **refractors**, while those with mirrors are called **reflectors**. The objective forms an image that is inspected through the second part of the telescope, the **eyepiece**. The eyepiece has a shorter focal length than the objective.

The principle of the astronomical telescope is simple. The objective forms a real, inverted image of the object being observed. If you were to place a transparent screen there, you would see the image and be able to photograph it.

THE LIGHT-GATHERING POWER OF A TELESCOPE

Inquiry 12–2 How many times more light is collected by a 10-cm-diameter objective than by one having a 1-cm diameter?

The objective's area collects light. The larger the area, the greater the **light-gathering power** of a telescope, which is its ability to collect light. In the Inquiry, the lens diameter has increased 10 times and the area 10^2 times. For this reason, 100 times as much light is collected by the larger lens.

The diameter of a telescope's objective is its most important single characteristic as far as an astronomer is concerned, because the diameter determines the light-gathering power. Even a 1-cm lens will collect about four times as much light as the human eye. The world's largest optical telescope, the 10-meter Keck telescope on Mauna Kea in Hawaii, has a diameter approximately 2000 times that of the human pupil and, as a consequence, it is able to gather roughly 2000^2 or 4,000,000 times as much light as the eye (**Figure 12–7**).

Lists of the largest telescopes in the world invariably are arranged in order of diameter. **Table 12–1** lists telescopes having diameters of 3.0 meters or greater.

Inquiry 12–3 If the dark-adapted eye has a diameter of 0.5 cm, how much more light can a lens of diameter D cm collect than the human eye?

The advantage of big telescopes over the human eye is even more dramatic than just light-gathering power, because at the focus of the telescope the astronomer can mount a camera or sensitive electronic detector and

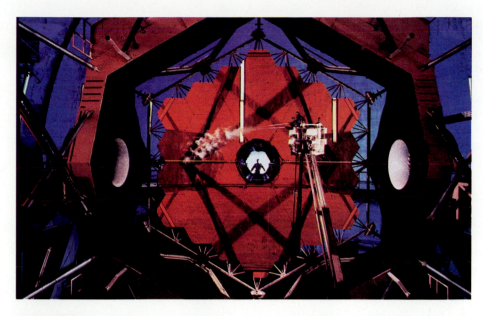

FIGURE 12–7. The 10-meter Keck telescope.

TABLE 12–1

Telescopes with Apertures Greater than 3.0 Meters

Observatory	Funding Country	Location	Diameter of Objective (meters)
European Southern Observatory*	Europe	La Silla, Chile	16-m Very Large Telescope (four 8-m mirrors)
Keck	USA	Mauna Kea, Hawaii	two 10.0-m Keck telescopes
NOAO*	USA	Mauna Kea, Hawaii	8-m (1998)
NOAO*	USA	Cerro Tololo, Chile	8-m (2000)
Mt. Graham International Obs.*	USA	Mt. Graham, Arizona	two 8-m telescopes (1998)
Special Astrophysical Obs.	Russia	Mt. Pastukhov	6-m Bol'shoi Teleskop Azimutal'nvi
Hale Observatories	USA	Mt. Palomar, California	5.08-m George Ellery Hale Telescope
Smithsonian Astrophysical Observatory/Univ. Arizona	USA	Mt. Hopkins, Arizona	4.5-m Multi-Mirror Telescope (MMT)
Royal Greenwich Observatory	United Kingdom	La Palma, Canary Islands	4.2-m William Herschel telescope
Cerro Tololo Inter-American Obs.	USA	Cerro Tololo, Chile	4.0-m
Anglo-Australian Observatory	USA	Siding Springs, Australia	3.9-m Anglo Australian telescope
Kitt Peak National Observatory	USA	Kitt Peak, Arizona	3.8-m Nicholas U. Mayall telescope
Royal Observatory Edinburgh	United Kingdom	Mauna Kea, Hawaii	3.8-m United Kingdom infrared telescope
Canada-France-Hawaii Telescope	Canada, France USA	Mauna Kea, Hawaii	3.6-m Canada-France-Hawaii telescope
European Southern Observatory	Europe	Cerro La Silla, Chile	3.57-m
WIYN*	USA	Kitt Peak, Arizona	3.5-m
Max Planck Institute (Bonn)	Germany	Calar Alto, Spain	3.5-m
Astronomical Research Consortium*	USA	Sacramento Peak, New Mexico	3.5-m
European Southern Observatory	Europe	Cerro La Silla, Chile	3.5-m New Technology telescope
Lick Observatory	USA	Mt. Hamilton, California	3.05-m C. Donald Shane telescope
Mauna Kea Observatory	USA	Mauna Kea, Hawaii	3.0-m NASA Infrared telescope

*Telescopes planned or currently under construction are noted by an asterisk.

make an observation over an extended length of time (e.g., a time exposure on film). The longer the observation, the fainter the objects that can be recorded. The human eye, on the other hand, is adapted to be efficient in detecting motion, so it "clears" itself many times a second; as a consequence, it cannot accumulate light as a photographic plate can. Our large telescopes are capable of detecting objects more than 100 million times fainter than the faintest stars detectable by the naked eye.

RESOLVING POWER

Because the edges of a telescope objective diffract the light passing around it, no image can ever be perfectly sharp. The larger the diameter of the lens, however, the sharper the image will be. Sharpness is measured in terms of the **resolving power**, or **resolution**, of the telescope, which is the angular size of the smallest detail that can be observed with it (**Figure 12–8**). Because short wavelengths diffract less than longer wavelengths, resolution is better at shorter wavelengths. The resolving power therefore depends on both the wavelength of the incoming radiation and the diameter of the telescope objective. Expressed mathematically,

$$\frac{\text{resolving power}}{\text{(in degrees)}} = 57.3° \times \frac{\text{wavelength of light}}{\text{diameter of objective}},$$

where the wavelength and diameter are measured in the same units.

For optical wavelengths (at about 5000 Å), a practical rule of thumb that measures resolving power in the more convenient unit of seconds of arc is given by the formula

$$\frac{\text{resolving power}}{\text{(in seconds of arc)}} = \frac{10}{\text{diameter of objective in cm}}.$$

Note that resolving power is expressed as an angle. A 1-inch (2.5-cm) telescope can distinguish objects separated by an angle of 4 seconds of arc as two distinct images. We say that these objects are *resolved*. Objects whose angular separation is less than 4 seconds of arc will not be resolved by a 2.5-cm telescope. In looking at a planet with this same telescope, only details separated by at least 4 seconds of arc will be distinguishable from one another. Using this formula, the 10-m Keck telescope is theoretically capable of resolving objects 0.01 second of arc apart, about four orders of magnitude better than the unaided human eye.

Unfortunately, the Earth's atmosphere plays a crucial, limiting role in ground-based observations under most situations. The atmosphere behaves like a lens in that it refracts the light passing through it. However, this lens is constantly moving. You have probably seen the

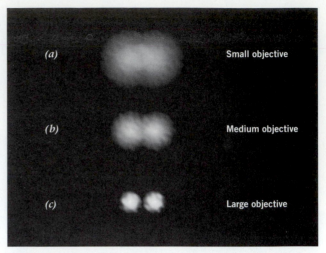

FIGURE 12–8. A double star as seen through telescopes with objectives of different diameters. (*a*) Small objective. (*b*) Medium objective. (*c*) Large objective.

effects of this: if you have ever observed stars carefully at night, even with the naked eye, you will have noticed that they twinkle. This is due to moving masses of air in the atmosphere through which light passes. Observed with a high-magnification telescope, the image of the star dances rapidly about and becomes blurred. A similar effect is seen when you look at scenery through a hot fire or through heat waves rising from a hot road.

The steadiness of the atmosphere can vary from minute to minute during the night, and from night to night. Astronomers refer to the amount of the variation as **seeing**. The seeing is expressed as an angular size and typically is 1 to 2 seconds of arc. On a night of good seeing, however, the atmosphere is stable enough that the blurring may be as small as a quarter of a second of arc from the best observatory sites.

The formulae given above for resolving power are the theoretically possible values. **Figure 12–9** compares the blurring effects of diffraction and seeing by showing the theoretically possible resolving power of telescopes of different objective sizes, and the seeing for a typical night. You will note that even modest-sized telescopes are more limited by seeing than by diffraction, and that only the very smallest telescopes are limited in their performance by diffraction. Seeing limitations are alleviated, however, when telescopes are placed above the Earth's atmosphere. They are then able to operate at their theoretically possible resolution limit.

Inquiry 12–4 At roughly what objective diameter does seeing rather than objective size become the

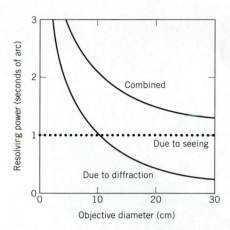

FIGURE 12–9. The blurring effects of seeing and diffraction in telescopes of different diameter objectives. The top curve gives the actual resolution achieved.

limiting factor in determining actual telescopic resolution? (Assume the best seeing is between 0.25 and 0.5 seconds of arc.)

MAGNIFICATION

We have said that a telescope forms a real image, and if a screen were placed at the focus, we would see an image. But that image would not be very large (as those readers who did the activities will recall). For this reason, an *eyepiece* lens is used to magnify the real image produced by the objective (see Figure 12–6). Because we are using the eyepiece as a magnifier, we actually observe a virtual image.

The overall **magnification** of a telescope depends on both the size of the real image formed by the objective, and the amount by which the eyepiece magnifies this real image. We said previously that the size of the image produced by the objective lens depends on its focal length. For this reason, the overall magnification of the telescope must depend on the focal lengths of *both* the objective and eyepiece. In fact, the magnification is expressed simply as

$$\text{magnification} = \frac{\text{focal length of objective}}{\text{focal length of eyepiece}},$$

where the focal lengths are in the same units.

Inquiry 12–5 If an amateur astronomer's telescope having an objective diameter of 15 cm and a 150-cm focal length is used with an eyepiece of 2-cm focal length, what would its magnification be?

Inquiry 12–6 Would the ability of this amateur telescope to resolve detail in an image be limited by diffraction or by seeing when the seeing is 0.5 seconds of arc?

It is possible in theory to make a telescope with as much magnification as desired simply by choosing eyepieces of shorter and shorter focal length. However, there are practical limits to the amount of magnification that can be used, because of diffraction in the telescope and unsteadiness in the atmosphere, both of which tend to blur the image. Beyond a certain point, one only magnifies the blur without seeing anything new. Astronomers rarely work at a magnification of more than about 500, and it would be an unusually good night when a small, amateur telescope could be used at a magnification greater than 10 times the diameter of the objective measured in centimeters.

Inquiry 12–7 Suppose you see a catalog that offers to sell you a 5-cm-diameter telescope with a magnification of 300. How do you respond? What would be a more reasonable magnification?

Inquiry 12–8 In what three ways can we talk about the "power" of a telescope. Is it meaningful to ask someone what the power of his or her telescope is?

> READERS HAVING THE ACTIVITY KIT
> SHOULD DO ACTIVITY 12–3,
> MAKING A TELESCOPE, NOW.

THE MOUNTING OF A TELESCOPE

All large telescopes, and many small ones, have special mountings that allow them to continue pointing toward the same part of the sky as it appears to turn overhead. The telescopes thus can point to objects for long periods and take time exposures with their cameras. Most telescopes are constructed so that they rotate about an axis parallel to the Earth's rotation axis; rotating the telescope about this axis, which is called the **polar axis**, compensates for the Earth's rotation. Such a mounting is known as an **equatorial mounting**. Figure 12–10 shows the equatorial mounting of the University of Texas 107-inch (2.7-m) telescope. The polar axis, from the lower left to the upper right, points toward the north celestial pole.

Mountings that allow the telescope to move in altitude and azimuth are much easier to construct and less expensive than equatorial mountings. However, such telescopes are more difficult to use, because keeping an object in the center of the field requires moving the tele-

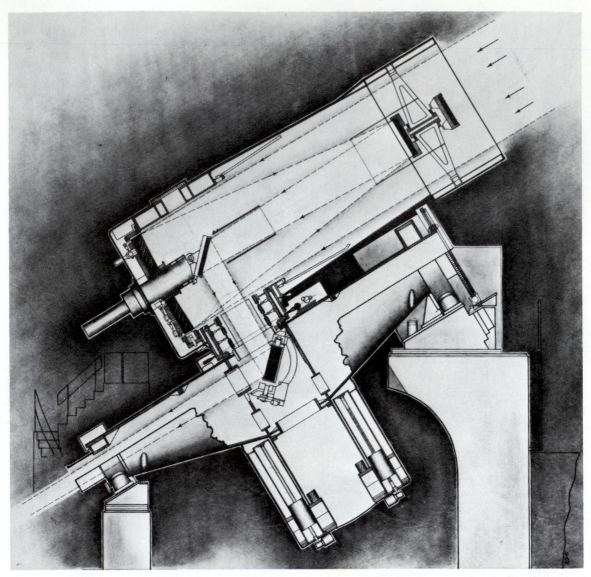

FIGURE 12–10.   A cross section of the 2.7-meter reflector at McDonald Observatory.   Notice the coudé focus.

scope in two directions instead of one. Because computers can perform these operations easily and without error, radio telescopes and large modern reflectors are built with such mounts.

12.3
COMPARING REFLECTING AND REFRACTING TELESCOPES

Refracting telescopes suffer from numerous disadvantages. In a refractor, the light passes through the glass.

Because, as we saw in the previous chapter, the amount of refraction depends on the wavelength, different colors are refracted in slightly different directions. This means that in a simple refracting telescope consisting of only two lenses, red and blue light focus at different points and produce colored halos around star images. This effect, known as *chromatic aberration*, is shown in **Figure 12–11a**. The resulting image lacks clarity. Chromatic aberration can be corrected by making compound lenses containing several lens elements of different materials and having different shapes (**Figure 12–11b**). Good lenses with excellent correction of various aberrations may have many lens elements—the standard

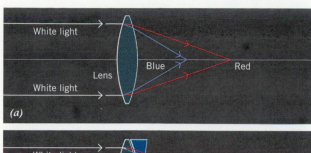
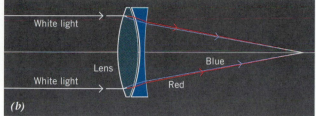

FIGURE 12–11. (*a*) Chromatic aberration. (*b*) The correction for chromatic aberration using a compound lens.

FIGURE 12–12. The 40-inch refractor of Yerkes Observatory, Williams Bay, Wisconsin.

55-mm lens on a good camera may have, for example, from five to nine elements. Such lenses are expensive and difficult to design. Furthermore, they lose light at each lens surface, which can seriously degrade their performance. On the other hand, as Sir Isaac Newton first pointed out, mirrors are inherently free of color effects.

Two additional drawbacks of lenses result from the fact that the light must pass through the lens. When the telescope moves to follow a heavenly object, the lens changes its orientation with respect to the center of the Earth and bends a little under its own weight. Because astronomical lenses and mirrors are ground to close tolerances—a millionth of a centimeter or so—even a miniscule amount of flexure can seriously distort the image. The lens can be made thicker, but it is difficult to cast a large, thick piece of glass that is so perfect it does not contain many bubbles and other imperfections. In addition, thicker lenses weigh more (requiring stronger, more expensive mountings to hold them) and absorb more light. The practical limit for refracting telescopes was attained in the 40-inch (1.02-m) lens at Yerkes Observatory, Williams Bay, Wisconsin (**Figure 12–12**). A mirror, on the other hand, does not suffer these limitations. Because it is supported from the back, it can be made quite stiff and can be cast with reinforcing ribs right in it.

A second drawback to refractors is that all lens surfaces must be perfect; in the case of a compound lens, that means at least four surfaces. Only the front surface of a mirror needs to be ground to high tolerances, and the presence of imperfections inside the mirror material does not degrade the performance because light does

not pass through it. For these reasons, per centimeter of objective diameter, reflectors are also much less expensive to build.

12.4
REFLECTING TELESCOPES OF VARIOUS TYPES

Reflecting telescopes are made with concave mirrors. While there are various possible curved shapes, *spherical* shapes are desirable because they are easy to attain. **Figure 12–13** shows a section of a spherical mirror with numerous parallel light rays from a distant source of light impinging on it. For each incident ray, the reflected ray is shown, following the law of reflection discussed in Chapter 11. Notice that rays of light from a star will *not* be focused into a single point by this mirror; this problem is called *spherical aberration*. One way to achieve a proper focus is to give the mirror a surface that is *para-*

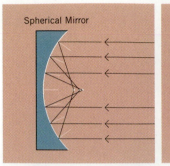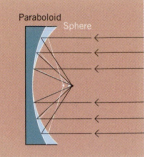

FIGURE 12-13. (*above left*) Spherical aberration, in which a spherical mirror will not bring parallel rays of light from a distant star to a single focus.

FIGURE 12-14. (*above right*) Spherical aberration, and the correction provided by changing the mirror's shape into a parabola.

bolic, in which the surface (shown in **Figure 12–14**) is not as steep as a sphere. In this case, the rays striking the outer part of the mirror focus at the same place as those closer to the mirror's center.

Another method of overcoming spherical aberration is found in the **Schmidt telescope**. This system (**Figure 12–15**) uses a spherical mirror because of its ease of manufacture. To overcome spherical aberration, a specially shaped, thin correcting lens is mounted in front of the mirror. It is thicker in the middle and around the edges and thinner in a ring-shaped region in between. The amount of correction required is small, and as a result the telescope is not nearly as sensitive to chromatic aberrations as an ordinary refractor would be.

The Schmidt design combines both refracting and reflecting elements in the objective. Its advantage is that it combines a wide field of view with a short focal length, so that short exposure times can be used. For this reason it is ideal for making surveys of large regions of the sky.

Because large telescopes are expensive, astronomers desire to make them as versatile as possible. Astronom-

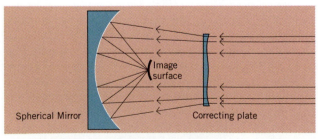

FIGURE 12-15. The Schmidt telescope. The aberrations of the spherical mirror are corrected by the thin correcting plate, producing an excellent wide-field image.

ical telescopes are usually built so that they can focus light in a variety of ways, which means a variety of observational goals can be accomplished with a single telescope.

One problem with a mirror is that the image forms in front of it. A method is required for observing the image without blocking most of the incoming light. If the mirror is large, this is not a major problem, and the telescope can be used at the mirror's focus, which is also called the **prime focus** (Figure 12–16*a*). A camera or other instrument can be mounted at this position. Although some light is blocked, most passes by and is focused by the main mirror. Until recently, in the largest telescopes the astronomer would actually ride in a cage placed at the prime focus. During the long period of observation throughout the night, the astronomer would adjust the image in the focal plane to obtain the best possible images. The observer's cage blocks only a small fraction of the light falling on the mirror. Nowadays, however, advances in electronics have moved the astronomer out of the cage and into a room where observing is done in front of a computer terminal.

In smaller telescopes, the amount of light blocked can be minimized by inserting a small, flat mirror into the optical path to redirect the image to a focus outside the telescope tube. This arrangement is common in amateur telescopes. **Figure 12–16***b* shows a common way of accomplishing this in the **Newtonian telescope**, named for its inventor, Sir Isaac Newton. The focal position in such an arrangement is called the Newtonian focus and is used in the least-expensive amateur telescopes.

In another arrangement, a convex mirror reflects the light back through a hole in the objective mirror. No additional light is lost on account of the hole, because it is in the shadow of the secondary mirror. Because the secondary mirror has some curvature, it actually increases the focal length of the system over either the prime or Newtonian focus. This arrangement is called the **Cassegrain focus** (Figure 12–16*c*). (The Cassegrain focus can also be seen in Figure 12–10 at the back end of the telescope.) High-quality amateur telescopes with a Cassegrain focus and a Schmidt correcting plate to alleviate spherical aberration are common.

Some telescopes can be set up to direct the light to yet a third focus, the *coudé* (from the French verb meaning "to bend"), and the word describes the rather complicated path taken by the light in this case. A sequence of mirrors brings the light down the hollow axis of the telescope mounting and into a room below (**Figure 12–16***d*). With this arrangement, no matter how the telescope moves, the light is brought to a focus at the same point in the room. The advantage of the coudé focus is that large and heavy instruments, which cannot be mounted on the moving telescope itself, can be

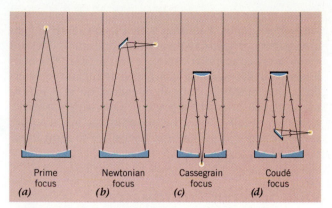

FIGURE 12–16. A variety of focal configurations are available with reflecting telescopes. (*a*) Prime focus. (*b*) Newtonian focus. (*c*) Cassegrain focus. (*d*) Coudé focus.

placed here for analysis of the light. Figure 12–10 shows the path of light taken to the coudé focus located off the figure to the bottom left.

Large telescopes are usually set up so that by moving mirrors any one of the foci—prime, Cassegrain, or coudé—can be selected. Why would an astronomer choose one focus over another? The focal length increases from prime focus to Cassegrain to coudé. Because, as we have seen, the image size depends on the focal length, the coudé focus produces the largest image of any of the foci. However, the larger the image size, the more spread out the photons are, making the image fainter and increasing the time required to obtain precise observations. Therefore, faint objects cannot be observed with the coudé focus. On the other hand, the short focal length of the prime focus is ideal for the faintest objects. The Cassegrain focus is in between. Thus the most appropriate focus to use for an observation depends on the brightness of the object and exactly what the astronomer is hoping to learn.

12.5
THE LARGE TELESCOPES OF THE FUTURE

During the last two decades, astronomers and engineers have experimented with radically new telescope designs to achieve reductions in the cost of building large instruments. The pioneering effort in this direction was the Multiple Mirror Telescope (MMT) located on Mount Hopkins south of Tucson, Arizona. This telescope originally had six 72-inch-diameter mirrors held in a common frame. All the mirrors were precisely aligned so that after some reflections they formed their images in

the same place. Thus the brightness of the image that was seen was equal to the sum of the light collected by all the separate mirrors. Telescopes of this design are capable of great light-gathering power at moderate cost.

One fascinating use of this telescope has been to combine the light from each mirror to produce interference. When used in this way the mirrors form an **interferometer**, an instrument that produces interference fringes. A complex analysis of the interference pattern provides astronomers with higher-resolution images than otherwise possible.

The European Southern Observatory, in planning its 16-meter Very Large Telescope (VLT) shown in the chapter opening photograph, is providing for the improvements allowed by interferometry. Once this state-of-the-art instrument is completed (1999), modern techniques should allow it to provide a resolution of 0.025 seconds of arc at a wavelength of 1 micron (10^{-4} cm) when the telescopes are separated by 130 meters. Such possibilities provide a dramatic improvement over the limits imposed by seeing and are tremendously exciting!

Inquiry 12–9 How large a single objective would be required *theoretically* to give the same possible resolution as the VLT at a wavelength of 1 micron?

Another important telescope is the Keck telescope built in Hawaii by the University of California and the California Institute of Technology. This telescope has a 10-meter diameter with a "segmented" mirror, which means that it has been ground in sections and assembled into one giant mirror like pieces of a puzzle. This type of assembly takes state-of-the-art computer and optics technology to keep all the various segments properly aligned. An identical telescope is being built 85 meters away and will allow astronomers to do interferometry. The National Optical Astronomy Observatory's Gemini project will place an 8-meter telescope in both the northern and southern hemispheres. Other such telescopes around the world are also in the planning stages.

The traditional design of telescopes contains limitations when you want to build them larger. For example, the mirror in the 240-inch Russian telescope is an enormous and heavy piece of Pyrex glass, the largest ever created. It requires an incredibly strong mounting to support and move it. This massive chunk of glass has other disadvantages. In particular, it takes a long time for it to reach a stable temperature as the surrounding air temperature changes. Consider the effect this has on an astronomer trying to observe during the night. As its temperature changes, the mirror responds by changing shape, but slowly, and as it does, the focus changes. All night long this huge massive mirror is continually changing its shape (and hence its focus); it has difficulty reaching temperature equilibrium and in forming a stable image.

Modern technology has made radical changes in telescope design, and it is now possible to make astronomical mirrors that are thin and lightweight. As the telescope points in different directions, a thin mirror will not maintain its desired shape, but this can be compensated for by computers that respond to the measured shape of the mirror. If the mirror is found to be deviating from the ideal shape required to focus the light exactly, a computer activates small motors to restore the mirror's shape. The advantages of building thin, lightweight mirrors are enormous, and in the end they translate into an ability to build very large ground-based optical telescopes at prices that are still reasonable.

Another development astronomers will be using is *active optics* (also known as adaptive optics). The idea of active optics is to change the shape of the mirror in response to changes in the Earth's atmosphere. By monitoring a star, or a laser beam reflected from the atmosphere, astronomers can make rapid small changes in the shape of the mirror to compensate for changes in the atmosphere. In this way, the effects of the atmosphere can be dramatically reduced.

12.6
DETECTORS AND INSTRUMENTS

Astronomers no longer use their eyes at the focus of a telescope. The eye is not as sensitive a detector of radiation as film or modern electronic detectors. Furthermore, the eye is connected neither to a perfect storage medium nor to one that allows for perfect duplication of the data. For these reasons, astronomers use a variety of detectors.

CAMERAS AND FILM

As important as the telescope itself is the radiation detector that is placed at the focus. We cannot overemphasize the dramatic boost given to astronomy by the invention of the photographic emulsion. The ability to take a time exposure at the telescope enabled astronomers to penetrate much more deeply into space, and it also created a permanent record for other scientists to study. If you realize that a photograph is composed of many microscopic picture elements called **pixels**, you can see that the data storage capacity of film is enormous. An 8-by-10-inch photograph has some hundred million pixels.

The major problem astronomers have always had with photographic emulsion is that its response to light intensity is not linear: twice as much brightness does not result in twice as much exposure on the film. This nonlinearity of photographic film makes it difficult to determine the intensity of the light falling on it. Furthermore, if you bombard a piece of film with too much light, eventually it stops responding at all. Another problem with film is that it is insensitive to most infrared radiation.

PHOTOELECTRIC PHOTOMETERS

The desire of astronomers for a more sensitive detector was answered by the discovery of the photoelectric effect discussed in Chapter 11. When incident light falls on certain metals (such as cesium), they respond by ejecting electrons; that is, they produce an electrical current. Such photoelectric surfaces are highly sensitive, and the amount of current given off is strictly proportional to the intensity of the incident light.

Astronomers found that such detectors could be used to register the brightness of stars to a high degree of accuracy. The light from a star is passed through a colored filter before being measured, so that by changing filters we can measure the amount of light the star gives off at various wavelengths. For example, by putting a blue filter (one that passes only blue light) in the light path, we can measure the amount of energy the star gives off at blue wavelengths only.

The photoelectric photometer described above has one dramatic disadvantage: It can measure only one star at a time because you must pass the light from the star through a small hole that excludes light from other stars that might be confused with it. The system then gives an accurate reading of the brightness. So although a photographic emulsion can register many stars at once, but with low accuracy, photoelectric systems can measure only one star at a time, but with high accuracy.

MODERN ELECTRONIC DETECTORS

Astronomers have always longed for an instrument that combines the sensitivity of the photoelectric detector with the two-dimensional data storage capacity of the photographic emulsion. For a few years they experimented with TV cameras mounted at the focus of the telescope. While not ideal detectors, they were used successfully in the International Ultraviolet Explorer satellite. In the 1980s a new type of detector called a **charge-coupled device** (**CCD**) was invented. When light strikes the detector surface, electric charges are produced in small regions of the material (called pixels, as

(a)

(b)

FIGURE 12–17. (*a*) A CCD image appears to be continuous and smooth but, as shown in part (*b*), is made up of distinct pixels.

FIGURE 12–18. An image of a globular cluster after various techniques of image processing have been applied.

in a photograph). Because the charge is directly proportional to the amount of light falling on the material, the amount of incident light can be determined by finding the electrical charge within each pixel. An image can be produced by displaying all the pixel data at once. A CCD image of one of the authors is shown in **Figure 12–17a**; part *b* shows the pixel structure around the eye. The CCD is found also in TV cameras and makes use of the same kind of technology that underlies transistors and microchips. With the advent of these devices, astronomers finally have detectors that approach the ideal of detecting virtually all the light from a celestial source.[2]

---

[2]Photography has a limited role in astronomy today. In fact, the last photograph ever to be taken with the Hale 200-inch telescope was made by Sidney van den Bergh on September 29, 1989.

Because the CCD provides us with a number proportional to the brightness of the radiation incident on each pixel, we describe the image in terms of numbers, referred to as *digital data*. Analysis of the millions of pixels in an image has given rise to what is called **image processing**. It is the mathematical manipulation of digital data to extract the maximum amount of information contained in the image. Such processing allows astronomers to view the universe in ways not previously possible. **Figure 12–18**, for example, shows a processed view of a globular cluster (see Figure 1-14 for an example). Numerous other examples will be presented throughout this book.

Such improvements come at a price, however. These new instruments collect data so rapidly that it is possible to become completely buried under it. Astronomers are having to devote considerable effort to developing efficient methods of data reduction, ways to get computers to take all the raw numbers, do all the corrections (e.g., for the absorption of the Earth's atmosphere), and complete the image processing automatically. Data reduction itself has become a career specialty for astronomers.

12.7

RADIO ASTRONOMY

In the early 1930s, while studying certain kinds of interference that affected transatlantic radiotelephone transmissions, Karl Jansky of Bell Telephone Laboratories discovered that some of the interference was coming from a region in the sky that moved in the same way the stars do. His antenna was rigidly mounted and could not compensate for the rotation of the Earth. A source of radiation in the sky would move across his antenna about once a day as a result of the apparent rotation of the sky. In fact, because of the difference between the solar and sidereal days, this source among the stars would be observed about four minutes earlier each day.

Jansky had discovered radio emissions from the center of our own galaxy. Strangely enough, his discovery did not generate much interest among professional astronomers. One of the few people to become seriously interested was an amateur astronomer and professional radio technician named Grote Reber, who set up a small radio telescope—the first one designed for the purpose—in his backyard in Illinois. He began to make systematic observations and was the first to make a map of the radio sky.

During World War II Reber was probably the world's only radio astronomer. The rapid growth of radar technology and the enormous amount of surplus radar equipment left over after the war soon triggered an explosion of postwar astronomical activity.

Radio astronomy is important because it enables astronomers to study an entirely different portion of the electromagnetic spectrum than that traditionally available to astronomy. Many objects that appeared to be uninteresting faint stars when examined optically turned out to be powerful emitters of radio radiation. An excellent example is a subset of the class of objects known as *quasars*. Although they are insignificant in optical telescopes, some quasars are among the most luminous objects in the sky when viewed by radio telescope. The *pulsar*, another type of object that has attracted great attention from astronomers and has provided much insight into the final stages of the lives of stars, might never have been detected using only optical techniques.

Radio telescopes have other advantages as well. Clouds of interstellar dust in the plane of our galaxy obscure most of our own stellar system from the view of optical telescopes. But radio waves pass through the dust without much dimming, and radio astronomy has become an important tool for the study of the Milky Way. Because the daytime sky does not appear as bright to radio telescopes as it does to optical telescopes, radio astronomers can make many observations during the day. Bad weather is not as much of a problem for the radio astronomers either, because radio waves readily penetrate clouds.

However, radio telescopes do have some limitations. Most serious is that their resolving power is generally poor. We saw earlier (Section 12.2) that the resolving power of any telescope depends on the wavelength of

FIGURE 12–19. The 1000-foot radio telescope at Arecibo, Puerto Rico.

radiation divided by the diameter of the objective. Therefore, the long wavelength of radio waves will produce poor resolution unless the objective diameter is extremely large. Even the world's largest radio "dish," the 1000-foot (300-m) telescope at Arecibo, Puerto Rico (**Figure 12–19**), can be seriously affected by diffraction and produce poorly resolved data.

Until recent years radio astronomy data were presented only as *contours*—lines that connect regions of constant intensity. **Figure 12–20** shows intensity numbers from the image in Figure 12–17b and contour lines connecting locations of constant intensity. **Figure 12–21** shows a series of such radio contours superimposed over an optical photograph. Because the cross-hatched oval indicates the degree of resolution of the telescope that obtained the radio data, apparent structure smaller than the oval is not fully resolved. As shown below, modern techniques of image processing allow radio data to be presented as images rather than numbers and contours.

Inquiry 12–10 Estimate the resolving power of the 1000-ft (300-m) dish if it is operating at a wavelength of 1 meter. How does this compare with your eye (with a resolution of about 1/60°)?

Figure 12–22 shows the basic components of a radio telescope. An antenna collects the electromagnetic energy and converts it to a weak electrical current. The antenna can be a dish, as shown here, or it could resemble a television antenna. If it is a dish antenna, it need not be polished to the great accuracy that is required for an optical telescope. In fact, the surface is often just a wire mesh, which is as efficient in reflecting radio waves as a solid surface would be, and is also lighter and less af-

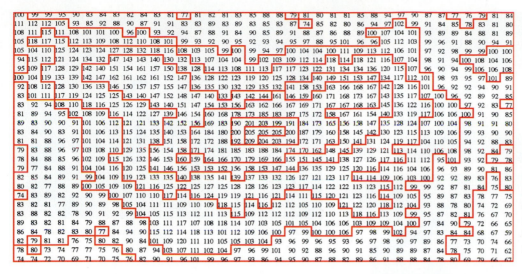

FIGURE 12–20. Raw data from part of Figure 12-17*b* is composed of numbers representing intensities. Contour lines connect regions of equal intensity.

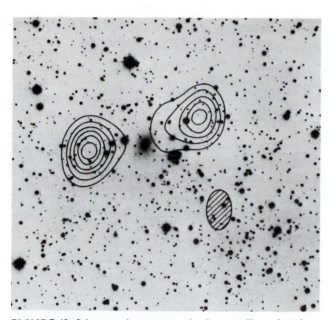

FIGURE 12–21. A galaxy as seen both optically and with a radio telescope. The radio contours have been superimposed on an optical photograph.

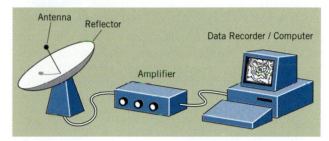

FIGURE 12–22. The basic components of a radio telescope are an antenna to collect the radiation, an amplifier to amplify it to detectable levels, and a recorder to make a permanent record.

drew a wiggly line whose height was proportional to the signal strength; modern radio telescopes record the intensity data (which are simply numbers) directly onto magnetic tape, so that it can be fed to a computer for further analysis.

INTERFEROMETERS

A simple way to overcome the problem of poor resolving power is to combine signals from two separated antennas, as shown in **Figure 12–23**. When this is done, a much-increased resolving power is obtained, equivalent to a radio dish whose diameter is equal to the distance between the two antennas. Such an arrangement is called a **radio interferometer**. For example, two 25-m-diameter radio dishes separated by 3 km will provide data having a resolution equal to a single telescope 3 km in diameter. Radio astronomers have even used antennas on opposite sides of the Earth as an interferometer to obtain highly precise information. In this case, the sepa-

fected by wind. It is required only that the imperfections in the surface of the dish be an order of magnitude smaller than the shortest wavelength at which the dish is to be used. Deviations of a few inches can often be tolerated.

The second major component of a radio telescope is a highly sensitive amplifier, which multiplies the strength of the signal millions of times. Finally, a device to record the amplified signal for further study is required. In the early days this was usually a chart recorder that simply

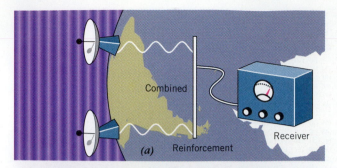

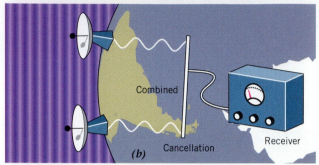

FIGURE 12–23. Principle of the interferometer. (*a*) When the waves reaching the two antennas are in phase, the output of the receiver is at a maximum. (*b*) After the Earth has turned a bit, the waves are out of phase, and destructive interference takes place. The output of the receiver is at a minimum.

FIGURE 12–24. The Very Large Array (VLA) in New Mexico.

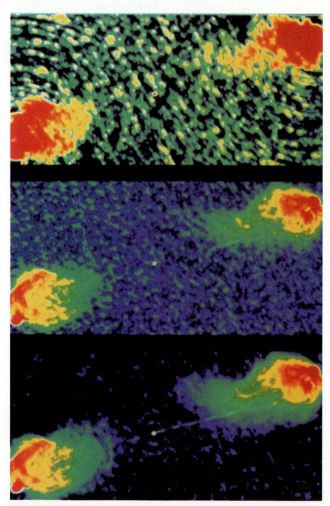

FIGURE 12–25. An image in various stages of image processing.

rate telescopes are not physically connected, but the data from each are stored on magnetic tape, along with accurate time signals from state-of-the-art atomic clocks, and then a computer combines the two signals. By putting antennas on opposite sides of the Earth, international collaborations of astronomers are able to simulate the resolving power of a telescope as large as the Earth (although the radiation gathering power is not enhanced to a comparable degree, of course). This field of astronomy is known as **very-long-baseline interferometry**, or VLBI. Ironically, after struggling with the notoriously low resolution of single-dish telescopes, radio astronomers using VLBI obtain higher resolution than any optical equipment under construction at this time.

One of the most powerful radio interferometers in the world, called the Very Large Array (VLA), has been constructed on the desert plains near Socorro, New Mexico (see **Figure 12–24**). It consists of 27 parabolic reflectors, each of which is 25 meters in diameter. The telescopes can be arranged along a giant Y pattern covering an area 17 miles in extent. Using this system, a radio astronomer can see radio-image details comparable to the resolution obtained by optical astronomers.

A large step forward for radio astronomy was the 1992 completion of the Very Large Baseline Array

(VLBA), a system of ten 25-meter dishes spaced out over the United States and the Caribbean. This system will provide both radiation-gathering power and resolution, with a resolution some 100 times better than any existing telescope system.

Other important radio observatories are located in England, the Netherlands, Australia, Germany, and the former Soviet Union.

Once the data have been obtained by each of the component telescopes in the interferometer array, the data are combined into one set of numbers containing the total information. The raw data are represented in an image in the top part of **Figure 12–25**. To convert these data into a final image requires extensive image processing. The middle section shows an intermediate stage in the production of a final image (shown in the bottom part of the figure) of what is called a double-lobed radio galaxy. Production of such images requires the largest, most sophisticated computers available.

Neither radio astronomy nor optical astronomy can stand by itself in modern research, because each technique is capable of providing information that the other cannot. Used together, the two techniques reinforce each other and give us a much clearer idea of the nature of celestial objects. In a real sense, the whole becomes greater than the sum of its parts.

12.8
OPTICAL AND RADIO OBSERVATORY SITES

Astronomical observatories are located on remote mountain tops for a variety of reasons. As discussed previously, the Earth's atmosphere absorbs radiation incident upon it. Even in the optical region where the transparency is relatively high, only some 80% of the incident optical radiation reaches sea level. This amount increases to 90% at 7000 feet. Not only does the atmosphere absorb radiation, but it emits its own faint glow. With less atmosphere above an observer on a high mountain, there is both less atmospheric emission and less turbulence to distort the image. Cloud cover above the observatory must be relatively infrequent because optical telescopes cannot see through clouds. In addition to these requirements, an observatory site must also be accessible for personnel and necessary power, water, sundries, and maintenance.

Because infrared radiation is absorbed by the Earth's atmosphere, ground-based infrared observations require a location also having the lowest possible amount of infrared-absorbing water vapor overhead. Such locations are found on high mountain tops in the arid regions of Arizona, Hawaii, and Chile.

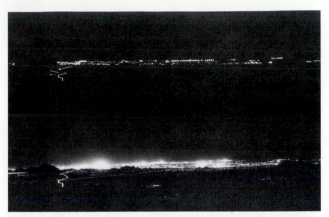

FIGURE 12–26. The growth of light pollution around Tucson, Arizona, between the years 1959 to 1980.

The final and perhaps the most important requirement for locating an optical observatory is to be away from city lights. Just as you see few stars when the full moon is up, city lights cause important details to be lost in astronomical images. This problem is known as **light pollution**. Encroachment of cities, both large and small, into once-dark areas can drown out the view of the universe not only to astronomers but to humanity as a whole. As an example, **Figure 12–26** shows the growth of lighting in the Tucson area as observed from Kitt Peak National Observatory between the years 1959 and 1980. Lights from cities as far as 100 miles from an observatory measurably increase sky brightness. The number of areas on Earth suitable for astronomy is small and getting smaller; the only remaining sites in the continental United States include limited parts of Arizona and New Mexico. **Figure 12–27** shows a satellite mosaic at night on which cities even as small as Lawrence, Kansas, appear. For this reason astronomers help com-

FIGURE 12–27. A photomosaic of the United States at night as seen from a satellite.

munities around observatories find ways to provide adequate lighting while still protecting the astronomical environment. They recommend the use of low-pressure sodium lamps. These lights use less energy and cost less to operate than incandescent, mercury vapor, or high-pressure sodium lamps, yet provide enough light to illuminate streets without washing out a large amount of the optical spectrum. Well-designed shielding around light fixtures directs light downward where it is useful, thereby reducing sky illumination.

Proponents of bright lighting argue that astronomy can be done from space. The expense of making all astronomical observations from orbit is prohibitive and impractical. Perhaps even more important is the loss of the night sky to citizens, for the night sky is a treasure everyone should be able to enjoy.

Radio astronomy also has a pollution problem; there is a limited number of radio frequencies and innumerable uses for them. Powerful radio and TV transmitters, along with lower-power CB radios, cellular phones, telecommunications via microwave, and so on "illuminate" the sky at radio wavelengths and make observations of faint astronomical sources impossible at certain wavelengths. While some astrophysically important radio frequencies are currently protected by international agreement, some others have been lost.

12.9

BEYOND OPTICAL AND RADIO ASTRONOMY

Especially since 1960, astronomers have learned to exploit the entire electromagnetic spectrum, from long radio waves to very short gamma rays. The impetus of this explosion of effort has been the tremendous growth of technological capability, including improvements in detectors, computers, and the ability to send instruments above the turbulent, obscuring atmosphere of the Earth. To the astronomer, the atmosphere is basically a nuisance because, as illustrated in Figure 11–7, it absorbs most of the radiation coming into it. Indeed, the most important reason for building observatories on mountain tops is to get above some of the atmosphere. Even mountain tops are not high enough to allow undisturbed observation of X-rays, gamma rays, ultraviolet, and infrared radiation. While the technology required to make instruments in these wavelength regions differs from that in the optical region, the fundamental principles of optics discussed previously still apply.

A surprising amount of successful research has been carried out by instruments carried up in balloon gondo-las, which can reach altitudes in excess of 20 miles. Prior to the use of satellites equipped with telescopes, such balloons provided the only technique to obtain some ultraviolet and infrared data. More recently, the Kuiper Airborne Observatory (KAO), a C-141 jet carrying a 1-meter infrared telescope, has provided astronomers with a routine view of the infrared sky. Because the most serious obstacle to astronomical observations at infrared wavelengths is the absorption by atmospheric water vapor, the KAO rises above about 99% of the water vapor by flying at about 40,000 feet.

The Infrared Astronomy Satellite (IRAS), which operated 1983–1984, contained a small telescope with advanced infrared detectors. Because the telescope was warm and radiated at infrared wavelengths, to obtain the highest quality data the entire telescope had to be cooled to 4 K in a bath of liquid helium. Once the liquid helium evaporated (after about nine months), the instrument became inoperable. However, during its short period of operation, IRAS made numerous important discoveries concerning star formation, galaxies, dust within the solar system, and many other subjects discussed elsewhere in this book.

Important ultraviolet observations were made by instrumented rockets, but they were aloft for only a few minutes per flight. Longer opportunities to observe were presented by the Copernicus satellite during the years 1972–1980. The International Ultraviolet Explorer (IUE) satellite, with a 45-cm mirror, was launched in 1978. Although its design lifetime was only a few years and it is partly crippled, IUE is still collecting new data.

The ability to observe in the ultraviolet was one of the original justifications for building the Hubble Space Telescope (HST). With its larger diameter and more modern and sensitive detectors, HST is providing a more detailed view of the ultraviolet universe than Copernicus or IUE.

The hottest and most energetic objects in the universe emit energy in the X-ray and gamma ray regions of the spectrum. For this reason, the X-ray telescope known as the Einstein observatory, which operated from 1978–1981, provided crucial information on a variety of objects and processes. The German satellite Roentgen (ROSAT), launched in 1990, and the Japanese Ginga satellite are the only X-ray instruments currently in orbit. They are monitoring X-rays from such objects as neutron stars, quasars, possible black holes, and galaxies. The Gamma Ray Observatory (GRO) is one of the United States' so-called Great Observatories, which includes HST. Launched in 1991, it is investigating the highest energy photons in the universe. Among its discoveries are incredible bursts of energy whose nature is not yet understood.

12.10

THE HUBBLE SPACE TELESCOPE

The **Hubble Space Telescope** (**HST**), launched in 1990, is providing astronomers with a long-desired observatory in space (**Figure 12–28**). This telescope has the largest diameter of any thus far sent into space—2.4 meters. More important, however, is that for the first time we have a complete observatory in space with the full array of observing instruments available in ground-based observatories. All the previous telescopes that have been sent into space have not only been smaller but also, more fundamentally, they have been special-purpose instruments designed for only a few specific types of observations. The Hubble telescope was designed to be able to do complete studies of objects many times fainter and almost 10 times deeper into space than our best ground-based systems; that would mean almost 1000 times larger a volume of space. In addition, without the degrading effects of the Earth's atmosphere, the telescope will obtain more highly resolved images than possible from the ground.

In recent years advances in the technology of active optics, detectors, and software have allowed dramatic improvements in ground-based astronomy. Therefore, the advances in optical astronomy provided by HST, although certainly strong, may not be as great relative to ground-based work as originally thought. In any case, the telescope is the largest ultraviolet telescope in orbit today, and is producing the highest-resolution images ever made.

The telescope consists of five main instruments. The wide-field planetary camera is the main instrument for obtaining images of planets, galaxies, clusters of galaxies, and quasars. The faint-object camera obtains images of the faintest objects observable with the telescope. The faint-object spectrograph can record ultraviolet and optical spectra of comets, galaxies, quasars, and more. The Goddard high-resolution spectrograph is an ultraviolet spectrograph for observing objects 1000 times fainter than previous space-borne detectors. The fine-guidance sensors, used to point the telescope, are designed to determine positions of stars more precisely than ever before. One of the authors (W.H.J.) is involved in this part of the project.

There will be a natural symbiosis between HST and the upcoming generation of new, giant ground-based

FIGURE 12–28. The Hubble Space Telescope.

telescopes. The Space Telescope, with its sharp and unblurred view of much of the electromagnetic spectrum, will doubtless be the *discovery* instrument that finds new and surprising objects and results that jerk science forward, sometimes in spite of itself. When a new object is discovered, then the giant ground-based telescopes, with their enormous light-gathering power and versatility, will take over and provide the detailed studies necessary to understand the new object.

Unfortunately, the primary mirror of the telescope suffered from spherical aberration due to a manufacturing error. However, astronomers found methods to improve the image quality so that, in fact, much new information has been obtained. The spherical aberration has been corrected by using the general technique described for the Schmidt telescope. Just as the Schmidt telescope has a corrector which exactly cancels the effects of spherical aberration, NASA has placed correcting mirrors in front of each instrument. As a result, the images the telescope produces now exceed the original specifications.

CHAPTER SUMMARY

OBSERVATIONS

- The main purpose of a telescope is to gather light and bring it to a focus. Secondary purposes are to resolve and magnify the images.
- The **light-gathering power** of a telescope depends on the area of the objective lens or mirror. The **resolving power** of a telescope depends inversely on the objective diameter. The **magnification** depends on the ratio of the objective focal length to the eyepiece focal length.
- Image size is determined by the focal length. The greater the focal length, the larger the image produced.
- **Seeing** measures the degree of motion of the Earth's atmosphere above an observatory site.

- **Refracting telescopes** suffer from chromatic aberration. Spherical aberration is corrected by shaping a mirror into a nonspherical shape or by using a correcting lens (**Schmidt telescope**).
- The properties of **interference** may be used to improve the resolution of a telescope. **Long-base-line interferometry** is necessary to obtain usable resolution with radio telescopes because of the long wavelengths of radio waves.
- Detectors used by astronomers include photographic film and various electronic devices, including the **charge-coupled device (CCD)**. These detectors are used with cameras, photometers, and spectrographs.
- Astronomical studies have been made from balloons, airplanes, rockets, and satellites.

SUMMARY QUESTIONS

1. How do both a pinhole camera and a lens form a real image of an object? Use a diagram in your explanation.

2. What effects do spherical and chromatic aberration have on an image?

3. What are the basic parts of a telescope? What is the function of each part? Use diagrams in your answers.

4. What is meant by resolving power as it applies to a telescope?

5. How does seeing affect the image produced by a telescope? How does the effect of seeing compare with the effects of diffraction?

6. What is the principal advantage that telescopes of large diameter have over those with small diameter?

7. What are the advantages and disadvantages of each of the principal types of reflecting telescopes?

8. What are the advantages and disadvantages of radio telescopes as compared with optical telescopes?

APPLYING YOUR KNOWLEDGE

1. What are the pros and cons of placing an optical observatory on the roof of a university building?

2. Why do astronomers need large telescopes on Earth? In space? Why is the word *large* important?

3. Why are radio telescopes so much larger than optical telescopes?

4. Why must the shape of the mirror of an optical telescope be more accurate than the shape of the antenna of a radio telescope?

5. Given the choice of a reflecting telescope at the prime focus, Cassegrain focus, or coudé focus, which would you choose to use to photograph a faint galaxy? To obtain a detailed picture of a planet? Explain your reasoning.

6. Giovanni Schiaparelli, who claimed to observe *canali* (later translated as *canals*) on Mars, used a telescope having a 22-cm-diameter objective. What is the resolving power of such a telescope? What would the true size of Martian "canals" have to be for Schiaparelli to have observed them? Assume Mars is 0.5 AU from Earth at the time of observation.

7. An 8-inch amateur telescope with a focal length of 2000 mm produces an image of the moon 1.7 cm in diameter. How large an image of the Moon does the 200-inch telescope (focal length = 16.7 meters) produce? What would be the image size for a 50-inch telescope having the same focal length as the 200-inch telescope?

8. Suppose you want the Moon to appear 45° across when viewed through a telescope of 125-cm focal length. What focal length eyepiece must you use?

◾ **9.** Compare the theoretical light-gathering and resolving powers of telescopes having the following objective diameters: 29-cm (8-inch), 127-cm (50-inch), 2.54-m (100-inch), 5.08-m (200-inch), 10-m (400-inch).

◾ **10.** Compare the resolving power of the 10-m Keck telescope with what will be possible at the completion of the Keck 2 telescope 85 meters away, when the combi-nation is used for interferometry at a wavelength of 2 microns (1 micron = 0.0001 cm).

◾ **11.** The European Southern Observatory's VLT will consist of four separate telescopes, each having 8-meter diameter mirrors. How large would a single mirror have to be to have the same light-gathering power? How much more light than the human eye will it collect?

ANSWERS TO INQUIRIES

12–1. Using ratios, 1 cm/ 50 mm = image size/150 mm; therefore, image size is 3 cm.

12–2. The explanation is in the following paragraph.

12–3. $(D/0.5)^2 = 4D^2$ if D is in centimeters.

12–4. Roughly 20 to 30 cm.

12–5. The magnification would be 150/2 = 75; the diameter is unimportant.

12–6. Diffraction

12–7. A 5-cm telescope has such small light-gathering and resolving power as to be not worth purchasing for astronomical purposes. From the rule of thumb given in the text, the maximum magnification you would want to use with such a telescope would be about 50.

12–8. Light-gathering power, resolving power, magnification. No, because the word "power" has no single meaning.

12–9. To get a resolution of 0.025 seconds of arc, which is 6.9×10^{-6} degrees, at a wavelength of 1 micron (10^{-4} cm) requires a telescope of diameter D = 57.3 λ/resolving power = 57.3×10^{-4} cm / 6.9×10^{-6} = 8.3 meters. (Remember, the VLT will work against the limits of the Earth's atmosphere, not the limits of diffraction.)

12–10. The resolving power is 57.3° × 1/300 = 0.2°. The naked eye can resolve detail more than 10 times finer than this.

13

SPECTRA: THE KEY TO UNDERSTANDING THE UNIVERSE

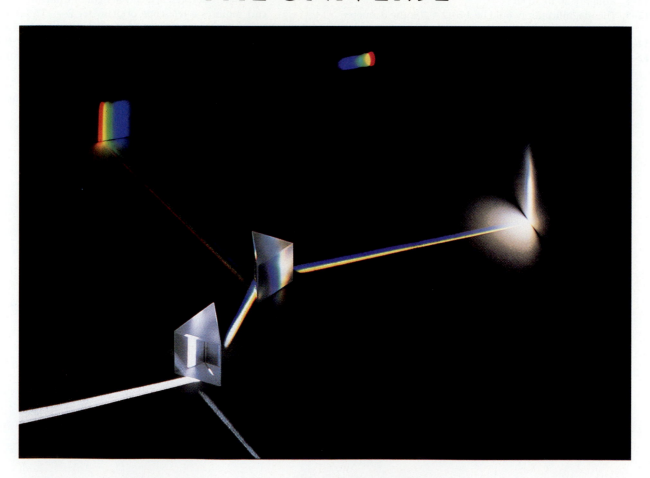

We understand the possibility of determining their [celestial bodies'] shapes, their distances, their sizes and motions, whereas never, by any means, will we be able to study their chemical composition.

AUGUSTE COMTE, *COURS DE PHILOSOPHE POSITIVE,* 1835

The techniques used to understand light derive from investigations begun by Newton in the mid-1600s. Newton used a prism to break up a beam of white light into a band of colors called a **spectrum** (the plural is **spectra**). He also showed that the colors could not be broken down further, and that they could be recombined to give white light. The existence of radiation beyond the two ends of the visible spectrum was unheard of at that time.

Because spectra of the light from stars do in fact reveal their chemical makeup, the chapter introductory statement by Auguste Comte is a classic example of the risk inherent in trying to set a priori limits to human knowledge. Though Comte could conceive of no method that could span the vast distances between the stars and determine their composition, a century later astronomers would carry out such analyses routinely.

During the nineteenth century, laboratory experiments by physicists showed that the spectra of light emitted from various luminous substances differed greatly from each other. Finally in the twentieth century, it became possible to obtain greatly detailed information about the stars by studying their spectra. It is no exaggeration to say that the overwhelming majority of modern astrophysical knowledge comes from a study of the spectra of celestial bodies. For this reason, this chapter is of fundamental importance.

In this chapter we investigate the three principal types of spectra and learn their significance. We will also examine the principles by which the chemical compositions of and physical conditions in stars can be measured, and apply these principles to analyzing various sources of radiation both in the laboratory and in the outside world.

13.1
OBSERVATIONS OF SPECTRA

Scientists made tremendous advances in understanding spectra toward the end of the nineteenth century. Many of these advances occurred because of scientists like Gustav Kirchhoff, who analyzed light by passing it through a prism and breaking it into its component parts. From his studies he observed that all light sources could be described in terms of three types of spectra. One of these had a continuous spread of color from violet to red (**Figure 13–1**) and is called the **continuous-emission spectrum**, or **continuous spectrum** for short. Such a spectrum, when photographed with black-and-white film, appears as continuously changing shades of gray. A second type of spectrum showed discrete, discontinuous bright lines, and is called the **bright-line spectrum** (Figure 13–2). The third type has discrete, discontinuous dark lines superimposed on

FIGURE 13–1. A continuous spectrum.

FIGURE 13–2. A bright-line spectrum.

FIGURE 13–3. A dark-line spectrum.

a continuous spectrum (**Figure 13–3**). This **dark-line spectrum** occurs only in conjunction with a continuous spectrum. While the reasons for and significance of these different types of spectra were not originally appreciated by scientists, Kirchhoff's research provided important clues, as will be discussed throughout this chapter.

The spectrum of light is studied with an instrument called a **spectrometer**. We detour briefly to examine this instrument before trying to understand astronomical spectra.

THE PRINCIPLE OF THE SPECTROMETER

READERS HAVING THE ACTIVITY KIT
SHOULD DO KIT ACTIVITY 13–1,
OBSERVING PATTERNS FROM RADIATION
SOURCES, AND KIT ACTIVITY 13–2,
ANALYZING LIGHT SOURCES WITH A
SPECTROMETER, NOW.

A spectrometer is a device to view the spectrum with the eye. A **spectrograph**, on the other hand, is a device for recording a spectrum; basically it is a spectrometer, but with the eye replaced by another type of detector (a camera or some type of electronic device that makes a permanent record of the spectrum). You could make a simple spectrograph by taping a diffraction grating over the lens of a camera and then using high-speed color film to photograph street lamps and other outdoor sources of radiation at night.

Astronomers sometimes use a similar technique to record the spectra of stars. They place a large prism over the end of a telescope and photograph stars directly (**Figure 13–4a**). The arrangement is called an *objective prism spectrograph* and the advantage it offers is that spec-

tra of many stars can be recorded on a single plate. Instead of a photograph full of star images, one obtains a photograph full of stellar spectra (**Figure 13–4b**).

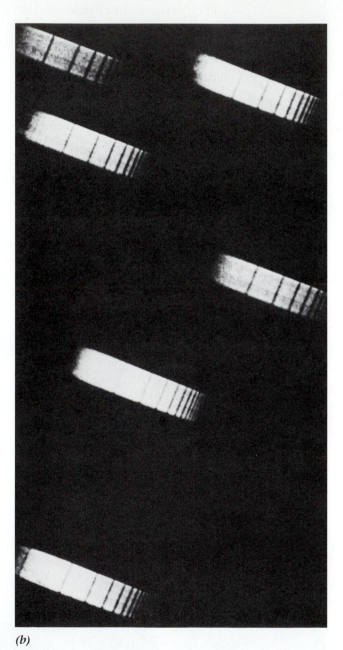

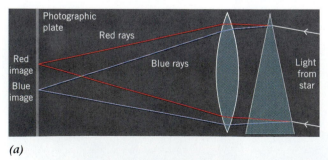

(a)

(b)

FIGURE 13–4. *(a)* An objective prism spectrograph can be used to obtain the spectra of many stars at the same time. *(b)* An objective prism plate showing the spectra of numerous stars.

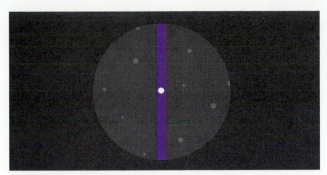

FIGURE 13–5. A spectrograph slit at the focus of the telescope limits the light entering the spectrograph to only that light coming from a particular star.

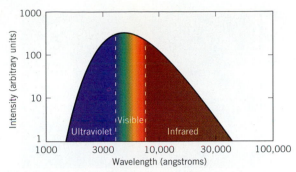

FIGURE 13–6. Energy distribution curve for a continuous spectrum.

Although the objective prism spectrograph is especially useful for making surveys of many stars rapidly, the spectra it produces show relatively little detail and cannot be made at all if the object is faint or is an extended source of radiation, such as a planet or galaxy. The typical spectrograph, therefore, has a number of additional parts that maximize both the amount of radiation that falls on the detector and the amount of detail in the spectrum. One of these is a slit placed at the telescope focus (**Figure 13–5**) to limit the light entering the detector to that of the star of interest. Because the spectrometer forms an image of the slit at each wavelength, the width of the slit is made small to maximize the ability of the spectrograph to distinguish fine detail; that is, to maximize the **resolution** of the spectrograph. Without high resolution, we are unable to distinguish the small details in the spectrum that are of importance in understanding the conditions in and around stars.

THE CONTINUOUS SPECTRUM: KIRCHHOFF'S FIRST LAW

We will now formulate the laws first elaborated by Kirchhoff. **Kirchhoff's first law** addressed the continuous emission spectrum.

> **Radiation given off by luminous solids, liquids, and opaque gases forms a continuous-emission spectrum.**

An incandescent bulb is an example of a solid giving off radiation; it contains a tungsten filament through which an electrical current flows, causing it to heat up and glow. The hot, molten lava that flows from an erupting volcano is an example of a hot liquid that, if observed with a spectrometer, would show a continuous spectrum. By an opaque gas, we mean a gas you cannot see through, such as the Sun.

A precise and quantitative description of a continuous spectrum is obtained by measuring the amount of energy emitted at each wavelength. This is done by passing the radiation through a diffraction grating or prism, which sends each wavelength in a slightly different direction. A photoelectric cell (e.g., a photographic exposure meter) placed at each wavelength in the spectrum then measures the amount of radiation at each wavelength. A graph of the resulting readings is then drawn—the horizontal axis giving the wavelength and the vertical axis the amount of energy at that wavelength. The resulting energy distribution curve for a typical continuous spectrum might look like **Figure 13–6**. Thus the word spectrum can refer either to a graph as in Figure 13–6 or a photograph as in Figure 13–3.

THE DEPENDENCE OF THE CONTINUOUS SPECTRUM ON TEMPERATURE

You probably have noticed that as an object is heated it begins to glow, and that the color of the glow changes slightly as the temperature of the object rises. For example, when you turn on an electric stove, you feel the heat, which is infrared radiation, before you see any visible changes. After a while, the burner glows red and then may take on an orange hue. The changing color as the temperature rises is caused by a change in the continuous distribution of energy with wavelength. In other words, as the object's temperature increases, the strength of its radiation shifts, from low in the visual spectral region and high in the infrared to relatively high in the visual and low in the infrared.

Inquiry 13–1 The hotter an object gets, the bluer the radiation from it becomes. Does a hotter object emit its most intense radiation at shorter wavelengths than a cooler object?

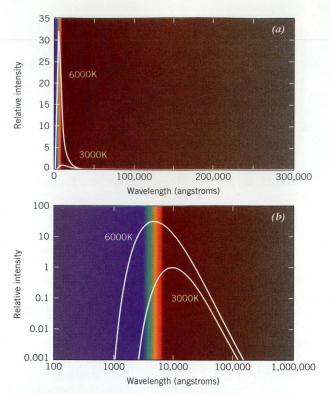

FIGURE 13–7. Radiation from a 3000-K body compared with that from a 6000-K body. *(a)* The scales are linear. *(b)* The scales are logarithmic.

A graph of the variation of intensity of radiation with wavelength—that is the spectrum of radiation—between wavelengths 0 Å and 300,000 Å is shown in **Figure 13–7a** for bodies of temperature 3000 K and 6000 K. Because both the wavelength and the intensity ranges in the graph are large, the graph is highly compressed. In particular, no detail can be seen in the 3000 K spectrum.

Figure 13–7b graphs the identical data as in part *a*. However, the scales of the axis are drawn differently so that the large ranges of wavelength and intensity can be incorporated on the one graph while making visible the differences between the curves for the two temperatures. Whereas Figure 13–7a has *linear* axes in which the interval between tick marks changes by a constant number of angstroms or intensity, the axes in part *b* are *logarithmic*; that is each tickmark increases by the same *multiplicative* factor. For example, in Figure 13–7a, the interval between each wavelength tick mark is 50,000 Å, while in part *b* each tick mark corresponds to an increase of wavelength by a *factor* of 10. Note that the lengths on the axis between the wavelengths of 1000 Å,

10,000 Å, and 100,000 Å are all the same. A similar logarithmic scale is used for the intensity axis. These scales allow more detail to be seen on one graph than could otherwise be seen. Because of the large range in numbers used in astronomy, logarithmic scales are used often by astronomers, as we will see in many chapters in this book.

Inquiry 13–2   What number comes next on the logarithmic scale?

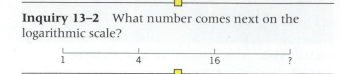

The one disadvantage of the logarithmic scale is that it is difficult to interpolate between divisions on the axis. On a linear scale, half the distance between 2 and 4 is 3; however, on a logarithmic scale, half-way between 10 and 100 is 31.62.

Any object whose continuous spectrum has the shape shown in Figure 13–7 is known as a **blackbody**. A blackbody (not to be confused with the term *black hole*) is a hypothetical object physicists find convenient when describing radiation emitted from any body whose temperature is greater than 0 K. Because, to a rough approximation, the spectra of some stars are similar to those of blackbodies, their characteristics can be used to determine temperatures of stars. This works because the spectrum of a blackbody depends only on temperature and not on chemical composition.

Figure 13–8 compares spectra of blackbodies having temperatures between 1500 K and 25,000 K. The higher

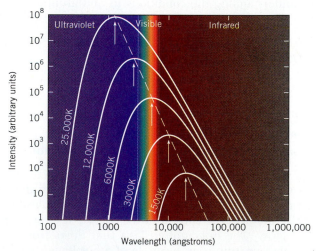

FIGURE 13–8.   Radiation from bodies of various temperatures is compared. The variation of the wavelength of maximum intensity in relation to the temperature is indicated.

temperature bodies emit more of their energy in the ultraviolet part of the electromagnetic spectrum. In fact, because the eye is insensitive to radiation whose wavelength is shorter than about 4000 Å, much of the radiation emitted by a body whose temperature is 12,000 K or more is not visible. The predominant visible radiation from this body is in the shorter wavelengths and appears bluish to the eye.

If we measure the total energy radiated, we find that the 3000-K blackbody radiates most of its energy in the red and infrared part of the spectrum, whereas the 6000-K blackbody has its maximum energy output in the yellow-green part of the spectrum. In addition, we see that all blackbodies of temperatures greater than 0 K radiate at least some energy at all wavelengths, from the shortest to the longest. From the spectral energy curves we see that hot bodies radiate more energy at each wavelength than cooler bodies.

WIEN'S LAW

Each spectrum shown in Figure 13–8 exhibits a maximum intensity at some specific wavelength. We denote the wavelength of the maximum intensity as λ_{max}. Laboratory experiments with approximate blackbodies, later confirmed by theoretical calculations, show that there is a relationship between λ_{max} and the temperature of the object. This relationship, first given by the German physicist Wilhelm Wien, is

$$\lambda_{max} = \frac{3 \times 10^7}{T},$$

where the wavelength λ_{max} is in angstroms and the temperature T is in kelvins. More energy is emitted at λ_{max} than at any other wavelength. Because a hot body emits most of its energy in the ultraviolet, the body will emit more blue light than red light in the visual spectral region and thus appear to be blue. The same type of argument for a cool body, which emits most of its energy at long wavelengths, shows that it will appear to be red. For these reasons, the color of an emitting body is determined by λ_{max}. However, because the wavelength of maximum emission is much more objectively and precisely defined than the subjective notion of color, we will use the wavelength frequently. To give an example of the use of this formula, let's ask a question: What will be the wavelength of maximum emission for a body whose temperature is 6000 K? The answer is

$$\lambda_{max} = \frac{3 \times 10^7}{6000} = 5000 \text{ angstroms.}$$

This wavelength is in the green part of the electromagnetic spectrum. In practice the question is turned

around, because the wavelength of maximum emission is the observed quantity and the temperature is the computed result.

A graph of the relationship between λ_{max} and temperature is shown in **Figure 13–9**. Instead of using the formula, you can read values directly from the graph if you prefer. Notice, however, that the scales are logarithmic rather than linear, so interpolation is not easy.

Inquiry 13–3 What is the wavelength of maximum emission for a body whose temperature is 12,000 K? Check your result on the graph in Figure 13–9 and compare it with the curve in Figure 13–8.

Inquiry 13–4 Although the temperature on the surface of Mars varies, let us assume that it is at approximately 300 K (about room temperature). At what wavelength is its peak emission of radiation found? (This is nearly the same as for your body.)

The curves shown in Figure 13–8 actually represent the amount of energy per second radiated by *each square centimeter*[1] of a blackbody having a temperature greater than 0 K. Notice that the area between the wavelength axis and the curve itself is greater for the hotter body; this means that the hotter body radiates more energy per second from each square centimeter than the cooler one does. How much more? This question was answered by the Austrian scientists Josef Stefan and Ludwig Boltzmann, who found that the energy radiated per second from one square centimeter of a body is propor-

---

[1]A more general way of saying *energy radiated by each square centimeter* is to say *energy radiated per unit area*. A unit area can be one square centimeter, one square inch, or one square kilometer. However, the energy per square centimeter will not be the same as the energy per square kilometer.

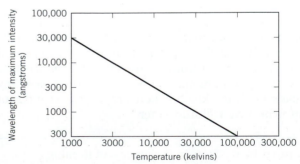

FIGURE 13–9. A graph showing Wien's law, the relationship between temperature and the wavelength of maximum intensity.

tional to the fourth power of the temperature. Symbolically, the **Stefan-Boltzmann law** is

$$E \propto T^4,$$

where T is the temperature in kelvins, and $\propto$ means "proportional to." The units of E will be energy/sec/cm^2.

Inquiry 13–5 How much more energy does each square centimeter of a 50,000-K star radiate than a 5000-K star?

In summary, there are three important points about blackbodies. First, Wien's law shows that the wavelength of maximum emission is shorter for hot bodies than for cool ones. Second, the Stefan-Boltzmann law shows that *every* object with a temperature above absolute zero emits radiation. Third, the amount of continuous radiation depends only on temperature, not on chemical composition.

OBJECTS CAN BE SEEN BY THEIR EMITTED LIGHT OR THEIR REFLECTED LIGHT

It is important to realize that the energy emitted by Mars and the other planets that results from their temperature is *not* the radiation by which we see these objects with our eyes. The reason is that such cool bodies radiate their energy at wavelengths to which human eyes are insensitive. We *see* the planets, and everyday objects such as tables and chairs, by light *reflected* from them, not by radiation resulting from their temperature. The visible colors of these objects are determined by their reflective properties. If a chair appears to be red, it is because it reflects light of longer, redder wavelengths more efficiently than shorter, bluer wavelengths, not because it has been heated to a temperature of several thousand degrees Kelvin. Similarly, the visible-light spectrum of Mars closely resembles that of sunlight because it *is* reflected sunlight.

Most astronomical objects, the stars in particular, have sufficiently high temperatures that we see them by the radiation they *emit* as a result of their high temperatures; therefore, their temperatures can be determined from their colors or from the wavelength at which they emit the largest amount of energy, λ_{max}, provided they radiate like a blackbody. There are some exceptions to this, which we will study later.

Inquiry 13–6 Stars in the process of forming (we call these objects *protostars*) emit most of their radiation in the infrared part of the spectrum. What would be the surface temperature of such a protostar if the wavelength at which it emits the greatest amount of energy were 30,000 Å?

Inquiry 13–7 In the 1960s, astronomers discovered a new type of object called an *X-ray source*. These objects show a continuous spectrum with the peak emission at very short wavelengths. What is the temperature of an X-ray source whose peak radiation falls at 0.5 Å?

DISCRETE EMISSION SPECTRA: KIRCHHOFF'S SECOND LAW

Kirchhoff's second law concerns the second type of spectral pattern, the **discrete-emission** or **bright-line spectrum**.

> **An emission or bright-line spectrum (one that emits energy only at specific wavelengths) is characteristic of transparent (i.e., rarefied, thin) gases that have been excited to glow by being heated or by other means.**

The energy distribution curve of an emission spectrum can also be graphed, as in **Figure 13–10a**, which shows the emission spectrum of low-density mercury vapors. Because energy is emitted only at certain wavelengths, the energy distribution curve consists of "spikes" at several wavelengths. A different amount of energy may be radiated in each of the spectral lines, so the heights of the spikes will, in general, be different. (When we speak of the strength of a bright spectral line, we mean the height of the peak.) The brightest spectral lines of mercury are in the blue/ultraviolet region of the spectrum; thus the mercury lamp appears bluish in color. Similarly, **Figure 13–10b** shows a spectrum for

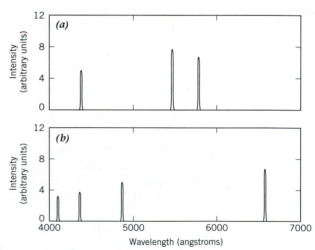

FIGURE 13–10. Energy distribution curve for a discrete-emission spectrum. *(a)* Mercury. *(b)* Hydrogen.

hydrogen. The light from hydrogen appears red because the structure of the hydrogen atom causes the brightest spectral line to appear in the red part of the spectrum. Notice that the separation of the hydrogen emission lines decreases toward shorter wavelengths; any useful model of spectra must explain this observed convergence of the spectral lines.

The properties of these transparent gases are not the same as those of denser, opaque gases that cause a continuous spectrum. For a denser gas, the spectrum of the emitted radiation is continuous and depends completely on the body's temperature. None of the laws governing radiation from dense bodies holds for transparent gases. In fact, a transparent gas does not need to be "hot" to emit energy; low-temperature gases can also produce emission lines.

The Spectral "Fingerprints" of the Chemical Elements

Early researchers noticed that each chemical element emitted a different pattern of wavelengths in the spectrum. Consequently, this pattern could be used as a guide to identify the chemical element. If several elements were mixed together in the same gas, the resulting spectrum would contain a mix of the emission features of each of them, so the presence of each element could be confirmed. Furthermore, the strengths of the bright spectral lines would give a measure of the proportions of the elements in the mixture.

In other words, the presence of discrete features in the spectrum of a celestial object, as opposed to continuous features, allows us to determine its chemical composition. We can identify not only the kinds of elements present in the stars but also their amounts. It is clear that Auguste Comte was wrong in stating that we would never know the composition of celestial objects!

Figure 13–11 shows emission spectra of several other elements found in common sources of radiation. Readers who have done Kit Activity 13–2 can compare these with the spectra of radiation sources that they observed with their spectrometer and attempt to identify the elements involved.

The Sun's Dark-Line Spectrum: Kirchhoff's Third Law

Kirchhoff observed a third type of spectrum when light having a continuous spectrum, characteristic of a blackbody at some temperature, was passed through a low-density, cooler gas, as shown in Figure 13–12. He found that under such circumstances the rarefied cool gas *absorbed* radiation at specific wavelengths. Most of the continuous spectrum passed through the cool gas unhindered, but at certain wavelengths, energy was removed from the beam. Significantly, the wavelengths absorbed by the cool gas were the *same* as those that would have been emitted if the gas had been heated. Such a spectrum is called an **absorption** or **dark-line spectrum**. (Those readers who were able to make the observation connected with Kit Inquiry 13–2f in Kit Activity 13–2 may have been able to observe a dark line in the yellow region of the spectrum of sodium.)

The energy distribution curve for the resulting beam of radiation would resemble Figure 13–13, a continuous

FIGURE 13–12. How an absorption spectrum is produced.

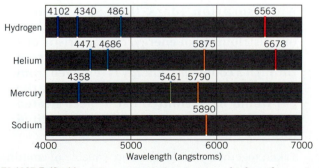

FIGURE 13–11. Discrete-emission spectra of selected elements in the visible-wavelength region. The colored lines marked at certain wavelengths are places where energy is being emitted.

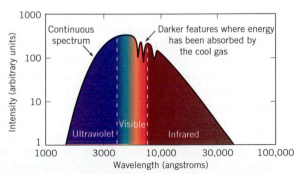

FIGURE 13–13. Energy distribution curve of a continuous-emission spectrum with energy removed at certain wavelengths to produce an absorption-line spectrum.

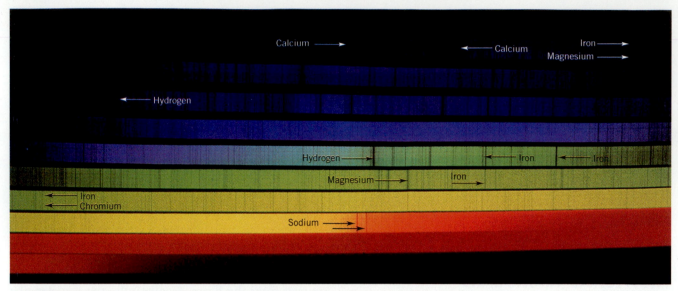

FIGURE 13–14. The spectrum of the Sun, showing many of the elements in the solar atmosphere.

distribution of energy whose basic features, determined by the star's temperature, are preserved, but with dips in the energy at specific wavelengths. The continuous part of the spectrum can still be studied to learn about the temperature of its source, but the presence of the cooler gas reveals itself by the absorption it produces. Note that when we say "cool" gas we mean it is cooler than the object that produced the continuous spectrum that flows through it.

Figure 13–14 shows a spectrum of the Sun, with the elements that produced many of the dark lines identified. Each dark line is at the same wavelength as would be emitted by the same element if it were vaporized into a transparent gas at a high temperature. In other words, each chemical element also exhibits its characteristic "fingerprint" pattern of lines even when it appears in dark-line form. This means that we can determine the chemical composition of the Sun and other stars, because they all show a dark-line spectrum.

A dark-line spectrum cannot exist without a background continuous-emission spectrum. For this reason, the Sun's dark-line spectrum has a continuous emission of energy between the absorption lines. We can extract a great deal of information from its total spectrum, namely, its temperature *and* its chemical composition. We will see in later chapters that when we examine the lines in a spectrum in magnified detail, we can determine even more properties of the object that emitted that spectrum—for example, its density, motion through space, rotation, turbulence, and even the presence and strength of a magnetic field.

Inquiry 13–8 The Sun is so hot that water vapor cannot exist there; it would be broken up into atoms of hydrogen and oxygen. Yet when astronomers observe the solar spectrum, they find absorption lines produced by water molecules. How can this be?

Inquiry 13–9 What kind of spectrum would you expect to observe from the Moon? What additional features might you expect to see in the spectrum of the planet Venus that you would not see in the Moon's spectrum?

Kirchhoff's third law on absorption spectra explains these observations, and can be summarized as follows:

An absorption (dark-line) spectrum is formed whenever radiation with a continuous distribution of energy, characteristic of a blackbody of some temperature, is passed through a cooler gas. Energy is removed from the beam at the same wavelengths that the gas would have emitted had it been heated to a higher temperature or otherwise excited into emission.

Taken together, the elements are capable of emitting and absorbing at probably millions of specific wavelengths. With the large number of possibilities, scientists have developed extensive tables that give the wavelengths and characteristic strengths of the various spectral lines of all elements. Astronomers refer to such tables in their analysis of the spectra of celestial bodies.

13.2

UNDERSTANDING SPECTRA

In Section 13.1 we examined observations of spectra with no consideration of their cause beyond Kirchhoff's nineteenth-century laws. In the early part of the twentieth century, scientists began to lay the groundwork for understanding spectra as they began to learn more about the structure of atoms. About all they knew was that atoms were composed of both positively and negatively charged particles. In fact, Ernest Rutherford proposed a model for an atom that had a heavy, positively charged **nucleus** surrounded by orbiting negatively charged electrons.

On the basis of experiments Rutherford constructed a model for hydrogen, the simplest atom, that consisted of a single positively charged particle, called the **proton**, at the center of the atom with an electron orbiting around it. Because the mass of the proton was found to be about 1800 times the mass of the electron, virtually all the mass of the atom resided in its nucleus. However, Rutherford showed that the nucleus was very small (approximately 10^{-13} cm in size) and hence the "size" of the atom depended on how far away the electron was from the nucleus. Because an average electron stays approximately 10^{-8} cm (1 Å) away from the nucleus, the diameter of the atom is about 10^5 times the diameter of the nucleus itself. Atoms, therefore, are mostly empty space.

The most common form of the helium atom has a nucleus that consists of two protons and two **neutrons** (chargeless particles of approximately the same mass as the proton). Its total **atomic weight**, the number of protons plus neutrons in the nucleus, is four. However, the electrical charge of the nucleus is only two (because of the two protons), so only two electrons are needed to orbit outside the nucleus to give the helium atom overall electrical neutrality.

The number of protons in the nucleus determines what element a nucleus is. Two nuclei having a different number of protons are different elements. In general, the nuclei of the chemical elements are designated with a concise notation in which the *charge* of the nucleus—the number of protons in it—is written before an abbreviation of the element's name as a subscript, and the total atomic weight is written as a superscript before the element name. Thus we designate hydrogen as $^1_1$H and helium as $^4_2$He. As a final example, consider the element carbon, which is designated by $^{12}_6$C. The subscript of 6 tells us that the carbon nucleus has 6 protons, whereas the superscript of 12 tells us that the nucleus

has a total atomic weight of 12, so it must have 6 neutrons in addition to the 6 protons. To electrically balance the 6 protons, it has 6 electrons as well. The atom $^{13}_6$C is a carbon atom having 7 neutrons rather than 6. Atoms having the same number of protons but different numbers of neutrons are called **isotopes** of an atom.

Inquiry 13–10 While the most common isotope of oxygen has 8 neutrons, isotopes having 7, 9, and 10 neutrons also exist. What is the chemical designation for each of these isotopes?

Rutherford's simple model agreed well with experiments in which atoms scattered fast-moving helium nuclei incident upon them, but it ran into difficulty when it was used to explain the observed spectra of atoms. The reason for this is that, according to the theory of electromagnetism, accelerated electrons ought to radiate away energy in the form of electromagnetic waves so rapidly that in less than a millionth of a second the electrons should collapse into the nucleus. This prediction of the model was in violent disagreement with experiments.

In 1913 the Danish physicist Niels Bohr proposed a radical modification to Rutherford's model. In the older model, electrons could orbit the nucleus at *any* distance; all possible orbital distances from the nucleus were allowed. Bohr's radical proposal was that electrons could exist only in specific, well-defined orbits around the nucleus. He put forth a model for the hydrogen atom that consisted of a single, heavy, positively charged particle, a proton, with a negatively charged electron in orbit around it. The electron could occupy orbits only at specific radii—other orbits were forbidden to it. In particular, in the Bohr model there was an orbit having minimum radius and energy; if the electron were in that orbit, it could not radiate away any more energy. This idea ruled out the catastrophic collapse of atoms that was marring the earlier theory. The resulting model resembles a miniature solar system in which the more massive proton plays the role of the Sun and the less massive electron the role of a single planet (**Figure 13–15**). Planets, however, can occupy any orbit; elec-

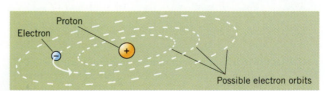

FIGURE 13–15. The Bohr model pictured the atom as a miniature solar system, with electrons orbiting the nucleus as planets orbit the Sun.

trons, only specific ones. In addition, the electron and proton are bound together by electromagnetic rather than gravitational forces.

It now became possible to explain the spectrum of an atom. Bohr realized that the closer the electron was to the nucleus, the more energy would be required to remove it from the atom because of the strong electromagnetic attraction between the oppositely charged nucleus and the electron. In this way, he was able to assign to each orbit a precise amount of energy equal to that required to just remove an electron in that orbit from the atom. Bohr then proposed a revolutionary idea; he said that an atom emits and absorbs energy by emitting and absorbing photons having *exactly* the energy required to move the electron from one orbit to another. Because the energy and wavelength of photons are related as previously discussed ($E=hc/\lambda$), once Bohr knew the energy he could calculate the possible wavelengths that could be emitted or absorbed by a hydrogen atom. When he did so, the agreement of the theory with the observed spectrum of hydrogen was excellent.

The absorption and emission of photons in the Bohr model of the atom are illustrated in **Figure 13–16**. Symbolically, we can write absorption as

atom in lower energy state + energy → atom in higher energy state.

and emission as

atom in higher energy state → energy + atom in lower energy state.

To produce an absorption line, an electron must be excited to a higher energy level. The energy that excites the electron can come from two places. One is photons incident on the gas. The photon's energy excites the electron to a higher energy state and is therefore removed from the incident beam. Thus an absorption line is produced in the spectrum. Second, energy to excite an electron can come from collisions with other particles, in which one particle's kinetic energy is decreased by

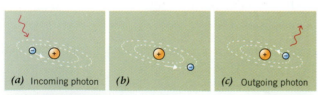

(a) Incoming photon   **(b)**   **(c)** Outgoing photon

FIGURE 13–16. Emission and absorption of light by a hydrogen atom. *(a)* A photon impinging on an atom whose electron is close to the nucleus. The photon is absorbed and the electron becomes excited as shown in *(b)*. After a while, the electron drops to a lower energy-level, with the emission of a photon, as in part *(c)*.

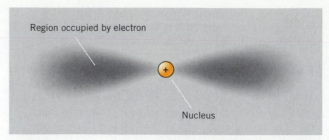

Region occupied by electron

Nucleus

FIGURE 13–17. The electron orbit closest to the nucleus in the hydrogen atom. The darkness of the "cloud" in the figure represents the probability of where the electron can be. The exact position of an electron in an atom cannot be precisely located.

exactly the amount of energy required to excite the electron to a higher energy level.

Once an electron is in an excited energy state, emission will occur whenever the electron undergoes a transition to a lower energy state. The emitted energy will appear as a photon, in which case an emission line will be produced.

Successful as it was in explaining the hydrogen atom, the Bohr model is too simple to explain the spectra of the other elements, which have more electrons. The theory of *quantum mechanics* (quantum theory for short), which replaced the Bohr theory in the 1920s, shares some of its main features, but recognizes that because of its wave nature the electron cannot be localized to simple circular orbits as in the Bohr theory. **Figure 13–17** shows a schematic representation of the electron cloud closest to the nucleus. Rather than being represented by a thin line as in Figure 13–16, the electron orbit is represented as a fuzzy "cloud," in which the darkest regions represent places where the electron has the highest probability of being at any given moment. Like the Bohr theory, quantum theory predicts that atoms can exist only in specific energy states and that to move from one energy state to another, the atom must emit or absorb a photon of *precisely* the right energy or wavelength. If the energy is not right, no absorption occurs. Mathematically, the relationship between the energies of the two states and the energy of the photon is

energy of photon = energy of higher energy state − energy of lower energy state.

In symbols, if E is the energy, then

$$E_{\text{photon}} = E_{\text{higher state}} - E_{\text{lower state}}$$

Inquiry 13–11 How does the relationship between the energies of the two states and the energy of the photon lead to the conclusion that the total amount of energy in the universe is not changed when an electron jumps from one energy state to another?

Inquiry 13–12 What physical principle is embodied in Inquiry 13–11?

ENERGY-LEVEL DIAGRAM OF THE HYDROGEN ATOM

Atoms having electrons in different orbits are said to be in different energy states or **energy levels** because each orbit corresponds to a different energy. These energy levels are best understood in terms of the **energy-level diagram** (shown in **Figure 13–18** for the hydrogen atom). Each numbered energy level corresponds to an allowed electron orbit in the Bohr model. The size of the interval between the levels labeled *1* and *2* in Figure 13–18 represents the amount of energy required to move the electron from its lowest allowed level (called the **ground state**) to the level labeled *2* (any level other than the ground state is called an **excited state**). The figure shows that a smaller amount of energy is required to move the electron from level *2* to level *3*; still less is required from *3* to *4*, and so on. Going to successively higher energy levels requires successively less energy because the electron occupies more and more distant orbits, where the attractive force between the nucleus and the electron becomes weaker. For this reason, the energy levels appear to converge toward the top of the diagram; this convergence of energy levels predicts the convergence of spectral lines observed in Figure 13–3.

The arrows in Figure 13–18 represent **transitions**, in which the electron jumps from one energy level to another, with the simultaneous absorption or emission of energy. The amount of energy absorbed or emitted is equal to the difference between the energies of the two levels. Several of the transitions that are responsible for producing lines in the hydrogen spectrum are shown. For example, when the electron makes a transition be-

tween levels *2* and *3*, an absorption line at a wavelength of 6563 Å is produced; the 4861-Å line is produced when the transition is between levels *2* and *4*.

Inquiry 13–13 The arrow in the diagram points from the initial state of the atom toward the final state. Which direction represents emission, an upward-pointing arrow or a downward-pointing arrow? Which represents absorption? (Hint: For each case, absorption and emission, which state has the higher energy—the initial or the final?)

The spectral lines at 6563 and 4861 Å belong to a series of lines called the **Balmer series**. These are lines formed when the hydrogen atom makes a transition between the second energy level and *any* higher energy level. A transition from level *2* to *4* will produce a Balmer *absorption* line at 4861 Å; a transition from level *4* to *2* will produce a Balmer *emission* line at the identical wavelength. The Balmer series is important because its wavelengths are in the visual region of the spectrum and are thus observable in spectra taken by telescopes located on the Earth's surface.

In a similar manner, **Figure 13–19** shows that the ultraviolet **Lyman series** of spectral lines of hydrogen is generated by transitions between the ground state and

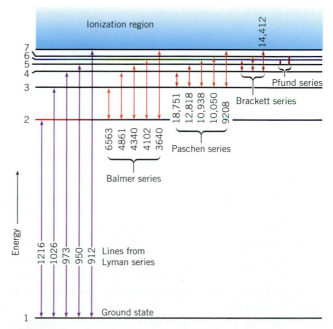

FIGURE 13–19. Energy-level diagram for the hydrogen atom, showing the principal transition in the Lyman, Balmer, Paschen, Brackett, and Pfund series.

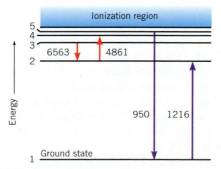

FIGURE 13–18. Energy-level diagram for the hydrogen atom, showing several transitions.

any higher level; the Paschen series by transitions between the third level and a higher level; the Brackett series between the fourth level and a higher level; and the Pfund series between the fifth level and a higher level. Series involving the sixth, seventh, eighth, and higher levels also exist but do not have special names.

Inquiry 13–14 In which regions of the electromagnetic spectrum will each of the hydrogen series produce spectral lines? Which can be observed from the Earth's surface? Which require instruments above the Earth's atmosphere?

The wavelengths of some of these high-numbered series are so long that they lie in the radio part of the electromagnetic spectrum. Some of them have been detected by astronomers.

Inquiry 13–15 Each of the following transitions gives rise to a line in one of the named series above. Which series does each line belong to? Does the transition give rise to an emission or absorption line?
(a) From level 4 to level 2
(b) From level 1 to level 2
(c) From level 5 to level 3
(d) From level 5 to level 1
(e) From level 5 to level 28

The shaded region at the top of the energy-level diagram represents what happens when the electron is given so much energy that it can no longer be held by the electrical attraction of the proton. The electron is removed from the atom, and the electron and proton go their separate ways. In such a case, we say that the atom has been **ionized**. We may represent this process, called **ionization**, by the upward arrow labeled *a* in **Figure 13–20**, and by the formula

hydrogen atom + sufficiently energetic photon → proton + free electron.

Any energy the photon has above that required to ionize the atom goes into the electron's kinetic energy of movement. Because the free electron can move with any speed (that is, can have any energy), there is a continuous range of ionized states that are distinguished from the specific energy states of the bound electron. After ionization, the proton is all that remains of this simplest of atoms.

The reverse process can also occur, in which the proton recombines with a free electron and gives off a photon. This reaction, which is called **recombination**, may

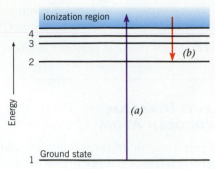

FIGURE 13–20.   Energy-level diagram for hydrogen, showing (a) ionization and (b) recombination.

be represented as a downward transition from the shaded region of Figure 13–20 into the lower region. The formula is

proton + free electron → hydrogen atom + emitted photon.

As represented in the diagram the only difference between emission and absorption is the direction of the arrow. It would thus appear that there is complete symmetry between emission and absorption. In fact, this is not so, because if an atom is in an excited energy state, it will spontaneously give off its excess energy in a short time (about 10^{-8} seconds) and make a transition into a lower energy state. But an atom in a lower energy state cannot make a transition to a higher state unless a photon of exactly the right energy arrives and is absorbed.

For example, referring to Figure 13–19, if the hydrogen atom is in its ground state (level *1*) and a photon of 1210 Å passes by, nothing will happen because the atom cannot absorb a photon of this energy. Similarly, a photon of wavelength 1220 Å will also be ignored. Only if the photon has just the right energy to lift the electron to a higher state—for example, if its wavelength is 1216 or 950 Å—can it be absorbed. Similar considerations hold for absorption to higher levels if the atom is already in one of the excited energy states.

Inquiry 13–16   Under which of the following situations is it possible for the photon to be absorbed by the hydrogen atom and produce a discrete absorption line? If the photon is absorbed, in what state will the atom find itself? Use Figure 13–19.
(a) Atom in ground state (level *1*); the photon wavelength is 1026 Å
(b) Atom in excited state (level *2*); the photon wavelength is 1026 Å

(c) Atom in excited state (level *2*); the photon wavelength is 4861 Å

(d) Atom in excited state (level *3*); the photon wavelength is 6563 Å

Once the atom is in the excited state, however, it will have excess energy and can rapidly and spontaneously make a transition into any lower state, if another absorption has not occurred first. If this lower state is also an excited state, then the atom will again make another transition into an even lower state, and so on. In other words, after *each* transition, the electron will cascade to lower energy levels with the emission of a photon having an appropriate wavelength.

Inquiry 13–17 What wavelength or wavelengths is it possible for an atom in level *3* to emit by spontaneous emission?

Inquiry 13–18 What are all the possible wavelengths an atom in level *4* could possibly emit by spontaneous emission as the electron cascades to the ground state?

WHY DIFFERENT ELEMENTS HAVE DIFFERENT SPECTRA

The spectra of most elements are much more complicated than that of hydrogen. Although the simple Bohr model of the atom was adequate to explain the wavelengths of the hydrogen spectrum, the development of quantum theory was required to give a satisfactory account of the other elements. The reason for this is that in the Bohr model the electrons are imagined to be localized as points orbiting the nucleus, whereas in fact the wave nature of the electron makes this picture inadequate if more than one electron is present.

For example, consider the neutral helium atom. It has two protons in its nucleus and therefore two electrons orbiting it; thus the atom is electrically neutral. In addition, there are two neutrons in the nucleus. The two electrons are located somewhere in a hazy cloud surrounding the nucleus, because their wave nature makes them impossible to localize, as illustrated in **Figure 13–21**. Because the two electrons occupy roughly the same region of space and are both negatively charged, they tend to repel one another, which distorts the regions each can occupy and affects the energy states that the atom can have. The resulting energy-level diagram is considerably more complex than the one for hydrogen.

Similarly, as one progresses through heavier and heavier elements, the number of protons in the nucleus

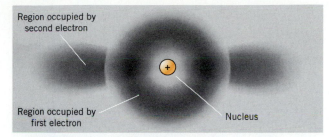

FIGURE 13–21. In a helium atom, the two electrons occupy overlapping regions.

increases. It follows that the number of electrons surrounding the nucleus must also increase. The electrons tend to arrange themselves into shell-like regions, each containing a certain number of electrons; the electrons in the outermost shell are the ones responsible for the visible spectrum of the particular element. With more and more electrons, the mutual repulsion between the various electrons becomes more and more complex, with a resulting increase in the complexity of the energy-level diagram and spectrum. Thus the spectra of two elements with almost the same number of protons will be completely different. In general, it is so difficult to calculate the spectrum of even simple atoms that physicists must make use of laboratory measurements of spectra to interpret astronomical observations.

Inquiry 13–19 Considering everything you have learned, would you expect the spectrum of singly-ionized helium to more closely resemble the spectrum of neutral helium, neutral hydrogen, or ionized hydrogen?

ATOMIC EXPLANATION OF KIRCHHOFF'S LAWS

Let's not forget Kirchhoff's laws, which provided us with information on the conditions under which spectra of various types are formed. We now need to merge his ideas with those of atomic structure discussed above.

You might well wonder at this point why a dark-line spectrum is ever seen. It might seem that the absorption of a photon will usually be followed by re-emission of a photon of the same wavelength so that *nothing* would be observed. If this is so, then why is a dark line seen?

Consider **Figure 13–22**, which shows light that has a continuous spectrum passing through a rarefied, cool gas. The atoms in the gas absorb photons of certain wavelengths from the passing beam of radiation. The atoms then re-emit photons at wavelengths characteristic of the

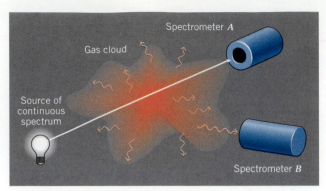

FIGURE 13–22. Light with a continuous spectrum passing through a gas. Both absorption *(A)* and emission *(B)* spectra can be seen, depending on the location of the observer.

gas. An important difference between the absorbed and emitted photons can be seen: Whereas the photons absorbed were all traveling in the same direction, the emitted photons are radiated uniformly into all directions in space. At the particular wavelengths at which absorption is taking place, there will be a net decrease in the number of photons in the beam moving toward spectrometer *A*, because most of the photons that are emitted at the wavelength will go in other directions. Observer *A* will therefore see a continuous spectrum with energy missing at the wavelengths that the gas is capable of absorbing. In other words, there will be a continuous spectrum with a superimposed dark-line spectrum.

Inquiry 13–20 Suppose an observer were to study the spectrum of the radiation received by a spectrometer pointed in the direction of spectrometer *B* in Figure 13–22. What kind of spectrum would be seen? Why?

This inquiry shows that the type of spectrum seen depends not only on the physical conditions in the source of light but also on the geometrical relationship between the source and the observer. We illustrate spectra further in the next section.

*13.3
EXAMPLES OF SPECTRA

In this section we present examples of spectra formed under a variety of physical conditions. These examples, which include the Sun, gas clouds found near hot stars, and the pervasive cool interstellar gas, contain no new fundamental concepts but will enhance your understanding of the concepts presented in this chapter. They

also serve to emphasize the overriding importance of spectra to the study of the universe. Because space contains extremes in the physical conditions of density, temperature, pressure, mass, and energy that have never been duplicated in a laboratory on Earth, these illustrations also reveal extraterrestrial space as a place that provides a unique laboratory for the study of nature.

A MODEL OF THE SUN
BASED ON ITS SPECTRUM

Observations of the solar spectrum, along with Kirchhoff's laws of radiation, can be used to construct a first approximation of a model for the Sun. First, the fact that the Sun exhibits a continuous spectrum with absorption features shows that the light-emitting portion of the Sun must be either a hot solid, a hot liquid, or a hot, dense, opaque gas. But we also know that the Sun is yellow in color, corresponding to a peak emission of energy at about 5000 Å or a bit less.

Inquiry 13–21 Adopting 5000 Å as the wavelength of peak emission for the Sun, roughly what is the temperature (in K) of the Sun's surface?

At such a high temperature (6000 K, or almost 11,000°F), matter cannot remain as a solid or liquid because it is quickly vaporized. It is clear, then, that the Sun must be composed of a hot, dense, opaque gas. This part of the Sun is called the **photosphere** (Greek for "sphere of light").

The Sun also has dark lines in its spectrum (see Figure 13–14) indicating that there must be a layer of relatively cool gas above the photosphere. So, in the first approximation, our model of the Sun is as follows: A ball of hot, dense, opaque gas with a temperature around 6000 K surrounded by a cooler outer atmosphere (by cooler we mean perhaps 4500 K) that produces absorption lines in the spectrum. Such a model of the Sun is shown in **Figure 13–23**. As we will see in Chapter 14, other stars, too, exhibit dark-line spectra and surface temperatures comparable to the Sun's; as a result, similar models can be used to explain the principal observed features of the stars.

Given the basic model, astronomers can refine it by careful measurement of the wavelengths and strengths of all the dark lines in the spectrum. By comparing these measurements with laboratory measurements, astronomers can determine not only the chemical composition of the Sun but also the *abundances* of the various chemical elements in the Sun and stars. In addition, other details about the structure of the stars can be determined, some of them surprising. These aspects

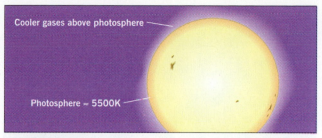

FIGURE 13–23. A simple model of the Sun. Light from the center of the Sun's disk comes from a deeper, hotter layer and appears brighter than the light from the Sun's edge.

of spectral analysis will be developed further in Chapter 14.

THE SUN'S FLASH SPECTRUM

Astronomers expected the temperature of the Sun to decrease with distance from its center. They therefore were surprised to find a spectacular confirmation of Kirchhoff's laws during a total eclipse of the Sun. As the Moon moves in front of the Sun, it progressively blocks more and more of the sunlight but does not change the character of the spectrum, which remains a dark-line spectrum, but of decreasing intensity. Eventually, however, the Moon completely covers the Sun's photosphere. Now, only those atmospheric gases surrounding the photosphere of the Sun and beyond the edge of the Moon are visible to us, and in an instant the spectrum changes from a dark-line to a bright-line spectrum. This change occurs so suddenly that the new emission spectrum has been named the flash spectrum (**Figure 13–24**). (The bright-line spectrum is, of course, always there, but is washed out by the bright photosphere.) Because the emission lines appear pinkish for this part of

the Sun, the emitting region is called the **chromosphere** (color sphere).

Inquiry 13–22 Accepting the idea that the Sun is composed primarily of hydrogen, how can you explain the pink color of the flash spectrum?

Each spectral line in the flash spectrum is curved because the source of light is the thin crescent of the radiating solar atmosphere. No slit is used in the spectrograph, so the spectrum photographed consists of many thin crescents, one for each wavelength that the gases are capable of emitting. In effect, the Moon is acting like the spectrograph slit.

From the observed flash spectrum, and from Kirchhoff's second law, we are forced to conclude that the region surrounding the photosphere, from which the flash spectrum comes, is hotter than the photosphere itself.

In the middle of a total eclipse, when the Sun's disk is completely covered by the Moon, the solar **corona** appears (see Figure 4–26b). This normally invisible tenuous envelope of the Sun emits a discrete-emission-line spectrum of highly ionized gases, indicating a temperature of over one million degrees.

Inquiry 13–23 The spectra of a certain class of stars (known as Wolf-Rayet stars) possess not only characteristics of the absorption spectra found in ordinary stars but also bright-line emission features. Suggest a model of a Wolf-Rayet star that might explain the observed spectrum.

Inquiry 13–24 The great galaxy in Andromeda, M31, has an absorption-line spectrum. What does this indicate about the types of objects that make up M31?

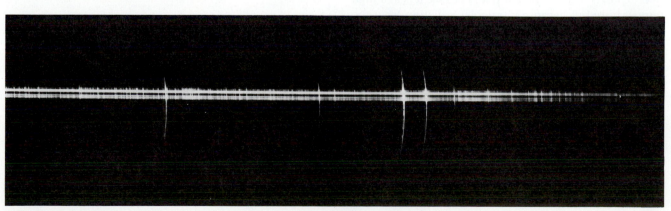

FIGURE 13–24. The Sun's flash spectrum. In this case we are positioned like the observer at point B of Figure 13–22.

IONIZED GAS BETWEEN THE STARS

Our next examples of spectra involve colorful clouds of gas. First we look at the spectrum of hydrogen from these clouds, then we see the spectrum from helium and heavier elements.

The Nature of Nebulae The term **nebula** comes from a root meaning *cloud*, and historically it referred to any fuzzy, indistinct, nonstellar object in the sky. Even as recently as the twentieth century the controversy about the nature of the the "nebulae" was still very much alive. The first progress in understanding nebulae was made when the science of spectroscopy began to develop. It was found that the spectra of some nebulae looked like the spectra of stars, indicating that they consisted of a group of stars so distant that individual objects could not be recognized (just as the Milky Way is simply a glow to the naked eye). Some prominent nebulae, on the other hand, had bright-line spectra due to the emission of energy only at certain wavelengths.

Inquiry 13–25 What conclusion about the nature of nebulae could be drawn from those that showed bright-line spectra?

A good example of a nebula having a bright-line spectrum is the Lagoon Nebula, M8, in Sagittarius (**Figure 13–25**). M8, which is just visible to the naked eye and quite satisfying to view in a small telescope, was the eighth object on a list of "fuzzy" objects compiled by the eighteenth-century French comet hunter Charles Messier. An **emission nebula** (or **diffuse nebula**) such as M8 is a cloud of rarefied gas found in association with a hot star or group of stars. In most cases, the star is not difficult to see, although in long-exposure photographs made to reveal the nebular gas, the star is usually overexposed. The gas radiates not because it is at a high temperature but because ultraviolet radiation from the hot star excites the gas to higher energy levels. The process can be described symbolically as

$$\text{hydrogen atom} + \text{ultraviolet photon} \rightarrow \text{proton} + \text{electron}.$$

To explain further, **Figure 13–26** illustrates the continuous spectrum of a hot star. Only those photons having wavelengths shorter than 912 Å have sufficient energy to ionize the hydrogen atom (compare with Figure 13–19). The surrounding region of ionized gas created by this reaction is known as an **H II region** (hydrogen that is not ionized is called H I). The ionization of hydrogen atoms creates a hot gas consisting of protons and electrons at a temperature of about 10,000 K.

FIGURE 13–25. The Lagoon Nebula in Sagittarius is a good example of an emission nebula—a region where stars, gas, and dust interact.

Even though the high temperatures cause the particles to move about at high speeds, a proton will occasionally recombine with an electron. When this occurs, the electron will cascade to lower and lower energy levels and emit photons, as described previously. Quantum mechanics shows us that an electron will almost always be involved in the transition from the second to the first excited state, with the emission of a photon of wavelength 6563 Å. Because this wavelength is in the red end of the visual spectral region, nebulae generally appear reddish. It is these photons resulting from recombination that make the nebula visible to us.

Recombination does not simply re-create a spectrum like that of the original star. The stellar spectrum in Figure 13–26 is a continuous spectrum, whereas recombination produces primarily an emission-line spectrum in the ultraviolet, visible, and infrared parts of the spectrum from the cascade process described earlier for the hydrogen atom.

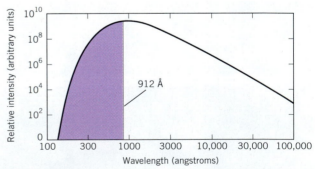

FIGURE 13–26. The spectrum of a star of surface temperature 30,000 K. The wavelength region containing photons energetic enough to ionize hydrogen atoms is shaded.

Forbidden-Line Emission Even after astronomers understood how the ionized gas and emission lines were created, questions remained, because there were many emission lines in the spectra of nebulae besides those of hydrogen, as **Figure 13–27** shows. In fact, in the early part of the twentieth century researchers had no success in identifying many of these emission features. Because helium was first observed in the solar spectrum and only subsequently found on Earth, frustrated astronomers looked for a new chemical element—dubbed "nebulium"—that did not exist on Earth. The explanation of these mysterious emission lines was provided by Ira S. Bowen in 1927. He showed that they were due to common chemical elements, such as oxygen, in unusual energy states called *metastable states*. Metastable levels are unlike normal energy levels in that once an atom finds itself in a metastable level it will stay there for a relatively long time before radiating. Although an ordinary transition downward typically takes place within one hundred millionth of a second after the electron arrives in the excited state, an atom may remain in a metastable level for a second or even more before radiating and dropping to a less excited state. As a result, emission lines not otherwise possible can be formed. Emission lines resulting from transitions from metastable states are called forbidden lines, not because they disobey laws of physics but because they had not been observed previously and were not understood originally.

Two forbidden lines that often occur in nebulae are at wavelengths of 4959 and 5007 Å. These lines are caused by transitions in oxygen that has had two electrons removed. Because these strong lines are in the green part of the electromagnetic spectrum, nebulae often have a greenish tint to them.

But even granted that these lines are unusual, why hadn't such transitions been seen in laboratories on Earth? The answer is that forbidden lines can form only when the density of gas is extremely low. Even the best vacuum that can be created on Earth is four to seven orders of magnitude more dense than the gas in diffuse nebulae. The observation of forbidden lines in nebulae immediately tells us that the density of material is low. Thus the study of nebulae provided scientists with objects under never-before-seen physical conditions.

NEUTRAL GAS BETWEEN THE STARS: INTERSTELLAR ABSORPTION LINES

The ionized gas in the emission nebulae is the easiest component of the gas between stars to observe because the hot star excites the gas cloud to radiate. It is more difficult to detect interstellar gas that is predominantly *neutral*; that is, not ionized. This gas is cool and does not emit energy efficiently; in fact, it was first found by virtue of its absorbing powers.

Shortly after the turn of the century, the astronomer Johannes Hartmann was studying the spectrum of the binary star δ Orionis (one of the stars in Orion's belt). He discovered that some spectral lines changed wavelength with time while others did not. (The reason why some lines change position will be discussed in Chapter 14.) He also noticed that the unchanging lines were extremely narrow in wavelength, distinctly different in appearance from the broad and constantly shifting absorption lines of the stars. He reasoned—correctly—that these absorption features were caused by a cloud of material between the star and himself (**Figure 13–28**). In other words, from an analysis of spectra he accidently discovered **interstellar absorption lines** and the fact that the space between stars is filled with gas. The narrowness of the lines also indicated to him that the cloud consisted of very cold material, with a temperature of only about 100 K (−173°C).

Interstellar absorption lines are difficult to study because there are not many of them in the visible part of

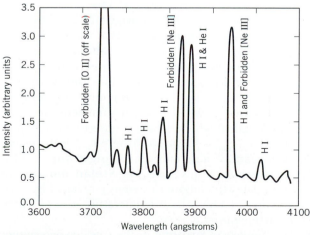

FIGURE 13–27. The spectrum of the Orion Nebula, showing the hydrogen lines and the forbidden-line emission.

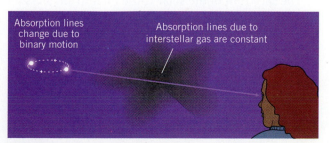

FIGURE 13–28. Interstellar gas clouds were first discovered by examining the spectra of binary stars.

the spectrum. The reason for this is worth noting, because it also explains part of the great interest astronomers have in observations made from space.

The schematic energy-level diagram of a typical atom is shown in **Figure 13–29**. Most atoms have energy levels spaced so that transitions from the ground state to an excited state involve a considerable amount of energy. Such transitions correspond to wavelengths in the ultraviolet part of the spectrum, typically in the range from 500 to 2000 Å. Because most atoms in cool interstellar space have their electrons in the ground state, transitions from the ground level are the strongest ones. Unfortunately, they are also the ones that are hardest to observe, because ozone in the Earth's atmosphere absorbs these ultraviolet wavelengths. We can therefore see them only with telescopes placed above the atmosphere.

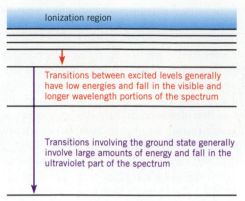

Ionization region

Transitions between excited levels generally have low energies and fall in the visible and longer wavelength portions of the spectrum

Transitions involving the ground state generally involve large amounts of energy and fall in the ultraviolet part of the spectrum

FIGURE 13–29. An energy-level diagram for a typical atom.

Inquiry 13–26 It has been stated that 90% of the universe is hydrogen, so presumably this applies also to interstellar gas. Consult the energy-level diagram of the hydrogen atom. Would we be likely to observe interstellar absorption lines of hydrogen gas in the visible part of the spectrum? Explain your reasoning.

The ultraviolet spectrograph on the Copernicus satellite, which gathered data between the years 1972–1980, surveyed the absorption-line transitions found in the ultraviolet region of the spectrum for a number of bright stars. Because it was able to observe the strong transitions beginning in the ground state, the satellite returned a wealth of data on the abundances of elements in interstellar gases. It found that many chemical abundances corresponded well to the values that had been discovered for most stars, but, interestingly, there turned out to be a set of chemical elements whose abundances were dramatically *depleted*. Elements found from both ultraviolet and optical observations to be depleted include carbon, nitrogen, calcium, potassium, nickel, aluminum, and iron. We will argue later in the book that these depleted elements have probably gone into forming dust grains in space.

CHAPTER SUMMARY

OBSERVATIONS

- The solar spectrum contains a continuous spectrum with dark absorption lines superimposed on top of it. Just as the Sun reaches totality during a total eclipse, the spectrum becomes an emission-line spectrum. The spectrum of the solar corona is also an emission spectrum of highly ionized atoms.
- The spectrum of a **gaseous nebula** shows emission lines.
- Cool gas in the space between stars absorbs background starlight and produces **interstellar absorption lines**.

THEORY

- **Kirchhoff's first law** tells us that radiation from an opaque solid, liquid, or dense gas will emit a **continuous spectrum**. His **second law** is that a low-density hot gas will emit a **bright-line (emission) spectrum**. His **third law** says that white light passing through a cooler gas will be absorbed and produce a **dark-line (absorption) spectrum**.
- The (continuous) spectrum of a **blackbody** depends on temperature but not on chemical composition.
- **Wien's law** relates a blackbody's temperature to the wavelength at which the body emits its most intense radiation: $\lambda_{max}T = 3 \times 10^7$, where λ_{max} is the wavelength of maximum radiation in **angstroms** and T is the temperature in kelvins. The **Stefan-Boltzmann law** relates the amount of energy radiated by each unit area of a body to its temperature: $E \propto T^4$ where T is temperature.
- The number of **protons** in the nucleus determines what element a nucleus is. Differing numbers of **neutrons** produce different **isotopes** of an element.

To achieve electrical neutrality, there are as many **electrons** as there are protons.

- The **energy-level diagram** describes the amount of energy required to excite an electron from one discrete orbit to another. Electrons in lower energy states may absorb a photon and move to more energetic, **excited states**. Electrons in excited states move to less excited states (closer to the nucleus) by emitting a photon. The frequency of the absorbed or emitted photon is determined by the energy difference between the two orbits. An electron in the **ground state** cannot emit a photon. An atom can become **ionized** if the electron is given enough energy to be removed from the atom.

- Each atom and isotope has a unique energy-level diagram, meaning that each atom's absorption-line and emission-line spectrum is unique.

- Quantum mechanics provides scientists with a model of the atom that offers a more complete description of the workings of nature on the atomic level.

Conclusions

- From the observed continuous spectrum, we infer that the Sun's interior consists of a hot, opaque gas.

From the observed absorption lines, we infer the presence of a lower-temperature atmosphere surrounding the photosphere. From the observed flash spectrum we infer the presence of a higher-temperature chromosphere. From observations of the highly ionized corona, we infer temperatures in excess of 10^6 K.

- The chemical composition of a gas can be determined by comparing an observed emission-line or absorption-line spectrum with that of a gas in a laboratory.

- The **Balmer series** of hydrogen occurs in the visual region of the spectrum as a result of transitions involving the first excited state. The **Lyman series** involving the ground state appears in the ultraviolet, while all other hydrogen series occur at infrared or longer wavelengths.

- The emission-line spectrum of a gaseous nebula comes from ionization of hydrogen gas by high-energy photons from a hot star near the hydrogen. Some strong emission lines come from transitions involving metastable states, states whose lifetimes are longer than normal.

- From absorption lines observed in the spectra of some stars, astronomers infer the presence of cool gas in the regions between stars.

SUMMARY QUESTIONS

1. What are the three principal types of spectra? Explain the conditions under which each is produced.
2. What effect does temperature have on the color and energy distribution of the radiation emitted by a hot object?
3. How can you describe the formation of the absorption spectrum of the Sun in terms of a simple solar model?
4. How do the interactions between photons and atoms produce emission and absorption spectra?
5. What is represented in an energy-level diagram?

Draw the energy-level diagram for a hydrogen atom, and indicate the principal transitions.
6. Why do different chemical elements have different characteristic spectral lines?
7. What are Kirchhoff's laws for dark-line and bright-line spectra? How does atomic theory explain the Sun's flash spectrum?
8. How can you use the spectra of nebulae to show that some are composed of stars and others are composed of hot gases?
9. How is the spectrum of an emission nebula formed?

APPLYING YOUR KNOWLEDGE

1. Consider the following experiment. A clear glass tube filled with sodium gas has light from an electric sodium lamp focused into it. (A sodium lamp is yellow in color.) The inside of the tube is observed to contain a cone of yellow light in the region in which the sodium lamp is focused. However, when the electric sodium lamp is replaced with an electric lamp filled with mercury gas, no cone of light is observed from inside the tube. Explain the experiment.

2. List all possible *downward* transitions an electron in the sixth energy level in the hydrogen atom can make. List their wavelengths when known.
3. Why does the Lyman series occur at shorter wavelengths than the Balmer series?
4. Given that most of the hydrogen atoms in the Sun's photosphere are in their lowest energy state, what spectral series of hydrogen would be produced most strongly in the Sun?

5. Suppose you observed the spectrum of some astronomical object and found it to have both absorption and emission lines. What might be a reasonable explanation for the nature of the object?

6. If you are star-gazing on a clear night and notice two stars, one of which appears bluish and the other reddish, what could you conclude about the relative characteristics of the two stars?

7. What is the wavelength of the strongest emission of energy for a body of temperature 3 K? (As you will see later in the book, this temperature is the current temperature of radiation still around from shortly after the universe began!)

8. What are the frequency and the energy of a photon having a wavelength of 21 cm?

9. Relative to the Sun, how much more energy per unit area does a star with a temperature of 22,000 K radiate?

10. Relative to the Sun, how much more energy per second comes from the entire surface of a star that has a radius one-third that of the Sun and a temperature of 22,000 K? (Hint: Remember that the energy per second *per unit area* depends on T^4.)

11. Consider the following logarithmic scale:

What number should be in the place of the question mark? Explain your reasoning.

12. Compute the temperature of a star that appears to be red. Do the same for a star that appears to be blue.

ANSWERS TO INQUIRIES

13–1. Yes.

13–2. In this example each succeeding tick mark is four times greater than the previous one. Thus the next number is 64.

13–3. $3 \times 10^7/12,000$ K = 2500 Å

13–4. $3 \times 10^7/300 = 10^5$ Å.

13–5. $(50,000/5,000)^4 = 10^4$

13–6. $3 \times 10^7/30,000$ Å = 10^3 K.

13–7. $3 \times 10^7/0.5$ Å = 6×10^7 K.

13–8. There is water vapor in the Earth's atmosphere through which the stars are observed.

13–9. The Moon's visible spectrum should be similar to the Sun's, because the Moon shines by reflected sunlight. We would also see CO_2 in Venus's spectrum.

13–10. $^{15}_{8}O$, $^{17}_{8}O$, $^{18}_{8}O$

13–11. If the relationship is rewritten, it becomes

> energy of higher energy state = energy of photon + energy of lower energy state.

Therefore, the total energy after the transition is the same as the total energy before the transition.

13–12. The principle of conservation of energy.

13–13. Emission is represented by a downward-pointing arrow; absorption by an upward-pointing one.

13–14. Lyman: ultraviolet; Balmer: optical; Paschen, Brackett, Pfund: infrared. Only the Balmer lines can be observed from the Earth's surface; all the others require instruments above the Earth's atmosphere.

13–15. (a) Balmer emission (b) Lyman absorption (c) Paschen emission (d) Lyman emission (e) Pfund absorption

13–16. (a) Photon may be absorbed and the atom will be in the third energy state. (b) Photon cannot be absorbed to produce an absorption line. (c) Photon may be absorbed and the atom will be in the fourth energy state. (d) Photon will be emitted, not absorbed, and the atom will be in the second energy state.

13–17. 6563 Å, 1026 Å

13–18. 973 Å, 4861 Å followed by 1026 Å, 18,751 Å followed by either 1026 Å or both 6563Å and 1216 Å.

13–19. Neutral hydrogen, because both have a small nucleus surrounded by a region containing one electron. But because the nuclear charge of helium is greater, all the energy levels will be higher, and the wavelengths will be shorter than the corresponding hydrogen wavelengths.

13–20. An emission spectrum. The observer would see the light being emitted by the atoms in other directions after being absorbed by the beam. There is no continuous spectrum because there is no higher temperature object behind the cloud.

13–21. $3 \times 10^7 / 5000$ Å = 6000 K

13–22. High temperature will ionize hydrogen. Recombination of protons with electrons will produce transitions between levels *3* and *2*, giving a strong emission line at 6563 Å, in the red.

13–23. This could be produced by a star (which gives the absorption spectrum) surrounded by a shell or ring of gas that is hotter than the outer surface of the star.

13–24. Because the stars have absorption spectra, we are presumably seeing the composite effect of a large number of stellar spectra superimposed when we observe the galaxy.

13–25. They must consist of rarefied gases (by Kirchhoff's laws) at some temperature.

13–26. We would not be likely to observe interstellar absorption lines of hydrogen gas in the visible part of the spectrum because most of the hydrogen would be in the ground state. For this reason, the only transitions that occur involve photons with wavelengths in the ultraviolet part of the spectrum.

PART FOUR

DISCOVERING THE NATURE AND EVOLUTION OF STARS

When we view galaxies with our eyes, we see stars, which appear to be the fundamental building blocks of galaxies. However, looking further, we find that stars are formed from clouds containing atoms, molecules, and dust. Some people therefore might describe these as the fundamental components of galaxies. The purpose of Part 4 is to describe these components and to obtain a basic understanding of them—their characteristics, formation, and evolution.

Most of what we know about the universe comes from the study of spectra; thus we open Part 4, with a study of stellar spectra. In Chapter 14, we study what astronomers can learn about stars from an analysis of their spectra and try to understand how conclusions about stars are made from spectral analysis.

In Chapter 15 we look at a large population of stars to see what general conclusions we can form about them. It is here we find that stars range in size from that of the Earth up to the diameter of Saturn's orbit, with masses ranging from a small fraction of the Sun's mass to nearly 100 times that. We also begin to determine the distances to stars that are too far away to have measurable stellar parallaxes. Finally, we look at binary stars, because more than half the stars in our solar neighborhood are in such systems.

Chapter 16 contains our study of the Sun, along with the beginning of our discussion of stellar evolution. We look at possible sources of energy in the Sun and examine the conditions required for it to generate the energy we observe. We examine the conditions of equilibrium that must occur within a star for it to survive for long periods of time. These conditions, along with the sources of energy, determine how long a star will "live." The final section of the chapter examines the Sun as a typical star. We learn about its various layers, and the range of processes that occur both within it and on its surface. Finally, we see how astronomers are studying the interior of the Sun by observing oscillations on the surface.

The evolution of stars is the subject of Chapters 17, 18, and 19. In Chapter 17 we learn about the formation of stars—where and how they form. In Chapter 18, we look at middle age and the approach to old age for stars of all masses. The death of low-mass stars is discussed here. Finally, in Chapter 19, comes our discussion of the deaths of massive stars, those that explode in the fury of a supernova explosion, leaving behind either a neutron star or a black hole. All along the way, particular attention is paid to observational evidence for or against our theoretical picture of star formation. These observations include those of star clusters as well as clues given by detailed examination of stellar spectra.

14

UNDERSTANDING STELLAR SPECTRA

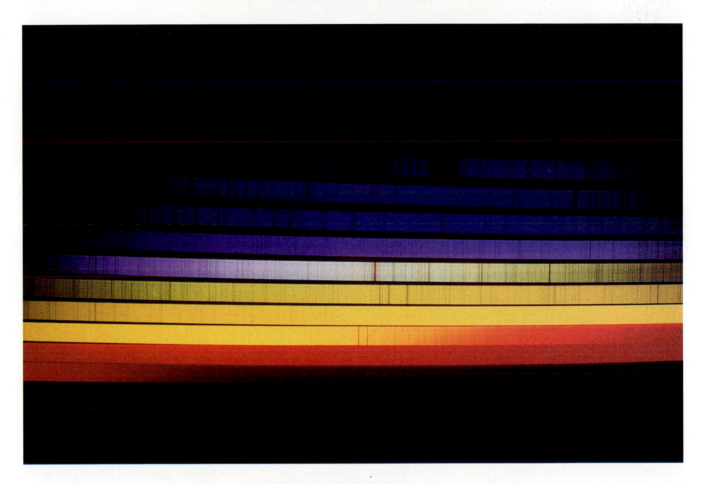

Oh Be A Fine Girl, Kiss Me! (The original mnemonic)
Oh Boy, An F Grade Kills Me!
Oven Baked Ants, Fried Gently, Kept Moist, Retain Natural Succulence
Our Best Aid For Gnat Killing—My Reindeer's Nose. Snort.
Only Boys Accepting Feminism Get Kissed Meaningfully

MNEMONICS

When faced with large quantities of seemingly random information, scientists respond by first classifying the data. For example, biologists classify animals into kingdom, phylum, subphylum, class, order, family, genus, species, and subspecies; geologists classify minerals in terms of form, cleavage, and hardness. Classifications are like boxes into which like objects are placed. Such classifications provide scientists with opportunities to see similarities and differences between objects. In astronomy, we classify types of galaxies (see Chapter 21) as well as types of stars. As we shall see, stars can be classified on the basis of their spectrum (this chapter), their variation in light output (Chapter 18), and their membership in groups of stars (Chapters 15 and 18).

14.1

SPECTRAL TYPES

The light we receive from a star is emitted by its outer layers. By carefully studying the spectrum of this light, astronomers try to learn not only about the star's surface but also about interior conditions that are not directly observable.

In the late nineteenth century, astronomers built up a scheme of spectral classification that simply designated the various types of spectra by letters of the alphabet. Thus, there were stars of **spectral type** A, B, C, and so on. But around 1900, when our story properly begins, the American Annie Jump Cannon of Harvard College Observatory (**Figure 14–1**) began a project in which she was eventually to classify the spectra of nearly 400,000 stars on the basis of their spectral appearance and the relative strengths of the dark-line features in them. Her classifications were published between the years 1918 and 1924 as the famous *Henry Draper Catalog,*[1] which is still in constant use by astronomers.

At the time Cannon did her work, astronomers did not know why the spectra of stars varied so greatly in appearance—that understanding had to wait until the full development of quantum physics in the 1920s. However, the classification scheme Cannon established led the way toward a proper understanding of stellar spectra, and it is still used today in modified form (and is currently being updated by Nancy Houk at the University of Michigan).

FIGURE 14–1. Annie Jump Cannon (1863–1941), classifier of stellar spectra.

The classification scheme now employed by astronomers is shown in **Figure 14-2**. To the side of each spectrum is the name of the star; each also has a letter designation for its **spectral class**. Each major class has been divided into 10 subclasses using the numbers 0 through 9. Thus, within the class of stars of type B, astronomers distinguish subtypes B0, B1, . . . , B9. Within each major class, the spectra are similar; yet the various classes merge smoothly into each other, so that a B9 star and an A0 star are more similar to each other than a B9 star and a B0 star, for example. At the top and bottom of Figure 14–2, the major spectral lines are identified. If you follow a particular spectral line down the figure, you will notice that it varies in strength as the spectral type varies. For example, examine the line of ionized

---

[1]The catalog is named for the person whose estate funded its publication. Henry Draper was the first astronomer to photograph the spectrum of a star.

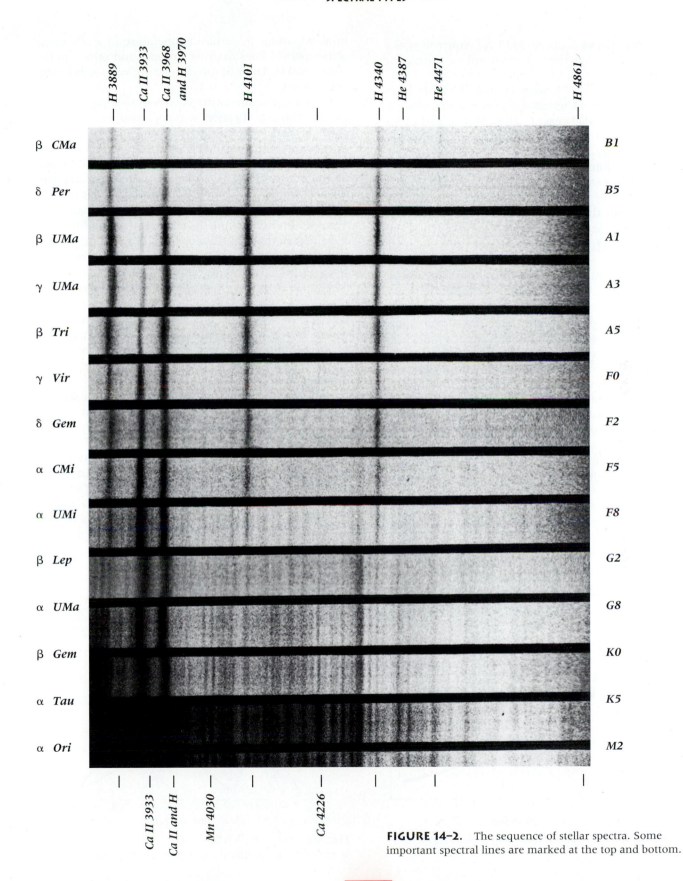

FIGURE 14–2. The sequence of stellar spectra. Some important spectral lines are marked at the top and bottom.

calcium, denoted as Ca II, at 3933 Å.[2] You will notice that as you go through the A stars toward the F stars, the strength of this line continually increases.

In the modern classification scheme, the letter types do not follow in alphabetical order. Furthermore, some letters of the alphabet (e.g., C, D, E, H, I, J, and L) are missing. In the original Harvard classification of Annie Cannon, the most important criterion was the strength of the Balmer lines of hydrogen. Notice that they are strongest in type A, but they are also strong in types B and F and faintly visible in most of the other types. It also turned out that there were some duplicate categories that had been given different letters; when this was understood, certain letters disappeared from the list.

More important, when the colors of the stars in various classes were examined, it was found that stars of a given spectral type had similar colors. For example, O and B stars were very blue, whereas M stars were quite red. We have already seen that the color of a hot, glowing mass, such as a star, is an indication of its temperature (Wien's law in Chapter 13). When these variations in color were studied by astronomers, they came to realize that each of the different spectral types belonged to stars of different temperature; type O stars had temperatures greater than about 25,000 K, type M stars had temperatures below 3500 K, and the other types were in between (**Figure 14–3**). This, then, must be the underlying reason for the variation of stellar spectra. Some-

how, variations in surface temperature of a star cause some spectral lines to grow stronger and others to become weaker. Later in this chapter we will see how this can happen.

Once these temperature variations were appreciated, the spectral classification sequence was rearranged in order of *decreasing* temperature, and the resulting sequence came out as OBAFGKM, omitting the spurious and duplicate classes. Three rarer categories (R, N, and S) include somewhat unusual spectra and are frequently named at the end of the list (although they do not really follow the temperature sequence). Generations of astronomy students have remembered the correct order of the spectral sequence using the mnemonic phrase first suggested in less enlightened times by Henry Norris Russell, "Oh Be A Fine Girl, Kiss Me," to which some add the phrase "Right Now, Smack" to account for the special classes R, N, and S. You can substitute "guy" for "girl" if you wish, but over the years various astronomers and students have had fun constructing their own pet mnemonics for the spectral sequence. A few that we have heard have been given as the chapter opening quote.

The variation in the observed strengths of the spectral lines for various elements is summarized in **Figure 14–4**. We can readily see from this figure which spectral features dominate at each spectral type.

The basic characteristics of stars having various observed spectral types are summarized in **Table 14–1**. Notice that the surface temperatures of these "normal" stars vary from roughly 50,000 K down to as little as 2000 K, and that the spectral features vary in a continuous manner. Also included in the table is the approxi-

---

[2]When discussing stellar spectra in the visible region, we will use angstroms (Å) because of historical and user preference, although some scientists prefer nanometers, where 3933 Å = 393.3 nm.

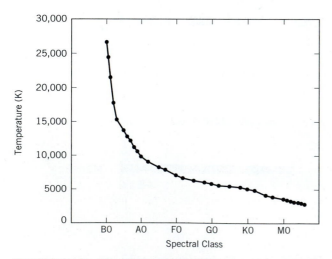

FIGURE 14–3. The relationship between spectral type and surface temperature.

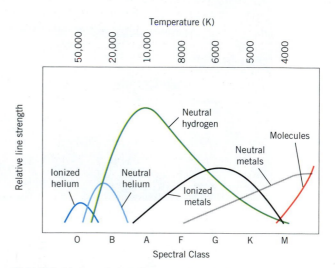

FIGURE 14–4. The variation of the relative strengths of spectral lines from different elements with spectral type.

TABLE 14–1

Characteristics of Spectral Classes for Normal Stars

Type	Temperature	Color	Prominent Spectral Features	Percentage	Examples
O	25,000–50,000 K	Blue	Lines of ionized helium; plus lines of multiply ionized elements, such as O III, N III, C III Si IV. Balmer lines of hydrogen rather weak.	0	Mintaka Meissa Alnitak Naos
B	11,000–25,000 K	Blue	Lines of neutral helium; plus lines of ionized elements, such as O II, N II, C III, Fe III. Balmer lines of hydrogen moderately strong.	0.1	Rigel Regulus Spica Bellatrix
A	7500–11,000 K	Blue-white	Balmer lines very strong; other features very weak or absent.	1	Vega, Sirius, Altair, Deneb
F	6000–7500 K	White	Balmer lines moderately strong; plus lines of some ionized elements, such as Ca II, Ti II, Fe II. (Ca II, at 3933 Å is about as strong as Hγ at 4340 Å.)		Polaris Procyon A
G	5000–6000 K	Yellow-white	Balmer lines present but weaker than in hotter stars. Lines of neutral elements strong (e.g., Fe, Ti, Mg). Lines of easily ionized elements (e.g., Ca II) are strong.	4	Sun α Cen A τ Cet Capella
K	3500–5000 K	Red-orange	Lines of neutral elements strongest. Ca II still present but weaker.	14	Arcturus Aldebaran Pollux
M	2000–3500 K	Red	Balmer lines weak. Many lines of neutral elements. Spectrum dominated by molecular features, such as TiO, C_2, and CH.	72	Betelgeuse Antares Mira Barnard's star

mate percentage of normal stars in each spectral class. Take the time to compare the descriptions in Table 14–1 with the spectra themselves shown in Figure 14–2. Remember that the classification scheme is based only on the *observed* appearance of the spectrum, especially the spectral lines. Only after the spectra have been classified on the basis of their appearance do we attempt to understand the physical meaning of the classes.

Inquiry 14–1 The spectra at the top of Figure 14–2 are shaded to the right side (meaning they are less intense), while those at the bottom are shaded to the left side. Why is there this difference in appearance?

WHY SHOULD SPECTRA VARY IN APPEARANCE?

Most stars show a continuous spectrum with absorption lines superimposed. This means that there is a hot,

opaque region (in the Sun, its **photosphere**) that emits a continuous spectrum, and above it a cooler, more transparent layer where the absorption lines are formed. This cooler layer is the star's atmosphere, and it is the part of the star on which we now focus our attention.

Scientists were able to understand the different appearances of stellar spectra once they understood two ideas. First, as discussed further in Section 14.2, most stars are chemically similar to the Sun, so that for most stars differences in chemical composition are unimportant. Stars consist of some 75% hydrogen, 23% helium, and only 2% everything else, which we will refer to as *metals*. (While 75% of the *mass* of a star is hydrogen, 90% of its *atoms* are hydrogen.) The second important idea is the understanding of the structure of the atom and the energy-level diagrams that we discussed in Chapter 13. Because of stars' similar chemical compositions of hydrogen, helium, and metals, we can understand spectra by examining simplified energy-level diagrams for these three types of atoms.

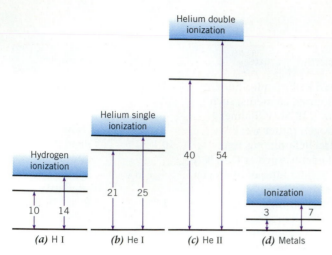

FIGURE 14–5. Energy-level diagrams for hydrogen, neutral and ionized helium, and a typical metal.

Figure 14–5 shows the ground state, first excited state, and ionization level for hydrogen, neutral and ionized helium, and a typical metal. In part *a* we see that it takes 10 units of energy to excite hydrogen to its first excited state. If 14 units of energy are available, the electron is given enough energy to leave the atom, and the result is an ion. The energy to excite and ionize usually comes from the collision of two atoms. Because the greater the temperature the greater the collision speed between particles in a gas, high levels of excitation, including ionization, occur only when the temperature is sufficiently high. In the case of hydrogen, the temperature must be about 6000 K to reach the first excited state, and 12,000 K to ionize.

Figure 14–5*b* shows a similar diagram for neutral helium. Because helium has two protons in its nucleus, the locations and spacings of the allowed orbits are different from those of hydrogen. Helium's greater nuclear charge also means that more energy is required to excite an electron. In fact, it takes more energy to *excite* helium (21 units of energy) than to *ionize* hydrogen. Helium ionization requires 25 units of energy. Such ionization releases one of helium's two electrons.

The one electron remaining in the helium atom can also undergo transitions. However, due to the loss of the first electron, the locations of the allowed orbits are different from that of the neutral atom. Therefore, as indicated in Figure 14–5*c*, the transitions that occur in ionized helium require different energies than was the case for neutral helium. From the figure we see that ionized helium becomes excited if there are at least 40 units of energy, and that the second electron will be dislodged only if 54 units of energy are available.

Lastly, we have the energy level diagram of a typical metal (*d*). Although metals have many electrons, only the outermost one(s) undergo transitions and produce spectral lines that are important to us at this time. As indicated in **Figure 14–6a** for sodium, the outermost electron is at a large enough distance from the nucleus that it feels little of the positive nuclear charge. In fact, the many electrons closer to the nucleus provide additional repulsion on the outer electron, which makes it even less strongly held. For these reasons, Figure 14–5*d* indicates that only some 3 units of energy are required to excite a typical metal. A total of 7 units of energy are sufficient to remove one electron from many elements.

With this understanding of the structure of the important atoms, we can attempt to predict what stellar spectra should look like for stars of different temperatures. We will confine ourselves to the visual region of the spectrum, because historically astronomers were limited to that region due to the Earth's atmosphere.

Low-Temperature Stars Low-temperature stars (spectral types K and M) have temperatures less than about 5000 K. Low temperatures are sufficient to excite electrons in most elements other than hydrogen and helium. Excitation readily occurs because the outer electron, which is the most important one in producing spectral lines, is only weakly held to the nucleus, for reasons discussed above. Therefore, electron transitions occur easily. The large variety of atoms present in a star, combined with the ease with which transitions can occur, produces a huge number of possible absorption lines. Therefore, cool stars are expected to have a plethora of lines, and to have highly complex spectra.

In cool stars, hydrogen and helium exist almost entirely in their ground states.

Inquiry 14–2 Do you expect the Balmer lines to be strong, weak, or absent from the spectrum of a low-temperature star? Explain why.

Inquiry 14–3 Do you expect absorption lines of helium to be strong, weak, or absent from the spectrum of a cool star? Explain why.

Medium-Temperature Stars Medium-temperature stars (spectral types A, F, and G) are those having temperatures between 5000 and about 12,000 K. Metals in a medium-temperature star are readily excited and sometimes ionized. The removal of electrons has two important effects on the spectrum. First, with distant electrons removed, the new outermost electrons are now closer to the nucleus, where the attractive electrical force is greater (**Figure 14–6b**). Second, there are

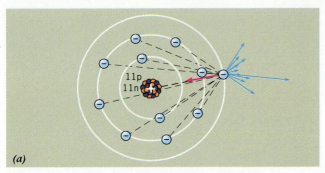
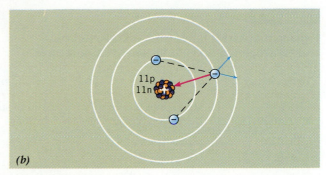

FIGURE 14–6. (*a*) The atomic structure of sodium, showing that the outermost electron feels a smaller attractive force from the positively charged nucleus than do the electrons closer to the nucleus. Furthermore, the distant electron feels a repulsive force from all the inner electrons. (*b*) When ionized, the outer electron is closer to the nucleus, and there is a smaller repulsive force from the few remaining electrons. (Drawing is not to scale.)

fewer electrons repelling the now-outer electron. The result is that the now-outer electron is held somewhat more tightly than before, and more energy is required to excite it. The greater energy means spectral lines from ions will generally occur at a higher frequency (because $E = hf$), or shorter wavelength, than for a neutral atom. In other words, spectral lines for ions are expected to be primarily in the ultraviolet spectral region. The result is that medium-temperature stars are not expected to have as many spectral lines in the visible region of the spectrum as lower-temperature stars.

In the medium-temperature range, much of the hydrogen is in the second energy level (the first excited state). This is so because higher temperatures provide additional energy to excite the electrons. Because transitions involving the first excited state produce spectral lines belonging to the Balmer series (Chapter 13), the Balmer lines of hydrogen will be strong in the spectrum of such a star.

Inquiry 14–4 Do you expect absorption lines of helium to be strong, weak, or absent from the spectrum of a medium-temperature star? Explain why.

High-Temperature Stars In a hot star (one of spectral type O and the hot B stars, with temperatures greater than about 20,000 K), the high temperature causes ionization of the metals. The situation is similar to but more extreme than that for medium-temperature stars, because more than one electron can be removed from each metal atom. For this reason, few if any spectral lines from metals are expected to be present in the visible part of the spectrum because they will all be in the ultraviolet.

Inquiry 14–5 In a hot star, hydrogen will be mostly ionized. What would be the appearance of the *absorption-line spectrum* from hydrogen in such a star? (Hint: There is a slight trick to the question!)

A large fraction of the hydrogen will be ionized in a hot star. Remember, spectral lines are formed by electron transitions. Because there are few hydrogen atoms still having electrons to make transitions, there will be only weak lines from hydrogen in the spectrum.

Helium in a hot star can be excited to higher energy levels. Therefore, absorption lines from neutral helium will be present. In the hottest stars, helium will be ionized. The one remaining electron will be able to undergo transitions and produce spectral lines at wavelengths characteristic of ionized helium.

A summary of the discussion of what is expected in spectra is in **Table 14–2**.

Molecules in Cool Stars The spectra of cool stars may be expected to have one additional attribute. In such stars, the low particle speeds allow two or more atoms to combine to form a molecule. Most of the absorption observed in the visible part of a cool star's spectrum comes from molecules of titanium oxide (TiO) (**Figure 14–7**), but in certain stars there is also noticeable absorption from molecules of C_2 (two carbon atoms bound together) and CN (cyanogen). The TiO, C_2, and CN molecules are exceptionally effective at absorbing light. Although a typical atom has only a relatively small number of absorption lines in the visible part of the spectrum, molecules are able to absorb at so many wavelengths that in certain parts of the spectrum they seem to block nearly all the light.

TABLE 14-2

Expected Spectral Lines from Metals, Helium, and Hydrogen in Stars of Various Temperatures

Temperature	Metals	Helium	Hydrogen
COOL (T<3500 K)	Metal lines dominate TiO molecule	Ground state None	Ground state Weak to none
MEDIUM	Some lines present	None	First excited state Strong
HOT (T>20,000 K)	Ionized None in visual	Yes, He I He II in hottest	Ionized Yes, but weak

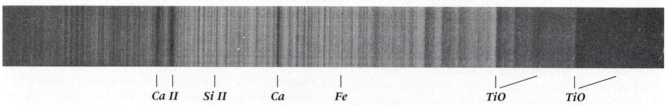

Ca II   Si II   Ca   Fe   TiO   TiO

FIGURE 14–7. The spectrum of an M-type star, with absorption bands from molecules indicated.

The reason molecules are so effective at absorbing light is not hard to understand. A simple molecule can be modeled as a dumbbell (see **Figure 14–8**) in which the bar represents the electrical forces holding the atoms together. This structure can vibrate along the line between the atoms in the molecule, as well as rotate about an axis perpendicular to this line. Just as the electrons in the atom are allowed to occupy only certain states, not all molecular vibrations and rotations are allowed. Those vibrations and rotations that are allowed exhibit slightly different amounts of energy and give rise to separate energy levels. Because each vibrational and rotational energy level is close to the next one, when transitions do occur, the lines are close together and blend to form a band of absorption rather than a line. (The individual lines can be seen if a high-resolution spectrograph is used).

The forces that hold a molecule together are quite weak compared with those that bind an electron to the nucleus of an atom. At the high temperatures found in stellar atmospheres, most molecules are decomposed into their component atoms, a process called **dissociation**. Dissociation can be caused either by the absorption of a photon or by the collision of a molecule with a high-speed atom, electron, or other molecule. The titanium oxide molecule is one of the hardiest, and enough of it survives at the temperatures found in the atmospheres of M stars for it to be prominent in their spectra.

Inquiry 14–6 On the basis of the explanations in the previous paragraphs, if you found a star whose spectrum contained only the absorption features of titanium oxide, would you be justified in concluding that the star was composed entirely of titanium oxide? Explain.

OBSERVATIONS OF STELLAR SPECTRA

Now that we know what to expect on the basis of atomic theory, we will examine what astronomers actually observe.

> YOU SHOULD DO DISCOVERY 14–1, CLASSIFICATION USING KNOWN SPECTRA, AT THIS TIME.

The M stars, shown at the bottom of Figure 14–2 and in Figure 14–7, have strong TiO bands as well as numerous lines from metals. On the basis of our previous discussion, they are clearly low-temperature stars.

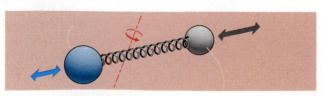

FIGURE 14–8. Atoms in molecules vibrate along the line of force holding them together. The molecule also rotates.

Moving upward in the sequence, we now consider the spectra of K and G stars. As Figure 14–2 illustrates, most of the observed spectral features in a K-type star are the absorption lines of neutral metals (such as iron and calcium), whereas only weak molecular absorption bands (chiefly CH) are seen. Evidently, the atmospheres of K stars are hot enough that most of the molecules have been decomposed into their constituent elements.

Inquiry 14–7 Raising the temperature of a gas causes the average velocity of the molecules and atoms in the gas to increase. How does this explain the smaller number of molecules in the atmosphere of a K-type star as compared to an M-type star?

The G stars have temperatures in the range of 5000 K to 6000 K. The Sun, for example, is a typical G2 star. In G stars, absorption lines due to ionized elements begin to appear. As we saw before, the spectrum of an ionized atom is expected to be quite different from that of a neutral atom, because the energy levels are completely rearranged by the change to the ionized state. Therefore, the different pattern of lines makes it possible to distinguish between lines due to ionized atoms and lines due to neutral atoms of the same element.

For example, the strongest single feature in the entire solar spectrum is the absorption line of ionized calcium (Ca II) at 3933 Å. This is easy to understand, because not only is the outer electron in the atom relatively far from the nucleus, but also the attractive force of the nucleus on this electron is nearly neutralized by the repulsive force of the inner electrons. This means that the outer electron is only weakly bound to the atom and can easily be dislodged from it either by absorbing a high-energy photon (photoionization) or by colliding with another atom (collisional ionization). These two ways of ionizing the atom can be written as *photoionization*:

Ca + high-energy photon → Ca II + free electron

and *collisional ionization*:

Ca + high-velocity particle → Ca II + free electron + lower-velocity particle.

Inquiry 14–8 Raising the temperature of a star's surface increases the average energy of the photons it emits. How does this affect the ionization of atoms in the star's atmosphere?

With F-type stars, few absorption features due to neutral elements are seen. From the earlier discussion, these stars must have higher temperatures than G-type stars. At the temperatures of F-type stars, about 7000 K, most elements have lost at least one electron, and many lines of multiply ionized elements are seen. In stars with surface temperatures around 7000 K, for example, the absorption features of Ca have faded and absorption by Ca II is seen instead.

The main observed spectral characteristics of A-type stars are the dominating Balmer lines. These strong lines arise because the temperature is such that the electrons are mostly in the first excited state as previously discussed. The temperatures of these stars are, therefore, about 10,000 K.

B-type stars show lines of neutral helium with a decreased strength of the hydrogen Balmer lines. According to our arguments based on atomic structure, the temperatures must be greater yet, between about 11,000–25,000 K.

The hottest stars—the O stars—show a high degree of ionization. Absorption lines due to ionized helium (He II) are observed in the spectra of O stars, indicating that a significant amount of ionized helium does exist in their atmospheres and that the temperatures must be high, in excess of 25,000 K.

Inquiry 14–9 Absorption features of neutral silicon (Si I) are seen in G stars. In F stars, these lines are weak, but lines of ionized silicon (Si II) are strong. In B stars, the only strong silicon lines are due to silicon atoms that have lost two electrons and are said to be doubly ionized silicon (Si III). Why is this so?

We should clarify our notation of atoms and ions. Using iron as an example, we refer to neutral iron as Fe I; that is, the Roman numeral I is attached to the element name. Singly ionized iron, which has lost one electron, is designated as Fe II, doubly ionized iron as Fe III and so forth. The spectrum of an iron atom with 15 electrons removed (as actually seen in the solar corona) would be designated as Fe XVI. This convention will be used throughout the text where appropriate.

Inquiry 14–10 Why should it be more difficult to remove a second electron from an ionized atom than it was to ionize the atom in the first place? (There are two main reasons.)

Inquiry 14–11 If you photograph the spectrum of the Sun from the Earth, you will find certain strong absorption bands due to molecular oxygen. How can this observation be explained?

Inquiry 14–12 Ionized helium has a Lyman and a Balmer spectral series like hydrogen, except that all the wavelengths are shorter. Which of these two series, the helium Lyman or Balmer series, would probably appear first in the ultraviolet spectrum of stars as the temperature is raised?

14.2

THE COSMIC ABUNDANCE OF THE CHEMICAL ELEMENTS

The abundance of a chemical element in a star does not follow immediately from its presence or absence in the star's spectrum. To understand what an observed spectrum is telling us, we must first know the star's surface temperature. For example, we have seen that the strengths of the hydrogen Balmer lines change with temperature, as do the line strengths for ionized calcium. Furthermore, the *relative* strength of the hydrogen line compared to the calcium line depends on the temperature. Once the temperature is known, it is possible (with much work) to interpret the spectrum and deduce the proportions of each of the chemical species appearing in the spectrum. This is not an easy task, because some substances (such as ionized calcium) exhibit very strong lines even though there may be only a small amount of the substance in the star. In the case of hydrogen, spectral lines are strong only because 90% of the atoms in stars are hydrogen.

Another complication has made it difficult to deduce precise chemical abundances. Some substances have no strong transitions (and therefore no dark or bright lines) in the visible part of the spectrum. For example, neutral carbon has almost all its significant transitions in the ul-

traviolet, at wavelengths shorter than 3000 Å. Earth's atmosphere, which absorbs ultraviolet light, has prevented us from observing these lines from the Earth's surface. In fact, many important chemical elements have their strongest transitions, which involve the ground state, in the ultraviolet. Astronomers who wish to study these elements at visible wavelengths are forced to observe transitions between two excited states, even though only a tiny fraction of the atoms are in those states. The resulting spectral lines are weak and difficult to observe.

Some other substances have all their lines in the infrared, microwave, and radio regions of the spectrum. For example, molecular hydrogen (H_2) and many other molecular species fall into this category. One benefit of the new astronomical observing techniques, including observations from space, has been to open these previously inaccessible regions of the spectrum for study.

Inquiry 14–13 The visible-light spectrum of Saturn is dominated by absorption bands of ammonia (NH_3) and methane (CH_4), yet we saw in Chapter 10 that the low density of Saturn can be explained only if the planet's atmosphere is composed of at least 90% hydrogen. How can so much hydrogen be present and not be seen in the visual spectrum?

Starting in the 1930s, when the structure of atoms had become reasonably well understood, astronomers finally began to draw some broad inferences about the abundances of various chemical elements in stars. A surprising finding was made by Cecilia Payne-Gaposchkin, the first woman ever to receive a doctorate from Harvard, and one of the more prolific astronomers of the century (**Figure 14–9**). She found that most of the stars in the galaxy had similar chemical compositions—in each star almost 90% of the atoms are hydrogen, nearly 10% are helium, and everything else in the periodic table accounts for the remaining 1 or 2%! Only one or two atoms in a hundred are the heavier elements we think of as being common on Earth—carbon, nitrogen, oxygen, silicon, and the rest. The small amount of heavy elements in the universe explains why we earlier considered spectra of only hydrogen, helium, and a typical metal.

The interstellar gas also turned out to have approximately the same chemical composition, as did individual stars in other galaxies. In fact, the only parts of the universe that did *not* have such a composition seemed to be the terrestrial planets in the solar system, meteorites, and interstellar dust grains. These three items make up only a small part of the universe as a whole. Therefore, it appears that Earth is an atypical sample of the material found in the universe.

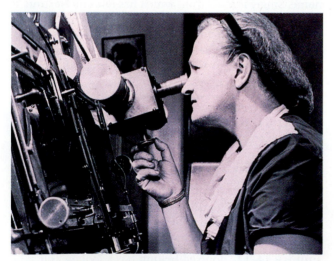

FIGURE 14–9. Cecilia Payne-Gaposchkin (1900–1979), who showed that stars are primarily hydrogen.

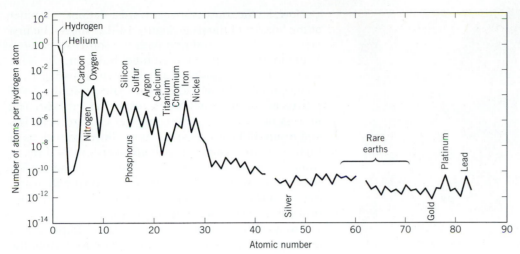

FIGURE 14–10. The relative abundances of the chemical elements in the universe. Note the peak at iron, due to the stability of the iron nucleus.

Overall, we can characterize the chemical composition of the universe by a broadly uniform **cosmic abundance**. Figure 14–10 illustrates the relative abundances of the chemical elements in the universe as they have been determined from studies of all available sources, such as the Sun, stars, interstellar gas, and meteoritic rubble that has landed on Earth. Note that the abundance scale is a logarithmic scale, where each tick mark represents a factor-of-10 change in abundance. The attempt to understand the reasons for the detailed shape of this element-abundance curve spawned the field of nuclear astrophysics and provided an understanding of the evolution of stars (which we will discuss in Chapters 16–19).

The existence of a common elemental composition for almost all the objects in the universe is a spectacular idea and certainly something that a successful theory of the origin and evolution of the universe should be able to explain. We will return to this idea later.

Inquiry 14–14 We have just seen that meteorites are *not* typical of the universe as a whole—they have little hydrogen and helium. How, then, can studies of meteorites help us understand the compositions of the stars, which are *mostly* hydrogen and helium?

14.3
THE DOPPLER EFFECT

Once the basic spectral types of stars have been recognized, and the normal wavelengths of their spectral fea-
tures have been measured, it is possible to look at fine details in spectra and acquire even more specific knowledge about individual stars. For example, there turns out to be a straightforward way to measure the speed at which a star is moving through space. The physical principle behind the measurement is based on the wave nature of light and is most easily introduced through an analogy with the familiar behavior of sound waves.

If you have ever stood alongside a highway or railroad track as a car or train moved past while blowing its horn, you have probably noticed a change in the pitch of the horn as the vehicle moves by. The pitch is high when the vehicle is approaching, but after it passes the pitch suddenly lowers. The changing frequency is called the **Doppler shift** or **Doppler effect**, and it is due to the wave nature of sound.

A sound wave is a disturbance that propagates through the air at a speed of about 300 meters per second (the speed of sound). The disturbance consists of waves of high and low pressure that follow each other in rapid succession, causing the eardrum to vibrate. (They are similar to the seismic P waves discussed in Chapter 8.) The more rapidly the eardrum vibrates, the higher will be the pitch of the sound we perceive.

In **Figure 14–11**, a sound wave is passing two observers, one stationary and the other moving toward the source of sound. In the figure, the lines correspond to the high-pressure areas while the low-pressure regions are halfway between. In any second of time, a larger number of high-pressure and low-pressure regions will pass the moving observer than will pass the stationary observer. As a result, the moving observer will hear a higher-pitched note than the stationary observer. Had

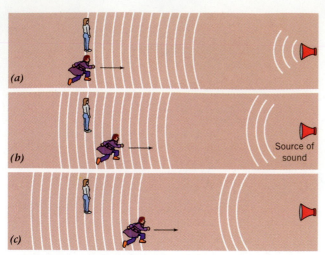

FIGURE 14–11. Sound wave passing by two observers (a) initially, (b) later, and (c) still later. More wave crests per second pass the moving observer, who detects a higher pitch than the stationary observer hears.

the moving observer been going *away* from the sound source, the effect would be the opposite: fewer regions of high and low pressure would pass by each second, and a lower-pitched note would be heard.

Consider a source of sound waves emitted in all directions. This is similar to a pebble being dropped into a pond and producing a circular wave (**Figure 14–12a**) moving outward in all directions from the point of contact with the water. After, say, one second the peak of the wave will be as shown. Now another pebble is dropped, producing wave number 2. During this time, the first wave has moved outward. Continue dropping pebbles until four waves have been produced and are at the positions shown in parts *a-d*. The space between each wave is the wavelength.

Now, let the source move toward a stationary listener at the bottom of the page. Figure 14–12e shows the first wave, which was produced with the source at position 1. Before the next wave is emitted, however, the wave source moves to position 2, as shown in part f. Each wave produced is circular about the position of the source at the time when the wave was emitted. The process of wave emission continues each second. The end result, in Figure 14–12h, shows how the regions of high pressure—called *wavefronts*—get bunched up in front of the moving source of sound and spread out behind it. The listener in the direction shown will hear a higher-pitched sound than a listener located in the opposite direction (toward the top of the figure).

From the discussion, you can see that a Doppler shift occurs only if there is motion toward or away from the observer. If the motion is across the line of sight between the object and the observer, there is no Doppler shift. If the motion is at an angle to the line of sight, only the component of motion toward or away from the observer contributes to the Doppler shift. Motion along the line the sight—that is, radial to the object—is known as the **radial velocity**. This term will be used frequently throughout the rest of the book.

Inquiry 14–15 What will be the pitch of sound perceived by a listener toward the right of the diagram in Figure 14–12, relative to observers at the top and bottom of the diagram? (Note that, at the instant shown, the source of sound is moving perpendicular to the line between it and the listener, so it is moving neither toward nor away from the listener.)

An analogous effect takes place with light. If an observer moves toward a source of light, or if the source moves toward the observer, an electromagnetic wave of

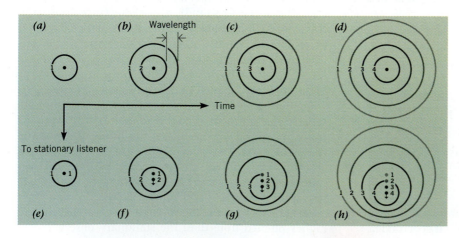

FIGURE 14–12. Sound waves bunch up in front of a moving source and are stretched out behind it, producing a Doppler shift.

a higher frequency (shorter wavelength) will be detected than if the relative motion between source and observer were zero. If the source and observer are moving away from each other, then the frequency detected will be lower and the wavelength longer. From Figure 14–12, it makes sense that the larger the relative speed between the source and the observer, the larger will be the resulting shift in wavelength. By observing the wavelength shift, we should be able to determine the relative speed of the source and the observer.

The amount of change in the wavelength produced by the relative motion of the source and the observer is easy to compute, using the *Doppler shift formula*

$$\frac{\text{relative speed between the source and observer}}{\text{speed of light}} =$$

$$\frac{\text{change in wavelength}}{\text{rest wavelength}}.$$

Expressed symbolically, we have

$$\frac{v}{c} = \frac{\Delta\lambda}{\lambda},$$

where the symbol $\Delta\lambda$ stands for the *change* in the wavelength (*not* some quantity Δ times λ), λ is the rest wavelength (the wavelength the wave would have if there were no motion), c is the speed of light, and v is the relative velocity of the source and the observer. The change in the wavelength, $\Delta\lambda$, is defined as the *observed* wavelength minus the *rest* wavelength. If the observed wavelength is greater than the rest wavelength, $\Delta\lambda$ is positive and the source and observer are separating. Because the wavelength has been shifted to the red, it is called a **redshift**. If the observed wavelength is less than the rest wavelength, $\Delta\lambda$ is negative and the source and observer are approaching each other, and the result is a **blueshift**.

The Doppler effect is readily seen in **Figure 14–13**, which shows the spectrum of a star that revolves about another, fainter, star. The figure is divided into four parts: the central part includes two spectra of the star taken at different times, while the top and bottom quarters include emission spectra from a lamp attached to the telescope. These spectral emission lines are at known, fixed positions and therefore define fixed reference positions. The upper stellar spectrum shows absorption lines to the blue (left) of the brightest reference lines, while the lower stellar spectrum shows the same lines redshifted. The changed positions are caused by the star's movement toward the observer when the upper spectrum was obtained and away from the observer at the time the second one was taken. (Each spectrum also includes a component of the Earth's orbital motion.)

The Doppler shift formula is accurate provided the speeds are not too close to the speed of light. Stars, in fact, move at only a small fraction of the speed of light. In words, the formula for the Doppler shift states that the fractional change in wavelength ($\Delta\lambda/\lambda$) is the same as the ratio of the relative speed between source and observer to the speed of light (v/c). For example, if the observed change in the wavelength were 1% of the rest wavelength, the relative motion of the objects would be 1% of the speed of light, or 3000 km/sec.

As an additional example, consider the case of an absorption line normally observed at 6000 Å. Suppose it is shifted to longer wavelengths (a redshift) by 1% to 6060 Å. The source and the observer therefore are moving apart at 1% the speed of light, or 3000 km/sec. If the 6000-Å line were shifted to shorter wavelengths (a blueshift) by 1% to 5940 Å, the source and the observer would be *approaching* one another at 1% the speed of light.

To use the Doppler shift formula to find speed requires knowing the rest wavelength. The rest wavelength is determined through measurements in a laboratory. Once the identity of a particular spectral line is known, its rest wavelength can be found in a book of wavelengths.

Here is another example. In the spectrum of M31, the great galaxy in Andromeda (see Figure 1–15), the line of hydrogen whose rest wavelength is 4861 Å is observed at 4856 Å. What is the speed of this galaxy relative to the Earth? Is it moving toward or away from us?

Because the observed wavelength is shorter than the rest wavelength, the light has been blueshifted, and the

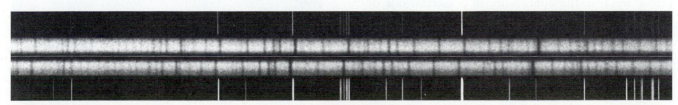

FIGURE 14–13. Spectra of α^1 Geminorum at two different times. The spectra are Doppler shifted relative to one another because of the orbital motion of the star.

galaxy must be moving toward us. The change in wavelength is 4856 Å − 4861 Å or −5 Å. Therefore, using the Doppler formula we find that

$$\frac{\text{relative speed between M31 and Earth}}{\text{speed of light}} = \frac{-5 \text{ Å}}{4861 \text{ Å}}$$

$$= -0.001.$$

The galaxy's speed relative to the Earth is therefore −0.001 times the speed of light or −0.001 × 300,000 km/sec = −300 km/sec, which is toward each other. At that speed, one could travel around the Earth in two and a quarter minutes! Yet this particular galaxy is a slowpoke compared with most of the galaxies in the universe. On the other hand, nearby stars are found to move somewhat more slowly than this, around 30 km/sec relative to the Earth. Note that the wavelength change produced by stars is so small that no noticeable change in the star's color results. A further numerical example using the Doppler shift formula is in Appendix A9.

Inquiry 14–16 In the spectra of certain stars, the wavelengths of all the lines shift from longer to shorter wavelengths and back to longer wavelengths in a regular fashion over the course of a few days. Can you suggest a possible explanation for this?

14.4
WHAT WE CAN LEARN FROM SPECTRAL LINES

We have seen that the strengths of lines in a stellar spectrum are affected by the temperature and chemical composition of the star, and how these physical parameters can be measured, in principle, by studying line strengths. We also have seen how the observed positions of lines can be used to measure the speed of the object toward or away from us. Not all spectral lines appear the same: some lines appear strong while others appear weak, some appear narrow while others appear broad. Because the appearance of a spectral line depends on a variety of a star's physical characteristics, much information may be learned through a detailed study of the *shape* of the spectral line. Such a detailed shape is known as a **line profile** (Figure 14–14). By studying the line profile, we obtain information about a variety of stellar properties: rotation, atmospheric density, atmospheric turbulence, and magnetic field. We discuss each of these next.

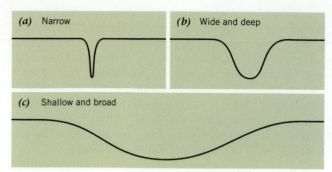

FIGURE 14–14. Line profiles of different characteristics.

THE ROTATION OF STARS

Although stars are too far away for us to see different parts of the actual surface, we can consider the light coming from different parts of the stellar surface. **Figure 14–15a** shows a spectral line formed by a nonrotating star. In **Figure 14–15b**, one side of the rotating star is approaching us; the radiation we receive from this side must be shifted by the Doppler effect to shorter wavelengths. The light from the star's other side must be similarly shifted to longer wavelengths, because that part of the star is moving away from us. Because the part of the star in the center moves perpendicularly to our line of sight, that light will not be shifted by the Doppler effect at all (unless, of course, the entire star is moving along the line of sight). When we observe the distant pointlike star, we observe a mixture of the light from all parts of the star, each shifted by its own amount. The net effect is to broaden the spectral lines compared with that for a nonrotating star. Thus, the greater the rotation speed, the broader the line. In principle we can reverse the

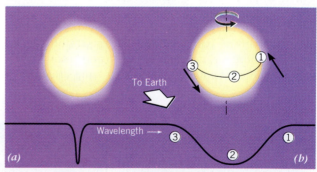

FIGURE 14–15. The non–rotating star on the left produces a narrow spectral line, while the rotating star on the right emits radiation that is Doppler shifted by different amounts across the stellar surface. The resulting spectral line is broadened.

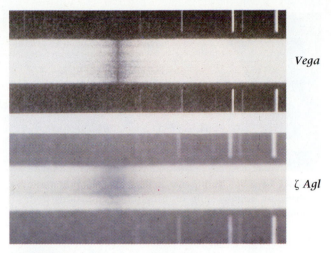

Vega

ζ Agl

FIGURE 14–16. The spectrum of Zeta Aquilae, a rapidly rotating star. The slowly rotating star Vega is shown for comparison.

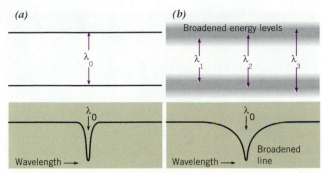

(a) *(b)*

FIGURE 14–17. Energy levels and line profiles of (a) an isolated atom and (b) an atom affected by its neighbors.

process; an observation of the width of the spectral line can provide information on the rotation speed (**Figure 14–16**).

DENSITY IN A STELLAR ATMOSPHERE

Thus far we have discussed absorption lines in stellar spectra as if they appeared at a single precise wavelength (which may, of course, shift position because of the Doppler effect). This description is actually oversimplified, because quantum theory tells us that spectral lines are not infinitely sharp.

An atom in a real gas (such as one in a stellar atmosphere) will continually interact with neighboring atoms. These collisions disturb the energy levels in the atom in random, unpredictable ways, partly as a result of the force of the collision itself and partly as a result of the electromagnetic effects of the charged particles in the neighboring atoms. **Figure 14–17** shows the effect of collisions on the atom. The energy level diagram on the left is for an isolated atom; it has sharp energy levels, and transitions between them produce a narrow line. The diagram on the right corresponds to an atom that is

strongly affected by its neighbors. At any time the energy levels might be farther apart than normal; at another time, closer. Transitions between the two disturbed levels shown can involve photons of greater or lesser energies and wavelengths than for undisturbed energy levels. Because a spectral line in a star is made up of transitions from a large number of atoms, instead of absorbing only photons of a wavelength close to the wavelength of an isolated atom (called λ_0 in the figure), the atoms can absorb photons of many different wavelengths in a range around λ_0. The result is a broadened line centered on λ_0 as a central wavelength.

This effect primarily depends on the *density* of a star's atmosphere. If the density is high, there will be many collisions between atoms and the energy levels will be much perturbed. In addition, the denser the material, the closer together the atoms are, on the average, and the greater the effect of one atom's electrical forces on another.

Inquiry 14–17 Figure 14–18 shows the spectra of two stars of the same temperature. One is for a dwarf star and the other is for a supergiant star (where the density of the atmospheric gases is thousands of times less than that in a dwarf star). Which is which?

Astronomers can distinguish between collisional broadening of spectral lines and the broadening caused

FIGURE 14–18. Spectra of two stars that have the same surface temperature but different atmospheric densities.

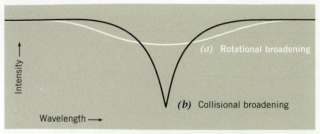

FIGURE 14–19. Line profiles for (*a*) rotational broadening and (*b*) collisional broadening. Note the sharper peak in the case of collisional broadening.

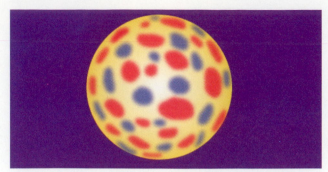

FIGURE 14–20. Turbulence in the atmosphere of a star is observed by Doppler shifts of rising (blue) and falling (red) regions.

by rotation discussed earlier because the detailed shape of the line is different in the two cases. **Figure 14–19** shows the theoretical shape of the line profile in each situation. Notice that the rotational broadening produces a smooth profile, whereas collisional broadening produces a profile with a sharp center.

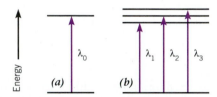

FIGURE 14–21. Zeeman effect. (*a*) No magnetic field. (*b*) Magnetic field present. Note the splitting of the upper energy level, giving rise to several closely spaced transitions.

TURBULENCE IN STARS

Figure 14–20 is a diagram that illustrates parts of a stellar atmosphere rising while others are falling. With such atmospheric turbulence, rising elements are moving toward the observer while falling ones are moving away. The result is that rising elements emit radiation that will appear at shorter wavelengths than normal, while falling ones will be at longer wavelengths. The end result of a large number of rising and falling elements, each moving with slightly different velocities, is a broadened spectral line. Because turbulent motions are relatively small, they are important only in highly detailed analyses.

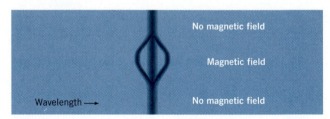

FIGURE 14–22. An absorption line in the Sun's spectrum. The upper and lower parts of the spectrum, where we observe a single narrow line, come from gas that is not in a magnetic field. The central area shows the splitting of lines because of the Zeeman effect discussed in the text.

MAGNETIC FIELDS IN STARS

If an atom is placed in a strong magnetic field, the energy levels in the atom will be split into several different, closely spaced levels (**Figure 14–21**). The larger the magnetic field, the greater the amount of splitting. Transitions can occur between the various levels, as shown in Figure 14–21*b*, which produces a corresponding splitting of the spectral lines. **Figure 14–22** shows the spectrum of a gas (the Sun) in which the top and bottom areas show no magnetic field and no splitting, while the central part shows the absorption line split into three parts. By measuring the amount of line splitting, astronomers can determine the strength of a star's magnetic field. (The splitting of spectral lines in a magnetic field is called the *Zeeman effect*.)

Except for those few bright stars that have strong magnetic fields and are observed with a high-resolution spectrograph, it is unusual to see the lines actually split. Instead, they usually merge together and appear broadened. Because the light from a Zeeman-split line is polarized (see Chapter 11) this kind of broadening can be distinguished from the types previously mentioned.

The Sun has a magnetic field, which changes with time, that is analyzed daily from observations of the sunlight's Zeeman effect and polarization. The observations show that regions of intense magnetic fields are associated with sunspots.

A FINAL EXAMPLE OF STELLAR SPECTRA

Let's pull many of the ideas of this chapter together. Suppose a star is in a state of rapid rotation. An example might be the star β Lyrae, whose equatorial regions are moving at hundreds of kilometers per second—so rapidly that material is actually flying away from the star's equator (**Figure 14–23**). Let's consider the spectrum of such an object.

Inquiry 14–18   What type of spectrum comes from the star itself?

Inquiry 14–19   What type of spectrum would you expect to see if you consider only the light from the edges of the gaseous ring?

Inquiry 14–20   What type of spectrum would you expect to see if you consider only that part of the ring directly between the observer and the star?

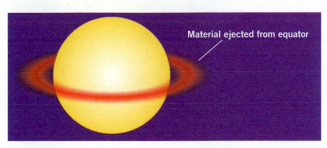

FIGURE 14–23.   Artist's conception of β Lyrae.

The resulting spectrum is diagrammed in **Figure 14–24**. Such a star and ring system would have a continuous spectrum with absorption lines from the star itself. Rapid rotation of the star would broaden the star's spectral lines. The radiation from the edges of the ring would be in emission, because the ring edge is "hot" compared with its background of cold space. Because of the ring's rotation, the emission line will be broader than if the ring were not rotating. Finally, the gas directly in front of the star will produce an absorption line in the middle of the emission line because this gas will be cool compared with the stellar photosphere behind it.

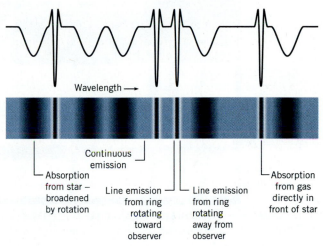

FIGURE 14–24.   A schematic spectrum resulting from a star surrounded by a rotating, nonexpanding shell.

DISCOVERY 14–1
CLASSIFICATION USING KNOWN SPECTRA

After completing this activity you will be able to:

- Classify an unknown spectrum to within one-half a spectral class, given a sequence of known comparison spectra.
- Discuss the characteristics of spectra of stars having different temperatures.

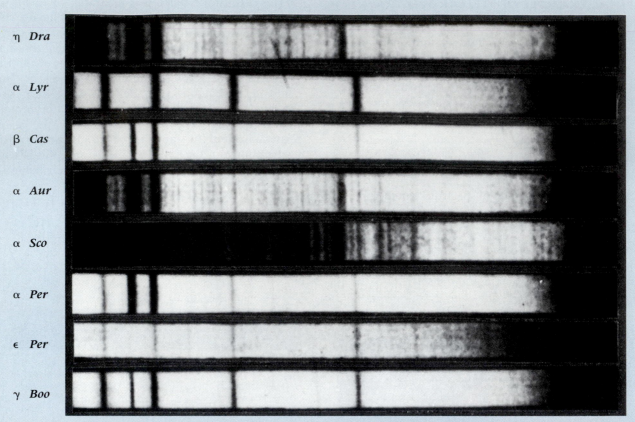

η *Dra*

α *Lyr*

β *Cas*

α *Aur*

α *Sco*

α *Per*

ε *Per*

γ *Boo*

DISCOVERY FIGURE 14–1–1.   Unknown spectra for classification, using the spectra in Figure 14–2 for comparison.

This exercise can be done using either photographic spectra or the more modern digital spectra in the Activity Kit, which were obtained with an electronic detector.

If you are using photographic spectra, compare each of the unknown spectra in **Discovery Figure 14–1–1** with the spectra of Figure 14–2. If you are using digital spectra, compare the unknown spectra in Kit Figure 14–1–1 with the standards in Kit Figure 14–1–2. None of the unknown spectra will be exactly like the comparison spectra, because they are for different stars; some will be somewhat similar and others will be quite different. Attempt to classify each spectrum by finding the standard spectrum that most closely resembles the unknown spectrum, feature for feature. Write down the spectral type of the spectrum that gives you the best match. If the unknown spectrum clearly lies about halfway between two known spectra, write down the spectral type that is about halfway between the two known spectra (e.g., if your unknown spectrum lies halfway between A0 and A7, then you could write down either A3 or A4).

Remember that you are looking for specific spectral features that characterize each spectral type. Are lines of neutral or ionized helium present when the hydrogen Balmer lines are weak or absent? Then the spectrum is an O or B type, if you identified the helium lines correctly. Are the hydrogen lines the dominant feature? Then the spectrum is probably type A. How about metals? Molecular bands? Then the star is cooler, probably type F, G, K, or M, depending on what spectral lines are present and how strong they are. In this way you can make a rough classification.

To match the spectra more exactly, you will need to look at more subtle characteristics. For example, in stars of types A to G, you can compare the strengths of the hydrogen Balmer lines at 4340 Å and the ionized calcium line at 3933 Å. The hotter the star, the weaker the calcium line will be in comparison to the hydrogen line. At spectral type A7, they are about equal in strength, whereas in F and G stars, the line of ionized calcium is stronger.

In a similar way, you can compare the strength of the neutral calcium line at 4226 Å and the group of iron lines at 4271 Å with the hydrogen line at 4340 Å. The iron and calcium lines will become more and more prominent as the star gets cooler. Similarly, the titanium oxide (TiO) bands near 4761 Å and 4944 Å can be compared with the calcium and iron lines if the star is very cool. In O and B stars, you can compare the strengths of the hydrogen, helium, and ionized helium lines.

Summarize your results. Include information about what features you found and the relative strengths of the various features that guided you to the classification you made.

- ☐ **Discovery Inquiry 14–1a** Summarize in your own words the characteristic features of each different spectral type, referring only to Figure 14–2. Then compare your answers with **Table 14–1** to check your understanding.
- ☐ **Discovery Inquiry 14–1b** What is the contradiction if a stellar spectrum exhibits both molecular lines and helium lines? How might such a contradiction be resolved?

CHAPTER SUMMARY

OBSERVATION

- Stars are observed to have a variety of different types of spectra. The types of spectra are given alphabetic names that we refer to as the **spectral type** or **spectral class** (OBAFGKM).
- Stars of spectral type O and B show lines of helium; type A stars have strong hydrogen lines; type F, G, and K stars have ever-stronger metal lines and ever-decreasing hydrogen lines. The M stars are dominated by absorption from molecules along with a large number of lines from neutral metals.
- Stars of spectral type O and B are observed to be hot stars, while stars of spectral type K and M are cool.
- Large stars with low atmospheric densities are observed to have lines that are narrow, while small stars with higher atmospheric densities have lines that are much broader. The broader lines in small stars come about because the atoms collide more frequently and disturb the energy levels.

THEORY

- The variety of spectra are interpreted with the help of atomic energy-level diagrams. The higher the temperature, the more electrons can be excited to higher atomic orbits, and the greater the **ionization** can be.
- The **Doppler effect** results from the wave nature of light; it is a shift in the wavelength of the observed radiation caused by the relative motion of a source and the observer. The motion that results in the wavelength shift is along a line connecting the source and observer; motion perpendicular to this line does not contribute to the observed shifting of the spectrum.
- Rotating stars produce broadened spectral lines.
- The turbulent motions of gases within the atmosphere of a star cause the line profile to broaden.
- Magnetic fields cause energy levels to split, thus splitting spectral lines into multiple lines.

CONCLUSIONS

- From a detailed analysis of the spectra of stars and galaxies, astronomers conclude that hydrogen and helium are the most abundant elements in the universe. All other elements in the universe, the metals, comprise only about 2% of the elements present.
- From a minute analyses of spectral **line profiles**, and comparison with theoretical models, we can de-

duce detailed information on the elemental abundances in stellar atmospheres.

- From an analysis of a star's spectrum, astronomers can infer many of the star's characteristics: its atmospheric pressure, whether or not it rotates and how rapidly, whether or not the atmosphere is turbulent, whether or not it has a magnetic field and its strength, and whether or not the star has a shell of gas surrounding it.

SUMMARY QUESTIONS

1. What are the standard spectral types in order from hottest to coolest? What are the main characteristics of each class (temperature range, color, prominent spectral features)?

2. Why do the principal features of spectra change from one star to another?

3. Why does relative motion between a source and an observer result in a change in the observed wavelength of emitted light?

4. What are three effects that alter the observed profile of a spectral line? Explain why each one changes the line profile.

APPLYING YOUR KNOWLEDGE

1. When lines of *ionized* helium are present in a spectrum, will the lines of *neutral* helium have the same or different strength when compared to a spectrum in which no ionized helium lines are present? Explain.

2. In what energy state must a hydrogen atom be for it to absorb light at a wavelength of 4861 Å? Describe how the atom might get to such an energy state.

3. What will happen to a photon of wavelength 1220 Å when it encounters a hydrogen atom?

4. Stars known as white dwarfs (studied in Chapter 18) have very high densities. How do you think spectral lines of a white dwarf might differ from that of a "normal," less dense star?

5. Consider a star with an expanding shell of hydrogen surrounding it. Make a drawing of the system, and then describe and draw the resulting spectrum by considering the radiation coming from the system's various parts.

6. It has been found that the lines of hydrogen in the spectra of galaxies tend to be observed at longer and longer wavelengths the farther the galaxy is from us. Interpret this observation in terms of what you have learned.

■ **7.** How rapidly would an object have to be moving

for its color to change from blue (4000 Å) to red (7000 Å) because of the Doppler effect? What (incorrect!) assumption are you implicitly making?

■ **8.** What would be the maximum contribution (in angstroms) of the Earth's orbital motion to the shift of a stellar absorption line? Assume a rest wavelength of 4000 Å, and that the Earth travels at some 30 km/sec around the Sun. What would be the maximum contribution from the Earth's rotational motion, which is 0.46 km/sec at the equator?

■ **9.** At what wavelengths will the "D" lines of sodium at a rest wavelength of about 5900 Å be observed in the spectrum of a star that is moving away from us at 90 km/sec? If it is moving toward us at 150 km/sec?

■ **10.** What would be the approximate spectral type of a star whose wavelength of maximum intensity was (*a*) 10^{-5} cm, and (*b*) 1.2×10^{-4} cm ?

■ **11.** A galaxy is observed to have two prominent lines in its spectrum at wavelengths of 5104 and 6891 Å. Assuming these are redshifted Balmer lines, what is the speed of the galaxy relative to the Earth? (Hint: If you have identified the lines correctly, you should get the same speed regardless of which line you see.)

ANSWERS TO INQUIRIES

14–1. The stars at the top are hot and thus emit more strongly in the blue; the stars at the bottom are cool, thus emit more strongly in the red and are deficient in blue radiation.

14–2. The Balmer lines involve the first excited state. In a low-temperature star, because few atoms will be in the first excited state, the Balmer lines will be weak if present at all.

14–3. There are no helium absorption lines.

14–4. The temperature is too low for helium to be excited.

14–5. Because absorption lines are caused by transitions of electrons, and because there are no electrons remaining attached to the nucleus, there will be *no* hydrogen absorption lines in this situation.

14–6. No. The presence of TiO means the temperature is very low; therefore, other elements may not become excited at such low temperatures.

14–7. The higher the temperature, the more frequent and violent are their collisions, which dissociate them.

14–8. It will increase the level of ionization, because more frequent and energetic collisions will ionize more particles.

14–9. The argument here is similar to the changing of the strengths of hydrogen lines explained in the text. The higher the temperature, the greater the amount of ionization and the stronger the lines due to ionized atoms.

14–10. (*a*) The second electron is closer to the nucleus and therefore more strongly bound. (*b*) There are fewer electrons to repel the second electron than to repel the first one.

14–11. These are lines from molecular oxygen in the Earth's atmosphere.

14–12. The helium Lyman lines should appear before the helium Balmer lines.

14–13. Saturn is cool, so most of the hydrogen is in molecular form. Because molecular hydrogen does not have a spectrum in the visual spectral region, we don't detect it.

14–14. Even though hydrogen and helium are lacking, the meteorites may give information about the proportions of the heavier elements in the universe.

14–15. There is no change in pitch.

14–16. Such stars are the so-called *spectroscopic binaries*, which consist of two stars in orbit around each other. The period is short enough and the relative velocities high enough that the change in Doppler shift is readily seen.

14–17. The spectrum on the bottom has broader lines and comes from the dwarf star.

14–18. A continuous spectrum with absorption lines from the stellar atmosphere; a "normal" stellar spectrum.

14–19. Because the gas would be a low-density gas that is hot compared with whatever is behind it, the spectrum would be an emission-line spectrum.

14–20. There would an absorption line from the ring superimposed on the stellar continuous spectrum.

Discovery Inquiry 14–1b. The presence of molecules requires a star with a low surface temperature, whereas helium lines require a star with high surface temperature. A single star cannot have both characteristics, so this apparently single star must be an unresolved binary system.

15

THE OBSERVED PROPERTIES OF NORMAL STARS

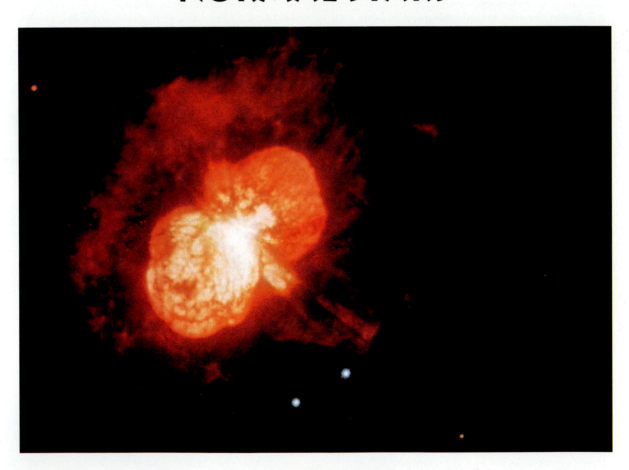

Twinkle, twinkle little star
How I wonder what you are
Up above the world so high
Like a diamond in the sky.

NURSERY RHYME

In this chapter we will discover some fundamental stellar properties by examining graphs of energy output versus surface temperature for a large number of stars. Such a procedure for comparing stars was first suggested early in this century by Danish astronomer Ejnar Hertzsprung and American astronomer Henry Norris Russell. These diagrams are so useful in exhibiting differences among the various types of stars that we will make constant reference to them from this point on.

We will see that from their positions on such a diagram, we can identify stars of vastly different sizes; this will lead us to an understanding of the interrelationships among the most fundamental characteristics of stars—their energy output, surface temperatures, radii, and masses. With this knowledge, we will then see how the distances of many stars can be estimated once their spectra have been studied.

THE HERTZSPRUNG-RUSSELL DIAGRAM

One of major goals of science is to make sense out of apparent chaos. Given the large number of stars in the sky, astronomers desire to understand them by searching for relationships between various stellar properties. This section examines perhaps the most important of all such relationships for stars.

A GRAPHING EXPERIMENT

To illustrate the procedure we will employ in this chapter, let us first discuss a graphing experiment that is analogous to the idea behind the Hertzsprung-Russell diagram and that is straightforward to interpret. In 1992, one of the authors requested all students in introductory astronomy courses at the University of Kansas to anonymously submit to him their sex, height, weight, and number of letters in their last name. In **Figure 15–1a** we have plotted the height of each person versus the number of letters in his or her last name. In the graph, each person is represented by a single point. In **Figure 15–1b**, we have plotted height versus weight.

In the first case, it is quite apparent that there is no systematic relationship between the two quantities plotted; such a diagram provides no useful information on the class population. In the second case, there is evidence of a systematic relationship. The taller a person is,

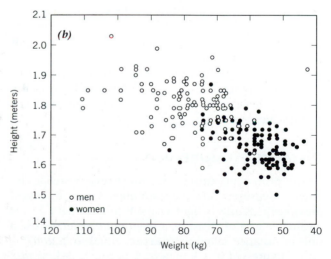

FIGURE 15–1. (*a*) Graph comparing the heights of a group of people with the number of letters in their last name. (*b*) Graph of the heights of the same group of people compared with their weights.

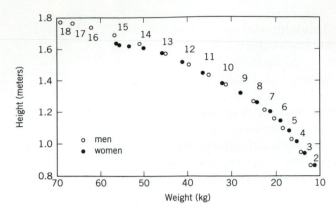

FIGURE 15–2. Graph that shows the evolutionary path of children's height and weight from age 2 to 18.

the heavier that person tends to be. There is, of course, some scatter; given any height, there will always be people of various weights. But the variations do not obscure the existence of the relationship shown in the diagram. In other words, the points do not simply scatter around the diagram but form a sequence of points within the graph that can provide useful information about the characteristics of people.

Furthermore, in Figure 15–1*b* we can even detect evidence of two different relationships, one for males and one for females. If you did not realize that there were any differences between the body types and structures of males and females, you could have discovered it by looking at this graph. (Where do you fall within the two graphs in Figure 15–1?)

More detailed analysis of such data shows a tendency for people's location in the diagram to change with time from areas of low weight (for a given height) to regions of higher weight. A graph of height versus weight for children from the ages of 2 to 18 is shown in **Figure 15–2**. This height/weight diagram demonstrates that children change as they age. We can say that as people change, they "move" in the height/weight diagram. This notion of a systematic change in the position within a diagram will be used for the discussion of stellar evolution in coming chapters.

DISTANCE MEASUREMENTS

Before proceeding further, we must review stellar distances. Measurements to distant objects were discussed previously in units of light-years. While the light year is a valuable unit of measure, astronomers generally use a unit of distance called the **parsec**, equal to 3.26 light-years. Expressed in kilometers, 1 parsec = 3.1×10^{13} km. The reason for defining the parsec, which we ab-

breviate pc, is that it is the distance to a star whose parallax angle is exactly 1 second of arc. (Rather than always saying parallax angle, astronomers generally speak of the parallax.) With this definition, the formula relating distance and parallax, which was presented in Section 5.6, is particularly simple:

$$D \text{ (in parsecs)} = \frac{1 \text{ AU}}{p \text{ (in seconds of arc)}}.$$

As an example, we measure the parallax of Alpha Centauri to be 0.75 seconds of arc. The distance is then 1/0.75 = 1.33 pc. Expressed in light-years we have 1.33 pc times 3.26 ly/pc = 4.3 ly.

Inquiry 15–1 If the smallest angle an instrument can accurately measure is 0.01 second of arc, what is the largest distance (in parsecs) for which this method can be applied?

Inquiry 15–2 If the instrument in the previous question were placed on one of Saturn's satellites, which is 10 AU from the Sun, to what distance could we accurately determine the positions of stars?

The parallax angle can be measured directly only for the nearest stars. Beyond about 100 parsecs (300 light-years) the distances become increasingly unreliable. Because the galaxy is some 100,000 light-years in diameter, it is evident that the distances to most of the stars in the galaxy must be determined by other means, which will be discussed later in this chapter and in subsequent chapters.

VARIATION OF STELLAR BRIGHTNESS WITH DISTANCE

When looking at the night sky, you cannot readily tell if the stars you see are relatively nearby or distant. All you can determine by looking is how bright a star appears to be; that is, what its apparent magnitude is. (Apparent magnitudes were first presented in Section 4.10.) It would be possible to determine a star's actual energy output, or **luminosity**, from its observed brightness if its distance were known. For nearby stars distances are known, because their parallaxes can be measured and the distances readily determined from the formula given above. Then their luminosities can be found.

The apparent brightness of a light source decreases with distance in accordance with the **inverse square law of light**. This law tells us that the amount of light received from a distant source decreases in proportion to the square of its distance from us. Suppose a star of luminosity L is at a distance d from Earth. It will have an

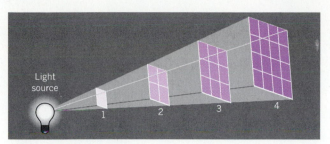

FIGURE 15–3. The inverse square law of light.

apparent brightness B. The inverse square law can then be expressed as

$$B \propto \frac{L}{d^2}.$$

An easy way to understand the inverse square law is through **Figure 15–3**. Although the light source emits photons in all directions, consider only those photons confined between the straight lines. Because photons generally travel in straight lines, no photons will enter or leave the confined region. A certain number of photons from the light source pass through an area at a distance of 1 unit from the source, as shown. At a distance of 2 units, the same number of photons pass through an area that is now twice as large on each side, or four times greater in area than at location *1*. Thus the apparent brightness, which is the number of photons passing through a unit area, is $(1/2)^2 = 1/4$ what it was before. At a distance of 3 units, the apparent brightness will be $(1/3)^2 = 1/9$ the apparent brightness at location *1*.

As an example of the inverse square law, suppose we observe two identical stars, one of which is 5 times closer to us than the other. The inverse square law of light tells us that the nearer star will *appear* to be 5^2 or 25 times brighter than the farther one. Or, to turn this around, again suppose we know from a detailed study of their spectra that the two stars have the same luminosity but we *observe* that one star appears 100 times brighter than the other. We then *deduce* that the apparently fainter star is actually 10 times farther away than the brighter one, because $\sqrt{100} = 10$.

Inquiry 15–3 Suppose two stars of equal luminosity are observed and one of them appears 50 times brighter than the other. If the nearer star is at a distance of 12 light-years, how far away is the distant star?

Inquiry 15–4 Suppose one star is 8 light-years from us and another 48 light-years away. If the two stars have the same luminosity, how many times brighter does the nearer star appear?

INTRODUCTION TO THE HERTZSPRUNG-RUSSELL DIAGRAM

You have already learned quite a bit about stars. You have seen that stars have a spectral type that can be determined by temperature, perhaps rotation and turbulence that can be determined from the Doppler effect, and perhaps a magnetic field that can be studied using the Zeeman effect. Furthermore, stars have radial velocities toward or away from the observer that can be studied with the Doppler effect. From the inverse square law, we can find a star's luminosity once its apparent brightness and distance are known. Are any of these characteristics related to any other? For example, **Figure 15–4** shows a graph of apparent magnitude plotted against spectral type. No relation is seen, and none is expected because a star's apparent brightness should not be determined primarily by its temperature.

Early in the twentieth century, Hertzsprung and Russell independently proposed a powerful method for comparing stars by plotting their luminosity against their surface temperatures. Now known as the **Hertzsprung-Russell diagram**, or simply the **H-R diagram**, this graph, which is perhaps the most important one in the field of astronomy, pulls together fundamental information on stellar temperatures, luminosities, radii, and masses. We will use the H-R diagram to find the distances to stars and to understand both stellar and galactic evolution.

We saw previously that the surface temperature of a star can be determined from Wien's law by measuring the star's color (as discussed in Chapter 13) or by determining its spectral type (as discussed in Chapter 14). Astronomers frequently use the spectral type rather than the temperature, because spectral type is what they determine directly when classifying a spectrum. In addition, the spectral sequence is easy to remember. On the other hand, temperature is so much more fundamental a quantity than spectral type that it is preferable to plot

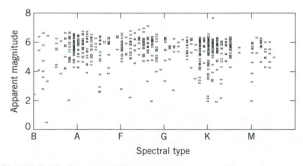

FIGURE 15–4. Graph showing an expected lack of correlation between apparent magnitude and spectral type for a large number of bright stars.

the H-R diagram using temperature. To emphasize the strong relationship between spectral type and temperature, in the diagrams that follow, temperature is on the bottom axis and approximate spectral type along the top.

We can find a star's luminosity when its distance is known. We often express it in terms of the Sun's luminosity. For example, the star Arcturus radiates approximately 100 times more energy than the Sun. Its luminosity in solar units is therefore equal to 100 $L_\odot$ where the symbol $L_\odot$ represents the Sun's luminosity. Because the Sun's luminosity in metric units is 3.8×10^{33} ergs[1] per second, it follows that the luminosity of Arcturus in metric units is 3.8×10^{35} ergs per second.

Inquiry 15–5   What would be the luminosity, in metric units, of a star of 5 $L_\odot$? Of a star of 0.2 $L_\odot$?

Inquiry 15–6   What would be the luminosity, relative to the Sun, of a star whose luminosity in metric units was 10^{31} ergs per second? Of a star whose luminosity was 8×10^{34} ergs per second?

THE NEAREST STARS

The nearest visible stars all have observable parallaxes. Because the distance in parsecs is given by $1/p$ (seconds of arc), the distances to these stars are known. Therefore, we can find their luminosities. **Figure 15–5** is the H-R diagram for the 51 stars *nearest* the Sun; it is a complete survey of all stars closer than about 15 ly. Many of these stars are double or triple star systems (i.e., several stars bound together by gravity), and in such cases each member of the system is plotted as a separate point. For each, the surface temperature (in kelvins) and luminosity (relative to that of the Sun) are plotted. Notice the point representing the Sun.

Notice, too, that in the H-R diagram the temperatures *decrease* from left to right. Furthermore, because this is a *logarithmic* graph (discussed in Chapter 13), the divisions shown are not equally spaced. Equal *distances* on the graph correspond to equal *ratios* of the quantity being plotted. For example, the interval between 3000 and 6000 K is equal to the intervals between 10,000–20,000 K and 20,000–40,000 K, because the temperature ratio in each case is two. Logarithmic graphs are required here because the luminosity axis in Figure 15–5 has a large range (10^{12}).

[1] An erg is a small unit of energy. It is the amount of energy you expend when lifting one milligram (0.001 gram) a distance of one centimeter.

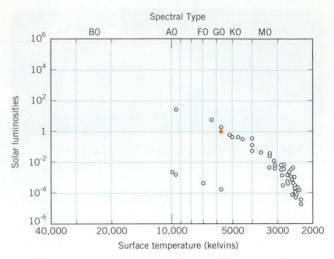

FIGURE 15–5.   The Hertzsprung-Russell diagram for the stars closest to the Sun. The Sun is denoted as an orange point.

Inquiry 15–7   Where on Figure 15–5 do most of the points seem to be found?

Inquiry 15–8   Can you identify a second group of stars that is distinctly separate from the first?

THE BRIGHTEST STARS

Most of the stars that appear brightest to the naked eye at night are close enough to the Sun to have observable parallaxes, and thus known luminosities. An H-R diagram for these stars is shown in **Figure 15–6**. In cases of multiple-star systems, we have selected and plotted only the brightest star of the group.

Inquiry 15–9   Can you identify any distinctly separate groups of stars among the ones plotted in Figure 15–6?

Inquiry 15–10   Can you identify one of these groups as being a natural extension of a group you discovered among the nearby stars?

Figure 15–7 combines the plots of both the nearest and the brightest stars.

Inquiry 15–11   Considering *all* the stars in Figure 15–7, approximately what is the ratio of luminosity

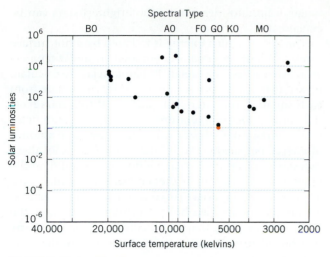

FIGURE 15-6. The Hertzsprung-Russell diagram for the brightest stars in the night sky. The Sun is denoted as an orange point.

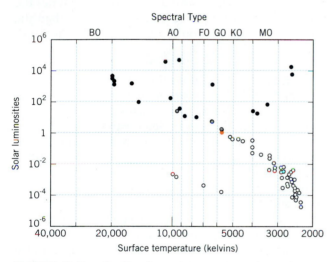

FIGURE 15-7. The data from Figures 15–5 and 15–6 have been plotted together in this diagram. Note the separation into distinct groups.

between the most luminous and least luminous stars plotted?

Inquiry 15–12 Are the stars nearest the Sun the ones that appear brightest in the sky? How do you explain this?

Inquiry 15–13 Compare the luminosities of the stars having temperatures near 3000 K. Within this group, approximately what is the ratio of luminosity between the most luminous and least luminous stars?

INTERPRETING THE HERTZSPRUNG-RUSSELL DIAGRAM

There is a wealth of information in the diagrams you have been inspecting, and now is a good time to isolate some of the important characteristics of the stars displayed. It is immediately apparent from the H-R diagrams that the two groups of stars, near and bright, are very different in nature; only three objects (Sirius, Procyon, and α Centauri) appear on both graphs.

Inquiry 15–14 Are the brightest stars we see in the sky that way because they are nearby?

The brightest stars in the sky are definitely *not* the nearby stars—they are, instead, extremely luminous objects, so intrinsically bright that even though they are not nearby they are still the most conspicuous objects in the night sky. From the diagrams you can see that the ratio of the luminosity of the brightest to the faintest star shown here is greater than 10^{10}—ten billion—and the actual ratio is even greater, because there are stars even more and less luminous than those shown.

Because most of the brightest stars in the sky are superluminous objects at various distances, it should not be difficult to convince yourself that the diagram of the *nearest* stars is more representative of the *average* population of stars. The plot of the nearest stars includes the statistics on *all* the known stars within a certain volume of space—a sphere centered on the Sun whose radius is 15 ly. It is a nearly complete sample.

By analogy, you would probably get a better idea of the characteristics of U. S. citizens by surveying everyone in your hometown rather than those people whose names appear in the headlines of the newspaper. The people in the headlines are generally atypical in some respect, which is why they got there in the first place.

Inquiry 15–15 There may still be stars within a 15-ly radius that have not yet been detected, so the plot of Figure 15–5 may not be complete. What do you think would be the most likely reason for such a star not to have been detected?

Inquiry 15–16 In arguing that the stars near the Sun are probably typical of the average stars in our galaxy, what fundamental (post-Copernican) assumption are astronomers making?

Inquiry 15–17 Make this assumption, and compare the number of stars in Figure 15–5 that are fainter than the Sun to the total number of stars. Roughly what proportion of stars in our galaxy are fainter than the

Sun? (In making your counts, count each component of a multiple-star system as a separate star.)

It is apparent from the H-R diagram that the super-luminous stars are relatively rare objects and that the majority of stars fall in the lower (less luminous) part of the diagram. In fact, the fainter types of star are more numerous (at least within the limits of the diagram, a point to which we will return).

15.2
STELLAR RADII: MAIN-SEQUENCE, GIANT, SUPERGIANT, AND WHITE DWARF STARS

In examining these H-R diagrams, we clearly see that some parts of the graph are heavily populated with stars, while other parts are empty. The most obvious feature of the H-R diagram is the band of stars running diagonally across it from the upper left to the lower right; this band has come to be known as the **main sequence**. It is evident that the majority of stars are found in this region of the diagram and thus are called main-sequence stars. Indeed, because the nearby stars are representative of average stars in our galaxy, and most of them fall on the main sequence, the obvious conclusion is that most stars in our galaxy are main-sequence stars. Later, when we consider the evidence that stars evolve and change their characteristics with time, we will also see an additional significance to this clustering in the diagram; namely, that the main sequence is where stars spend most of their lives.

There is, however, a second group of stars that makes up a fair proportion of the nearby stars and that therefore must be a fairly common type of star in the universe. This is the group below the main sequence in the diagram, called **white dwarf** stars (for reasons that will soon become apparent). These stars are the most important and populous category of non-main-sequence stars.

Inquiry 15–18 What percentage of stars in our galaxy do you estimate are white dwarf stars? Use information in Figure 15–5.

When we consider the *brightest* stars, we find that many of them are *not* main-sequence stars. Notice that among the stars with temperatures near 4000 K, there are some with luminosities around 100 times the Sun's and others with luminosities only 1/100 that of the Sun. We need to understand how this ten-thousandfold range of luminosities among the brightest stars can be explained.

The rate at which energy is radiated by a hot surface is related to its temperature. For a given stellar temperature, we can use the Stefan-Boltzmann law to calculate how much energy per second the star will radiate from each square centimeter of its surface. If two stars have the same surface temperature, then a square centimeter on the surface of each star will radiate the same amount of energy per second. So it follows that if one of these two stars radiates more *total* energy per second than the other, the star with the higher energy output must have a greater surface area. To consider a more "earthly" analogy, compare two hot plates that are each at the same temperature; that is, equally red. The only way one hot plate will radiate more total energy than the other is if it is larger.

In comparing stars with a 4000-K temperature, we see that a square centimeter on the surface of each of these stars radiates the same amount of energy per second. The stars that have a luminosity 100 times the solar value must simply be much larger stars than the other group that had only a hundredth of the Sun's luminosity.

The stars in the luminous group have come to be called **giants** to distinguish them from the more common stars on the lower part of the main sequence. Arcturus is a familiar example of a giant star.

This same argument can demonstrate the existence of even larger stars in the diagram. Notice that there are some stars with surface temperatures about 3000 K that are over a million times more luminous than main-sequence stars of this same temperature. Betelgeuse and Antares are two of them. It immediately follows that these stars must have over a million times the surface area of the main-sequence stars. Because of their enormous size, these stars have come to be called **supergiants**. If the supergiant Antares were to replace the Sun in our solar system, it would swallow up the Earth and extend all the way out to the orbit of Mars!

Stars on the main sequence are also generally called **dwarfs**. Dwarf stars should not be confused with the dramatically different white dwarfs. Thus we have divided stars into three **luminosity classes**: dwarfs, giants, and supergiants. These distinctions will help us find the distances to stars.

The relationship between the luminosity, temperature, and size of a star can be expressed in a simple way: The luminosity is the total energy per second radiated by the star and is given by the formula

luminosity = (energy per second radiated by a square centimeter of the star's surface) × (the total surface area of the star).

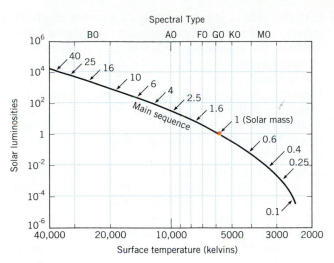

FIGURE 15–10. The main sequence, showing the locations of stars of various masses.

limits. Surprisingly enough, there are theoretical reasons to believe that neither case is possible. In particular, an object much more massive than the upper limit would be unstable; it would eject enough matter to bring it within the observed range. An object much less massive than 0.1 solar mass will not become self-luminous; that is, it cannot produce enough energy to become a star. Given that there are some uncertainties in the measurement of both the highest and lowest masses, we can say that the range of stellar masses runs from approximately 0.1 to 100 solar masses.

There is a sound physical reason for the existence of the main sequence. For normal main-sequence stars mass determines a star's temperature and radius, and thus its luminosity: the more massive the main-sequence star, the more luminous it is. In Chapter 16, where we consider the internal structure of stars, we will see that this relationship exists because of nuclear

reactions taking place in the centers of stars. In the meantime, we will simply take it as an observed fact. If we use the masses and luminosities from Figure 15–10 and plot a graph of these quantities, the result is **Figure 15–11**, which is called the **mass-luminosity relationship**. Like the Hertzsprung-Russell diagram, it is a logarithmic graph, and it shows that the increase of luminosity with mass is rapid. For example, a star only twice as massive as the Sun is roughly 10 times as luminous. Mathematically, astronomers find $L \propto M^4$ where the mass and luminosities are expressed in units of the Sun.

It follows that the stars in the upper part of the main sequence are more luminous than the Sun because they are more massive. Their radii are all roughly three times the Sun's, so they lie along a constant-radius line just above the line that runs through the Sun. As we travel down the main sequence, the stars become less and less massive, and hence less luminous. At roughly the mass of the Sun, the decrease of luminosity with mass becomes more gradual; we will see the reason for this in a later chapter. At the same time, the radii of the stars on the main sequence start to decrease, so the main sequence dips below the line for one solar radius and into the region of cool stars called **red dwarfs**.

Inquiry 15–24 About how many times more massive is Plaskett's star than the least massive observed stars?

Inquiry 15–25 Which are most numerous in our galaxy: stars of high mass or stars of low mass?

DENSITIES OF STARS

We have just seen that the range in the masses of stars is relatively modest, whereas the range in their diameters is large. It follows that the range in stellar *densities* (i.e., mass per unit volume) is enormous. For example, the diameter of a typical white dwarf is roughly 1/100 that of the Sun's (that is, the size of the Earth), yet the star may have the same mass as the Sun. Because the volume of a spherical star is proportional to the *cube* of its diameter, the volume occupied by a white dwarf is only $(1/100)^3$, or one-millionth that occupied by the Sun. If the same amount of mass is packed into only one-millionth the volume, the density must be a million times as great. Because the average density of the material in the Sun is about that of water, the density of the material in a white dwarf must therefore be roughly one million times that of water. A cubic centimeter, if brought to Earth, would weigh over a ton! Similarly, the material in a red supergiant is, on the average, only one-millionth as dense as water.

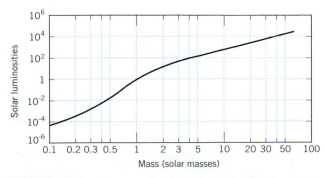

FIGURE 15–11. The mass-luminosity relationship for main-sequence stars.

15.4

STELLAR DISTANCES AND THE H-R DIAGRAM

Hertzsprung and Russell took data supplied by others and drew conclusions about stars from the data. Their work was possible only because accurate parallaxes using photographic techniques were becoming available, and Annie Cannon was publishing her spectral types in the *Henry Draper Catalog* of stellar spectra.

One of the first applications of the Hertzsprung-Russell diagram was to determine the distance to stars that were too distant to exhibit an observable parallax. Although parallaxes (and therefore distances) can be directly observed only for the nearest stars, if it were possible to estimate a star's luminosity by some means, then from its observed brightness one could infer the distance using the inverse square law of light.

We can see from the formula in Section 15.2 that to determine the luminosity of a star we must first be able to estimate both its surface temperature and its size.

Inquiry 15-26 Assuming that a star is a main-sequence star and that its temperature has been measured to be 10,000 K, approximately what would be its luminosity relative to the Sun? Use one of the H-R diagrams discussed earlier to determine your answer.

We cannot assume, as done in Inquiry 15–26, that a given star is necessarily a main-sequence star. Fortunately, it is possible to distinguish among luminosity classes (main-sequence, giant, and supergiant stars) by carefully studying their spectra. Distinguishing between the groups is possible because of the enormous range in the densities of stellar atmospheres, which is accompanied by a similar range in the atmospheric pressures. Thus, as we saw in the previous chapter, the low-density stars that we now call supergiants have narrow spectral lines, while the high-density ones, the dwarfs, have broadened lines (see Figure 14–18). This technique works well when hydrogen lines are present; the situation is more complex in stars having weak or no hydrogen lines in the spectrum.

A star's spectrum thus tells us both the star's surface temperature and its luminosity class. With these two pieces of information, we can tell approximately where the star falls on the Hertzsprung-Russell diagram. From its location on the H-R diagram, we infer the star's luminosity. While it is true that there will be some uncertainty in this inference, an imperfect result is better than none at all. Then, by comparing its inferred luminosity

to its observed brightness (which is a quantity that is relatively easy to get), we can estimate the distance to the star from the inverse square law of light. A distance determined in this way, by using the *observed* spectral type and luminosity class, and the H-R diagram to infer the luminosity, is known as a **spectroscopic parallax**. The name is something of a misnomer, because a parallax is an angle, but the word is used here as a synonym for distance. The spectroscopic-parallax technique is summarized in **Figure 15–12**.

Spectroscopic examination of a star and the use of the H-R diagram to infer its luminosity are the keys to this method. In using the H-R diagram, we are assuming that the properties of the star whose distance we wish to find are similar to those of the vast majority of other stars. Although direct parallax measurements are valid only to, at most, 300 light-years from the Sun, the spectroscopic method can be used on any star for which a good spectrum can be obtained and classified with accuracy, excluding those stars that are "peculiar" in some way. Astronomers are therefore able to determine distances to a large number of stars.

As an illustration of the spectroscopic-parallax method of finding distance, consider the star α Centauri. Spectroscopic observations of its absorption lines show that it is similar to the Sun (a main-sequence, G-type star) and, in fact, is only slightly more luminous. Direct measurements, however, show that the much-closer Sun *appears* much brighter—almost 9×10^{10} times

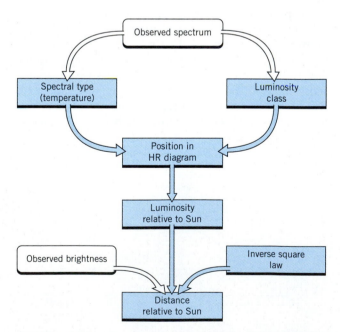

FIGURE 15–12. The process of determining stellar distances using the spectroscopic-parallax method.

brighter than α Centauri. From the inverse square law, the distance is given by the square root of this brightness ratio. Therefore, we find that α Centauri is nearly 3×10^5 times farther away from us than the Sun. This distance of 3×10^5 AU may be expressed in other units as follows:

$$\frac{(3 \times 10^5 \text{ AU})}{(2 \times 10^5 \text{ AU/parsec})} = 1.5 \text{ parsecs}$$

$$(1.5 \text{ parsecs}) \times (3 \text{ light years/parsec}) = 4.5 \text{ ly}$$

$$(3 \times 10^5 \text{ AU}) \times (1.5 \times 10^8 \text{ km/AU}) = 4.5 \times 10^{13} \text{ km}.$$

Inquiry 15–27 How many orders of magnitude are there between the numbers used to express a distance in parsecs and the same distance expressed in astronomical units?

On the other hand, measurements of Canopus show it to appear just about as bright as α Centauri, and yet its location in the Hertzsprung-Russell diagram tells us it is about 1200 times as luminous. Again we use the inverse square law: because the square root of 1200 is about 35, it follows that Canopus is about 35 times farther from us than α Centauri, or about 160 light-years. (Readers desiring a more general mathematical treatment of spectroscopic parallax may refer to Appendix A, part 10.)

15.5
BINARY STARS

William Herschel was the first astronomer to demonstrate from continued observations that there were such things as stars orbiting other stars under their mutual gravitational attraction. In fact, we now know that more than half the stars in the vicinity of the Sun are members of multiple-star systems. Binary systems are particularly valuable to astronomy, because it is from them that virtually all *direct* measurements of stellar masses, and many stellar-diameter measurements, are made. For these reasons, we now examine three types of binary-star systems.

VISUAL BINARIES

Binary systems in which both components can be seen are called visual double or **visual binary stars** (Figure 15–13). The discovery of such pairs gave astronomers their first clue that stars had a large range in luminosity. For example, because both of the stars on the left side of each photograph in Figure 15–13 are at the same distance, the brightness difference they exhibit is their actual ratio of luminosities. Not all stars that appear close together are gravitationally connected, however. Some pairs of stars, which we call *optical doubles*, appear to lie near each other in the sky but are actually at different distances from Earth.

Visual binaries give astronomers their best information about the masses of stars. Kepler's third law (Chapter 5), as modified by Newton to account for the masses, is used for this purpose:

$$(M_1 + M_2)P^2 = A^3.$$

Remember that the masses are in solar mass units; P is the orbital period in years, and A is the semi-major axis of the elliptical orbit in astronomical units.

The period of the two stars in their orbit is observed directly, and the average separation between the stars can be worked out from the observed angular size of the

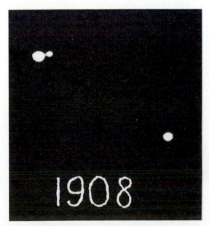  

FIGURE 15–13. The visual binary Krüger 60, photographed in 1908, 1915, and 1920.

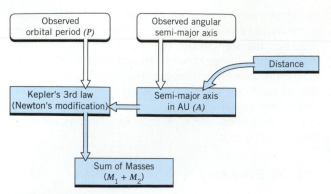

FIGURE 15–14. The process of obtaining information on stellar masses using observations of binary stars.

SPECTROSCOPIC BINARIES

In a binary system that is too distant or in which the stars are too close together to be resolved, direct viewing of each star is not possible. The presence of a binary system may be inferred, however, from a periodic Doppler shift observed for the spectral lines as the stars orbit one another. Such a system is known as a **spectroscopic binary**. Sometimes, when the spectrum of what appears on a photograph to be a single star is analyzed, two spectra are seen to be superimposed, showing two stars to be present. The spectrum of the star Mizar at two different times is shown in **Figure 15–15**.

orbit and the distance to the star system.[2] Therefore, if the distance to the binary is known, we can compute the sum of the masses of the two stars. **Figure 15–14** summarizes this procedure for obtaining information about stellar masses.

Finally, observations of the binary system over many years can, in some cases, provide the ratio of the masses of the two stars. The mass ratio, when combined with the sum of the masses from Kepler's third law, allows us to solve for the mass of each star.

> YOU SHOULD DO DISCOVERY 15–1, VISUAL
> BINARY STARS, AT THIS TIME.

[2]The separation of the stars in astronomical units is related to the observed angular separation and the distance through the angular-size–formula introduced in Chapter 3. An extremely useful form of this equation is

$$\text{angular separation (seconds of arc)} = \frac{\text{linear separation (AU)}}{\text{distance (parsecs)}}.$$

To understand these spectra, we consider two such stars in various positions in circular orbits as shown in **Figure 15–16a. Figure 15–16b** contains schematic spectra at each position. At the bottoms of Figures 15–16b are comparison spectra, which are emission spectra of a lamp attached to the telescope to provide lines of known wavelength for reference. In the stellar spectra, the spectral lines for each star are noted.

At time 1, the stars are moving along the line of sight from the observer to the stars. Because star A is moving away from the observer, its spectrum will be Doppler shifted toward longer wavelength while the lines for star B will be blueshifted, as shown in position 1 of Figure 15–16b. When the stars have moved in their orbits to position 2, the stars are moving across the line of sight; neither star shows a Doppler shift, and the spectral lines from both stars coincide. At position 3, the situation is the reverse of position 1; star A shows a blueshift and star B, a redshift. Finally, the stars will again move across the line of sight (time 4), and there will again be no Doppler shift of the spectral lines.

$\lambda 4415.1\text{Å}$ $\lambda\,4481\text{Å}$ $\lambda 4526.6\text{Å}$

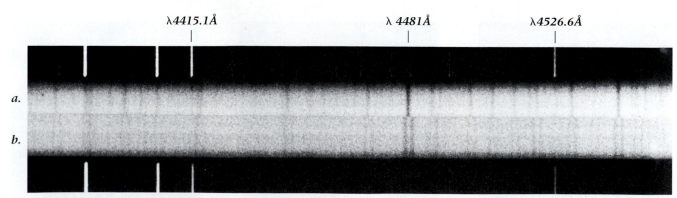

a.

b.

FIGURE 15–15. Spectra of the spectroscopic binary Mizar. (a) The spectral lines are superimposed when the stars are moving across the line of sight. (b) The spectral lines are separated when the stars move toward or away from the observer. Note the comparison spectra above and below the stellar spectra.

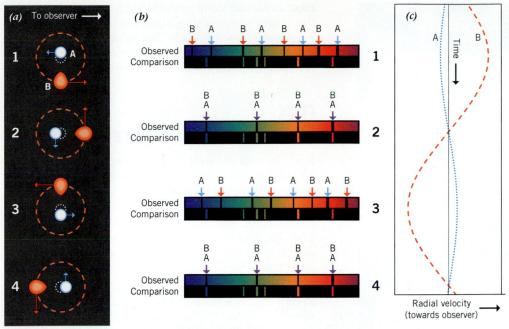

FIGURE 15–16. (a) Two stars in circular orbits in a binary system at different times. (b) Schematic spectra of the stars. The symbols for the spectral lines for each star are the same as for the stars in part a. (c) The radial velocity curves resulting from the spectra in part b.

A star system like the one just described is called a double-lined spectroscopic binary—one in which the spectral lines of both components can be observed. From the observed Doppler shift, the radial velocity of each star at every position in the orbit is computed using the Doppler shift formula. The results are then graphed, producing what astronomers call a **radial velocity curve** (Figure 15–16c). Because the stars remain opposite the center of the orbits at all times, both stars have the same period. For this reason, because the outermost star has the greater distance to go during the orbital period, its velocity is larger, as shown in the radial velocity curve. Furthermore, the less massive star moves faster than the more massive one. From these considerations we find that the ratio of the observed velocities of the stars equals the ratio of the masses of the stars. Thus spectroscopic binaries are important to astronomers because of the information they provide about masses.

Inquiry 15–28 Refer to Figure 15–16c. Suppose that this radial velocity curve showed one star (call it star A) to have a maximum radial velocity of 25 km/sec while the other star (B) has a maximum velocity of 125 km/sec. What is the mass of star A relative to that of star B?

Whenever one star in a spectroscopic binary system is more than about 3 magnitudes (15 times) fainter than the other, only one spectrum will be visible. However, because both the visible and invisible objects still move about a common point, the spectral lines of the brighter star may be seen to shift back and forth periodically. In such a single-lined spectroscopic binary, detailed analysis allows us to infer the mass of the *unseen* star. Astronomers use such techniques to search for black holes.

> YOU SHOULD DO DISCOVERY 15–2,
> SPECTROSCOPIC BINARY STARS,
> AT THIS TIME.

ECLIPSING BINARIES

Sometimes, one star will pass in front of the other, producing what we call an **eclipsing binary**. The archetype of this kind of star is β Persei, or Algol, meaning "the demon star" in Arabic. The brightness of an eclipsing binary is normally quite constant, but every few days it suddenly dims for a short time and then brightens again (**Figure 15–17**). The variation of the brightness with time produces a **light curve**. The shape of the light curve of an eclipsing binary is distinctive and is easily explained by the fact that periodically one of the stars moves in front of the other, partially or completely

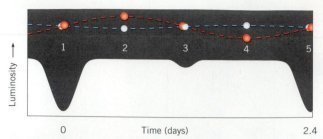

FIGURE 15–17. The light curve of an eclipsing binary. Notice that the light curve is constant except during the eclipses.

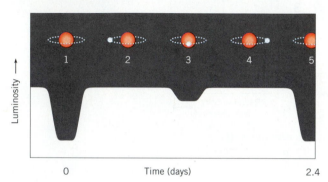

FIGURE 15–18. The cause of light variations in an eclipsing binary.

cutting off the light of the second star (**Figure 15–18**). Approximately half an orbit later, the second star eclipses the first one in a similar fashion, producing a secondary drop in brightness.

From a detailed study of the light curve of an eclipsing binary, we can deduce many details about the orbit and the eclipsing stars themselves. In particular, we can determine the diameters of the stars, in terms of the average distance between them.[3] Stellar radii measured in eclipsing binary systems are in good agreement with those inferred from the formula that relates radius, luminosity, and temperature; this agreement gives us confidence that the radii determined by this indirect means are reliable.

For stars that are both spectroscopic *and* eclipsing binaries, it is possible to find the mass of each star in the binary system. Because the mass of a star is the single

---

[3]Sharp-eyed readers may have noticed the differences between the light curves in Figures 15–17 and 15–18. These differences are due to different relative sizes of the two stars in the systems that produced the light curves shown. The stars that produced the light curve in Figure 15–17 are of nearly equal size whereas there is a substantial difference for the stars in Figure 15–18.

most important thing we need to know to understand it, the binary systems that allow us to determine stellar masses are prized finds in astronomy. It was from the masses of binary stars, for example, that the mass-luminosity relation was originally discovered.

15.6
RECENT STUDIES OF STELLAR SIZES

Knowledge of the sizes of stars is necessary for complete knowledge of stellar properties. We have seen that some direct information on stellar sizes can be obtained from binary stars, and some indirect information from the H-R diagram. Technology has provided some additional methods, two of which we now mention.

LUNAR OCCULTATIONS

Recent developments in instrumentation for observing stars have opened up new channels for obtaining information on stellar sizes. In the 1970s, considerable effort was put into the study of events called *lunar occultations* in which the Moon passes in front of a star. From observations of these occultations, direct measurements of stellar diameters are sometimes possible.

For years it was assumed that when the sharp edge of the Moon passed between us and a star, the light from the star would extinguish instantly. But the advent of modern photometric instruments and electronics has made it possible to take an accurate reading of stellar brightnesses every thousandth of a second or so. As a consequence, we can measure the actual duration of time it takes for the Moon's edge to move across a tiny stellar disk. Because the Moon is outside our atmosphere and has no atmosphere of its own, there is no blurring of the data involved anywhere along the line of sight.

Figure 15–19 illustrates what was seen when astronomers at the University of Texas applied these superfast photometric techniques to a lunar occultation of the star λ Aquarii. The dotted curve is what we would expect to see if the star really were an infinitesimally small point source; the ups and downs are the *diffraction pattern* associated with a ray of light coming from the star passing near a straight obstacle (the lunar edge). This is exactly analogous to your observations of diffraction in Chapter 11.

In Figure 15–19, notice the curve that is actually seen for the star λ Aquarii; the pattern is somewhat smoothed and broadened. The light does not fall off as rapidly as it would for a point source. In fact, the stellar trace is exactly what we would expect to see from an occultation of a star with an apparent diameter of 0.0074 seconds of

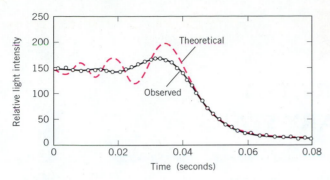

FIGURE 15–19. Occultation light curve of λ Aquarii. Shown are a theoretical curve (dashed line) for an infinitesimally small point source, and the observed light curve (solid line).

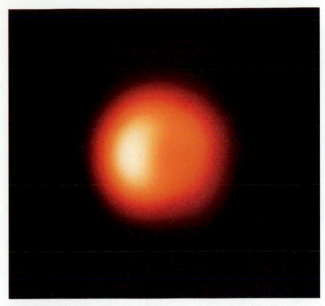

FIGURE 15–20. The resolved disc of Betelgeuse, showing a hot spot.

arc—the diameter of a penny seen from 300 km away. An independent determination of the distance of the star (say, measuring a parallax for it) would allow us to convert this apparent size into a linear size in kilometers.

OPTICAL INTERFEROMETRY

We explained the principles of radio interferometry in Chapter 12; these same principles can be applied at optical wavelengths, although the instruments are more difficult to use and the data harder to interpret. For a few very large, nearby stars the measurements have been successful. With a radius 640 times that of the Sun, Betelgeuse is the largest star that has been measured in this way. Recent work has provided a remarkable image of Betelgeuse and revealed a disc 0.05 second of arc in diameter (**Figure 15–20**). Furthermore, Betelgeuse is the first star other than the Sun on which surface markings have been seen. The figure shows a "hot spot" or bright area that is hypothesized to be caused by a small, hot companion star. The smallest star that has been observed in this way is Arcturus, with a radius 23 times that of the Sun. The only reason Arcturus could be measured in this way is that it is relatively near the Sun (only 35 light-years), so that its angular size is relatively large.

Technological advances in recent years, and those expected in the near future, will allow optical interferometric determination of the diameters of many more stars. Anticipation of these advances is the reason for the layout of the European Southern Observatory's telescope shown in the photograph at the opening of Chapter 12, and for construction of the 10-meter Keck 2 telescope nearby the Keck 1. Not only will these telescopes directly measure radii for many more stars, but changes in stellar diameter will be observable for certain stars whose sizes vary.

DISCOVERY 15–1
VISUAL BINARY STARS

When you have completed this Discovery, you should be able to do the following:

- Determine the period of stars in a visual-binary system from a series of photographs.
- Determine the scale of a photograph, given the angular separation of two objects in the photograph.
- Determine the separation of two stars in a binary system given their observed angular separation and the distance of the stars from the Sun.
- Determine the sum of the masses of the stars in a visual binary system.

Figure 15–13 contains photographs of the binary-star system Krüger 60 taken in 1908, 1915, and 1920. We will refer to the brighter star in the binary system as the *primary*, and the fainter one as the *secondary*. The star in the lower right corner is simply a star within this field of view. The motion of the secondary relative to the line between the primary and the field star is clear.

Note: This activity calls upon many pieces of knowledge from earlier in the book. You may want to review Section 3.3, the end of Section 5.6, and the discussion of distance at the beginning of this chapter. A drawing may help you visualize the relationships of the stars in the system and yourself.

You should make a photocopy of Figure 15–13 or use tracing paper rather than drawing over the original.

You will first determine the period of the binary system. Start by drawing a line through the primary star parallel to the left edge of the photograph. This line is your reference line. Repeat for each photograph. Place a pencil mark at the center of each primary and secondary star image and draw lines connecting the primary and secondary stars. Your lines should extend well past the secondary star. Use a protractor to determine the angle at the primary between your reference line and the line to the secondary. Tabulate the angle for each of the dates. Determine the angle through which the star has moved between each date.

☐ **Discovery Inquiry 15–1a** In a complete cycle the secondary star will travel through an angle of 360 degrees. Using the information about the motion of the secondary derived above, determine the period of the binary star system. (We are making the valid assumption that all the photographs are from a single period.)

To determine the scale of each photograph—that is, the number of seconds of arc on the sky in each millimeter on the photograph—use the fact that in 1908 the field star to the lower right was 30 seconds of arc from the primary. (Although the scale in each photograph is the same, it appears different because the stars themselves have moved.)

☐ **Discovery Inquiry 15–1b** Using this number and a ruler, determine the angular separation *in seconds of arc* between the primary and secondary components for each photograph.

We will assume that the average of your measured angular separations equals the semi-major axis of the elliptical orbit of the secondary around the primary. Although this assumption is not necessarily true, it is sufficient for our purposes here.

Krüger 60 has a parallax of 0.254 second of arc.

☐ **Discovery Inquiry 15–1c**   What is the distance of Krüger 60 from the Earth in parsecs? In AU?

☐ **Discovery Inquiry 15–1d**   Use the distance you just computed, along with the average angular separation you found, to determine the length of the semi-major axis of the orbit in both astronomical units and kilometers. A drawing that includes the observer and the two stars may help; then use the form of the angular-size formula given in the chapter.

☐ **Discovery Inquiry 15–1e**   Using all the information you now have, determine the sum of the masses of the stars in the Krüger 60 binary system.

While we have not determined the masses of the individual stars in this system, knowledge of their sum is important. In some systems, the motion of the primary and the secondary relative to background stars can be determined. In these cases, we can determine the *ratio* of their masses. Then, given the ratio, and the sum as determined above, we can obtain the masses of the individual stars.

DISCOVERY 15–2
SPECTROSCOPIC BINARY STARS

When you have completed this Discovery, you should be able to do the following:

- Determine the scale of a spectrum, given the wavelengths of two lines in a comparison spectrum.
- Measure the shift between spectral lines in millimeters and determine their shift in Å.
- Apply the Doppler shift formula to these measurements to determine the relative radial velocity of the two stars.

If a binary system is composed of two stars that have about the same brightness, the spectrum of the system will consist of the spectrum of each star superimposed on one another. If we see the orbital plane edge-on, there will be times at which one star approaches the observer and the other moves away. At these times the Doppler shift will cause two spectral lines rather than one to be seen. From the separation of the lines, we can determine the relative speeds of the stars in the system.

The spectrum of ζ Ursae Majoris (also known as Mizar) is shown in Figure 15–15. The top spectrum (*A*) was taken when the stars were moving across the line of sight so there was no Doppler shift of the stellar spectrum; the bottom one (*B*) was taken when one star was moving toward the observer and the other was moving away, thus producing the observed change in spectral-line position.

Above and below the two stellar spectra are emission spectra of a standard lamp that is attached to the spectrograph. Because this lamp is motionless with respect to the spectrograph, its lines are at their rest wavelengths and may be used as reference markers.

Use a millimeter ruler to determine the scale of the photograph by finding the number of angstrom units in each millimeter measured on the figure. Next, use your ruler to measure the separation of the two lines centered on 4481 Å in the bottom spectrum; then, use the value of the scale you computed to find the separation of the stellar lines in Å. This separation is due to the relative motion of the stars.

☐ **Discovery Inquiry 15–2a** What is the value of the scale of the spectrum in Å/mm?
☐ **Discovery Inquiry 15–2b** What is the value of the separation of the lines in mm? In Å?
☐ **Discovery Inquiry 15–2c** What is the value of the relative radial velocity of the two stars?

In practice, measurements such as this are made at many times throughout a complete orbit. When this is done we have a graph of the radial velocity versus time—a radial velocity curve as shown in Figure 15–16c. Further analysis of this graph gives detailed information about the orbit and the masses of the stars.

CHAPTER SUMMARY

OBSERVATIONS

- The **Hertzsprung-Russell diagram**, the most important graph in astronomy, plots **luminosity** and **temperature**. The majority of stars lie on the **main sequence**, with higher-luminosity groupings called **giants** and **supergiants**, and lower-luminosity **white dwarfs**.
- The brightest stars in the sky appear bright because they have high luminosities, not because they are close to the Sun.
- For stars on the main sequence there is a **mass-luminosity relation** showing that stellar luminosity increases strongly with stellar mass.
- There are three types of binary star systems: **visual binaries**, **spectroscopic binaries**, and **eclipsing binaries**. Such systems are most important in determining stellar masses and sometimes their diameters.

THEORY

- **Luminosity** is proportional to the 4th power of the surface temperature and to the star's surface area (which is proportional to the square of the radius).
- The **inverse square law** of light states that the apparent brightness varies inversely with the distance squared.

CONCLUSIONS

- Most stars in our galaxy are main-sequence stars.
- The **spectroscopic parallax** method of distance determination requires the star's **spectral type** and **luminosity class**, both of which are readily obtained from the spectrum. The star's luminosity is then inferred from the H-R diagram. When inferring the luminosity, we implicitly assume the star whose distance is to be found is similar to the other stars in the H-R diagram. The determination of luminosity from the H-R diagram is uncertain because we do not know if the star in question actually has the average luminosity of its luminosity class, or if it is more or less luminous. The distances determined using these luminosities, therefore, contain uncertainties.
- Stellar masses are inferred from observations of the period and semi-major axes of stars in binary systems. If a star's luminosity is known, and if the star is a main-sequence star, its mass can be inferred from the mass-luminosity relation.
- Stellar diameters may sometimes be obtained from analysis of binary stars. Studies that use the techniques of interferometry and occultations also give astronomers information on stellar sizes.

SUMMARY QUESTIONS

1. Sketch and label the Hertzsprung-Russell diagram for all stars and identify each of the various groups of stars on it. Explain the terminology "main sequence," "red dwarf," "white dwarf," "giant," and "supergiant."

2. Why is the Hertzsprung-Russell diagram for the nearest stars different from that for the brightest stars in the sky?

3. Why are a star's radius, temperature, and luminosity interrelated?

4. How do the various groups of stars on the Hertzsprung-Russell diagram differ in terms of their radii?

5. Where on the main sequence do stars of different masses lie? Draw a schematic H-R diagram and indicate such locations.

6. What is the range of orders of magnitude for stellar mass, luminosity, radius, surface temperature, and density?

7. What is the method of spectroscopic parallax? What are its advantages and disadvantages?

8. What are the three main types of binary star systems? Describe them.

9. How can astronomers determine the masses and radii of stars directly?

APPLYING YOUR KNOWLEDGE

1. Suppose you have an eclipsing system in which the stars have the same diameter. Assuming the eclipses are central—that is, during the eclipses the centers of the stars coincide—draw the resulting light curve.

2. Discuss the process by which astronomers determine a spectroscopic parallax. In your discussion be sure to specify what is observed and what is inferred.

3. Make up a table that shows the maximum and minimum observed values for the following properties of stars: temperature, mass, diameter, density, and luminosity. Express your results both in absolute numbers and in terms of the Sun's value.

4. Rewrite the formula relating distance in parsecs and the parallax angle to be valid for astronauts observing from an orbit whose average distance from the Sun is A astronomical units.

5. What would be the distance to a star identical to the Sun but whose apparent brightness is 10^{16} times less than that of the Sun?

6. What would be the approximate period of revolution for a binary system composed of two stars like the Sun for which the semi-major axis is (a) 1 AU, (b) 10 AU, and (c) 40 AU?

7. To explain a *possible* period of 26 million years in the extinction of some biological species, scientists have suggested that the Sun might have a companion whose period is the interval between extinctions. What would be the distance to such a star (which is known as Nemesis)? How does this distance compare with the distance to the nearest star?

8. The star α Canis Majoris (Sirius) is a binary of period 50 years. Its semimajor axis is 7.5 seconds of arc, and the system's parallax is 0.378 second of arc. What is the sum of the masses of the two stars in solar mass units?

9. What is the distance to a blue supergiant star whose apparent brightness is 10^{14} times less than the Sun's brightness, assuming a true luminosity of 10^4 the luminosity of the Sun? As seen in the H-R diagram in Figure 15–9, such supergiants do not all have the same luminosity. If the true luminosity is actually 2×10^4 that of the Sun, what is the star's distance? Compare the two distance estimates, and discuss how uncertainties in the inferred luminosity produce uncertainties in the derived distance. (Hint: Appendix A10 may be helpful.)

10. How many times less massive is Jupiter than the least massive star shown in the mass-luminosity relation in Figure 15–11?

ANSWERS TO INQUIRIES

15–1. Distance = 1/parallax angle = 1/0.01 = 100 parsecs.

15–2. The distance is directly proportional to the length of the baseline. Because Saturn is 10 AU away, the distance will increase to 1000 parsecs.

15–3. From the inverse square law, $\sqrt{50} \approx 7$, so the more distant star is about $7 \times 12 = 84$ ly away.

15–4. The ratio of the distances is 6. From the inverse square law, the luminosity depends on the distance squared, or 36 times brighter.

15–5. 2×10^{34} ergs per second; 8×10^{32} ergs per second.

15–6. Approximately 1/400 $L_{\odot}$; 20 $L_{\odot}$

15–7. On a rough diagonal from the center of the plot to the lower right corner.

15–8. There is a group of stars below the main sequence which, as we will see, are called white dwarfs.

15–9. The bright stars divide into two main groups: those along a diagonal line in the upper left part of the diagram, and a group to the right of the first set of points.

15–10. Those on a rough diagonal from the upper left corner to the center of the plot.

15–11. Roughly 10^9 to 10^{10}.

15–12. The nearest and brightest stars form two virtually distinct groups. From the H-R diagram we see that the nearest stars are intrinsically faint, while the brightest-*appearing* ones are also intrinsically bright. The nearest ones are typical of the vast majority of stars, while the brightest ones are seen only because they are so luminous. The latter group is atypical.

15–13. The ratio between the brightest and faintest stars on the graph approaches 10^6 to 10^7.

15–14. The brightest stars in the sky are that way because they are intrinsically very luminous, not because they are nearby.

15–15. Its faintness.

15–16. We are located in a part of the universe that is typical of the universe as a whole.

15–17. It is often stated that the Sun is a "typical" star. Insofar as it is outshined by many other stars, this is true; however, only 3 of the 51 stars in the graph of nearby stars are more luminous than the Sun. Therefore, the Sun is more luminous than roughly 90% of the stars in our galaxy.

15–18. The data in the figure indicate that roughly 10% of the stars are white dwarfs.

15–19. $2^2 = 4$ times as luminous.

15–20. $2^4 = 16$ times as luminous.

15–21. The luminosity ratio is 2×10^6. Because luminosity depends on the square of the radius, the size depends on the square root of the luminosity. Antares is $\sqrt{(2 \times 10^6)} = 1.4 \times 10^3$ times, or 3 orders of magnitude larger in radius than CD $- 46°11540$.

15–22. Over 100 times as large, about the same as the Earth's orbital size. Only 1/100 as large, or about the size of the Earth.

15–23. White dwarfs are 100 times smaller or roughly 15,000 km (about the diameter of the Earth). Supergiants are roughly 300 times greater or 5×10^8 km in diameter (a few astronomical units).

15–24. Plaskett's star is roughly 700 times more massive than the least massive star.

15–25. Stars of low mass.

15–26. From Figure 15–8, a main sequence star with a temperature of 10,000 K would have a luminosity of 50–60 $L_\odot$. (Remember, the luminosity scale is a logarithmic scale, so interpolation is not easy.)

15–27. The number representing the distance in parsecs is 5 orders of magnitude larger than the number of astronomical units.

15–28. The mass ratio equals the velocity ratio, with the more massive star moving more slowly. Therefore, because $V_B = 5V_A$, we have $M_A = 5M_B$.

16

THE SUN AND STARS:
THEIR ENERGY SOURCES
AND STRUCTURE

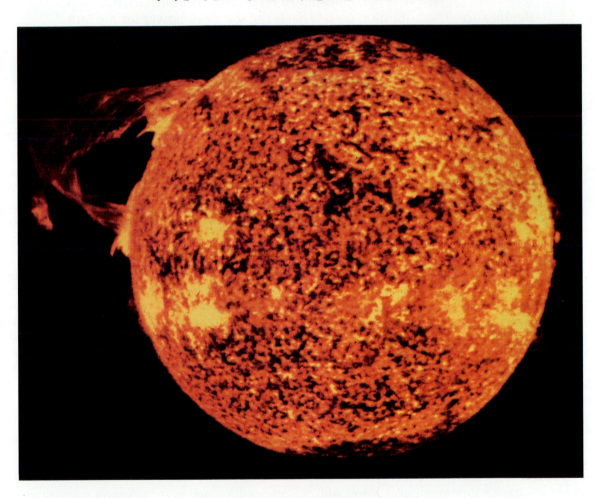

If the radiance of a thousand suns
Were to burst into the sky,
That would be
The splendor of the Mighty One

BHAGAVAD-GITA

When scientific research depends heavily on government support, as it does at the present time, debate over funding priorities is inevitable. Many feel that research should be directed primarily toward the solution of practical problems such as improvement of crop yields, discovery of new ways to produce energy, cures for disease, and the like. Others point out that many of the most important practical discoveries of science have come from basic research conducted solely for the purpose of unlocking nature's secrets, without any practical objectives in mind. In fact, both kinds of research are important.

The discovery of nuclear fusion is one of the most dramatic examples of the importance of basic research. For good or ill, few discoveries have had more practical implications for the survival of humanity, yet the basic concepts were first discovered by astronomers and physicists doing research that, at the time, appeared to be utterly without any practical significance. Their question was: What is the source of power for the Sun and other stars?

16.1

THE POWER PRODUCED BY THE SUN

As soon as the approximate distance to the Sun was known, it became possible to determine its luminosity, which we can also think of as the amount of power it produces. Measurements show that 1400 watts—enough power to light 14 standard 100-watt light bulbs—falls on a square meter at the Earth's distance from the Sun. Averaged throughout the year, some 13% of this prodigious amount of energy strikes the ground in the United States. For example, in the course of a year approximately 10^{14} kilowatt-hours of sunlight fall above the state of Kansas. If it were possible to recover it all, at seven cents per kilowatt-hour it would be worth about $7 *trillion* and would amount to roughly 13 times the world's entire electricity consumption.

The Sun supplies abundant energy to the Earth. Unfortunately, even with the best technology only a small portion of this power could actually be converted into electricity. The "energy crisis" might well be called an energy *cost* crisis, because at present it is not *economically* feasible to harness most of this solar energy. But if and when the cost of conventional types of energy rises high enough, there will come a time when solar power will indeed be economically competitive, and we can look forward to a day when much of the energy we use will be derived directly from the Sun.

Given the power received on Earth, it is possible to calculate that the luminosity of the Sun is about 4×10^{26} watts. Expressed in ergs, the appropriate unit of energy when mass is measured in grams and length in centimeters, the Sun radiates about 4×10^{33} ergs per second of energy into space.

Inquiry 16–1 What principle or law would you use to compute the energy radiated by the Sun, given an observation of the amount of energy falling on the Earth?

The amount of power radiated by the Sun is enormous, and it was obvious to nineteenth-century astronomers that ordinary combustion simply could not supply this amount of energy for more than a few thousand years. At the same time, because geologists were providing convincing evidence that the Earth was millions of years old, it was clear that some other method of producing energy in the Sun must exist.

16.2

ENERGY SOURCES

In our discussion of stellar energy sources, we consider a variety of mechanisms, some that work and some that do not. In this way can we understand the energy requirements of stars. We begin with the conversion of the Sun's gravitational potential energy into radiation and then consider a medley of mechanisms involving the atomic nucleus.

GRAVITATIONAL COLLAPSE

One answer to the mystery of the Sun's energy source was proposed by the physicists Lord Kelvin and Hermann von Helmholtz, who pointed out that a large mass such as the Sun will, under the action of its own gravity, arrange itself into a spherical shape. Because each particle of matter will tend to be drawn toward the center of the Sun (**Figure 16–1a**), the effect of gravity will be to cause the Sun to contract and shrink in size (**Figure 16–1b**).

The arguments of Kelvin and Helmholtz are easy to understand. Recall from the discussion of energy in Chapter 6 that bodies in motion have kinetic energy, while bodies having the *possibility* of motion have potential energy. A star has potential energy by virtue of its gravity. Gravity's pull on the outer parts of a star causes those parts to move towards its center. This slight decrease in the size of the star decreases the amount of potential energy present. The law of conservation of energy, however, requires that the decrease in potential energy result in an increase in the kinetic energy within the star. The end result is that half the loss of potential energy produces an increase in the temperature of the star, while the other half produces energy that is radiated away.

Let's hypothesize that the energy of the Sun comes from a continuous but slow decrease in its size through the conversion of gravitational potential energy into radiation. While this *possibility* makes good physical sense, is it *reasonable* as the energy source in the Sun?

To check this hypothesis, let's see how long the Sun could last if its luminosity came from the loss of gravitational potential energy. The lifetime of a star depends on two quantities: the amount of energy available, and the rate at which the star emits this energy. In other words,

$$\text{lifetime} = \frac{\text{energy available}}{\text{rate of loss of energy}} = \frac{\text{energy}}{\text{energy/sec}}.$$

If the Sun's energy were from gravitational contraction, it would have lost some 10^{49} ergs of potential energy since its birth. The Sun radiates 4×10^{33} ergs/sec; this is, of course, its luminosity. Thus the lifetime over which gravitational collapse could supply the observed luminosity would be

$$\text{lifetime} = \frac{10^{49} \text{ ergs}}{4 \times 10^{33} \text{ ergs/sec}}$$

$$= 2.5 \times 10^{15} \text{ sec}$$

$$\approx 10^{8} \text{ years}.$$

Throughout most of the nineteenth century this **Kelvin-Helmholtz contraction time** appeared to be a sufficiently long time, and the question of the source of the Sun's power was not considered troublesome.

However, by the end of the nineteenth century, geological evidence had increased the estimated age of the Earth to *several* hundred millions of years, and the discovery of radioactivity at the close of the century made it possible to measure the Earth's age with even greater certainty at around 4.5 *billion* years. Because it is hard to imagine how the Earth could be much older than the Sun, the source of the Sun's energy once again became one of the great puzzles of astronomy.

Inquiry 16–2 The conclusion that gravitational contraction is not the source of the Sun's energy depends on what assumptions about the constancy of the Sun's energy output over time? Explain.

Although the Sun does not shine from gravitational collapse, starting in the next chapter we will see that this mechanism plays an important role for certain stars at a number of points in their lives.

There are objects that do shine primarily by the Kelvin-Helmholtz process, although one hesitates to call them stars because their luminosities are so low. They are known as **brown dwarfs**. One object, LHS 2924, which has a surface temperature of only 1950 K, is significantly cooler than M-type dwarf stars; it is not a confirmed brown dwarf, however. In 1984 there was a flurry of excitement when some astronomers announced that they had detected another brown dwarf in orbit around the star Van Biesbroeck 8. However, other astronomers were unable to find the object and the original announcement of the discovery was withdrawn. This incident shows how important it is for scientists to confirm the results of others. Many research groups in the 1990s are searching for brown dwarfs because of the important role they play in studies of both star and planetary formation.

(a)   *(b)*

FIGURE 16–1. (*a*) Gravity acting on a volume of material pulls the matter toward the center. (*b*) Gravity causes the gas from which the Sun formed to collapse slowly into a sphere.

RADIOACTIVITY

Is it possible that the energetic particles released when heavy elements, such as uranium and radium, undergo radioactive decay provide the energy of the Sun? While this hypothesis perhaps was never given serious consideration, the famous British physicist Sir Arthur Eddington did discuss it. The answer to the question is no, for a variety of reasons. The phenomenon of natural radioactivity is almost entirely restricted to heavy atoms—those with an atomic mass number greater than 206 (corresponding to the element lead). As we have seen, spectroscopic studies of the Sun show elements other than hydrogen and helium to be exceedingly rare. Furthermore, calculations show that a Sun composed entirely of uranium and its products would produce only half the observed luminosity.

NUCLEAR REACTIONS: FISSION VERSUS FUSION

Nevertheless, the attention of astronomers and physicists had been drawn to the nucleus of the atom as a possible source of power for the stars and the Sun. Theoretical investigations showed that there were two basic types of nuclear reactions that might be a source of large quantities of energy. In the first of these, **fission** reactions, heavy nuclei split into lighter ones. An example is the fission of uranium into nuclei of krypton and barium, along with neutrons, as illustrated in **Figure 16–2**. The other type of reaction, **fusion**, takes place when several light nuclei collide and combine (that is, fuse) into a heavier nucleus, releasing energy in the process. Because the Sun is composed almost entirely of light elements, Eddington in around 1920 reasoned that fusion might well provide the Sun's energy. We now look more closely at fusion reactions.

THE PROTON-PROTON CHAIN

One possible fusion reaction was shown in the 1930s to be workable. A young graduate student, Charles Critchfield, who worked for the famous physicist George Gamow, calculated that at the temperatures and pressures expected to exist at the center of the Sun, it would be possible for two hydrogen nuclei (which are protons) to collide and form a nucleus of deuterium, or heavy hydrogen (**Figure 16–3**). Because the process involves two protons, it is called the **proton-proton chain**. Additions and refinements were made by Hans Bethe, who was awarded the 1967 Nobel Prize in physics for his important work. While we refer to such reactions as **hydrogen burning**, it is "burning" not in the usual chemical sense but, as we will see, in the sense of consuming fuel. Symbolically, the reaction can be written

$$\,^{1}_{1}\mathrm{H} + \,^{1}_{1}\mathrm{H} \rightarrow \,^{2}_{1}\mathrm{H} + e^{+} + \nu \,.$$

The deuterium nucleus, called a *deuteron* and written $\,^{2}_{1}\mathrm{H}$, consists of one proton and one neutron; therefore, its charge is one unit and its atomic mass number is two units. In addition, two other particles are produced: a positron, a positively charged electron written e^{+}, and a massless, chargeless particle known as a **neutrino** and designated by the Greek letter ν (nu). We will have more to say about these particles in Section 16.4.

Inquiry 16–3   Assuming the mass of the positron is negligible in comparison with that of a proton or a neutron, is the total atomic mass number conserved in the fusion of two protons? The electrical charge? Justify your answer.

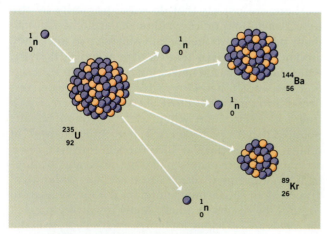

FIGURE 16–2.   The fission of uranium into barium and krypton.

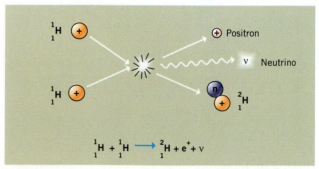

FIGURE 16–3.   The fusion of two hydrogen nuclei to form a deuterium nucleus also releases energy.

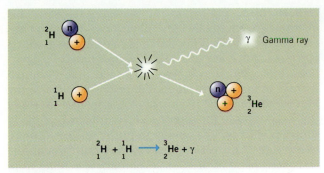

FIGURE 16-4. The fusion of hydrogen and a deuterium nucleus to form a nucleus of helium-3.

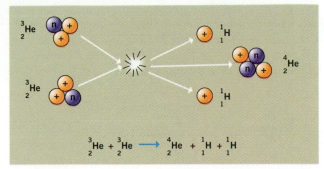

FIGURE 16-5. The fusion of two helium-3 nuclei to form a helium-4 nucleus. To get the two helium-3 nuclei requires cycling through the previous steps two times.

Once created, the deuteron can collide with yet another proton, producing a light isotope of helium that has two protons and one neutron. At the same time, a high-energy photon, or gamma ray (symbolized by the Greek letter γ), is produced (**Figure 16-4**). Symbolically, the reaction is written

$$^2_1H + ^1_1H \rightarrow ^3_2He + \gamma .$$

Inquiry 16-4 Because the photon is massless and chargeless, show that this reaction conserves atomic mass number and charge.

Inquiry 16-5 Nuclear reactions must conserve charge and total atomic mass number. On the basis of such considerations, which of the following reactions is impossible and why?

$$(a) \ ^2_1H + ^1_1H \rightarrow ^3_1H + e^+ + \gamma$$

$$(b) \ ^2_1H + ^1_1H \rightarrow ^3_1H + \gamma$$

The final step in the proton-proton chain takes place when a second nucleus of light helium, created in the same way as the first, collides with another nucleus of light helium, forming a nucleus of ordinary helium and ejecting two protons (**Figure 16-5**). In symbols,

$$^3_2He + ^3_2He \rightarrow ^4_2He + 2^1_1H$$

Inquiry 16-6 Considering the entire sequence of the three reactions in the proton-proton chain, how many protons are involved in the production of one nucleus of helium-4 (^4_2He)? How many are left over?

The first two parts of the cycle occur twice for every one occurrence of the final step. Counting all the protons that go into the proton-proton chain, we find that the net result is to transform four hydrogen nuclei into one nucleus of helium-4. But if we compare the *masses* of the four protons with the resulting helium nucleus, we find that there is a discrepancy. In other words, mass appears to be missing because the mass of four protons is not precisely equal to the mass of a helium nucleus. In fact, the mass of a proton is 1.67252×10^{-24} gm, whereas the mass of a helium nucleus is 6.64258×10^{-24} gm. If we subtract the mass of one helium nucleus from that of four hydrogen nuclei, we find that

mass of four hydrogen nuclei = 6.69008×10^{-24} gm
mass of one helium nucleus = $\underline{6.64258 \times 10^{-24} \text{ gm}}$
difference in mass = 0.04750×10^{-24} gm.

It appears that a small fraction of the original mass ($0.04750/6.69008 = 0.007$) is missing. What has happened to it? The answer was given by Albert Einstein (**Figure 16-6**) in the year 1905. As a consequence of his

FIGURE 16-6. Albert Einstein (1879-1955). This great physicist was also a humanitarian and an accomplished violinist.

theory of special relativity, Einstein showed that mass and energy are but different forms of the same thing and that each can be converted into the other. His famous formula $E = Mc^2$ gives the relationship between the amount of mass and an equivalent amount of energy. In words,

$$\text{energy} = \text{mass} \times (\text{speed of light})^2.$$

If we express energy in ergs, the speed of light is 3×10^{10} cm/sec and its square is 9×10^{20} cm²/sec², a large number. This means that even a small amount of mass is equivalent to a large amount of energy. Just one gram of matter, if converted entirely into energy, is equivalent to nearly 10^{21} ergs, or about 3×10^7 kilowatt-hours. At a cost of seven cents per kilowatt-hour, this would be worth about $2 million! And so, the missing mass was not missing but was converted into energy.

Inquiry 16–7 How much energy (in ergs) is produced by fusion of four hydrogen nuclei into one helium nucleus? Assume that 0.05×10^{-24} gm is entirely converted into energy.

Inquiry 16–8 Each second the Sun radiates about 4×10^{33} ergs of energy into space. How many grams of matter are consumed each second, assuming that the source of the Sun's energy is the conversion of mass into energy?

The rate at which mass is converted to energy in the Sun, as computed in Inquiry 16–8, is huge by Earthly standards. How long would it take the Sun to convert only 10% of its mass of hydrogen into helium plus energy at the current rate? Remember from earlier in the chapter that the lifetime of an energy source is given by the energy available divided by the rate of loss of energy. Accounting for the fact discussed above that in hydrogen burning only 0.007 of the mass becomes energy, we can compute the lifetime as follows:

$$t = \frac{0.007 \times (\text{mass to be converted}) \times (\text{speed of light})^2}{\text{energy radiated per sec}}$$

$$t = \frac{(0.007)(0.1)(2 \times 10^{33} \text{ grams})(3 \times 10^{10} \text{ cm/sec})^2}{4 \times 10^{33} \text{ ergs/sec}}$$

$$t \approx 10^{10} \text{ years!}$$

Even at the prodigious rate of conversion of mass to energy implied by Inquiry 16–8, the Sun would be able to shine for 10 billion years and still have radiated away less than one-tenth of one percent of its mass. So nuclear reactions do not strictly conserve mass (because a small amount is used up), but they do conserve mass *plus* energy.

Both fission and fusion are capable of releasing vast quantities of energy by the conversion of mass into energy. In both cases, the combined mass of the products of the reaction is less than the combined mass of the particles that went into it. We have seen that, although there is an insufficient amount of heavy elements in the Sun for fission to be its source of energy, the fusion of the Sun's abundant reserves of hydrogen is a likely source of energy. The energy from the nuclear furnace in the core of the Sun diffuses its way out through the enormous mass of the Sun, eventually emerging from its outer surface as its visible luminosity.

Does the fact that fusion is physically plausible and abundant for a long time scale *prove* its existence in the Sun? No. A discussion of observations concerning fusion in the Sun will come in Section 16.4.

THE CARBON-NITROGEN-OXYGEN CYCLE

Astrophysicists have good reason to believe that the proton-proton chain is the main source of the energy generated by the Sun, and that it is also the power source for stars less massive than the Sun (which means the majority of stars). But there is another important set of hydrogen-burning reactions, called the **carbon-nitrogen-oxygen cycle**. Although it contributes only a small amount to the Sun's luminosity, it dominates in stars that are more massive than a few times the mass of the Sun.

In the carbon-nitrogen-oxygen, or **CNO**, cycle, hydrogen is also converted into helium. However, nuclei of the elements carbon, nitrogen, and oxygen are also involved, although they are constantly renewed in the cycle and are not used up permanently. The sequence of nuclear reactions in the CNO cycle is

$$^{12}_{6}\text{C} + ^{1}_{1}\text{H} \rightarrow ^{13}_{7}\text{N} + \gamma$$

$$^{13}_{7}\text{N} \rightarrow ^{13}_{6}\text{C} + e^+ + \nu$$

$$^{13}_{6}\text{C} + ^{1}_{1}\text{H} \rightarrow ^{14}_{7}\text{N} + \gamma$$

$$^{14}_{7}\text{N} + ^{1}_{1}\text{H} \rightarrow ^{15}_{8}\text{O} + \gamma$$

$$^{15}_{8}\text{O} \rightarrow ^{15}_{7}\text{N} + e^+ + \nu$$

$$^{15}_{7}\text{N} + ^{1}_{1}\text{H} \rightarrow ^{12}_{6}\text{C} + ^{4}_{2}\text{He}.$$

You should be able to verify the conservation of charge and atomic mass number in each of these reactions. Notice that, at the reaction's end, the original carbon-12 nucleus is returned unchanged. Here, carbon functions as a catalyst—a substance that helps a reaction to take place without itself ultimately being changed. If a star contained no carbon atoms, it would not be able to generate energy using this cycle, even if a star of its mass would normally do so.

The CNO cycle, in addition to producing helium, also produces high-energy photons (gamma rays) and neutrinos.

Inquiry 16–9 How many hydrogen nuclei enter the CNO cycle? How many helium nuclei and positrons are produced? How is the net effect of the CNO cycle similar to that of the proton-proton chain?

16.3
THE CONDITIONS REQUIRED FOR FUSION

Fusion will occur only where the physical conditions are right. In this section we will discuss what these conditions are and where within a star they are met.

TEMPERATURES AND DENSITIES

Fusion cannot take place spontaneously; the right conditions must be present in a gas or nothing will happen. Unfortunately, the required conditions are difficult to produce on Earth. If it were not so, humanity's energy crisis could be solved easily.

In the Sun, the temperatures and densities are high compared with those on Earth. Models tell us that the Sun's central temperature is about 15 *million* kelvins, and that the central density is about 150 times that of water. Under such extreme conditions, atoms are flying about at such great velocities and colliding so frequently and violently that their electrons are stripped away from the atomic nuclei. A gas such as this, composed of charged particles, is known as a **plasma**. Although it is a somewhat uncommon state of matter on Earth, it is extremely common in the cosmos. In fact, most of the matter in the universe probably exists in the form of a plasma.

For nuclear fusion to take place, two conditions are required: high temperatures and high densities. Because the particles that fuse together in the various reactions are all positively charged, the electrical forces acting on the particles cause them to repel one another. The repulsive force is a strong barrier to incoming particles— the only stronger force known in the universe is the nuclear force that binds protons and neutrons in the nucleus. But if two particles can be brought sufficiently close together for the nuclear force (which is attractive and functions only over extremely short distances) to

act, the particles will be pulled together and fusion will take place.

High density also helps the fusion process. Although some reactions will certainly take place at low densities if the temperature is high enough, only at a sufficiently high density will collisions be frequent enough for substantial amounts of energy to be produced.

Inquiry 16–10 Would you expect the rate of energy generation to *increase* or *decrease* as the temperature is raised? How about if the density is raised?

The rate of energy generation per gram of material depends strongly on the temperature. For the proton-proton reaction, the energy generation rate per gram depends on T^4. This means that a star with a central temperature twice that of the Sun would generate 16 times as much energy. The CNO cycle is even more strongly dependent on the temperature; the energy generation rate is proportional to T^{17}. Even a small increase or decrease in the temperature will produce a large change in the amount of energy produced: a 1% change in temperature will produce nearly a 20% change in energy. Because more massive stars will in general have higher central temperatures, it is apparent why the CNO cycle will dominate in their energy generation.

CONTROLLED THERMONUCLEAR FUSION— AN ASIDE

Due to its large mass, the Sun has no difficulty attaining the temperatures and densities necessary for nuclear fusion. On Earth, however, it has proved difficult to achieve these conditions in a laboratory. The hydrogen bomb accomplishes fusion, but in an uncontrolled manner. For the reliable generation of power on Earth, we would want a reaction that could be completely and dependably controlled. An ultimate power source would also be free of the undesirable side effects of today's *fission* reactors, such as poisonous and radioactive by-products.

The generation of power by fusion contains the promise of all these benefits. Suppose, for example, that a reactor could employ the second reaction in the proton-proton chain. Both hydrogen and deuterium are virtually limitless on Earth, and the reaction produces no odious by-products. These factors have led to intense efforts by physicists to discover ways to produce controlled fusion reactions in the laboratory. If successful, these efforts could point the way to a long-term solution of energy problems.

It turns out to be relatively simple to generate the necessary high temperatures; this simply means that the reacting particles must be moving at high rates of speed. But it has proven to be difficult to contain the hot particles long enough for them to react together. One tactic for doing this involves the construction of a "magnetic bottle"—a region of magnetic lines of force that hold the hot plasma particles inside until they can react. Unfortunately, however, the magnetic bottles that have been constructed to date are highly unstable and break down in a fraction of a second. When this happens, the hot particles run into the walls of the container and cool down, and all possibility of a nuclear reaction is lost.

Another method is to fire a set of powerful laser beams onto a small frozen pellet of deuterium and tritium, the heaviest hydrogen isotope. The laser radiation causes a sufficient compression of the material, to a density about 20 times that of lead, for fusion to occur. Research in both areas is occurring, and, barring unexpected difficulties, many experts feel that safe and dependable power from nuclear fusion could be coming on line at about the middle of the twenty-first century.

16.4
ANTIMATTER AND NEUTRINOS

Recall that the first reaction in the proton-proton chain produced two unfamiliar particles, the positron and the neutrino. The positron, denoted by the symbol e^+, was predicted theoretically by the great physicist Paul Dirac several years before it was actually discovered. A positron is an antielectron—the same as an ordinary electron except with an opposite electrical charge. If a positron and an electron meet, they will annihilate each other totally, their mass being converted into energy in the form of two gamma rays. In symbols,

$$e^+ + e^- \rightarrow \text{energy.}$$

The positron, or antielectron, was the first of a whole range of antiparticles to be detected. **Antimatter**, which consists of these various antiparticles, is the mirror image of ordinary matter. Every particle has a corresponding antiparticle. The negatively charged antiproton corresponds to the proton, the antineutron corresponds to the neutron, and so on. Whenever a particle meets its own antiparticle, the two annihilate each other in a flash of energy. (The photon is somewhat special; it is its own antiparticle, and photons do not annihilate each other.)

Like the positron, the neutrino was also first predicted theoretically and only later observed. The neutrino is chargeless and, like the photon, is generally believed to be massless and to move at the speed of light.

Unlike the photon, however, it interacts with matter only rarely. This makes it difficult to detect. The neutrinos produced in the center of the Sun, for example, stream freely outward. When they reach Earth, most of them pass right through as if it were not there.

Inquiry 16–11 It would be of great value to detect neutrinos that have traveled from the center of the Sun. Why? (Hint: Do we receive any of the photons that are created in the center of the Sun?)

Although the probability of a given neutrino being absorbed by any intervening matter is extremely small, we believe that so many neutrinos are produced in the Sun that a certain number must inevitably be absorbed when passing through matter. Since the late 1960s, Raymond Davis, an American chemist, has been trying to detect solar neutrinos. His apparatus, located deep underground in a gold mine to shield it from as much stray radiation as possible, consists of a 100,000-gallon tank filled with perchloroethylene (C_2Cl_4), a common commercial dry-cleaning fluid (**Figure 16–7**). The idea behind the experiment is that there is an extremely

FIGURE 16–7. The Davis solar neutrino experiment.

small but non-zero probability that a solar neutrino will be absorbed by one of the chlorine atoms in the tank and be turned into an atom of radioactive argon. The reaction can be written symbolically (using an asterisk to indicate the radioactive argon) as

$$^{37}_{17}\text{Cl} + \nu \rightarrow (^{37}_{18}\text{Ar})^* + e^-.$$

Inquiry 16–12 Is charge conserved in this equation? Show that your answer is correct.

Periodically, helium gas is bubbled through the tank and any argon atoms that have been created are swept up. When the radioactive argon decays after a time, the number of decays are counted and the number of captured neutrinos is determined.

It would be nice to be able to say that Davis has detected enough solar neutrinos to observationally confirm nuclear fusion as the source of the Sun's power. Some neutrinos have been detected, but less than a third as many as detailed computer models of the Sun predict. Because the lack of agreement between the models and the experiment is crucially important, much effort has gone into trying to account for it. However, the reason for the discrepancy has not yet been found. Some scientists have suggested that our models of the Sun might be wrong, and that the temperature at the Sun's center might be less than we calculate, although theoreticians feel that an error of this magnitude seems unlikely.

A more exotic possible explanation for the lack of agreement between the expected and observed neutrino counts comes from recent experiments that indicate that the neutrino, contrary to previous belief, may actually have a tiny mass, much smaller than the mass of an electron. If so, the neutrino would not be a stable particle but instead would turn into other particles to which Davis's detector is not sensitive. If this occurs, it would explain the failure of his experiment to detect significant numbers of neutrinos. So far, other experiments have failed to confirm the existence of the neutrino's mass, so this idea is still speculative.

Another possibility is that our understanding of the theory of fusion is incomplete. Davis's experiment detects neutrinos from a rather rare reaction that accounts for only 2% of the Sun's energy. The more common reactions produce neutrinos whose energy is too small to convert chlorine into argon. For a number of years proposals have been made for a new, more sensitive experiment that would monitor a large mass of the element gallium. Gallium is so expensive that the experiment must be international in scope. A project known as **SAGE**, the **S**oviet-**A**merican **G**allium **E**xperiment, lo-

cated in the Caucasus Mountains, uses 30 tons of liquid gallium. Interactions of gallium with neutrinos produce the element germanium. The number of interactions is so low that after one month of running the experiment, scientists were looking for 36 atoms of germanium out of 10^{29} atoms of gallium! Much to everyone's consternation, this experiment has detected only one-sixth of the expected neutrinos, in agreement with the Davis experiment.

Understanding the neutrino discrepancy is so important that another group of countries, including Italy, Germany, France, the United States, and Israel, have organized an experiment known as Gallex, inside a mountain at Gran Saso, Italy. Preliminary results announced in mid-1992 had detected about half the neutrinos expected. The differences in the results between this experiment and those of SAGE are yet to be understood. Finally, Japan has an experiment at Kamiokande using 680 tons of pure water. This experiment has not yet provided results.

In 1986 two Soviet nuclear physicists suggested yet another possible solution, one that does not require any disturbances to accepted theory. They proposed that neutrinos escaping from the Sun are scattered by encounters with the particles of matter making up the solar material. This interaction tends to transform the incident neutrino into other types of neutrinos not detectable by the Davis experiment. This suggestion explains the discrepancies in terms of known physical effects; it simply took a while for someone to think of it. A definitive observational test of this explanation will use results of the gallium experiments mentioned earlier.

16.5
STELLAR STRUCTURE

In addition to having a source of energy, a star must satisfy a variety of nature's laws to remain stable for a long time. In this section we discuss the ideas that permit stars to live long lives. We will then put these laws together to produce a model of the interior of a star.

PRESSURE AND ENERGY EQUILIBRIUM

At the high temperatures found in the interior of a star like the Sun, all matter is gaseous. This greatly simplifies the study of stars because, unlike the physical laws that govern the behavior of solids and liquids, those that govern gases are quite simple.

At first it might appear impossible for hydrogen to exist as a gas at the densities found in the interior of the Sun. The hydrogen *atom* (as opposed to the hydrogen nucleus) is mostly empty space because the electrons are far from the nucleus. Therefore, if hydrogen *atoms* were packed as closely as possible, they would form metallic hydrogen, with a density of about 1 gram per cubic centimeter. Yet the density estimated for the Sun's interior is much greater—about 150 grams per cubic centimeter. How can this be?

The answer is that because of its high temperatures the Sun is composed of a plasma, which as we saw means that the electrons have been stripped away from the protons that form the nucleus. The diameter of a hydrogen *nucleus* is much less than that of a hydrogen *atom*, and the smaller nuclear particles can be compressed further than for a gas composed of the larger atoms. For this reason, the density of a plasma consisting of protons and electrons can be much greater than that of an equivalent mass of atomic hydrogen.

The force of gravity, acting on each atom in the Sun, wants to collapse it. Two principal factors prevent collapse. The first is that any gas, such as that composing the star, has an internal **gas pressure** caused by the motions of atoms. This outward motion resists compression just as the gases in an inflated balloon resist compression. At a temperature of absolute zero, this pressure is nearly zero. The hotter or denser the gas is, the greater this pressure will be. In other words,

$$\text{gas pressure} \propto \text{temperature} \times \text{density}.$$

This relationship between pressure, temperature, and density is known as the **perfect** or **ideal gas law**. As the interior of the star becomes hotter and more compressed, the gas law tells us that the outward pressure increases until the star is capable of resisting further collapse.

The second contributor to preventing collapse is a consequence of the radiation that streams outward from the center of a star, where vast quantities of energy are being generated by the fusion of hydrogen into helium. Because photons have momentum, as they impinge on the gaseous material they tend to push the gas outward. This effect is called **radiation pressure**.

Stars of low to moderate mass, such as the Sun, are supported primarily by gas pressure, whereas in stars of high mass radiation pressure is more important because of their higher temperature. In fact, it is believed that no stars larger than about 70 solar masses exist because in such a large star the radiation pressure would be so great as to blow away the outer layer of the star, until its mass is reduced sufficiently.

A star such as the Sun is said to be in a state of equilibrium if changes in it take place slowly. In such a state, at every point within the star the inward force of gravity is balanced by the outward force of gas and radiation pressure, so that a blob of gas does not rise or fall (Figure 16–8). Without this **gravity-pressure equilib-**

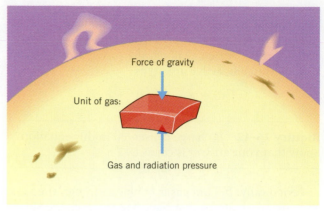

FIGURE 16–8. Equilibrium between the inward force of gravity and the outward force of gas and radiation pressure in a star.

rium (the technical term astronomers use is *hydrostatic* equilibrium) the Sun would change rapidly. For example, if the outward pressure maintaining the Sun in equilibrium were suddenly to disappear, the Sun would collapse in roughly 1000 seconds!

Another aspect of equilibrium is that the amount of energy lost at the surface of a star is precisely compensated by an equivalent amount of energy generated in its interior. We refer to this aspect of equilibrium as **thermal equilibrium**.

It is instructive to ask what would happen if a star were not precisely in equilibrium (**Figure 16–9**). For example, suppose the interior temperature were to increase slightly. Due to the ideal gas law, the pressure in the interior of the star would then increase slightly and become larger than the inward force of gravity. The star would expand a little. This slight expansion would lower the temperature and the density. Because, as we have seen, the star's energy-generation rate depends strongly on temperature, the slightly lowered temperature would lower the rate at which the star produces energy. This small decrease in energy production would cause the temperature to decrease slightly and, due to the ideal gas law, cause the pressure to decrease slightly. The star would then return to a new state of equilibrium in which it is slightly larger and cooler than before. This natural "thermostatic" action is constantly in operation within a star, always tending to bring it back into a state of equilibrium should any minor imbalances occur.

Inquiry 16–13 Suppose the energy-generation rate at the center of a star were to become slightly smaller than its equilibrium value. How would the star regain its state of equilibrium?

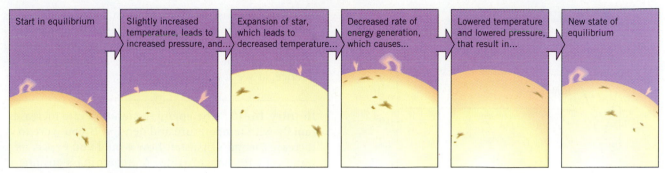

| Start in equilibrium | Slightly increased temperature, leads to increased pressure, and... | Expansion of star, which leads to decreased temperature... | Decreased rate of energy generation, which causes... | Lowered temperature and lowered pressure, that result in... | New state of equilibrium |

FIGURE 16–9. How a star adjusts from one equilibrium state to another.

ENERGY TRANSPORT

Once energy has been generated in the core of a star, it must get to the surface to be radiated away. The high-energy photons produced by the fusion of hydrogen into helium travel only a short distance (typically 1 cm) before they are absorbed by slightly cooler material above. This material then radiates the energy it has acquired in the form of new photons. Because the material is slightly cooler than that at the center, the new photons emitted will have a slightly lower energy, and thus a slightly longer wavelength. These photons in turn are absorbed by even cooler material, and again new photons are emitted with even longer wavelengths. This process continues until the energy finally reaches the surface in the form of visible radiation, where it is then radiated into space. Because in the core the photon travels only a short distance before being absorbed, and because the re-emission of a new photon can be in *any* direction, it takes energy hundreds of thousands of years to escape (**Figure 16–10**). The process just described, known as **radiation transport** or **radiative transport**, occurs throughout most of the Sun's interior.

In the outer regions of the Sun, radiative transport does not work well because the cooler gases are more efficient at absorbing radiation. The energy trapped by the highly absorbing gas heats it more than would otherwise be the case, much as a blanket keeps us warm by trapping the heat of our bodies. Due to this effect, a blob of warmed gas may become slightly lighter than the surrounding material and tend to rise, like a hot air balloon. After rising a certain distance, it gives off its heat to the cooler material that was above it. Cooled off, it sinks again to the region below, where the process starts all over again. This process, called **convection**, is rather familiar in everyday life. For example, in a pot of boiling water, the energy is transported from the bottom of the pot to the surface by convection. Another example is the "waves of heat" that are seen over a hot fire or rising

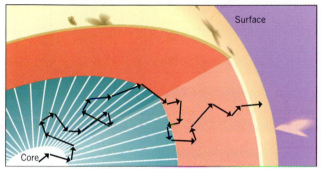

FIGURE 16–10. The path followed by photons from production in the solar interior until they reach the surface. The original gamma-ray photon is not the one that is emitted from the surface; different photons are constantly absorbed and reemitted.

from a highway on a hot summer day; these actually consist of rising currents of hot air. Convection is highly efficient at moving energy from one point to another.

The ideas we have been discussing—gravitational and thermal equilibrium, and energy generation and transport mechanisms—can be expressed in complex mathematical terms. The equations include such quantities as temperature, pressure, density, chemical composition, luminosity, and mass. Solution of these equations, which requires a large, fast computer, yields values of temperature, pressure, density, chemical composition, luminosity, and mass at each point throughout a star in equilibrium. These quantities, as they vary throughout the star, provide astronomers with a **stellar model**. Such a table of numbers, or a graph, gives us a picture of the interior of a star. The results of one such calculation for the Sun are displayed in **Figure 16–11**. The graph shows rapid drops in temperature and density with distance from the center. This rapid decrease in temperature and density is why energy generation oc-

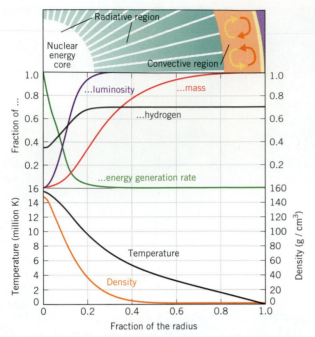

FIGURE 16–11.   A theoretical model of the interior of the Sun. Note the different axes on both the left and right sides. The top part of the figure shows the location of the nuclear energy core, and the radiative and convective regions.

curs only in the core region, as shown by the appropriate curve in the figure.

Inquiry 16–14   What fraction of the mass of the Sun is in the inner 50% of the Sun's radius? The inner 70%? What fraction of the Sun's mass is in the outer 30% of the radius?

Inquiry 16–15   What fraction of the luminosity of the Sun is produced in the inner 30% of the radius?

Inquiry 16–16   What fraction of the original amount of hydrogen in the Sun's core has now been converted into helium?

We might expect stars of different masses to reach different states of equilibrium, with different central temperatures and densities, and this is indeed so. In fact, we can go further and state that if a star obtains its energy from the conversion of hydrogen into helium, as do main-sequence stars, then the star's equilibrium state is determined uniquely by its mass, its chemical composition, and its age. This idea is known as the **Russell-Vogt theorem**. It tells us that for stars of the same chemical composition and age, the different equilibrium states that we see along the main sequence of the

Hertzsprung-Russell diagram exist because the stars have different masses. Therefore, because stellar spectroscopy tells us that most stars have the same chemical composition, the main sequence is actually a *mass* sequence, as we saw in Figure 15–10.

Inquiry 16–17   The energy-generation rate increases rapidly with temperature and density. If you were to increase the mass of a star, how would you expect its luminosity to change? (Hint: A greater mass pushing down on the central regions of the star produces higher pressures, temperatures, and densities there. This is the explanation for the mass-luminosity relationship discussed in Chapter 15.)

16.6
THE LIFETIMES OF STARS

The more massive a star, the more hydrogen it has to burn. The greater the luminosity, the more rapidly the star loses energy. As we saw earlier in the chapter, the lifetime of a star depends upon the amount of energy available divided by the rate at which energy is lost. Because the amount of energy available depends on the star's mass (through $E = Mc^2$), and the rate of energy loss is the luminosity, symbolically we have

$$t \propto \frac{M}{L}.$$

The existence of the mass-luminosity relationship has important consequences for the lifetimes of stars of differing masses. Both theory and observation indicate that a star of about the Sun's mass will remain on the main sequence as a hydrogen-burning star for about 10 billion years. Consider a star that is twice as massive as the Sun. Although it has twice as much fuel to burn, it consumes it at a rate nearly 16 times as fast as the Sun, because its luminosity is almost 16 times greater than the Sun's. As a consequence, its larger supply of fuel does not last nearly as long.

Inquiry 16–18   Using the above information, how long do you estimate a star of two solar masses can be expected to remain on the main sequence?

The variation of stellar lifetime with mass is dramatic. From the above discussions, astronomers determine that the lifetime in years of a star on the main sequence is given by

$$t = 10^{10} \frac{M}{L}.$$

TABLE 16–1

Main–Sequence Lifetimes

Mass (in solar masses)	Main-sequence lifetime (in years)
60	2 million
30	5 million
10	25 million
3	350 million
1.5	1.6 billion
1.0	9 billion
0.1	thousands of billions

where M and L are in units of the Sun's mass and luminosity. In the previous chapter, we saw that the mass-luminosity relation tells us that $L \propto M^4$. Substituting for luminosity in the equation, we find

$$t \propto 10^{10} \frac{M}{M^4} \propto \frac{10^{10}}{M^3}.$$

This expression says that the lifetime of a star on the main sequence varies inversely with the cube of the mass. Thus, the greater the star's mass, the shorter its main-sequence lifetime. Although the most massive stars can be expected to live for only a few million years or so, the least massive red dwarfs consume their fuel so slowly that they can be expected to live for thousands of billions of years—so long, in fact, that astronomers are not even sure whether there will still be a universe when they finally use up their fuel. **Table 16–1** gives the **main-sequence lifetimes** (i.e., the time spent on the main sequence burning hydrogen) for stars of various masses.

The variation of stellar lifetime with mass has many important consequences. To mention just one, we can seek out regions of active star formation in our galaxy by looking for luminous blue stars; that is, the high-mass O and B stars at the top of the main sequence. Because these stars have such short lifetimes compared to the age of the galaxy itself, the fact that they exist at all means they must have formed relatively recently and are probably not far from the place where they were born. On the other hand, when we observe a star of low mass—say, one-tenth of the Sun's mass—we cannot tell whether it formed recently or not, because its main-sequence lifetime is so long that it could have formed at almost any time in the history of the galaxy.

Our discussion up to this point has examined the theoretical ideas governing the structure and evolution of stars. These ideas are summarized in **Figure 16–12**, which illustrates the interior structure of the Sun as determined from our models. The illustration shows other phenomena on the Sun's surface, which we are able to investigate because of its nearness. Detailed observa-

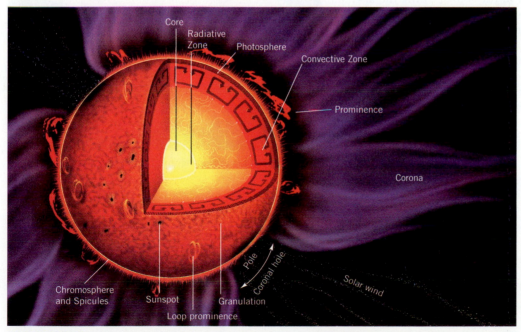

FIGURE 16–12. A cross-sectional drawing of the Sun, showing the energy-generating core, the surrounding radiative and convective zones, and the solar atmosphere.

tions then allow us to make comparisons with theory. We study the Sun's observable outer regions for the remainder of this chapter.

16.7
THE SUN—A TYPICAL STAR

Much of what we know about the stars comes from our study of the Sun. Because it is so close, we know its distance, and thus its mass and luminosity, accurately. We are able to see details of its surface and surrounding gases that we cannot see for other stars. For these reasons, we have been able to learn more about it than about any other star. We can expect that if good models of the Sun can be constructed that agree with the detailed data we have observed, then similar models for stars not much different from the Sun are probably not far off the mark. Of course, the more different a star is from the Sun, the more uncertain our models will be. The rest of this chapter looks at some of our detailed knowledge of our star. **Table 16–2** summarizes some basic data about the Sun.

THE PHOTOSPHERE

The bright, visible surface of the Sun that we see is called the photosphere. More precisely, it is the shell of hot, opaque gas several hundred kilometers thick, within which the Sun's continuous spectrum is produced. **Figure 16–13** is a photograph of the Sun showing the appearance of the visible "surface," and illustrating that the center appears considerably brighter than its edge or limb. This phenomenon, known as *limb darkening*, is easily observed by projecting the Sun's image with a small telescope (see the Activity Book). The reason for limb darkening is that when we look at the edge of the Sun, we are looking at the upper, cooler, and hence fainter levels of the Sun's atmosphere. But when we look toward the center of the disk of the Sun, our line of sight actually reaches to a deeper, hotter, and brighter

region (**Figure 16–14**). Even this simple observation provides direct evidence for the fact that the temperature of the Sun increases toward its center, confirming numerous theoretical calculations.

Above the photosphere is a layer of cooler gas a few thousand kilometers thick. When light from the photosphere passes through this layer, which begins the solar atmosphere, photons are absorbed at wavelengths characteristic of the atoms in the gas. This produces the observed *absorption spectrum* of the Sun. The boundary between the photosphere and the various atmospheric layers is not well defined; it is difficult to say exactly where one ends and the other begins. For this reason, astronomers have a difficult time defining a "surface" of the Sun.

Inquiry 16–19 Observations of the Sun show that its radiation is most intense at a wavelength of about 5000 Å. What is the photospheric temperature?

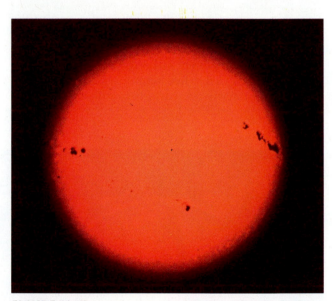

FIGURE 16–13. A photograph of the Sun, showing limb darkening.

FIGURE 16–14. The cause of limb darkening. The observer's line of sight penetrates to a deeper, hotter layer when looking at the center of the Sun's disk than when observing the limb.

TABLE 16–2
Properties of the Sun

Diameter	1,391,980 km (109.3 $D_\oplus$)
Mass	1.99×10^{33} gm
	(332,943 $M_\oplus$)
Average density	1.41 g/cm³
Luminosity	3.90×10^{33} ergs/sec
Photospheric temperature	5780 K
Equatorial rotation period	25.04 (Earth) days

In addition to finding the temperature from Wien's law, we can also use the Sun's G2 spectral type to determine the temperature from the degree of excitation and ionization, as discussed in the previous chapter. From these considerations we obtain a temperature for the absorbing gases of 5870 K.

THE CONVECTIVE ZONE

Other interesting details about the photosphere are revealed in photographs made with a telescope above much of the Earth's turbulent atmosphere. **Figure 16–15** is a photograph taken with Princeton University's Project Stratoscope, for which a telescope was suspended from a balloon about 100,000 feet above the Earth's surface. At this altitude, much sharper pictures revealing great detail can be made.

In this photograph, you will notice a pattern of light and dark areas known as **granulation**. The bright areas are slightly hotter than the darker ones. Photographs taken in rapid succession show that this pattern changes with time, with a typical bright area lasting for only a few minutes. If we look at a moving picture of this granulation, the impression we get is of a region of boiling gas, with material bubbling out of the bright areas and falling back into the darker ones.

Inquiry 16–20   Figure 16–16a illustrates photospheric granulation with the slit of the spectrograph across a number of the bright and dark areas. **Figure 16–16b** shows one spectral line at 8542 Å. Wavelength, increasing to the right as shown, indicates that light from the bright areas is blueshifted whereas that from the dark areas is redshifted. How should this observation be interpreted?

Spectra of the granulation show the individual granules to be in motion. The bright regions are blueshifted and thus rising, while the darker regions in between are redshifted and thus falling. The observed Doppler shifts of the granules tell us that the observed granulation is simply the visible tops of the convection cells. Thus we have observational evidence that the energy transport mechanism of convection occurs in the outer regions of the Sun, just as predicted by theory.

Further characteristics of the outer convection regions are revealed by detailed computer models of the Sun. This convection occurs in a shell that extends from the Sun's surface down to a depth of nearly 150,000 km, or about one-tenth of its diameter (Figure 16–12). In the interior, the energy is transported by the radiative transport mechanism. A star less massive than the Sun is similar, except that the region of convection is proportion-

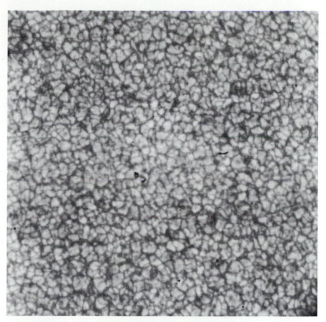

FIGURE 16–15.   Solar granulation, as seen from Project Stratoscope. The region shown is some 24,000 km on a side.

Spectrograph slit allows light through

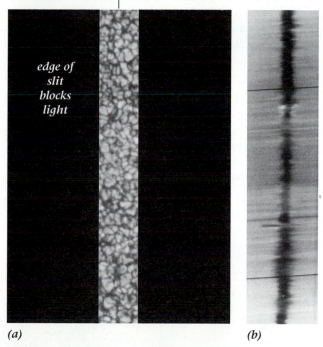

edge of slit blocks light

(a)   *(b)*

FIGURE 16–16.   (a) A spectrograph slit over solar granulation; only that light coming through the slit forms the spectrum. (b) The Calcium line at 8542 Å of the solar spectrum examined in great detail. Parts of the solar surface are rising (blueshifted) and parts are falling (redshifted portions of the line), giving rise to the wiggly appearance of the line as the slit samples different points on the Sun.

ately larger; in low-mass stars, in fact, the entire star is convective. Calculations indicate a different story for high-mass stars; in these, models show that the envelope is transporting energy by radiative transport, whereas the core is convective.

> YOU SHOULD DO DISCOVERY 16–1, SOLAR GRANULATION, AT THIS TIME.

SUNSPOTS

Chinese and Japanese astronomers occasionally observed sunspots thousands of years ago, but systematic observations have been made only for several hundred years. Large sunspots sometimes appear that are visible to the naked eye, especially when the Sun's glare has been dimmed near sunrise or sunset. With a small telescope, the easiest (and safest!) way to observe sunspots is to project the image on a screen (see the Activity Book; **do not look directly at the Sun with a telescope**).

Sunspots are complex entities. For example, they generally occur in elaborate groups (see Figure 16–13). Furthermore, they are not uniformly dark. **Figure 16–17** shows a high-resolution view of a sunspot. The dark region in the center of the spot is known as the *umbra* (shadow) and the lighter region around it is called the *penumbra*. Although they seem dark on the face of the Sun, sunspots are actually quite bright; their dark appearance results from the contrast with the even brighter solar disk.

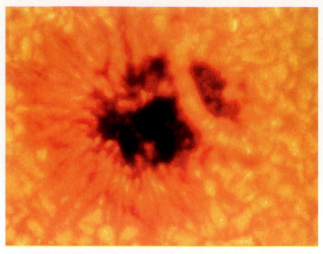

FIGURE 16–17. High-resolution photograph of a sunspot.

Inquiry 16–21 The spectrum of a typical sunspot resembles that of a star of spectral type K2. From this information, what is the temperature of such a sunspot? By how many degrees does it differ from the surrounding 5800-K photosphere? How much less energy does the sunspot radiate per unit area?

Inquiry 16–22 Suppose a rotating star had a large "starspot" on one side. How might the presence of the spot be detected by an observer on Earth?

The number of sunspots visible at any given time, and their location on the solar disk, varies. Astronomers have been counting sunspots for many years; a graph of the number of sunspots with time is shown in **Figure 16–18.** You can see an obvious periodicity of 11 years, more or less. The number of spots decreases from a strong maximum until, at the end of the cycle, few if any sunspots exist. The number of spots at any given maximum is not the same. In the late 1600s the number of sunspots was lower than at other times. This so-called **Maunder minimum** indicates a possible important change in the Sun's activity. The time of the Maunder minimum corresponds to a cold period known in Europe as the "Little Ice Age." While still controversial, a correlation between sunspot activity and climactic conditions on Earth is inferred by some scientists. The possibility of an important connection provides practical motivation for understanding the Sun so that accurate predictions of future solar behavior can be made.

Any given sunspot has a finite lifetime; a spot may form, only to dissolve a few days or weeks later. Spots

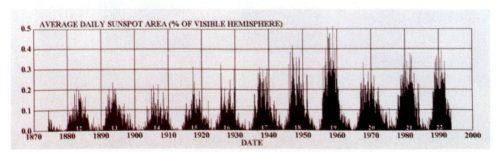

FIGURE 16–18. Variations of the sunspot numbers with time.

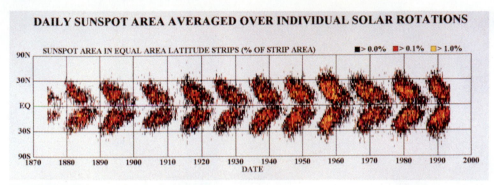

FIGURE 16–19. The butterfly diagram, showing the location of newly formed sunspots on the solar disc throughout the 11-year cycle.

continually form as others disappear over the course of the sunspot cycle. As time passes, however, definite trends become apparent. Early in the sunspot cycle, spots are formed in modest numbers at high latitudes; that is, far from the Sun's equator in both hemispheres. As the years within a cycle pass, spots form and dissolve at lower latitudes, while the number of spots visible at any time increases to a maximum. **Figure 16–19** shows the positions at which sunspots form throughout the 11-year cycles; because of its shape, this figure is known as the **butterfly diagram**. A given spot does not change latitude during its short lifetime; it is the location of formation that changes throughout an 11-year cycle.

THE SUN'S MAGNETIC FIELD

When astronomers place the slit of a spectrograph across a sunspot (**Figure 16–20a**), they observe the spectral lines to be split into two or more parts inside the sunspot but not outside it (**Figure 16-20b**). Such a splitting is caused by the Zeeman effect, discussed in Chapter 14, produced by magnetic zones in the photosphere. A solar magnetogram is made with a special instrument that uses the Zeeman effect. Such magnetograms show astronomers how solar magnetic fields change with location and time.

Sunspots and magnetic fields on the Sun's surface are intimately related. It has long been known that sunspots generally come in pairs, with one leading spot and one trailing. Observations of their magnetic fields revealed that the two spots in a pair have opposite polarities; that is, if the leading spot is a north magnetic pole, the trailing spot is a south magnetic pole, and vice versa. Furthermore, the polarities of the leading and trailing spots in the southern hemisphere are reversed from those in the northern hemisphere (**Figure 16–21**). After an 11-year cycle has run its course, the new spots that appear

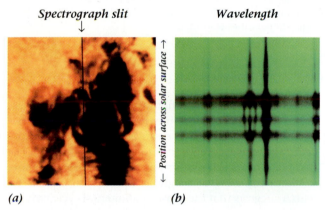

Spectrograph slit | *Wavelength*

(a) | *(b)*

FIGURE 16–20. (a) The solar photosphere with a spectrograph slit across the photosphere and a sunspot. (b) The spectrum of the photosphere and a sunspot, showing Zeeman splitting of radiation from the sunspot. Each point on the spectrum in the vertical direction corresponds to a unique point of light on the Sun coming through the slit.

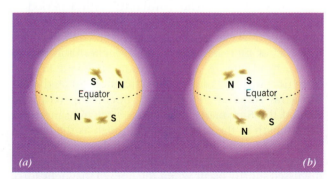

(a) *(b)*

FIGURE 16–21. Sunspot polarities in the northern and southern hemispheres. (a) Notice that the polarities of the leading and trailing spots are reversed in the northern and southern hemispheres. (b) Eleven years later, the polarities in each hemisphere are reversed.

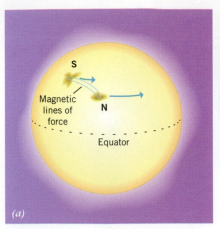
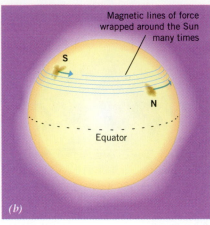
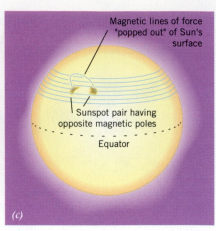

FIGURE 16–22. Differential rotation of the Sun tends to stretch out the magnetic lines of force (*a*) until they are wrapped many times around the Sun, beneath its surface (*b*). If a loop of magnetic lines of force breaks through the surface, as shown in (*c*), a sunspot pair having opposite polarities is formed.

at high latitudes in each hemisphere will have the opposite polarities from the previous cycle. If this reversal in magnetic polarities is counted, the actual sunspot cycle averages 22 years in length rather than 11. The sunspot cycle is simply a visible manifestation of a more fundamental 22-year magnetic cycle. More recent research suggests that the most fundamental cycle may instead be 44 years. Verification and further study are required before the final answer is in.

Astronomers believe that the mechanism that produces sunspots works something like this: Intertwined within the plasma of the Sun's surface is a magnetic field. In any plasma there is a tendency for the magnetic field to become "frozen in"—that is, to be dragged along by the charged particles of the plasma. However, the Sun shows **differential rotation**. With a period of about 25 days at the equator, 28 days at mid-latitudes, and even longer at higher latitudes, the equator of the Sun rotates more rapidly than the higher latitudes, as shown in **Figure 16–22**. As a result, magnetic lines of force near the Sun's surface are stretched out by the differential rotation, eventually becoming wrapped around the Sun many times. This stretching increases the strength of the magnetic field, much as the tension in a rubber band increases when it is stretched. Astronomers believe sunspots and other solar activity to be discussed shortly occur when the lines of magnetic force erupt through the surface and disturb the normally quiescent atmosphere. Exactly why the magnetic polarities reverse themselves every 11 years is more complex and less well understood.

THE CHROMOSPHERE

One of the examples of spectra that we examined in Chapter 13 was a solar spectrum at the moment during a solar eclipse when the Moon exactly covered the photosphere and revealed a reddish glow peeking around the rim of the Moon. This region was referred to as the chromosphere, or color sphere. We saw that the spectrum, referred to as the flash spectrum because it appears suddenly, was an emission-line spectrum characteristic of a hot, rarefied gas. We also saw that the red color comes from the excited hydrogen gas in this layer, which emits much of its energy at 6563 Å, the red (Hα) line of the Balmer series.

The chromosphere also plays a role in the formation of the solar absorption-line spectrum. In the lower portions of the chromosphere, where the gases are still cooler than those that cause the continuous spectrum, most of the absorption lines in the solar spectrum are formed (although some are formed in the upper photosphere).

Although the chromosphere is best studied during an eclipse, it is also studied routinely by using an instrument called a coronagraph, which creates an artificial eclipse by blocking out the light from the Sun's bright photosphere with an occulting disk inside the instrument. This permits the fainter chromospheric emissions to be examined. The chromosphere has also been studied from spacecraft.

Figure 16–23 is a photograph of the edge of the Sun. It displays finer details of the chromospheric region. Notice particularly the needlelike filamentary structures known as *spicules*. Detailed study has shown that the material in these regions is rising and falling at speeds of about 30 km/sec, carrying heat upward. The spicules rise to heights of 5000 to 20,000 km above the photosphere.

Inquiry 16–23 How does the typical height of a spicule above the photosphere compare with the diameter of the Earth?

The atmosphere of the Sun attains its lowest temperature, perhaps as low as 3000–3500 K, at the base of the chromosphere some 500 km above the photosphere. Through the 1500-km-thick chromosphere, the temperature rises to some 30,000 K. Thereafter, temperature rises rapidly through a thin transition region until it eventually merges with the solar corona above it (**Figure 16–24**).

THE CORONA

Above the chromosphere is the beautiful and tenuous **corona**, from the Latin word meaning *crown*. **Figure 16–25**, taken during two eclipses, shows the corona as a pearly light extending far beyond the solar disk. The

FIGURE 16–23. A photograph of the Sun showing spicules.

(a)

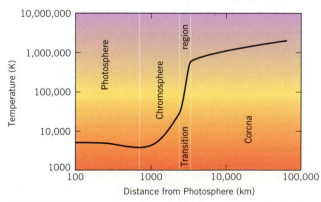

FIGURE 16–24. The variation of temperature with depth in the chromosphere.

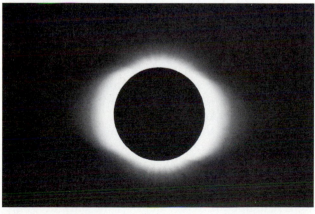

(b)

FIGURE 16–25. The solar corona at (a) sunspot minimum and (b) maximum.

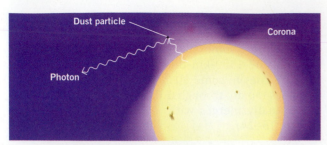

FIGURE 16–26. Scattering of light by particles in the solar corona.

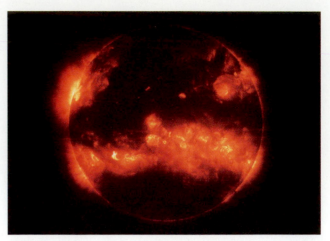

FIGURE 16–27. X-ray image of the Sun.

total brightness of the corona is about a millionth the brightness of the Sun itself—about as bright as the full Moon.

The corona changes from month to month; sometimes it exhibits long streamers and at other times is virtually featureless, as shown in Figure 16–25b. Detailed studies show that it consists of several components. Part of the light from the corona is nothing more than sunlight scattered in our direction by electrons and dust particles, as illustrated in **Figure 16–26**. In addition, there are a few bright emission lines, such as might be expected from a region of hot, rarefied gas. These emission lines were a source of much mystery for many years after their discovery because they did not appear to be the result of any known element; they were attributed to a hypothetical element known as "coronium" that presumably was found in the Sun but did not occur naturally on the Earth.

It seemed natural to speculate the existence of a new element, because the element helium was first detected in the Sun (the name comes from the Greek word for Sun) before it was found on Earth. But in 1941 it was discovered that two of the most prominent emission lines were really due to highly ionized iron atoms. For example, the strongest line, at 5303 Å, was shown to be the result of iron atoms that had lost 13 of their electrons. Because the amount of energy required to remove 13 electrons from iron is large, the conclusion was inescapable: the temperature of the corona must be much higher than that of the surface of the Sun, exceeding 1 million kelvins.

Figure 16–27 shows the Sun imaged in the short wavelengths of X-rays. The fact that the Sun emits strongly at these short wavelengths provides additional evidence for the extreme temperature of its outer regions. The bright regions in this figure are the areas of highest temperature and density; the dark areas are regions known as **coronal holes** where the density is much lessened. Such images taken over a long span of time provide astronomers with further data for

and against various hypotheses about the workings of the Sun.

The corona's incredibly high temperature was a great surprise to astronomers. Why didn't the Sun's outer atmosphere just grow cooler and cooler as it moved progressively outward, as might be expected? A hypothesis accepted as reasonable since the 1940s was overthrown in about 1980 on the basis of new data from spacecraft. The corona shows a structure dependent upon the degree of sunspot activity and thus the Sun's magnetic field. Figure 16–25a shows the corona at sunspot minimum, while Figure 16–25b shows its appearance at maximum. Our understanding of these differences is still slight. About all we can currently say regarding the corona's high temperature is that it undoubtedly has to do with magnetic fields and their motions through the coronal gases.

THE SOLAR WIND

When the first satellites were launched into orbit in the late 1950s, in addition to discovering the Earth's Van Allen belts (Chapter 8), they also confirmed an earlier hypothesis, based on the changing appearance of comet tails, of a flow of particles away from the Sun. Called the **solar wind**, this flow of particles consists of protons, electrons, and helium nuclei. Somehow, the outer atmosphere of the Sun is ejecting these particles and accelerating them to high velocities, several hundred kilometers per second at the distance of the Earth. It would not be inaccurate to think of them as having simply "boiled off" of the hot corona. In a way, the solar wind is really the outermost extension of the Sun's atmosphere. Although the actual amount of mass lost by the Sun is small, there is evidence that some stars have strong stellar winds, which in some cases are much stronger than the Sun's and involve the ejection of much larger amounts of mass.

The particles in the solar wind do not come equally from all parts of the Sun, but rather travel along open magnetic field lines through the coronal holes discussed above (see Figure 16–12).

FLARES AND OTHER SURFACE ACTIVITY ON THE SUN

Sunspots are not isolated phenomena; they are often associated with other surface activity, such as **flares** and **prominences**. Solar flares are sudden brightenings on the surface that are thought to occur when broken magnetic field lines reconnect with a subsequent burst of energy. For years scientists have felt that particles produced during these bursts brought about changes in the Earth's ionosphere, thus disrupting radio communications, which depend on the reflective power of the ionosphere. It has also been thought that solar flares play a major role in producing aurorae on Earth. However, recent work questions this paradigm and instead relates aurorae and communications problems to another phenomenon called coronal mass ejections. These coronal mass ejections discharge substantial amounts of material from the corona at average speeds of 400 km/sec. Thus, our understanding of these phenomena and their complex effects on Earth may be undergoing a fundamental change.

Prominences, like the ones shown in **Figure 16–28**, are among the most dramatic phenomena seen on the Sun. They are enormous filaments of excited gas arching above the surface and usually stretching hundreds of thousands of kilometers into the corona. They frequently run between active sunspot regions, and it is generally assumed that their form and structure are strongly influenced by the Sun's magnetic field. The loop prominence in Figure 16–28b has a shape similar to that of a horseshoe-shaped magnet. Like a horseshoe magnet, for which each side has opposite polarity, sunspots at the foot of each side of the loop would have opposite polarity, as observed. When seen at the edge of the Sun, prominences can be spectacular, as Figure 16–28 attests. When seen against the solar disk they appear as dark, filament-like structures. At times they appear to move rapidly; in time-lapse photography they seem to blow away from the Sun at high speed and fly off into space. The origin of prominences is unclear. They are not nearly as energetic as solar flares.

Although the Sun is the only star on which we can observe surface phenomena in detail, the presence of

(a)

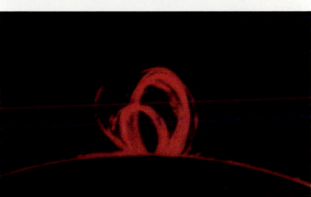

(b)

FIGURE 16–28. Solar prominences. (*a*) A large prominence. (*b*) A loop prominence.

starspots on other stars has been inferred from observations. In addition, stars called **flare stars** are known to show flares that are much more energetic than those on the Sun; these stars are located at the lower right end of the main sequence. Because other stars are so far away, it has not been possible to see directly surface details that probably are present. There is spectroscopic evidence that some stars have a strong **stellar wind**, analogous to the solar wind; yet others show evidence of very large spots on their surface, proportionately much larger than those of the Sun.

Figure 16–29 summarizes the appearance of the Sun in a number of wavelength regions. The various sections of this mosaic look different because the radiations shown emanate from different layers, each having its own temperature and density. This illustration emphasizes the importance of viewing the universe in all spectral regions.

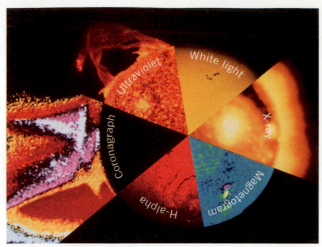

FIGURE 16–29. Mosaic of the Sun at various wavelengths and in a magnetogram.

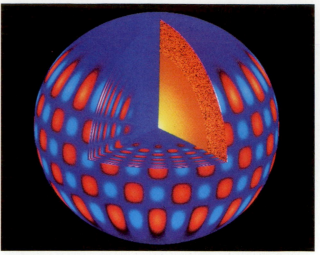

FIGURE 16–30. Theoretical calculations show what solar oscillations would look like on the surface of the Sun.

SOLAR OSCILLATIONS

One of the more interesting and important solar observations in recent decades is that some spectral lines move back and forth periodically. Interpreting these cyclical wavelength shifts as periodic Doppler shifts suggests that the solar photosphere oscillates rhythmically. One important oscillation has a five-minute period and is known as the five-minute oscillation. Oscillations with other periods have also been measured. These observations have given rise to the new research field of *helioseismology*, the study of oscillations of the solar surface.

The oscillations can be likened to those on the surface of a ringing bell. The details of the oscillations, such as period and amplitude, are determined by the variations of temperature and density in the solar interior. Furthermore, the rotation of the Sun's interior influences the photospheric oscillations. Theoretical calculations that include a range of physical conditions and internal rotations produce models such as that shown in **Figure 16–30**, which shows regions participating in the photospheric oscillation. Figures such as this provide a new method of studying the solar interior by making observations of surface oscillations and comparing them with theoretical models. Such comparisons provide information on the solar interior that is not otherwise obtainable, and will be pivotal in choosing between competing hypotheses of the solar structure.

The current interpretation of the results is complex and controversial. The results do not appear consistent with the standard stellar models discussed in this chapter, in that they require the solar interior to be cooler than previously thought. Furthermore, they indicate that the bottom of the convection zone rotates more rapidly than the top; this leads to the possibility that the core rotates more rapidly than the outer regions.

To better understand the Sun requires that it be continuously monitored. For this reason, solar telescopes have been placed at the South Pole so that six-month long continuous observations can be made. A worldwide network of specially designed solar telescopes, known as the Global Oscillation Network Group (GONG) is being established by the United States National Solar Observatory. The GONG will permit year-round observations of the Sun for many years to come.

SOLAR POSTSCRIPT

Our entire discussion of the Sun in this chapter is based on data obtained from within the ecliptic plane. How do the Sun's polar regions behave? Are there differences in the properties of the solar wind coming from the polar regions compared with the rest of the Sun? These and other questions are to be answered by the **Ulysses** spacecraft, which arrived below the solar south pole in August 1994 and will arrive above its north pole in July 1995. While the spacecraft will still be some 2 AU away, the data obtained will provide a never-before-seen view of our star, the Sun.

DISCOVERY 16–1
SOLAR GRANULATION

In this activity you will determine some of the basic properties of solar granulation.

☐ **Discovery Inquiry 16–1a** If the region of the Sun shown in Figure 16–15 is some 24,000 km on a side, determine the average size of a single granule. Find the average size by measuring five granules and finding their average. Determine the uncertainty in your result.

☐ **Discovery Inquiry 16–1b** If a typical granule lasts for about eight minutes, and if it disappears after traveling a distance approximately equal to its diameter, how fast does a typical granule move?

☐ **Discovery Inquiry 16–1c**   Refer to Figure 16–16*b*. In this figure, there is approximately 0.023 Å for each millimeter on the photograph. Because the width of the line is produced by the motion of the solar granules you can determine the speed of the granulation by measuring the width of the spectral line. Measure the spectral line width in millimeters and convert the line width to angstroms. Then use the Doppler shift formula to determine the velocity with which the granules move up and down. Is this result consistent with the answer you computed previously?

CHAPTER SUMMARY

OBSERVATIONS

- The observable outer regions of the Sun include the visible **photosphere**, the **chromosphere**, and the transition region into the **corona**.
- Convection is observable in the Sun as **granulation** in the solar **photosphere**.
- **Sunspots** are cooler regions on the photosphere that are strongly associated with solar magnetism, which varies with a 22-year cycle. Solar **flares** and **prominences** are other phenomena observed to be associated with solar magnetic activity.
- The **solar wind** is made up of charged particles emitted by the Sun through **coronal holes**.

THEORY

- Sources of energy in stars include gravitational collapse and nuclear fusion at various times in a star's lifetime. Gravitational collapse produces energy from the conversion of potential energy into kinetic energy. Two mechanisms can produce nuclear energy. **Fusion** is when two nuclei collide with enough energy to overcome the electrostatic repulsion of the positive charges of the colliding particles; as a result the nuclei merge. **Fission** is when a massive nucleus breaks into less massive ones. The amount of energy released in a nuclear reaction is given by the difference between the total mass of all the fusing particles and the mass of the resulting particle, multiplied by the speed of light squared ($E = Mc^2$).

- In the **proton-proton chain**, four hydrogen nuclei are transformed into a helium nucleus plus energy plus additional particles. Higher-mass stars on the main sequence obtain their energy from the **carbon-nitrogen-oxygen (CNO)** cycle, which also converts four hydrogen nuclei to helium, with carbon acting as a catalyst.

- The proton-proton reaction produces both **antimatter** and **neutrinos**. Neutrinos are thought to be massless or to have very small mass. Because neutrinos interact only weakly with matter, they exit the Sun without absorption. The experimental search for neutrinos from the Sun in the amount predicted by theory has the capability of providing a test of theories of the solar interior. Thus far, neutrinos have not been observed to come from the Sun in the expected amount.

- The rate at which fusion proceeds depends strongly on temperature, which determines the particles' ability to overcome the repulsive forces, and density, which determines how often particles will interact.

- A star is in **equilibrium** if its gravity and outward pressure (**gas pressure** and **radiation pressure**) are in balance (**gravity-pressure** or **hydrostatic equilibrium**), and if the amount of energy radiated equals the amount of energy generated in the core (**thermal equilibrium**).

- Energy produced in a star's core by fusion moves toward the surface by radiative transport and/or convective transport. In the Sun, radiation is important throughout most of the volume, but convection is important in the outer regions.

- The complex equations describing equilibrium conditions along with energy production and transport laws are solved by a computer to give a **stellar model**, which describes how temperature, pressure, density, chemical composition, mass, and luminosity vary throughout a stellar interior.

- The **Russell-Vogt theorem** says that a star in equilibrium has a structure determined by its mass, chemical composition, and age.

- The lifetime of a star on the main sequence depends on the mass and luminosity. From the **mass-luminosity relation** we find the lifetime is shorter for more-massive stars.

Conclusions

- The Sun is too old for gravitational collapse to have been the main source of energy over its lifetime. There is sufficient nuclear energy available within the Sun, however, to provide the observed energy for an adequately long time.

- The proton-proton chain occurs in the Sun and other stars of similar and lower mass.

- We infer that the corona is hot from the presence of spectral lines of highly ionized atoms and from the emission of X-rays.

- Although not understood in detail, the complex magnetic field structure observed in the Sun apparently result at least partly from **differential rotation**.

- The Sun is observed to oscillate. The details of the oscillations depend on the interior structure and rotation. Astronomers have used observations of oscillations to model the Sun's interior rotation.

SUMMARY QUESTIONS

1. What is meant by Kelvin-Helmholtz contraction? What is its significance insofar as providing the Sun's energy?

2. What do we mean by nuclear fission and by nuclear fusion? Why is nuclear fusion a viable source for the Sun's power, while fission is not?

3. What are the steps involved in the proton-proton chain? What is the net result in terms of what goes in and what comes out?

4. What are the general features of the CNO cycle? What is the end result in terms of particles going into the reaction and those coming out?

5. What are the conditions of density and temperature required for fusion to take place? Why is each variable important? What would happen to the total amount of energy if the variables were altered?

6. What is antimatter? What would be the result if antimatter and matter came into contact with one another?

7. What do we mean by a neutrino? What is the importance of the solar neutrino experiment for astronomy, and what are the current results?

8. What is equilibrium? What forces in a typical star maintain a state of equilibrium? What happens to a star if it deviates only slightly from equilibrium?

9. What is meant by a stellar model? How does the study of a stellar model allow astronomers to understand the interior structure of a star?

10. Why does the mass-luminosity relationship exist? What are its implications for the lifetime of a star?

11. What are the various layers of the Sun? What type of spectrum does each layer produce? Explain.

12. What is the meaning of the term convection? What direct evidence is there that convection occurs in the Sun?

13. How is it known that the solar corona is very hot?

14. What is the sunspot cycle? How are sunspots believed to arise?

APPLYING YOUR KNOWLEDGE

1. Hypothesis: The Sun's energy comes from a slow contraction and conversion of gravitational potential energy. Question: Present evidence for or against the hypothesis.

2. Suppose the energy generated in the core of a star were to increase suddenly, and that a new state of equilibrium could not be attained rapidly. What would occur and why?

3. What would happen if a balloon filled with air at room temperature (300 K) were submerged in liquid nitrogen (78 K)? Why?

4. Why is radiation pressure more important in hot stars than in cool ones?

5. Why is it not possible for you to compress an inflated balloon with your hands? (Assume your hands are large enough to encompass the balloon.) Explain the source of the forces involved.

6. Why is blueshifted granulation brighter than the surrounding darker regions that are redshifted?

■ **7.** Verify that charge and atomic mass number are both conserved in each of the reactions in the CNO cycle.

■ **8.** How large is the largest sunspot visible in Figure 16–13? How large is the smallest? Express your answers in kilometers, and compare your answers with the diameter of the Earth.

■ **9.** Use Figure 16–18 to determine the percentage variation of sunspot number from cycle to cycle.

■ **10.** Determine the height above the photosphere of the prominence shown in Figure 16–28a in both kilometers and Earth diameters. Similarly, estimate the apparent length of the prominence.

■ **11.** Use Figure 16–18 to predict when the next sunspot maximum will occur. Use the figure to predict the part of the cycle the Sun is currently in.

■ **12.** Although a photon is a "massless" particle, it behaves as though it has mass. Use Einstein's equation to find the apparent mass of a photon whose wavelength is 5000 Å. (Hint: How is wavelength related to a photon's energy?)

■ **13.** If the typical human eye is able to resolve one minute of arc, how large would a sunspot have to be for it to be seen with the naked eye? Express your answer both in kilometers and in Earth diameters.

ANSWERS TO INQUIRIES

16–1. The inverse square law.

16–2. The assumption is made that the Sun's energy output has not varied significantly in roughly five billion years. This is reasonable, in view of the fact that the earliest known precursors to life on Earth are roughly four billion years old. For life to have survived, the Earth's temperature could not have varied greatly since that time.

16–3. *Atomic mass number:* There are two protons on the left and one proton and one neutron on the right, for a total atomic mass number of two (as shown by the superscript 2 for deuteron). *Charge:* On the right, the deuteron and the positron each have one positive unit of charge.

16–4. *Atomic mass number:* The sum of the superscripts—i.e., the atomic mass number—is 3 on each side. *Charge:* Each hydrogen nucleus has 1 charge for a total of 2. The helium nucleus has a charge of 2.

16–5. Reaction (b) does not conserve charge. Reaction (a) actually does take place in the Sun, although only infrequently. The nucleus designated ^3_1H is that of an extra-heavy form of hydrogen known as **tritium**.

16–6. Six protons are involved in the creation of each helium nucleus, with two protons left over.

16–7. 0.05×10^{-24} gm $\times (3 \times 10^{10}$ cm/sec$)^2 = 5 \times 10^{-5}$ ergs, roughly.

16–8. $M = E/0.007c^2 = 4 \times 10^{33} / 0.007(3 \times 10^{10})^2 = 6.3 \times 10^{14}$ gm, or almost 700 million tons each second.

16–9. Four hydrogen nuclei enter. One helium nucleus and two positrons are produced. Both convert four protons to helium plus energy.

16–10. Raising temperature increases the rate of energy generation in two ways: by making it easier for reactions to occur, and by increasing the frequency of collisions between particles. Raising density will increase the rate of energy generation by making collisions more frequent.

16–11. Detection of neutrinos allows us to "look" into the solar interior. They would confirm the theory of nuclear fusion as the Sun's energy source, and would offer a tool for measuring the temperature and pressure at the Sun's center by providing a method of inferring the reaction rate from the number of neutrinos produced per second.

16–12. The charge of the chlorine atom is 17; that of the argon atom is 18, plus the −1 charge of the electron, gives a total charge on the right of 17.

16–13. If the energy-generation rate were too low, the interior of the star would not be hot enough to support its weight. The star would collapse slightly, increasing the internal density, temperature, and pressure. The energy-generation rate would increase, and a new equilibrium would be attained.

16–14. Using Figure 16–11, we see that the inner 50% of the radius has 88% of the mass; 97% of the mass is inside 70% of the radius. Only 3% of the mass is in the outer 30%.

16–15. From Figure 16–11, 99% of the luminosity comes from the inner 30% of the radius.

16–16. From Figure 16–11, the original amount of hydrogen in the core is the same as currently present in its outer layers. Then, given an original amount of 70% hydrogen and a current amount of only 35%, (70 − 35)/70 = 50% of the original hydrogen in the core has changed to helium.

16–17. We would expect more massive stars to attain higher internal temperatures, densities, and pressures. There-fore, their energy-generation rate would be higher. Thus more-massive stars have greater luminosities (at least on the main sequence).

16–18. It burns twice as much fuel at 16 times the rate of the Sun, so it can be expected to live only about one-eighth as long—about a billion years.

16–19. From Wien's law, 3×10^7 / 5000 Å = 6000 K.

16–20. The bright areas are rising and the dark areas are falling. This is direct evidence for the existence of con-vection in the outer layers of the Sun.

16–21. A K2 star has a temperature of about 4800 K (Chapter 14), so the sunspot is 1000 K cooler. Because the en-ergy per unit area depends on T^4, we would have $(4800/5780)^4 = 0.48$. The sunspot radiates about half as much energy per unit area.

16–22. The light from the star will vary with time.

16–23. A height of 5000 to 20,000 km is 0.4 to 1.6 times the diameter of the Earth.

17

STAR FORMATION AND EVOLUTION TO THE MAIN SEQUENCE

When we reflect . . . , we may conceive that, perhaps in progress of time, these nebulae which are already in such a state of compression, may still be farther condensed so as to actually become stars.

SIR WILLIAM HERSCHEL

The study of star formation is a lively subject in which "accepted" ideas change rapidly. Much of the new information in these areas results from the development of new instrumentation that is sensitive to electromagnetic radiation in regions other than the visible spectrum. For example, observations at millimeter, centimeter, and meter wavelengths have been crucial in providing evidence for and against a variety of hypotheses, as have observations in the ultraviolet and infrared parts of the spectrum, made possible by the space program.

In this chapter we study the material from which stars are made. Then we examine some of the theories about how stars form. Finally, we consider a variety of observations that show that the theoretical picture we have painted is essentially correct in its general outline.

17.1

MATTER FOR STAR FORMATION

Stars must form from *something*! The most likely material is what we call the **interstellar medium**, which consists of atoms, mostly in the form of neutral and ionized hydrogen gas, as well as molecules and dust located between the stars in the Milky Way. **Figure 17–1** is a visual-light photograph showing ionized hydrogen (the reddish regions) and dust (the bluish and the black regions). Some aspects of the ionized and neutral components of the interstellar medium were presented in Chapter 13 as examples of what we can learn about the universe with the use of spectroscopy. The roles played by neutral and ionized gas, and dust, in star formation will be presented in this chapter. (Further aspects of dust will be described in Chapter 20.) We will also discuss the important role molecules play both in forming stars and in providing us with observational tools with which we can better understand star formation.

A little background on the interstellar medium is useful here. In Chapter 13 we learned that cool, low-density neutral gas produces interstellar absorption lines in the spectra of stars lying beyond the absorbing gas. Often, the interstellar lines consist of many components, thus indicating that the interstellar medium consists of numerous **diffuse interstellar clouds** of gas along any given line of sight. A typical diffuse gas cloud is 50 light-years across and contains some 400 solar masses of material. This typical cloud has a density of some 10 atoms per cubic centimeter—an extremely good vacuum—at a low temperature of 100 K. (For comparison, the density of air at sea level on Earth is some 10^{19} atoms/cm³; a high vacuum in a laboratory on Earth still has 10^5 atoms/cm³.) These properties of the

FIGURE 17–1. The Trifid Nebula, showing ionized hydrogen in red, light reflecting off dust in blue, and opaque dust in black.

average interstellar cloud are useful for comparison with the regions in which stars form.

MOLECULES IN THE INTERSTELLAR MEDIUM

Astronomers have known for many years that molecules exist in the atmospheres of some stars. For example, as discussed in Chapter 14, M-type stars show titanium oxide (TiO) in their spectra. Other molecules, such as C_2, CH, CH^+, and CN have also been seen in stellar spectra for many years. These molecular lines are attributed to absorption by cool gases within the extended and diffuse stellar atmosphere. In spite of these observations, astronomers generally argued that the interstellar medium was an unfavorable location for the formation of molecules.

Inquiry 17–1 The densities and temperatures of many interstellar clouds are quite low. Why might these facts have led astronomers to believe that the formation of molecules in such clouds would be highly unlikely? (Hint: How would the collision rate between single atoms be affected by these conditions?)

Some astronomers pointed out an additional reason why molecules were not expected in interstellar space. Even if a molecule were to form, high-energy ultraviolet radiation flooding space from the many hot stars in the galaxy would soon break it apart, because only a small amount of energy is needed to hold a molecule together.

Observations using new technology have proven these early theoretical lines of reasoning to be entirely wrong. In Chapter 14 we found that molecules are capable of radiating energy in many different parts of the spectrum. When a molecule goes from one state of rotation to a state of slightly slower rotation, it will give off the excess energy in the form of a low-energy photon with a wavelength on the order of a millimeter or so, in the radio region of the spectrum. When a molecule goes from one vibrational state to another, it typically emits photons in the infrared spectral region. When electrons within a molecule make transitions, the photons may appear in the ultraviolet, visible, or near-infrared part of the spectrum. Radiation at infrared wavelengths has been difficult to observe, however, because the Earth's atmosphere obscures much of it. Millimeter-wavelength radiation, however, does penetrate the atmosphere, but only recently has technology developed to the point where it can be detected easily. For example, the radio-wavelength transitions of the hydroxyl (OH) molecule were first observed only as recently as 1963. Thus technological innovations are critical to the advancement of astronomical knowledge.

New interstellar molecules continue to be discovered. **Table 17–1** shows a partial list of some of the interstellar molecules that have been identified, ranging from simple molecules such as NH_3 (ammonia), H_2O (water), and H_2CO (formaldehyde, or embalming fluid) to more complex compounds such as CH_3CH_2OH (ethyl alcohol, the stuff of hangovers). Numerous spectral features have been observed in the millimeter region of the spectrum but are not yet identified with a specific molecule; in all probability, even more complex molecules than these exist.

We mention these molecules in interstellar space because they play two important roles in star-formation studies. First, without molecules star formation might not be able to occur; they are a necessary part of the process. Second, molecules make their presence known through their characteristic emissions in various spectral regions. They provide information about the sites where star formation occurs. Where and how these molecules form is the subject of the next section.

THE FORMATION OF MOLECULES IN DUST CLOUDS

Why did interstellar molecules turn out to be so abundant in space when theoretical predictions suggested just the opposite? The answer to this question comes from an examination of the regions in space where molecules occur. Our basic observation is that molecules are located inside clouds; the molecules are not simply scattered uniformly throughout interstellar space. Furthermore, they are not located within the diffuse clouds discussed above, but in clouds that contain extensive amounts of light-absorbing dust. For this reason such clouds are known as **dark clouds**.

TABLE 17-1

Some Observed Interstellar Molecules

H_2	Molecular hydrogen	NH_3	Ammonia
CN	Cyanogen	H_2CO	Formaldehyde
CO	Carbon monoxide	H_2CS	Thioformaldehyde
CS	Carbon monosulfide	HNCO	Hydrocyanic acid
SO, SO_2	Sulfur monoxide and dioxide	H_2CNH	Methylenimine
SiO	Silicon monoxide	HC_3N	Cyanoacetylene
OH	Hydroxyl	H_2CO_2	Formic acid
CH, CH^+		CH_3OH	Methyl alcohol
H_2O	Water	CH_3CN	Methyl cyanide
H_2S	Hydrogen sulfide	CH_3CH_2OH	Ethyl alcohol
HCN	Hydrogen cyanide	CH_3CH_2CN	Ethyl cyanide

Inquiry 17–2 Using what you know about spectra, how do you think spectral lines formed in molecular clouds might appear different from the lines formed in less dense diffuse clouds? Explain your reasoning.

The strengths of molecular spectral features frequently make it possible to determine the density and temperature of the regions where molecules are found. Astronomers have discovered that the dark clouds containing molecules are, in general, *more* massive, denser, and cooler than the diffuse interstellar clouds that cause optical absorption lines (Chapter 13).

A specific example is the dense molecular cloud Sagittarius B2, which lies in a direction from the Sun toward the center of our galaxy. **Figure 17–2** shows a radio contour map[1] of this enormous cloud, which may have a mass greater than one million times the Sun's. It contains every molecule listed in Table 17–1. Typical dark clouds have densities as high as 10^6 particles per cubic centimeter and temperatures below 100 K, with values from 10 K to 30 K being common. These dense, cold molecular clouds, which may well be the most massive objects in the galaxy, are called **giant molecular clouds**.

Their high mass, relatively high density, and cool temperatures make giant molecular clouds a likely place for star formation to occur. Their large masses and densities produce strong gravitational forces, and their low temperatures mean that the pressure that would normally work against gravitational collapse is small. The possibility (and hope) of seeing the very early stages of star formation inside these clouds is one of the principal motivations for continuing to observe them.

The presence of large quantities of dust in the clouds is the key to understanding how molecules can exist there. The dust is highly effective at absorbing optical and ultraviolet radiation. This means that molecules deep within a dust cloud are almost totally shielded from the ultraviolet radiation that tends to break up, or dissociate, molecules. Once a molecule in such a cloud is formed, it is not easily destroyed, and a concentration of molecules can build up.

But the dust does pose serious difficulties when we attempt to observe the visible radiation emitted by stars that might be forming inside the clouds. The scattering and absorption of visible light by the dust in a dense cloud can be so great that less than one-millionth of the light escapes. It is probable that there are stars born inside clouds containing so much dust that the stars live out their entire lives inside the dense layer of dust and never reveal themselves optically to the outside universe.

[1]Contour maps were introduced in Chapter 12.

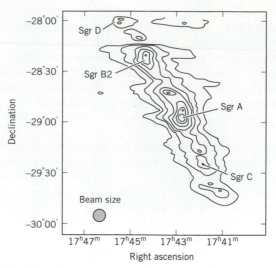

FIGURE 17–2. An infrared map of the galactic center. The infrared contours show concentrations that coincide with the strong radio sources Sgr A, B2, C, and D.

Note, however, the following fortunate circumstance: Because dust is not effective in absorbing or scattering long-wavelength radiation, the millimeter-wave emission that is given off by molecules embedded in the dusty cloud can freely escape from a cloud and be observed. Thus the same dust that is such a villain to the optical astronomer is a virtual blessing to the astronomer studying millimeter-wave emission. By shielding the centers of the dense clouds from stray radiation, the dust particles allow the concentration of molecules to build up. The long-wavelength radiation emitted by these molecules passes through the dust unhindered on its way to a distant observer.

THE HYDROGEN MOLECULE

The most abundant molecule in the giant molecular clouds is the hydrogen molecule, H_2. For complex reasons related to its symmetrical shape, the hydrogen molecule cannot be detected at millimeter wavelengths like most molecules. We must instead rely on observing it in the ultraviolet part of the spectrum. The Copernicus satellite carried an ultraviolet spectrometer capable of measuring these wavelengths, and it was extremely successful in detecting molecular hydrogen in the galaxy. The satellite looked at bright stars and detected H_2 in their absorption spectra. **Figure 17–3** shows a high-resolution spectrum of the star ζ Ophiuchi. The cloud in front of ζ Ophiuchi has a relatively low density for a dark cloud; to some extent, radiation can reach the interior regions and destroy molecules there. The ob-

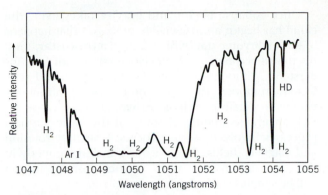

FIGURE 17–3. Measurements from the Copernicus satellite of the ultraviolet spectrum of the H_2 molecule, as seen in absorption in the star ζ Ophiuchi.

served presence of H_2 in this relatively low-density dark cloud suggests that in extremely dense and dusty molecular clouds there must be a good deal of H_2. From this we infer that most of the mass of many clouds is probably molecular, rather than atomic, hydrogen.

The relationship between H_2 and dust particles is especially intimate, because dust grains are responsible for the formation of the hydrogen molecule. If two hydrogen atoms happen to collide, it is unlikely that they will combine into a molecule. However, if a hydrogen atom collides with a dust grain, it has a high probability of sticking there. If two hydrogen atoms have become stuck to a particular grain, they will migrate slowly and randomly around the grain until they encounter each other; in this situation, they will be able to combine. When the molecule forms, there is a small amount of energy released that ejects the H_2 molecule from the grain and forces it back into the cloud. Thus the grains are the vehicles that allow the molecules to form, and the grains protect the molecules from energetic photons that might destroy them.

Now that we know that matter for star formation exists, we will examine some of the theoretical questions concerning the collapse of the giant molecular cloud into one or more stars.

17.2
THEORETICAL STAR-FORMATION STUDIES

Beyond the general assertion that star formation is taking place, we would like to know more about *how* star formation actually proceeds. For example, what causes stars to form from gas clouds? What do they look like

before they become stars? How long does it take them to form? Did star formation proceed more rapidly in the past, or has it occurred at a uniform rate over the history of the galaxy? How do the temperature and luminosity change as the star evolves (in other words, what path in the H-R diagram does an evolving star take)? What observational evidence is there that stars are born, live a long life, and then die out? Astronomers are only now beginning to answer some of these difficult questions.

Inquiry 17–3   Dark clouds and hot, luminous, massive (therefore young) main-sequence stars are associated with each other far more often than chance would predict. What conclusion do you draw from this fact?

INITIATION OF STAR FORMATION

Star formation requires not only the presence of material from which the stars will form, but a mechanism to cause the cloud to begin to collapse on itself. The physical association of luminous stars with dark clouds provides evidence of a causal relationship. A gas cloud has mass and therefore gravity. A cloud will begin spontaneously to collapse if its gravitational pull upon itself (often referred to as **self-gravity**) is greater than the outward pressure caused by the motions of the particles in the cloud. That outward pressure, of course, is determined by the density and temperature of the gas through the ideal gas law discussed in Chapter 16. The more massive the cloud, the more likely it is to collapse.

The famous British astrophysicist, Sir James Jeans, used these ideas in the early 1900s to derive a simple equation that shows that a typical interstellar cloud having a temperature of some 50 K will spontaneously collapse if it is at least two parsecs in radius; such a cloud would contain 1000 solar masses of material. Because this so-called **Jeans mass** is far greater than the mass of any known single star, astronomers believe some additional mechanism(s) must be active to allow stars of smaller mass to form.

To find additional mechanisms that might cause a cloud to begin to collapse, we need to find a force that could squeeze a cloud to the point where its gravitational attraction for itself could take over and continue the collapse. For example, if a cloud finds itself near a hot star (**Figure 17–4a**), the star's photons will provide a pressure (called radiation pressure) that will tend to compress the cloud. Or, as shown in **Figure 17–4b**, clouds might collide and initiate a collapse.

In general, the motions of the gas in a cloud are turbulent. The gas in different parts of the cloud will swirl

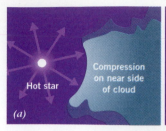

FIGURE 17–4. Two mechanisms that have been suggested for starting the collapse of interstellar clouds. (*a*) Radiation pressure from a nearby hot star. (*b*) The collision of two clouds.

in different directions. It is possible that the motions of gases could produce condensations within the cloud—regions of high density that could start to collapse (perhaps like in Figure 6–11, which refers to planet formation). In fact, we infer that something like this must occur, because many stars are seen grouped together in star clusters. This presumably means that they formed from a large, massive cloud that fragmented and condensed into many stars. Little is understood about the fragmentation process, however.

In recent years, astronomers have begun to understand that a gas cloud may be given a "push" by shock waves produced by a supernova explosion. Such an event occurs, as we will see in more detail in Chapter 19, when a massive star explodes, not only spewing mass and energy into space, but also producing a compression, or shock, wave. Should such a shock wave encounter a gas cloud, it may induce the cloud's collapse. Because we now know that interstellar space is filled with remnants of supernovae explosions, large numbers of such supernova-induced collapses must have happened in the past. In Chapter 7 we saw observational evidence from the abundance of certain elements in meteorites that the solar system may have begun in this way.

There is one additional mechanism astronomers consider that might initiate cloud collapse. As we will see in Chapter 20, our galaxy's spiral-shaped structure may have been determined by a spiral-shaped wave that rotates relative to the gas, dust, and stars. When a rapidly moving gas cloud overtakes a slower moving density wave, compression of the cloud occurs. This compression might initiate cloud collapse so gravity can take over and star formation can begin.

Inquiry 17–4 What are six mechanisms that astronomers believe can initiate the collapse of a cloud to form a star? Explain how each causes collapse to occur.

Once collapse begins and gravity takes over, the cloud has begun an irreversible process. What happens to this embryonic star, which we call a **protostar**, until the time it becomes a fully developed star is not well understood. Although astronomers have computed star-formation models, the results do not agree with observations in detail as well as we would like because we lack an understanding of some of the basic physical processes involved. Some processes, such as rotation and magnetic fields, are difficult to include properly in the computations. For these reasons, our description will be simplified and far from complete.

THEORETICAL MODELS OF EARLY STELLAR EVOLUTION

Our approach to understanding stellar evolution will be to examine how the location of the point representing the star changes in the H-R diagram as time progresses. Our theoretical models predict changes in the star's luminosity and surface temperature at each stage throughout its lifetime. Connecting the resulting successive points in the H-R diagram will produce an evolutionary path in the diagram, much as we did in Chapter 15 for aging children in the height-weight diagram. We will then be able to compare the results of these theoretical models with observations. Agreement means we understand the physical conditions and processes that occur. Lack of agreement might mean a number of things, the most probable of which is our lack of detailed understanding of the physical processes included in the model calculation. In such a situation the researcher would consider the problem further, change the computer program to incorporate new ideas, compute a new model, and again compare the model with observations.

Because a large, cool gas cloud will have a low luminosity, it will be located on the H-R diagram at the far lower right. During collapse, part of the gravitational potential energy will be converted into heat and part into radiation. The density of the gas is low, so radiation will escape from the cloud rather than be absorbed. Because this radiation loss is an energy loss, which cools the interior, the collapse is able to continue because the interior temperature and pressure are kept from rising. In addition, the processes of dissociation of molecular hydrogen and ionization of atomic hydrogen take up energy, thus preventing the temperature from rising much.

While the entire cloud collapses, the inner region collapses more rapidly than the outer ones (**Figure 17–5***a*). A pre-stellar core is thus formed (Figure 17–5*b*). Eventually, the collapsing core builds up such a pressure (Figure 17–5*c*) that the contraction stops, reverses direction, and expands (Figure 17–5*d*). This phase is often referred

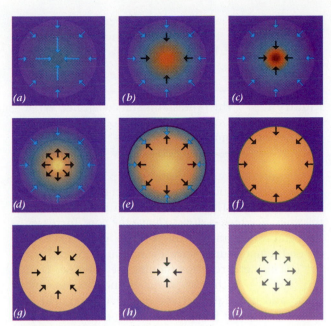

FIGURE 17–5. A series of diagrams showing the dynamical collapse of a cloud into a protostar.

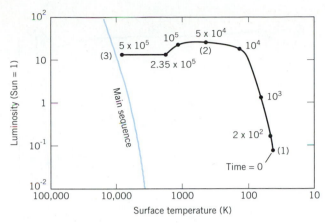

FIGURE 17–6. The path in the H-R diagram of a protostar during dynamical collapse.

to as the core "bounce." The expanding core then interacts with the still-collapsing outer regions (Figure 17–5e) until the entire mass is once again collapsing (Figure 17–5f). The core density now increases (Figure 17–5g) until photons are no longer free to escape and the core becomes opaque. The renewed collapse process continues again but with increasing energy (Figure 17–5h) until there is yet another bounce of the core (Figure 17–5i). After this second core bounce, the temperature and luminosity place the object in an observable part of the H-R diagram. Because the entire process occurs relatively rapidly, perhaps in only thousands of years, astronomers refer to it as the *dynamical phase* of pre-main-sequence evolution. The final bounce (Figure 17–5i) and increase in luminosity are thought to take only some 200 days!

The variation in temperature and luminosity in the H-R diagram during the rapidly changing dynamical phase is shown in **Figure 17–6.** The protostar evolves from a pre-stellar core at point *1* to point *2* in 5×10^4 years. Still contracting, it reaches point *3* in an additional 4.5×10^5 years. All stars, no matter what their mass, begin in this general region of the H-R diagram.[2]

Our protostar at point *3* is not yet in equilibrium because it is still collapsing and producing energy by the

slow conversion of gravitational potential energy into radiation. This energy is transported from one place to another entirely by convection. In other words, the protostar is completely turbulent at this point.

We do not observe visible radiation during this phase of collapse because dust surrounding the protostar absorbs the radiation. These dust grains will absorb some of the energy of the collapse and heat up a bit. Such warmed grains will radiate in the infrared part of the spectrum. Thus an expected observational signature of star formation will be the presence of infrared sources within dusty and gaseous regions of space. In fact, because the dust will usually hide the protostar from view, this infrared radiation provides the main method of observing star formation in progress.

Eventually the protostar may break out of its dusty cocoon and be seen in visible light. Even now the still-contracting star is probably accreting (adding) hydrogen from its surroundings. The situation is complex because, while mass **accretion** (growth by externally added material) is expected to occur, astronomers have not yet observed it for sure. On the other hand, extensive mass loss is observed to occur through an intense stellar wind. What is dominant, mass loss by the stellar wind, or mass accretion? How do the degrees of mass loss and mass accretion vary with time?

THE APPROACH TO THE MAIN SEQUENCE

We now continue our discussion of stellar formation from point *3* in Figure 17–6. While the following discussion is for a star of one solar mass, we will see that it is basically correct for stars of other masses, also. To properly portray the evolution in the H-R diagram, we begin a new diagram in **Figure 17–7.** The protostar continues

---

[2]The reason is that temperature depends on the ratio of mass to radius. Since the ratio remains nearly constant during this part of the evolution, all stars will have nearly the same temperature.

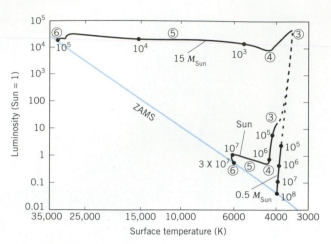

FIGURE 17–7. The path in the H-R diagram of protostars of different masses.

to undergo changes in its density and surface temperature, which makes its position in the H-R diagram change. While the protostar is still collapsing and still obtaining its energy from gravity, its luminosity begins to decrease (points 3 to 4 in Figure 17–7), so the evolutionary path in the diagram is nearly vertical. During this time, the changing temperatures and densities cause the energy transport to change from convection to radiation. Eventually, at point 4, all the energy transport is by radiation, which halts the decrease in luminosity. The surface temperature begins to increase, changing the direction of the evolutionary path in the H-R diagram. All the while, of course, the star has been collaps-

ing and the core temperature has been increasing. At point 5, the core temperature reaches the 10 million kelvins required for the proton-proton reaction to commence.

Inquiry 17–5 The protostar is still collapsing. What changes in physical conditions must occur to halt the collapse and bring it into equilibrium?

The onset of nuclear fusion near point 5 has an important consequence that results from the events shown in **Figure 17–8.** The onset of fusion produces energy; this energy, which is slow to leave the core, increases its temperature slightly (recall the ideal gas law). The increased temperature increases the outward pressure in the collapsing core, thus halting the collapse. During this time, in which nuclear fusion increases in importance and gravitational collapse decreases, the star moves from point 5 to point 6 in Figure 17–7. Our protostar has become a star.[3]

This description is basically correct for stars of all masses; details differ depending on the mass. For example, for a more massive star, the changes will occur more rapidly. A general rule in stellar evolution is that the more massive the star, the more rapidly the changes occur. The path in the H-R diagram for a 15-solar-mass star is shown in Figure 17–7. The more massive star reaches equilibrium at both higher temperature and luminosity, as shown. For a 0.5-solar-mass star, the vertical distance moved in the diagram is longer, and the time is considerably lengthened, as the star achieves equilibrium at low temperature and luminosity.

When stars of intermediate mass reach equilibrium, they are located at intermediate points in the H-R diagram. Connecting those points for stars in equilibrium that cover the range of observed stellar masses determines a line known as the **zero-age main sequence**, or **ZAMS**. The ZAMS is the initial main sequence along which stars exist. Because the pre-main-sequence lifetime is short compared with the main-sequence lifetime, the star's age is taken as zero at this point.

The results just discussed came from the calculation of theoretical models. These models provide us with the theoretical basis for stating that the main sequence is a mass sequence for stars having hydrogen fusion as the source of energy. Furthermore, these models also provide a theoretical reason for the observed main-sequence mass-luminosity relation presented in Chapter 15.

> Hydrogen-burning ignition
> ↓ produces
> Energy generation
> ↓ which produces
> Higher temperature
> ↓ which produces
> Higher pressure
> ↓ which produces
> Slight expansion of core
> ↓ which produces
> Slight cooling of core
> ↓ which results in
> Thermal equilibrium
> ↓ which means
> Collapse ceases

FIGURE 17–8. The chain of events after the onset of hydrogen fusion in a stellar core.

[3]There is no agreed-upon definition of when a protostar becomes a star. We will take it to be around the time when core hydrogen burning becomes important.

TABLE 17–2

Time Scales for Pre-Main-Sequence Evolution

Mass (solar masses)	Spectral type	Time to reach ZAMS (in years)
30	O6	30,000
10	B3	300,000
4	B8	1,000,000
2	A4	8,000,000
1	G2	30,000,000
0.5	K8	100,000,000
0.2	M5	1,000,000,000

The time scales for collapse to the main sequence of stars of different masses are shown in **Table 17–2**.

What about objects less massive than those in Table 17–2? In a protostar less massive than 0.08 solar mass, the temperature never reaches the 10^7 K required for ignition of the proton-proton reaction. Therefore, such an object never becomes what we might call a star, but instead becomes a **brown dwarf** or, in the case of a very-low-mass object, a planet.

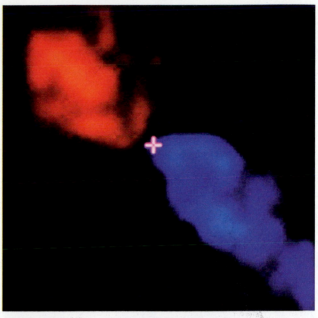

(a)

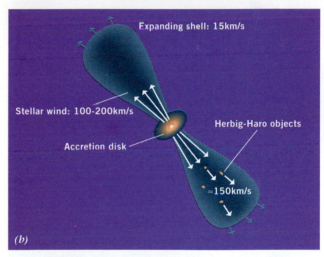

(b)

FIGURE 17–9. (*a*) An image of L1551 as observed by a radio telescope in radiation from the CO molecule. Note the bipolar nature. (*b*) A model to explain the observations in part *a*.

17.3

OBSERVATIONAL STAR-FORMATION STUDIES: THE SEARCH FOR STELLAR PRECURSORS

Our discussion of star-formation theory provides us with the ability to interpret new observations, and to make a number of predictions that may be checked observationally. Unfortunately, astronomers have not yet found a clear-cut example of an object whose sole energy source is contraction. Because the search for one is so intense, such an object has been described as "the Holy Grail of infrared astronomy." In their search astronomers have found a variety of different types of objects that appear to play some role in the overall picture of star formation. These are often referred to as **young stellar objects**, and are the subjects of the following sections. A major goal of astronomers studying young stellar objects is to tie all the pieces together into a coherent picture.

COMPLEXES OF MOLECULAR CLOUDS

Young stellar objects of all kinds are observed by radio astronomers to be surrounded by symmetrically placed molecular clouds (**Figure 17–9a**). This figure shows blueshifted molecular gas on the lower right and redshifted gas on the upper left. These observed gas flows are called **bipolar flows** (*bi* meaning two, *polar* meaning from opposite extremes). They are symmetrically placed with respect to a bright infrared source centered between them. The bipolar flows are thought to result when strong stellar winds, moving at 100–200 km/sec, drill holes through gas accreting onto the collapsing protostar. Due to rotation the gas surrounding the object forms into an **accretion disc** in the equatorial plane

TABLE 17-3

Properties of Molecular Clouds

Property	Giant Molecular Clouds		Small Molecular Clouds	
	Core	Envelope	Core	Envelope
Temperature (K)	20–100	10–20	10–20	10
Mass (solar)	100–5000	10^3–10^5	1–100	10–1000
Size (parsecs)	1–3	10–200	0.05–1	1–10
H_2 density (per cm^3)	10^4–10^7	10^2–10^3	10^4–10^6	10^2–10^3
Number in Galaxy	5000		25,000	

(review Figure 6–8 to understand why the disc is flat). Because the disc is thinner along the rotation axis than in its equatorial plane, the intense stellar wind can more easily penetrate the gas along the rotation axes. Two holes are punched in the gas in opposite directions, thus explaining the bipolar nature of the young stellar objects (**Figure 17–9***b*).

Radio observations have identified two types of molecular clouds associated with young stellar objects: the giant molecular clouds introduced previously, and small molecular clouds. These molecular clouds have sufficient mass for gravity to prevent them from dissipating. It is these clouds, whose characteristics are given in **Table 17–3**, that eventually collapse to form stars. The warm and massive giant molecular clouds form high-mass pre-stellar cores, while the cooler and less massive small clouds form low-mass cores. These cores then fragment into stars. The giant molecular clouds produce both large-mass and small-mass stars, but the small clouds form low massive stars almost entirely.

OB ASSOCIATIONS

The giant molecular clouds are the birth places of massive stars (those greater than three solar masses). Such massive stars have high surface temperatures and are therefore of spectral types O and B. Because a large number of such stars will form from a single cloud and are observed to be associated with each other, such stellar groupings are called **OB associations**. Remembering that the very existence of luminous O and B stars reveals their youth (because they burn their fuel so rapidly), we know the stars cannot have strayed far from their birthplace. The observation of remnant gas within OB associations is consistent with their status as groupings of newly formed stars.

One of the early observations made of the OB associations was that they were expanding—that is, they were not held together by gravity. In fact, the observed stellar motions were so large that the associations would re-

main together for only a few million years. But how can the OB associations be unbound when the giant molecular clusters from which they formed were bound? This is a long-standing problem for which an answer was only recently discovered. We describe this problem here because it presents us with a beautiful, consistent picture of star formation involving giant molecular clouds.

The first panel in **Figure 17–10** shows a giant molecular cloud some 50 parsecs across in which about 1% of the mass forms low-mass young stellar objects (YSOs). Each of these young stellar objects has its own random motion relative to every other object. The huge mass of the giant molecular cloud provides enough gravity to prevent the young stellar objects from escaping. Eventually, some massive stars may form. These massive stars, which make up a small subgroup of OB stars (panel *2*), emit large amounts of ultraviolet photons that dissociate the H_2 molecules and ionize hydrogen, producing an expanding region of ionized hydrogen called an **H II region**. Some of the dissociated and ionized gas then disperses and is lost to the giant molecular cloud (panel *3*). The expanding H II region may stimulate another episode of star formation farther inside the cloud. As before, more gas will be dissociated, ionized, and lost, and yet another subgroup of OB stars may form that has its own expanding H II region.

What is happening is that some of the mass of the giant molecular cloud is being converted into stars, and some is being dispersed. In a relatively short time, 90–99% of the original gas that gravitationally bound the giant molecular cloud and stars together in the first place is lost. The random motions of the low-mass stars formed earlier are now large enough for the stars to leave the site of their birth. The end result, then, is an expanding association of OB stars, some of which are older than others (panel *4*). The less massive stars that were formed originally are then able to expand into the surrounding field. This model, which fits the observations well, is an example of how a useful model provides a complete, consistent picture.

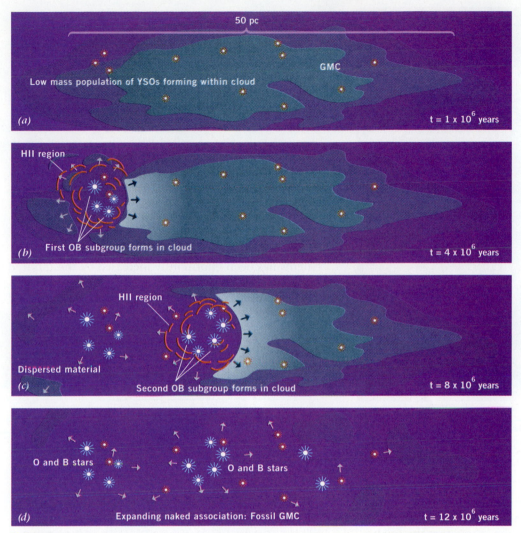

FIGURE 17–10. The formation of an expanding OB association from a giant molecular cloud.

Inquiry 17–6 Stars of type O and B are often seen in regions where star formation is thought to be taking place. How do we know that such OB associations must be young?

Inquiry 17–7 Observationally it has been found that the youngest OB associations are also the smallest in size. What conclusion can you draw from this fact?

BOK GLOBULES

Because star formation begins in the coldest, dustiest regions of the galaxy, the earliest stages of star formation are hidden from view by the thick dust. Astronomers have long been fascinated by the existence of small, dark regions often seen projected against bright nebulae, as in **Figure 17–11**. These regions are known as **Bok globules**, after astronomer Bart Bok, who first studied them in detail and speculated they might be protostars. Observations show that the smaller of these regions are the cores of the small, cold clouds discussed earlier. Recent research has found that a quarter of them are actually observed to have a young star inside; in reality, perhaps all of them do.

However, it should be noted that there are regions in our galaxy and in other galaxies where star formation seems to be taking place *without* globules. Some astronomers therefore argue that, although globules may indeed be self-gravitating objects that are on their way to becoming stars, some stars may *not* begin this way.

FIGURE 17–11. Globules in the diffuse nebula NGC 2244, believed to be the cradles of newly forming stars.

are interested in studying the infrared portion of the spectrum?

In the 1970s it became possible to observe moderately bright point sources of infrared radiation accurately, and it has become clear that these infrared point sources are frequently associated with star-forming regions. A good example of such an object is found in the Orion Nebula, the so-called BN source, named for its codiscoverers Eric Becklin and Gerry Neugebauer. Spectra of this object indicate the presence of dust grains composed of water ice and silicates. Because BN is so small that it appears as a point source of radiation, it may be a protostar developing in a dust cloud visible only at infrared wavelengths.

More recently, the amount of data at infrared wavelengths has been enormously increased by the Infrared Astronomy Satellite (IRAS), which was launched in 1983 and operated for nearly a year before the liquid helium that kept its detectors near a temperature of absolute zero completely evaporated. During its brief lifetime, IRAS detected hundreds of thousands of previously unknown point sources of infrared radiation and discovered new evidence for the formation of stars.

Inquiry 17–10 Why do you think the detectors of IRAS were cooled to such a low temperature?

INFRARED SOURCES

Suppose a condensation does form and grow within a dusty interstellar cloud and increases in temperature and energy output. Calculations show that the condensation will develop a denser, hotter core that is surrounded by a cooler, more rarefied envelope. While the hotter core temperatures may vaporize nearby dust particles, the surrounding *envelope* will remain heavily laced with obscuring grains. However, even though the energy produced by the central object is absorbed by the dust grains, it must reappear somewhere—the law of conservation of energy says that the energy cannot simply disappear. Therefore, the dust grains are heated up and reradiate this energy as heat.

Inquiry 17–8 The temperature of the dust grains in many interstellar clouds is about 300 K. At about what wavelength would one expect most of the radiation from such dust grains to be emitted? In what portion of the electromagnetic spectrum does this wavelength lie? (Hint: Use Wien's Law.)

Inquiry 17–9 What is one reason why astronomers

T TAURI STARS

There are many stars that are called **T Tauri stars** located in the H-R diagram above the ZAMS. The light output of these stars varies irregularly, in a way that suggests they are not completely stable. This observation proves nothing; however, a large number of these stars emit radiation in the infrared, indicating the presence of dust particles, which are necessary for star formation to occur, in the immediate region. T Tauri stars also have spectra showing lines of the element lithium. Lithium is not generally present in older stars because it is easily destroyed at the temperatures present in main-sequence stars. Thus lithium is often taken as a sign of stellar youth. In addition, T Tauri stars show signs of intense mass loss from a stellar wind, just as expected for a protostar. Finally, many T Tauri stars are associated with the bipolar flows discussed previously.

A dark cloud complex important to star formation studies is the Taurus-Auriga complex, which contains a large number of T Tauri stars. Their locations in the H-R diagram are superimposed over theoretical evolutionary paths in **Figure 17–12.** As expected, the stars are located above the ZAMS. Not only that, the stars are mostly on or

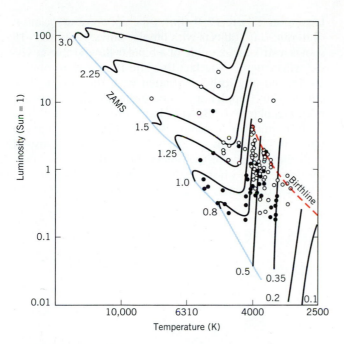

FIGURE 17–12. H-R diagram showing T Tauri stars in the Taurus-Auriga cloud. Also included are theoretical evolution tracks for stars of various masses. Note that most of the stars fall on or below the theoretical birthline.

below a theoretical *birthline*, which shows from theoretical model calculations where stars should first appear on the H-R diagram. The open circles are the so-called classical T Tauri stars, while the filled circles are the naked T Tauri stars, those not having dust surrounding them.

All these data taken together show that T Tauri stars must be stars still collapsing on their way to the zero-age main sequence. They provide an important observational confirmation of the entire stellar-formation picture thus far presented.

Inquiry 17–11 How would you expect intense mass loss in T Tauri stars to make its presence observable?

A summary of the formation of these low mass T Tauri stars is presented in **Figure 17–13**. In part *a*, cores form within the molecular cloud because collapse is more rapid in the central regions. The core, which has some rotation, accretes material onto the protostar and its disc (part *b*). Eventually, the strong stellar wind from the protostar punches holes in the surrounding material along the rotation axes, producing a bipolar flow of gas away from the system (part *c*). Next, the infall of material ceases and the protostellar system becomes a T Tauri star (part *d*). Finally, there is a naked T Tauri or post–T Tauri phase in which there is no dust left within the system (part *e*).

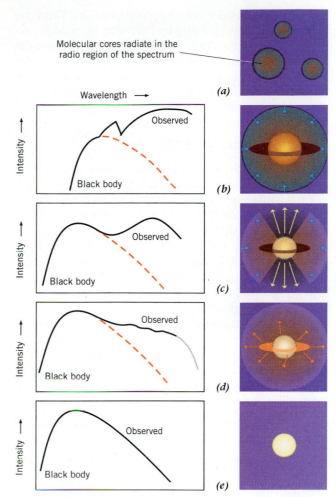

FIGURE 17–13. Various stages in the formation of a low-mass star; the predicted infrared spectrum at each stage is also shown.

This figure also furnishes an example of how observations and theoretical models are used to provide an understanding of phenomena. Each of the final four phases of evolution presented in Figure 17–13 produces a different infrared spectrum. The results of calculations of these infrared spectra are shown in the left side of the figure. Thus, by comparing the observed spectrum with a series of model spectra, astronomers can learn where in the evolutionary cycle a given pre-main-sequence star is.

FU ORIONIS STARS

Comparisons of photographs of the region around the star FU Orionis taken a few years apart revealed the star to have suddenly appeared where no bright star previously existed. Other examples of this phenomenon have also been found and are referred to as **FU Orionis stars**.

The event is thought to be associated with a T Tauri star whose brightness increases by 10 to 100 times due to accretion of material from the surrounding disc onto the star. Such outbursts appear to be repetitive and may occur as many as 100 times for a given T Tauri star. Astronomers do not know whether all T Tauri stars go through this phase.

HERBIG-HARO OBJECTS

In the early 1950s, several objects whose spectrum showed Balmer emission as well as emission of several other elements were discovered. Due to their location within or near dark clouds, their association with star formation was natural. Named after their discoverers, George Herbig and Guillermo Haro, these **Herbig-Haro objects** are now known to be the by-products of star formation, not part of the initial star-formation process itself. **Figure 17–14** shows a Herbig-Haro object within a dark cloud region.

Recent observations reveal Herbig-Haro objects moving away from some young stellar objects, indicating ejection of material producing the Herbig-Haro objects and therefore a direct link between them and star formation. Observations associating Herbig-Haro objects with dusty infrared sources embedded within dark clouds add further evidence of a connection with star

formation. Figure 17–9 shows the apparent association of Herbig-Haro objects with bipolar flows. **Figure 17–15** shows a striking example of a bipolar flow visible at visual wavelengths. The fact that the Herbig-Haro objects are indeed physically associated with the bipolar flow comes from Doppler shift measurements that show the northern object (HH34N) to have a redshift of 120–220 km/sec, and the southern object (HH34S) to have a blueshift of 130 km/sec. These velocities are consistent with the motions of the bipolar flows themselves. Furthermore, this photograph shows an infrared source with a jet of material feeding the Herbig-Haro object

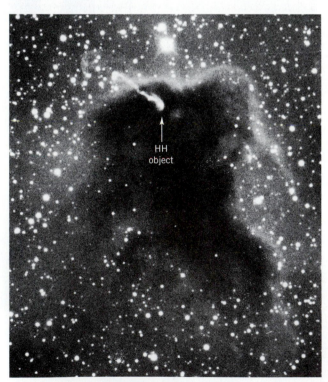

FIGURE 17–14. Herbig-Haro object.

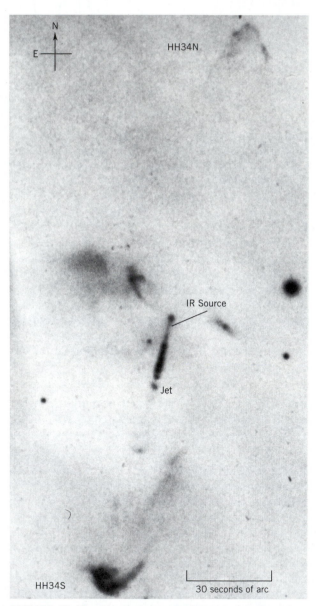

FIGURE 17–15. The region around HH34, showing a jet from a central source toward HH34S.

HH34S to its south; while there is no evidence of a northern jet feeding the northern Herbig-Haro object, it may be there and hidden by thick dust. We readily conclude that a variety of data show Herbig-Haro objects to be intimately related to star-formation processes.

The details of how Herbig-Haro objects form are not fully understood. It has been suggested that the outburst that produces FU Orionis stars ejects material that becomes Herbig-Haro objects. Whatever the details, all the theories include the complexities of both stellar winds and shock waves.

STAR CLUSTERS

We are unable to observe the evolution of a single star due to the long time scale for change relative to our short human lifetime. Even the longer time interval since Galileo first pointed a telescope toward the sky is much too short to allow us to observe the normal evolution of even the most massive stars. The way for us to make headway, observationally, is to study a large number of stars at different stages of their lifetimes. But how can we decide which star is where in its history? Here is where star clusters play an important role.

A star cluster is a gravitationally bound grouping of stars. One of the more famous is the easily observable Pleiades, or Seven Sisters, in the constellation of Taurus (**Figure 17–16**). Another important one, also in Taurus and visible with the naked eye, is the Hyades; it forms the V-shaped head of the Bull. These and other clusters are important to us because of three (assumed) properties: (1) all the stars are at the same distance; (2) all the stars formed out of the same material; (3) all the stars formed at the same time. The presumption of equal dis-

FIGURE 17–16. The Pleiades star cluster.

tance means that differences in star brightness are real differences, not manifestations of unequal distances. From the second assumption of forming from the same material comes the premise of similar chemical composition. Finally, the assumption of formation at one time means that each star in the cluster is the same age.

The Russell-Vogt theorem (Chapter 16) says that the structure of stars is determined by mass, chemical composition, and age. Because of assumptions 2 and 3 above, the only difference between stars in a cluster is the mass. But we know that stars of different masses evolve at different rates. Thus, by the time a cluster reaches a certain age, the high-mass stars will have evolved more than the low-mass stars.

For example, **Figure 17–17** shows the expected H-R diagram for an imaginary cluster. In part *a*, which occurs at an age of zero years, all the stars are just reach-

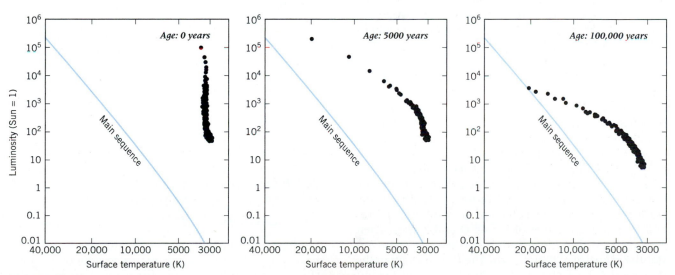

FIGURE 17–17. The evolution of an imaginary cluster in the H-R diagram.

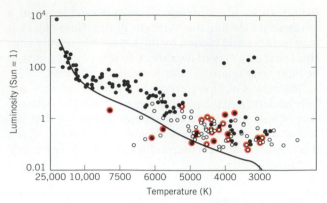

FIGURE 17–19. The Hertzsprung-Russell diagram for NGC 2264, a very young cluster, showing collapsing stars that have not yet reached the main sequence.

FIGURE 17–18. The region around the star cluster NGC 2264. Note the large amounts of both gas and dust.

ing the beginning of their protostellar stage at the end of the dynamical phase of pre-main-sequence evolution; the more massive stars are somewhat more luminous. After 5000 years (part *b*), the more massive stars have evolved toward the left, toward the ZAMS. The less massive stars have not made any significant changes. At 10^5 years (part *c*), the process continues, with the more massive stars evolving more quickly than the less massive ones. The appearance of the diagram at this point presents us with something we can look for observationally: an H-R diagram for a cluster in which the upper main sequence is populated with stars, but with the less bright stars *above* the projected ZAMS.

Some real clusters present evidence of stars that are not yet completely formed. **Figure 17–18** shows the cluster NGC 2264, which contains large amounts of gas and dust, and a number of young O and B stars. Their association with the gas and dust is excellent circumstantial evidence of their youth. The H-R diagram of the stars in NGC 2264, shown in **Figure 17–19**, matches our

expectations from the imaginary cluster, and is therefore an example of a young cluster. Furthermore, symbols outlined in red represent T Tauri stars, and are just where they are expected to be.

A final and important observation related to star formation is that astronomers have never found an example of an *isolated* young star. All the stars that are recognizably young are in associations or young clusters. This strongly suggests that all stars form initially in clusters, and that the isolated stars we see in the galaxy today were created by the disruption or evaporation of star clusters over time. Gravitational interactions with their neighbors and gravitational tidal effects in the galaxy may be sufficient to explain the dissolution of star clusters.

Inquiry 17–12 How would you explain the fact that only the stars in the lower part of the main sequence in Figure 17–19 have not yet reached the main sequence, whereas the stars in the upper part of the diagram have already reached the main sequence?

Inquiry 17–13 Draw an H-R diagram showing two young clusters of different ages. Which one is older, and why do your cluster diagrams differ?

17.4
A PROMINENT REGION OF STAR FORMATION: THE ORION NEBULA

To illustrate and pull together many of the ideas we have discussed, we conclude with a description of one of the most prominent star-forming regions in our

FIGURE 17–20. The Orion Nebula, M42, the nearest and brightest of the diffuse nebulae. This picture shows both the Trapezium stars and the surrounding nebula.

galaxy—the Orion Nebula. **Figure 17–20** shows a photograph of the Orion Nebula, M42, the most famous diffuse nebula in the sky. (See Figure 1–11 for a different view.) It is relatively close to the Sun, at a distance of approximately 1500 light-years, and is bright enough to be seen with the naked eye. The Orion Nebula is a center of active star formation, and it is interesting to study this photograph in some detail, because it can help make our

rather theoretical discussions of star formation earlier in the chapter more concrete.

At the center of Figure 17–20 we see four bright stars known as the Trapezium. The brightest of them is the powerful star θ^1C Orionis, a main-sequence star of perhaps 50 solar masses. Its spectral type is O4.5, which means that it is extremely hot, with a spectrum rich in ultraviolet radiation. The energy from this star is almost entirely responsible for the bright, ionized H II region we see as the Orion Nebula.

θ^1C Orionis is the brightest star in an extensive cluster of stars in the vicinity of the nebula. This cluster contains literally hundreds of stars, all of them young. Other stars in the Orion Nebula have extremely high velocities; apparently they are rapidly leaving the region where they were created. Although it is not obvious why they are moving so fast, their very existence gives another means of estimating the age of the Orion Nebula, because we can easily determine how long ago these stars were in the center of the cluster.

Inquiry 17–14 If you observed a star in Orion to be located 0.1 degree from the center and to have a velocity of 100 km/sec, how long ago would it have been ejected from the center?

Infrared emissions from the Orion region are shown in **Figure 17–21**. The region around the Trapezium stars is very bright at these wavelengths, with numerous discrete sources. Near the Trapezium is a source discovered by Edward Ney and David Allen (NA in the figure) that shows an emission feature at 100,000 Å that is characteristic of heated silicate dust. But the main concentration of infrared emission is to the north and west of the Trapezium stars, as indicated in the figure. In this region there are a number of discrete sources that emit at wavelengths characteristic of OH and H_2O, as well as the sources labeled BN and KL. BN is the Becklin-Neugebauer source discussed earlier, and KL designates an extended region of far-infrared emission called the Kleinman-Low nebula. This cluster of infrared sources is located in a dense region of one of the dark molecular clouds in the Orion region of the sky.

To summarize star formation in the Orion Nebula, the bright diffuse nebula visible to the eye appears to be located on the front edge of this giant molecular cloud. The Trapezium is on the near side of the cloud, ionizing and illuminating the gases in that region (**Figure 17–22**). There may be many other stars in the region embedded deeply in the cloud and invisible optically. Our understanding of Orion has expanded because of the new, unfamiliar view that was produced from data

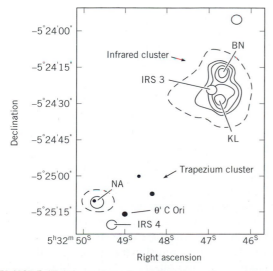

FIGURE 17–21. Infrared emission from the region of the Orion Nebula.

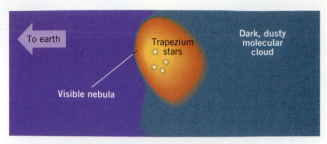

FIGURE 17-22. A model of the visible Orion Nebula and its surroundings.

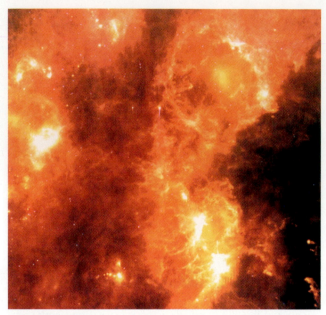

FIGURE 17-23. A view of the region around the Orion Nebula as viewed in the infrared by the IRAS satellite.

gathered by the IRAS satellite (**Figure 17–23**). This image shows the Orion Nebula at the lower right and a huge dust ring around λ Orionis in the upper right (λ Orionis is the top star in Orion's head). In fact, all evidence suggests that an impressive new cluster of stars is forming in this region. One day, when the dust is dispersed and we see this cluster in all its glory, the Orion region will be even more spectacular and beautiful to the naked eye than it is now.

CHAPTER SUMMARY

OBSERVATIONS

- A variety of molecules are found in interstellar clouds located inside dust clouds. The largest molecular clouds are the **giant molecular clouds**, which have high mass, relatively high density, and low temperature. The most abundant molecule is molecular hydrogen, which is best observed in the ultraviolet spectral region. Many molecular clouds exhibit **bipolar flows**, which show that material is ejected from clouds in opposite directions.
- **OB associations** are expanding groups of O and B stars.
- **Bok globules** are small dark regions in dark clouds.
- **T Tauri stars** vary irregularly, contain the element lithium, and are located in the H-R diagram above the ZAMS. Many show bipolar flows of gas away from the star. **FU Orionis stars** are associated with T Tauri stars that suddenly flare up. **Herbig-Haro objects** are emission regions observed to move away from young stellar objects in a bipolar flow.
- H-R diagrams of young clusters show T Tauri stars still collapsing onto the zero-age main sequence.

THEORY

- Cloud collapse can be caused by gravity in a massive cloud, by an external push from radiation pressure from nearby massive hot stars, by turbulence, by shock waves produced in a supernova explosion, or by clouds colliding with density waves in the Galaxy.
- Clouds collapse faster in the central regions than in the outer ones. As the pressure builds up, there is a core bounce. Eventually the protostar becomes visible in the H-R diagram, at which time energy is transported by convection. Continued collapse increases central pressures and temperatures until hydrogen burning ignites in the core, and the star settles onto the zero-age main sequence.
- High-mass stars evolve more rapidly than low-mass stars.
- Dust surrounding protostars is expected to radiate in the infrared spectral region. Protostars in different stages of their evolution are expected to produce differently shaped infrared spectra, which can then be compared with observations.
- Model stars of different masses that are burning

hydrogen in the core array themselves in the H-R diagram in order of mass along the **zero-age main sequence (ZAMS)**. The arrangement has the least-massive stars at the bottom of the ZAMS and thus produces a relationship between mass and luminosity.

- Rotation causes collapsing clouds to form a disc.

CONCLUSIONS

- From observations of bipolar flows we infer the presence of an **accretion disc** surrounding the protostar.
- From observations of infrared sources astronomers infer the presence of heated dust around protostars.

- **Bok globules** are deduced to be the cores of small, cold molecular clouds. From the observations of **T Tauri stars** we conclude that they are young stars still collapsing toward the zero-age main sequence. **FU Orionis stars** are inferred to be T Tauri stars that suddenly accrete matter from their accretion disc. Astronomers conclude that **Herbig-Haro objects** result from bipolar flows during the star-formation process.
- Although there are a number of uncertain details, a variety of consistent observations lend credence to the overall theoretical picture of star formation.

SUMMARY QUESTIONS

1. What is the picture that astronomers have of the formation of stars? Include in your answer a description of the path that a protostar takes to the main sequence, the conditions under which stars are believed to form, and the significance of the fact that O and B stars are often found associated with gas and dust.

2. What is the significance of the main sequence in terms of the results of stellar model calculations?

3. Why are the interiors of dust clouds favorable for the formation and continued existence of interstellar molecules?

4. What are the physical conditions that exist in giant molecular clouds? Include density, temperature, and masses. Why are these conditions favorable for star formation?

5. What is the observational evidence that the variety of objects discussed in the chapter are indeed the precursors to stars?

6. How do observations of one young star cluster differ from observations of a somewhat older cluster? Explain your reasoning.

APPLYING YOUR KNOWLEDGE

1. How would the spectrum of a protostar having both substantial mass loss from a strong wind and substantial accretion from surrounding material differ from that of a star having neither of these traits?

2. What characteristic of the objects shown in Figure 17–13 causes the spectrum in the infrared to deviate from that of a blackbody?

3. Given that a high-mass star has more hydrogen than a low-mass star, why does a high-mass star remain on the main sequence for a shorter time?

4. Why might it be possible to simply *look* at Figure 17–16 and deduce that the bright stars in it are young?

5. In what ways do observations of molecular clouds, bipolar flows, OB associations, Bok globules, T Tauri stars, FU Orionis stars, and Herbig-Haro objects discussed above all provide data relating to star formation?

6. Consider an object similar to that shown in Figure

17–15 in which two Herbig-Haro objects are ejected from a central object. Suppose the central source has a spectral line at 6000 Å. Suppose further that one component has the same spectral line at 6002 Å, while the other component has the same line at 5997 Å. How fast is each component moving with respect to the source, and with respect to each other? What implicit assumption about the motion are you making?

7. Figure 17–2 is an infrared contour map of the central region of our galaxy. If the distance from the Sun to the center of our galaxy is 8,500 parsecs, what is the size in kilometers of the infrared region shown? (Hint: Think about the angular size formula.)

8. Compute what temperature the protosun had when its luminosity and radius were each 1000 times the present values? (Hint: In Chapter 15 we saw a relationship between the luminosity, temperature, and radius.)

ANSWERS TO INQUIRIES

17–1. Both low densities and low temperatures tend to decrease the number of collisions between single atoms. This would decrease the rate at which one might expect molecules to form.

17–2. Lines formed in more dense regions are broader than those formed in less dense regions, because the higher the particle density the more particle interactions occur.

17–3. Their association cannot be due to chance, so it must be causal; that is, the stars must form from the clouds.

17–4. Gravity, radiation pressure from a nearby hot star, cloud collision, turbulence, shock wave compression from a supernova explosion, cloud collision with a density wave within the galaxy.

17–5. To halt the collapse requires increasing the outward pressure; according to the ideal gas law, that can be done by increasing the temperature or the density.

17–6. The O and B stars have a very short lifetime.

17–7. The idea of young associations being smaller than older ones is consistent with their observed expansion.

17–8. $\lambda_{max} = 3 \times 10^7 / 3 \times 10^2 = 10^5$ Å, which is in the infrared spectral region.

17–9. Infrared radiation can pass through dust, so observing it allows astronomers to look into the interiors of dust-gas clouds and acquire information about star formation. Infrared radiation also allows the study of cool objects that radiate very little in the visible portion of the spectrum.

17–10. IRAS was designed to detect infrared radiation; that is, heat. If the detectors are not cooled to a very low temperature, the satellite would detect *itself* and not the desired astronomical object.

17–11. Blueshifted absorption lines would be present.

17–12. The stars in the upper part of the H-R diagram are more massive than the stars in the lower part of the diagram. Because they are more massive, the force of gravity is stronger, and they tend to contract more rapidly.

17–13. Your drawing may be an H-R diagram in which the lower right part of the main sequence for both clusters coincides. At the upper end of the main sequence, one cluster's data should turn off toward the right; this is the older cluster. The diagram appears this way because the more massive stars in the older cluster have evolved away from the main sequence; the stars in the younger cluster have not had enough time to evolve away from the main sequence.

17–14. We are given a velocity and angle; the distance is given in the text as 1500 light-years. From the angular-size relation in Chapter 3, we find the distance of the star from the center of Orion is 0.1° × 1500 ly/57.3° = 2.6 ly, or 2.5×10^{13} km. Because time is distance/velocity, we find t = $(2.5 \times 10^{13}$ km$)/[(100$ km/sec$)/(3 \times 10^7$ sec/year$)]$ = 8300 years.

18

STELLAR EVOLUTION AFTER THE MAIN SEQUENCE

This is the way the world ends
Not with a bang but a whimper.

T. S. ELIOT, *THE HOLLOW MEN,* **1925**

The development of computers has made it possible for astronomers to calculate the changes expected to take place in stars as they age. Although the actual evolution of a star is slow by human standards, taking millions or even billions of years, a modern computer, in only a few hours, can do the calculations necessary to follow the evolution of a star over most of its lifetime. In this chapter, we will first describe the theoretical picture of stellar evolution that results from extensive computer modeling, and then compare it to a variety of observations.

18.1

THE MID-LIFE EVOLUTION OF SUN-LIKE STARS

Stars cannot remain on the main sequence forever. The reasons why stars leave the main sequence is the subject of the next section. Thereafter, we will examine evolution up to the giant phase and beyond.

WHY STARS LEAVE THE MAIN SEQUENCE

We have seen that a stable main-sequence star constantly maintains a state of equilibrium between the inward force of gravity and the outward force of gas and radiation pressure. The energy required to maintain the outward pressure is generated by the fusion of hydrogen in the star's core, the only place where the temperature and density are high enough for fusion to take place.

The star cannot remain on the zero-age main sequence indefinitely, because eventually most of the hydrogen in the star's core will be converted into helium. At this time the star's structure will be as shown in **Figure 18–1**. The changed chemical composition within the core will have changed the star's internal structure. With the core's hydrogen mostly depleted, fusion must cease because, although the core is a hot 15 million kelvins, it is not hot enough to fuse helium, a process that requires temperatures of about 100 million (10^8) kelvins. At least temporarily, therefore, the star will have lost its source of energy.

Inquiry 18–1 With its internal energy source gone, what will happen to the core of the star?

How long does it take for a star to burn its core hydrogen? We saw in Chapter 16 that the lifetime of a star on the main sequence is determined by $10^{10} M/L$, where M and L are the mass and luminosity in solar units. Thus

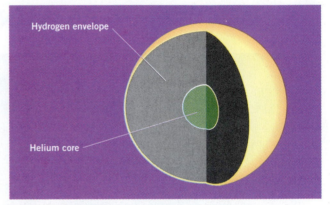

FIGURE 18–1. Once a star converts its core hydrogen into helium, as shown here, it is ready to leave the main sequence.

a star of the Sun's mass will remain relatively stable in diameter and luminosity for around 10 billion years. Thereafter, it will find itself undergoing relatively rapid changes, as we now discuss.

The star does not darken immediately when fusion stops, for two reasons. First, as discussed in Chapter 16, it takes a considerable length of time for the photons in the core to leak out and reach the surface; the outward diffusion of energy that has already been produced will take tens of thousands of years. Even more important, the gravitational contraction that must set in after fusion ceases will tend to heat the star further by releasing gravitational potential energy. Paradoxically, it seems, the cessation of fusion actually causes the interior of the star to heat up! In a star of one solar mass, such as the ones we are considering, the temperature of the region immediately surrounding the helium core will rather quickly reach 15 million kelvins and cause the hydrogen around the core to fuse. The outward flood of photons will resume, and contraction of the stellar core will halt. The interior structure will look like that in **Figure 18–2**. At this point the star is said to have a **hydrogen-burning shell**.

FIGURE 18-2. Hydrogen burning in a shell around an inert helium core in an evolved star of one solar mass.

Why do the changes just described occur? The basic reason is that as hydrogen fuses into helium in the hot core, the chemical composition of the core changes. You might think that the increasing proportion of helium in the star would be detected in its spectrum, but it is not. The reason is that the convective zone in the outer stellar envelope does not extend deep enough to scoop up helium and move it outward. However, we know there is a helium core because the composition change produces a structural change that modifies the star's surface temperature and luminosity, and thus its location in the H-R diagram. Such changes are shown in **Figure 18-3**. This highly schematic figure is not meant to be correct in detail but to provide a general picture of events. The one-solar-mass star's initial position is on the ZAMS at point *1*. As the core helium abundance increases during the first 4.5 billion years, the point representing the star's position moves to number *2*, which corresponds to the location of the Sun in the H-R diagram today. Further hydrogen burning moves the location to point *3*, at which the core hydrogen is depleted. The core is enriched by an ash of helium where once there was mostly hydrogen. Hereafter, the core collapses, heats, and ignites the hydrogen-burning shell.

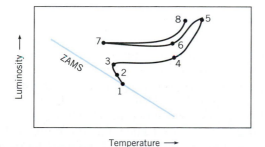

FIGURE 18-3. A schematic diagram of the evolutionary track of a star of one solar mass in the Hertzsprung-Russell diagram. The events occurring at each numbered point are discussed in the text.

Recall from Chapter 15 that the luminosity of a star is determined by its surface temperature and surface area, so that an increase in the star's luminosity must be due to an increase in its surface temperature, its surface area (or, equivalently, its diameter), or both.

Inquiry 18-2 You will notice that at point *3* in Figure 18-3 the star has a slightly *larger* luminosity and nearly the same temperature as at its ZAMS location at point *1*. What does this imply about the star's diameter?

As hydrogen fuses in the shell, the helium that is produced is added to the central helium core, which slowly shrinks and increases in density. However, the temperature of the core does not increase much at this point, because it is able to radiate away most of the heat liberated by the compression of its gases.

The photons generated in the hydrogen-burning shell do not readily flow from the star, but rather are absorbed and cause the outer layers to expand (although the core is contracting). Because the shell photons are not getting out, the star's luminosity does not increase but remains approximately constant as the atmosphere expands and the star's point moves toward number *4* in the H-R diagram.

BECOMING A RED GIANT

As the star expands and its point approaches number *4* in Figure 18-3, models show that the entire envelope of the star outside the hydrogen-burning shell becomes convective. Energy generated in the shell will be transported outward by the turbulent motions of the envelope gases. Convection is much more efficient than radiation in transporting the energy of the star, and the star becomes more luminous. The shell's high temperature causes the reactions to proceed rapidly and thus supply the additional energy required to maintain equilibrium. The increased flood of energy, in turn, causes the envelope of the star to expand rapidly; as the envelope expands, it cools—just as the gases expanding out of an aerosol can cool (making the nozzle cold to the touch). Our star of one solar mass has become a **red giant** star at point *5*. The red giant Aldebaran in Taurus is one of the brightest stars in the winter sky; Arcturus in Boötes is a bright red giant star in the summer sky.

Inquiry 18-3 The decreased surface temperature and greatly increased luminosity of the star move its position in the Hertzsprung-Russell diagram to point *5*. What kind of star has resulted?

Inquiry 18–4 At point 5, the Sun will be about 100 times more luminous and 30 times larger than it is now. If the Earth is now about 4.6 billion years old, how long will it be before this event takes place? What consequences will it have for life on Earth?

EVOLUTION AFTER THE GIANT PHASE

Even though most stars eventually become red giants, relatively few stars are red giants at any one time. This is so because the evolution of a star in this stage is rapid compared to its lifetime on the main sequence. (For an analogy, return to the photographs taken by the visiting astronaut in Chapter 2. These photos would show relatively few people getting their hair cut at any given instant, even though nearly everyone has their hair cut sooner or later.)

We have seen that even as the stellar core contracts, the star's envelope expands. The two parts of the star are moving toward opposite extremes—the core becoming hotter and denser as the envelope becomes cooler and more rarefied. As more and more helium is added to the core, its temperature will rise until at a temperature of 10^8 K helium fusion will commence. The process consists of two basic steps, which must take place in rapid succession (**Figure 18–4**).

$$^4_2\text{He} + ^4_2\text{He} \rightarrow (^8_4\text{Be})^*$$

$$^8_4\text{Be} + ^4_2\text{He} \rightarrow ^{12}_6\text{C} + \gamma.$$

Because the helium fusion reaction involves *three* helium nuclei, or **alpha particles**, it is often called the **triple-alpha reaction**. (The term *alpha particle* for a helium nucleus goes back to the early days of nuclear physics.) Three helium nuclei have been converted into a single carbon nucleus, with an excited beryllium nucleus as an intermediate product. The star is producing new elements, in a general process called **nucleosynthesis**. The reason the two reactions must take place in rapid succession is that the beryllium nucleus is unstable (indicated by the asterisk) and will quickly break down into two helium nuclei if the density is not great enough to make the second reaction take place immediately.

Inquiry 18–5 The three helium nuclei together have a greater mass than the resulting carbon nucleus. What has happened to the rest of the mass?

An interesting thing now happens. In a normal gas, the energy released by helium fusion would heat the gas and, because the pressure in the gas depends on its temperature, the core would expand and cool, slowing

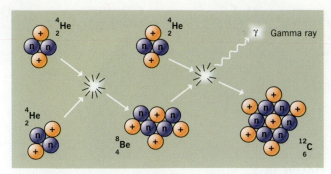

FIGURE 18–4. The triple-alpha process, which burns helium into carbon in the interior of stars.

down the rate of energy production. This process was described in Figure 17–8 in the case of hydrogen burning, in which a natural thermostatic safety valve regulates the temperature and energy-production rate and prevents it from running out of control.

Here, however, the gas at the center of our hypothetical star is not an ideal gas. It is in a special state in which the electrons are packed together so closely that the ideal gas laws do not apply. Physicists call such a gas a **degenerate gas**. In the core of a star whose electrons are degenerate, increases in the temperature do not increase the pressure, and so no subsequent expansion takes place. This means that when the helium core heats to the point where the triple-alpha reaction starts, the energy generated by the fusion raises the temperature, which increases the energy-generation rate, which raises the temperature further, which increases the energy-generation rate even more, and so on. Without expansion to cool the central core, the star becomes involved in what we call a **thermal runaway**. The process we have just discussed is summarized in **Figure 18–5**, where we compare the events after nuclear burning in an ideal gas with those after burning in a degenerate one.

The thermal runaway occurs rapidly because the energy generation depends so strongly on temperature; the energy-generation rate depends on T^{40}! With such a strong dependence on temperature, a 1% increase in temperature produces nearly a 50% increase in energy generation. Changes occur so quickly that the star is no longer in equilibrium. The energy within the core rapidly becomes so great that the core undergoes a violent explosion, an event known as the **helium flash**. During the few seconds of the helium flash, an incredible amount of energy—equal to the visible radiation from an entire galaxy—is produced!

Following the helium flash, a number of important events occur: (1) the core rapidly expands and cools,

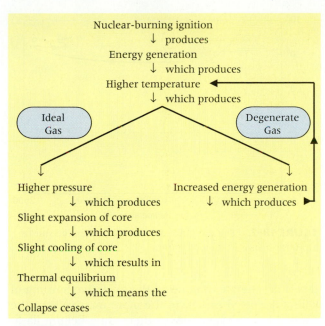

Nuclear-burning ignition
↓ produces
Energy generation
↓ which produces
Higher temperature
↓ which produces

Ideal Gas   Degenerate Gas

Higher pressure   Increased energy generation
↓ which produces   ↓ which produces
Slight expansion of core
↓ which produces
Slight cooling of core
↓ which results in
Thermal equilibrium
↓ which means the
Collapse ceases

FIGURE 18–5. Events after the onset of fusion in an ideal gas and in a degenerate gas.

soaking up a great deal of the energy and thus keeping the stellar luminosity from increasing greatly; (2) the core's degeneracy ends so it becomes an ideal gas again; (3) the hydrogen-burning shell is disrupted so that its importance as a source of energy decreases; (4) the star's luminosity decreases because the energy from the flash is absorbed internally; (5) the envelope of the star contracts. These events cause the location of the star in the H-R diagram to move from point 5 down to point 6 (Figure 18–3), where the structure is like that shown in Figure 18–6.

By the time a star's point reaches number 5 in Figure 18–3 and becomes a red giant, its structure is already rather complex, and it becomes more difficult to model with a computer. The evolutionary track indicated in the figure depends on a number of assumptions about

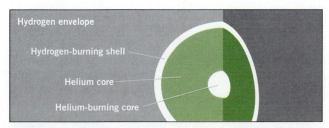

Hydrogen envelope

Hydrogen-burning shell

Helium core

Helium-burning core

FIGURE 18–6. The core of a star of one solar mass, after core helium burning is established. The hydrogen envelope cannot be illustrated to scale here, as it extends out to many times the diameter of the core region.

the star, including its chemical composition and how rapidly the star begins to lose mass. We have already seen that neutrinos thought to be produced in the Sun present problems for astronomers; the poorly understood role of neutrinos in red giant evolution dramatically complicates the computation of the models. The track indicated in the figure is only one of many possibilities that the Sun might follow in its movement leftward across the Hertzsprung-Russell diagram.

Eventually, gravity causes the core's expansion to halt and core contraction to ensue. Because the core is again an ideal gas, core contraction causes the temperature to increase. Once the temperature reaches 10^8 K, the triple-alpha reaction begins again (point 6 in Figure 18–3). This time, however, helium ignition is nonviolent, because the core is not degenerate. The situation is similar to that when hydrogen burning first began: the core temperature increases, the core expands and settles into an equilibrium state, and core expansion ceases. From points 6 to 7, then, the star is burning helium into carbon.

At point 7 in Figure 18–3 the core is an ash of carbon, just as it was an ash of helium at point 3. As before, because the temperature is not high enough for nuclear fusion to begin, the core begins to contract and heat up. Schematically, the star then moves in the direction of point 8. A solar-type star will never reach temperatures sufficient for further nuclear fusion to occur. This, then, marks the end of middle life for stars having masses near that of the Sun. Hereafter, the star enters its death stages discussed later in this chapter.

18.2

THE MID-LIFE EVOLUTION OF LESS MASSIVE STARS

As we saw in the previous chapter, if an object contains less than about 0.1 solar mass, it will never become a star at all; that is, it will not have a main-sequence phase. The force of gravity impelling contraction is simply insufficient to generate enough central heat to fire up a hydrogen fusion reaction. The object never becomes self-luminous. It will become a degenerate brown dwarf and ultimately (over an extremely long time period) just cool off to become a dead object. Brown dwarfs that circle other stars could become planets.

On the other hand, stars of mass greater than about 0.1 solar mass will burn hydrogen during a main-sequence phase. Such stars will go through some of the processes described for the Sun. Helium burning will occur only for stars having masses greater than about 0.2 solar mass. Stars of very low mass will not produce red giants. In fact, they are not able to evolve upward from the main sequence at all. If a star has less than about 0.2 solar mass of material, the gravitational contraction that begins after the core has used up all its hydrogen will not be able to heat up the central regions of the star enough to fire up any other nuclear reactions. After the main-sequence phase, these stars will then go directly to a long phase of "cooling off," fading away from their main-sequence position down and to the lower right in the H-R diagram.

18.3

THE MID-LIFE EVOLUTION OF MORE MASSIVE STARS

Although there is some uncertainty about where the line should be drawn between "ordinary" and "massive" stars, we adopt 5 solar masses as a convenient dividing point. The structure of more massive stars is considerably different from that of stars lower down on the main sequence. Because more massive stars produce most of their energy by means of the carbon-nitrogen-oxygen (CNO) cycle, rather than the proton-proton cycle, the core transports energy by convection. On the other hand, their outer regions transport energy by radiation. In other words, the convective and radiative regions are switched compared to their locations in low-mass stars. In general, the more massive the star, the greater the fraction of its energy that is carried by radiation transport. That is why, as we noted earlier, a star more massive than 60 to 70 times the Sun's mass would generate so much radiation that its pressure would cause mass in excess of 60 solar masses to be lost rapidly. In further contrast to the less massive stars, the radiation pressure in massive stars is much more important than gas pressure in supporting the star against the inward force of gravity.

In spite of the structural differences we have enumerated, the early stages in the life of a massive star are similar to that of the less massive ones. A massive star spends a relatively quiet and stable life on the main sequence (**Figure 18–7** between points *1* and *2*), though for a much shorter time owing to its rapid consumption of fuel. When its core has been depleted of hydrogen (Figure 18–7, point *3*), a massive star begins to contract. Just as for the lower-mass stars, a point is rapidly reached where hydrogen burning begins in a shell around the helium core. This core eventually contracts to the point at which helium burning (by the triple-alpha reaction) can begin.

There is a crucial difference between the initiation of helium burning in the more massive stars and in the less massive stars. Although the material in the less massive

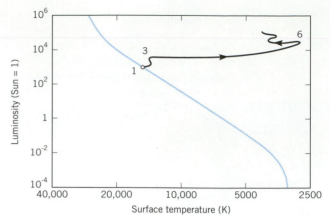

FIGURE 18–7. The evolutionary track of a massive star on the Hertzsprung-Russell diagram.

stars is degenerate, producing the explosive burning of helium discussed earlier, in the more massive stars the material is *not* degenerate, so there is no helium flash. When the helium ignites, the star is at about position *6* in Figure 18–7. The core expands, because it is not degenerate; this pushes the hydrogen-burning shell outward, cooling it off. Helium burning then becomes the source of most of the star's energy.

As carbon accumulates in the core of the star, further nucleosynthesis can take place to produce oxygen from a collision between carbon and an alpha-particle:

$$ {}^{12}_{6}C + {}^{4}_{2}He \rightarrow {}^{16}_{8}O + \gamma. $$

As a result, the star forms a carbon-oxygen core. Eventually, of course, most of the helium is burned up; when this happens, the fusion reactions in the core cease and the core begins its collapse once again. Soon, the conditions in the helium surrounding the carbon-oxygen core become right for the formation of a **helium-burning shell**. At this point the star's energy comes from a **double-shell source** (Figure 18–8). One shell is the new helium-burning shell, and the other is the old hydrogen-burning shell that surrounds it and is slowly disappearing. The surface temperature of the star decreases, and

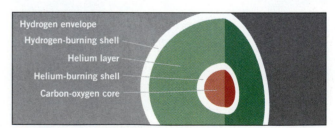

FIGURE 18–8. The double-shell structure of the core of a massive evolved star. The hydrogen envelope extends out to thousands of times the diameter of the core.

the star moves to the right in the Hertzsprung-Russell diagram, along a roughly horizontal path, and into the red giant region (Figure 18–7). Because at point 6 the star's position in the diagram is not far from that occupied by low-mass stars during parts of their evolution, it is difficult to distinguish among red giants of various masses from their positions in the diagram.

The subsequent evolution of massive stars is not understood as well as the evolution described thus far. In contrast to stars like the Sun, the more massive stars eventually generate interior temperatures and pressures sufficiently high for carbon and oxygen burning to take place with the synthesis of heavier elements. Such burning involves the addition of helium nuclei in a series of **alpha-capture reactions**, in which helium nuclei react with successively heavier nuclei. In the most massive stars, even the burning of elements such as magnesium, silicon, sulfur, and so on, up to but not including the heavier element iron, may occur. As each heavier nucleus begins to fuse into still heavier nuclei, a shell is formed where this reaction takes place. The temperature at which various nuclear reactions occur, and the time it takes for each energy source to become exhausted, is shown in **Table 18–1** for stars of about 20 solar masses.

The structure of the star becomes more and more complex. The core consists of many shells, much like an onion, each burning the elements in it into still heavier elements (**Figure 18–9**). At the same time, the star shuttles back and forth on the Hertzsprung-Russell diagram, following a path that is difficult to calculate, even on the most sophisticated computers. The final evolutionary phases of massive stars are complex and varied, and it will take an entire chapter (Chapter 19) to do the subject justice.

18.4
PULSATING STARS

The Hertzsprung-Russell diagram shown in **Figure 18–10** includes a shaded region known as the **Cepheid instability strip** because stars in this part of the diagram vary in brightness over periods of hours, days, or weeks. It appears that when a star enters this region of the diagram, its internal structure becomes less stable, and it begins to pulsate—actually expanding and contracting in a more or less regular way. At the same time, the surface temperature varies. The star becomes a **variable star**.

The properties of variable stars are determined by their internal structure. For this reason, by observing the properties of variable stars and relating the observed properties to theoretical models describing the pulsation, astronomers can attempt to understand the internal structure of the stars and predict their future behavior.

Inquiry 18–6 What happens to the luminosity of a star whose diameter and temperature vary regularly?

TABLE 18–1

Temperature Required for Nuclear Burning and Time for Exhaustion of Nuclear Fuel in Massive Stars (about 20 Solar Masses)

Energy source	Temperature (K)	Time for exhaustion (years)
Hydrogen burning	1.5×10^7	10^7 years
Helium burning	1.7×10^8	10^6 years
Carbon burning	7×10^8	10^3 years
Neon burning	1.4×10^9	3 years
Oxygen burning	1.9×10^9	1 year
Silicon burning	3.3×10^9	1 day

FIGURE 18–9. A multishelled star. Each shell converts the element in it into the heavier elements in the shell beneath. Again, the hydrogen envelope is too extensive to illustrate.

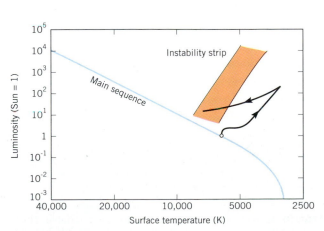

FIGURE 18–10. The instability strip in the Hertzsprung-Russell diagram, showing the evolutionary track of a star through it.

A number of different types of variable stars are known, the differences depending primarily on mass and their stage of evolution. Usually they are named for the first star of their type to be discovered. Thus **Cepheid variables** are named for δ Cephei, a bright star in the constellation of Cepheus. Cepheids are distinguished from other variable stars by their periods, which range from about a day to some weeks in length, and a distinctive brightness variation or **light curve** (**Figure 18–11**), in which the star rises to maximum brightness somewhat more rapidly than it fades to minimum brightness. The Cepheid stars are fairly massive (3–10 solar masses) and quite luminous; some of them are as much as 10,000 times as luminous as the Sun. Their surface temperatures range between 4000 K and 8000 K, so they might be called "yellow giants," because their colors are similar to the Sun's. They evolve through the Cepheid stage in tens of thousands of years.

Stars having masses similar to the Sun also cross into the Cepheid instability strip. One such group is the **RR Lyrae stars**, named for a variable star in the constellation Lyra. These stars are somewhat similar to the more massive Cepheids, but are observed to have pulsation periods of less than a day. In addition to their lower mass, they are less luminous than the Cepheids and all have the same average luminosity. At still lower luminosity is a group of variable stars known as δ Scuti stars. These variables, which have low amplitudes, have periods of only about two hours and are found in the area where the instability strip and the main sequence intersect. All these types of variable stars form a sequence within the Cepheid instability strip.

A highly distinctive group are the **long-period variable stars**, also called Mira variables after the first one discovered. They vary in visual light by hundreds of times over periods of 100 to 1000 days. These stars have the double-shell sources of hydrogen and helium discussed previously.

We will return to Cepheid and RR Lyrae variables when we discuss stellar distances in Chapters 20 and 21.

18.5

MASS LOSS, BINARY STARS, AND STELLAR EVOLUTION

Our discussion thus far has considered only isolated stars having a constant mass. Reality, however, is more complex. Evidence indicates that many stars lose substantial amounts of mass while they are red giants. Our lack of understanding of the cause of this mass loss is the biggest obstacle in the way of our further understanding of stellar evolution.

Inquiry 18–7   What would you expect to happen to the rate at which a star evolves if the star were to lose a substantial amount of its mass?

Inquiry 18–8   How might mass loss be observable in the spectrum of a star?

Mass loss is detected by its effects on a star's spectrum. We observe absorption lines from elements such as hydrogen and calcium that are blueshifted relative to their usual wavelengths (**Figure 18–12**). The blueshifted gas is interpreted as gas expelled from the star, traveling with a considerable velocity toward us. Such observations have been made around the supergiant star Betelgeuse. Consistent with these Doppler shift observations are infrared images of Betelgeuse that show an extensive envelope of gas surrounding the star. This giant shell of gas strongly suggests rapid mass loss, which has been estimated to take place at the rate of 10^{-5} to 10^{-6} solar masses per year. Depending on the amount of time that a star remains a red giant, the total amount of lost mass

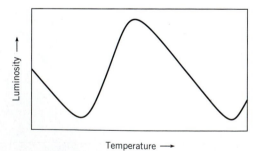

FIGURE 18–11.   The light curve of a Cepheid variable. The shape of the light curve is distinctly different from that of an eclipsing variable, as can be seen by comparing this figure with Figures 15–17 and 15–18.

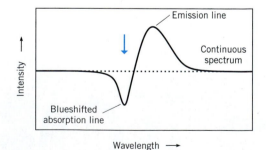

FIGURE 18–12.   Spectrum of a star that is losing mass. The arrow points to an absorption feature on the shorter (blue) side of a line in the star itself, indicating the rapid motion of mass from the star toward us.

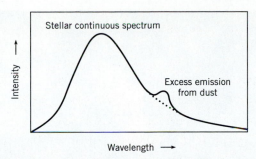

FIGURE 18–13. Infrared emission from dust adds to the continuous spectrum of a star to produce a hump at infrared wavelengths.

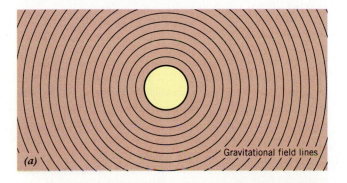

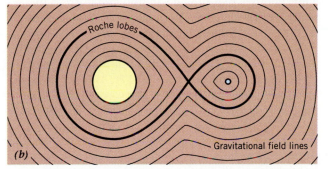

FIGURE 18–14. (a) The gravitational surface of a single spherical star. Along each surface, gravity is constant. (b) In a binary system, the gravitational surfaces are distorted. The surface forming a figure 8 defines the two Roche lobes.

could be significant. For example, a star initially as massive as six solar masses may lose as much as four to five solar masses through a strong stellar wind before dying.

The loss of gas from a red giant is often accompanied by the ejection of dust particles into a **circumstellar shell** surrounding the star. These dust particles, which are heated by radiation from the star, emit in the infrared spectral regions and are detected as an excess emission over and above that from the star itself (**Figure 18–13**). Furthermore, because the circumstellar shell is often a disc or an elliptical cloud, the light may be polarized. Studies of how the polarization changes with time and wavelength provide information on the size and composition of the dust particles, and on the nature of the circumstellar shell itself.

Our discussion has considered mass loss from individual stars. Binary star systems exist in which the component stars are so close to each other that they strongly interact. Each component of the binary system distorts the gravitational field of the other. The gravity of a single isolated star may be represented as shown in **Figure 18–14a**, in which each circle represents a three-dimensional surface on which gravity is constant. (Gravity's strength decreases with the inverse square of the distance.) Gas inside each of these surfaces is bound to the star. However, when the star has a nearby companion the total gravitational surface will be distorted and appear as in Figure 18–14b. One particular gravitational surface appears like a figure 8; each part of the figure 8 is called a **Roche lobe**. Because gravity is the same all along the surface, gas inside one Roche lobe can readily transfer to the other lobe through the crossing point. Gas inside more distant surfaces forms a common envelope surrounding both stars.

Suppose one star in the binary system has 6 solar masses, while the other has 20 solar masses (**Figure 18–15**). Because the more massive star will evolve more rapidly, it will become a red giant before the 6-solar-mass star has left the main sequence. The red giant may completely fill its Roche lobe. As the giant star continues

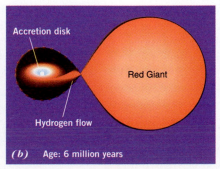

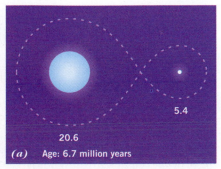

FIGURE 18–15. The evolution of a close binary system composed of two massive stars. As the 20-solar-mass star evolves rapidly, it becomes a red giant and expands to fill its Roche lobe, transferring material to the less massive star.

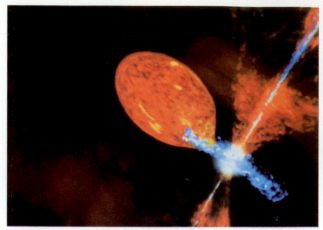

FIGURE 18–16. An artist's conception of the binary system R Aquarii, consisting of a cool giant and a white dwarf.

FIGURE 18–17. NGC 6852, the "dumbbell" nebula.

to try to expand, less energy is required for matter to pass through the crossing point to the other side than it would take to expand beyond the Roche lobe. Therefore, the matter from the more massive star comes under the gravitational influence of the less massive one and transfers to it.

Inquiry 18–9 What happens to the evolution of each star due to the transfer of mass between them?

Depending on the exact situation, it is possible for the less massive star to gain enough mass so it becomes more massive than its companion. The originally less massive star speeds up its evolution, and eventually it may be possible for mass to transfer back to the original star. Such a process can continue back and forth for a while. **Figure 18–16** shows an example of a system that has gone through some of the processes just described. R Aquarii is a bloated red giant with a white dwarf companion. Matter from the red giant moves through the Roche lobe and accumulates in an accretion disk surrounding the hot white dwarf before detonating in a nuclear explosion. The process described above occurs in a variety of stellar systems, some of which are so important we will discuss them in further detail in Chapter 19.

18.6
THE DEATH OF LOW-MASS STARS

While astronomers have confidence in their understanding of *what* happens to low-mass stars after their red giant phase, the theoretical details of *why* are not firmly understood. Remember that at the end of the red giant

stage the star contains a burned-out core of carbon along with a helium-burning shell. The star is so bloated that its radius approaches an astronomical unit. A number of events are thought to work together: a series of explosive events within the helium-burning shell produce instability in the star's outer regions. Because the outer regions are far from the core, they are cool enough for electrons and ions to recombine with one another. Recombination emits photons that provide an outward push to the extended stellar envelope. The end result of these events is that the star's outer envelope is ejected.

Once the star sheds its outer envelope, the hotter inner regions are exposed. The resulting object is called a **planetary nebula** (Figure 18–17). Another example, known as the Ring Nebula in the constellation Lyra, is shown in Figure 1–12. Note the star in the center; it ejected the observed nebula. The central star eventually cools to the point where we recognize it as a **white dwarf** star. White dwarf stars are discussed further in a later section of this chapter.

PLANETARY NEBULAE

Planetary nebulae received their name because in a small telescope they somewhat resemble the pale disk of a planet. (You can easily see the Ring Nebula using an amateur-sized telescope.) Actually, they are not at all like planets. Their spectra show a series of bright emission features at specific wavelengths.

Inquiry 18–10 What specifically can you conclude about the properties of planetary nebulae given that they are observed to have emission lines?

The temperature of the gas in these objects is typically about 10,000 K, whereas their densities are extremely low, ranging from a few hundred to a few thousand atoms of gas per cubic centimeter. (For comparison, a cubic centimeter of air has nearly 3×10^{19} atoms.) In the center of a planetary nebula is a very hot star, and it is the ultraviolet radiation streaming from this star that heats the surrounding gas and excites it to glow.

Inquiry 18–11 Why does the previous sentence specify ultraviolet radiation rather than, say, visible or infrared radiation as the cause of the nebular glow?

The ringlike appearance of the nebula is only an illusion. Recent observations show, at least in the case of the Ring Nebula, that the star has ejected bipolar lobes of gas, and that the nebula looks the way it does because we are viewing it along the lobe axis (**Figure 18–18**). A bipolar lobe model helps to explain the variety of the observed shapes of planetary nebulae.

If the light from the nebular ring is analyzed with a spectrograph, the shell of gas is found to be expanding, as expected for a shell of gas that has been ejected from the central star. Typically, it moves away from the central star at speeds of around 25 to 30 kilometers per second. Though this may seem fast, it is actually quite slow when compared to the more violently ejected gases observed around exploded stars, as we will see in the next chapter.

Because the gas in a planetary nebula is excited to glow by the central star, the nebula can be used to study the central star itself. These stars are somewhat small, but they are luminous and have high surface temperatures. In fact, several have surface temperatures in excess of 100,000 K. When the positions of the central stars are plotted on the Hertzsprung-Russell diagram, as shown in **Figure 18–19**, they are found to lie just below the upper end of the main sequence.

Inquiry 18–12 What do the small size of the central stars in planetary nebulae, their location in the Hertzsprung-Russell diagram, and the fact that they have recently shed considerable amounts of mass indicate about the probable place of planetary nebulae in the story of stellar evolution?

There is a great variety in the appearance of planetary nebulae; a regular, uniform nebula, such as the Ring Nebula, is actually the exception rather than the rule. It is possible that the matter is not expelled uniformly from the surface of the star, or the material may run into interstellar gas and dust as it expands away from the star. Multiple shells have been observed for some planetary nebulae.

While no one has yet observed a star form a planetary nebula around it, astronomers have discovered a number of candidate objects for planetary nebulae in formation. There are many good reasons to suppose that the precursor star was a red giant (actually a long-period variable star consisting of both helium- and hydrogen-burning shells) before entering the planetary nebula phase. We know that red giant stars have dense core regions, surrounded by tenuous atmospheres. Such an object is consistent with theoretical predictions about the death of low-mass stars.

Evidence in favor of the scenario of red giant to planetary nebula to white dwarf comes from a comparison of the number of planetary nebulae existing at any time and their estimated lifetimes. It is estimated that approximately two to three planetary nebulae are born

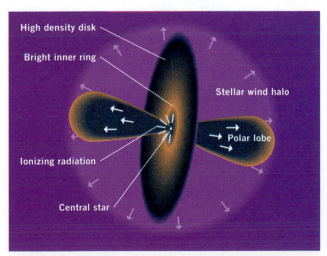

FIGURE 18–18. Explanation of the ringlike appearance of the shell of gas surrounding the central star of a planetary nebula.

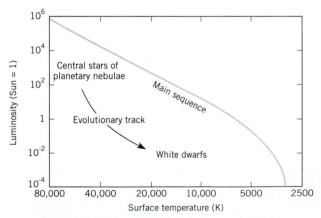

FIGURE 18–19. The positions of the central stars of planetary nebulae plotted on the Hertzsprung-Russell diagram. The evolutionary path to the white dwarf region is also shown.

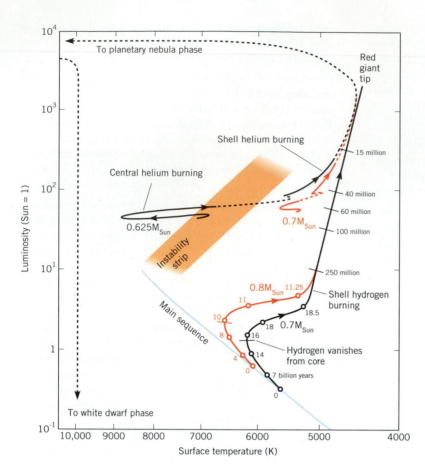

FIGURE 18–20. Evolutionary tracks for stars of 0.7 and 0.8 solar mass. The times shown are the times since the ZAMS. The times along the giant branch are the times from the point indicated until the star reaches the red giant tip.

each year in our galaxy. Because each of these nebulae is supposed to produce a white dwarf, about two to three white dwarfs would be expected to form in the galaxy each year. On the other hand, there are about 10 billion white dwarfs in the galaxy. Because the galaxy is about 15 billion years old, this means that an average of a little less than one white dwarf per year has been produced in the galaxy over its entire history; this number agrees tolerably well (to the same order of magnitude) with the other estimate. This agreement has convinced astronomers that the majority of stars do indeed end their lives as white dwarfs.

Inquiry 18–13 The estimate of the present production rate for white dwarfs, obtained from the number and lifetimes of planetary nebulae, is about three times the average rate over the entire history of the galaxy. Assuming that this difference is real, and not just due to inaccuracies in our estimates of the numbers, what would it imply about the present white dwarf production rate compared to the rate in the past?

The post-main-sequence evolution of stars is summarized in **Figure 18–20**, which shows the theoretical evolutionary tracks for stars of 0.8 and 0.7 solar mass. As each star evolves away from the main sequence, the times since the ZAMS are indicated. Notice the difference that a change of just 0.1 solar mass makes in the evolutionary times. The times along the giant branch are the times from that point to the red giant tip. (As an aside, the 250 million years up the red giant branch is approximately the time it takes the Sun to revolve once around the galaxy.)

OBSERVATIONAL PROPERTIES OF WHITE DWARFS

We previously indicated that our understanding of the evolution of stars having nuclear-burning shells is lacking many details. The difficulty relates to the complex structure of stars having helium- and hydrogen-burning shells. To finish our story of the evolution of low-mass stars, therefore, we will replace our computer models with evidence based on observations.

One way to proceed would be to look for stars that *appear* to be either dead or dying. If we wanted to make a list of the characteristics that such a star might have, we might make the following inquiries.

Inquiry 18–14 Would a dying star have a larger or smaller proportion of hydrogen, relative to heavier elements such as helium and carbon, than main-sequence stars?

Inquiry 18–15 Considering that it would have no source of internal energy to provide the high temperatures and pressures required to support it against the force of gravity, would such a star be large or small?

Inquiry 18–16 Assuming that the mass of the star has not changed much, would the star's density be low or high? How might this density affect the appearance of the spectrum?

Inquiry 18–17 In view of the size and lack of energy source for the star, would it have a high or a low luminosity?

A class of stars that appears to have all the expected characteristics of dying stars is the group of white dwarf stars lying below the main sequence. About 150 years ago, the German astronomer Friedrich Wilhelm Bessel (who also successfully measured the parallax of a star) discovered that the bright star Sirius exhibited a "wobble" in its apparent motion across the sky. Sirius is a bright star primarily because it is close to Earth; as a result, its motion through space can be measured more easily than the motions of most stars. The wobbles in the path of Sirius indicated the presence of a faint, unseen companion orbiting around it that was affecting its motion (see Figure 6–18 and 6–19). In 1862, the famous telescope maker Alvan Clark actually saw the faint companion, now known as Sirius B, in a newly built telescope. (Even amateur astronomers can sometimes catch a glimpse of it with a 10-inch telescope.) Sirius B is noted in **Figure 18–21**, in which the difference in image sizes between Sirius A and the white dwarf is caused by its brightness being some 10,000 times less than that of Sirius A. Thus Sirius B became the first white dwarf to be discovered. Over a period of time, the orbit of the object was determined and, when the mass of the star was calculated, it turned out to be similar to that of the Sun.

Inquiry 18–18 Describe what specific information astronomers had to have to determine the mass of Sirius B.

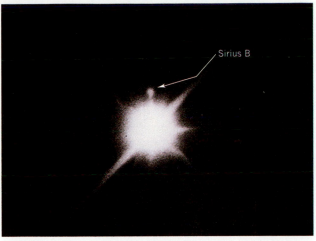

FIGURE 18–21. Sirius A and its white dwarf companion, Sirius B. The difference in the apparent sizes of the images of the two stars is caused by their great difference in luminosity.

Inquiry 18–19 Does the companion of Sirius B obey the mass-luminosity relationship discussed in Chapter 15? Suggest a reason for this.

From the surface temperature and luminosity of this object, it was simple to calculate its diameter. Imagine the astonishment of astronomers when they found that the diameter of Sirius B is only a hundredth that of the Sun: a mass equal to the entire Sun is crammed into a volume similar to that of the Earth (**Figure 18–22**). Its density must be extraordinarily high—because its diameter is only a hundredth of the Sun's, its volume must be only a millionth $[(10^{-2})^3 = 10^{-6}]$ of the Sun's, making its density about a million times greater than the Sun's. In earthly terms, a *cubic centimeter* of white dwarf material, brought here and weighed, would weigh as much as a pickup truck! Sirius B is indeed an extraordinary object.

FIGURE 18–22. A comparison between the sizes of the Earth, the Sun, and Sirius B.

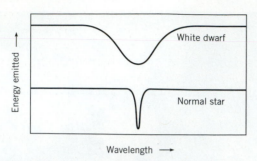

FIGURE 18–23. The broad absorption lines of a white dwarf compared with the absorption spectrum of a main-sequence star of the same spectral type.

The spectra of white dwarfs add additional details to the picture we are constructing of small, dense objects. For one thing, the absorption lines are observed to be very broad (**Figure 18–23**). This breadth can be attributed to the high density of the star's atmosphere. (Remember why the spectral lines in main sequence stars are broader than those in supergiants?).

Because the white dwarf is small, the inverse square law of gravity tells us that the strength of gravity on the star's surface must be large. A photon leaving the strong gravitational field at the surface of a white dwarf loses considerable energy, meaning that its corresponding wavelength must become longer than usual. In other words, light leaving a strong gravitational field will have its wavelength shifted toward redder wavelengths. This redshift is not caused by the Doppler shift, but by the gravitational field itself. This effect, known as a **gravitational redshift**, was first predicted by Albert Einstein as a consequence of the theory of relativity. The observed gravitational redshift of light from a white dwarf is one of many pieces of data that show the theory of relativity is a correct description of nature.

Inquiry 18–20 To be detected, the gravitational redshift of a white dwarf must be distinguished from any *velocity* redshift (Doppler shift) due to the motion of the star in space. How might this be done in the case of Sirius B? (Hint: Sirius B is a member of a binary system.)

If the Sun were collapsed to the size of a white dwarf, its magnetic field would also collapse and strengthen. White dwarfs can be expected, therefore, to have intense magnetic fields. Such strong fields are capable of polarizing the emitted radiation. Therefore, observations of polarized light from magnetic white dwarfs tell astronomers about the magnetic conditions in the star

and the properties of the compressed white dwarf material itself.

How do our observations of white dwarfs compare with our theories of them? We expect a white dwarf to be deficient in hydrogen because it is the remnant of a star whose core hydrogen has been depleted. It is difficult, however, to make observations that confirm the expected hydrogen deficiency. Unless material from the core were to become mixed with the surface layers of the star, the products of nuclear fusion would not be detectable in the spectrum. Furthermore, the extreme broadening of the spectral lines makes them harder to detect. Thus, although models lead us to believe that white dwarfs have cores of helium and heavier elements, we have no *direct* observations of it.

THE STRUCTURE OF WHITE DWARFS

When the laws governing degenerate gases were discovered, theoreticians became seriously interested in white dwarfs, because it soon became evident that these were stars composed almost entirely of degenerate material.

The laws of degenerate gases are actually simple, and the structure of stars composed of degenerate material can be worked out with pencil and paper—a computer is not needed. Whereas "normal" stars are held up by the outward gas and radiation pressure, the electrons of a degenerate gas provide a pressure known as **degenerate electron pressure**. This pressure comes about because the laws of nature will not allow electrons to be compressed further. It is this pressure, therefore, that prevents a white dwarf from collapsing further. Because temperature and pressure in a degenerate gas do not depend on each other, a star composed of degenerate gas cannot shrink, even if the temperature inside it becomes very low. White dwarfs are therefore deprived not only of a source of nuclear energy but also of gravitational energy; with no source of new energy, they can do nothing but slowly cool off, radiating into space the remnants of energy left over from their younger days. Because of their small surface area, this takes billions to trillions of years.

In the early 1930s, the Indian-American astrophysicist S. Chandrasekhar (**Figure 18–24**) made a remarkable discovery about degenerate stars: There is a relationship between the mass and the radius of white dwarfs (**Figure 18–25**). The white dwarf **mass-radius relation** shows that the more massive the star, the *smaller* it is. Not only that, and even more surprising, white dwarfs of mass greater than about 1.4 times the mass of the Sun cannot exist.

The significance of this **Chandrasekhar limiting mass** for stellar evolution will be seen in the next sec-

FIGURE 18–24. S. Chandrasekhar, who discovered the mass-radius relationship for white dwarfs. He was awarded the Nobel prize in Physics in 1983 for his work related to studies of stellar evolution.

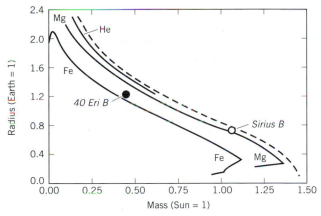

FIGURE 18–25. The mass-radius relationship for white dwarfs. The particular curve depends on the chemical composition. The observed locations of two white dwarfs are noted.

tion. Figure 18–25 shows that the mass-radius relation differs for stars of different chemical composition. Accurate observations allow astronomers to determine the overall chemical composition of a white dwarf. Knowing the chemical composition allows astronomers the opportunity to deduce the mass of the star that formed the white dwarf. The figure shows the locations in the mass-radius diagram of two white dwarfs. Note the uncertainty in both mass and radius.

Inquiry 18–21 Statistics on the proportion of stars of various masses and the number of white dwarfs that are seen today indicate that most stars end up as white dwarfs, including all stars up to at least several times the Sun's mass. What must happen during the evolution of these more massive stars for them to end up as white dwarfs?

18.7
THE OBSERVATIONAL EVIDENCE FOR STELLAR EVOLUTION

Most of our discussion thus far has looked at the evolution of stars on the basis of the predictions of computer models, because a compact and logical presentation can be made in this way. Actually, many of the important stages in stellar evolution had been worked out from observational evidence before large computers became available. It is now time to see what the observations have to say about stellar evolution, and whether any of the theory discussed in this chapter has anything to do with reality.

THE HERTZSPRUNG-RUSSELL DIAGRAMS OF STAR CLUSTERS

Some of the most dramatic and conclusive evidence for stellar evolution comes from a study of the Hertzsprung-Russell diagrams of the stars in star clusters. Astronomers have long been aware of the existence of stars that are close to each other and move with a common motion through space. Two examples are the Pleiades (Figure 17–16) and the Hyades, both of which can be seen with the naked eye in the constellation of Taurus, the Bull. The Pleiades is especially pretty when observed with a good pair of binoculars or a small telescope. Because of the "open" appearance of these clusters, they are often called **open clusters**.

When the cloud from which the cluster forms collapses and fragments into stars, different stars will have different masses and, therefore, different luminosities. In Chapter 17 we saw how the H-R diagram of an imaginary but typical cluster changes as the stars evolve onto the main sequence (Figure 17–17) during the first 10^5 years. The rest of the evolutionary picture is shown in **Figure 18–26**. Between 10^5 and 3 million years (Figure 18–26a), the more massive stars evolve onto the upper end of the ZAMS. The less massive stars evolve toward the bottom end of the ZAMS but at a significantly slower pace. Many star clusters are observed to have

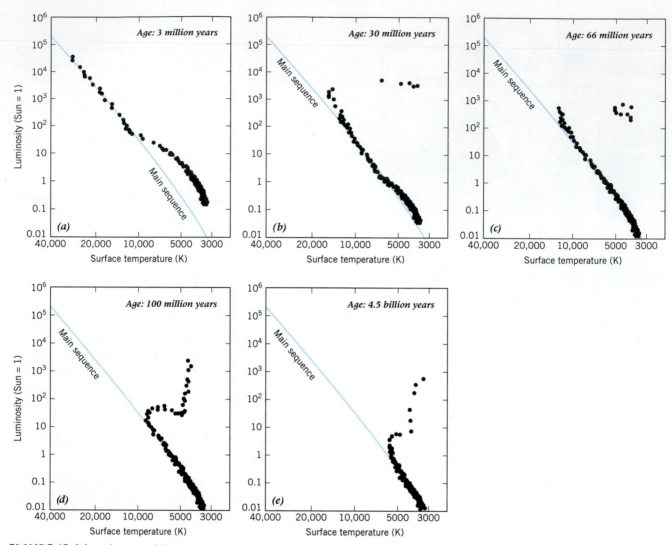

FIGURE 18–26. The rest of the evolution of an imaginary open cluster. The earlier evolution is shown in Figure 17–17.

Hertzsprung-Russell diagrams of this form, with stars on the upper end of the ZAMS and stars above the lower ZAMS. Such clusters have formed recently.

After only a few tens of millions of years, the most massive stars in the cluster will begin to exhaust the hydrogen fuel in their cores and evolve, moving away from the main sequence and into the red giant region, as shown in **Figure 18–26b**. As more time passes, stars of lower and lower mass farther down the main sequence will also have time to complete their hydrogen-burning phase, and they will also begin to leave the main sequence and move toward the giant region (**Figure 18–26c-e**).

We call the top of the main sequence, from which stars have begun to evolve, the main-sequence **turnoff**

point. The older the cluster, the fainter and redder is its location. By observing the location of the turnoff point, astronomers can determine the age of the cluster. The ages corresponding to the locations of the main-sequence turnoff point are shown in **Figure 18–27**.

Inquiry 18–22 How is the age of a cluster related to the position of its turnoff point?

Inquiry 18–23 The turnoff point provides a method for estimating the age of a cluster using a plot of its Hertzsprung-Russell diagram. What crucial assumption is being made about the formation of the cluster for this to be a valid method?

Inquiry 18–24 Figure 18–27 shows the Hertzsprung-

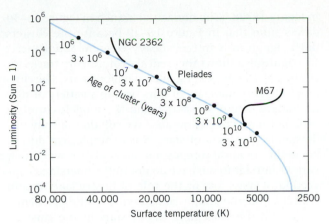

FIGURE 18–27. The Hertzsprung-Russell diagram of three clusters. The dots on the main sequence indicate the position of the turnoff point for clusters of the indicated age.

FIGURE 18–28. The globular cluster 47 Tucanae in the southern hemisphere.

Russell diagrams of three clusters plotted on the same graph. Estimate the ages of (*a*) the Pleiades, (*b*) NGC 2362, and (*c*) M67.

For isolated stars not found in clusters, it is not possible to estimate ages with any degree of certainty. However, in some circumstances we can draw useful conclusions. When we see a single star of one solar mass in the galaxy, it is in general difficult to tell whether it formed recently or billions of years ago, because such stars spend about 10 billion years in their stable, main-sequence phase. But if we see a massive, luminous star near the top of the main sequence, the short lifetime of such stars assures us that it must have formed recently.

Current observations raise doubts about the assumption that all the stars in a cluster formed at the same time. For example, the open cluster NGC 2214 appears to have two distinct supergiant branches corresponding to ages of *both* 50 and 126 million years. While we have no definitive explanation, it appears that this cluster had two periods of star formation. Other recent studies indicate that some clusters have four or five distinct periods of star formation.

There is yet another type of star cluster, which has an appearance dramatically different from that of the open clusters: the so-called **globular cluster** (named for its round shape and dense appearance). Figure 1–14 shows the bright cluster known as M13, while **Figure 18–28** shows the bright southern cluster 47 Tucanae.

Globular clusters provide us with further confirmation of the evolutionary picture we have been constructing. The H-R diagram of a typical globular cluster is shown in **Figure 18–29**. Whereas open clusters exhibit a wide variety to their diagrams (because of the

variety in their ages), *all* globular clusters have similar-appearing main sequence and giant branches. They all have a short main sequence with a turnoff point at both a low luminosity and a low temperature, thus indicating that they are old. In fact, these clusters appear to be somewhere between 10 and 16 billion years old. All the massive stars in such a cluster have had plenty of time to go through their entire lifetimes. By now they have evolved into dead stars (i.e., nonluminous objects) or dying stars (feeble white dwarfs of such low luminosity that we cannot detect them when we photograph the cluster). In such an old cluster, even stars with as little mass as the Sun will have died and become faint white dwarfs.

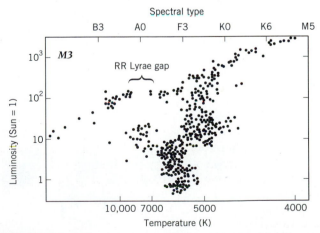

FIGURE 18–29. The H-R diagram of the globular cluster M3. Only RR Lyrae stars inhabit the region noted.

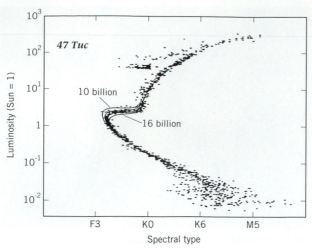

FIGURE 18–30. The H-R diagram of the globular cluster 47 Tucanae along with evolutionary tracks for clusters of age 10, 12, 14, and 16 billion years. The spread at the bottom of the main sequence is caused by increasing observational uncertainty as the stars become fainter.

The stars we see near the turnoff point in Figure 18–29 are just beginning to evolve away from the main sequence. These stars have a mass of approximately 0.85 to 0.90 solar mass. The stars that are a little farther along in their evolution into the giant region will have masses slightly above this value (having left the main sequence slightly earlier). In addition to the shortened main sequence and the giant branch, globular cluster diagrams also show a **horizontal branch** of stars. These stars are burning helium in the core and have either single or double shell sources. The RR Lyrae stars mentioned previously are located along the horizontal branch within the Cepheid instability strip. Because there are no stable stars located here, this part of the horizontal branch is known as the **RR Lyrae gap**.

Figure 18–30 shows some of the best modern data, obtained by Jim Hesser, for the cluster 47 Tucanae, a cluster located in the southern hemisphere and pictured

in Figure 18–28. (The horizontal branch in Figure 18–30 differs from that in Figure 18–29 because the clusters differ significantly in their heavy-element abundance.) The figure also shows theoretical evolutionary tracks for clusters of ages 10, 12, 14, and 16 billion years. The cluster models are made up of stars having a variety of stellar masses. The observations and the models correlate well, but not perfectly because we still do not fully understand all aspects of stars. For example, we do not know the chemical composition precisely (or whether it varies from star to star); we do not fully understand convection or mass loss or the role of binary stars within clusters. But the rough coincidence of the calculated track with the placement of the stars in the cluster's H-R diagram gives us some confidence that our theory of stellar evolution as we have presented it is basically correct.

EVIDENCE FROM SPECTRA PERTAINING TO STELLAR EVOLUTION

In the 1940s, it was discovered that absorption lines due to a newly discovered and rather exotic heavy element, **technetium**, existed in the spectra of some stars. An example is the star ST Hercules, which shows lines of the technetium isotope having 99 protrons plus neutrons (^{99}Tc) (**Figure 18–31**). This isotope of technetium is radioactive with a half-life of only about 200,000 years. Its presence in a steller atmosphere tells us that the nuclear processes required to produce it were operating when the light we see left the star. Thus the observation of technetium in the spectra of certain stars provides a *direct* observational verification that nuclear processes actually take place in stars and that the resulting elements are mixed into the material of the star's surface, where the spectrum is formed.

Some red giant stars show the element barium in their spectra. How are these so-called **barium stars** formed? This question plagued astronomers for many years until it was found that all stars showing large

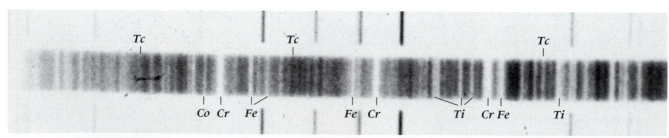

FIGURE 18–31. The spectrum of the ST Hercules shows the element technetium, indicating current nuclear reactions in the star's interior.

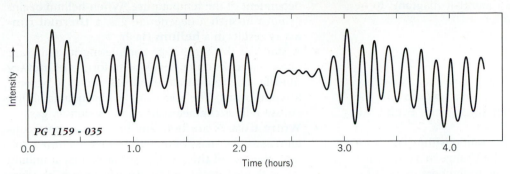

FIGURE 18–32. The variation in brightness with time of the pulsating white dwarf PG1159–035.

abundances of barium are in fact binary stars. The barium observed in the red giant spectrum is now understood to have formed in a white dwarf companion *when it was a red giant*. The barium arrived at the surface by convection, and then was transported to the binary companion by transfer through the crossing points of the Roche lobes, as described earlier.

A DIRECT DETECTION OF STELLAR EVOLUTION

Because stars live out their lives on such long time scales, making observational progress in stellar evolution studies means finding stars at different phases of their evolutionary cycles and fitting these various pieces together in sequence. New technology has allowed direct observational verification of the fact that stars do change their characteristics with time.

A white dwarf will cool and decrease in luminosity along the path shown in Figure 18–19. As the star cools, its surface oscillates. The pattern of these small oscilla-

tions will change in a well-determined way as the star cools. In studies over time of a hot white dwarf star called PG 1159–035, Don Winget and his collaborators from the University of Texas were able to observe a changing pulsation period in the star, just as expected (see **Figure 18–32** for a sample of their data). Enough data have been obtained to show that the oscillation is speeding up about 1 second in 1,270 years, about the rate that cooling theory predicts. In other words, the predicted movement of a white dwarf star along the cooling track illustrated in Figure 18–19 has been detected.

We end this section on observations important to stellar evolution with the observational search for brown dwarfs. Locating such low-mass, faint objects is difficult. While many research groups are trying to find brown dwarfs, and some have claimed success, the problem is so difficult that the astronomical community has yet to agree that even one brown dwarf has been discovered. The search will continue either until they are found, or until astronomers find a good reason why they should not exist after all.

CHAPTER SUMMARY

OBSERVATIONS

- A variety of types of stars are observed to vary periodically in their light output. Such **variable stars** include the **Cepheids**, **RR Lyrae stars**, δ Scuti stars, and long-period variables. Observationally, these stars differ in their period and amount of light variation. Light variation occurs when evolving stars pass through a part of the H-R diagram known as the **Cepheid instability strip**.

- Luminous stars exhibit strong stellar winds that cause the stars to lose a large fraction of their mass in a relatively short interval of time.
- The central star in a planetary nebula is observed to be hot, often up to 100,000 K. Planetary nebulae shells expand with speeds of about 25 km/sec.
- White dwarf spectra contain highly broadened, gravitationally redshifted hydrogen lines.
- Observations of **open clusters** show them to exhibit a variety of types of H-R diagrams. Observations of

globular clusters show their H-R diagrams to be nearly the same.

THEORY

- **Figure 18–33** summarizes the post-main-sequence evolution of a star of one solar mass by showing the internal structure at each stage of its evolution, which is shown in Figure 18–3.
- The changing chemical composition within a star's interior provides the impetus for change in its structure. These structural changes lead to temperature and luminosity changes that trace an evolutionary path in the H-R diagram.
- Stars similar in mass to the Sun have their energy produced by the proton-proton reaction, transported out of the core by radiation, and through the envelope by convection. In more massive stars, energy is produced by the CNO cycle, transported through the core by convection, and through the envelope by radiation transfer.
- Once the core reaches a temperature of 10^8 K, **helium burning** in the **triple-alpha reaction** begins.
- A **degenerate gas** is one in which the pressure is independent of the temperature. When helium begins to burn in such a degenerate gas, a **thermal runaway** results in a **helium flash**.
- A star like the Sun, after completing its red giant phase, ejects its outer envelope to form a **planetary nebula**. The central star eventually cools to become a white dwarf. The spectrum of the nebula shows emission lines characteristic of a low-density gas.
- **White dwarfs** are hot, low-luminosity stars containing a solar mass of material squeezed into a volume the size of the Earth. With a density a million times that of water, white dwarfs cannot be more massive than the **Chandrasekhar limiting mass** of 1.4 solar masses. Stars above this mass limit must lose substantial quantities of mass if they are to become white dwarfs. The degenerate electron gases that compose white dwarfs are subject to a **mass-radius relation** that says that the greater the mass (up to the 1.4-solar-mass limit), the smaller the radius. White dwarf spectra often show a **gravitational redshift**, which is caused by a photon's loss of energy as it escapes from the strong gravitational field at the white dwarf's surface. All these theoretical ideas are consistent with observations.

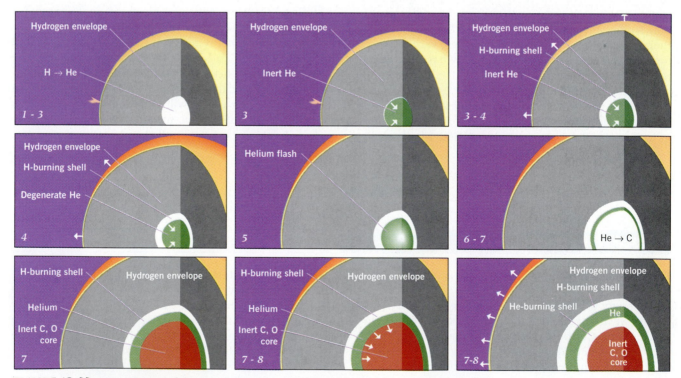

FIGURE 18–33. A summary of the structure of a star of one solar mass during each part of its midlife phases. The numbers refer to the positions in Figure 18–3. The drawings are not to scale.

- Depending on the relative masses of each star in a binary system, they may transfer mass back and forth between the **Roche lobes**. The star gaining mass speeds up its evolution, while the star losing mass slows down its evolution.
- Objects whose masses are less than about 0.1 solar mass are expected to produce **brown dwarfs**.

CONCLUSIONS

- From the agreement between the theory of stellar evolution and observations, astronomers conclude that elements heavier than hydrogen are synthesized inside stars. Observations of the presence of certain elements in stellar spectra (technetium and barium) provide observational evidence that nuclear reactions occur in stars.

- A planetary nebula forms from a red giant. The central star that remains eventually becomes a white dwarf.
- Oscillations of the light emitted by white dwarfs are observed to change in the manner predicted by theory, thus providing observational evidence that stellar evolution actually occurs.
- From the variety of types of H-R diagrams exhibited by **open clusters**, astronomers conclude that they have a range of ages. Cluster ages are determined from **main-sequence turnoff points**. The similarity of H-R diagrams for **globular clusters** shows them to have similar ages of 10–16 billion years. Such diagrams agree well, but not perfectly, with those predicted by theoretical stellar model calculations. These observations provide evidence that our general understanding of stellar evolution is valid.

SUMMARY QUESTIONS

1. Why must a star inevitably leave the main sequence? How is the length of time a star spends on the main sequence affected by such things as the star's mass and luminosity?

2. What are the internal and visible changes that take place in a star after its main-sequence evolution? Answer in terms of the physical processes that are taking place. Why does the internal temperature of a star rise when its core nuclear fuel is exhausted?

3. What does the evolutionary track of a 1-solar-mass star on the Hertzsprung-Russell diagram look like? Relate the luminosity, surface temperature, radius of the star, internal structure, and physical processes taking place in the star to its position along the evolutionary track. Do the same for a 10-solar-mass star.

4. What unusual properties distinguish a degenerate gas from a normal gas? How are these properties responsible for the helium flash? How are they important for the structure of white dwarf stars?

5. What are the nuclear reactions in the triple-alpha reaction? What are the physical conditions under which the reaction takes place? When in the lifetime of a star is the triple-alpha reaction important?

6. Where is the Cepheid instability strip in the Hertzsprung-Russell diagram? What is its significance?

7. What characteristics might a dying star be expected to exhibit? How do these compare with the known properties of white dwarfs? What is the role of the Chandrasekhar mass limit in the theory of white dwarfs?

8. What evidence is there that planetary nebulae are an intermediate stage between red giants and white dwarfs? How does mass loss affect the possibility that a given star may become a white dwarf? What is the approximate maximum initial mass of stars that eventually become white dwarfs?

9. What observational evidence is there for stellar evolution? How would the Hertzsprung-Russell diagram of a star cluster be expected to evolve over time?

10. What observations of stellar chemical compositions have a bearing on studies of stellar evolution?

APPLYING YOUR KNOWLEDGE

1. Hypothesis: The Sun is hollow. Use all your astronomical knowledge to argue against this hypothesis.

2. How do observations of the presence of certain elements in a star's atmosphere provide information about stellar evolution?

3. Why is the observed main sequence not an "infinitely thin" line on the H-R diagram? (Two main reasons are sought: one observational and one theoretical.)

4. Stars do *not* evolve up or down the main sequence. From what you know of the properties of stars on the

main sequence, what would have to happen to a star to make it evolve up the main sequence?

5. Summarize the events occurring at each point in Figure 18–3.

■ **6.** What percentage of its mass would a 10-solar-mass

star lose in 10^5 years if it were losing mass at a rate of 10^{-5} solar mass per year?

■ **7.** Use the formula for the main-sequence lifetime, along with the mass-luminosity relation, to compare the lifetimes of stars of 0.1, 1, 10, and 20 solar masses.

ANSWERS TO INQUIRIES

18–1. The core of the star must start to contract, because the source of thermal energy that was supporting the star against gravity is gone.

18–2. It is bigger than it was.

18–3. The star is a red giant.

18–4. It will be another 4.4 billion years or so before this happens. But when it does the consequences of the Sun's becoming a red giant will be catastrophic for life, which will no longer be able to exist on our planet.

18–5. The remaining mass is converted into energy.

18–6. The luminosity of the star would also vary in a regular fashion.

18–7. The speed of evolution would slow down.

18–8. Emission lines, probably Doppler shifted, would be observable.

18–9. The star gaining mass speeds up its evolution while the one losing mass slows its evolution.

18–10. From Kirchhoff's laws planetary nebulae must consist of a rarefied gas at a high temperature.

18–11. Ultraviolet photons have enough energy to excite hydrogen, whereas neither visible nor infrared radiation do.

18–12. It is believed that they are intermediate stages between red giants and white dwarfs.

18–13. The present rate of white dwarf production is greater than that in the past; perhaps more stars are now reaching the stage in their lifetimes when they become white dwarfs. This is reasonable, because the kind of stars that become white dwarfs are low-mass stars, which take a long time to evolve.

18–14. It would have a deficiency of hydrogen, because it has converted much of it to helium and carbon.

18–15. Small; it would collapse to a point where some other force—specifically, degenerate electron pressure—would be sufficient to support the star.

18–16. Because its diameter must be small (see the previous Inquiry), its density must be high. High density would produce highly broadened spectral lines.

18–17. Its luminosity would be low because it has only a small surface area from which to radiate energy.

18–18. Because the mass would come from Kepler's third law, the period and semi-major axis were needed. Determination of the semi-major axis required the observed angular semi-major axis and the star's distance.

18–19. No; the mass-luminosity relationship is valid only for main-sequence stars.

18–20. The other star is a main-sequence star and should have very little gravitational redshift. Because the two stars are moving together through space, their Doppler shift due to their common motion through space would be the same. Any difference between the redshifts of the two stars, therefore, must be the result of orbital motion (which can be calculated and allowed for) and the gravitational redshift.

18–21. The stars must lose mass, otherwise they would be more massive than the Chandrasekhar limit of 1.4 solar masses and could not become white dwarfs.

18–22. The lower the turnoff point, the older the cluster.

18–23. It is assumed that all the stars in the cluster were formed at the same time, so they are the same age.

18–24. According to the diagram, (*a*) the Pleiades is a little less than 10^8 years old, (*b*) NGC 2362 is about 5×10^6 years old, and (*c*) M67 is somewhat less than 10^{10} years old.

19

CATACLYSMIC ASTRONOMY

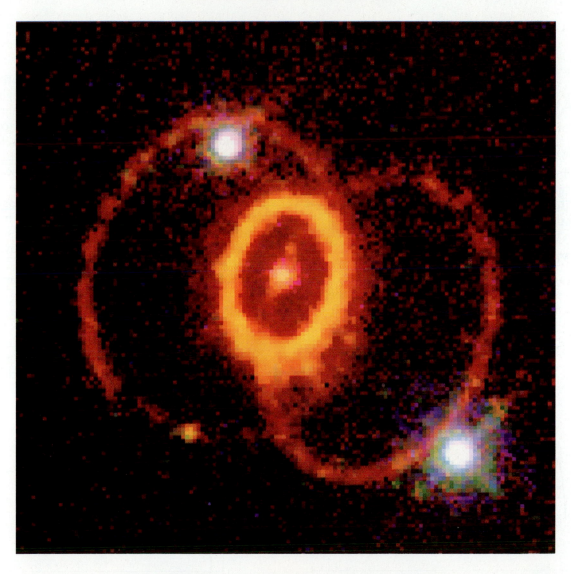

A leaf of grass is no less than the journey work of the stars.

WALT WHITMAN, *SONG OF MYSELF*

407

In Chapter 18, we studied the evolution of the overwhelming majority of stars—those of modest to moderate mass. We have seen how they evolve from the main sequence into the giant region, become variable stars, and then eject planetary nebulae. As the nebula dissipates, the central star becomes a white dwarf and slowly cools.

The previous chapter left several important questions unanswered. In particular, we did not finish discussing the life cycle of stars of more than a few solar masses because the evolution of these stars is different from that of low-mass stars. In addition to a discussion of the death of high-mass stars, we will also further examine the evolution of binary stars. Such stars often experience violent changes as they age, and they play an important role in our discussion.

19.1
NOVAE

Since ancient times people have observed stars to brighten suddenly and then fade slowly. Such events were called *nova stella* (Latin for *new star*) or **nova** (plural **novae**) for short. Only in the twentieth century has it become apparent that these objects are actually old stars passing through an explosive phase of their evolution, and that the term nova is therefore a misnomer.

OBSERVATIONS

In late August of 1975, when the sky had darkened sufficiently, many amateur and professional stargazers noticed that a bright nova had suddenly appeared in the constellation of Cygnus, approaching the brightness of the blue giant Deneb at the tip of the Northern Cross. Novae are given names that include both the constellation and the year in which they occur. Therefore, this nova is known as Nova Cygni 1975.

Nova Cygni 1975 faded rapidly in only a few days. On the other hand, Nova Delphini 1967, after reaching its maximum brightness, declined in luminosity very slowly, taking about a decade to return to its former state. These two novae, the fastest and slowest of modern times, represent the extremes for most novae; some nova light curves are shown in **Figure 19–1**.

Several kinds of evidence suggest that novae eject mass.

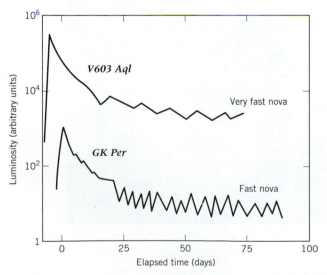

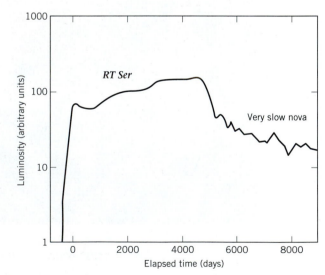

FIGURE 19–1. The change in brightness over time of a slow nova and a fast nova.

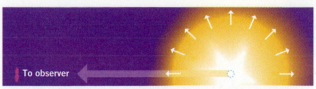

FIGURE 19–2. The early spectrum of a nova exhibits blueshifted features because the expanding gases on the near side of the star are moving toward the observer.

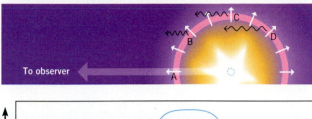

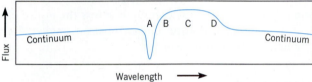

FIGURE 19–3. A schematic spectrum produced by a cloud of gas expanding around a nova.

FIGURE 19–4. A photograph of Nova Persei 1901 taken many years later, showing the expanding nebulosity around the star.

Inquiry 19–1 In what ways might the loss of material in a nova explosion make its presence known?

During its early phases, the spectrum of a nova shows normal stellar absorption lines, but displaced to shorter wavelengths. **Figure 19–2** illustrates that this occurs because the near side of an expanding shell of gas around a star consists of absorbing material moving toward the observer, so the light is Doppler shifted toward shorter wavelengths. The speed with which the gas moves may also be found from observations of the widths of the broad emission lines. **Figure 19–3** shows a star with an expanding ring about it, and a schematic diagram of the spectrum it forms. The greater the expansion speed, the greater the width of the line. The expansion velocities of the ejected gas are considerably greater than those found in planetary nebulae; the speeds typically range from a few hundred to more than a thousand kilometers per second.

If one waits long enough after the nova outburst, the ejected material may become visible on photographs as an expanding nebulosity (**Figure 19–4**).

A great deal of energy is involved in a nova outburst. Increases of over 100,000 times in a day or two have been found by comparing the star's maximum brightness with its pre-explosion value. At their maximum brightness, novae are temporarily among the most luminous objects in the galaxy—up to 1 million times more luminous than the Sun.

We cannot yet predict which stars are going to "go nova"; we can only hope that when one does it has been observed previously, so that we can study the changes in its properties. From such observations astronomers have concluded that during the early rise to maximum brightness, the surface temperature of the star does not change by much. The increase in luminosity, therefore, primarily comes from the increased surface area of the rapidly swelling star immediately after the explosion.

Inquiry 19–2 Suppose a star were to increase its radius by 10 times so that its surface area increased by a factor of 100 times. Suppose also the temperature of the star did not change. How many times brighter would the star become?

Inquiry 19-3 The situation in Figure 19–2 must eventually be followed by a stage at which some material escapes from the object. What kind of spectral changes would you expect to witness in this ejected material with time? How would the expansion of the gases affect the spectrum?

THE PLACE OF NOVAE IN STELLAR EVOLUTION

When we plot the positions on a Hertzsprung-Russell diagram of stars that undergo nova explosions, we find that they fall into the same region of the diagram as the hot central stars of planetary nebulae. Apparently, both types of objects are late stages in the lifetime of a star that is trying to adjust its mass to reach a stable condition. But while the nova outburst is much more vigorous, ejecting material at high speeds and increasing the star's luminosity greatly, a typical nova will lose only about 10^{-4} solar masses, about 1000 times less than that lost by a planetary nebula. So exactly what is the difference between the two? To answer this, let us briefly consider the current theories about novae, and the evidence that led to the development of these theories.

DO NOVAE INVOLVE LARGE OR SMALL STARS?

Observations of the brightness of novae, made with accurate equipment that can determine stellar brightness every thousandth of a second, show that small and rapid brightness fluctuations continually take place. The rapidity with which a light source can vary its intensity is related to the size of the emitting source. In other words, an object cannot make a substantial change in its radiation output in a time shorter than the time it takes light to travel across the object's emitting region. For this reason, we conclude that the nova event occurs in a small space.

To understand why, consider **Figure 19–5**, which shows an object with various points on it noted. If the object is uniformly bright across its surface, an increase in brightness at point A will have little effect on the object's overall brightness. Furthermore, if the change were random, an increase at point A is likely as not to occur during a decrease at point B. To have a substantial change in brightness requires that a change at point A be followed by changes at points B, C, D, and so on across the object. In other words, it is necessary for a change at one point to be causally related to changes at other points. But a change at point B cannot be caused by a change at point A sooner than the time light would take to travel from A to B because nothing can travel faster

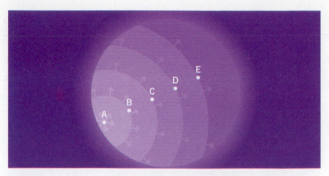

FIGURE 19–5. If a change in brightness occurs at point A, a coordinated change at B cannot occur before light from A can reach point B. Extending the argument across the object means it must be smaller than the light-travel time across it.

than light. Carrying this argument across the entire object, we conclude that an object must be smaller than the light-travel time across it.[1]

HOW A NOVA EXPLODES

In 1954 the astronomer Merle Walker discovered that the nova DQ Herculis is actually a close binary system with an orbital period of 4 hours and 39 minutes. In subsequent years it has become apparent that novae exist in binary systems where the stars are perhaps a solar diameter apart. The current models of a nova explosion depend on this fact.

Spectra of novae may sometimes reveal the presence of more than one star, but most often the brighter of the pair so dominates the light from the other that the existence of the second star is not easy to detect.

The model of a nova that best fits most observations consists of a white dwarf star and a low-luminosity main-sequence star in orbit around each other. If the two stars are relatively close to each other, it is possible that mass may transfer from one to the other as discussed in the previous chapter. Such a system is shown in **Figure 19–6**. Material from the outer parts of the main-sequence star is forcibly removed by a shrinking Roche lobe,[2] just as toothpaste is forced from a tube whose volume shrinks when it is squeezed. The white dwarf star, which in spite of its small diameter may have a mass equal to or greater than that of the Sun, has a strong

---

[1] Just as we can speak of distance in terms of light-years, it makes sense to speak of the size of an object in terms of the time it takes light to travel across it. This is the light-travel time.

[2] The reason why the Roche lobe is shrinking is unclear, but it may be due to magnetic braking, discussed for the Sun in Chapter 6, which helps the system lose angular momentum.

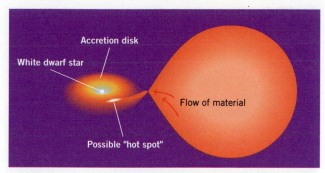

FIGURE 19–6. Currently accepted models of novae involve the interaction between stars in a binary system.

gravitational field and will attract material from the atmosphere of the main-sequence star. As material flows between the stars, the rotation of the system forces it to spiral around the star, forming an **accretion disc** around the white dwarf. The accretion disc is continuous all the way to the white dwarf's surface. Observations suggest that there is actually a dancing "hot spot" where the spiraling stream of material from the main-sequence star collides with the gaseous disk around the white dwarf, liberating large amounts of energy.

The white dwarf has a high surface temperature, and if large amounts of gas from the main-sequence star are dumped into the accretion disc and eventually onto the white dwarf, the infall of material will generate even more heat. The infalling gas is rich in hydrogen. As time progresses a layer of hydrogen builds up on the star's surface. Eventually the temperature at the base of the layer becomes sufficiently high for explosive hydrogen fusion to take place. In effect, the surface of the white dwarf becomes a giant hydrogen bomb and the blast blows off some mass. The ejected mass probably comes from the thin envelope of the white dwarf star. One reason for studying novae now becomes clear: by studying observable novae ejecta, astronomers are studying the outer layers of a white dwarf star which, we hope, will tell us about the later stages of stellar evolution.

The observation that many novae are recurrent was crucial to the development of this model because it showed astronomers that a nova explosion was not a one-shot, catastrophic event. It showed that if a nova results from some sort of instability, the unstable conditions are able to recur some time later. Further evidence for a recurring event comes from observations that after the ejected material has been blown off the star and dissipated, an object very similar to the prenova star remains. Not only has it survived the cataclysmic event, in many respects it seems virtually unscathed and ready to begin the process leading to the explosion all over again.

Inquiry 19–4 After an outburst as just described, the companion star may start to dump mass onto the white dwarf star all over again. How might this fact be used to explain the existence of recurrent novae?

Inquiry 19–5 The answer to Inquiry 19–4 indicates that we may expect many and perhaps all novae to be recurrent. If this is so, how would it affect our conclusions as to how many stars in the galaxy are involved in the nova phenomenon?

To interpret the significance of novae in stellar evolution, we should correct the observed nova frequency for the fraction that are recurrent. But this fraction is not known; nor is it known how often or how many times a typical recurrent nova explodes. Thus it is difficult to determine how important novae are for stellar evolution. Present evidence suggests that most normal stars having masses less than several times the Sun's mass eject planetary nebulae if they are single stars or widely separated binaries, or become novae if they are close binaries.

Recent observations indicate that there is a difference in chemical composition between "fast" and "slow" novae, with fast novae showing an unusually high abundance of carbon, nitrogen, and oxygen, and slow novae showing abundances closer to that of the Sun. This suggests that perhaps these two types of objects have a different stellar history, that their predecessors were stars of different internal structure, and that the mechanisms causing their explosions are different.

Inquiry 19–6 In a nova system, a companion star dumps mass onto the surface of a white dwarf star. But the white dwarf must have originally been a normal star itself, possibly more massive than the companion star, because it arrived at the white dwarf stage sooner. Might there have been a stage at which the present white dwarf star was a giant, dumping mass onto its companion? What effect might this have had on the evolution of the system?

Novae and recurrent novae belong to a class of objects often referred to as **cataclysmic variables**. Because the variety of possible stellar systems is so large (due to variations in mass, radius, magnetic field, mass loss rate, and stage in evolution) these stars can produce bewilderingly complex systems. Full understanding of their nature will come only with advances in observations over the entire electromagnetic spectrum, and in our knowledge of the applicable laws of nature.

19.2

SUPERNOVAE

As their name suggests, **supernovae** are even more dramatic and energetic than novae. The supernova shown in **Figure 19–7** is nearly as bright as all the rest of the galaxy in which it is found. Since ancient times, observers have occasionally noticed that a star would suddenly appear in the sky, becoming so bright that it might even be visible during the day. Such an object was the supernova of 1572 that Tycho Brahe observed in Cassiopeia; he studied it carefully and demonstrated that it must be at a greater distance from Earth than the Moon, thus showing the universe beyond Earth to be changeable. Another such object, appearing only 32 years later in the constellation of Ophiuchus, was observed by Johannes Kepler. Since this 1604 event, no supernovae have been observed in our galaxy, a fact that makes the study of these enigmatic objects more difficult. Until recently, astronomers were forced to grapple with evidence from supernova events in distant galaxies, from which much less information is available.

OBSERVATIONS

This situation changed dramatically on February 24, 1987, when astronomers around the world were thrilled by a discovery made by a young Canadian observing assistant named Ian Shelton. Shelton was fortunate enough to be the first person to report the appearance of a supernova in the Large Magellanic Cloud, a nearby galaxy in the southern hemisphere that can be seen with the naked eye (**Figure 19–8**). While not within our galaxy, SN 1987A (so named because it was the first supernova of 1987) is the next best thing because it is only some 170,000 light years away. Furthermore, it is the first bright supernova since the invention of the telescope! Because the star was readily observable with the naked eye, there was no problem studying it with even small telescopes. Large telescopes allowed astronomers to study it in more detail than any other supernova. It was also the first supernova to be studied with a variety of spaceborne instruments: the International Ultraviolet Explorer, the Japanese Ginga X-ray satellite that had been launched shortly before the explosion, and rocket and balloon experiments. Before we discuss the important findings of these studies, however, we need to learn more of the fundamentals of supernovae.

Figure 19–9 shows a photograph of the famous Crab Nebula in Taurus (see also Figure 1–13). This object is the remnant of a supernova explosion that was witnessed by Chinese astronomers in the year A.D. 1054, when it was at least as bright as Venus. The photograph

FIGURE 19–7. A supernova explosion in a distant galaxy.

gives a visual impression of a violent explosion, and detailed observations confirm this. Doppler-effect measurements of the filaments of gas have shown that they are moving at a high rate of speed, on the order of 1450 km/sec. When studied spectroscopically, the gas itself exhibits a high degree of excitation, indicating that there is a considerable amount of energy in these expanding filaments. Photographs taken of the Crab Nebula over an interval of years clearly reveal the outward motion of the gases. In fact, this expansion, combined with the measured Doppler shifts, provides us with the means to measure the Crab's distance, about 6000 light-years. If we take the presently observed motion and extrapolate it backward in time, it confirms the Chinese observation that the original explosion occurred almost 1000 years ago.

> READERS HAVING THE ACTIVITY KIT
> WILL WANT TO DO ACTIVITY 19–1,
> THE CRAB NEBULA, AT THIS TIME.

There is a considerable amount of mass in the gaseous filaments of the Crab Nebula and it seems as if a star obliterated itself in the explosion. Could it be that

the mass of the exploding star, and will be discussed later in this chapter.

Computer models suggest that an implosion followed by an explosion should be exhibited by stars with masses greater than approximately 10 solar masses. The supernova formation rate also tends to agree with the star formation rate in this mass range. To put that another way, if only the most massive stars formed supernovae, we would not see as many as we do. If low-mass stars formed supernovae, we should see many more than we do.

Inquiry 19–10 In what ways do the formation of Type I and Type II supernovae differ? How do these different formation mechanisms produce differences in the presence of hydrogen in the spectrum?

PRODUCTION OF HEAVY ELEMENTS BY MASSIVE STARS

Although heavier elements cannot be produced by fusion reactions as they occur throughout a star's lifetime, nucleosynthesis of elements beyond ^{56}Fe can occur by a process that adds neutrons to elements. Because a neutron has no charge, it can enter a nucleus—for example, that of iron—at a low speed without disrupting it. The most stable elements are formed by the slow addition of neutrons to atoms, in a process called the **s-process** (s for slow). When a neutron is captured, the nucleus stabilizes by emitting high-energy electrons. This process of neutron capture can synthesize elements that go as high in the periodic table as bismuth (^{209}Bi), beyond which elements become radioactively unstable. Technetium, discussed in the previous chapter, is formed by the s-process. Its presence in stars proves that the s-process of neutron addition operates.

Inquiry 19–11 In general, we would expect that the addition of helium nuclei to such elements as ^{12}C, ^{16}O, and so on, would produce nuclei whose atomic weight is a multiple of 4. Yet other nuclei, such as ^{17}O and ^{18}O, are found in nature. How might the existence of these nuclei be explained?

Inquiry 19–12 The most common form of oxygen is ^{16}O. Keeping this fact in mind, what conclusion can you draw about how most of the oxygen in our universe was created?

A supernova explosion involves extremely high temperatures and densities. We saw that the collapse of the iron core produces neutrons. In fact, a huge number of neutrons is produced quickly. These neutrons can be absorbed rapidly by the other elements, thus producing heavy elements that can be manufactured only under the conditions present during a supernova explosion. Production of these heavy elements takes place because of the rapid addition of neutrons in a nuclear process known as the **r-process** (r for rapid). One of the elements built up during these catastrophic moments of stellar evolution is ^{235}U, famous on Earth for its use in the first atomic bombs and in nuclear power plants.

Heavy-element production during a supernova explosion, along with the elements' subsequent ejection into interstellar space, place supernovae in an important position in the ecology of the galaxy. Eventually, the turbulence of the interstellar medium will incorporate these supernova-produced heavy elements into giant molecular clouds. When these GMCs produce new stars, they will be formed from materials of an earlier generation of stars, in the ultimate recycling process.

SUPERNOVA 1987A

The most exciting observation connected with Supernova 1987A was the simultaneous detection by Japanese and American neutrino telescopes of bursts of neutrinos even before the supernova was detected optically. From the 21 neutrinos detected on Earth, astronomers calculated that neutrino emission from the supernova carried 100 times as much energy as all other forms of energy produced in the explosion. This, together with the observation that this supernova was a Type II, is a dramatic confirmation of the general theoretical model of the Type II supernova explosion that we have described.

If astronomers happened to have made observations of the star that became SN 1987A, the observations would go a long way toward identifying the type of star that undergoes a supernova explosion. While there was a strong debate as to which star actually blew up, we now know it was one designated as SK −69°202. The surprise here is that SK −69°202 was a blue supergiant rather than a red supergiant, as expected from current theory. The fact that it was a blue supergiant, however, accounts for the lower than expected luminosity of the supernova, because a blue supergiant is less massive than a red one. The data show the star had a mass of some 20 solar masses and a radius of slightly more than 40 solar radii.

Light reaches Earth not only from the supernova itself, but also from photons scattered by dust in interstellar space. Because the path this reflected light travels is longer than the direct one from the supernova itself, it

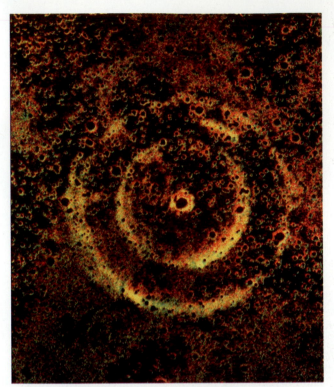

FIGURE 19–17. SN 1987A light echoes from interstellar matter.

arrives later. Such "light echoes" from interstellar material are shown in **Figure 19–17.** Analysis of such data provides us with new information on the distribution of interstellar clouds in the general direction of the star.

Observations made at radio wavelengths three and one-half years after the explosion of Supernova 1987A revealed the beginning of formation of a supernova remnant. By observing this remnant over time, we will be better able to understand the development of such objects and to place older supernova remnants into their proper context. The fact that it took more than three years before such radio energy was seen is important, because it shows that gas in the vicinity of the star had been swept away by the star's strong stellar wind, just as expected. The radio emission was produced when the ejected material finally encountered dense clumps of gas in the interstellar medium.

19.3
NEUTRON STARS AND PULSARS

The model of a Type II supernova explosion that we have described predicts different effects on different parts of the star. Although the outer envelope of the star

is blown off, the central core undergoes a catastrophic collapse. In the last rush of nuclear reactions in the collapsing core, electrons and protons combine to make neutrons. This occurs even within the nuclei of the heavier elements. For example, a silicon nucleus in the central regions of the star might find its protons converted to neutrons. If this process continues long enough, the silicon nucleus will become unstable and decay into two smaller nuclei. Thus the process—electron + proton → neutron + neutrino—not only destroys individual protons but also tends to break down heavier nuclei and convert them into neutrons.

These reactions had been known in theory for a number of years and led Walter Baade and Fritz Zwicky in 1934 to suggest that after a supernova explosion had taken place, we might be left with a remnant star composed entirely of neutrons—a **neutron star.** The theory suggested that such an object would have a mass less than about 3 solar masses, just as a white dwarf's mass must be less than 1.4 solar masses. Furthermore, the theory said that the mass would be contained in a sphere only about 20 km in diameter. The density would therefore be enormous, equal to the density of a thimble that had been stuffed with a billion automobiles! This is 10^8 times the density of a white dwarf.

The difficulty with this idea was to demonstrate that a neutron star was indeed left over after the explosion. Although such an object might have a high surface temperature, its extremely small surface area would make it a very faint object. It took about 30 years for astronomers to confirm this prediction observationally.

Astronomers had examined the various stars within the confines of the Crab Nebula and had not been able to find any clues that any of them might be a remnant star, although Zwicky did suggest that one of them was a likely candidate (he was right!). Most of them were foreground or background stars not associated with the nebulosity at all. Perhaps the explosion had left nothing behind. Also, it was not difficult to show that the Crab Nebula should have disappeared long ago, because the nebula was radiating energy so furiously that it should have cooled off quickly. Nevertheless, the nebular gases remained extremely turbulent, hot, and excited. This implied that, although no visible energy source was present, somewhere within the nebula there was a source of energy that continually excited the gas to glow.

Observations of the radio spectrum of the Crab Nebula show it to differ from that of a normal blackbody. Bodies that emit energy due to their temperature have a continuous spectrum whose intensity beyond a peak decreases with increasing wavelength (Chapter 13). For the Crab Nebula, however, the intensity is observed to *increase* with increasing wavelength (**Figure 19–18**). Furthermore, astronomers observe the radiation to be

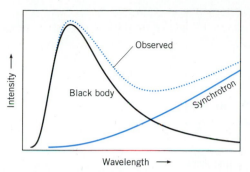

FIGURE 19–18. A schematic comparison of the spectrum of the Crab Nebula compared with that of a blackbody. The increase toward longer wavelengths is characteristic of synchrotron radiation.

strongly polarized (see Figure 11–12 for photographs in polarized light). These observed characteristics are completely understood if the radiation from the Crab Nebula is due to the emission of energy by charged particles moving near the speed of light in a magnetic field. Such radiation is called **synchrotron radiation**; it is often called **nonthermal radiation** to distinguish it from that produced by a hot (thermal) body.

Synchrotron radiation from an accelerated electron is not radiated in all directions, but into a cone along the direction of the electron's motion (**Figure 19–19**). This directionality results in its characteristic spectrum and its polarization. Synchrotron radiation was first observed coming from synchrotrons, atom smashers built to study nuclear reactions by accelerating charged particles to near-light speeds and allowing them to collide with other particles at high energies. Synchrotrons use magnetic fields to accelerate charged particles to high velocities and to confine their motion. In the laboratory, physicists found that a charged particle moving at a high rate of speed through a magnetic field radiates away co-

pious amounts of polarized energy whose intensity increases with increasing wavelength, exactly as now observed for the Crab Nebula.

PULSARS

The big breakthrough that tied many of the pieces in the supernova puzzle together took place in 1967, with the discovery of the objects that have come to be called **pulsars**. In that year a graduate student at Cambridge University, Jocelyn Bell, was working for astronomer Anthony Hewish and conducting a radio survey of the sky, searching for fluctuations in the emission from radio sources (**Figure 19–20**). She noticed a peculiar object (now known as CP 1919) that appeared to be "pulsing" at radio wavelengths in an extremely regular way. **Figure 19–21** shows one of the records of the object; it was distinctive in that the interval between the pulses of radiation was constant to an extremely high degree of precision. The interval was determined to be an unchanging 1.33730113 seconds. Nothing of this regularity had ever been seen in nature before. In fact, the repeatability of the pulse interval meant that the pulsars

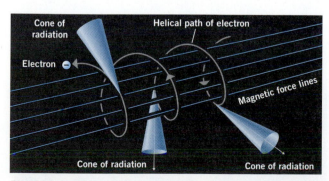

FIGURE 19–19. An electron follows a helical (spiral) path about a magnetic line of force, emitting synchrotron radiation in a narrow cone along the direction of motion.

FIGURE 19–20. Jocelyn Bell Burnell, the discoverer of pulsars.

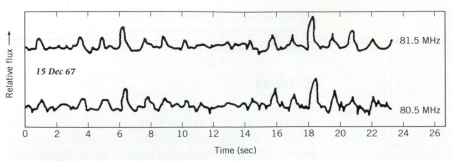

FIGURE 19–21. A graph of radiation intensity over time for CP 1919, which first drew astronomers' attention to the remarkable objects known as pulsars. Note the periodic nature of the intensity.

were providing more accurate time-keeping capabilities than any clocks on Earth. The question arose: was the source terrestrial or extraterrestrial?

Inquiry 19–13 Because the radio telescope Jocelyn Bell used did not move, it detected objects as they passed over the telescope while the Earth rotated. The telescope observed the radio emission to occur nearly four minutes earlier on each successive night of observation. On the basis of this piece of information, what might you conclude about the location of the emitting source? Explain.

The time intervals between pulses repeated so well that astronomers working on the data later confessed that they could not prevent themselves from considering the possibility that it might actually have been a coded message from an extraterrestrial civilization (the LGM theory, for Little Green Men). The obvious first step, however, was to look for other objects of this type; once astronomers knew what to look for, many others were found. Furthermore, pulsar periods were not constant but were observed in many cases to increase by tiny but measurable amounts. A natural cause now be-

came a more plausible hypothesis. As a result of this and further work, Hewish was awarded the Nobel Prize in physics in 1974.

How large are pulsars? We have seen that an object's brightness cannot change faster than the length of time it takes light to travel across it. [In the case of a pulsar, whose intensity changed over a time of 0.00007 seconds, the emitting region could not be larger than 0.00007 light-seconds across.] This is about 20 km—in agreement with what Baade and Zwicky had predicted! Regardless of what was causing the variation in the light signal, the conclusion was still valid. Whatever a pulsar was, it was considerably smaller than a white dwarf star. Suddenly theoretical speculation about the possibility of incredibly compact neutron stars was revived and made very real. **Figure 19–22** emphasizes the small sizes of white dwarfs and neutron stars by showing them relative to the Earth and Central Park in New York City.

One early hypothesis about the nature of pulsars described them as pulsating white dwarfs or neutron stars. Models of such pulsating stars, however, showed that they would not be able to pulsate as rapidly as the 0.033-second rate observed for the pulsar found in the Crab Nebula. This is a perfect example of observations falsifying a hypothesis.

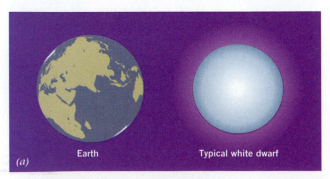

FIGURE 19–22. (a) The relative sizes of Earth and a white dwarf star. (b) The sizes of a neutron star and Manhattan Island, New York.

Another hypothesis was that a pulsar is a rotating white dwarf or neutron star. Because a white dwarf rotating with the period of the Crab pulsar would break apart, that hypothesis cannot be valid.

The hypothesis that pulsars are rotating neutron stars makes several predictions. A rotating neutron star should have an intense magnetic field, as would be the case if the Sun were shrunk to a neutron star's size. In addition, just as magnetic braking was seen to slow the Sun's rotation, a similar mechanism would slow the neutron star's rotation. Both these predictions are verified by observations. In addition, the change in rotation period would mean that rotational energy is being lost. The law of conservation of energy would require that the lost energy reappear elsewhere. When Thomas Gold computed the amount of energy loss represented by the slowdown of the Crab Nebula pulsar, he discovered it to be almost exactly the amount required to explain the observed luminosity of the Crab Nebula. In other words, using the observed slowdown of the Crab pulsar and the assumption that the pulsar is a neutron star, the long-standing problem of the energy source for the Crab Nebula was solved, thus affirming the hypothesis that pulsars are indeed rotating neutron stars. On the order of 200 pulsars are now known, and the hypothesis that they are rotating neutron stars has been well confirmed.

Do supernovae remnants contain pulsars? As we have seen, the Crab Nebula does. In the constellation of Vela, where a cloud of expanding gas had also been suggested as a supernova remnant, another pulsar was found near the center of the expanding gases. The Vela pulsar is 10 times closer to the Earth than the Crab Nebula, and its explosion must have been an awesome sight. If it was an explosion of the same force as the Crab, the star would have appeared approximately 100 times brighter than the supernova of A.D. 1054, as seen from Earth.

Inquiry 19–14 What are the arguments that the Crab pulsar is a rotating neutron star?

Inquiry 19–15 The Vela pulsar is much slower than the Crab pulsar. What conclusion do you draw about the age of the Vela pulsar?

The discovery of the pulsar in the Crab Nebula challenged observational astronomers to identify the remnant star and observe its pulsations in visible light as well as in the radio region of the spectrum. By synchronizing their observations with the observed pulse time of the object, astronomers were able to detect optical pulsing for one of the stars in the Crab (**Figure 19–23**). Finally, the star that exploded has been identified.

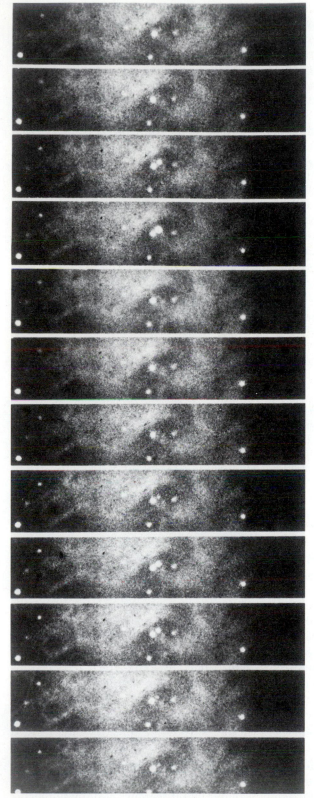

FIGURE 19–23. The pulses from the central star of the Crab Nebula, seen at optical wavelengths.

Many speculations concerning supernova explosions and neutron stars now have been tied together. However, not all supernovae remnants have observable pulsars, nor do all pulsars have observable remnants. For these reasons, astronomers continue to anxiously await the discovery of a pulsar at the location of SN 1987A. Understanding why requires discovering the mechanism by which a rotating neutron star produces the observed pulsar pulses; we do this next.

WHY PULSARS "PULSE"

How does a rotating neutron star produce the observed pulsations of a pulsar? The most successful models appear to be those in which a rotating neutron star is enclosed in a region of plasma (ionized gas). Consider, for example, **Figure 19–24**, in which a rotating neutron star is shown together with the magnetic lines of force that are supposed to surround the star. If lines of magnetic force are threaded through a region of ionized gas (such as one that composes a star) then the gas and the magnetic field lines tend to move together. If the gas moves, the lines of magnetic force are dragged along with it.

This means that when the central core of a supernova collapses, the magnetic field lines will collapse with it. As the magnetic field is compressed together into the collapsing core, its strength increases, building up to an enormous value. It is not known exactly what form the magnetic field lines might take around a collapsed star, but a simple pattern has been drawn in Figure 19–24. Notice that the rotational axis of the star and the axis of symmetry of the magnetic field do not necessarily coincide. (Remember, the magnetic and rotation axes are not coincident for Earth, Jupiter, Uranus, or Neptune.)

The outer surface of a neutron star is expected to consist of nondegenerate ionized gas. What would happen if some of these charged particles were to evaporate from the hot surface of the neutron star? As the charged particles flow away from the star, they are caught up by the magnetic field lines and forced toward the magnetic poles, where the field is stronger. The particles are accelerated by the rapidly rotating magnetic field, until at a certain distance from the star their speed approaches the speed of light. When this occurs, the material begins to radiate synchrotron radiation near the poles, like two searchlights sweeping around through space. If one of these rotating beacons happens to sweep past the Earth, we will see a pulsar, as in **Figure 19–25**.

Inquiry 19–16 What does the searchlight mechanism of pulsar emission imply about the completeness of our pulsar searches and the number of stars that might become pulsars?

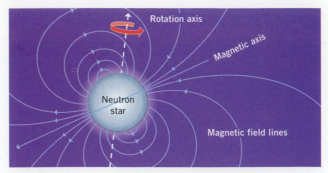

FIGURE 19–24. The central regions of a pulsar, showing the rotating neutron star and its magnetic field.

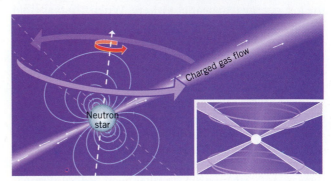

FIGURE 19–25. Viewed from a distance, the material that flows out along the magnetic axis of a neutron star is accelerated and generates a sweeping beam of radiation that gives the effect of a rotating searchlight.

Even if a pulsar's radio beacon is aimed away from us, there are still observational techniques that can detect its presence. In particular, all strong pulsars apparently are immersed in a cloud of electrons that move near the speed of light and that radiate X-rays. X-ray surveillance satellites, which can detect the X-rays from the electron clouds, provide indirect evidence of the presence of a pulsar. Even so, only about 20% of supernova remnants show evidence of neutron stars inside. At present, the significance of this fact is uncertain.

Inquiry 19–17 How might you explain the fact that no pulsars are known whose periods are more than a few seconds long?

A MODEL OF A NEUTRON STAR

We have already talked about the unbelievably high density of neutron stars, but we wish to illustrate further what truly odd objects they are. Because we have at present no way of creating material of such tremendous densities in a laboratory, theory is the only guide we have to understand these stars. A neutron star may have an atmosphere, but because of its strong surface gravity it would be only a few centimeters thick. Underneath it is the star's outer envelope, which would actually be *crystalline*; that is, a solid rather than a gas, with the particles composing it arranged in a rigid, regular structure.

Evidence for the crystalline outer layer of neutron stars is provided by a surprising event occasionally seen in observations of pulsars. Although in general the rotation of pulsars slows gradually with the passage of time, in some cases they speed up suddenly, after which the pulsations gradually slow down again. These "glitches" have been interpreted as the result of adjustments of the crystalline shell of the star, accompanied by a slight collapse of the star. These episodes in the life of a neutron star have been called "starquakes," by analogy to earthquakes.

Inquiry 19–18 What physical law causes a pulsar's rotation to speed up if, as a result of a starquake, the star collapses slightly?

BINARY PULSARS

The first discovery of a **binary pulsar**, in which a pulsar is found together with another star in a close binary system, was exciting for astronomers because they could now determine the masses of neutron stars from observations. In this binary pulsar, PSR 1913+16, the sum of the masses is about 2.8 solar masses. Thus it is possible that both stars are neutron stars of 1.4 solar masses each. Only two such binary pulsars are currently known.

Another extremely important observation of these binary systems is that the orbital period of the two stars is observed to decrease in exactly the manner predicted by Einstein's theory of general relativity for an object emitting gravitational waves, which are extremely feeble waves predicted to be emitted whenever a mass is accelerated. The binary pulsars have turned out to provide an unexpected but important test of the theory of relativity.

MILLISECOND PULSARS

Another unusual group of pulsars contains the "millisecond pulsars," one of which is PSR 1937+214. With a period of 0.001558 seconds, this pulsar rotates on its axis at the phenomenal rate of 642 rpm! If it were spinning only 10% faster, in fact, the forces generated would tear it apart. It appears that this pulsar is very

old—an age of 10^9 years has been suggested—but its magnetic field is so small that it loses energy slowly. Because the picture we gave above was one of pulsar periods lengthening over time, why is this old pulsar rotating so rapidly? The explanation is that millisecond pulsars are members of binary systems. The low-mass companion finally reached a stage where it could dump mass into an accretion disc which, when transferred to the neutron star surface, increased the rotation speed. (Think of the way you can increase the speed of a spinning basketball by hitting it periodically.) Many such systems are turning up in the centers of globular clusters, where the density of stars is high and where binary star systems should be abundant.

ARE THERE PLANETS AROUND PULSARS?

Recently, astronomers observed pulsars for which the pulses arrived some 0.003 seconds later than expected. This short delay time is significant. These data have been interpreted in terms of two small planets orbiting a neutron star. (These observations were mentioned in the concluding section of Chapter 6.) One idea as to how planets might form in this unfriendly environment suggests that energetic particles from the neutron star erode the companion's gas, which eventually condenses into one or more planets. More data are required, however, before more refined models can be produced.

19.4
BLACK HOLES

The objects known as black holes have captured the imaginations of many people. Our investigation will first describe what they are and their properties. Then we will see how astronomers might go about detecting objects that are thought to emit no light. During this discussion, we hope to dispel some popular misconceptions about black holes.

THE THEORETICAL PREDICTION OF BLACK HOLES

We saw that theory had suggested the possibility of neutron stars some 30 years before their existence was verified. The simple fact that these objects were theoretically possible was a stimulus to thought and observation and aided in their ultimate discovery. Similar predictions and later discoveries had occurred earlier in physics with the neutrino and with antimatter. For these reasons when the solution of an apparently valid equation suggests that a certain type of physical behavior *can* take place, scientists are conditioned to think that somewhere in the vast universe it actually *may* take place. In a sense, the universe is so large that if something can happen, it probably will.

If you take that last statement with a grain of salt, but still keep it in mind, you will see why astronomers were intrigued with another idea that kept popping up in calculations from time to time—the possibility that under some circumstances a dense gaseous configuration could begin to collapse and there might be nothing to stop the collapse.

We have seen that in the center of a supernova explosion there is a dense imploding remnant, but at some point various sources of outward pressure (high internal temperature and neutron degeneracy pressure) combine to stop the collapse and restore equilibrium, thus producing a neutron star. Yet the possibility remains that there might be circumstances in which an imploding star could either be pushed past this equilibrium point or else simply have so much mass that nothing could stop the tremendous inward crush caused by the gravitational forces. Remember that if a white dwarf gains mass beyond its maximum 1.4 solar masses, it collapses into a Type I supernova. What happens in the collapse of an object of mass greater than the neutron-star limit of 3 solar masses?

As a star collapses during its implosion, its surface gravity will increase rapidly as its size decreases. Eventually, the speed required for anything to escape from the star's surface (its escape velocity) will become greater than the speed of light. Because, according to the theory of relativity, nothing can travel faster than the speed of light, not even light would be able to escape. Astronomers call such a collapsed object a **black hole**. Because no known forces can stop the implosion, it continues until the mass becomes a point of zero radius and infinite density called a **singularity**. Surrounding the singularity is a region of empty space from which light cannot escape. The radius of this region is determined by the object's mass and is known as the **Schwarzschild radius**.[3] As an example, a one-solar-mass black hole will have a 3-km Schwarzschild radius. Light emitted inside this radius will be trapped. However, light emitted a few Schwarzschild radii away will be able to escape. Because communication with the universe outside a black hole is impossible, the Schwarzschild radius is often referred to as the **event horizon**—outside observers are unable to see events occurring inside.

To understand the unusual properties of a black hole, it is necessary first to see how a gravitational field can af-

[3]The Schwarzschild radius is given by $2GM/c^2$ where G is the gravitational constant, M is the object's mass, and c is the speed of light. Thus doubling the mass doubles the size of the Schwarzschild radius.

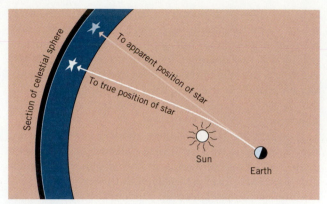

FIGURE 19–26. The bending of light rays from a distant star when the light passes near the Sun alters the apparent position of the star from its true position.

fect radiation that is passing through it. The modern theory of how radiation and gravity interact is Albert Einstein's general theory of relativity. In its most complete form, it is highly mathematical, but some of its consequences can be discussed in a simple way.

One of the first predictions Einstein made from his theory was that when radiation passes very close to a one-solar-mass star, its path should be bent 1.75 seconds of arc at the Sun's edge. A total eclipse of the Sun offered astronomers the first opportunity to verify the prediction. When the Moon covers the bright disk of the Sun, the sky near the Sun darkens and it is possible to see stars close to the Sun's edge. **Figure 19–26** illustrates what should happen if the light rays from these stars are bent as they pass near the Sun during an eclipse. By comparing a photograph taken during the eclipse with one taken with the same telescope when the Sun has moved away from that part of the sky, we should be able to measure a change in the apparent positions of the stars. During the total solar eclipse of 1919, a research team led by English astronomer Sir Arthur Eddington (**Figure 19–27**) performed this extremely difficult experiment, and they indeed observed the predicted effect. This experiment has been refined and repeated successfully numerous times.

Another manifestation of this same phenomenon is seen at the surface of white dwarf stars, as discussed in the last chapter. As photons try to leave the white dwarf, they must expend considerable energy to escape the strong pull of gravity. As a consequence they lose energy. When a photon loses energy, its wavelength shifts toward the red, producing a gravitational redshift.

The cause of the bending of light can be understood from two points of view. Classically, using Newton's law of gravity, a photon will be attracted to a massive body,

FIGURE 19–27. Sir Arthur Eddington, one of the more prominent astrophysicists of the 20th Century.

FIGURE 19–28. (a) A ball traveling in "flat" space always travels in the same direction. (b) If the table top is made of rubber, a brick in the center "warps" the space through which the ball travels. In this way, the path changes as it passes through the warped area.

just the same as anything else with mass. (A photon, which moves at the speed of light, behaves as though it has mass. When a photon stops moving, its so-called *rest mass* is zero. That is why a photon is called a *massless particle*.) From the viewpoint of the theory of relativity, the photon travels through a space that itself has been changed, or "warped," by the presence of mass. As an analogy, the path of a billiard ball on a table made of a stretched sheet of rubber would be different from the path if the rubber top were warped by a brick placed on it (**Figure 19–28**). Einstein predicted a bending of light that agrees perfectly with what astronomers observe and is twice that predicted by the Newtonian gravitational law. (We will expand upon Einstein's ideas of the geometry of space in the final chapter of the book.)

The popular idea of a black hole is that of a cosmic vacuum cleaner, something that sucks in everything near it. This concept is true only for matter within or near the event horizon. Thus transforming the Sun into a black hole would not cause the Earth to be sucked in because there would be no change in the gravitational attraction felt by the Earth at its distance of 1 AU.

OBSERVATIONS OF BLACK HOLES

If an object cannot radiate energy, then we must be quite clever to observe it. Obviously a photograph is useless, and it is unlikely that we would be able to detect the effect a black hole might have in bending radiation passing near it (as in an eclipse), because we would have no way to make a comparison photograph without the black hole present. But there are some other possibilities; for example, we may ask what effect a black hole might have on mass that is near it, but not yet inside the Schwarzschild radius.

Consider **Figure 19–29**, in which a black hole is associated with a giant star in a binary system. Such a configuration could be formed if a massive star in a binary

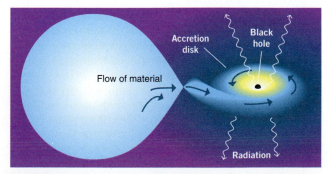

FIGURE 19–29. The most likely way to detect the presence of a black hole is by observing X-rays emitted by the hot gas that falls into such an object.

evolved into a black hole, leaving its less massive companion to live out its life normally. Much of the mass of the original star would be in the black hole singularity, so it still would be a gravitationally functioning member of the binary system, and the two objects would continue to orbit each other. However, if the companion were to become a giant, or for any reason spill mass over onto the black hole, a situation similar to that of a nova would develop. Mass would be dumped onto the black hole (through an accretion disc) and, upon falling in, would disappear. However, before it disappeared from our sight, it would be compressed and heated to extreme temperatures by friction with other particles.

Inquiry 19–19 If friction heated gases to a temperature of 10^6 K, in what wavelength region should astronomers look for the most intense radiation?

Due to the high temperatures of the gas falling into a black hole, considerable interest has been focused on X-ray surveys of the sky that continue to be carried out by satellites, and astronomers have turned their telescopes toward those parts of the sky where strong X-ray sources have been seen. One of the brighter X-ray objects known is Cygnus X-1 (the first X-ray source located in the constellation Cygnus). Its X-ray radiation fluctuates both randomly and also with a period of 5.6 days, indicating its membership in a binary system in which the X-rays are periodically eclipsed. Although it is often difficult to associate a visible object with an X-ray source, Cygnus X-1 turned out be at the same location as a known star (called HDE 226868) that happens to be a single-lined spectroscopic binary of period 5.6 days. The fact that only one set of spectral lines is observed is important because it means the companion star is fainter than the observed one, which is a blue supergiant.

If Cygnus X-1 consists of an invisible object and a blue supergiant, how is it possible to tell whether the invisible companion is a black hole? Perhaps it is merely a neutron star, for matter falling onto a neutron star will also probably generate a high enough temperature to give off X-rays. The one way to tell is to find the mass of the invisible star. If it could be demonstrated that the companion star had a mass greater than about three solar masses, theory says it must be a black hole.

Inquiry 19–20 Why couldn't the invisible object with a mass greater than three solar masses be a white dwarf or a neutron star?

It is not possible to solve for the mass of both stars in a binary system if only one of the stars shows spectral lines. However, it is possible to set some limits on the mass of the invisible object. The blue supergiant star in Cygnus X-1 should have a mass of around 20 solar masses (assuming that it is not some kind of peculiar object). This mass, along with the observed Doppler shifts in its spectrum, implies that the mass of the invisible star must be at least five to eight solar masses. Neither a neutron star nor a white dwarf can be anywhere near this massive, so the only remaining possibility is that the collapsed object in Cygnus X-1 is a black hole.

Cygnus X-1 is neither the only nor the best black hole candidate. The star V404 Cygnus is an X-ray source that emits more energy in X-rays than a million stars like the Sun do in all spectral regions. The observed orbital motion of 420 km/sec means that the collapsed object must have a mass at least greater than 6.3 solar masses and probably between 8–15 solar masses. V404 Cygnus is a stronger black hole candidate than either Cygnus X-1 or another candidate, A0620-00, because it requires us to make fewer assumptions about the mass of the observable component.

At present, we have only theoretical speculations as to how a black hole might actually form. It is possible that during a supernova explosion, a region from the central core of the star might be able to implode into a black hole in one quick transition. It is also conceivable that a neutron star near the upper limit of mass allowed for neutron stars could gravitationally accrete more material from its vicinity (perhaps from a companion star) and collapse into a black hole once the mass exceeds the neutron star limiting mass. If a star survived the red giant and planetary nebula mass-loss phases and still had several solar masses worth of material, it might be able to continuously contract into a black hole under the action of its own self-gravitational attraction.

Do black holes exist? While intriguing evidence points toward their existence, the astronomical community as a whole does *not* agree that stellar black holes have been observed. There are simply too many other ways to explain the observations, although many of these explanations are ad hoc and contrived (for example, that Cygnus X-1 is a multiple star system made up of the supergiant and two or three neutron stars; or that such objects are so-called Q stars, objects composed of protons, electrons, and neutrons held together by the strong nuclear force rather than by gravity). The black hole explanation is the simplest and therefore, by Occam's razor, preferred.

Our theoretical understanding of the formation of black holes is far from complete. A few scientists doubt they can form. Because making the statement "black holes exist" is tantamount to making an important statement about how nature works, scientists are required to be skeptical and take a conservative approach until all the evidence is overwhelmingly in favor. (The possibility that million-solar-mass black holes lie at the centers of certain galaxies will be discussed in Chapter 22.)

*19.5
EXAMPLES OF BEASTS IN THE COSMIC ZOO

In recent years the "zoo" of objects that appear to involve a compact object (white dwarf, neutron star, or black hole) in a binary system has continued to grow. In addition to the novae and Type I supernovae, the ranks have been augmented by objects known as X-ray bursters, gamma-ray bursters, and the enigmatic object known as SS433. Some of these objects are found within globular clusters. X-ray and gamma-ray bursters are sources of highly variable, high-energy electromagnetic radiation. It is suspected that these also involve an evolved star transferring mass onto a neutron star or black hole.

SS433

SS433 exhibits a variable energy output in the optical, radio, and X-ray regions. It also shows two sets of emission lines at displaced wavelengths that indicate the existence of clouds of material in the system streaming at speeds as much as 20% of the speed of light. Also, the velocities vary periodically over a 164-day period, from a blueshift of 8% the speed of light to a redshift of 16%

the speed of light. The best model for this bizarre behavior is shown in **Figure 19–30**. A normal star forms a binary pair with a compact object, probably a 0.8-solar-mass neutron star and quite possibly a star that has in the past been a supernova. Material from the normal star streams toward the compact object, forming an accretion disk. Two jets of material are ejected perpendicular to the accretion disk at approximately 27% the speed of light. The accretion disk itself precesses, or changes its plane of rotation, under the gravitational influence of the normal star, much as the Earth's axis of rotation precesses under the influence of the Moon. This causes the direction of the two jets of material to change, which in turn causes the velocities we observe to vary periodically.

CYGNUS X-3

The object Cygnus X-3 is observed from the longest radio wavelengths to the most energetic gamma rays. It might contribute strongly to the high-energy cosmic rays observed in our galaxy. Its location at some 30,000 light years behind a dust cloud makes it invisible in optical wavelengths. With a period of 4.8 days, it was assumed to be a neutron star in a binary system. Only recently, however, has its true nature been determined through infrared observations capable of penetrating the obscuring dust. They revealed broad emission lines of both neutral and ionized helium but a deficiency of hydrogen. These observations are reminiscent of the type of star known as a Wolf-Rayet star, an extremely hot star with such a strong wind that the outer layers have been blown away, thus exposing the helium-rich core. This star will probably explode as a supernova; because it is a binary system, it will produce another binary pulsar.

GEMINGA

Geminga is a gamma-ray source, the second brightest in the sky. The amount of γ-ray energy is 1000 times the X-ray energy, which itself is nearly 2000 times the optical energy. Finding an optical counterpart to the γ-ray source, however, has proven to be difficult and is the subject of controversy among astronomers. Some researchers have interpreted the data as indicative of an isolated neutron star, others as showing a pair of orbiting neutron stars less than 33 light-years from Earth, and yet others as a young pulsar at a distance of 3000 light-years. Recent observations made with the Rosat X-ray satellite found Geminga to be a strong source of pulsed X-ray emission with a 0.237-second period, and thus an X-ray pulsar.

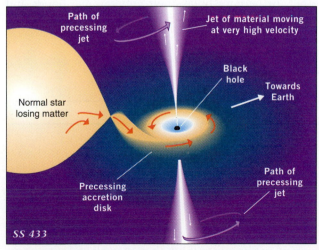

FIGURE 19–30. A model of SS433.

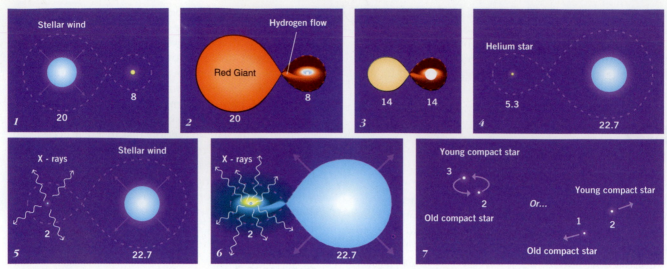

FIGURE 19–31. An evolutionary model for Centaurus X-3, showing how mass transfer changes a binary system and produces a variety of observable effects.

A GENERAL MODEL

These three examples of "weird beasts in the astronomical zoo" serve to emphasize two points. First, the universe is filled with many strange and exotic objects whose comprehension provides astronomers with an understanding of concepts not otherwise fathomable. Second, many of these objects involve mass-transferring, interacting binary systems. While details differ from one object to another, mass transfer between members of a binary system is the overriding concept that ties them all together. An example of the process is shown in **Figure 19–31**, which presents a scenario for the evolution of the binary X-ray source Centaurus X-3. It begins with two stars having 20 and 8 solar masses, respectively. The 20-solar-mass star evolves more quickly and begins to eject a stellar wind (*1*). Eventually it becomes a red

giant, fills its Roche lobe (*2*), and begins to dump matter onto the companion until the masses are equalized (*3*) and a slower rate of mass transfer occurs. When enough mass has been transferred (*4*), a helium-rich core is exposed, and we have a Wolf-Rayet star that eventually becomes a supernova and leaves behind a neutron star. The evolution of the secondary star now speeds up so that material from its stellar wind falls onto the neutron star, which appears as an X-ray source (*5*). The secondary expands to fill its Roche lobe (*6*) and then rapidly transfers matter back to the primary; this is Centaurus X-3 today. Mass may be lost from the entire system at this state. Finally, the system will evolve into a compact object (*7*) in which the stars may orbit each other, or perhaps break their gravitational bonds and separate from one another.

CHAPTER SUMMARY

- **Figure 19–32** summarizes our entire discussion of stellar evolution from the main sequence onward. The figure shows the post-main-sequence evolution and final stages for stars of various masses. Development of this summary consumed the careers of numerous astronomers over many decades; more complete understanding of these evolutionary phases is the job of future generations of astronomers.

OBSERVATIONS

- **Novae** are stars that increase in brightness as much as 10^5 times in a day or two. Their spectra show expansion velocities of hundreds up to a thousand km/sec. Some novae are observed to occur in binary star systems. While only some appear to be recurrent, astronomers think they may all be.

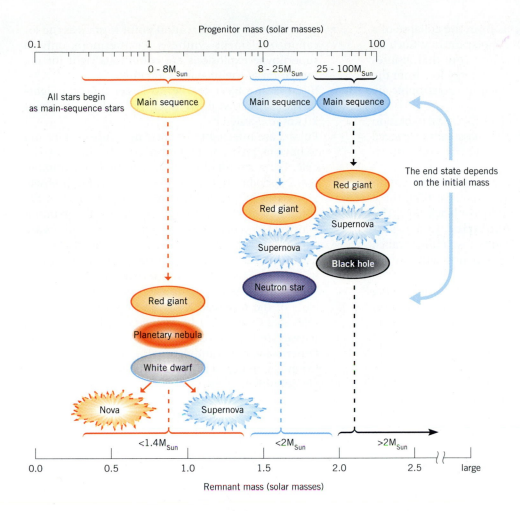

FIGURE 19–32. A summary of stellar evolution from the main sequence through stellar death.

- Supernovae are objects whose brightness increases by billions of times over the pre-outburst brightness. **Type I supernovae** have no hydrogen in their spectra, whereas **Type II supernovae** do show hydrogen. The spectra of gas in supernovae remnants show the presence of heavy elements, which were formed by nuclear burning during the explosion.
- **Pulsars** are objects for which astronomers observe precisely timed emissions of radio energy. The periods between pulses generally increase with time. Observations show some pulsars to have a sudden change in the pulse period.
- The observed increase of a pulsar's continuous spectrum with increasing wavelength, along with the radiation's polarization, is interpreted as synchrotron radiation from charged particles accelerating near the speed of light in a magnetic field.
- X-ray and gamma-ray telescopes in orbit about the Earth have made observations of a variety of objects.

Many of these objects change brightness rapidly, in a fraction of a second.

THEORY

- Nova explosions are thought to occur when matter from one star in a binary system collects in an **accretion disc** surrounding a companion **white dwarf**, with the subsequent buildup of hydrogen on its surface. Once the temperature and pressure at the bottom of this accreted hydrogen layer reach high-enough values, a thermonuclear explosion occurs, causing the ejection of matter from the white dwarf.
- A **Type I supernova** results when a white dwarf accretes enough mass to reach the Chandrasekhar limiting mass of 1.4 solar masses. Because all Type I supernovae are formed from the same mechanism, they are all similar.

- A **Type II supernova** results from the collapse of a massive star containing 10–20 solar masses. Such an object has produced an inert iron core that disintegrates and emits neutrinos. The energy loss from the star's interior causes it to collapse catastrophically and then to explode.

- For the collapsing remnant of a supernova explosion to produce a neutron star, the mass cannot exceed about three solar masses. During the collapse, its magnetic field intensifies.

- A massive star that undergoes a supernova explosion that leaves behind a core more massive than three solar masses is thought to produce a black hole. The object collapses into a **singularity**, and is surrounded by an **event horizon** whose radius (called the **Schwarzschild radius**) depends only on the remnant's mass.

- Matter falling into a black hole from a surrounding accretion disc will be heated to high temperatures and emit radiation in the X-ray and gamma-ray spectral regions.

CONCLUSIONS

- Single stars get rid of mass by ejecting a planetary nebula, while binary stars eject matter by (recurrent) novae explosions.

- Both theory and observation point to stars as the location of element synthesis. Furthermore, only in supernovae explosions are conditions right for the formation of the heaviest elements.

- Because an object can be no larger than the light-travel time across it, we infer that pulsars are about 20 km in diameter.

- Pulsars are inferred to be rotating neutron stars and to have magnetic and rotation axes that do not coincide. The star's rotation causes a beam of radiation to point periodically toward the observer. From observations of binary pulsars, neutron star masses are computed and found to be less than the maximum allowed by theory. Observations of binary pulsars are consistent with Einstein's theory of relativity.

- Cygnus X-1 is an X-ray emitting object with a period of 5.6 days. Its optical counterpart is a single-lined spectroscopic binary with a 5.6-day period. On the basis of the observed X-ray and optical spectra, we infer that Cygnus X-1 has a mass of at least five solar masses and therefore is likely to be a black hole. Other black hole candidates also exist. The evidence for the existence of black holes is compelling but not yet definitive.

SUMMARY QUESTIONS

1. What do we mean by the term "nova"? What are two pieces of evidence that novae eject mass?

2. Why do astronomers conclude that objects must be smaller than the light-travel time across them?

3. What model for novae phenomena best allows astronomers to understand them?

4. What are the observed features of the Crab Nebula? What evidence exists that it is indeed the object observed about 1000 years ago by Chinese astronomers?

5. What are the observed differences between ordinary novae and supernovae? Explain them in terms of the violence of the explosion, the increase in brightness of the object, and the amount of mass expelled.

6. In what ways do Type I and Type II supernovae differ? Your answer should be based on their observed properties and the models we have to explain them.

7. How do supernovae provide us with direct evidence for the synthesis of heavy elements within stars? Name and describe three basic processes by which heavy elements are synthesized.

8. Describe the interior structure of a neutron star.

9. What evidence is there that pulsars are actually rotating neutron stars? What observational evidence is there for the existence of pulsars?

10. How do light and matter behave near a black hole? Why can nothing escape from a black hole?

11. How might astronomers detect a black hole? What evidence is there that the Cygnus X-1 system actually consists of a giant star and a black hole, rather than some other collapsed object such as a white dwarf or a neutron star?

12. What is the basic model that explains most of the objects in the astronomical "zoo"?

APPLYING YOUR KNOWLEDGE

1. Outline the steps leading to the inference that Cygnus X-1 and V404 Cyg might or might not be black holes.

2. What evidence might Tycho Brahe have used when he said that the supernova of 1572 was farther away than the Moon? (Remember, this was prior to the telescope!)

3. Your body consists of numerous chemical elements, including carbon, oxygen, and iron. Discuss the origin of these elements in terms of their formation in stars.

■ **4.** Suppose the spectrum of a nova contains the Hβ spectral line centered at a wavelength of 4856 Å. Remembering that Hβ is normally at a wavelength of 4861 Å, calculate how rapidly the gaseous envelope expands.

■ **5.** Novae usually show broad emission lines caused by the expanding gas. Suppose such a spectral line is observed to be 10 Å wide. Using the observed width of the emission line, determine the speed of expansion of the gaseous envelope relative to the star itself.

■ **6.** What would be the size of a pulsar whose brightness changes substantially in 0.00007 seconds?

■ **7.** What would be the mass of a black hole whose binary companion is a 20-solar-mass star separated by 0.175 AU and whose period is 5 days?

ANSWERS TO INQUIRIES

19–1. Presence of emission lines; emission lines that are Doppler shifted relative to absorption lines; presence of gaseous nebula visible on photographs around the star.

19–2. The star would be 100 times more luminous.

19–3. As the gas expands away from the star, it will get thinner, and we might expect the line in the spectrum due to the gas to get weaker. We might also expect the gravitational force of the star to slow down expansion of the gases. This would cause a change in the amount of the Doppler shift of the lines that are due to the gases.

19–4. Recurrent novae might be explained by a periodic process of mass dumping from the companion to the compact object, followed by an outburst, followed by a repeat of this process for an indefinite number of times.

19–5. If many novae are recurrent, then the number that we count overestimates the number of stars that "go nova" in our galaxy.

19–6. It is certainly true that the white dwarf star must have started out with the larger amount of mass. It probably did undergo a stage of mass transfer onto the other, initially smaller star. The evolution of that star then would have been speeded up. In addition, the dumping of material consisting of heavier elements onto the surface of the initially smaller star might alter the chemical composition of its atmosphere.

19–7. Because supernovae are very unusual events, we might expect that if there is a remnant star in the Crab Nebula, it should appear to be peculiar or unusual compared to most stars. In fact, Fritz Zwicky, applying this principle, actually identified the remnant star of the Crab Nebula in the 1930s.

19–8. Heavy elements are believed to have been produced in the interiors of massive stars. The Crab Nebula confirms this production, because we see these elements in the remnant, and because of the elements' dispersal into space.

19–9. We would expect heavier elements to be less abundant than lighter elements.

19–10. Type II supernovae result from the collapse of a single massive star as a natural part of its evolution. Theory tells us that abundant neutrinos will be produced during the core collapse, and that they will aid in blowing the star apart. Because the outer layers of the precursor were hydrogen, the spectrum will include hydrogen. Type I supernovae, on the other hand, result when matter from one star is dumped onto a white dwarf, whose mass is increased above the Chandrasekhar limiting mass. Because the white dwarf contains little hydrogen, the spectrum will not contain hydrogen.

19–11. Some process other than the addition of helium nuclei must be at work in the production of ^{17}O and ^{18}O. The addition of neutrons may be one such process.

19–12. Probably most of the oxygen is produced by reactions between helium nuclei and the ^{12}C produced by the triple-alpha process.

19–13. The four-minute difference is the difference between the solar and sidereal day. It is reasonable to conclude that the emission is from an extraterrestrial source.

19–14. The Crab pulsar is observed to be slowing, and the

Crab Nebula itself emits more energy than can be accounted for. Using the observed slowdown of the pulsar, and the expected size of a neutron star (say, 20 km), astronomers could compute the amount of energy lost as the pulsar rotation slowed. This energy loss was nearly identical to the luminosity of the Crab Nebula.

19–15. Presumably the Vela pulsar is much older than the Crab pulsar.

19–16. If the Earth is not in the region "swept out" by the beam from a pulsar, we may not see or notice the beam. This in turn would mean that we would *undercount* the number of pulsars. It might be possible to determine if this is the case by comparing the ages and number of known pulsars with the expected number of supernovae per century in our galaxy. Regrettably, this is difficult to do because the supernova rate itself is very uncertain.

19–17. As a rotating neutron star slows down, it loses rotational energy. Eventually it will reach a point at which it has too little energy to accelerate the charged particles to near-light speeds. When this happens, the pulsar would no longer emit radio emission.

19–18. Conservation of the angular momentum.

19–19. Wien's law would tell us the X-ray region.

19–20. Theory indicates that the limiting mass for white dwarfs is about 1.4 solar masses. The neutron star limit is between 2 and 3 solar masses; it is more uncertain because it depends on assumptions that are less well understood.

PART FIVE

DISCOVERING THE NATURE AND EVOLUTION OF GALAXIES AND THE UNIVERSE

The stars discussed in Part 4 are grouped together into galaxies, the building blocks of the observable universe as a whole.

Our study of galaxies begins at home in our own galaxy, the Milky Way. In Chapter 20 we learn how astronomers have determined our location within our galaxy, and we examine the optical and radio evidence that our galaxy has a spiral shape. In discussing how the galaxy's mass is determined, we find there are good reasons to suspect that invisible matter may play an important part in the overall makeup of the Milky Way. Discussions of the cause of spiral structure, of cosmic rays, and of different populations of stars lead us to an examination of the formation and chemical evolution of the Milky Way. Because gas and dust between the stars play an important role in the galaxy's structure, and because they profoundly affect observations, the interstellar medium is further discussed here.

The study of galaxies beyond our own is the subject of Chapter 21. After learning about the different shapes of galaxies, we investigate how their distances are found. This discussion ends with the all-important Hubble relation, which connects the observed motions of distant galaxies to their distances. The study of clusters of galaxies, beginning with the Local Group of which we are a part, comes next. By examining galaxies in distant clusters, we are able to learn more about the properties of galaxies as a group. Such knowledge allows us to discuss galaxy evolution.

Because not all galaxies are as "normal," as are those discussed in Chapter 21, Chapter 22 examines peculiar galaxies. We find that their peculiarities come about from galactic collisions. We also find galaxies that emit tremendous amounts of energy. Much of Chapter 22 concerns our attempts to understand these vast amounts of energy. There is, however, a minority of astronomers who believe the generally accepted picture to be incorrect. Some of their arguments, and the evidence for and against their ideas, are also discussed.

In a sense, all discovery in astronomy is aimed at understanding the origin and evolution of the universe, the subject with which we close the book in Chapter 23. The purpose of this chapter is to present the generally accepted cosmological ideas and the evidence in favor of them. We also look at the steady state cosmological theory, because, even though it is passé, it is worth considering in any treatment of how science makes critical decisions. An important goal of this chapter is to provide you with evidence that the general idea of the big bang theory of cosmology is valid. The concepts of the geometry of curved space are also presented, along with ideas as to how we can predict the fate of the universe.

20

THE MILKY WAY: OUR GALAXY

Be it ever so humble, there's no place like home.

JOHN HOWARD PAYNE
(FROM THE OPERA *CLARI, THE MAID OF MILAN*, 1823)

Thomas Wright is generally credited with the first published statements, in 1750, describing our galaxy as a lens-shaped, or disklike, system of stars. He also suggested that the Sun and other stars travel in extended orbits about the "Universal Center of Gravitation," which we would later call the *galactic center*. His thinking was both bold and correct, for although Galileo's telescopic observations over a century earlier had shown the Milky Way to be composed of many faint stars, further progress had been slow. By the 1780s, William Herschel had confirmed, using systematic star counts, that the Galaxy[1] was shaped like a lens, and he popularized the notion in his writings.

We have already examined various component objects within our own stellar system, which we call the **Milky Way**, because when viewed from a dark location the stars are so bright they look like milk spilled across the sky. We see these objects through a filter of dust that permeates the Milky Way. The growth of technology since World War II has given astronomers tools to penetrate the dust—detectors and telescopes sensitive to radiation at radio, infrared, and millimeter wavelengths. Technology has also made it possible to assemble bits and pieces of knowledge into a coherent overall picture of our galaxy.

But even today many questions about the Galaxy are difficult to answer because of our location inside it. Therefore, astronomers also study other galaxies in the hope of understanding our own galactic system better. (The study of other galaxies is the subject of the next chapter.)

Can we make an educated guess as to the shape of our galaxy? Looking at other galaxies, we find that they consist mostly of two types: spirals (**Figure 20–1a**) and ellipticals (**Figure 20–1b**). Spirals appear flattened, with a bulge in the center and extensive regions of gas and dust. Ellipticals, on the other hand, appear to have a more uniform appearance and to lack extensive dusty regions. Because photographs of our night sky show it to be filled with dark, dusty regions (**Figure 20–2**), we may hypothesize that our galaxy is a spiral. Affirming this hypothesis and understanding our location within the Galaxy are among our goals in this chapter.

20.1

OUR PLACE IN THE MILKY WAY

Where are we located within our galaxy? How big is it? How rapidly are we moving? How long does it take for us to complete one orbit? Do other stars near us have identical motions? What is the Galaxy's mass? The answers to these questions, which we now seek, provide fundamental information about our home galaxy.

THE DISTRIBUTION OF GLOBULAR CLUSTERS

In Chapter 3 we introduced attempts to determine the structure of the Milky Way by means of star counts, and we discussed how the dust in the Galaxy made those results thoroughly misleading. These studies indicated that the Galaxy was about 10,000 light-years in diameter, with the Sun in the center. In 1917, before astronomers had figured out a way to correct for the effects of dust, Harvard astronomer Harlow Shapley was already pursuing a different line of research. He was studying globular clusters, the giant spherical star clusters containing approximately 100,000 stars (and introduced in Chapter 18 in the context of stellar evolution). Figure 18–28 shows a photograph of the globular cluster 47 Tucanae, which is easily observed in the southern sky.

[1]Whenever the word "galaxy" refers specifically to the Milky Way, it is capitalized.

436

(a)

(b)

FIGURE 20-1. (*a*) A spiral galaxy. (*b*) An elliptical galaxy.

Shapley believed that these enormous star clusters must be an important component of the Galaxy, and he was interested in studying their distribution in space. To determine the distances to these clusters, Shapley made use of RR Lyrae stars located in them. These objects are readily identified in a given cluster by comparing photographs taken at different times and identifying stars whose brightness has changed in a characteristic manner and with a period of less than a day.

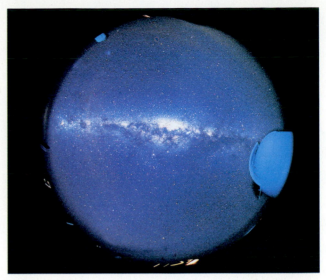

FIGURE 20-2. A fisheye view of the sky showing the Milky Way with its strong absorption by dust along the galactic plane.

How, then, do we find the distance to an RR Lyrae star? Remember, to find the distance to *any* star requires knowing two quantities: the apparent brightness and the luminosity. These quantities are then used with the inverse square law to find the distance. The apparent brightness of a star is readily obtained from observation. The luminosity of all RR Lyrae stars is about the same,[2] at maximum brightness about 100 times that of the Sun. Thus, if you know the luminosity for one RR Lyrae star, you know it (approximately) for every other one. Therefore, Shapley could find the distance to any star identified as an RR Lyrae star and thus the distance to the cluster.

Once we have the distance, a measurement of the cluster's angular diameter allows us to compute its linear size. We can then repeat the procedure for all clusters having RR Lyrae stars and obtain an average linear size for a typical globular cluster. The techniques for finding distance and linear size are summarized in parts *a* and *b* of **Figure 20-3.**

Inquiry 20-1 What assumptions are made in determining the distance to a cluster in this way?

Inquiry 20-2 What is the linear size of a globular cluster at a distance of 5000 ly if its observed angular size is 0.057 degrees?

[2]The truth of this statement can be seen by remembering that RR Lyrae stars are found in the horizontal branch of the globular cluster H-R diagram (Chapter 18). All objects on the horizontal branch have nearly the same luminosity.

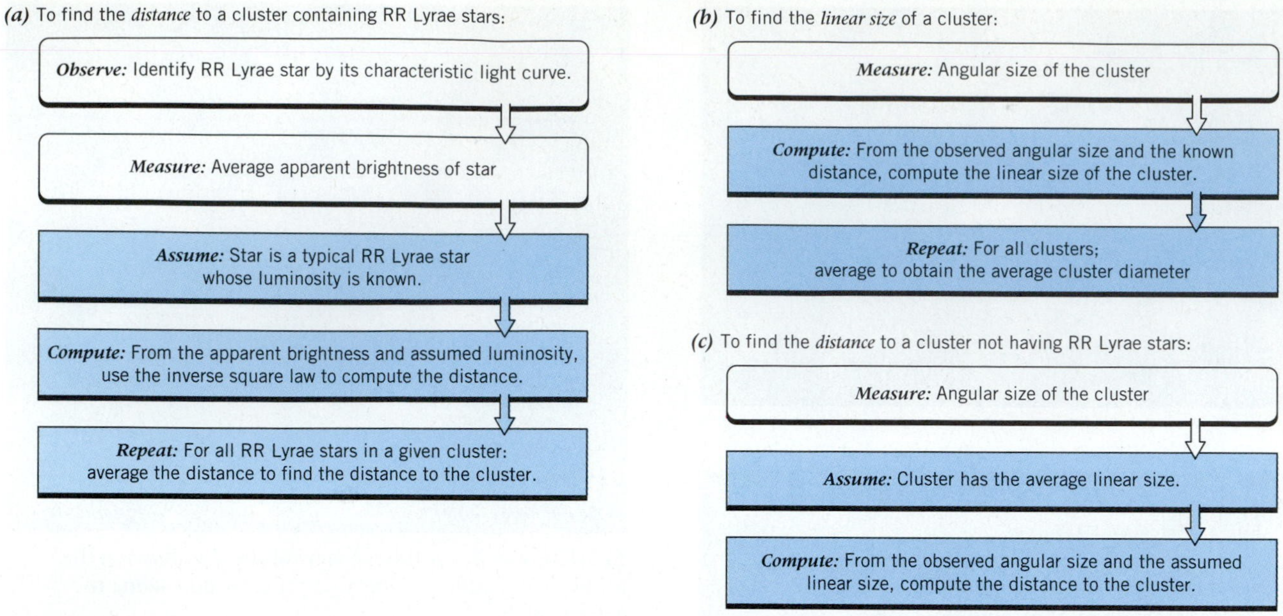

(a) To find the *distance* to a cluster containing RR Lyrae stars:

> *Observe:* Identify RR Lyrae star by its characteristic light curve.

> *Measure:* Average apparent brightness of star

> *Assume:* Star is a typical RR Lyrae star whose luminosity is known.

> *Compute:* From the apparent brightness and assumed luminosity, use the inverse square law to compute the distance.

> *Repeat:* For all RR Lyrae stars in a given cluster: average the distance to find the distance to the cluster.

(b) To find the *linear size* of a cluster:

> *Measure:* Angular size of the cluster

> *Compute:* From the observed angular size and the known distance, compute the linear size of the cluster.

> *Repeat:* For all clusters; average to obtain the average cluster diameter

(c) To find the *distance* to a cluster not having RR Lyrae stars:

> *Measure:* Angular size of the cluster

> *Assume:* Cluster has the average linear size.

> *Compute:* From the observed angular size and the assumed linear size, compute the distance to the cluster.

FIGURE 20–3. Summary of distance and size determination for globular clusters. (*a*) For clusters having RR Lyrae stars. (*b*) Finding a cluster's linear size. (*c*) Whenever no RR Lyrae stars are observed in a cluster.

Many globular clusters, however, either lacked RR Lyrae stars or were too far away for individual variable stars to be photographed. To estimate distance for these clusters, Shapley used the familiar concept that angular size decreases as distance increases. We assume the linear diameter of the unknown cluster is the same as the diameter of the average cluster. A measurement of the unknown cluster's angular size then results in an estimate of its distance. The technique of finding the distance for these clusters is summarized in part *c* of Figure 20–3.

Inquiry 20-3 What assumption did Shapley make when he estimated the distance to the farthest globular clusters from their apparent sizes?

Shapley found that some of the globular clusters are very far from the Sun, and that the Galaxy must therefore be much larger than had been previously thought. More significantly, he found that these clusters fall into a spherical distribution centered on a location far from the Sun (**Figure 20–4**). He reasoned that globular clusters were such massive and important components of the Galaxy that it was unlikely that they would be concentrated on one side of it. In other words, Shapley suggested that the center of the distribution of the globular cluster system was also the center of the entire Galaxy,

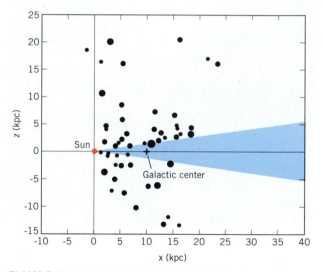

FIGURE 20–4. The distribution of globular clusters as determined by Harlow Shapley. Note that the center of the distribution is not at the Sun.

and that the Sun was actually quite far from this center. It is now estimated that the Sun is between 25,000 and 30,000 light-years from the galactic center. We will use a value of 30,000 ly. Shapley is thus a modern Copernicus in that he removed the Sun from the center of our stellar system.

Inquiry 20–4 What principle would have guided Shapley in placing the center of the Galaxy at the center of the observed distribution of globular clusters rather than at the center of the observed distribution of stars?

THE MOTION OF THE SUN AROUND THE GALAXY

The Sun orbits around the center of the Galaxy. Its motion can be determined by observing the Sun's movement relative to something stationary. The problem is, nothing in space is motionless! However, by observing the apparent motions of many globular clusters and distant galaxies distributed around the sky, we find that the Sun moves with a speed of about 250 km/sec around the center of the Milky Way.

Knowing the distance of the Sun from the galactic center we can find the distance it travels in one rotation. Because we now know how fast the Sun moves, we can find how long it takes. The resulting galactic year is about 200 to 250 million years.

THE MASS OF OUR GALAXY

We now have enough information to estimate the Milky Way's mass. We know that the period of the Sun around the galactic center is about 200 million years (2×10^8 years). We also know that the distance from the Sun to the center of the Galaxy is 30,000 light-years, or about 2×10^9 astronomical units (there are about 63,000 astronomical units in a light-year). Neglecting the mass of the Sun relative to the mass of the Galaxy, Newton's version of Kepler's third law applied to the Milky Way gives

$$\text{mass} = \frac{\text{distance}^3}{\text{period}^2}$$

$$= \frac{(2 \times 10^9)^3}{(2 \times 10^8)^2}$$

$$= 2 \times 10^{11} \text{ solar masses.}$$

If we assume that an average star has 1 solar mass, then our result translates into approximately 200 billion stars. To make a more precise estimate, we would have to allow for the numbers of stars of different masses, as well as the fraction of mass in the form of interstellar gas and dust. However, for an order-of-magnitude estimate, this number will do.

There are two important approximations in our calculation. First, we assume the orbit of the Sun is an el-lipse, which it isn't because matter in the Galaxy is not concentrated into a point. Second (and more important), the mass outside the Sun's orbit, and much of the mass outside the galactic plane, has been neglected in using Kepler's third law. Taking these factors into account, the entire Galaxy is probably a few times more massive than our simple estimate gives.

THE MOTION OF OTHER STARS

Because we know the size of the Milky Way and the speed of the Sun around its center, and we have an educated guess about its shape, we can now ask: How are the other stars near the Sun moving? The stars whose motions around the Galaxy are similar to that of the Sun are moving along with the Sun in the **galactic disk**, the flattened plane, about 1000 ly thick, containing most of the observable matter in the Galaxy. The galactic disk is the fundamental plane of the Milky Way, just as the ecliptic is for the solar system.

The Doppler effect gives us a means to determine the traffic patterns obeyed by stars in different locations of the Galaxy. Because the motions of objects reflect to some extent their origins, studies of stellar motions using the Doppler effect may provide information to help us unravel the story of how the Galaxy evolved.

If the Galaxy rotated like a solid wheel, then no star would ever change its position relative to any other; like wooden horses on a carousel, they would remain at constant distances from each other. However, we observe Doppler shifts in the light of other stars. Remember, a Doppler shift provides information about relative motion *toward* or *away from* the observer. It follows that if the Sun and stars are moving relative to one another, their distances must also be changing. A natural question to ask is whether the stars move around the center of the Galaxy in a manner analogous to the motion of the planets around the Sun, with the stars farthest out moving the slowest.

Figure 20–5 illustrates how we might answer this question for stars in the solar vicinity. For simplicity, the stars are shown as having circular orbits. We will hypothesize that stars closer to the center of the Galaxy move faster in their orbits, as do planets in the solar system. A star at position A would then show a Doppler shift to longer (redder) wavelengths because the Sun would be leaving it behind. A star at F, on the other hand, would show a Doppler shift to shorter (bluer) wavelengths, indicating that the Sun and the star are approaching one another. This is so because the star at F moves more rapidly than the Sun and overtakes it. Stars in the direction of the galactic center, toward point G, show no radial velocity because they move parallel to the Sun, neither approaching nor receding from it. A

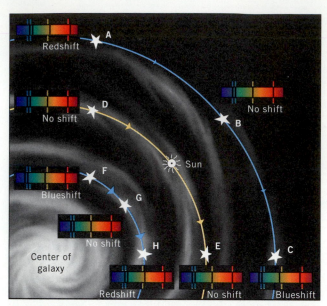

FIGURE 20–5. The motions of stars near the Sun produce various Doppler shifts in stellar spectra, depending on the object's location relative to the Sun. Note the position of the absorption lines relative to the comparison lines.

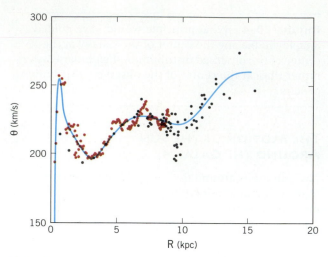

FIGURE 20–6. The rotation curve of the Milky Way.

star at *E* will also exhibit no radial velocity, because it is at the same distance as the Sun from the galactic center and therefore orbits the Galaxy with the same velocity.

Inquiry 20–5 What radial velocity, relative to the Sun, would a star have if it were located at point *B* in the diagram? Point *C*? Point *D*? Point *H*? Explain your reasoning.

Observations of the Doppler effect for stars in the immediate vicinity of the Sun agree well with the hypothesis just described. We are, therefore, able to conclude that the motion of stars close to the Sun is similar to the Keplerian motion of planets around the Sun (that is, planets farther from the Sun move with lower velocities). Just as the motions of planets are controlled by a dominant mass at the center of the solar system, so too is the motion of the Sun around the Galaxy controlled by the mass interior to its orbit. Likewise, the observed motions of the stars can occur only if the central regions of the Galaxy contain a large concentration of mass that dominates and controls the motions of stars and gas clouds. Because we are far outside this mass concentration, we may assume as a first approximation that all the mass is concentrated at a point at the center of the Galaxy.

> YOU SHOULD DO DISCOVERY 20–1,
> GALACTIC ROTATION, AT THIS TIME.

To compare observations with the predicted Keplerian motions, astronomers graph the rotation speeds of stars and gas clouds at various points throughout the Galaxy. The resulting **rotation curve** for the Milky Way is shown in **Figure 20–6**. The first feature we notice is that in the central regions, the velocity change with distance is a straight line. Such a straight line is exactly what you would get if you drew a rotation curve for a rigid body such as a rotating plate: the farther from the center, the greater the speed. Thus the interior regions of the Milky Way rotate like a solid body. That does not mean the nuclear region is solid, but that the numerous stars interact strongly with one another.

Beyond the location of the Sun the orbital motions are observed to remain nearly constant, rather than decreasing with distance as expected from Kepler's laws. This observation is highly significant and tells us that the mass of the Milky Way is *not* concentrated at the galactic center as we previously had assumed, but is spread out over large distances beyond the Sun's orbit. Astronomers have concluded that the Milky Way galaxy is surrounded by an extensive **galactic halo**. However, the total amount of material visible in the halo, such as globular clusters and some stars, is insufficient to explain the observed rotation curve. For this reason astronomers hypothesize the existence of unseen material we call **dark matter**.

DARK MATTER

Dark matter is not confined to the halo defined by the globular clusters but is part of an extended halo. What does this dark matter consist of? Because we have yet to detect it directly, it must be in a state that is difficult to recognize. A variety of suggestions have been made, from black holes, brown dwarfs, and other stellar remnants to a cloud of massive neutrinos. There is the possibility that it could simply be large numbers of faint stars that are difficult to discern individually at great distances. A generic term for such objects is MACHO, standing for **MA**ssive **C**ompact **H**alo **O**bject. A few astronomers reported an observation in 1993 that they interpret as caused by MACHOs. Before the general astronomical community accepts their conclusion, however, a great deal of additional data must be obtained by astronomers around the world. Other possibilities for dark matter include WIMPS, or **W**eakly **I**nteracting **M**assive **P**articles. These are theorized and have never been detected, not even on Earth. There are good reasons for hypothesizing such particles. They would solve not only the dark matter problem but also the solar neutrino problem discussed earlier. In constructing models that fit the observed rotation curve, astronomers find the total mass of the Galaxy may exceed a trillion solar masses.

HIGH-VELOCITY STARS

The majority of stars near the Sun move around the galactic center in nearly circular orbits at more or less the same velocity of about 250 km/sec. Relative to the Sun, however, the neighboring stars appear to be moving slowly. The situation is similar to cars speeding around a racetrack. While cars at the Indianapolis 500 race are going 185 miles per hour relative to their pit crews, their speed relative to each other is small.

Globular clusters, on the other hand, move around the Galaxy in elliptical orbits inclined at random angles to the galactic disk. Such orbits cause globular clusters to appear to be moving with high speeds relative to the Sun. For this reason, the globular clusters are called high-velocity objects. Similarly, stars that belong to the halo and whose tilted, elliptical orbits happen to carry

them through the disk near the Sun, will also have high speeds, similar to those of the clusters. Astronomers call these stars **high-velocity stars** because *relative to the Sun* they appear to move at high speeds.

How is it possible to determine whether a given star belongs to the halo or the disk? By observing its motion relative to the Sun. An object observed to be a high-velocity star belongs to the halo (**Figure 20-7**). We have been able to distinguish several thousand halo stars from disk stars on the basis of their apparent motions relative to the Sun. The ability to distinguish which part of the Galaxy a star belongs to is a prerequisite to understanding the formation and subsequent evolution of the Milky Way.

20.2
INTERSTELLAR GAS AND DUST

The space between the stars, the **interstellar medium**, is not empty but contains atoms, molecules, and dust grains. The interstellar medium provides the material from which stars form, and into which they lose matter by means of stellar winds, planetary nebulae ejection, and supernovae explosions.

We are discussing the interstellar medium in this chapter because the interstellar medium strongly affects our observations of the Galaxy. If we were to neglect the effects of the interstellar medium, our interpretations of observations would be subject to systematic errors, as was the case for early star counts. The known interstellar medium amounts to 10% the mass of the Milky Way. A study of the interstellar material itself tells us much about the conditions present in the Galaxy today and at times in the distant past.

INTERSTELLAR DUST: EXTINCTION OF STARLIGHT

Photographs of emission nebulae frequently reveal the existence of dust along with the gas. In **Figure 20-8** of the Trifid Nebula, the dark lanes across the middle as well as the mottled structure are caused by dust particles that block radiation coming from behind. A famous example of obscuration by dust is illustrated in **Figure 20-9**, a dense cloud of dust called the Horsehead Nebula.

One effect that dust has on passing radiation is illustrated in **Figure 20-10**. Some of the photons passing through dust clouds are scattered or reflected from the dust grains. These photons generally leave the dust in a direction different from the radiation's original direction, thus the star's intensity is diminished. Other pho-

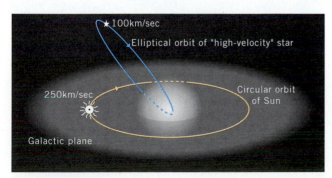

FIGURE 20-7. The orbit of a typical high-velocity star is highly elliptical and tends to lie outside the galactic plane.

FIGURE 20–8. The Trifid nebula in Sagittarius is a good example of an emission nebula—a region where stars, gas, and dust interact.

FIGURE 20–9. The Horsehead nebula, a region of dust in the constellation of Orion.

tons are absorbed. Their energy goes into heating the dust grains slightly.

Inquiry 20–6 What would be the effect on our estimate of the distance to a star if its light was affected by an unknown amount of intervening dust?

Inquiry 20–7 If our estimates of distances to stars were adversely affected by an unknown amount of dust, our knowledge of what stellar properties would be affected?

INTERSTELLAR DUST: REDDENING

The amount of light scattering depends on wavelength. Although radiation of all wavelengths is scattered, shorter (bluer) wavelengths are more readily scattered than longer (redder) wavelengths (see Figure 20–10). You see the effect of the scattering of blue light by dust and molecules every clear day—scattering is what

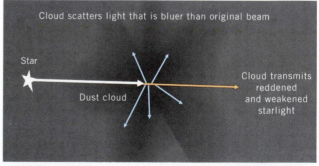

FIGURE 20–10. Scattering of light by dust particles both dims and reddens starlight as it passes through the interstellar medium.

makes the sky blue. Removal of blue photons by the scattering process is enhanced at sunset because sunlight moves through a greater path length, thus causing the Sun's reddish color.

Likewise, as starlight passes through interstellar dust, blue photons are scattered away, thus causing the light reaching an observer to be redder than it would have been had there been no dust. The reddening of starlight gives us a clue to the size of the dust grains, because the particle size determines the extent to which light of different wavelengths is scattered. For example, a beam of light passed through an assembly of particles the size of bowling balls would have its intensity lessened, but the decrease would not depend on wavelength.

On the other hand, when the size of the scattering particles is about the same size as the wavelength of the light, there are important effects on the light's color caused by diffraction and interference. For this reason, astronomers deduce that the dust particles in interstellar space must average about 1500 Å in diameter.

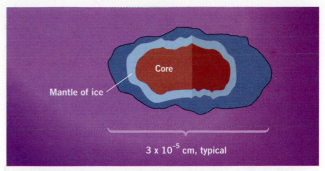

FIGURE 20–11. A suggested model for a typical interstellar dust grain.

Inquiry 20–8 What would be the effect on our estimate of a star's temperature if its light was reddened by dust, compared to a similar estimate for an unreddened star?

Early in the twentieth century, astronomers could not tell whether a star was red because it was cool or because its light had passed through interstellar dust. It was not possible to unscramble these two effects until a thorough understanding of stellar spectra was developed, which allowed astronomers to deduce the temperatures of stars from the absorption lines in their spectra rather than from their color. Knowing the star's temperature, astronomers could then predict what its *true* color should be. If its *measured* color did not match its predicted true color, the difference in the colors could be attributed to reddening by interstellar dust.

From the amount by which the starlight was reddened, astronomers could estimate the total amount of absorption and scattering, and therefore the star's true brightness. The correct distance could now be determined. In this manner, astronomers began to account for the dust and to gain a truer picture of the distribution of stars in space.

INTERSTELLAR DUST: REFLECTION NEBULAE

The bluish "haze" near the Trifid Nebula in Figure 20–8 results from starlight reflected from dust surrounding a star. Light reflected from dust produces a **reflection nebula**. Astronomers know the light is reflected from dust because the light's spectrum is that of the bright illuminating star and because the light is polarized.

An important piece of information about interstellar dust particles is available only from reflection nebulae. By studying the reflected light, we discover that the par-

ticles are highly reflective. In fact, a dust grain reflects approximately 90% of the light that falls on it. This demonstrates that the particles have smooth surfaces, and that whatever their chemical composition, they are probably covered with an outer coating of some type of ice (see **Figure 20–11**). Note that "ice" does not have to mean water ice. A number of simple compounds could form ices at the temperatures we see in these nebulae.

Scattering by dust also polarizes light. By analyzing the polarization we can determine more about the particles' sizes and chemical composition. The polarization results confirm the idea that the dust grains have an icy surface with, perhaps, graphite (a crystalline form of carbon) in the interior.

INTERSTELLAR DUST: POLARIZATION OF STARLIGHT

Astronomers obtain additional information about interstellar dust particles by studying the polarization exhibited by starlight passing directly through the dust. Starlight that has been reddened after passing through long paths of dust is also generally polarized to some degree, indicating that the polarization is indeed caused by the interstellar dust grains. To produce polarization, the dust particles must be somewhat elongated in shape and must also have their long axes aligned with one another. Only then will the light become polarized.

Inquiry 20–9 Why would this mechanism not work if the dust grains were spherical?

This whole picture requires something that "lines up" the dust grains over large areas of interstellar space. Such an agent can be found in the interstellar magnetic field, to be discussed more fully later in this chapter. Because our focus now is on the dust grains themselves, we can say that the existence of the polarization tells us

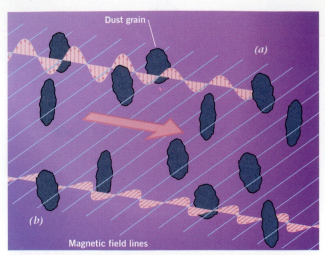

FIGURE 20–12. Interstellar grains tend to be aligned perpendicular to magnetic field lines. Radiation with its electric-field vibration lined up with the long axis of the grains (*a*) tends to be removed from the beam. Radiation not so aligned (*b*) tends to pass through undiminished. This absorption produces polarization.

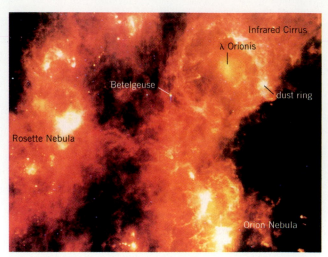

FIGURE 20–13. An infrared view from IRAS of the Orion region. Bright areas are where star formation occurs. Note, too, the wispy infrared cirrus.

something about the shape of the grains (they are elongated) and something about their chemical composition (at least some of the atoms in the grain must be iron or an element that can be magnetized). **Figure 20–12** summarizes the production of polarized light by magnetically aligned grains.

INTERSTELLAR DUST: THERMAL RADIATION

Dust heated by the starlight it absorbs radiates energy in the infrared spectral region. By observing the distribution of infrared radiation throughout the Galaxy, astronomers are able to determine the distribution of dust in the Galaxy. In addition, information on the properties of dust itself is obtained. **Figure 20–13** shows a false-color image of the region containing the constellation Orion as viewed by the Infrared Astronomy Satellite (IRAS). The brightest areas, corresponding to star formation regions, are bright because hidden protostars heat the surrounding dust. Also shown are filaments of dust that are reminiscent of cirrus clouds; for this reason, these features are called infrared cirrus.

INTERSTELLAR DUST: CHEMICAL COMPOSITION AND FORMATION

What is interstellar dust made of? While the scattering and absorbing properties of small dust grains are somewhat dependent on the dust's composition, scattering

studies do not provide unambiguous answers to the question. By merely examining a beam of light that had passed through a cloud of dust, for example, you would have difficulty distinguishing whether the light had gone through chalk or flour dust. We must realize that a dust grain is an aggregation of several hundred to perhaps several thousand atoms bound together into some sort of solid matrix. Such a particle does not produce simple spectral features at well-defined wavelengths, like an atom or a molecule. Yet knowledge of the composition of the dust particles in the Galaxy is crucial to understanding how and where they are formed, and to determining the true chemical composition of the material in the interstellar medium. Furthermore, the chemical elements contained in the dust grains must be included in any tally of chemical abundances.

An important clue about the composition of the grains came from an early series of small space telescopes known as the Orbiting Astronomical Observatories. These telescopes, which were designed to view ultraviolet radiation, discovered a spectral feature at 2200 Å, which laboratory studies showed matched that of graphite, which is made of carbon. It was natural to assume that graphite was at least one of the components of interstellar dust. This result is consistent with the polarization data discussed previously.

More information comes from observations made in the infrared. Some stars have been found to be surrounded by dust shells that heat up and then reradiate in the infrared. While the radiation emitted by the dust generally has a continuous spectrum, on some occasions the spectrum shows a distinct hump above the stellar spectrum at around 100,000 Å (see Figures 17–13 and 18–13 for examples). Comparison with laboratory spectra suggests that this is due to the presence of silicate minerals in the dust cloud. An example of a common silicate is SiO_2 (quartz).

How could grains containing carbon and silicon be formed? It has been suggested that graphite grains are formed in the carbon-rich atmospheres of certain giant pulsating stars (the long-period variable stars mentioned in Chapter 18). During part of their pulsation cycle, these stars expand and their outer atmospheres cool off. Theory suggests that at some point carbon atoms begin to stick together and form graphite flakes. Later on, the star contracts and its outer regions heat up. Increased radiation pressure from the star exerts pressure on the carbon flakes and literally pushes them out of the stellar atmosphere and into interstellar space. The fraction of interstellar grains that might be graphite flakes depends on how many of these long-period variable stars exist and how efficiently they manufacture grains. This is difficult to estimate.

Silicate grains have been discovered in the atmospheres of long-period variables that have more oxygen than carbon, thus demonstrating that interstellar dust probably has no single, uniform composition or origin. This is, in all probability, the reason why it has proved so difficult to make a definitive model of an interstellar dust particle.

Inquiry 20–10 The binding energy that holds atoms together in a dust grain is not very large, and it may easily be exceeded by the random energy of collisions in a gas that is a thousand degrees or more in temperature. What does this imply about the fate of dust grains in a hot cloud?

INTERSTELLAR GAS

The interstellar medium also contains large amounts of gas. Most of the gas is hydrogen, but other elements are also present. For example, sometimes stellar spectra exhibit lines of sodium and calcium that are formed in interstellar gas clouds. The interstellar lines are distinguishable from those formed within the star itself because the lines are extremely narrow.

The coldness of interstellar space means that much of the hydrogen is not atomic but molecular. These molecules are not broadly scattered around the Galaxy but form giant molecular clouds. Other molecules, too, are present in these clouds. A variety of different molecules are observed in the interstellar medium, including formaldehyde and ethyl alcohol. Table 17–1 lists others, which were discussed when we looked at star formation.

When interstellar hydrogen is close to a hot star, the hydrogen is ionized and forms an H II region.

Inquiry 20–11 Why would spectral lines formed in interstellar gas clouds be narrow? (Hint: Why are lines of dwarf stars broader than those formed in supergiant stars?)

Inquiry 20–12 Why does hydrogen gas near a hot star produce an H II region?

Inquiry 20–13 We frequently find several interstellar absorption lines in a spectrum, all due to the same atomic transition, but at slightly different wavelengths. What conclusion do you draw from this observation?

Interstellar lines often show multiple components occurring at slightly different wavelengths. Each component is produced by a gas cloud at a different distance and moving with its own velocity. Radiation from each cloud is Doppler shifted by a different amount. Such gas clouds are often called **diffuse interstellar clouds**.

Because of absorption by dust, we cannot satisfy our goal of determining the overall structure of the Milky Way when looking only at visible wavelengths. Fortunately, the vast amount of neutral hydrogen within the Galaxy has some properties we can exploit to study the Milky Way further.

21-CM HYDROGEN EMISSION

While pacing the beaches of occupied Holland during World War II, the young Dutch astronomer Hendrik Van de Hulst had ample time for reflection. He wondered whether there was any way to observe the cold neutral hydrogen gas that must be so abundant in the Galaxy. After analyzing the details of the hydrogen-energy-level diagram, he concluded that the ground state of neutral hydrogen was not a single level, as we drew it in Chapter 13, but was in fact split into two closely spaced energy states.

Researchers in the 1920s and 1930s had been able to show that both the proton and the electron behave as if they were spinning like miniature tops (**Figure 20–14a**). As a result, each of these charged particles also has a tiny magnetic field. The magnetic fields of the proton and electron interact and affect the energy levels of the hydrogen atom. For example, if the proton and the electron are both spinning in the same direction—a situation we describe by saying the spins are "parallel"—the atom has a slightly higher energy level than if they are

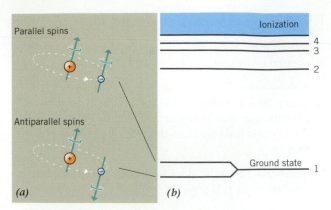

Parallel spins

Antiparallel spins

Ionization

4
3

2

Ground state 1

(a)   *(b)*

FIGURE 20–14. *(a)* The proton and electron can have their spins in the same (parallel) or opposite (antiparallel) direction. *(b)* Each situation has a different energy, so the energy level of the ground state is split in two. A transition between these states produces a photon corresponding to a wavelength of 21 cm.

spinning in opposite directions, which we describe as "antiparallel" (Figure 20–14*b*). This gives the hydrogen atom another way to radiate, because if an atom in the antiparallel ground state collides with another atom, it can absorb a small amount of energy from the collision and end up in the parallel ground state. This is a metastable energy level, not unlike those discussed earlier (Chapter 13, Section 3), and the atom stays at the high energy state for a long time before ultimately making a transition downward.

Because the energy levels for the parallel and antiparallel states are close to one another, the emitted photon has a low energy. A low-energy photon has a low frequency or a long wavelength—in this case 21 cm, or about 8.3 inches—in the radio part of the electromagnetic spectrum. The significance of van de Hulst's work was his prediction that all hydrogen atoms absorb and emit at the radio wavelength of 21 centimeters.

Observation of 21-cm radiation revolutionized our knowledge of the Milky Way and other galaxies for several reasons. First, it is a spectral feature of the most abundant element in the universe. Second, it indicates the presence of *neutral* hydrogen gas, which had previously been difficult to study. Third, this radiation is not absorbed by interstellar dust but passes easily through it.

Inquiry 20–14 Why should 21-cm radiation pass so easily through interstellar dust, while visible light is so strongly absorbed and scattered? (Hint: Recall the discussion of diffraction in Chapter 11.)

The passage of 21-cm radiation through interstellar dust is enormously significant because it means that 21-cm radiation emitted anywhere in the Galaxy can reach Earth and be detected by radio telescopes. The discovery of 21-cm radiation made it possible for the first time to study the structure of nearly the entire Galaxy, undeterred by the presence of absorbing interstellar dust particles. Later in this chapter we will see how it is used to study the overall structure of the Milky Way.

THE STRUCTURE OF THE INTERSTELLAR MEDIUM

To complete our interlude on the interstellar medium, we examine its structure. An analysis of interstellar absorption lines shows that diffuse interstellar gas clouds come in a range of sizes, with masses varying from just a few solar masses up to thousands of times the mass of the Sun. But for simplicity, it is frequently useful to speak of an average, or "typical," cloud, which can be characterized as having a diameter on the order of 50 light-years and a total mass of perhaps 400 solar masses. The gas density averages about 10 atoms per cubic centimeter—a very good vacuum using Earth's standards, though more dense than the average of 1 atom per cubic centimeter for interstellar space as a whole. These clouds have temperatures of about 100 K. There are many such clouds; a volume in the Galaxy equivalent to a cube 1000 light-years on each side contains approximately 2000 such clouds, occupying approximately 10% of the volume of the cube.

Inquiry 20–15 The masses of typical star clusters in the Galaxy are several hundred to several thousand solar masses. These masses are similar to the masses of interstellar gas clouds. What does this suggest about the origin of clusters?

Observations of 21-cm radiation tend to show emission profiles similar to that sketched in **Figure 20–15**. The sharp emission spike comes from a cold diffuse cloud. The broader "shoulder" of emission covering a greater range of wavelengths comes from atoms at temperatures of about 1000 K. The hotter atoms move with higher speeds, which then produce larger Doppler shifts that broaden the spectral line. From these line profiles we therefore conclude that the space between the diffuse clouds is filled with a hotter, lower-density gas, which we call the **warm intercloud gas**.

In other words, the cold clouds appear to be immersed in a tenuous medium of warm gas. This suggests

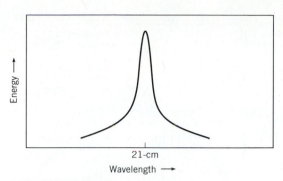

FIGURE 20–15. The narrow core of the line comes from clouds of cold gas, but the broad "wings" must come from hotter material.

that we should model the interstellar medium as a "two-component" system, with regions of sharply different properties coexisting.

In truth, however, the interstellar medium is probably even more complex than this. For example, the Copernicus satellite detected interstellar absorption lines due to O VI (oxygen five times ionized) in the spectrum of some stars. Because the removal of five electrons from oxygen atoms takes a great deal of energy, the observations suggest that at least these particular oxygen atoms were sitting in a region of several hundred thousand degrees Kelvin.

Rockets and satellites above the atmosphere have surveyed X-ray emissions received from space. These observations show that some X-ray emission comes from discrete sources (probably the quasars we will discuss in Chapter 22), but at least one component of X-ray emission is highly extended in structure, suggesting the possibility of emission from interstellar gas with temperatures in excess of a million degrees! This material is called "coronal" gas because of its high temperature.

Why does coronal gas have such a high temperature? Supernovae explosions, as we have seen, release tremendous amounts of energy in the form of photons and shock waves. With an average of one supernova exploding every 50 years or so, a continuous supply of energy goes into the interstellar medium. In addition, high-speed winds with velocities as great as 3000 km/sec come from the hottest stars. All this energy churns the interstellar medium. Furthermore, these events produce what astronomers call interstellar bubbles and superbubbles, cavities filled with hot, rarefied gases scattered around the Galaxy. Thus, the interstellar medium is not a quiet place but one filled with great activity.

Now that we have some understanding of the interstellar medium and its effect on observations, we can return to our study of the Milky Way's stellar components.

20.3 THE STRUCTURE OF THE MILKY WAY SYSTEM

Herschel and Kapteyn counted stars to determine the structure of the Milky Way. Their methods are basically valid even today, as long as we account for interstellar absorption.

> YOU SHOULD DO DISCOVERY 20-2,
> THE DISTRIBUTION OF DIFFERENT OBJECTS
> AROUND THE GALAXY,
> AT THIS TIME.

From photographs of other galaxies we see that the most luminous stars, as well as gas and dust, appear to be located in spiral-shaped arms. Given the evidence of extensive gas and dust within the Milky Way, we will hypothesize that ours is a spiral galaxy; that is, a galaxy with spiral arms, rather than a dustless elliptical galaxy. To detail the spiral structure of our galaxy, and to determine our location relative to the spirals, we must observe objects that we believe lie within the spiral pattern. Furthermore, we require objects that may be seen at relatively large distances. Finding a variety of such objects is the subject of the next section.

Inquiry 20–16 Considering all the types of objects studied in the past few chapters, which types of objects might you expect to be (a) observable at large distances, and (b) located within the spiral arms?

OPTICAL EVIDENCE FOR THE SPIRAL STRUCTURE OF THE GALAXY

Objects that trace out the spiral structure of a galaxy are known as **spiral tracers**. Our immediate goal is to identify a number of spiral tracers that optical astronomers can use. The detailed study of the spiral nature of the Milky Way was pioneered by the Dutch astronomer Jan Oort, who was also famous for his studies of the origins of comets.

The first type of optical spiral tracer is the OB association because O and B stars are young and expected to be in the thin galactic plane near the dust and gas from which they formed. Similarly, young open clusters observed within both the Milky Way and other galaxies are located only in the thin galactic plane, where material for star formation is plentiful. (Because open clusters are confined to the plane of the galaxy, they are often referred to as **galactic clusters**.) But because these stars are located near dust, their observed bright-

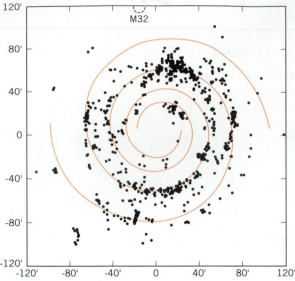

FIGURE 20–16. The distribution of H II regions in M31.

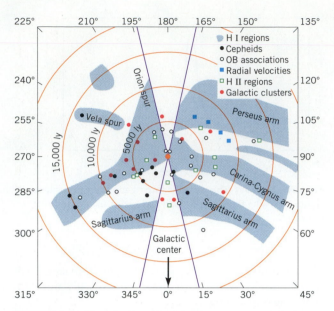

FIGURE 20–17. The distribution of various types of objects near the Sun, showing evidence for spiral structure (after Bok). The center of the Galaxy is located about 1.5 cm below the bottom of the figure.

nesses must be corrected for the reddening and absorbing effects of dust. In this way, the spectroscopic parallax technique discussed in Chapter 15 was now able to produce distances that were free of systematic errors caused by absorption of light by dust.

Because H II regions are associated with star-forming regions present in the galactic plane, they might be expected to be located in spiral arms. The locations of H II regions in the spiral galaxy M31 (**Figure 20–16**) agree with this view. We can therefore, assume that H II regions in the Milky Way will be valid tracers of spiral structure.

Dark nebulae, which also are associated with star-forming regions, may be observed against bright stellar backgrounds at large distances. Therefore, they, too, may be useful tracers of spiral structure.

Finally, some of the luminous Cepheid variables are found within the spiral pattern. Their distances may be calculated using an important technique discussed in the next chapter. When a picture of the locations of all these types of objects was made, a striking result began to emerge: these objects appeared to be distributed along pieces of a spiral pattern.

Figure 20–17 is a diagram of the locations of the types of objects just described. The view is that of an observer above the Milky Way. Objects are not randomly placed but appear to fall into pieces of spiral arms. However, these spiral fragments are far from being simple or well defined. Furthermore, because only objects close to the Sun can be observed at visual wavelengths due to

absorption by dust, only that part of the spiral pattern located near the Sun can be discerned.

Inquiry 20–17 Astronomers have concluded that the primary location of star formation in our galaxy at the present time is in the spiral arms. What reasoning must have led them to this conclusion?

Inquiry 20–18 What type of astronomical object discussed in this chapter would you expect to see in the spiral arms as a result of the simultaneous presence of hot O and B stars along with the considerable amount of hydrogen gas?

RADIO EVIDENCE FOR THE SPIRAL STRUCTURE OF THE GALAXY

We have learned that neutral hydrogen emits and absorbs radio radiation at a wavelength of 21 cm, and that this radiation penetrates the light-absorbing dust present within the plane of the Milky Way. Observations at 21 cm are therefore useful for studying those parts of the Galaxy containing neutral hydrogen. Because 21-cm observations of other galaxies show the hydrogen to be located in a spiral pattern (**Figure 20–18**

FIGURE 20–18. Radio emission from hydrogen at 21-cm from M51 overlayed on an optical photograph.

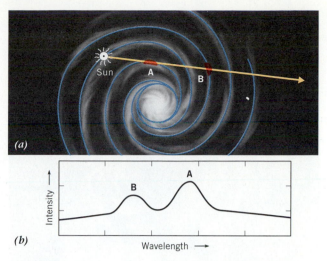

FIGURE 20–19. (*a*) Plotting the location of hydrogen gas in the Galaxy from 21-cm observations. The gas in different parts of the Galaxy will have different Doppler shifts and different intensities, enabling positions within the galaxy to be estimated. (*b*) The composite emission line detected by a radio telescope.

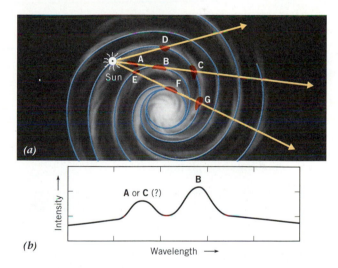

FIGURE 20–20. Resolving ambiguities concerning the distribution of gas requires an assumption that the gas is distributed in a regular way in the galaxy. (*a*) Plotting the location of hydrogen gas. (*b*) The composite emission line.

shows 21-cm contours for the galaxy M51), we can expect the hydrogen in the Milky Way likewise to lie in a spiral pattern. Mapping the distribution of 21-cm emission within the Milky Way is clearly the way to study the spiral arm structure.

Figure 20–19*a* indicates how the 21-cm-emission feature can be used to map out the spiral structure of the Galaxy. Two different clouds of neutral hydrogen, at positions *A* and *B*, will both emit 21-cm radiation. However, the spectral line emitted by the gas at *A* will be more intense because we are looking *along* the length of a spiral arm and therefore seeing through more gas than we do for cloud *B*. Cloud *A* will also show a larger Doppler shift to longer wavelengths, not only because *A* has a greater orbital velocity than *B* (being nearer the center of the Galaxy), but also because *A*'s motion carries it directly away from the Sun. The variation of intensity with wavelength of the spectral line that would be seen by the radio telescope is shown in Figure 20–19*b*.

There are some problems in applying these ideas to the real Galaxy. For example, the spectral lines formed by clouds *A*, *B*, and *C* in **Figure 20–20***a* are shown in Figure 20–20*b*. The large "bump" in the spectrum at the

longest wavelength is due to gas cloud *B*. However, it is not possible to tell just from the spectrum whether the shorter wavelength "bump" is due to gas at point *A* or at point *C* (although it would have to be at one of those two points because its velocity relative to the Sun is

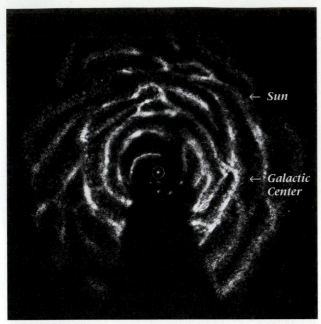

FIGURE 20–21. A map of the galaxy based on 21-centimeter emissions. The blank sector is a region where accurate data cannot be obtained because the radial velocities are so close to zero.

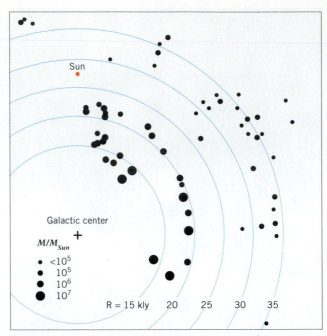

FIGURE 20–22. The distribution of molecular clouds in the Milky Way.

uniquely related to its distance from the center of the Galaxy). Gas at either point A or point C would show the *same* Doppler shift. In practice, ambiguities of this type must be resolved by assuming that the gas is distributed in a regular way in the Galaxy. For example, other observations in other directions (say in the direction of point D) might establish the existence of a continuous spiral arm going through points D and C.

Inquiry 20–19 Consider the line of sight EFG in Figure 20–20a. Notice that points E and G both lie in the arms and are the same distance from the galactic center. What problems might arise in interpreting the spectrum along this line of sight? What would the spectrum look like?

Figure 20–21 illustrates the distribution of neutral hydrogen in our galaxy, based on 21-cm observations. The positions of the Sun and the galactic center are indicated. When drawn like this, the results are a little bit confusing. Rather than the obvious and straightforward spiral pattern that some external spiral galaxies exhibit, we see a much more complex maze of smaller arms. Is our galaxy really a spiral like the others we see, or is it fundamentally different? To what extent is this complex picture created by the ambiguities that were discussed earlier?

We have no way to resolve the ambiguities of our model near the galactic center, because all the spiral arms in that direction contain material that is moving tangentially to our line of sight and hence has no Doppler shift. There is no way we can tell how far away the material is or how many clouds there are along such a line of sight. We know there is hydrogen in this direction, but we cannot tell where it is located. If our galaxy had shown a clearer pattern of spiral arms, we could have extrapolated that pattern into this unknown region; but given the observed distribution of gas there is no way to do this.

Further observational evidence of spiral structure comes from the distribution of molecular clouds. Because they emit at radio wavelengths, their emissions also pass through the dust. The observed distribution of molecular clouds, shown in **Figure 20–22**, is also a spiral pattern consistent with that of neutral hydrogen.

THE CAUSE OF SPIRAL STRUCTURE

We have ample evidence for spiral structure in our galaxy. The young, hot main-sequence stars and their H II regions arrange themselves in a spiral pattern;

neutral hydrogen and molecular emissions both show evidence of spiral structure. The question we have not yet asked is: Why?

One early hypothesis was simply that the approximately Keplerian motion of the material in the Galaxy tended to create spiral structures. Consider the extreme case shown in **Figure 20–23**, in which a set of stars distributed in a straight line moves about the center of the Galaxy. The inner stars go around rapidly and the outer ones slowly; this tends to stretch the line of stars out into "arms." However, on closer examination this simple model fails to explain the spiral structure that we presently observe.

Why does this simple model fail? Because the older stars in our galaxy exceed 10 billion years in age, the galactic system itself is at least that old. We know the Sun is about 4.6 billion years old and orbits the center of the Galaxy in a mere 200 million years or so. This means that since it was formed the Galaxy has rotated about 50 times and the Sun has revolved around it some 20 times. If this kind of motion (with the faster rotation

being closer to the galactic center) actually created the spiral arms out of matter wound like spaghetti on the tines of a fork, then the arms of the Galaxy would be wound tightly around the galactic nucleus. In fact, however, the arms of spiral galaxies are rather loosely wound, which contradicts the predictions of this hypothesis. What we have just described is often referred to as the "wind-up problem." Spiral arms that formed from such a model would not persist for long.

THE DENSITY-WAVE THEORY

In the mid-1960s, C. C. Lin and Frank Shu, then at MIT, refined earlier ideas that the spiral pattern of stars in the Galaxy could be explained by the existence of a spiral-shaped compression wave moving through the material of the galactic disk. This compression wave is similar to the seismic P-waves described in Chapter 8. As the wave moves through a certain part of the Galaxy, it compresses the gas it passes, just like a sound wave compresses the air it passes through. The increased density from the compression might cause interstellar clouds to begin to collapse and trigger star formation.

Figure 20–24 illustrates very schematically the simple form of such a wave. The material of the galactic disk is imagined to be rotating clockwise. The mathematics of the situation implies that the wave travels *more slowly* than the gas, so that a gas cloud approaches the wave *from behind*, enters into the wave, and is compressed. The simple model illustrated here is called the *two-armed spiral shock model*, because the passage of the wave causes a shock wave in the gas, much like the sonic boom produced by a jet plane.

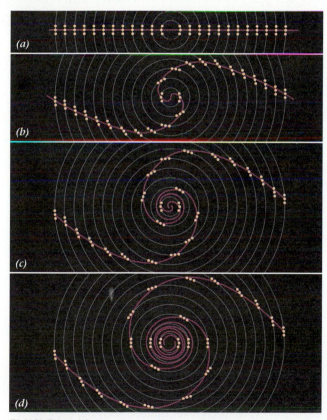

(a)

(b)

(c)

(d)

FIGURE 20–23. A long strip consisting of stars (a) would begin to wind up (b) and show spiral structure after some time if the inner part of the galaxy rotates more rapidly than the outer part (c, d).

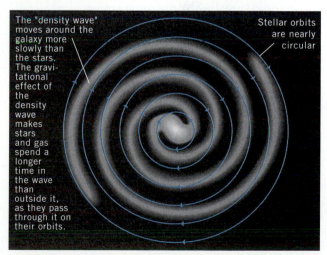

The "density wave" moves around the galaxy more slowly than the stars. The gravitational effect of the density wave makes stars and gas spend a longer time in the wave than outside it, as they pass through it on their orbits.

Stellar orbits are nearly circular

FIGURE 20–24. The density-wave theory of spiral structure. Because of the higher density within the wave, stars tend to spend more time there than between the arms.

Lin and Shu's work is important because it showed that a rotating disk of gas is susceptible to the formation of such spiral waves, and that if a spiral wave is started in a rotating disk of gas, it can persist for a long time.

Like any good theory, the density-wave theory makes a number of predictions. If a gas cloud (such as a giant molecular cloud) passes through the density wave and has star formation triggered, stars of all masses will form. However, remember that the luminosity of a main-sequence star depends approximately on the fourth power of its mass, and that the O and B stars, which are massive, will be by far the brightest objects that form. Thus, when we look at a galaxy, the light in the spiral arm regions will be dominated by the O and B stars. An easy observational confirmation of this idea is that the observed colors of the spiral arm regions of other spiral galaxies are indeed bluish.

Another prediction results when you remember that the lifetimes of these O and B stars are short by cosmic standards, and certainly short compared to the 200–300 million years it would take the wave to make one pass around the Galaxy. Thus, the O and B stars flare up, live their lives, and die on such a short time scale that the wave has hardly had time to move on at all. On a galactic time scale, the O and B stars are like "flashbulbs" that shine briefly and tell us where the wave *has just been*. The density-wave theory then predicts that the regions where we see bright O and B stars should be concentrated along the *inside* edges of spiral arms. Observations of numerous spiral galaxies show this to be the case.

Finally, the theory predicts that regions where the wave is currently compressing the gas and dust will be somewhat darker than other regions. Again, observations are in agreement with this prediction.

We should not think of the spiral arms as always being composed of the same stars. University of Texas astronomer Frank Bash has compared the locations through which a density wave passes to a hospital maternity ward; that is, a place where women and babies are always found and birth is constantly taking place. But the women and babies are not always the same individuals— some leave to be replaced by others. Similarly, the O and B stars that light up the arms so dramatically have short lifetimes and, although the arms are always outlined by such stars, new stars are constantly being born to replace the O and B stars that have died out. Thus a "spiral arm" is ephemeral—a region of a galaxy where interstellar gas is being illuminated by recently formed O and B stars that are strung out along the path through which the density wave has just passed, like a string of pearls, as astronomer Bart Bok has so aptly described it.

More recently, it has become apparent that even this theory is too simple. Although the two-armed spiral shock model does indeed produce long-lived spiral arms, they do not live as long as the galaxy does. Some additional (unknown) mechanism must be continually generating new density waves as the old ones die out.

Various researchers have noted that many spiral galaxies simply do not exhibit the clean, straightforward pattern of long, easy-to-trace spiral arms that the density-wave theory predicts. Indeed, Figure 20–21 shows that even our own galaxy does not. In the next chapter, we will discuss another possible mechanism of spiral arm formation based upon supernova explosions.

We have not discussed what produces the spiral density wave in the first place. The answer is not really known, but it certainly involves the nonuniform distribution of mass within the galaxy, the galaxy's tendency to collapse from gravity, and the transfer of angular momentum outward.

THE GALACTIC HALO AND THE NUCLEAR BULGE

We previously defined the galactic halo in terms of the spherical distribution of globular clusters. We have seen that stars passing near the Sun and having high velocities relative to it are also part of the halo. Astronomers have found RR Lyrae stars that are not associated with any globular cluster to be halo members. Because only a small fraction of stars are RR Lyrae stars, we conclude that there are probably large numbers of faint stars in the halo. Radio emission from the halo further suggests that there is some gas there, although the exact amount is a subject of considerable current controversy among astronomers.

The extent of the galactic halo is also a subject of debate. There is evidence that it extends at least to our neighboring galaxy, the Large Magellanic Cloud, some 170,000 ly away. This result is consistent with the conclusions drawn from the galactic rotation curve, that a far larger proportion of the mass of the Galaxy may be tied up in an extended galactic halo than had previously been thought. Later we will see that the subject of galactic halos is also an area of active research in connection with other galaxies.

In 1989 the COBE satellite produced a wide-angle image of the Milky Way in infrared radiation (**Figure 20–25**). The image covers the entire sky; the view is similar to spreading the globe of the Earth on a flat piece of paper. This view shows the concentration of old and cool stars in the galactic disk and the presence of a flattened **nuclear bulge** in the direction of the constellation Sagittarius. While the halo and bulge merge with one another, we do not yet understand how they are related to each other. For example, we do not know if they formed at the same time or as separate entities. In addition to observable stars, the bulge region contains more than 10^8 solar masses of molecular gas, hot X-ray-

FIGURE 20–25. The Milky Way observed towards the galactic center as observed by the COBE satellite in the infrared.

emitting gas, and the galactic center itself, to which we now turn our attention.

THE CENTER OF THE GALAXY

The central regions of our galaxy are hidden from the view of the optical astronomer because of interstellar dust. The development of radio and infrared observing techniques since World War II has allowed us to pene-

trate the dust to the very center of the Milky Way. Furthermore, technological advances have greatly improved the ability of the instruments observing at these long wavelengths to resolve detail.

The galactic center is a region of great complexity. **Figure 20–26** shows an early low-resolution contour map of the galactic center. The strongest peak is known as **Sagittarius A**. Other strong features are also noted. Sagittarius A radiates strongly in both radio and infrared wavelengths and is one of the brightest radio and infrared sources in the sky, in spite of its great distance. Its brightness indicates that events involving tremendous amounts of energy, which we have yet to comprehend, are occurring in the galactic center.

A higher-resolution radio image of the region known as Sagittarius A West shows the intricate nature of the galactic center region (**Figure 20–27**). A feature called Sagittarius A* (abbreviated Sgr A* and pronounced "Sag A star") is the brightest radio source in the galactic center. It is located very close to, and is perhaps at, the dynamical center of the Galaxy. About one second of arc from Sagittarius A* is an object referred to IRS 16, a complex of at least 15 components, all of which exhibit

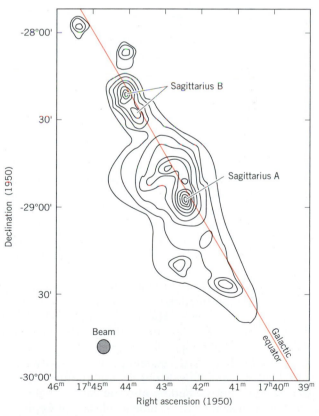

FIGURE 20–26. A radio contour map of the galactic center region.

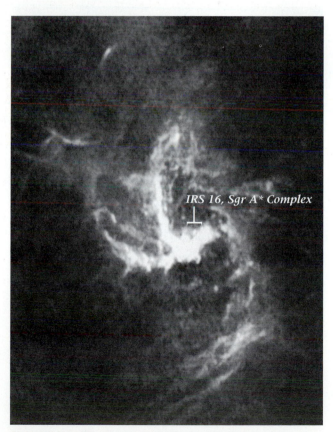

FIGURE 20–27. A radio image of the galactic center. Important features described in the text are indicated.

strong stellar winds. What is the nature of these objects, and how do they interact with one another?

An examination of the spectrum is necessary to answer these and other questions. The strong radio emissions from Sagittarius A form a continuous spectrum, but it appears to be a **nonthermal source**; that is, its brightness is *not* related to its temperature. The observed spectrum's increase in intensity into the radio wavelengths, along with its polarization, indicates that it is the same synchrotron radiation that we discussed in connection with the Crab Nebula.

The energy from Sagittarius A* is strongly concentrated in a small region of space, less than 20 astronomical units in size. Yet it radiates an amount of energy tens of millions of times greater than that emitted by the Sun. Whatever the nature of Sagittarius A*, it must truly be exceptional.

Attempts to explain the high concentration of so much energy in such a small volume of space have produced a proliferation of competing hypotheses. For example, a dense group of hot stars surrounded by dust shells is one possibility, and IRS 16 may be such an object. There is also the possibility that matter-antimatter reactions are the source of the energy concentration. This theory was given a surprise boost in 1977, when a high-altitude balloon flight, looking toward the direction of the galactic center, detected an emission line at a gamma-ray wavelength of 0.024 Å. This is a wavelength characteristic of matter-antimatter annihilation between positrons and electrons. These highly energetic photons coming from the center of the Galaxy indicate that whatever is happening there, it apparently is producing some antimatter. This emission has now stopped. Using our previous arguments relating brightness changes to the size of an emitting region, we conclude that the region producing the gamma rays cannot be very large; the best estimates suggest that it is less than a light-year in size.

The leading hypothesis today is that there is a black hole, perhaps containing more than a million solar masses, in the center of the Milky Way. Its energy may come from accretion of matter being lost by the nearby IRS 16. Such a hypothesis explains a wide range of observations both in the Milky Way and other galaxies as well. For this reason, black hole hypotheses will be discussed in more detail in Chapter 22, in connection with external galaxies whose nuclei emit even more energy than that detected from our system.

Figure 20–28 summarizes the Milky Way as we now believe it looks. The principal components are the nucleus, the halo region, the galactic disk, and the spiral arms.

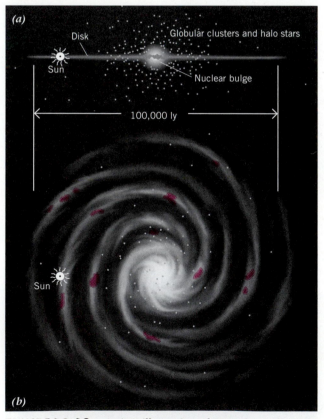

FIGURE 20–28. Artist's illustration of our galaxy, showing the locations of the globular clusters and halo, the nuclear bulge, the disk, and the spiral arms.

STELLAR POPULATIONS

In studies culminating during World War II, Mount Wilson astronomer Walter Baade, noticed an apparent division of stars into two strikingly different classes, which he called **population I** and **population II**. Population I objects are those confined most strongly to the plane of the galaxy, and, in particular, those found in the spiral arm regions. Population II objects are those less strongly confined to the galactic plane, and, in particular, include the objects in the nuclear and halo regions of the galaxy.

For example, the bright O and B stars that recently formed from molecular clouds in the spiral arm regions are population I objects. Also strongly concentrated in the spiral arm regions and grouped together as population I are the bright supergiant stars, young open clusters, some Cepheid variables, and the low-velocity stars. Examples of population II objects not confined to the galactic plane but extending into the galactic halo and into the nuclear regions include a subclass of Cepheids called W Virginis stars, the globular clusters, RR Lyrae variables, and the high velocity stars.

TABLE 20–1

Summary of Properties of Stellar Populations

Population	Typical Examples	Location In Galaxy	Motion	Age	Heavy-Element Abundance
I	Hot, luminous main-sequence stars, young open clusters, H II regions, OB associations, Sun dark clouds, giant molecular clouds, Type II supernovae	Spiral arms, disk	In the plane, orbits are almost circular. Low-velocity objects.	Young to medium	1–2%
II	RR Lyrae stars, globular clusters, high-velocity stars, planetary stars, planetary nebulae, Type I supernovae	Nuclear region, halo, extended galactic disc	More eccentric, less angular momentum about galactic center	Old	Less than 1% typically 0.1–0.01%

Significant differences in chemical compositions exist between the two stellar populations. We have stressed that most stars (and indeed virtually all objects that have been detected in the universe, with the exception of terrestrial-type planets) have a somewhat similar composition, with approximately 70% of the mass being hydrogen and approximately 28% helium. All the rest of the atoms in the periodic table contribute only about 2% of the mass of the universe.

When we examine the abundances of heavy elements in stars more closely, we find that, although the heavy elements are a minority component of all stars, there are significant differences between stars. For example, population II stars are extremely deficient in the heavy elements, with an abundance as much as 10,000 times smaller than in the Sun and other population I stars.

Inquiry 20–20 Where did the heavy elements found in population I stars form?

Modern studies have shown that this two-population model is only a first approximation to the truth, and that there are other populations of stars intermediate in properties between these two extremes. For example, many astronomers believe there is observational evidence that the Milky Way has a thick disk component, 7000 ly thick, between the disk (whose thickness is about 1000 ly) and the halo.

It is apparent that the Galaxy—a complex entity of some 10^{12} stars interacting with extensive amounts of gas and dust—is evolving with time. Star formation is going on now, and it clearly went on in the past. Our goal is to try to construct a picture that explains how the Galaxy was born and how it has evolved from that time to the present. Many of the most important clues are contained in **Table 20–1**, which summarizes the differences between the two extreme stellar populations. We will next attempt to use these differences to help us obtain a coherent picture of the formation of the Milky Way.

20.4
THE FORMATION AND EVOLUTION OF THE GALAXY

An understanding of the formation and evolution of the Milky Way requires us to explain why there is a disk, halo, and bulge. It must explain why population I objects contain more heavy metals than population II objects contain. In the following sections we present a general picture based on our current understanding.

THE CHEMICAL EVOLUTION OF THE GALAXY

As many stars end their lives, a portion of the heavy elements that they have manufactured by nuclear fusion in their cores is returned to the interstellar medium by means of a variety of mass-ejection mechanisms: stellar winds, planetary nebulae, and novae and supernovae explosions. Thus the next generation of stars forms out of gas clouds that have been "enriched" in heavy elements. These in turn make more heavy elements, and so we are led to expect that the percentage of the Galaxy's mass that is bound up in heavy elements increases with time.

The differences between population I and II stars tend to confirm this evolutionary picture. Population I stars are younger and composed of a significantly higher percentage of heavy elements than population II stars. The older average age of population II objects shows they were formed at an earlier time when the abundance of heavy elements was not as great. Although there have certainly been local variations in the rate at which star formation and heavy-element enrichment have gone on throughout the Galaxy, the general trend seems clear.

Inquiry 20–21 Some of the following objects are definitely population I, some are definitely population II, and some could be either. Which is which?

(a) An O-type main-sequence star.

(b) An M-type main-sequence star.

(c) A star whose velocity, relative to the Sun, is very high.

(d) A K-type star with weak metallic lines in its spectrum.

(e) A star in a globular cluster.

(f) An H II region.

It is a fairly simple thing to note that the heavy-element concentration in the Galaxy should be increasing with time, but it is much more difficult to build a model of the Galaxy that predicts the details of such enrichment. When we examine the chemical compositions of the various ejecta that we can see (e.g., planetary nebulae, stellar winds, supernovae remnants), we find considerable variations in the details of their heavy-element composition. Such variations must be due to detailed differences in the chemical evolution of the precursor stars.

For example, we have talked about the development of convective zones in the outer parts of stars as they evolve toward the red giant region (Chapter 18). If a convective zone is extensive enough, it may reach deeply into the star and actually bring up material from near the core. Such material could have been recently processed in nuclear fusion reactions, and it would be "scooped" out of the central regions of the star and mixed into the atmospheric gases, later to be ejected. This process is called "dredge-up," and the chemical composition and isotopic abundances of the material that is brought up to the surface of the star would depend on the exact details of the entire nuclear history of that particular star. Thus, until we have detailed calculations for the entire evolutionary track of stars of all masses, we will be unable to predict all the details of the chemical enrichment of interstellar gases.

THE FORMATION OF THE MILKY WAY

When we also take into account the motions of the different stellar populations, we begin to learn something about the formation and early life of the Galaxy. For example, studies of the H-R diagrams of globular clusters show them to be among the oldest objects in the Galaxy, with ages certainly in excess of 10^{10} years. Another way of saying this is that they were among the earliest objects formed. But if we examine their distribution in space and their motions, we can infer how the gases that formed them were moving at the time of their first appearance. This indicates that, as the globular clusters were forming, the Galaxy was much more spherical in shape than it is now. Apparently, the material in the Galaxy collapsed into a disk *after* the globular clusters were formed. The gas in this disk would have quickly established itself into circular orbits around the center of the Galaxy.

Inquiry 20–22 Stars continued to form after the material in the Galaxy collapsed into a disk. What must their orbits have looked like?

The wealth of observations we have described now allow us to construct a unified theory for the formation of the Galaxy. We emphasize that the number of unknowns and uncertainties is large. The steps are illustrated in the sequence of diagrams in **Figure 20–29**.

In the beginning, a gigantic cloud of gas somehow began to collapse under its own gravity. It must have contained at least as much mass as we presently see in the Galaxy. At some point, a large number of giant condensations, which would become the globular clusters, formed in the cloud. They each contained at least 100,000 solar masses of material, perhaps more. These condensations themselves collapsed under their own gravity, more rapidly than the Galaxy as a whole. Eventually they fragmented into a number of smaller condensations, each of which became a star. Thus the process leading to the formation of the globular clusters must be at least twofold—first, the total cluster mass must condense; second, the individual stars must form.

Because the Galaxy was roughly spherical when the globular clusters formed, the clusters have a spherical distribution in space. As the rest of the gas in the Galaxy collapsed into a disk, the globular clusters were no longer affected by the gas motions, and they continued to move through space in the orbits they had when they formed.

The rest of the gas collapsed even further, but the most rapid collapse would have occurred in the nuclear regions where the density was highest. One would ex-

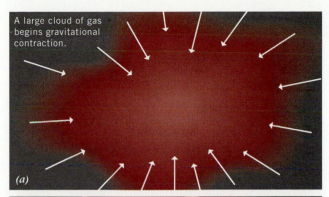

A large cloud of gas begins gravitational contraction.

(a)

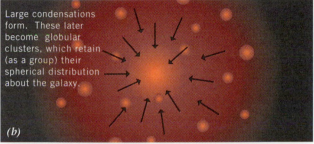

Large condensations form. These later become globular clusters, which retain (as a group) their spherical distribution about the galaxy.

(b)

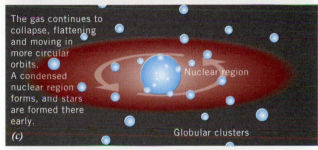

The gas continues to collapse, flattening and moving in more circular orbits. A condensed nuclear region forms, and stars are formed there early.

Nuclear region

Globular clusters

(c)

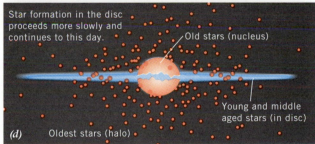

Star formation in the disc proceeds more slowly and continues to this day.

Old stars (nucleus)

Young and middle aged stars (in disc)

(d) Oldest stars (halo)

FIGURE 20–29. A hypothesized sequence of events in the formation of the Galaxy.

pect star formation to have occurred in that region next. The rate at which stars formed is not known, although there are indications, especially in some other galaxies, that the rate of star formation at this stage could have been extremely high.

As the gases collapsed, they would have begun to spin more rapidly as a consequence of the conservation of angular momentum. In a region as extensive as the

Galaxy, the system would flatten into a disk. Any star formation that took place during this stage would have produced a group of stars intermediate between population I and population II.

Finally, when the gases concentrated themselves strongly in the galactic plane, star formation would proceed within the flattened, pancake-shaped distribution of gases left over at the end. Such a flattened distribution of gas is susceptible to the formation of spiral density waves, which can then move through the gas and trigger star formation. As the Galaxy aged, the gases must have become increasingly rich in heavier elements that were manufactured within earlier generations of stars, although this constituent of the interstellar gases remains minor even today.

This description is simplified. It neglects the effects of nearby companions, and it does not explain any of the details of the bulge, the patterns of gas flow into the Galaxy, the energetics of the core, or the formation of spiral density waves. Also, it did not consider the possible role of magnetic fields, whose presence we describe next. We will have a somewhat more general discussion of the evolution of galaxies in the next chapter.

20.5
THE GALACTIC MAGNETIC FIELD

We have mentioned the existence of magnetic fields in a variety of objects within the Galaxy. For example, the radio source Sagittarius A* at the center of the Galaxy appears to be radiating by means of the nonthermal synchrotron mechanism in which energy is radiated by charged particles accelerated in magnetic fields to speeds near that of light. This emission mechanism was discussed in connection with the continuous emission from the filaments of the Crab Nebula, and as an important component of the emission from pulsars. In these cases, we were dealing with magnetic fields associated with individual radiation sources. However, it now appears that magnetism is widespread throughout the structure of the Galaxy itself. This widespread magnetic field is important because it might have played a significant role in the Milky Way's formation.

There are two direct lines of evidence for a pervasive magnetic field in our galaxy: the 21-cm Zeeman effect, and the polarization of starlight. Indirect evidence comes from cosmic rays arriving at Earth. Let us look at each.

THE 21-CM ZEEMAN EFFECT

In discussing magnetic fields in stars (Chapter 14) we pointed out that when an atom finds itself in a magnetic field, the atomic energy levels are split because of the

Zeeman effect. What would normally be observed in the spectrum as a single line at a specific wavelength is replaced by several closely spaced lines, each representing a transition between two of the split energy levels. In the 21-cm emission from neutral hydrogen clouds, astronomers have looked for and detected the Zeeman splitting of the individual energy levels that produce the 21-cm emission. From these observations astronomers conclude that the general interstellar magnetic field appears to have a strength that is about 10^{-5} times the average magnetic field at the surface of the Sun. This is large enough to have observable effects, but not so large as to enable the lines of magnetic force to affect the motions of gas clouds, as some astronomers had earlier thought possible.

INTERSTELLAR POLARIZATION

We often observe starlight arriving at Earth to be polarized as a result of its passage through interstellar dust. The only satisfactory explanation of this polarization assumes that interstellar dust particles are elongated and of such a composition that they are aligned by magnetic fields. Organized magnetic fields in the Galaxy will align interstellar dust particles over wide reaches of space, resulting in polarization (Figure 20–12).

20.6
COSMIC RAYS

Experiments in the early part of the twentieth century revealed that at high altitudes an *electroscope*—a simple device for measuring static electricity—lost its static charge more rapidly than at lower altitudes. The rate of loss was greater than could be explained by solar ultraviolet radiation, and at first it was supposed that high-energy photons from the cosmos were responsible. These photons were assumed to have wavelengths even shorter than gamma rays. But then it was found that the loss of charge also depended on the magnetic latitude. Because we do not expect the Earth's magnetic field to affect uncharged particles such as photons, it follows that the phenomenon must be due to energetic charged particles impinging on the Earth. The paths of such charged particles would be curved by the Earth's magnetic field and would tend to be concentrated toward the Earth's magnetic poles. Scientists making these observations referred to the phenomenon they were observing as **cosmic rays**.

OBSERVATIONS OF COSMIC RAYS

Cosmic rays rain down on Earth from all directions. We now know that the name cosmic "ray" is a misnomer,

dating from the early days of the study of radioactivity, well before photons and high-energy material particles had been distinguished clearly. We have also learned that these particles contain as much energy as all the radiation emitted by all the stars in the Galaxy!

Inquiry 20–23 Where on the Earth's surface would you expect a charged electroscope to lose its static charge most rapidly—near the equator or near the poles? Explain.

The discovery of these energetic particles is significant not only because it opened up the field of high-energy particle physics and advanced Einstein's theory of special relativity, but because it opened a window to the distant universe by allowing us to examine particles of matter from far-flung parts of the Galaxy.

Cosmic ray particles arrive at Earth traveling at speeds close to the speed of light. To study them before they interact with the Earth's atmosphere requires observing from altitudes greater than 100,000 feet, often using balloons. Studies show that the majority of cosmic ray particles are protons, the nuclei of hydrogen atoms. Approximately 15% are alpha particles—helium nuclei—whereas only 1% or so are nuclei of heavier elements. Overall, it appears that the abundances of various nuclei in cosmic rays are in agreement with the cosmic abundances that have been determined by other means.

ORIGIN OF COSMIC RAYS

Although there are such things as bursts of cosmic rays, one of the most striking and significant properties of them is that the incoming flux of particles is *isotropic*; that is, equal numbers of particles impinge on the Earth from all directions in space. The incoming material comes from no preferred direction in space.

Inquiry 20–24 Can the Sun be the source of most of these cosmic rays? Why or why not?

Cosmic rays appear to be a property of the Galaxy as a whole. We can rule out the Sun as the cause of these rays by direct observations of the types of particles that stream toward us from it. We find, for example, that the average energy of particles from the Sun is much less than the average energy of a cosmic ray particle, and that under no circumstances does the Sun produce particles as energetic as some seen in cosmic rays. Also, the cosmic ray particles show high abundances of lithium, beryllium, and boron that bear no resemblance to the solar abundances of these elements. We are led to feel that, although the Sun might be the source of a small percentage of the cosmic ray flux, it certainly cannot be an important contributor to it. A model that does account for the isotropy of the cosmic ray particles assumes that they are significantly affected by the presence of the interstellar magnetic field. Although this field has been accurately measured only in a few places, it is easy to imagine that the magnetic forces pervade the entire Galaxy, including the halo region. If this is true, energetic charged particles would be deflected by the magnetic lines of force, so that when they arrive at Earth the direction from which they come would be completely random and unrelated to where the cosmic rays were produced.

Yet this is not a complete explanation, because it does not account for the origin of these particles, nor how they acquired their great energies. An occasional cosmic ray particle enters the Earth's atmosphere moving at a speed of 99.999999999% the speed of light; its vast energy comes from its enormous velocity.

How can particles acquire such velocities? Two suggestions have been put forth, neither entirely satisfactory. One is that they are created in the blast wave of a supernova explosion, which is one of the most energetic events that can happen in a galaxy. However, the rate of supernova explosions is such a puzzle to us that it is difficult to tell whether there are enough of them to have created all the cosmic rays we see.

Another possibility, suggested by the famous physicist Enrico Fermi, is that interstellar particles could be accelerated to higher energies by being bounced back and forth between two interstellar clouds. The magnetic field lines in the clouds can act like mirrors for charged particles, and if two clouds happen to be approaching each other, ions and electrons between them could be accelerated just as a tennis ball is accelerated when it meets a moving racket.

DISCOVERY 20–1
GALACTIC ROTATION

When you have completed this discovery you will be able to do the following:

- Describe the rotation curve for a rotating solid body and for a differentially rotating body.
- Describe the rotation curve of the Milky Way and explain why astronomers believe the Galaxy contains dark matter.

To better understand the observed rotation of the Milky Way, you will now discover how bodies that follow Kepler's laws move, as well as the rules that govern the rotation of solid bodies.

KEPLERIAN MOTION

A. Using data on the planets from Appendix C, make a graph of orbital velocity on the vertical axis and distance (in au) along the horizontal axis. Draw a smooth curve through the points.

B. Describe in detail the appearance of your graph. You have discovered how velocity changes with distance for bodies that move according to Kepler's laws. The shape of curve you found results whenever bodies orbit a more massive one, such as planets orbiting a star.

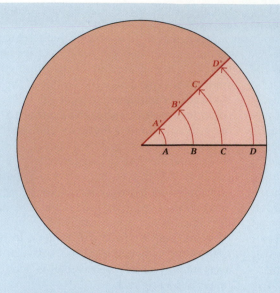

DISCOVERY FIGURE 20–1–1. The rotation of a solid object, such as a plate or frisbee. In the time point *A* rotates to point *A′*, all the other points will rotate through the same angle.

SOLID-BODY ROTATION

Consider a rotating plate or frisbee (see **Discovery Figure 20–1–1**), and in particular the points *A*, *B*, *C*, and *D*. Because the object is a solid body, when point *A* reaches *A′*, *B* will reach *B′*, and so forth.

C. Draw a graph showing the speed of rotation as the distance from the center increases. (Your axes should be the same as in part A above.) Describe the graph.

THE ROTATION OF THE MILKY WAY

A graph of rotation speed with distance from the galactic center is called a **rotation curve**, and is shown in Figure 20–6. You will now apply what you have discovered about rotation to the observed galactic rotation curve.

D. Describe the shape of the rotation curve of the Milky Way in the region from the center out to about 1 parsec, from a few parsecs out to the Sun, and beyond the Sun. Does the shape of the rotation curve in any of these parts follow either the rotation law for solid bodies or that for bodies following Kepler's laws?

E. How would the distribution of mass in a galaxy have to be modified in order to change a Keplerian rotation curve to one that looks like the Milky Way's observed rotation curve at distances greater than that of the Sun?

You have just discovered the reasons astronomers consider the Milky Way to contain a massive halo of matter surrounding the Galaxy.

DISCOVERY 20–2
THE DISTRIBUTIONS OF DIFFERENT OBJECTS AROUND THE GALAXY

When you have completed this discovery you will be able to do the following:

- Describe the distribution of open clusters, globular clusters, dark nebulae, HII regions, and stars within the Milky Way.

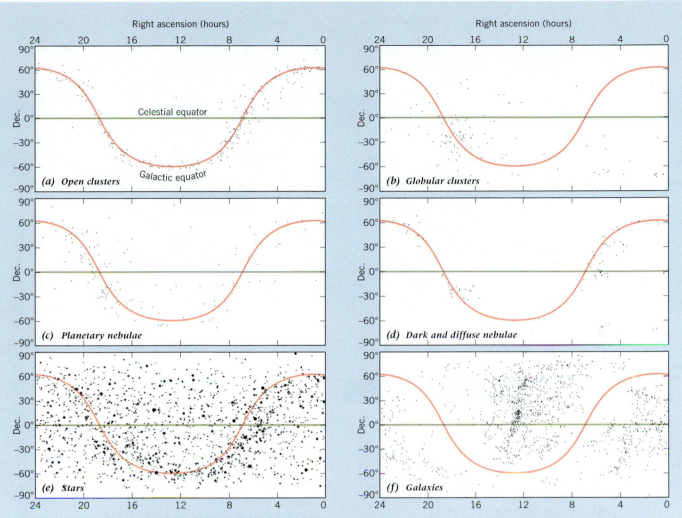

DISCOVERY FIGURE 20–2–1. The distribution on the sky of a variety of astronomical objects. The ecliptic and galactic planes are indicated. Produced by *Voyager* from Carina Software.

■ Describe the apparent distribution of galaxies on the celestial sphere.

Astronomers study the structure of the Milky Way by analyzing the way in which the various types of objects in the Galaxy are distributed around it. For example, the location throughout the Galaxy of open clusters is shown in **Discovery Figure 20–2–1a**. Also shown in the figure are the celestial equator and the galactic equator, which lies along the plane of the Galaxy. We see that the open clusters are not distributed randomly but are confined along the galactic plane. (The few objects that seem to deviate from the trend appear to do so because they are relatively close to the Sun.)

□ **Discovery Inquiry 20–2a** What is the distribution of the *globular* clusters shown in Discovery Figure 20–2–1*b*? Compare and contrast the distribution with that of the open clusters.

❑ **Discovery Inquiry 20–2b**   What is the distribution of the planetary nebulae shown in Discovery Figure 20–2–1c? What conclusions do you draw relative to the open clusters and globular clusters?

❑ **Discovery Inquiry 20–2c**   What is the distribution of the dark and diffuse nebulae shown in Discovery Figure 20–2–1d? What conclusions do you draw relative to the open clusters and globular clusters?

❑ **Discovery Inquiry 20–2d**   What is the distribution of the stars shown in Discovery Figure 20–2–1e. What conclusions do you draw relative to the open clusters and globular clusters?

❑ **Discovery Inquiry 20–2e**   What general conclusions can you draw about the structure of the Milky Way from the distribution of the objects studied above?

❑ **Discovery Inquiry 20–2f**   The distribution of the galaxies differs from that of objects belonging to the Milky Way. How might you explain the distribution of galaxies shown in Discovery Figure 20–2–1f?

If you had further knowledge about the properties of these objects, you would have been able to draw more thorough conclusions. The rest of the chapter provides you with this additional information.

CHAPTER SUMMARY

OBSERVATIONS

■ From studies of **globular cluster** distances and their distribution in the sky, **Harlow Shapley** found the Sun to be located some 30,000 light-years from the center of the **Milky Way**. He was able to determine cluster distances by finding **RR Lyrae** stars in the clusters. This method worked because all RR Lyrae stars have essentially the same luminosity.

■ The motion of the Sun around the center of the Galaxy is found from studies of the motions of globular clusters and distant galaxies. From these studies we find the Sun moves with a speed of about 250 km/sec.

■ From the observed motions of stars astronomers determined the galactic **rotation curve**, a figure showing how an object's speed depends on distance from the galactic center.

■ **Interstellar dust** affects starlight by **extinction, reddening**, and **polarization** of light waves. In addition, dust is observable by its scattering of light in **reflection nebulae**.

■ **Interstellar gas** is observable in the form of **H II regions** and **molecular clouds**, and by means of the emission by hydrogen of **21-cm radiation**, which has the ability to pass through dust without being absorbed.

■ The spiral structure of the Milky Way can be studied by means of **optical spiral tracers (OB associations, galactic clusters, H II regions, dark nebulae**, and **Cepheid variables)** and **radio spiral tracers (21-cm radiation** from neutral hydrogen, **molecular clouds)**.

■ The **galactic center** is a strong emitter of infrared and radio emission, in addition to gamma rays.

■ **Cosmic rays** consist of high-energy particles, usually electrons, protons, and the nuclei of some elements. They are observed to come to the Earth from all directions in space. Their intensity is related to the latitude on Earth from which they are observed.

■ Stars fall into two general populations. **Population I stars** move in the galactic plane, are young, have relatively high amounts of heavy elements, and move with low velocities *relative to the Sun*. **Population II stars**, on the other hand, are in the halo, are old, are metal poor, and are high-velocity objects.

THEORY

■ Spiral structure is generally understood by means of **spiral density waves** that are hypothesized to move with respect to the gas and dust in the galactic plane.

- The hydrogen atom's ground state consists of two close energy levels, which occur because when an atom's proton and electron both spin in the same direction, the energy is different from when they spin in opposite directions.

Conclusions

- It takes 200 to 250 million years for the Sun to complete one revolution about the Galaxy. From this period and the distance of the Sun from the galactic center, astronomers have determined a mass for the Galaxy of 10^{11} to 10^{12} solar masses.
- The interstellar medium has a complex structure consisting of **diffuse clouds** with temperatures of about 100 K, surrounded by a **warm intercloud gas** at a temperature of 1000 K, surrounded in turn by hot coronal gas at a million kelvins.
- From the observation that the **rotation curve** does not decrease with increasing distance, astronomers infer the presence of an extended **galactic halo** containing substantial amounts of **dark matter** beyond the Sun's orbit.

- Gamma rays from the galactic center appear to come from **matter-antimatter annihilation**.
- From the observation that radio emission from the galactic center increases in intensity with increasing wavelength and is polarized, astronomers infer that the radiation is produced by **synchrotron emission**.
- The presence of a magnetic field in the Galaxy is deduced from observations of the polarization of starlight, the 21-cm **Zeeman effect**, synchrotron radiation, and the distribution of **cosmic rays** in the Galaxy.
- The Galaxy is thought to have formed through the collapse of a rotating cloud of material. Because the globular clusters formed first, their orbits were determined by the material's motion at the time of formation; these orbits are therefore elliptical at great distances from the galactic center. After the Galaxy collapsed into a flat disk, newly formed stars moved into the galactic plane. Such a picture is supported by the observed differences between stellar populations.

SUMMARY QUESTIONS

1. Why did early star counts lead to an erroneous picture of the size of the Galaxy and the location of the Sun within it? How was Harlow Shapley able to improve this picture?

2. What is the modern picture of our galaxy? Include the Galaxy's dimensions and the location of the Sun, halo, disk, and nucleus.

3. What are several kinds of evidence for the existence of interstellar dust? How can the size of individual interstellar dust grains be estimated? What are the various lines of evidence that tell us about the composition of interstellar dust grains?

4. How do interstellar absorption lines give us information about the interstellar medium? What conditions (temperature, mass, density, diameter) are present in interstellar clouds? Why are interstellar absorption lines generally difficult to observe from the ground?

5. What is the observational evidence for a multicomponent interstellar medium?

6. What is the process by which neutral atomic hydrogen produces 21-cm emission? Why is 21-cm emission so important in astronomy?

7. What evidence is there for the existence of spiral structure in the Galaxy? Why do astronomers believe star formation is taking place in the spiral arms?

8. How can study of the motions of stars using the Doppler effect be used to study the structure of the Milky Way?

9. How are observations of neutral hydrogen (H I) used to study the rotation of the Galaxy and the distribution of gas within it? What are the main results of these observations, and what are the difficulties in interpreting them?

10. What evidence do astronomers have that unusual events occur near the galactic center?

11. What are the old (incorrect) ideas about spiral structure and why don't they work? Describe the density-wave theory of spiral structure.

12. What are three kinds of evidence for the existence of an overall galactic magnetic field?

13. What is the significance of cosmic rays? Where are they thought to form?

14. What is the difference between population I and population II objects? You should distinguish them on the basis of age, chemical composition, association with gas and dust, location within the Galaxy, motions, and orbital characteristics. How did these two populations come to be, and how are the characteristics of each related to the origin and evolution of our galaxy?

APPLYING YOUR KNOWLEDGE

1. The star in the center of a reflection nebula is generally a B star, typically B3 or cooler in spectral type. Why is it unlikely that a reflection nebula would be caused by a hotter star?

2. Show your understanding of how we analyze the spiral structure of the Milky Way by explaining Figures 20–19 and 20–20.

3. Astronomers believe there are still undiscovered globular clusters within the Milky Way. Where might such clusters exist? Explain. (Hint: Make a drawing of the Milky Way as seen edge-on, showing the Sun and the globular clusters.)

4. Explain how two stars can have identical spectra but different colors.

5. Examine Figure 20–5. For which positions (*A*, *B*, etc.) will stars appear to move through the greatest angular distance relative to distant background stars during one year? The least angular distance?

6. Explain why the caption in Figure 20–15 says that the "wings" must come from hotter material than the core of the line.

7. Explain how the methods used by Herschel and Kapteyn might be applied by a person lost in a forest to find the way out. What assumptions would the lost person be making?

8. What is the distance to a globular cluster inside of which an RR Lyrae star is observed that appears 3×10^{-17} as bright as the Sun? (Hint: You may want to refer to Section 15.4 as well as the H-R diagram for a globular cluster.)

9. Using the distance of the Sun from the galactic center, and the speed with which it moves, what is the length of the galactic year?

10. Consider two stars having identical spectra but different colors although they are at the same distance from Earth. Suppose the redder star appears two times fainter. How much farther away will the redder star appear to be? Explain why the redder one appears to be more distant.

11. If a dust particle consists of gas atoms that stick together during collisions, and a typical atom is 1 to 2 Å in diameter, what is the number of atoms in an average-sized dust grain? (Make an order-of-magnitude estimate, not a precise calculation.)

12. The smallest feature that can be resolved in Figure 20–26 is determined by the beam size shown. From the angular size of the beam in the figure, and the known distance to the galactic center, compute the linear size of the smallest resolvable object in the galactic center region. Express your answer in light-years, astronomical units, and kilometers. (Hint: Use the scale on the declination axis.)

13. If distances within the Milky Way were in error by a factor of two (i.e., if the Galaxy were actually twice as large as we thought), what would be the effect on the computed mass? (Assume the observed periods of observed objects remain the same.)

14. If a star at a distance of 15 light-years has a radial velocity of 150 km/sec, by what percentage does its distance from the Sun change in 100 years?

ANSWERS TO INQUIRIES

20–1. The principal assumption is that the stars identified as RR Lyrae variables in the cluster have the same properties as those known elsewhere, so we can assume that we know their luminosity correctly.

20–2. Angular size = 57.3° linear size/distance. Therefore linear size = 0.057 × 5,000 ly / 57.3° = 5 ly.

20–3. An important assumption is that all globular clusters have nearly the same size, and that the nearby clusters are similar to the distant ones. Because the first assumption, strictly speaking, is not true, Shapley's use of it is valid only in a statistical sense, much as one can talk about the height of an "average" adult human being.

20–4. Shapley was guided by the Copernican principle that the Earth must not be considered to be located in a favored location in the Galaxy.

20–5. Point *B*: no Doppler shift or relative velocity. Point *C*: blueshift, as the Sun overtakes the star. Point *D*: no shift because it has the same orbital speed as the Sun. Point *H*: redshift because, it moves faster than the Sun and is pulling away from it.

20–6. Because the star would appear fainter than it should be, we would estimate its distance to be greater than it really was.

20–7. Errors in distance would produce errors in, for example, luminosity, size, and mass.

20–8. We would estimate a temperature that would be too low.

20–9. There would be no long axis to preferentially absorb the radiation. Polarization is an effect that depends upon asymmetry.

20–10. High temperature will destroy grains over time.

20–11. They would be narrow because few if any collisions would occur to cause the lines to broaden.

20–12. Hot stars produce ultraviolet photons that are able to ionize the hydrogen and produce the H II region.

20–13. Several absorption features result when clouds of interstellar gas between the star and the observer move with slightly different velocities.

20–14. The wavelength of 21-cm radiation is much longer than the size of interstellar dust particles. Therefore, it is not effectively scattered or reflected by them.

20–15. It suggested that clusters form from large interstellar gas clouds. Furthermore, evidence suggests that virtually all stars originally form in clusters; for example, there is not a single known example of a young, isolated star. (Presumably, the isolated stars we see today, such as the Sun, came about when the cluster dissolved over a period of time, probably in response to tidal forces from the gravity of objects in the Galaxy.)

20–16. (*a*) Supergiants, OB associations, young open clusters, Cepheids, dark clouds, H II regions. (*b*) Same as part *a*, although the Cepheids might not be strictly in the plane.

20–17. The presence of very young stars, known to live only for short periods of time, near the spiral arms could not be explained otherwise. They cannot move more than a small fraction of the distance between spiral arms before they die.

20–18. Regions of ionized gas (H II regions)

20–19. Because *E* and *G* have the same distance from the Galaxy's center, the relative radial velocities of the two arms are the same. Therefore, the peaks in the spectrum would coincide. Only by observing in slightly different directions would the two peaks be seen to separate so that the ambiguity could be resolved. The spectrum would contain two peaks, one (from *E* and *G*) that was not shifted because the objects at *E* and *G* have the same orbital speed as the Sun, and a redshifted peak from *F*.

20–20. The heavy elements found in Population I stars (which are more recently formed) were produced in the interiors of previous generations of stars. The heavy elements thus formed were later ejected into the interstellar medium.

20–21. (*a*) Population I. (*b*) Could be either, because not enough information is given. The star could be young or old. (*c*) Population II. (*d*) Population II. The low metal content makes it possible to make a choice. (*e*) Population II. (*f*) Population I.

20–22. The stars that formed after the collapse of the galactic disk would have more circular orbits, confined to the galactic plane.

20–23. Near the poles, because the magnetic field of the Earth channels cosmic rays to move along the lines of force. This is also the reason why aurorae appear near the poles.

20–24. No. The paragraph following the inquiry provides an answer.

21

GALAXIES

[The] analogy [of the nebulae] with the system of stars in which we find ourselves . . . is in perfect agreement with the concept that these elliptical objects are just [island] universes—in other words, Milky Ways. . . .

IMMANUEL KANT, *UNIVERSAL NATURAL HISTORY AND THEORY OF THE HEAVENS, 1755*

21.1

THE HISTORICAL PROBLEM OF THE NEBULAE

In the eighteenth century astronomers observed large numbers of fuzzy (nebulous) objects, which were known as *nebulae*, and asked: What is the nature of these objects? Do they consist of gas or are they stars? In the mid-eighteenth century, the philosopher Immanuel Kant, influenced by the astronomer Thomas Wright of Durham, proposed that many of the nebulae that had

been observed were actually stellar systems similar to our own but external to it—in Kant's words, "island universes," at great distances from the Milky Way galaxy. It took almost 170 years of patient and careful research by both theoretical and observational astronomers to verify the validity of this hypothesis, and another 50 years before realistic models of galaxies could be constructed.

Probably the single most important factor in resolving the problem of the nature of the nebulae was the development of photography. By the 1880s, photographic emulsions were sensitive enough to permit time exposures of faint objects, allowing astronomers to penetrate more deeply into space than they could with the telescope-aided eye. **Figure 21–1a** is a drawing of the nebula known as the Whirlpool, M51, made by the Earl of Rosse, who observed the heavens with a giant 72-inch telescope. **Figure 21–1b** shows a photograph of the same object. The photograph shows much more detail than the drawing, and many of the details in the drawing are incorrect. Photography made it possible to produce a permanent record of the light from distant objects, a record that later could be analyzed at leisure and in more detail.

Even with photography, it was frequently impossible to distinguish between different types of objects solely on the basis of their appearance. Figure 18–28 shows the globular cluster 47 Tucanae; **Figure 21–2** is the elliptical galaxy NGC 4486 (meaning the 4486th object in the *New General Catalogue*, a catalogue of galaxies). 47

(a)

(b)

FIGURE 21–1. (a) The Whirlpool nebula as drawn by the Earl of Rosse. (b) The same galaxy in a modern photograph.

FIGURE 21–2. NGC 4486, an elliptical galaxy.

Tucanae consists of several hundred thousand stars and is about 20,000 light-years away, whereas NGC 4486 consists of approximately 100 billion stars and is 40 million light-years away. It would be difficult to tell them apart from their appearance alone.

Inquiry 21–1   Is there any evidence in Figure 18–28 and Figure 21–2 that indicates to you that the first object is actually much closer? What is it?

We can see many small bright spots in the outer parts of the elliptical galaxy shown in Figure 21–2; each of them is an entire globular cluster. This photograph dramatically shows the difference in size between a globular cluster and an elliptical galaxy.

When spectroscopic observations were begun at the end of the nineteenth century, it was found that objects such as the Orion Nebula and Ring Nebula had bright-line spectra of an excited, rarefied gas. But objects such as the great nebula in Andromeda, M31, (Figure 1–15) showed dark-line spectra similar to those of stars, indicating that they were in fact made up of stars. As we have already seen, the difficult questions were "How many stars?" and "How far away?" Were those nebulae that were made of stars only modest clusters in our own galaxy, or were they enormous and very distant stellar systems?

The first step in understanding these stellar systems is to discuss the observed differences in shape and form; that is, to set up a classification system that describes the variety of objects presented to us.

> READERS DOING ACTIVITIES SHOULD DO ACTIVITY 21–1, CLASSIFICATION OF GALAXY TYPES FROM SKY SURVEY PHOTOGRAPHS, AT THIS TIME.

21.2
THE MORPHOLOGY OF GALAXIES

When confronted with a variety of newly discovered types of objects, astronomers first establish a system for classifying them. Biologists do the same with the bewildering array of plants and animals present on Earth; geologists, too, have established comprehensive systems for classifying rocks and minerals. Classification is a technique for trying to make sense out of the apparent chaos nature presents to us. Although galaxy classification schemes have been around for nearly 100 years, astronomers are still refining them as modern instrumentation makes better data available. The most obvious division of galaxy types is between spirals and nonspirals, but within each category significant variations

exist. These galaxies form what we will often refer to as "normal" galaxies, to distinguish them from the "peculiar" galaxies discussed in the following chapter. We will look at spiral galaxies first.

SPIRAL GALAXIES

The classification of spiral galaxies has been subdivided primarily on the basis of the size of the nucleus relative to the entire galaxy, and the openness of the spiral arms. In **Figure 21–3** the main categories of spirals are shown. The designation S is employed for spirals; the letter following S designates the subclass. Sa galaxies are those with a large, dominant nuclear bulge and tightly wound spiral arms. Sc spirals have a small nucleus (relative to the entire galaxy) and loosely wound spiral arms. Between Sa and Sc spirals are the Sb spirals, similar in form to our own Milky Way galaxy, with a less prominent nucleus and more-open spiral arms than that of the Sa spirals.

Other spirals, shown along the bottom row in Figure 21–3, have a large "bar" through their central regions. They are called **barred spirals** and are designated SBa, SBb, and SBc. We do not yet fully understand the significance of the bar in these galaxies, but it may be a response of the system to some type of periodic gravitational disturbance (such as an orbiting companion) or simply a consequence of an asymmetric distribution of matter in the disk of the galaxy. Some astronomers suspect that the bar phenomenon is more widespread than present statistics indicate, and that these bars may be responsible in part for the generation of spiral density waves that we observe as spiral structure. There are indications that the Milky Way has a barlike structure.

ELLIPTICAL GALAXIES

The majority of nonspiral galaxies are classified as **elliptical galaxies** and are designated by the letter E, followed by a number that indicates how flat the galaxy appears. Type E0 galaxies are spherical in shape, whereas E1 galaxies are slightly flattened. E3 galaxies are shaped much like a football, whereas the most flattened of the elliptical galaxies are designated E7. Some of the various types are illustrated in **Figure 21–4**.

The elliptical galaxies shown have a more uniform appearance than the spirals. Their apparent lack of gas and dust produces this uniform look. Whereas catalogs of bright galaxies consist mostly of spiral galaxies and about 15% ellipticals, in actuality elliptical galaxies are more abundant. The apparent discrepancy occurs because spiral galaxies are easier to pick out on a photograph, and because there are more faint elliptical galaxies than there are bright ones. The misleading

FACE-ON

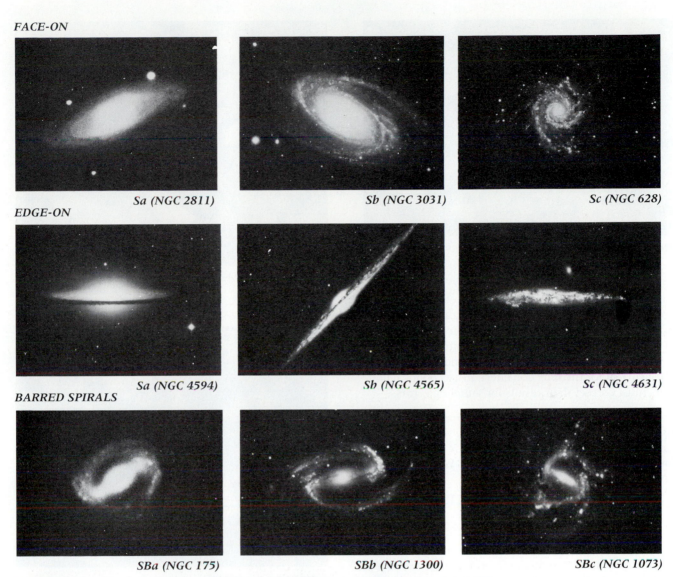

Sa (NGC 2811)   *Sb (NGC 3031)*   *Sc (NGC 628)*

EDGE-ON

Sa (NGC 4594)   *Sb (NGC 4565)*   *Sc (NGC 4631)*

BARRED SPIRALS

SBa (NGC 175)   *SBb (NGC 1300)*   *SBc (NGC 1073)*

FIGURE 21–3.   The main types of spiral galaxies, seen from various angles. The first two rows are "normal" spirals, while the bottom row shows barred spirals.

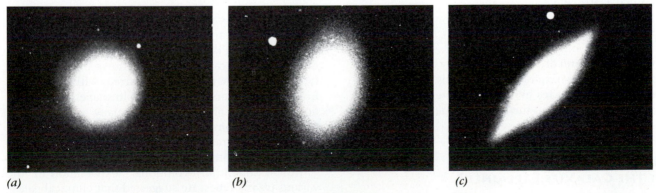

(a)   *(b)*   *(c)*

FIGURE 21–4.   The main types of elliptical galaxies. (*a*) An E0 galaxy, NGC 4278. (*b*) An E3 galaxy, NGC 4406. (*c*) An E7 galaxy, NGC 3115.

FIGURE 21–5. An irregular galaxy, NGC 6822.

FIGURE 21–6. Edwin Hubble.

conclusion about the relative numbers of spirals and ellipticals resulted from what scientists call a **selection effect**, in which a conclusion is influenced by the choice of objects studied. Hence, before reaching a conclusion about the relative numbers of spirals and ellipticals, we must correct for the selection effect, just as political pollsters must correct their results for sampling errors.

Inquiry 21–2   Is there any way to tell whether an E0 galaxy is really spherical or whether it is actually a football-shaped galaxy seen "end-on"? What other possibilities exist?

IRREGULAR GALAXIES

No matter how detailed the classification scheme, there are always objects that do not fit into it. In galaxy classification, these objects are called **irregular galaxies**. By definition, there is no "typical" irregular, but **Figure 21–5** shows an example. About 3% of all galaxies are classified as irregulars.

S0 GALAXIES

There is also a kind of "transition" galaxy that appears to be halfway between spirals and ellipticals. Such galaxies are called **S0 galaxies** and have a flattened disk form like the spirals, but no indication of spiral arms. They also have a large nucleus.

HOW ARE THE FORMS OF THE GALAXIES EXPLAINED?

The modern classification of galaxies was initiated by Mt. Wilson astronomer **Edwin Hubble** in the 1920s

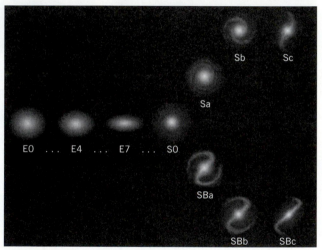

FIGURE 21–7. Hubble's tuning-fork diagram. Note the parallel sequences of spirals.

(**Figure 21–6**). As a way of displaying the types of galaxies classified, he arranged them in a diagram known as the **tuning fork diagram** (**Figure 21–7**). This diagram is interesting because it clearly shows the two major types of spirals (barred and nonbarred) with types going from a to b to c, and the variation in shapes of the ellipticals. It also emphasizes the S0 galaxy as a transition type between spirals and ellipticals. Modern modifications to this diagram place irregular galaxies on the end opposite the ellipticals.

Hubble sought an evolutionary explanation for the various galaxy types. He suggested that elliptical galaxies are in the early stages of formation and that they later flatten out and become spirals; eventually, he thought, they break up and become irregulars. After studying other properties of galaxies—most importantly, variations in their spin and stellar content—we will see that this idea cannot be correct. In fact, the visual appearance of galaxies is not sufficient to answer many of the questions one can ask about them. For example, we cannot even determine their intrinsic size from their appearance, nor can we estimate their distance.

21.3
THE DISTANCES OF THE GALAXIES

Much of our knowledge of the universe depends on having a reasonable understanding of the distances to galaxies. Only when we have such an understanding can we truly appreciate our place in the cosmos.

THE SHAPLEY-CURTIS DEBATES OF 1920

Whether the nebulae were nearby or distant objects was such an important question to the entire scientific community that in April of 1920, two astronomers debated it before the National Academy of Sciences, the most prestigious body of scientists in the United States. Harvard astronomer Harlow Shapley, whose pioneering studies had shown us the true size and shape of our galaxy, argued that the nebulae discussed at the start of the chapter were *not* separate island universes like the Milky Way because they were not far enough away. He had two principal pieces of evidence for his contention that they were relatively nearby. First, astronomers had occasionally discovered in these nebulae stars that had increased greatly in brightness, like the galactic novae. By assuming that they were similar to the novae in our galaxy, they could then compute the distances to these novae (and hence the nebulae that contained them). Most computations suggested that the galaxies were indeed far away—well outside our galaxy. However, two novae indicated the contrary, that the galaxies containing them were actually nearby. This made the distance scale highly uncertain.

Second, as soon as it became possible to detect individual stars in the photographs of these nebulae, astronomers began to look for evidence of the stars' motions across the sky. If such motions could be seen at all, it would indicate that the object was relatively nearby, just as a jet plane that appears to be moving rapidly cannot be very far off. Studies by the acknowledged expert in this area, Dutch astronomer Adriaan van Maanen, suggested that stars within M31 showed such motion. This meant that M31 had to be quite close to the Sun.

Inquiry 21–3 Why would M31's nearness have been a logical conclusion?

Shapley's opponent in the debate was Heber D. Curtis of Lick Observatory. His arguments were somewhat more intuitive and less grounded in observational "facts." However, Curtis turned out to be right, and it is valuable to examine, in retrospect, why Shapley's arguments were incorrect.

It was eventually realized that Shapley's novae were unusually bright. They were in fact what we would today call supernovae, objects whose existence was not suspected in 1920. Therefore, the distances that had been calculated on the assumption that they were ordinary novae were incorrect.

Inquiry 21–4 Why would mistaking a star for an ordinary nova instead of a more luminous supernova cause the nova's distance to be *underestimated*?

In addition, the data indicating that M31 exhibited a visible rotation turned out to be incorrect, and even today no one has been able to explain why this work was wrong. Van Maanen was one of the world's foremost experts in this branch of astronomy, and yet, for some reason, his results were incorrect.[1]

Still other factors undermined Shapley's arguments. The amount of obscuration by interstellar dust was not well understood at that time, and the correction was not properly applied. As a result, Shapley's calculations indicated that our own galaxy was 300,000 light-years in diameter, three times its currently accepted value.

At the same time, the distances to the nearest of the external galaxies—the Magellanic Clouds—had been determined to be only 75,000 light-years, by study of the apparent brightness of its Cepheid variable stars. How the Cepheids are used to find distance is important, so we now look at the method.

Cepheid variable stars turned out to possess an important property that made them highly useful as distance indicators. During a study of Cepheids in the Small Magellanic Cloud in 1912, Henrietta Leavitt, an astronomer working for Shapley, made a discovery that would rank among the most important in the history of modern astronomy. She discovered that Cepheids obey what is called a **period-luminosity law**, as illustrated in **Figure 21–8**. The Cepheids that vary in brightness

---

[1]A controversial explanation is presented in *Science and Objectivity: Episodes in the History of Astronomy* by Norris Hetherington, Iowa State University Press, 1988.

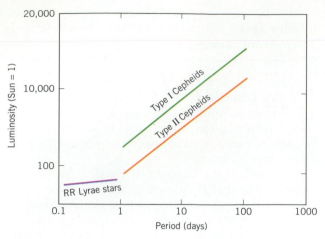

FIGURE 21–8. The period-luminosity law. Notice that there are two different laws for the two types of Cepheids.

over a time scale of just a few days have relatively low luminosities (on the order of 100 times the Sun's luminosity), whereas long-period Cepheids have luminosities on the order of 10,000 times that of the Sun. This means that if we take enough photographs to be able to determine the period of the brightness variation of a Cepheid in a galaxy, we can find out what its actual luminosity is by reading it from Figure 21–8. It is then a simple calculation, using the inverse square law of light, to compare the luminosity with the observed apparent brightness and see how far away the star is, and hence the distance to the galaxy it is in.

For example, suppose we observe a galaxy in which a Cepheid varies with a period of 10 days. Furthermore, suppose we observe the average brightness of the star to be 10^{-18} times that of the Sun. The first step in the solution is to assume (for this example only) that the period-luminosity relation for a Type I Cepheid applies to this star. We then use the period-luminosity relation to determine that its luminosity is 10,000 times that of the Sun. Using B, L, and d for the observed brightness, luminosity, and distance, respectively, from the inverse square law we find

$$\frac{B}{B_{sun}} = \frac{L/L_{sun}}{(d/d_{sun})^2}$$

$$10^{-18} = \frac{10^4}{(d/d_{sun})^2}$$

$$(d/d_{sun})^2 = 10^4 / 10^{-18} = 10^4 \times 10^{18} = 10^{22}$$

from which we determine that $d/d_{sun} \times 10^{11}$; then, $d = 10^{11}$ AU $= 93 \times 10^{17}$ miles $= 1.6 \times 10^6$ ly.

When the period-luminosity relationship was first discovered, however, it was not realized that nature has

provided us with two different types of Cepheids. Those seen in the Magellanic Clouds turned out to be the Type I objects that are significantly more luminous than the Type II objects they had been thought to be. Assuming that the Magellanic Cloud Cepheids were Type II objects made the calculated distance to the Magellanic Clouds much too small, giving the impression that they were part of our own galaxy. When it was realized that they were actually Type I objects, astronomers understood that the clouds were really farther away.

Thus we see that every error was in the wrong direction, and that Shapley's very convincing arguments were incorrect. However, only four years after the debate, Edwin Hubble's magnificent observational studies completely reshaped our picture of the universe.

Inquiry 21–5 What kind of error had astronomers made in assuming the Cepheids were of Type II rather than Type I—random or systematic?

THE DISTANCE OF THE GREAT GALAXY IN ANDROMEDA

Hubble had patiently studied the variable stars in M31 in Andromeda and its companion galaxies. When he discovered that some of these stars were Cepheid variables, he was able to calculate their distances. Although Cepheids are very luminous, they appeared faint on photographs of M31. Despite uncertainty about their absolute luminosities, Hubble estimated that M31 was at least 1 million light-years away and certainly far outside our own galaxy. In fact, when the period-luminosity relation of the Cepheids was finally modernized, it turned out that the Andromeda Galaxy is in fact 2×10^6 light-years away from us. The great extent of the universe was now comprehended for the first time.

OTHER DISTANCE INDICATORS FOR GALAXIES

Cepheid variables can be used to determine the distance to any galaxy that is near enough for individual Cepheids to be detected and measured. Using large ground-based telescopes, Cepheids are useful to a distance of about 20 million light-years.

Cepheids are examples of what astronomers call a **standard candle**, an object whose observed characteristics can be used to determine distance. To study systems more distant than we can locate with Cepheids, other standard candles must be found and their luminosities determined. That is, we must find objects that can be clearly recognized at great distances and whose absolute luminosities can be estimated. Then, when we see such an object in a galaxy, we can use its luminosity to calculate its distance.

Objects brighter than Cepheid variables include the most luminous of normal stars—the bright blue supergiants at the top of the main sequence, which are approximately 15 to 20 times brighter than the brightest Cepheids. Just as we saw that the maximum mass for a star is less than about 60 solar masses, so too there is a maximum possible luminosity for such stars. Assuming the brightest star in a distant galaxy is identical to the brightest star in a nearby galaxy allows us to compute a distance. Other standard candles include novae at maximum, and globular clusters, both of which are about as luminous as a blue supergiant.

Note some of the assumptions and perils of this process. Suppose we find an object that resembles an object in our own galaxy. We *assume* that this object is equivalent to the similar-appearing object in the Milky Way, and then we must locate examples of it that are close enough to the Sun for us to be able to determine their absolute luminosities. In the Cepheid example discussed above, even though they appeared similar, the Cepheids in the other galaxy were not the same as those in our own galaxy, and the difference led astronomy into a serious systematic error.

A safer process would be to determine how bright supergiants, novae, and globular clusters actually are in other galaxies. To this end, we study nearby galaxies containing these objects as well as Cepheids. The Cepheids give the distance to the galaxy. Using this distance we can then calculate the luminosity of the supergiants, novae, and globular clusters in the system. Because the luminosities of these objects are now known, we can use this information to infer distances of even more distant objects. Unfortunately, we do not always have the necessary overlap of distance indicators.

Inquiry 21–6 A key project of the Hubble Space Telescope is the detection and measurement of Cepheid variables in galaxies much more distant than those now being measured from the ground. What effects might this have on measuring the distances of even more distant galaxies?

By using supergiants, novae, and globular clusters, we can extend the extragalactic distance scale beyond that possible with Cepheids by perhaps 4 or 5 times, to about 80 to 100 million light-years.

The brightest H II regions in a galaxy are 15 to 20 times more luminous than supergiants, novae, and globular clusters. This is no surprise because an H II region may be excited by an entire cluster of bright stars. Once their luminosity is known, H II regions can be used to extend the distance scale by yet another factor of 4 or 5.

So far the brightest individual objects discovered are supernovae. On occasion, supernovae have even become brighter than the galaxy containing them! If the absolute luminosity of these objects can be found, we should be able to extend the distance scale by another factor of 25 or so. Unfortunately, however, the range of supernova luminosities is large, and we have no easy way to tell whether a given supernova is intrinsically bright or faint. Furthermore, supernovae are rare events, so even an average result is not a very accurate estimate. If the suggestion that all Type I supernovae have similar luminosities is confirmed, however, this situation could change, and Type I supernovae could become highly reliable indicators of distance.

Eventually, astronomers turned their attention to galaxies so far away that individual objects could not be detected in them. To extend the distance scale this far requires even more assumptions and, as a consequence, the estimated distances that result are even more uncertain. For example, we might suppose that all spiral galaxies have the same diameter. We could estimate this size by studying nearby spirals. From the smaller apparent (angular) size of more distant objects, we could then estimate their distances (see Figure 3–7).

THE DISTANCES TO CLUSTERS OF GALAXIES

Galaxies usually occur in clusters. For example, the Milky Way is part of the cluster we call the **Local Group** (discussed further in Section 21.6 below). The Virgo Cluster is close enough that we can study its members and structure in some detail. Its distance has been determined from the brightness of the globular clusters in one of its giant elliptical galaxies, NGC 4486 (also known more popularly as M87; this galaxy is illustrated in Figure 21–2). The Virgo Cluster has a complicated structure, and it has been suggested that it may really be two clusters at different distances that happen to lie in the same direction from us.

Studies of the Virgo Cluster give us our best information about the relative proportion of spiral to elliptical galaxies in the universe, because it is near enough that the smaller and fainter galaxies can be seen. As it turns out, ellipticals are more numerous than spirals. Distant clusters give a different result, showing more spirals than ellipticals, but this is so because many elliptical galaxies are small and faint, and thus cannot be seen in distant clusters. The misleading conclusion results from a selection effect, because faint elliptical galaxies can be seen only in nearby galaxy clusters. Hence, simple counts of galaxies may be misleading unless correction is made for such selection effects.

Obviously, the assumption that all galaxies of a certain type are the same size would not give us a very good method for estimating distances. But most galaxies are members of clusters of galaxies, which points the way to a partial solution of the problem. In a distant cluster of galaxies having a large number of members, it is proba-

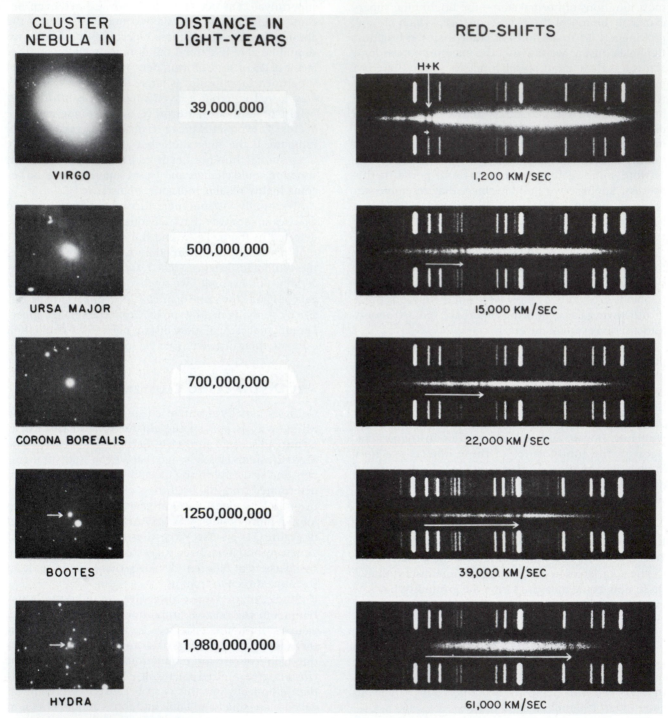

CLUSTER NEBULA IN	DISTANCE IN LIGHT-YEARS	RED-SHIFTS
VIRGO	39,000,000	1,200 KM/SEC
URSA MAJOR	500,000,000	15,000 KM/SEC
CORONA BOREALIS	700,000,000	22,000 KM/SEC
BOOTES	1,250,000,000	39,000 KM/SEC
HYDRA	1,980,000,000	61,000 KM/SEC

FIGURE 21-9. Photographs of various galaxies and their spectra, showing the increasing amount of redshift with increasing distance. The two close spectral lines indicated by "H + K" are caused by calcium in the stars composing the galaxy. Because the comparison spectrum is not calcium, the rest wavelengths of the galaxy lines do not coincide with lines in the comparison spectrum. The left end of the arrow indicates the approximate location of the rest wavelength of the shorter-wavelength calcium line.

bly somewhat safer to assume that the largest spirals are the same size as the largest spirals in *nearby* large clusters of galaxies. To be even more careful, we might take the average diameter of the 10 largest spirals and assume that this value would not vary much from cluster to cluster. A similar assumption could be used for the ellipticals. Clearly, there is uncertainty here, yet the approximate distances obtained in this way are much more useful than no distances at all. Put differently, "we do the best we can with what we have."

Inquiry 21–7 How might this discussion be affected if the cluster whose distance is being estimated has relatively few galaxies?

We might also use the brightness of the brightest galaxies in a cluster as a distance indicator rather than the sizes of the galaxies. With very distant clusters so far away that individual galaxies in a cluster cannot be recognized, we might even estimate the distance from the apparent brightness of the *entire* cluster, assuming a value for the absolute luminosity of a "typical" cluster of galaxies. Again, such estimates contain obvious uncertainties.

Of these methods, which are used? They all are, whenever possible. Astronomers look for consistency among their findings, all the while understanding the uncertainties of each technique. The final answer for a given galaxy is, then, an average that accounts for the uncertainties.

> READERS DOING ACTIVITIES SHOULD DO ACTIVITY 21–2, THE VARIATION IN SIZES OF GALAXIES, AT THIS TIME.

THE VELOCITY-DISTANCE RELATIONSHIP

We have not yet discussed one of the most important distance-measuring methods. In the early part of the twentieth century, as part of a program for studying the mysterious nebulae, U.S. astronomer V. M. Slipher photographed their spectra in order to measure their radial velocities. While some of the nebulae were known from their emission-line spectra to be gaseous, others were known from their dark-line spectra to contain large numbers of stars. From these spectra, Slipher was able to measure the wavelengths of the absorption lines, which he found to be shifted from their usual positions. Assuming these shifts were produced by the Doppler effect, Slipher showed that some of the objects were moving at high rates of speed. The largest speeds he meas-ured were in excess of 1000 km/sec, considerably greater than the highest speeds of any stars in our galaxy, and well in excess of the escape velocity from the Galaxy.

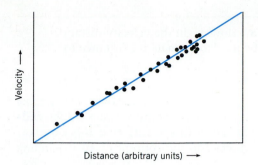

FIGURE 21–10. The Hubble velocity-distance relationship.

Hubble expanded upon these observations after Cepheid variable stars had established that the nebulae whose spectra contained absorption lines were external stellar systems we now call galaxies. With measured distances for some galaxies, Hubble and Milton Humason[2] were able to demonstrate by the year 1929 that there was a correlation between the speed at which a galaxy was moving and its distance from us. They found that the more distant the galaxy, the more rapidly it was moving away from us. **Figure 21–9** illustrates this correlation for objects at different distances. Notice that the more distant galaxies appear smaller and fainter, on the average; also, in their spectrum, as judged by the Doppler shift of common spectral lines, progressively greater Doppler shifts are seen. A graph of this relationship is shown in **Figure 21–10**. This relationship, which appears to be linear, may be expressed mathematically as

$$V = H d$$

where V is the velocity in km/sec determined from the observed Doppler shift, d is the distance in millions of light-years, and H is called the **Hubble constant**. This important relationship may be referred to as the **Hubble law**, or the **velocity-distance relation**. While we will return to a discussion of the Hubble constant in Chapter 23, for now we will just mention that the determination of its value is one of the most important studies in all of astronomy (for example, it provides a measure of the age of the universe). For now, we will take a value of 20 km/sec/Mly. This says that for every million light-years (Mly) farther we view, the speed will increase by 20 km/sec.

[2]Although Humason was a full staff astronomer at Mt. Wilson Observatory, he began as a mule driver helping ferry supplies to the observatory. He later became a night assistant, helping astronomers make observations and develop their photographic plates.

Inquiry 21–8 How might the velocity-distance relationship be used to estimate the distance to a galaxy?

Figure 21–10 shows that the galaxies are *receding* from us; that is, the spectra of most galaxies exhibit **redshifts**. We will see in Chapter 23 that this observation leads us directly to the concept of an expanding universe.

THE TULLY-FISHER RELATION

In discussing the steps leading to the Hubble law, we emphasized that uncertainties enter into distances all the way along. As the distance increases, so does the amount of uncertainty. For this reason, astronomers would like to have at least one method for determining distance that is independent of the standard candles discussed previously. Such an independent method will provide us with additional security in our knowledge of these vast distances.

Just as mass is the important property that determines all characteristics of stars, so too is mass important for galaxies (mass determination is discussed in Section 21.4). The greater the mass of a galaxy, the greater its luminosity. Similarly, the greater the mass, the greater the gravitational force at each point within the galaxy. This gravitational force determines the speed with which stars revolve around the galaxy.

Galaxies will have their 21–cm hydrogen emission broadened by rotation, just as rapidly rotating stars have broadened spectral lines. The end result is that the greater the galaxy mass, the broader its spectral lines, as summarized in the upper half of **Figure 21–11**. Therefore, we expect a relationship between rotational velocity and luminosity. Now, using this relationship and the rotational velocity that is computed from the observed widths of the 21–cm line, the galaxy's absolute luminosity can be found. Finally, by comparing this absolute luminosity with the observed brightness, we readily find the distance using the inverse square law of light. The flow of ideas just presented is summarized in the bottom half of Figure 21–11.

The above arguments were made by astronomers Brent Tully and Richard Fisher. The relationship they found between rotational velocity and luminosity, now known as the **Tully-Fisher relation**, provides a crucial *independent* distance-determination technique.

Figure 21–12 summarizes the determination of distances in the universe in what is often called the **distance pyramid**. The pyramid structure emphasizes the firm foundation of distances to nearby objects, and the

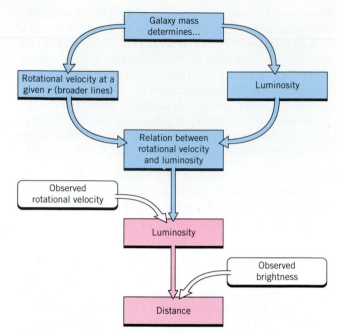

FIGURE 21–11. The flow of logic for finding galaxy distances using the Tully-Fisher method. The top half of the figure relies on theory, while the bottom half uses observations, which are placed in uncolored rectangles.

increasing uncertainty as distance increases. The standard candle at each level of the pyramid is on the left side; the distance to which it is valid is on the right.

Beginning with our knowledge of the Earth's diameter, radar is used to find most accurately the length of the astronomical unit. Once this is known, stellar parallaxes are observed to find distances to the nearest stars and to draw the H-R diagram. Once we have the H-R diagram, we can use it with the technique of spectroscopic parallaxes to determine distances to variable stars and star clusters. With these distances known, the properties of variable stars (Cepheids, RR Lyrae stars, and novae) and supergiant stars allow us to find distances to the nearest galaxies, those in the Local Group. From here, the properties of supernovae, the sizes of H II regions, and the properties of globular clusters and the brightest stars in these nearby galaxies are used to find distances to more distant (but still relatively nearby) clusters of galaxies. Finally, the properties of galaxies in these clusters, along with the Hubble law and the Tully-Fisher relation, are used to find distances to the most remote objects in the universe.

> TO REVIEW A NUMBER OF THE STEPS IN THE DISTANCE PYRAMID, YOU SHOULD DO DISCOVERY 21–1, DISTANCES THROUGHOUT THE UNIVERSE, NOW.

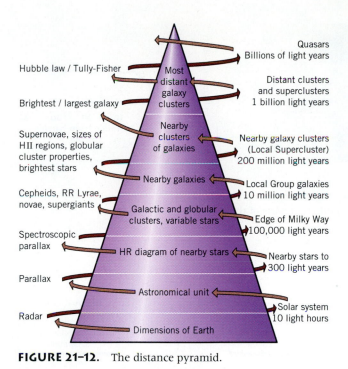

FIGURE 21–12. The distance pyramid.

tainties in size come directly from uncertainties in the angular size and distance. Here, too, the fuzziness of the edge of a galaxy makes measurement difficult.

Our standard of galaxy dimension is the Milky Way's diameter of 100,000 ly. Spiral galaxies have diameters that range from about 1/3 to 3 times that of the Milky Way. Irregular galaxies are all smaller, with sizes from 0.25 down to 0.05 times the diameter of the Milky Way. Ellipticals have a large range in size, from about 0.02 to 4 times the diameter of the Milky Way.

MASS

The determination of galaxy mass is simple in principle but complicated in practice. If you were to count all the stars you could see and assume some average mass per star, you could determine a mass for all the *visible* material. What we need is a method that includes the nonluminous matter, too.

Because Kepler's third law as modified by Newton is applicable in any situation in which the gravitational field of one body controls the movement of another, we can apply it to stars moving around a galaxy, or to one galaxy revolving about another. Just as we saw that the mass of a planet having a satellite can be found from Kepler's third law as modified by Newton, for a star traveling about a galaxy we have

$$(M_{star} + M_{galaxy})P^2 = A^3.$$

The masses (M_{star}, M_{galaxy}) are in units of the Sun, the period P is in years, and the distance from the galaxy's center A is in astronomical units. We will now apply this equation in a variety of ways to find the masses of galaxies.

We can find galactic masses using a method similar to that used to find the mass of the Milky Way. To find the mass of a single galaxy requires observing the motion (radial velocity) of individual stars orbiting the galaxy. Also, observing the angular distance of a given star from the galaxy's center, and knowing the galaxy's distance, we can find the distance of the star from the galaxy's center. With this distance, and the star's observed radial velocity, we can determine the star's orbital period around the galaxy. Putting the star's period and distance from the galactic center into Kepler's third law allows us to compute the galaxy's mass.

21.4

GENERAL GALAXY ATTRIBUTES

Just as when we studied stars we considered their luminosity, radii, mass, and magnetic fields, we do the same now for galaxies.

LUMINOSITY

Combining the distance with the observed brightness of a galaxy gives its luminosity, the total energy it radiates. But finding the total luminosity is not that easy because a galaxy does not have an abrupt edge. How far from the center of the galaxy is there energy to be observed? Do our observations include it all? How much of the radiation is absorbed by dust within the galaxy itself?

The observed range of galaxy luminosity is large and depends on the type of galaxy considered. For example, spirals typically range from 100 times less luminous than the Milky Way to twice as luminous. Ellipticals have a large range in luminosity, from 0.00005 to 5 times the Milky Way's luminosity.

DIMENSIONS

Galaxy dimensions are found from observations of the angular size along with the computed distance. Uncer-

Inquiry 21–9 How do you find the distance of a star from the center of a galaxy if you observe the star's angular distance from the center and know the galaxy's distance?

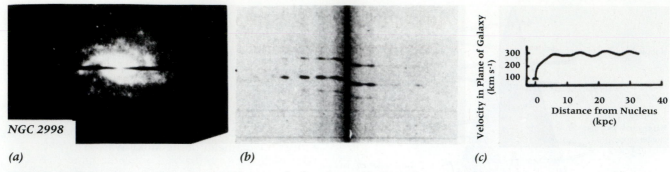

(a)   *(b)*   *(c)*

FIGURE 21–13. Galaxy rotation curve. *(a)* Photograph of a galaxy with the spectrograph slit across it. *(b)* The actual spectrum showing various Doppler shifts for different parts of the galaxy. *(c)* Rotation curve derived from the Doppler shifts.

Inquiry 21–10 How do you find a star's period around a galaxy if you know the star's velocity and its distance from the galaxy's center?

The above technique is crude and limited only to nearby galaxies. Uncertainties in the distance propagate into uncertainties in the mass.

Inquiry 21–11 Why is the above technique limited only to nearby galaxies?

Galaxies sometimes occur in binary systems. Just as binary star systems are used to determine stellar masses, binary galaxies can provide information about their masses. The determination of a galaxy's mass is complicated because we do not actually see the position of one galaxy change relative to the other, and because we do not know if the galaxies are at their closest or farthest separation, or at some point in between, at the time of observation. We also do not know whether we are seeing the orbital motion from the top or the side. These difficulties mean that our estimates of the masses of binary galaxies are uncertain. But remember, some mass information is better than no mass information.

A rotation curve, as we saw in Chapter 20, is a graph of the speed of galaxy rotation at each point within the galaxy. The rotation curve can be determined for spiral galaxies by making Doppler shift measurements of the galaxy spectra. **Figure 21–13** shows how the spectrograph slit is placed, and what the resulting spectrum and rotation curve look like for a spiral galaxy. The mass is then determined through comparison of the observed rotation curve with galaxy models having various masses. This method provides an estimate of the total mass of both luminous and nonluminous material.

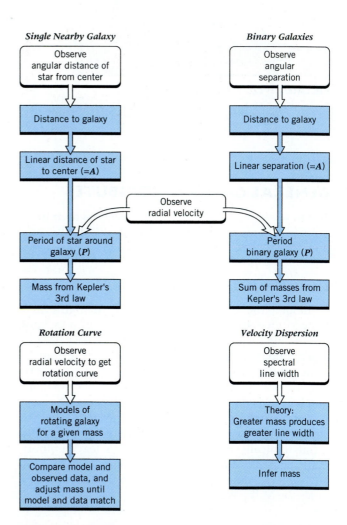

FIGURE 21–14. Summary of the logic behind four techniques for finding the masses of galaxies. The places in the logic where observations enter the picture are shown by uncolored rectangles.

FIGURE 21–15. A photograph of the spiral galaxy M83 indicating the flow of the galactic magnetic field as determined from radio polarization observations.

Elliptical galaxies rotate slowly; if they didn't, they probably would not be elliptical. Therefore, it is not possible to find a rotation curve for ellipticals. However, the larger the mass of a galaxy, the stronger its internal gravity, and the greater the range of velocities of its stars. The range of velocity is called the **velocity dispersion**. In looking at any region of an elliptical galaxy we see numerous stars, some of which are moving toward us and others away from us. The larger the velocity dispersion, the wider the spectral lines from the elliptical galaxy. Therefore, the mass of the galaxy can be computed from observations of the velocity dispersion. This is the basis for the Tully-Fisher method described above. The four methods we discussed for finding galaxy masses are summarized in **Figure 21–14**.

Typical spiral galaxies have masses ranging from double that of the Milky Way down to 1/100 of it. Irregular galaxies contain little mass, typically 0.5 to 0.0005 that of the Milky Way. The largest range is that of elliptical galaxies, which can go from a paltry 10^{-6} up to a whopping 50 times the mass of the Milky Way. Not surprisingly, the smaller ellipticals are called **dwarf elliptical galaxies** while the larger ones are **giant elliptical galaxies**.

MAGNETIC FIELDS

We have found that the Milky Way has a magnetic field, because the light of stars within the galaxy is often polarized. From polarization studies, we know that a magnetic field must align dust grains that have elongated shapes. The magnetic fields of other galaxies manifest themselves by polarizing the light of their stars. In addition to using radio observations, astronomers have studied the magnetic field of M31 by analyzing the polarization of light from the galaxy's globular clusters as the radiation passes through the galaxy's magnetic field.

Figure 21–15 shows a polarization map determined from radio observations for the galaxy M83. The lines superimposed over the photograph indicate the direction of the magnetic field around the galaxy. The magnetic field is ordered over large distances and follows a spiral pattern. Similar data that examine the small-scale structure find the magnetic field to be chaotic in regions where star formation is rampant and supernovae occur. This is probably due to shock waves that produce bubbles in the interstellar medium, which then disturb the ordered magnetic field.

21.5

THE COLOR AND STELLAR CONTENT OF OTHER GALAXIES

Are all galaxies made up of similar stellar populations? Are the stellar populations distributed uniformly throughout them? The fact that population I stars are younger and bluer than redder population II stars gives us a technique for studying the stellar content of galaxies through an examination of their colors.

SPIRAL GALAXIES

Spiral galaxies other than our own show many similarities to the Milky Way, but also some striking differences. Consider again Figure 1–15, the spiral galaxy in Andromeda.

Inquiry 21–12 Where within the Milky Way do blue stars exist? Red stars? What do you conclude from this?

In all spiral galaxies, the disk region is where current star formation is actively going on. As we have pointed out, stars generally form in clusters. When a cloud collapses and fragments into a cluster of stars, objects of different masses will be formed. However, when we view this star cluster from a distance, the light will be dominated by the most massive stars, the O and B stars from the top end of the main sequence. But these are also stars with high surface temperatures, so they are blue in color. The blue light from such stars dominates the light we receive from the spiral arm regions. In other words, spiral galaxies have population I objects located along the spiral regions.

Star formation in the bulge regions of spiral galaxies is thought to have taken place long ago. The stars we see there now tend to be population II; that is, somewhat older and redder than the population I objects. The massive (and blue) stars that were once in the bulge have all burned out and gone to their graves, and so the light from the galactic bulge region is lacking that blue component. Further, many of the stars in the central region would have had time to evolve into red giants. Both these factors combine to make the colors of the bulge regions of spiral galaxies somewhat reddish, in sharp contrast to the spiral arm regions. Thus we can see a correlation between age, color, and location of objects in a spiral galaxy. The make-up of a typical spiral galaxy is shown in **Figure 21–16**, which is an "exploded" figure separating the component parts.

ELLIPTICAL GALAXIES

Elliptical galaxies resemble the nuclear regions of spirals in that they are reddish in color and are dominated by Population II stars. It is apparent that there is little current star formation going on in elliptical galaxies. For many years astronomers felt that this was so because there was no loose gas or dust between the stars, but recent research has shown that elliptical galaxies do indeed have interstellar gas. However, it is fairly hot gas, probably material ejected by red giant stars in stellar winds. High-temperature gas also has a higher pressure, and as a consequence it is resistant to condensing and forming new stars.

IRREGULAR GALAXIES

A striking variety of galaxies fall into the catch-all group of irregular galaxies. Some contain large amounts of gas and dust, as well as O and B stars. Thus we know that star formation is currently taking place in these systems composed of population I stars. Galaxies in the other group of irregulars are amorphous in both shape and texture. They often contain dust and have the appearance of having undergone either a collision or an explosion. A number of irregular galaxies of this type will be discussed in Chapter 22.

Table 21–1 summarizes some of the most important features of galaxies of various types.

Whole galaxy

Open clusters, Population I stars

Dust clouds

Optical gas clouds

Neutral hydrogen

Globular clusters, Population II stars

FIGURE 21–16. An exploded view of the components of M31.

21.6

OUR NEIGHBORHOOD: THE LOCAL GROUP

Galaxies do not occur alone in space but rather as members of galaxy clusters. The Milky Way belongs to a cluster we call the **Local Group**, which we now examine.

GALAXIES GRAVITATIONALLY BOUND TO OURS

Once reliable methods for determining distances to nearby galaxies were developed, astronomers proceeded to map the universe, starting with the neighborhood of the Milky Way. It was during this mapping that we learned we belong to a cluster. The diameter of the

TABLE 21–1

Properties of Galaxies

Property	Irregular	Spiral	Elliptical
Gas	Substantial	Substantial	Little
Dust	Yes	Substantial	Very little
Population I	Yes	Yes	No
Population II	No	Yes	Yes
Oldest Stars	10^{10} years	10^{10} years	10^{10} years
Youngest Stars	Newly formed	Newly formed	10^{10} years
Rotation	Yes and no	High	Low or none
Color	Bluish	Bluish disc Reddish halo, bulge	Reddish
Spectral Type	A–F	A–K	G–K
Range in Diameter[1]	0.05–0.5	0.3–3	0.03–10
Range in Luminosity[1]	10^{-4}–10^{-2}	10^{-2}–2	5×10^{-5}–10
Range in Mass[1]	10^{-3}–0.5	10^{-2}–2	10^{-5}–50

[1]Values are in units of the value for the Milky Way.

Local Group is approximately 3 million light-years. Its two largest members are familiar galaxies—our own and M31 in Andromeda. These large "sister spirals" are the dominant masses in the Local Group, and associated with each are a number of smaller, less massive stellar systems (**Figure 21–17**). Because the Local Group is a gravitationally bound group of galaxies, each galaxy moves with its own motion relative to the others. For this reason, some of them show blueshifts rather than redshifts. M31, for example, is approaching us at a speed of around 275 km/sec.

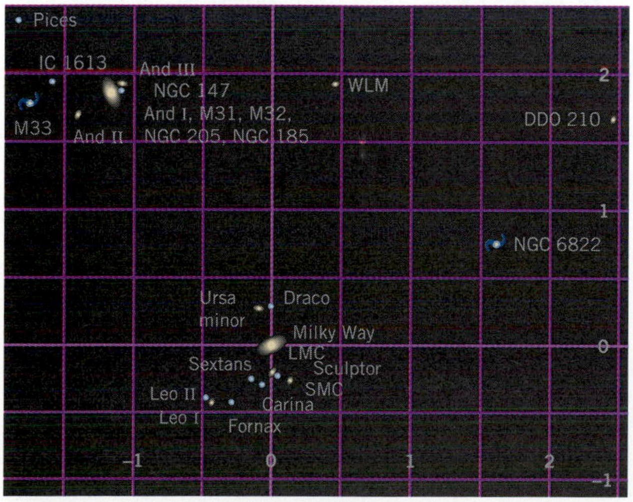

Because the Local Group galaxies are all relatively nearby, we are able to gain insight into some of the smaller, fainter galaxy types that exist but do not show up in distant clusters.

Two external galaxies close to our own are the **Large Magellanic Cloud** and the **Small Magellanic Cloud**. In fact, these are actually *three* galaxies, for it has recently been found that the Small Magellanic Cloud is not one but two galaxies superimposed in the sky. The Large Cloud is shown in **Figure 21–18**. These galaxies are naked-eye objects in the Southern Hemisphere sky; because both are less than 200,000 light-years from the Sun, it is not difficult to study individual stars in them. In fact, you can see as much in the Magellanic Clouds

FIGURE 21–17. Artist's conception of the appearance of the Local Group, as seen from the outside.

FIGURE 21–18. The Large Magellanic Cloud, an irregular galaxy.

with a 16-inch telescope as you can in M31 with a 200-inch telescope.

The Magellanic Clouds are irregular galaxies that appear chaotic, with no obvious symmetry (although the Large Cloud does show some suggestion of barred spiral structure). In them, we find many star clusters, gaseous nebulae, and supergiant stars. These young objects indicate that star formation is still important in the Magellanic Clouds, and we find that irregular galaxies generally have a large fraction of their mass, approximately 50%, in the form of interstellar gas. Active star formation in a few regions of dense clouds gives them their irregular appearance.

Inquiry 21-13 Do the globular clusters in the Milky Way all have nearly the same age or is there a substantial range of ages? What do you conclude about their formation from your answer?

The Magellanic Clouds also have globular clusters; however, there are two populations of them: one old like those in the Milky Way, and the other relatively young. These galaxies continue to confound astronomers and make it more difficult to fully understand stellar and galaxy evolution.

The nearby galaxies also provide information about interactions between galaxies. Radio astronomers have observed hydrogen gas flowing between the Magellanic Clouds and the Milky Way. This **Magellanic stream** shows that these galaxies have interacted with each other in the past. In fact, recent data indicate that the Milky Way's halo extends as far as the Magellanic Clouds. Once we understand the interactions of these bodies, we will have answers to many questions about both galactic systems.

Irregular galaxies, and the Magellanic Clouds in particular, are considerably less massive than large spiral systems like the Milky Way and M31. Although our galaxy undoubtedly contains more than 2×10^{11} solar masses, the Magellanic Clouds are 10 times smaller. They are also considerably less luminous.

Also associated with the Milky Way are several dwarf elliptical galaxies such as the Sculptor, Fornax, Leo I, Leo II , and the recently discovered one in Sagittarius that might be merging with the Milky Way. All these are modest systems, a few thousand light-years in diameter, and contain only a few million solar masses. They are somewhat like giant globular clusters. Elliptical galaxies, both large and small, are considerably redder than irregulars and spirals, which indicates that the stars in them are, on the average, older and more evolved.

The Andromeda Galaxy also has several companions. NGC 205 and M32 (NGC 221, seen in Figure 1-15) are small elliptical galaxies, somewhat larger than typical dwarf ellipticals, and they are comparable in mass to the Magellanic Clouds. Therefore, it is apparent that the Local Group is basically a binary system dominated by two large spirals—ours and M31—each of which has several satellite galaxies.

21.7
THE FORMATION AND EVOLUTION OF GALAXIES

How may the different types of galaxies be understood? What causes the differences in form and stellar content that we have seen? When studies of the various types of galaxies were first carried out, the population I nature of the spirals and the population II properties of the ellipticals led some astronomers to suggest that the difference was due to evolution, with young galaxies looking like spirals and later evolving into ellipticals.

However, it is apparent that the terms "older" and "younger" must be used with some care in describing galaxies. If we determine the ages of the oldest stars in both spiral and elliptical systems, we find that both types of galaxies contain equally old stars, around 10^{10} years old (Table 21-1). The stars in the halo region of our spiral galaxy are approximately the same age as the stars in an elliptical galaxy. Thus all the galaxies that we see began to form at about the same epoch in the history of the universe, so in that sense all galaxies are equally "old." However, the spiral and irregular galaxies have managed to continue star formation up until the present time.

THE IMPORTANCE OF DENSITY

The speed with which star formation takes place in a galaxy appears to be one significant difference between spirals and ellipticals. Ellipticals seem to be galaxies in which star formation took place rapidly early in their evolution. The central regions of spiral galaxies apparently underwent early and rapid star formation, but the arms of spiral galaxies and most parts of irregular galaxies are regions in which star formation proceeded more slowly, so that even today much gas is left over and stars are still being born.

An important factor in the initial rate of star formation is the gas density in the original, protogalactic cloud that collapsed and condensed to become the galaxy. A higher initial density would mean more rapid early star formation. If this were the dominant factor, it would argue that elliptical galaxies resulted from the densest clouds. Early star formation in the centers of spiral galaxies might be expected also, in that the central region of a collapsing cloud would be the densest part of the protogalactic cloud.

ANGULAR MOMENTUM

On the other hand, some elliptical galaxies show a certain amount of "flattening" in their shape, and certainly all spirals show a significant amount of material flattened out into a disk region. Thus it would appear that the various galaxy types can also be partly understood in terms of the amount of angular momentum (spin) that was contained in the original cloud of gas from which they formed. In discussing the origins of the solar system and the Milky Way galaxy, we pointed out that conservation of angular momentum means that the rotational speed of a collapsing gas cloud increases as the cloud collapses. This tends to flatten it out into a disk shape (see Figure 6–8).

The more angular momentum the cloud had initially, the more flattened the final shape would be. The evidence suggests that in the Milky Way, the globular clusters and the halo stars were formed first, when the galaxy was still roughly spherical in form, whereas the stars that formed later are in the flattened disk region.

If rotation were the dominant factor, it would argue that the ellipticals were formed from clouds that had a slower spin when they began to collapse. The clouds that began with more angular momentum would eventually form a disk region and end up as spirals. If the disk region had a lower density, it could explain why early star formation did not take place there.

However, there is evidence against the idea that angular momentum considerations are always dominant in determining the shape and form of galaxies. In particular, the most recent studies of the spin of elliptical galaxies suggest that their angular momentum distribution is much more complicated than the pattern that would result from a simple "collapse-and-spin-faster" hypothesis.

THE EFFECT OF MAGNETIC FIELDS

An early hypothesis that tried to explain spiral structure argued that magnetic fields channeled the radiating material into that shape. When the magnetic field in the interstellar material was eventually measured, however, it proved to be too weak to dominate matter in that way. Nevertheless, it is still possible that a magnetic field could have some effect during the collapse phase of a giant cloud. For example, ionized gas would be able to flow freely along magnetic lines of force but would be retarded if it tried to flow across magnetic field lines.

INTERACTIONS BETWEEN GALAXIES IN A CLUSTER

In light of the fact that most galaxies are found in clusters, astronomers have begun to tackle the complex problem of modeling the collapse of a huge cloud of gas into a cluster of galaxies. Such calculations are extremely difficult for a number of reasons. The situation is so complex that it taxes the abilities of even the fastest and most sophisticated computers; as a result, simplifications that are not physically realistic must sometimes be made in the calculations. Further, the gas in the original cloud would probably be turbulent, and we do not really have any satisfactory theory that describes the motions of turbulent gases.

A final complexity is revealed by inspecting Figure 1–18, which shows the Hercules cluster. It is apparent that galaxies in a cluster are *relatively close together*; that

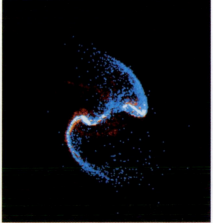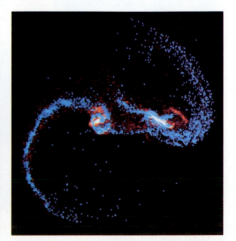

FIGURE 21–19. Computer simulation of a galaxy collision and subsequent merger.

is, their separations are not large in relation to their sizes. This means that these giant star systems will probably be in frequent interaction with each other, undergoing collisions and near-misses and strongly affecting each other with their respective gravitational fields. Close inspection of many of the galaxies in these photos shows some evidence of recent interaction, because we sometimes see irregularities and distortions of form. We must assume that the clouds that formed those galaxies interacted constantly during the time when galaxy and star formation were taking place.

The star formation history in a galaxy can be drastically altered by collisions. A galaxy undergoing a collision may be swept clean of interstellar gas, effectively ending star formation in that system. These scenarios will be considered in more detail in Chapter 22. Such collisions, by removing dust, gas, and loosely bound stars in spiral arms, can change a galaxy from a spiral to an elliptical.

The effects of one galaxy colliding with another can be studied using computer simulations. These calculations show that collisions and near-collisions between galaxies can dramatically modify their form; in fact, colliding galaxies can even merge (**Figure 21–19**). Collisions between the clouds that will become the individual galaxies can also cause changes in shape and form. For example, it is not difficult to show that a spherical cloud of gas that is moving at high speed through a

gaseous medium will tend to be flattened into a pancake form by the pressure of the gas. Thus flattening does not necessarily imply a significant spin.

Observational evidence of galaxy mergers can be seen in **Figure 21–20** in which the merged galaxy shows the nuclei of both colliding galaxies. The 1992 discovery by the Hubble Space Telescope that the Andromeda Galaxy has a double nucleus was a great surprise (Figure 21–20*b*). It is from such galactic cannibalism that massive galaxies become more massive and more dominating of the cluster volume. The actual galaxy merger simulated in Figure 21–19 would require over 2 billion years to be completed.

Calculations of galaxy collisions are so complex that they are still in their infancy, and it is not yet possible to summarize any general conclusions from them. However, it seems clear that collisions will be found quite relevant to the subject of the shapes and forms of galaxies. We will have occasion to consider galactic interactions again in the next chapter, when we consider unusual and peculiar galaxies.

DIFFERENCES BETWEEN SPIRALS: SUPERNOVA-INDUCED STAR FORMATION

The preceding discussion dealt with the formation of a spiral galaxy as if all such objects were similar. For ex-

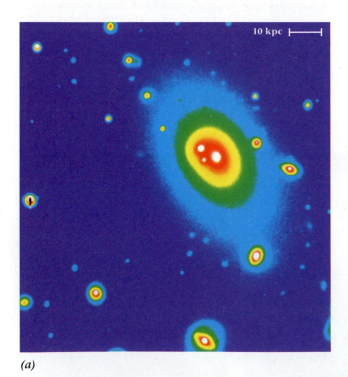

(a)

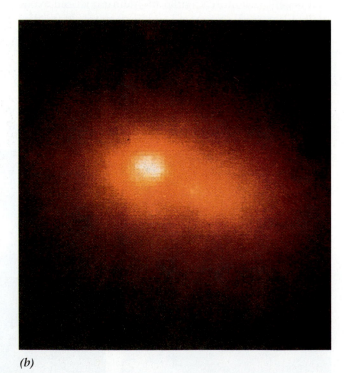

(b)

FIGURE 21–20. Galaxies having multiple nuclei. (*a*) NGC 6166. (*b*) M31 observed by HST.

ample, M51 in Figure 21–1 and NGC 628 in Figure 21–3 show galaxies with straightforward patterns of two well-defined spiral arms that can be followed out from the nucleus for a considerable distance. However, we find a much more complex spiral pattern in M101 (**Figure 21–21**). The arms seem to be made of "bits and pieces," and it is not possible to trace one spiral arm for any significant distance. It appears that that these two objects are responding to different influences.

It has been suggested that although a spiral density wave may explain the star formation pattern in galaxies such as Figures 21–1 and 21–3, it fails seriously for galaxies such as Figure 21–21. The density-wave theory suggests that spirals should have two cleanly defined arms that can be traced to great distances, and it seems fairly clear that not all spirals fit this description. Numerous astronomers have proposed a different mechanism of star formation in the more complex spirals, called *supernova-induced stochastic star formation*.

This theory begins by pointing out that a supernova is an incredibly energetic event. It releases so much energy in such a short period of time that it cannot be considered to be merely a disturbance in the structure of a galaxy; it must be a dominant force, at least for a certain period of time. A supernova has the power to disrupt an entire galaxy, as it pushes the interstellar gases around almost at will. Figure 19–10 is a photograph of the Veil Nebula, a remnant of an exploding star. The emission of light that we see is a consequence of the explosion's blast wave propagating through the interstellar gases and exciting them to radiate.

This blast wave is actually a shock wave not unlike the sonic boom caused by supersonic aircraft. It is a compression wave, and as it forces its way through the interstellar material it has a "snowplow" effect, piling up and compressing matter ahead of itself. Star formation may be initiated by the gas clouds being squeezed by the shock wave, just as we believe happened when the solar system began to form.

If a supernova triggers more star formation, some of those stars that are formed will surely be massive stars, and some fraction of those will evolve rapidly to a supernova event of their own. Thus we have a scenario somewhat like a chain reaction, with an explosive event rapidly triggering other explosive events, which in turn do the same again.

It is in general difficult to predict *where* on the perimeter of the supernova shock wave the next supernova will form, so the theories just imagine it to be something that can randomly take place anywhere around the supernova. It is this randomness that gives the process a statistical or stochastic nature. It is difficult to visualize what the outcome of this chain of events taking place in the disk of a galaxy might look like, but it is not difficult to program the scenario into a computer and display the pattern of star formation at certain time intervals (say every few million years). The results of such calculations are shown in **Figure 21–22**, where the pattern of bright stars is shown at different times. Remember that the disk of gas is rotating differentially; that is, the inner parts are rotating more rapidly than the outer parts. Thus the regions of star formation are

FIGURE 21–21. The spiral galaxy M101.

FIGURE 21–22. A computer calculation that simulates the spiral structure of M101 caused by exploding supernovae that trigger new star formation.

"stretched out" by the rotation, reproducing the pattern of bits and pieces of spiral arms. In a galaxy forming stars in this way, computer simulations show that the spiral pattern tends to form and fade, being more defined at certain times and less defined at other times.

Recall that in Section 7.5 we indicated that the chemical isotope abundances found in certain meteorites suggested that the formation of our own solar system was triggered by a supernova event. We also saw in Chapter 20 that the spiral pattern inferred from 21–cm emission is rather choppy and irregular. Our own galaxy is in many ways the most difficult of all for us to study, immersed as we are inside the dust layer in its central plane. It is not inconceivable that our own Milky Way might in fact be a "bits and pieces" type of spiral.

21.8
THE MASSES OF GALAXY CLUSTERS

The most obvious way to determine the mass of a cluster of galaxies is to determine the mass of each member galaxy and sum them. However, substantial uncertainties exist in the determination of the masses of individual galaxies. Furthermore, a large number of faint cluster members could be neglected, as well as non-luminous matter in the halo of a single galaxy or between galaxies.

A better technique extends the idea we used for finding the mass of an elliptical galaxy—that of velocity dispersion. The greater the mass of a cluster, the greater will be the dispersion in the motions of individual member galaxies. This technique has the advantage that its results include *all* the mass present in the cluster, not only the visible matter. Unfortunately, the method requires using the distance, uncertainties and all, but it always gives larger masses than the summing of individual masses, indicating the presence of nonluminous matter within the cluster.

INTERGALACTIC MATTER

Does matter exist between galaxies in a cluster? Yes. One piece of evidence was provided when astronomers accidentally discovered a large cloud of intergalactic neutral hydrogen at a distance of about 30 million light-years. This cloud has a diameter of 300,000 light-years and contains at least 10^9 and perhaps as much as 10^{11} solar masses of material, equivalent to quite a respectable galaxy. If such clouds turn out to be common, the finding would have a profound impact on our picture of the overall structure of the universe, as we will see in Chapter 23. Unfortunately, no other such clouds have yet been found.

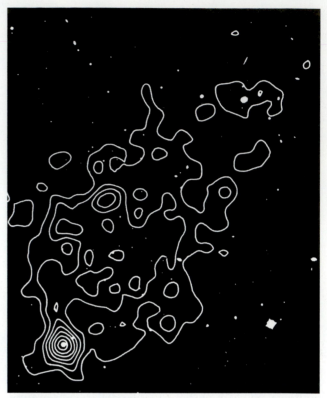

FIGURE 21–23. X-ray contours superimposed on an optical photograph of the galaxy cluster Abell 1367.

Further evidence of intergalactic matter came from observations made with X-ray telescopes pointed at galaxy clusters. Astronomers were surprised to find intense emission from regions of space that had appeared empty. **Figure 21–23** shows an optical image of the cluster Abell 1367 with contours superimposed outlining the region of X-ray emission. These results not only confirm the presence of intergalactic gas, they also tell us that it is hot, with temperatures in excess of a hundred million degrees. The energy source that produces these high temperatures is not known.

Inquiry 21–14 Why does the presence of X-rays tell us that intergalactic gas has such a high temperature?

DARK MATTER

One of the unresolved puzzles in our understanding of galaxy clusters is the fact that the observed matter, both in the form of galaxies and intergalactic material, is insufficient to hold the galaxy clusters together. However, given their great ages (10^{10} years) and their observed

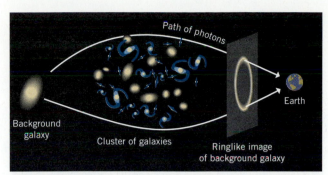

FIGURE 21–24. Gravitational lensing producing luminous arcs in clusters of galaxies.

large internal motions (typically 1000 km/sec or more), there must in fact be enough matter to bind clusters or they would no longer be observable as separate entities. The answer appears to be that there is additional unseen material known as **dark matter**. The question of what dark matter consists of is one of the more hotly debated topics of modern science today. Suggestions range from neutrinos to the most exotic of elementary particles.

Recent work spearheaded by Tony Tyson, an astronomer with Bell Laboratories, has actually mapped the distribution of dark matter within some galaxy clusters. His technique uses the idea that substantial amounts of mass deflect light passing close to distant objects. The deflection is similar to the deflection of light by the Sun or a black hole, as discussed in Chapter 19. In this case, the deflecting mass acts like a lens, as shown in **Figure 21–24**, and is therefore called a **gravitational lens**. If the source, gravitating mass, and the observer are on a straight line, the gravitationally lensed image will be a circle. If they are not on a straight line, the

gravitational lens will produce luminous arcs of light concentrically placed with respect to the galaxies in the cluster. **Figure 21–25a** shows a cluster of galaxies in which there are luminous blue arcs caused by this gravitational lensing effect. Tyson then produced computer models for a variety of dark matter distributions. Models are computed until the calculated results match the observed ones. The distribution of mass that produced the observations in Figure 21–25a is shown in part *b*.

The presence of dark matter plays an important role in the past and future history of the universe. For this reason, we will have more to say about it in Chapter 23.

21.9
CLUSTERS OF CLUSTERS: SUPERCLUSTERS

Some authorities have pointed out that within 70 million light-years of Earth there are a number of small to medium-sized clusters of galaxies, but that there is a considerable gap of empty space between these clusters and the next significant collection of galaxies. University of Texas astronomer Gerard deVaucouleurs suggested that all these small clusters of galaxies, our own included, are actually part of a larger "cluster of clusters" we call the **Local Supercluster** of galaxies. With a diameter of about 100 million light-years, it contains a total mass perhaps 5000 times that of the Milky Way.

Mt. Palomar astronomer George Abell, along with Harold Corwin and Ron Olowin, listed 4076 rich clusters in both hemispheres, each of which may contain several thousand galaxies out to a distance of about 3 billion light-years. These galaxy counts do not, of course, include those rich clusters that may be hiding

(a)

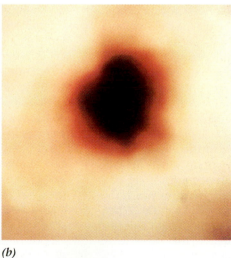

(b)

FIGURE 21–25. Dark matter in a cluster of galaxies. (*a*) Observations of blue arcs caused by foreground gravitational lenses. (*b*) The inferred distribution of dark matter required to give the observations in part *a*.

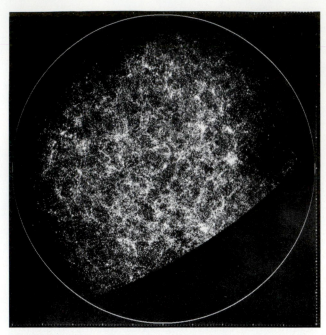

FIGURE 21–26.   The distribution of a million galaxies.

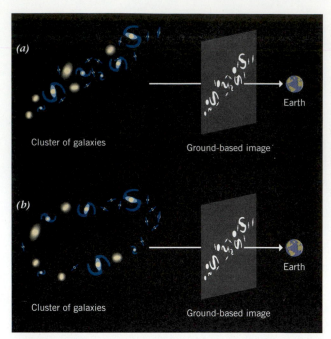

FIGURE 21–27.   Different actual distributions of galaxies as shown in (*a*) and (*b*) can produce identical apparent galaxy distributions.

behind the dust plane of the Milky Way, nor those clusters that are poorer in member galaxies. The rich clusters of galaxies and their spatial relationship to each other gave us our first inkling of large-scale structure in the universe. As our survey techniques are extended deeper and cover the entire sky, astronomers estimate that we will detect more than a billion galaxies and obtain a clearer picture of the structure they outline.

Figure 21–26 shows the results of a survey of relatively bright galaxies observable with moderate-sized telescopes from the Lick Observatory south of San Francisco. There are more than a million galaxies shown here. The galaxies are not distributed uniformly over the sky but fall into clumps and chains. While on a small scale the appearance changes from place to place, large areas in one location are basically similar to large areas in another. The importance of this statement will become clear in Chapter 23. The sharp curved edge is the limit of what could be observed from only the Northern Hemisphere.

Interpreting this figure is not straightforward, for reasons demonstrated in **Figure 21–27.** Suppose, for example, an observer examines a part of the sky in which galaxies are distributed in a line (part *a*). When this observer photographs the sky the galaxies *appear* to be in a straight line (part *b*). In part *c*, we see a case in which the galaxies are distributed in a plane around a circle. When photographed, they will *appear* to be in a straight line, just as in part *b*. We must always remember that we observe the positions of objects *projected onto the plane of the sky.*

Inquiry 21–15   What observations can astronomers make to determine the distribution of galaxies along the line of sight; that is, the distribution with distance?

We can calculate the true distribution of galaxies in space by obtaining spectra, measuring redshifts, and applying the Hubble law to determine distances.

Our problem is to present the three-dimensional reality on a two-dimensional piece of paper. What we can do is observe galaxies contained in a narrow slice of the celestial sphere, as shown in **Figure 21–28.** Here we are looking at galaxies contained in a wide strip of right ascension but a narrow region of declination (see Section 4.6). Once we have the spectrum of each object in the strip, we can graph its distance from us against its angular position on the sky (i.e., the right ascension).

Astronomers have obtained spectra for nearly 14,000 galaxies in two such slices in the northern and southern hemispheres (**Figure 21–29**). This remarkable illustration shows that galaxies are not uniformly distributed away from us, but that they fall along strings and circles, giving a "clumpy" or "bubbly" appearance. It is easy in such a diagram to see the clumping of galaxies into clusters, and of clusters into superclusters.

One large line of galaxies that covers the entire slice at a velocity of about 7500 km/sec in the northern part

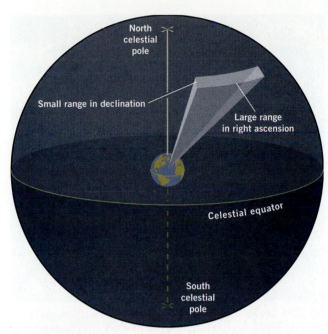

FIGURE 21–28. Defining a slice of the celestial sphere for use in finding the distribution of galaxies with distance.

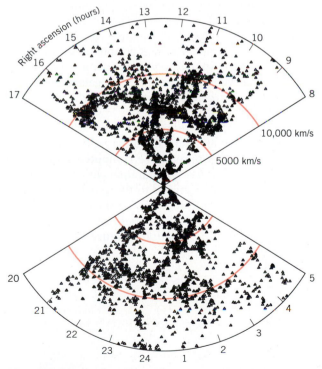

FIGURE 21–29. This figure shows two thin slices of the sky, as defined in the previous figure, that reveal the large-scale bubble structure of the universe. The declination is between 22° and 40° in both hemispheres.

has been named the *Great Wall*, because it appears to divide the slice in two. This structure is at least 500 million light-years long, but only some 15 million light-years thick. A similar southern wall extends several hundred light-years. Furthermore, there are large regions that galaxies appear to shun. These so-called **voids** are regions up to 600 million light-years in size containing few, if any, galaxies. A soapy froth of spherical voids with a characteristic diameter of about 50 megaparsecs (150 million light-years) appears to be a fundamental characteristic of our universe.

Although theorists have found some explosive-type physical mechanisms that can create a void-dominated structure in space, they have not succeeded in finding a way to produce voids that are as large as those observed. Explaining structures like these may give us important clues about the way the universe itself was formed. Indeed, it has been suggested that such giant bubbles are fundamental to the structure of the universe and that the galaxies form on the surfaces of the bubbles. We have found that most of the universe is empty space.

In studying the motions of the Local Group and other nearby galaxy clusters in the direction of (and closer than) the Virgo Cluster, astronomers nearly a decade ago found them all to have speeds somewhat greater than expected from Hubble's velocity-distance relation. Furthermore, galaxies in the same direction but beyond the Virgo Cluster appeared to be moving somewhat more slowly than expected on the basis of their distance (**Figure 21–30**). It is as though the Local Group and the

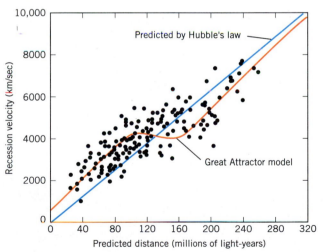

FIGURE 21–30. The velocities of galaxies in the direction of the Virgo supercluster differ from those expected from the Hubble law. Galaxies closer than about 150 million light-years are moving away faster than expected while those more distant than 150 million light-years are moving away slower than expected.

other nearby clusters are being accelerated by something, while the more distant ones are being decelerated. In other words, all these distant galaxies appear as though they are being pulled backward toward the Virgo Cluster. The interpretation of these observations has been that there is a large unseen mass, dubbed the **Great Attractor**, that is drawing us all toward it. However, in April 1994, as this book was in its final stages of preparation, strong doubt on its very existence was cast by possible errors in some important assumptions. This emphasizes the importance of verification to progress in science.

21.10

CLUSTER AND SUPERCLUSTER FORMATION

What formed first, individual galaxies that then came together to form clusters, or clusters that then fragmented to form individual galaxies? How rapidly after the beginning of the universe did they form? What conditions of temperature and pressure are required for galaxies and galaxy clusters to form? How important are magnetic fields in galaxy formation? What conditions produced the vast voids containing few if any galaxies? These and similar questions are among the most complex and least understood of any that astronomers are asking today. Our theoretical and observational understanding of galaxy formation is far behind our knowledge of star formation.

One suggestion for the formation of galaxy clusters is the so-called **top-down theory**, in which galaxy formation occurs after the formation of larger structures. Larger structures would form before the component galaxies if the temperature of the matter from which they formed was high. Just as a cloud forming a star must have a mass whose gravity exceeds the outward pressure of the moving gas atoms, a similar situation holds true for galaxy formation. A high temperature would require a mass of some 10^{15} solar masses for collapse to occur—the mass of a cluster. If this theory is

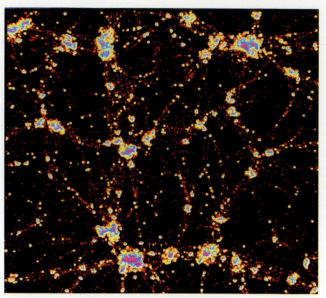

FIGURE 21–31. The results of computations showing that random gravitational clumping can produce structures whose appearance is not unlike that observed in the universe.

correct, the large structures would have formed within the first billion years or so, while the temperature of matter was still high.

If, on the other hand, the matter was cold, smaller concentrations of matter would be able to form readily. In this **bottom-up theory**, clustering occurs after galaxy formation. If this idea is correct, galaxy formation would have occurred more recently, "only" some 10 billion years ago.

Must one of these theories be correct? Not necessarily. Adrian Melott and Sergei Shandarin, cosmologists at the University of Kansas, have computed model universes in which *random* gravitational clumping appears to produce structures similar to those observed. While such models prove nothing, they do show that randomness plus gravity will produce large-scale structures that are consistent with observations (**Figure 21–31**).

DISCOVERY 21–1
DISTANCES THROUGHOUT THE UNIVERSE

When you have completed this activity, you will be able to do the following:

- Discuss how astronomers find distances from the nearest stars to the most distant galaxies.

Throughout this book you have read about a number of techniques astronomers use to determine distances. In many cases, the same basic technique is modified to apply to different objects at greater distances. While doing the Discovery Inquiries that follow, you may want to look at previous chapters for help in remembering the techniques to be applied.

NEAREST STARS

☐ **Discovery Inquiry 21–1a** If an astronomer measures the parallax to a star to be 0.025 second of arc, what is the star's distance?

☐ **Discovery Inquiry 21–1b** Suppose this star is observed to be 10^{-18} times as bright as the Sun. What is its luminosity relative to the Sun?

If we repeat what we have done in the previous two Discovery Inquiries for a large number of stars, and if we also know the spectral types for each of the stars, we can then draw an H-R diagram. Furthermore, we will assume that the H-R diagram for these stars is the same as the H-R diagram for any average sample of stars. In other words, we will assume that a large sample of distant stars is the same as a sample of nearby stars.

FAINTER STARS WITHIN THE GALAXY

We now observe a fainter star, whose apparent brightness is observed to be 10^{-20} times the brightness of the Sun. A spectrum of this star shows strong but extremely narrow spectral lines of hydrogen.

☐ **Discovery Inquiry 21–1c** What is the spectral type of the star?

☐ **Discovery Inquiry 21–1d** What is the luminosity class of the star?

☐ **Discovery Inquiry 21–1e** What is the star's luminosity relative to the Sun?

☐ **Discovery Inquiry 21–1f** What is the star's distance in parsecs?

The previous technique, the spectroscopic parallax, allows us to find the distance to any star for which we have a spectrum from which we can estimate the luminosity class.

VARIABLE STARS

The spectroscopic parallax also allows us to find the distance to star clusters, some of which contain variable stars. For example, globular clusters contain RR Lyrae variable stars, which are about 100 times as luminous as the Sun. If an RR Lyrae star is observed in a cluster whose distance is unknown, we can determine its distance.

☐ **Discovery Inquiry 21–1g** A faint RR Lyrae star in a distant cluster is observed to be 10^{-19} as bright as the Sun. What is the distance to the cluster?

A Cepheid variable is one whose period of light variation is determined directly by its luminosity. Thus by observing the apparent brightness and period of light variation for a Cepheid, and using the period-luminosity relation to find the luminosity, we can determine the distance.

☐ **Discovery Inquiry 21–1h** A Type I Cepheid variable having a period of 10 days is observed to be 10^{-20} as bright as the Sun. What is the distance to the star?

STAR CLUSTERS

Once we have the distance to a star cluster, we can find its intrinsic properties, such as linear diameter and luminosity. These properties will then be available for use further up the distance pyramid.

☐ **Discovery Inquiry 21–1i** If the angular diameter of the cluster in Discovery Inquiry 21–1g is 2 minutes of arc, what is the cluster's linear diameter?
☐ **Discovery Inquiry 21–1j** When looking at the light from the entire cluster, we find it to be 10^{-14} times as bright as the Sun. What, then, is the luminosity of the cluster in terms of the Sun's luminosity?

GALAXIES

The distance to a galaxy having Cepheid variables can be found using the same technique described above. Once the distance is found, the galaxy's linear size can be calculated. This linear size can be used to find distances to farther galaxies.

☐ **Discovery Inquiry 21–1k** What is the distance to a galaxy that appears to be identical to M31, but whose angular diameter is 100 times smaller?

CLUSTERS OF GALAXIES

Similar techniques can be used to find distances to clusters of galaxies.

☐ **Discovery Inquiry 21–1l** If the largest spiral in the Leo Cluster of galaxies has an angular diameter three times smaller than the largest spiral in the Coma Cluster, whose distance is 113 megaparsecs, what is the distance to Leo?
☐ **Discovery Inquiry 21–1m** If the brightest galaxy in the Leo Cluster is eight times fainter than the brightest galaxy in the Coma Cluster, what is the distance to Leo? What is the percentage difference from the previous determination?

HUBBLE LAW

Once the distances to relatively nearby galaxies are determined, and their Doppler shifts are found from their spectra, we can graph the velocity-distance relation. We can then use this relation to find the distance to more distant galaxies.

☐ **Discovery Inquiry 21–1n** A galaxy is observed to be receding from Earth with a velocity of 55,000 km/sec. Compute the galaxy's distance for a Hubble constant of 20 km/sec/Mly.

CHAPTER SUMMARY

OBSERVATIONS

- Galaxies were classified by **Edwin Hubble** into **spiral, elliptical**, and **irregular**. Furthermore, the spirals are divided into normal and **barred spirals**. The **S0** galaxies are transition types between spiral and elliptical galaxies. Observations of nearby galaxy clusters show there to be more elliptical than spiral galaxies.

- The **Hubble law**, also known as the **velocity-distance relation**, relates the distances of galaxies and their velocities determined by the Doppler effect.

- The Milky Way is part of a cluster of galaxies called the **Local Group**. It contains two to three dozen known galaxies, most of which are elliptical. The **Large** and **Small Magellanic Clouds**, which are observable with the naked eye, are also members.

- The properties of the various types of galaxies are summarized in Table 21–1.

- **Intergalactic matter** is observed by means of X-rays characteristic of a hot (million degree) gas between galaxies.

- Clusters of galaxies form into **superclusters**. We are in the **Local Supercluster**, which is some 10^8 ly across.

- Observations of the distribution of galaxies show a "clumped" or "bubbly" appearance. The **Great Wall** is such a clumped structure. Observations also show the presence of large regions of space in which few if any galaxies exist; these are called **voids**, and may be hundred of millions of light-years in extent.

THEORY

- The factors that determine the type of galaxy formed are only poorly understood. They probably involve rotation, magnetic fields, and the rate at which star formation occurred when the galaxy formed. The latter is determined by the density of the protogalactic cloud.

- The **Tully-Fisher relation**, which connects a galaxy's rotation rate and its luminosity, provides a means of determining distance that is independent of the Hubble law.

- Galaxies interact gravitationally when they pass near each other. Computer models indicate that such interactions can strip material from a galaxy and cause a spiral-shaped galaxy to eventually become an elliptical galaxy.

- Not all spiral galaxies are alike; some have well-defined two-armed spirals while others are more chaotic. The shapes of the two-armed spirals may be explained by the density-wave theory, and the more chaotic arms are better understood as having arisen from supernova-induced star formation.

- Theories of supercluster formation include the **bottom-up theory** (individual galaxies form and then clump together) and the **top-down theory** (clusters form and then individual galaxies fragment from the cluster).

CONCLUSIONS

- The understanding that the spiral nebulae are in fact "island universes" beyond the Milky Way galaxy came about when individual **Cepheid variable stars** were identified in the Andromeda Galaxy. Using the known **period-luminosity relation**, a distance of 2 million light years was computed.

- The methods used to determine the distances of galaxies are summarized in the **distance pyramid** in Figure 21–12.

- Galaxy luminosities, which are readily determined once the distance is known, range from nearly a million times less luminous than the Milky Way to some five times greater. Their sizes are found from the observed angular diameter and the computed distance, and range from less than a tenth the size of the Milky Way to about four times greater. Masses, which are found from examination of the **rotation curve** of a spiral, the **velocity dispersion** of an elliptical, or the application of Kepler's third law, range from a million times less massive than the Milky Way (**dwarf elliptical galaxies**) to some 50 times greater (**giant elliptical galaxies**).

- From the colors, shapes, and amounts of gas and dust present in galaxies, astronomers infer that elliptical galaxies had a high star formation rate early in their lives, while the rate within spirals was considerably lower. This means that elliptical galaxies may have formed from higher density concentrations of material than the spirals did.

- The amount of matter in a galaxy cluster may be found from the velocity dispersion of galaxies within the cluster. Such observations, which always indicate the presence of more mass than is luminous, provide evidence for **dark matter** in galaxy clusters. Luminous arcs around galaxy clusters are caused by **gravitational lensing** of distant galaxies by intervening dark matter.

SUMMARY QUESTIONS

1. How was the extragalactic nature of the galaxies discovered? Why were astronomers led astray by some observations?

2. What galaxy characteristics are used for their classification?

3. What are the population I and population II characteristics of spiral, elliptical, and irregular galaxies? What is the role of star formation rate on the determination of galaxy type?

4. What is the chain of observations necessary to estimate the distances to galaxies, from the nearest to the farthest? What is the impact of different types of errors on the distance pyramid?

5. What is the velocity-distance relation and how is it used to find distance?

6. What specific types of galaxies inhabit the Local Group?

7. What is the role of angular momentum in explaining differences among galaxy types?

8. How do astronomers determine the fundamental properties of mass, size, and luminosity for galaxies and galaxy clusters?

9. How do astronomers study the distribution of galaxies, galaxy clusters, and superclusters in space? What are some current results of such studies?

10. What is the evidence that indicates the presence of nonluminous matter in galaxies and galaxy clusters?

11. What are the various ideas astronomers currently have concerning the formation of galaxies and galaxy clusters? In your discussion include the effects of density.

12. What effects might supernovae have had on the formation of spiral arms?

APPLYING YOUR KNOWLEDGE

1. If the Cepheid in the example in the text were actually a Type II Cepheid rather than the assumed Type I Cepheid, would its distance be larger or smaller than the originally computed distance?

2. Look at the photograph of M31 in Figure 1–15. What assumption might you make to allow you to determine the inclination of the galaxy to our line of sight?

3. Is it possible that the drawing of the Whirlpool galaxy in Figure 21–1*a* was completely accurate when the Third Earl of Rosse made it in the middle of the nineteenth century, and that the galaxy changed to the appearance shown in Figure 21–1*b* in the intervening 150 years? Present explicit arguments based on material discussed in this book.

4. Classify everything you have in your own room into, at most, 10 categories. How might this classification scheme be useful to you?

5. Explain how our modern understanding of spectra allows us to say that there are fundamentally different classes of objects within the group of objects astronomers used to refer to as "nebulae."

6. Explain the distance pyramid in your own words without a drawing.

7. How do you implicitly use the idea of a standard candle to judge the distance to an oncoming car at night?

8. Solve the example of the Cepheid distance given in the text assuming the star is a Type II Cepheid rather than the assumed Type I. By what percentage does the distance change?

9. Suppose the brightest blue star in a galaxy is 20 times brighter than the brightest Cepheid. How much farther can astronomers see the blue star than a Cepheid?

10. What would be the angular size of a galaxy identical to the Milky Way if it were at a distance of 10 million light-years? If such a galaxy were observed to have an angular diameter 1/100 this size, what would its distance be? What assumptions did you make in finding this distance?

11. Look at the photograph of M31 in Figure 1–15. Compute the inclination of the galaxy to our line of sight. What assumption are you making in finding this angle? (Note: This problem requires basic knowledge of trigonometry.)

12. If a supernova in a galaxy is observed to be 10^{18} times fainter than the Sun, what is the galaxy's distance, assuming the supernova's luminosity is 10^9 times that of the Sun?

13. Suppose a galaxy is observed whose spectral lines of ionized calcium that normally would be at 3968 Å are actually at 4268 Å. What is the distance to this galaxy? What are you assuming in finding the distance?

ANSWERS TO INQUIRIES

21–1. You might have noted that individual stars can be seen in the globular cluster.

21–2. The answer is neither easy nor obvious. For example, there might be differences in the expected spread in stellar velocities for each case. Another possibility is that it could be flat like a plate, but seen "face-on."

21–3. The farther away an object is, the harder it is to see motions in it across the line of sight. Assuming velocities much less than that of light, the "motions" observed by van Maanen would have had to indicate that the galaxy was nearby.

21–4. An intrinsically fainter object would have to be closer to appear equally bright.

21–5. A systematic error.

21–6. We should be able to improve the distance scale by (a) measuring the distances of many more galaxies by the use of Cepheids, and (b) thereby increasing the accuracy of the calibration of supergiants, novae, and globular clusters.

21–7. A cluster that has relatively few galaxies would have fewer large galaxies, and its brightest galaxy would probably be less luminous than we would assume. Its distance would therefore tend to be overestimated.

21–8. First measure the galaxy's velocity away from us. Then, assuming the velocity-distance relationship applies to the galaxy in question, find its velocity on the velocity-distance graph and read the corresponding distance from the graph.

21–9. As first described in Chapter 3, linear size is proportional to angular size times distance.

21–10. Just as you know a car traveling 60 mph will go 60 miles after one hour, the distance the galaxy will travel in one orbit will be the circumference of the circle it travels around the galaxy's center. The period will be the distance traveled (circumference) divided by the observed velocity.

21–11. Only for nearby galaxies can you resolve individual stars.

21–12. Blue stars are in the disc, in particular within the spiral arms. Red stars, on the other hand, are in the bulge and the halo. These regions probably had much different densities in the early days of the Milky Way's formation.

21–13. As discussed in Chapter 20, they are all 10–16 billion years old. Thus, most were formed at about the same time, when the Milky Way was young.

21–14. Wien's law relating temperature and the wavelength of maximum intensity.

21–15. Obtain a spectrum and measure the Doppler shift. Interpreting Doppler shift in terms of galaxy motion away from us and applying the velocity-distance relationship gives the distribution along the line of sight.

22

PECULIAR AND UNUSUAL EXTRAGALACTIC OBJECTS

Eventually, we reach the dim boundary—the utmost limits of our telescopes. There, we measure shadows, and we search among ghostly errors of measurement for landmarks that are scarcely more substantial. The search will continue. Not until the empirical resources are exhausted, need we pass on to the dreamy realms of speculation.

EDWIN P. HUBBLE, *REALM OF THE NEBULAE, 1936*

For many years, astronomical theories emphasized the majestic, unchanging calm of the universe as we came to appreciate the enormous time scales over which cosmic changes took place. Sophisticated technology, however, has revealed that dynamic activity and incredibly energetic phenomena are occurring in many extragalactic systems, leading to a new discussion of a "violent" universe. But at the limits of our optical systems, even the largest telescopes cannot collect enough photons to answer all our questions. When the available data are limited, the theoretical possibilities are multiplied. This chapter will push against the limits of our knowledge and discuss many objects that we have only begun to appreciate. We begin with a look at some galaxies whose visual appearance is unlike the normal galaxies studied in the previous chapter. We then extend our observations to other wavelengths, at which time we will see how they have changed our view of the universe.

22.1
PECULIAR-LOOKING GALAXIES

Astronomers have observed many galaxies whose optical appearances are clearly peculiar. Even among the irregular galaxies, these galaxies stand out as highly unusual. **Figure 22–1** shows an old photograph of an object located in the constellation of Cygnus and known at one time as VV 72. Various suggestions have been made regarding the proper interpretation of this photograph. One early model suggested that it consists of two galaxies in collision at speeds of possibly thousands of kilometers per second. Because the average distance between stars in a galaxy is large compared to the sizes of the stars themselves, even if two galaxies were in collision, the two stellar systems would pass through each other relatively undisturbed. However, the extended interstellar gas in the two galaxies would collide violently. Shock waves from the collision would excite the gas and cause it to radiate furiously, possibly explaining strong emission at radio wavelengths. In such a collision, the two stellar systems would eventually go their own way, leaving all the excited gas behind (**Figure 22–2**).

FIGURE 22–1. The galaxy VV72 may be the result of a collision between two galaxies.

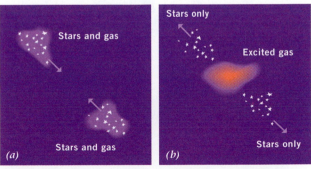

FIGURE 22–2. (a) Collisions of two galaxies containing both stars and gas. (b) In a highly supersonic collision, the stars pass through each other with little effect, but the gas clouds collide and merge. The galaxies are thus swept clean of gas, halting future star formation.

At present it is an open question whether such a collision would trigger star formation in the abandoned gas or whether the gas would just disperse. But it is clear that star formation in the two colliding systems would terminate, because they would be stripped of gas. An effect like this has probably operated in the globular clusters of our own galaxy where, as their orbits take them through the plane of our galaxy, they are swept clean of any gases they contain and continued star formation is prevented.

However, many astronomers dispute the notion that VV 72 is a pair of colliding galaxies, pointing out that a visual impression is hardly proof. The great distance of this object makes it difficult to observe optically.

We should beware of invoking collisions to explain all "funny-looking" systems, because we have no guarantee that a collision necessarily produces a strange appearance. The availability of ever faster computers has made it possible for astronomers to compute the details of complex gravitational interactions between simulated galaxies consisting of many stars. The most sophisticated computers are required to keep track of the positions, speeds, and directions of movement of some 10^4 stars during the billion or so years involved in an interaction. What computer simulations of close encounters show is that an enormously wide variety of appearances can be produced; the results depend critically on the details of how the two galaxies interact.

As one specific illustration, consider **Figure 22–3a**, which shows the object known as the Antennae. Figure 22–3b shows a time-sequence of computer models stemming from the effects of gravity during an encounter between large and small galaxies. Distortions and streamers of all types can result from the tidal interactions of passing galaxies.

Another visually interesting galaxy is the **ring galaxy** in Figure 22–4. Numerous examples of such

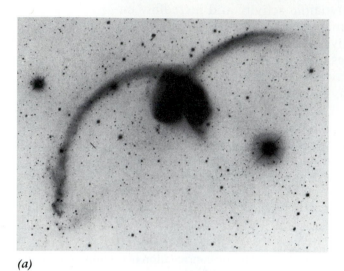

(a)

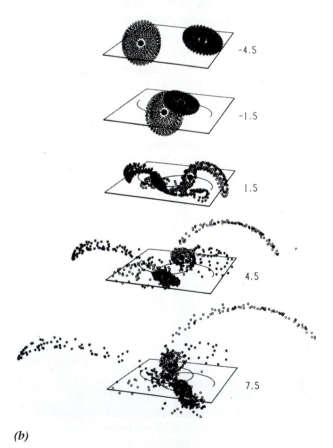

(b)

FIGURE 22–3. (a) The Antennae. (b) Computer simulations of colliding galaxies whose end result appears like the Antennae.

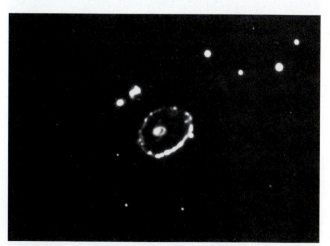

FIGURE 22–4. The Cartwheel galaxy. This is a ring galaxy, which is thought to have formed when a spiral galaxy (now forming the ring) collided with another galaxy.

galaxies exist. What mechanism could possibly produce a galaxy with a perfect detached ring around its nucleus? Computer models of a collision in which one galaxy collides face-on with another produce a perfect ring galaxy.

Optically peculiar galaxies show a large variety of appearances. While two classes of explanation, collision and explosion, were often invoked, only collisions are nowadays considered important.

22.2
RADIO GALAXIES

The first galaxy discovered to emit radio waves was the Milky Way itself; it was the first cosmic radio source to be detected, simply because it is so close. In 1931, a Bell Telephone engineer named Karl Jansky, who was testing antennas for long-distance telephone communications, discovered that every day his equipment detected a strong source of radio static. Each day when the interruption came, it occurred four minutes earlier—exactly what one would expect of an astronomical source, which followed the sidereal rather than the solar clock. In actuality, Jansky was observing radio signals from the gas in the plane of the Milky Way. By accident, he had founded the science of radio astronomy.

After World War II, when surplus radar equipment became available, astronomers began to survey the sky for radio sources. As discussed in Chapter 12 when we introduced the ideas of radio astronomy, an amateur astronomer named Grote Reber built a radio telescope and systematically mapped the sky in radio wavelengths.

Inquiry 22–1 If most radio sources are extragalactic, what does this imply about the amount of energy they emit?

Inquiry 22–2 Why would a galaxy composed of a mixture of ordinary stars like those studied in this book be expected to emit at least some energy in the radio region of the electromagnetic spectrum?

THREE BRIGHT RADIO GALAXIES: CYGNUS A, CENTAURUS A, AND M87

Naturally, the brightest objects were found first, and the first *extragalactic* radio object discovered was in the constellation of Cygnus. It was designated Cygnus A (Cyg A). Astronomers wanted to see whether there was an optically observable object coincident with the location of Cyg A. Finding an optical counterpart to a radio

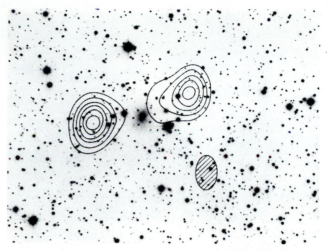

FIGURE 22–5. The typical double-lobed structure of a radio galaxy. In this photograph, the radio contours of the radio galaxy known as Cyg A are shown superimposed on a photograph of the optical galaxy between the radio lobes. Notice that the region of radio emission is much larger that the visible galaxy.

source was not easy in the early days of radio astronomy, however, because the resolving power of radio telescopes was poor and astronomers could not obtain accurate positions. Once it was done, however, astronomers found that Cyg A and the peculiar-looking VV 72 were the same object.

In the late 1960s and early 1970s, the resolving ability of radio telescopes had improved to the point where contour maps of the radio emission from Cygnus A (**Figure 22–5**) showed it to come from two lobes placed symmetrically on either side of the optical galaxy. These radio lobes appeared to be disconnected from the optical galaxy, and to extend many light years from it.

The distance to the optically visible galaxy was found using the observed Doppler shift to compute the galaxy's velocity, and then applying the Hubble law. From its distance of 750 million light-years and the observed intensity of its radio radiation, astronomers used the inverse square law of light to find that Cyg A emits about as much radio energy as the Milky Way does over *all* wavelengths. The large amount of energy emitted at radio wavelengths compared with the energy at visible wavelengths defines what we call a **radio galaxy**.

Another strong radio source, Centaurus A (Cen A), appears to coincide with the elliptical galaxy NGC 5128 in Centaurus (**Figure 22–6**). Optically, this elliptical galaxy appears to be laced with dust, which is unusual for elliptical galaxies, but its visual appearance gives no obvious indication of why it is such a strong radio emit-

FIGURE 22–6. NGC 5128, a strong radio galaxy known as Centaurus A (Cen A).

ter. It, too, has the symmetrical double-lobed structure of Cygnus A. Furthermore, the double-lobed structure of Cen A occurs on many size scales. As you view Cen A at successively higher resolution, the double-lobed structure continues into the central regions, as shown in the contours of **Figure 22–7**. Finally, in its innermost region, the galaxy shows a single, intense, elongated piece of material off to one side, referred to as a **jet**, that is also aligned with the lobes.

While not all doubled-lobed radio sources have optically identified galaxies associated with them, speculation is that as our ability to observe fainter and fainter galaxies improves, central galaxies will be found in all these cases. The double-lobed structure occurs in the majority of strong radio sources.

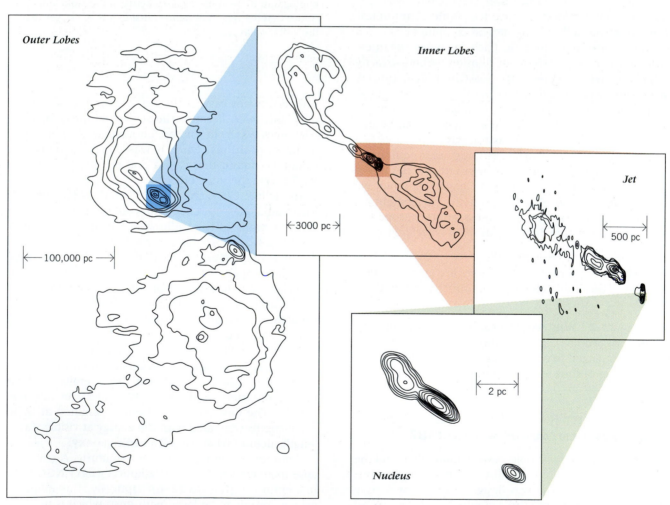

FIGURE 22–7. The double-lobed structure of Cen A is shown on a variety of scales, from the outermost lobes to the innermost jet.

At radio frequencies, the intensity of the radiation from these galaxies *increases* with increasing wavelength, opposite to the behavior of ordinary thermal emission (see Figure 19–18). In addition, the radiation is observed to be polarized. This strongly suggests that the radiation is synchrotron emission similar to that observed in the Crab Nebula. However, even knowing that the radiation is synchrotron emission, we must continue to try to understand how the gases and magnetic fields in these objects could have acquired as much energy as their great luminosity indicates.

Inquiry 22–3 What conditions are required to produce synchrotron radiation?

The radio lobes generally radiate energy most intensely at their outermost edge. The high intensity arises when particles from the central galaxy slam into the surrounding intergalactic material at high speeds and become heated by friction. From observations such as these, astronomers can determine the density of material in the intergalactic medium.

The galaxies discussed above, and indeed all galaxies having synchrotron sources, are known as **active galactic nuclei** or **AGN**. Their common properties can include high energy, polarized synchrotron radiation from a rapidly variable source, often a peculiar optical appearance, jets, and emission lines that are often broad, indicating highly turbulent motions within the gas.

Even if some of the strong radio sources we see are colliding galaxies, most clearly are not. An example is the giant elliptical galaxy NGC 4486, more commonly known as M87, that dominates the Virgo Cluster of galaxies (see Figure 21–2). This giant elliptical galaxy is one of the most luminous known galaxies at optical wavelengths, giving off more energy than the brightest known spiral galaxy. It also has strong radio emission, which is concentrated into a jet that protrudes from the central regions of the object. A photograph of NGC 4486 using a short exposure to reveal just the bright core of the galaxy (**Figure 22–8a**) shows the jet more clearly. Figure 22–8b is a high resolution image of the jet showing its many components.

Inquiry 22–4 What is one possible hypothesis for a cause of the jet in NGC 4486?

One additional class of radio galaxy is that shown in Figure 22–9. This galaxy is an example of what is called a **head-tail galaxy**. Its form appears to come from its rapid motion through the intergalactic medium, in

(a)

(b)

FIGURE 22–8. (*a*) A short-exposure photograph of NGC 4486 showing the jet. (*b*) A high-resolution image of the jet.

which its magnetic field is swept back by its interaction with intergalactic gas. The more rapid the motion, the more strongly the magnetic field is swept back, and the straighter the tail.

One of the surprising aspects of the peculiar radio galaxies is their sheer size. **Figure 22–10** shows the scale of some double-lobed radio sources. Note the comparison with the Milky Way!

HIGH-RESOLUTION OBSERVATIONS OF RADIO EMISSION FROM GALAXIES

Modern radio interferometers have made possible greatly improved resolution over the coarse resolution

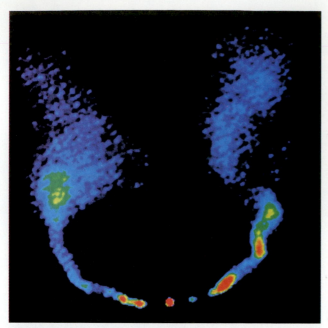

FIGURE 22-9. The head-tail radio galaxy NGC 1265.

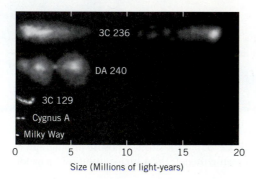

FIGURE 22-10. A schematic diagram showing the extent of some radio galaxies compared with the Milky Way.

FIGURE 22-11. A high-resolution image of Cyg A.

increased our ability to display the data as a picture rather than only as contour lines.

Astronomers have long been puzzled as to how material flowed from the central source to the outlying radio lobes. Clues leading to an answer come from high-resolution observations of Cyg A, which show conduits carrying material from the optical galaxy out to the distant lobes (Figure 22–11). What might cause the ejection of material is not clear.

Technological advances, along with placement of telescopes on mountain tops at high elevation, are making it possible to obtain higher-resolution images of galaxies from ground-based optical telescopes than could previously be obtained. A new era in high-resolution optical observations began with the launch of the Hubble Space Telescope. With its corrected optics, this instrument should be able to distinguish details that are 20 times closer together than has been possible from ground-based pictures. **Figure 22–12** shows images of the spiral galaxy M100 taken by the Hubble Space Telescope both before and after the repair mission. The image prior to the telescope's repair resolved more detail than ground-based images had obtained, thus indicating the overall gain provided by having a large telescope in space.

of early radio astronomy. Astronomers now carry out observations with resolution often greatly exceeding that of optical observations. Recent images of radio galaxies are amazing, as **Figure 22–11** of Cyg A shows. The greater resolution has occurred thanks to technology that has improved clocks and computers. Improved clocks allow us to time the arrival of energy at telescopes spread over the entire Earth in order to combine the collected signals later. Improved computer technology allows us to handle the torrent of data that radio interferometers produce. Finally, computer technology has also

22.3
SEYFERT GALAXIES AND OTHER ACTIVE GALAXIES

During World War II, Carl Seyfert, an astronomer at Vanderbilt University, noted some galaxies that were distinctive enough in appearance and spectra to merit an entirely new category. Most of these so-called **Seyfert galaxies** look like normal spirals, and some look like ellipticals. The Seyfert galaxy NGC 1068 is

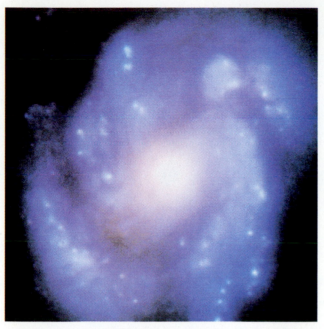

Wide Field Planetary Camera 1

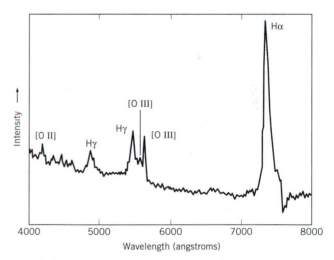

Wide Field Planetary Camera 2

FIGURE 22–12. M100 observed with the Hubble Space Telescope both before and after the servicing mission. The earlier image was better than what could be obtained with ground-based telescopes.

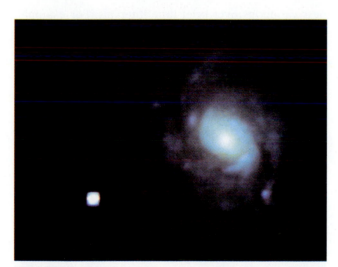

FIGURE 22–13. The Seyfert galaxy NGC 1068.

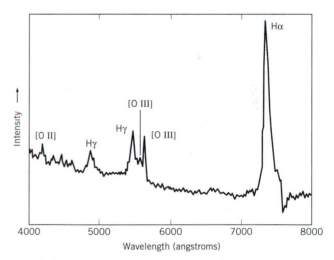

FIGURE 22–14. The spectrum of the nucleus of a Seyfert galaxy.

shown in **Figure 22–13**. Seyfert galaxies are characterized by two observational features. First, the nucleus of the galaxy is bright compared to the rest of the galaxy, and in the short-exposure photograph the nucleus even appears starlike. Second, the spectrum of the nucleus in

a Seyfert galaxy shows both continuum emission and bright-line emission (**Figure 22–14**), quite unlike that of, say, M31. The emission lines show that there is a large amount of hot gas in the nuclear region; the widths of the lines indicate that the gases in the nucleus are in a state of violent agitation with speeds of thousands of km/sec.

Inquiry 22–5 Why do broad lines indicate violent agitation in a Seyfert galaxy?

Distances to Seyfert galaxies can be found from the Doppler shift of the spectrum. From their distances and observed brightnesses we find that Seyfert galaxies are more luminous than normal spiral and elliptical galaxies. Furthermore, their radiation appears to come from an extremely small region, perhaps some 5–100 times the size of the solar system. The generally accepted explanation is that there is a massive black hole in the nucleus; the radiation we observe comes from material falling toward the black hole and interacting with an accretion disc before crossing the event horizon. One piece of evidence for this model comes from HST images of the Seyfert galaxy NGC 1068 (**Figure 22–15**), which show clouds of ionized gas in the innermost regions. These clouds appear to trace out a cone of radiation coming from an obscured area thought to contain the supermassive black hole. The model to produce this cone of clouds implies the presence of a disc. Ionizing radiation is emitted perpendicular to the disc, analogous to the way in which bipolar nebulae were produced in the star formation processes discussed in Chapter 17. While we know that about 1% of all galaxies have active nuclei, we don't know whether all galaxies spend a short time passing through an active phase or whether only a minority of objects ever become active.

Another example of violent activity in extragalactic objects is a class of objects that appear to be exploding. **Figure 22–16** shows the galaxy M82, long classified as an irregular (it clearly is irregular in form). It turns out to be a strong source of radio radiation, and its gases show violent agitation and high velocities. However, the obvious interpretation that this object is exploding has been challenged by an alternative model that suggests that we are only seeing dust filaments that scatter light to us from the galaxy's nucleus—something like a gigantic reflection nebulosity. The high velocities observed in the system would then come from violent, turbulent agitation of the gases in it. Furthermore, this galaxy is also classified as a **starburst galaxy**, one in which active star formation is booming. Such short but intense periods of star formation can be initiated by gravitational interactions or collisions with a passing galaxy.

BL Lac objects, named after the object BL Lacertae that was once thought to be an ordinary variable star, lack emission lines but are observed to have redshifts larger than most normal galaxies. They exhibit bright-

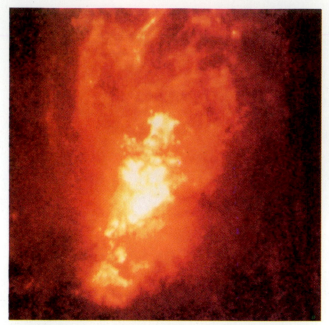

FIGURE 22–15. The center of NGC 1068 as observed by HST after the optics were corrected. HST reveals previously unknown detail. Compare with the ground-based photograph in Figure 22–13.

ness variations over brief time intervals. For example, some BL Lac objects change in brightness by 100% in less than a month or even 20% in a day. Because the emitting region of an object cannot be larger than the time it takes light to travel across it (see Chapter 19), we

FIGURE 22–16. The peculiar galaxy M82.

find them to be only a light-day across. If we block the light of the BL Lac object with a small obscuring spot at the telescope's focus, we are able to see faint radiation coming from surrounding areas, just as we are able to see the solar corona during an eclipse of the Sun. When we analyze this faint radiation, we find its spectrum to be similar to that of an elliptical galaxy. We conclude that in BL Lac objects we are looking directly into the core of an active galaxy.

22.4
QUASARS

The first step in the development of any new field of science is to catalog the observations. During the 1950s, large numbers of new sources of radio radiation were discovered and cataloged. The next step was for astronomers to attempt to find an optical object corresponding to every radio source. Some identifications were fairly obvious, but in many cases the low resolution of early radio telescopes frustrated these efforts. A high-resolution interferometer, built at Cambridge University in England in the 1950s, made it possible to identify many of the newly discovered radio sources with optical objects. Some turned out to be external galaxies, but one of them, 3C48 (the 48th object in the *Third Cambridge Catalog*), had no obvious optical counterpart other than what appeared to be a faint blue star.

The discovery of a strong radio source that looked like a star created a stir of interest in the astronomical community, because most previous radio sources had turned out to be galaxies. Normal stars were not expected to emit strongly at radio wavelengths. Other radio "stars" were soon found. The brightest known, 3C273, was found to be a point source by monitoring its brightness very carefully as the sharp edge of the Moon passed in front of it. So starlike did these objects appear that the name **quasar**, for "quasi-stellar radio source," was soon coined to apply to them.

Because quasars are stellar in appearance, photographs of them rarely look particularly interesting. **Figure 22–17a** shows a photograph of 3C48. Figure 22–17b shows a photograph of the brightest quasar, 3C273, which does have a structure in the form of a small jet emerging from one side.

Inquiry 22–6   What does the jet in 3C273 remind you of?

The first quasars were discovered by means of their radio emission. Further research, however, has shown that only 10% of the quasars are strong radio sources.

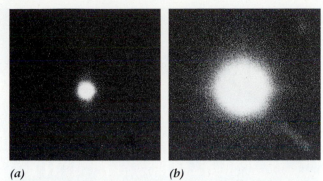

FIGURE 22–17.   Two quasars. (*a*) 3C48. (*b*) 3C273, showing the jet.

The name quasar is, therefore, a misnomer that arose from a case of observational selection, in which objects to be investigated optically were selected strictly on the basis of their radio emission.

THE SPECTRA, DISTANCES, AND LUMINOSITIES OF QUASARS

Once a radio object has been identified optically, the next step is to obtain its spectrum. These radio "stars" interested Maarten Schmidt of Mount Palomar Observatory, home of one of the few telescopes that was powerful enough to photograph their spectra. The spectrum of 3C273 (**Figure 22–18**) was found to consist of a continuous background and several bright emission lines, not unlike the spectrum of a Seyfert galaxy. However, the wavelengths of the emission lines did not seem to coincide with those of any known chemical elements.

Schmidt finally realized that there was a simple interpretation—the bright lines were not due to some mysterious element unknown on Earth but were simply considerably redshifted Balmer lines of hydrogen. Significant redshifts are associated with extragalactic objects, so this discovery elevated quasars from the category of star (i.e., potentially nearby objects of only modest luminosity) to the category of extragalactic object with a correspondingly large increase in energy output.

> YOU SHOULD DO DISCOVERY 22–1,
> MEASURING A QUASAR'S VELOCITY,
> AT THIS TIME.

The redshift of 3C273 was found to be almost 16%; that is, the object is receding from us at a speed of 16% the speed of light. *Assuming that the relationship between velocity and distance discovered by Hubble is applicable to 3C273*, it was possible to determine the distance.

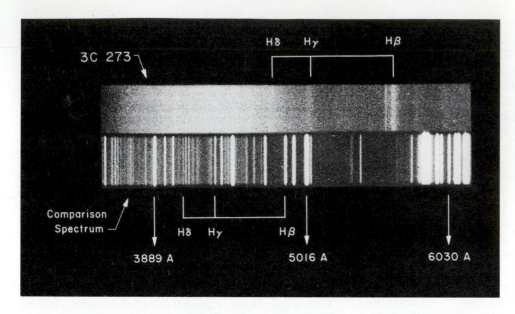

FIGURE 22–18. The spectrum of the quasar 3C273, which has a redshift 16% the speed of light.

Inquiry 22–7 What is the distance to 3C273? Use all the information you have and a Hubble constant of 17 km/sec/Mly.

You have just discovered that this object was found to be nearly 3 *billion* light-years away! From its distance and its observed brightness of 10^{-16} that of the Sun, it was possible to calculate the object's *intrinsic* luminosity. The result is that we now know it radiates between 10 and 100 times more energy than the entire Milky Way galaxy. Apparently, an important new kind of object had been discovered—superluminous, and with associated radio sources.

> YOU SHOULD DO DISCOVERY 22–2,
> DISTANCES IN THE UNIVERSE,
> AT THIS TIME.

3C48 was found to have a redshift of 37% of the speed of light, placing it over twice as far away as 3C273. The spectrum of one of the most distant quasars detected to date, known as PC 1247+3406, is shown in **Figure 22–19.** The narrow peak marked "H" is normally at a wavelength of 1216 Å in the ultraviolet. Because this spectrum shows the line at about 7200 Å, the quasar is moving away from us at some 94% the speed of light.

THE VARIABILITY OF QUASARS

At first, the quasars seemed to be interesting, even if their significance was unknown. Because they are stellar in appearance, it was reasoned that the brighter ones

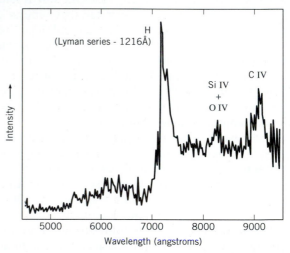

FIGURE 22–19. The spectrum of the high-redshift quasar PC 1247+3406, which has a redshift of more than 90% the speed of light.

must have gone unrecognized on old photographs of the sky, looking just like uninteresting faint stars. However, some major observatories have permanent photographic collections spanning long time intervals, and when the brightest quasar, 3C273, was found on them, it soon became evident that it varied significantly in brightness in only a few months. Other quasars showed similar irregular light variations, sometimes changing brightness by 50% a day. Indeed, more recently the HEAO-2 satellite discovered a quasar that varies its X-ray brightness significantly in as little as 200 *seconds*. Quasars soon became a tantalizing puzzle.

Inquiry 22-8   What does the rapid variability of these objects indicate about their size? If an object varies its brightness significantly in only 3.7 days (a hundredth of a year), what is its maximum size? Express this size in terms of the size of our planetary system (40 AU) and also in terms of the distance to the closest star, α Centauri.

We can now see why quasars are perplexing to astronomers. From the observations, we infer that quasars emit from 10 to 100 times as much energy as the entire Milky Way, but from a region that is as small as one-millionth the diameter of the galaxy (less than 0.1 ly). As you can imagine, these discoveries left astronomy in something of a state of shock.

OTHER PERPLEXING ASPECTS OF QUASARS

The spectra of quasars revealed other curiosities that were difficult to explain. They showed both absorption and emission lines, and on frequent occasions the two different types of spectral lines showed *different* redshifts. Even more puzzling, some quasars were found with multiple sets of redshifts in both their emission and absorption spectra. This really has only one simple interpretation—that the quasar has a complex structure in which various parts are moving at higher rates of speed than other parts. For example, multiple redshifts observed in absorption might indicate clouds of material moving at different rates of speed between us and the quasar's power source. Unfortunately, such high velocities between various parts of the quasar are difficult to reconcile with the relatively narrow widths of some of the sharp emission lines.

QUASAR HYPOTHESES AND SPECULATIONS

The tremendous distances to quasars, and thus the exceptionally high luminosities deduced, hinge entirely on the assumption that the Hubble law can be applied to the quasars. Observations tending to validate the assumption include the following:

1. Many quasars are found in or near galaxy clusters having similar redshifts. Given that over 7000 quasars are known, a small number compared with the number of known galaxies, the probability of such a quasar/cluster coincidence occurring by chance is small. The grouping is probably significant. **Figure**

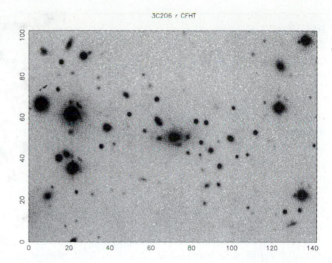

FIGURE 22–20.   A galaxy cluster with a quasar nearby.

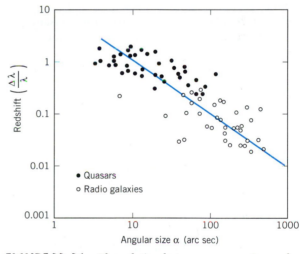

FIGURE 22–21.   The relation between a quasar's angular diameter and its redshift.

22–20 shows the coincidence between a quasar and a distant galaxy cluster.

2. If redshift is indeed an indicator of distance, a good correlation should exist between redshift and angular size. Such a correlation is observed in **Figure 22–21**.

3. High resolution images of some quasars show them to be located in the center of a fuzzy region of light that has the spectrum of a galaxy with the same redshift as the quasar. In other words, the quasar and a surrounding galaxy have the same redshifts.

Inquiry 22–9 Accepting that quasars are at great distances, are astronomers observing old or young objects?

If quasars are indeed at the large distances derived from the Hubble law, then what is the source of their immense energy? Finding viable energy sources has not been easy, and early suggestions included some wild speculations. One idea involved chain reactions of supernova explosions. According to this idea, many stars in a quasar are on the verge of becoming supernovae and are somehow triggered to explode by other supernovae explosions. It was reasoned that during the early stages of a galaxy's collapse there could be an era of extremely rapid star formation. After a brief period of rapid evolution by massive stars, the stage would be set for a supernova fireworks display. It was soon found that several different lines of argument seemed to work against this model—it even had difficulty generating enough energy.

Another proposed energy source for quasars was matter-antimatter reactions, which, as we have seen, appear to occur in the nucleus of the Milky Way. However, this model has been discredited by high-altitude balloon observations showing that quasars have little of the expected gamma-ray emissions.

SUPERMASSIVE BLACK HOLES

Theoretical calculations indicate that a cluster or galaxy of stars might be able to form a large black hole in its center. Gravitational theory puts no limits on the mass of a black hole, and the quasars have prompted astronomers to consider the possibility that a black hole of several million solar masses could form in the center of a galaxy.

Such a black hole would be an efficient and powerful source of energy. How? Mass from the galaxy's central regions would fall toward the black hole, become compressed, heat up, and radiate prodigious amounts of energy before being swallowed beyond the event horizon.

Most astronomers now think that the power source of quasars and Seyfert galaxies consists of an extremely active nuclear region containing a black hole of enormous mass (10^9 solar masses) that consumes large quantities of material to produce the observed luminosity. Current theoretical research suggests that a supermassive black hole model can indeed explain the enormous energy output of most quasars. Thus the idea of a supermassive black hole as the energy source seems to provide a consistent picture that others do not supply.

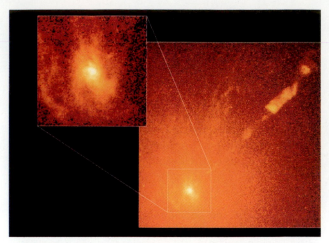

FIGURE 22–22. An HST image of the nucleus of M87 showing gas that appears to be spiraling about a massive central object, believed by astronomers to be a black hole.

This model suggests that many galaxies might have a massive black hole in the center. It also predicts that some of the most violent quasars should also be strong X-ray emitters because the large mass would provide efficient heating of infalling material. Observations are in agreement with this prediction.

The black hole model for AGN received what appears to be confirming observations with data obtained by HST in spring 1994. An image of the nucleus of the giant elliptical galaxy M87 (**Figure 22–22**; pre-HST images are shown in Figures 21–2 and 22–8) shows gas that appears to be spiraling around something. Spectra of the gas show it to have different velocities on opposite sides, indicating rotational motion of 450 km/sec at 60 ly from the center. From these observations, astronomers infer that the central object must have a mass of 2.5–3.5 billion solar masses in a region the size of the solar system. A massive black hole is the most reasonable conclusion.

M87 shows a relativistic jet (Figure 22–22), as do some quasars. How are these high-energy jets produced? Theoreticians imagine that material might spiral into a black hole in a manner somewhat similar to the way water spirals down a drain. The spiraling matter will in general form itself into a flattened-out accretion disk. The spinning black hole can also shoot out matter at high energy along the spin axes (**Figure 22–23**). This ejected matter becomes the narrow jet of energetic particles.

Some quasars appear to be putting more energy into their jet than they exhibit in their main body. Where the energy appears may depend on the details of the accretion process; that is, how the material flows into the black hole. However, the existence of these jets may

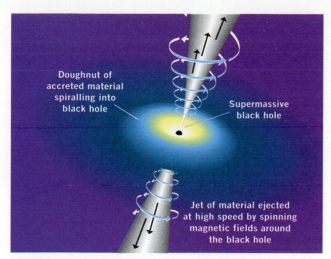

FIGURE 22–23. A black hole accretion disk of a quasar. Note the jet of gas squirted out at near the speed of light by the violent processes near the black hole.

FIGURE 22–24. An HST image, obtained with the Wide Field Planetary Camera, of the quasar 1229+204. Faint light surrounding the quasar comes from the galaxy that surrounds the luminous quasar.

also soften the severity of the basic energy problem of the quasars. For example, if a quasar happens to be pointing an energetic jet our way, we might think that it is more luminous than it really is, because we would assume that it is radiating uniformly in all directions when it is not.

If quasars are indeed at great distances, then we see them as they were many billions of years ago, and it is tempting to think that they are young galaxies. Evidence that quasars are galactic nuclei came when a photograph of 3C273 was made in which the light from the quasar was blocked during a long exposure, as we discussed for BL Lac objects. The photograph showed a "fuzz" around the quasar, the spectrum of which was later found to be similar to that of a normal galaxy.

The idea that quasars are galaxies received further confirmation from studies carried out by the Infrared Astronomy Satellite (IRAS). In 1985, IRAS discovered a number of objects that were radiating as brightly as quasars but were releasing most of their energy in the infrared rather than at visual wavelengths. Detailed spectral investigations of these objects with ground-based telescopes revealed starlike dark-line spectra, thus showing that they were indeed galaxies. They were also found to be rich in gas and dust, but most significantly there was evidence that each object was actually two spiral galaxies in the process of colliding.

Such collisions would send streams of matter to the central regions, where they would either create a black hole or fall into one if it were already there. In other words, the galaxy collisions might actually be triggering the birth of a quasar, but the large quantities of dust would hide the usual signatures of a quasar at visible wavelengths. The enormous energy generated in the region around the black hole would heat up the dust, causing it to emit huge quantities of infrared radiation. Should this cloud of dust ever burn away or be blown away, the theory suggests, a visible quasar would then emerge into view.

Observational verification that quasars are the cores of galaxies has been provided by ground-based and HST observations by Canadian astronomer John Hutchings. **Figure 22–24** shows his HST observation of the quasar 1229+204. What these results show is that the majority of the nearby quasars are in peculiar-looking galaxies. They appear to be tidally distorted by collisions and to result from galactic mergers between a high-mass and a low-mass galaxy. These observations show the importance of making observations from high-altitude, dark observing sites.

Finally, some astronomers have suggested a unified model in which radio galaxies, BL Lac objects, and some quasars are really the same type of object, but observed from different angles. **Figure 22–25** shows the suggested model. Observations explained by the model include the fact that quasars are strong at both visible and radio wavelengths; that radio galaxies are strong in the radio but weak in visual wavelengths; and that some radio galaxies show two jets while quasars never show

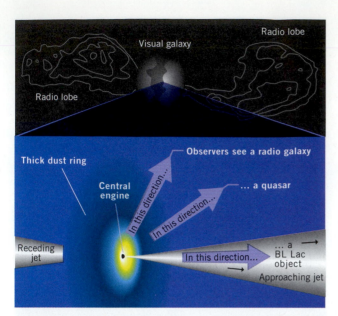

FIGURE 22–25. A model suggesting that radio galaxies, BL Lac objects, and quasars are the same type of object observed from different angles.

more than one. In the model, the central "engine" is surrounded by a thick ring of dust that causes jets to be emitted along the ring axis (this is similar to the bipolar nebulae observed in star-forming regions). Observers along the jet axis see a BL Lac object. Observers more than 45° from the jet axis see a radio galaxy because the ring of dust hides the nucleus. Observers less than 45° from the jet axis see a quasar, because the central source is no longer hidden. Such a unified model pulls together a great variety of observations and has great advantages over separate models.

A Dissenting Viewpoint

A small minority of astronomers object to the application of the Hubble law for finding quasar distances. They argue that the observed quasar redshifts are caused not by the expansion of the universe but by something else, discussed below. (Redshifts caused by the overall expansion of the universe are said to be **cosmological redshifts**. Distances found from the Hubble law are said to be **cosmological distances**. Thus the minority opinion is that quasars are not at cosmological distances.) They have pointed out a number of objects that appear on photographs to be physically associated but yet have vastly different redshifts. An example is the compact galaxy known as Markarian 205, which has a redshift of 21,000 km/sec, and appears physically connected to the larger galaxy NGC 4319, which has a redshift of 1700

km/sec and is presumably much closer. Further, one astronomer has claimed that he can observe the spectrum of the larger galaxy *through* the dust in the outer parts of the smaller one.

The astronomer Halton Arp has searched for such systems and claims that there are more close associations than chance alignments would predict. Other astronomers have presented calculations refuting this claim, and argue that the apparent "bridges" sometimes seen between high-redshift and low-redshift objects are simply artifacts of the photographic process. Observational astronomers are keenly aware of the fact that photographic emulsions can be tricky and unreliable at ultra-low-intensity levels. The most recent studies of the galaxies mentioned in the previous paragraph suggest that the bridge that appears to run between the two objects is indeed real, but that it is not in fact a physical connection between them. Instead, it appears to be a "tail" produced in the material surrounding the smaller galaxy by an unseen companion. The two objects in the picture are simply seen along the same line of sight by coincidence.

If the dissenting view were correct and quasars were not at cosmological distances, what then would be the source of their observed large redshift? One of the more radical suggestions is that quasars are not distant objects at all but are actually quite close to the Milky Way galaxy, having been ejected from it by a violent event in the recent past. Their high redshifts then are indicative only of their violent ejection and not of large distances. According to this hypothesis, they would actually be relatively faint, nearby objects. This theory has the added difficulty that no one has been able to find any other corroborating evidence from studies of our galaxy to indicate that there has been such a violent event at some time in the past.

Might quasars have been ejected by other galaxies that are relatively nearby? In this hypothesis, the Doppler shift comes from high ejection speeds, and we would expect ejection in all directions so that at least some objects would have been ejected toward us and show a blueshift. None are observed. The question of where the energy came from to eject objects with speeds near that of light is unanswered by the dissenting ideas.

Gravitational Redshifts?

Another interesting proposal is that the redshift of the quasars might be due only partly to the distance, and that the rest of it might be caused by an intense gravitational field in or near the object. If part of the redshift is gravitational, then the quasars need not be as far away as we think; hence they need not be as luminous as we think, so the energy dilemma might be eliminated.

To create a large gravitational redshift, the region of

gas in the quasar where the spectral lines are formed would have to surround an object of immense gravitational force. A supermassive black hole could do the job. However, photons from different regions of the object would have different redshifts and this would *broaden* the spectral lines, an effect that has not been observed.

The quasar controversy is a complicated one with many aspects that cannot be easily summarized here. It can be said, however, that none of the various alternatives to the cosmological interpretation of the quasar redshifts has achieved the coherence of the cosmological redshift model. At present, a poll of astronomers researching quasars would find most of them supporting the cosmological interpretation, but it is important to remember that majorities have been wrong before.

GRAVITATIONAL LENSES

During 1980, astronomers became increasingly puzzled by the properties of a mysterious object dubbed "the double quasar"—two quasars closer together in the sky than could be expected from statistics. Besides their angular proximity to each other, the two objects seemed to be virtually identical in observed characteristics, leading to the suspicion that what was actually being seen was some kind of double image of a single object. Finally, observations with a sensitive CCD detector revealed the presence of a third object between the two images. This object turned out to be a massive galaxy that could have a strong enough gravitational field to bend radiation passing near it, as predicted by Einstein's general theory of relativity.

Figure 22–26a shows a system, known as 2237+0305, that appears to contain four or perhaps five quasars. In Figure 22–26b, the four quasar images are subtracted, leaving the image of a foreground galaxy, which acts as a gravitational lens by focusing light from a single quasar located farther away into multiple images. An example of gravitational lensing is shown in Figure 22–27, in which a single object produces three

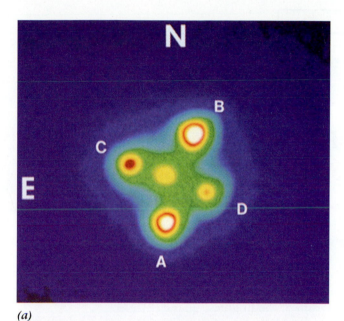

(a)

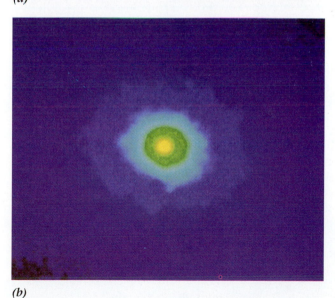

(b)

FIGURE 22–26. The gravitational lensing system 2237+0305. (a) Original image. (b) After subtraction of the quasar images, the lensing galaxy remains.

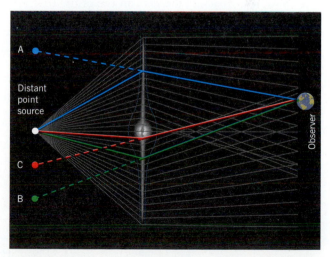

FIGURE 22–27. How a nearby galaxy produces three images of a quasar. The observer sees light from the quasar that appears to come from the three directions labeled A, B, and C.

additional images as the light passes through the mass of a closer, disk-shaped galaxy. Exactly what type of image is formed depends on the relative locations of the imaged object, the "lens" galaxy, and the observer. In some cases multiple images appear, and in other cases the image will be smeared out; in yet others, a perfect ring, called an Einstein ring, can be formed. The observation of these lensing systems provides further evidence that Einstein's general theory of relativity is a valid description of nature.

As is often the case in astronomy, once it was realized that this behavior was taking place, researchers began to look for other examples of it around the sky, and numerous examples of gravitational lenses have been discovered. Gravitational lensing of quasars by galaxies now appears to be a common phenomenon. An important finding from the observations of 2237+0305 shown in Figure 22–26 is that the system can be modeled exactly, but *only* by assuming that quasar redshifts are caused by the cosmological expansion of the universe.

DISCOVERY 22–1
MEASURING A QUASAR'S VELOCITY

When you have completed this Discovery, you will be able to do the following:

- Measure the redshift of a quasar.
- Compute the velocity with which a quasar is moving.

Once Maarten Schmidt realized that the spectral lines he was observing were Doppler-shifted hydrogen lines, he was able to determine the quasar's velocity. You will follow similar procedures using the spectrum of the brightest quasar, 3C273, given in Figure 22–18.

The first step in finding the velocity is to determine the scale of the spectrum using the comparison spectrum. The scale gives the number of angstrom units in each millimeter on the print. To find the scale, measure the number of millimeters between the comparison lines at 3880 Å and 6030 Å.

☐ **Discovery Inquiry 22–1a**   What is the number of angstroms per millimeter on the photograph?

Next, choose one of the hydrogen lines (Hβ at a rest wavelength of 4861 Å, Hγ at 4340 Å, or Hδ at 4101 Å). The rest positions of these lines are marked under the comparison spectrum. Measure the spectral line shift in millimeters from the rest position of one of the hydrogen lines to the emission line in the quasar spectrum. (Make your measurement parallel to the edge of the spectrum.) Then, use the scale determined above to compute the shift in angstroms.

☐ **Discovery Inquiry 22–1b**   What is the shift of the hydrogen line in millimeters? In angstroms?

The Doppler shift is given by $\Delta\lambda/\lambda$, the change in the wavelength divided by the rest wavelength. Compute this quantity for the hydrogen line you measured.

☐ **Discovery Inquiry 22–1c**   What is the measured Doppler shift, $\Delta\lambda/\lambda$, for 3C273?

☐ **Discovery Inquiry 22–1d**   Repeat the above for the other two hydrogen lines. What is the average Doppler shift? Estimate the error in the measured redshift by computing $\dfrac{\text{(maximum value – minimum value)}}{1.4}$.

☐ **Discovery Inquiry 22–1e**   Because quasar redshifts are large, the usual Doppler shift formula must be replaced by one that accounts for the theory of relativity. If $Z = \Delta\lambda/\lambda$, the expression to find the velocity becomes

$$\frac{V}{c} = \sqrt{\frac{(1 + Z)^2 - 1}{(1 + Z)^2 + 1}}.$$

Use this formula to compute the quasar's velocity relative to the speed of light, V/c.

☐ **Discovery Inquiry 22–1f**   To see how the measured uncertainty in the redshift affects the computed velocity, compute the maximum and minimum redshifts, $(Z + \text{error})$ and $(Z - \text{error})$, where error is the number found in Discovery Inquiry 22–1d. Use each of these numbers in the formula to find the maximum and minimum possible velocity consistent with your measured redshift.

DISCOVERY 22–2
A SCALE MODEL OF DISTANCES IN THE UNIVERSE

Now that we have studied the most distant objects yet observed, we can summarize distances with a scale model. If needed, refer to the distance pyramid in Chapter 21. This activity may be done with at least one other person.

Obtain a standard-size roll of toilet paper. The length of this roll will represent the distance from Earth to the most distant object known. The package label should tell you the number of tissues on the roll. Use this information to determine the distance (in light-years) represented by each sheet. Then, using a long hallway if available (if such a hall is not available, you can go back and forth within the longest area available), begin unrolling the paper, and mark the location of typical representatives of each of the following objects: a quasar, the edge of the Local Supercluster, a distant cluster of galaxies, M31, the center of the Milky Way, the nearest star, and the edge of the solar system.

☐ **Discovery Inquiry 22–2a**   How many light years does each sheet represent?
☐ **Discovery Inquiry 22–2b**   On the scale of your model, how wide a line represents the size of the Milky Way? The size of the solar system?

CHAPTER SUMMARY

OBSERVATIONS

- Peculiar-looking galaxies often turn out to be **radio galaxies**, objects that emit vastly more energy than would be expected from the stars composing them. The radio emission from these galaxies increases with increasing wavelength and is polarized; the emission is therefore produced by **synchrotron radiation**.

- Most radio galaxies have two radio-emitting lobes located symmetrically on either side of the optical galaxy. **Head-tail galaxies** appear the way they do because the galaxy's motion through the intergalactic medium and magnetic field causes material to stream behind, forming the tail.

- Some galaxies appear to be in collision.

- Radio images that show jets of material from the optical galaxy out to the radio-emitting lobes provide evidence of a physical connection between the galaxy and the distant lobes.

- **Seyfert galaxies** are most often spiral galaxies with bright nuclei and spectra that show gases in turbulent motion.

- **Quasars** appear to be starlike objects whose spectra show highly redshifted emission lines. Quasars often show multiple sets of absorption lines, vary over intervals of days to months, and are surrounded by a faint galaxy.

THEORY

- Computer models of collisions between galaxies are able to reproduce the appearance of real galaxies, including **ring galaxies**.

- Einstein's theory of general relativity predicts that a quasar's radiation, if passing by a massive galaxy, can be bent and formed into multiple images of the quasar. A number of quasars clearly show evidence of such **gravitational lensing**.

CONCLUSIONS

- If observed quasar redshifts are interpreted in terms of a redshift caused by the expansion of the universe, then Hubble's law can be used to find their distances. The computed distances are large. The luminosity inferred shows the quasars to have luminosities 10–100 times that of the Milky Way. Furthermore, variations over times as short as a few hours show the emitting regions to be small, only a few light-hours in size.

- The most generally accepted model of a quasar is the nucleus of a young galaxy in which there is a massive black hole surrounded by an accretion disc.

- A unified model that explains the variety of radio-emitting objects is given in Figure 22–25, in which a central energy source is surrounded by a thick ring of dust. The type of radio object observed is then determined by the angle from which the observer sees the disc.

SUMMARY QUESTIONS

1. What are some of the principal types of radio galaxies? Why is it sometimes difficult to associate a radio galaxy with an optical counterpart?

2. What are the observations that distinguish Seyfert galaxies from other galaxies?

3. What are the distinctive observations that distinguish quasars from other objects? (You should include redshift, spectrum, brightness variations.)

4. What are the characteristics of quasars inferred from their observations (distance, luminosity, size, energy source, their true nature)?

5. What are several theories for quasars? Evaluate them on the basis of the observational evidence.

6. What is meant by a gravitational lens? What can astronomers learn from studying lensed quasars?

7. What is the single model that gives a unifying picture of the various radio-emitting galaxies?

APPLYING YOUR KNOWLEDGE

1. Hypothesis: Quasars are at cosmological distances. Question: What evidence have astronomers collected in favor of this hypothesis? What types of evidence do some astronomers present against it?

2. How do astronomers infer the size of a quasar's emitting region from observations of the variability of its brightness?

3. Summarize the evidence that active galaxies contain massive black holes in their nuclei.

4. What effects do uncertainty in the value of the Hubble constant have on our understanding of quasars?

5. Consider four classes of objects: normal galaxies like the Milky Way, Seyfert galaxies, radio galaxies, and quasars. Arrange these objects in order of increasing energy output. Sketch a diagram that shows the relationship.

■ **6.** The following are observations for a fictitious quasar: redshift is 0.6 the speed of light; observed brightness is 10^{-17} that of the Sun. Compute the distance to the quasar and its luminosity in terms both of the Sun and of the Milky Way galaxy.

■ **7.** Suppose astronomers observe hydrogen lines in the spectrum of a quasar to have three components at wavelengths of 5200 Å, 5201 Å, and 5205 Å. The usual wavelength is at 4861 Å. If the three components are due to absorption along the line of sight to the quasar, what is the distance between each of the absorbing clouds? What assumption are you implicitly making in obtaining your answer?

■ **8.** Consider a simplified galaxy having only 10 stars. If F_{12} is the force between stars numbered 1 and 2, write down all the other forces that a computer program must consider in analyzing the movements of stars in a collision.

ANSWERS TO INQUIRIES

22–1. It must be large in order to be detected at such great distances.

22–2. Because all hot bodies emit some radiation at all wavelengths, such a galaxy would emit in the radio, too. However, the amount would be small compared with that in the visual spectral regions.

22–3. As discussed in Chapter 19, charged particles must be accelerated in a magnetic field to speeds near that of light.

22–4. One hypothesis is that it results from a gigantic explosion at the center of the galaxy.

22–5. Broad lines are produced by rapidly moving atoms, some of which are moving toward the observer while others are moving away. The faster they move, the greater the Doppler shifts and the broader the lines.

22–6. It is reminiscent of the jet in NGC 4486 (M87).

22–7. The Hubble law is $V = HD$, so $D = V/H$. Using $V = 0.16c$, we have $D = 0.16(3 \times 10^5)/17$ km/sec/Mly $= 2.8 \times 10^3$ Mly $= 2.8 \times 10^9$ ly.

22–8. They must be small. In the example given, the size would be only 1/100 of a light-year. Because α Cen is 4.3 light-years away, the size is about 0.04 light-year. Expressed in other time units, we have a size of about $365/100 = 3.7$ light-days $= 90$ light-hours $= 5400$ light-minutes. Because light takes 8 minutes to travel 1 AU, the size of the quasar's emitting region will be roughly $5400/8 = 675$ AU.

22–9. We observe the objects as they were when the light was emitted billions of years ago. Thus we see the objects as they were when they were young.

23

THE ORIGIN
AND EVOLUTION OF
THE UNIVERSE

The astronomers said:
"Give us matter, and a little motion, and we will construct the universe."

RALPH WALDO EMERSON

The universe is everything. As scientists, we cannot deal with everything but only those things we can detect and measure. For this reason, in this chapter we will be concerned primarily with the observable universe, because reasonable scientific models must be based on observation.

Consider the following idea that has received much discussion, especially in popular literature. Some have suggested that black holes are a kind of connecting bridge between universes and that, if a black hole is rotating, it might be possible to use it to travel from our universe to another universe (or to travel to some other time or place in our own universe).

But if we remember that the only possible meaning we can give to the word "universe" is what we can *detect* and *measure*, then it is impossible to say what we mean by *another* universe. If we cannot measure it, in what sense does it have any meaning? Moreover, such suggestions lead to ghastly paradoxes. For example, if we could use a black hole to travel backward in time, we might be able to alter events that have already happened. If such things were possible, there would be no way in which we could make sense out of our surroundings. (For example, you cannot meet your parents at a time before you were born!) But we can make sense out of them, because our universe obeys certain guiding principles of cause and effect. By keeping in mind that we are ultimately concerned with that which we can perceive and measure, we derive a principle to use in distinguishing between fact and fantasy. We should always remember that the purpose of science is to predict observations; that is, to answer the question: What will we see?

On the other hand, when we expand our horizons to include the entire universe, physics and astronomy can become highly speculative. This is inevitable, because at the limits of our telescopic view, objects are faint and data are scarce, so there are few observational constraints on the many theoretical possibilities. But speculations are fascinating (and fun!), and we should consider them, because they might lead us to new and powerful ways of looking at ourselves and the cosmos.

23.1

THE EXPANSION OF THE UNIVERSE

Astronomers take the concept of an expanding universe for granted. In this section, we see what the evidence is and the consequences of it.

THE VELOCITY-DISTANCE RELATIONSHIP

In Chapter 21 we saw that Hubble found a linear (straight-line) relationship between the velocity at which a galaxy moves away from the Earth and its distance. The **Hubble law**, illustrated in **Figure 23–1** can be expressed as

$$V = Hd,$$

where the constant of proportionality between the velocity V and the distance d is designated by the letter H, called the **Hubble constant**.

To determine the numerical value of H, we must measure the distances to galaxies using all the ideas discussed in Chapter 21 that are applicable to galaxies, and which were summarized in the distance pyramid of Figure 21–12. Such a project was undertaken at Mt. Palomar by astronomers Allan Sandage and Gustav Tam-

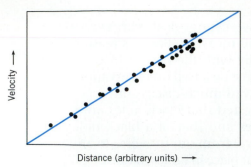

FIGURE 23–1. The Hubble velocity-distance relationship.

mann using the 200-inch telescope, which was then the largest telescope in the world. After many years of work obtaining velocities and distances, they derived a value for H of about 15 km/sec for each million light-years—that is, for each million light-years of its distance from us, a galaxy recedes by an additional 15 km/sec. For example, a galaxy 1 million light-years from Earth would recede at 15 km/sec, while one at 2 million light-years would recede at 30 km/sec, and so on.

Many astronomers believe that research from the last few years shows this value to be too low, and they suggest a figure somewhat higher, perhaps 25 or 30 km/sec per million light-years, a position that has long been advocated by University of Texas astronomer Gerard de-Vaucouleurs. While the most recent work by Sandage using the Hubble Space Telescope has reaffirmed his earlier, lower value, the subject is still highly controversial.

What difference does it make if the value of the Hubble constant is 15 or 30 km/sec/Mly? The numerical value of H is important not only because its size tells us the distances to the farthest objects in the universe, but also because it is a measure of the *age* of the universe. The association of the Hubble constant with age is not difficult to see. We again write the Hubble law:

$$V = Hd.$$

Furthermore, we also know that in the absence of deceleration the distance an object moves depends on its velocity and the length of time: $d = Vt$. Solving this for velocity, we find

$$V = \frac{1}{t}d.$$

These two equations for the velocity *must* be identical. For that to be the case, the Hubble constant must be equal to $1/t$. If H is 15 km/sec per million light-years, then the universe is about 20 billion (2×10^{10}) years old. However, if H is 30 km/sec/Mly, the universe is only half as old.

YOU SHOULD DO DISCOVERY 23–1, THE AGE OF THE UNIVERSE, AT THIS TIME.

In deriving the age of the universe from the Hubble constant, we assume the expansion velocity has been constant. However, astronomers have two reasons to believe that the expansion rate decreases with time. First, the mutual gravitational attraction of every body on every other body should slow the expansion. Second, astronomers have made some observations that indicate a possible slowing. The uncertainties in the measurements are so large, however, that a definitive statement cannot yet be made. If the expansion is slowing, then the true age of the universe would be somewhat less than the value indicated by H. The value of H therefore determines an upper limit to the age of the universe.

Inquiry 23–1 The distance of the Hercules Cluster of galaxies is about 300 million (3×10^8) light-years. At what velocity is it traveling away from us?

Inquiry 23–2 The redshift of the quasar 3C273 is about 16% the velocity of light. Approximately how far away is it? (Assume H to be the value proposed by Sandage and Tammann.)

Inquiry 23–3 If H gets smaller as the universe becomes older, would the value of H as measured from very distant galaxies be *larger* or *smaller* than that determined from nearby galaxies?

The galaxies in the Local Group move like bees in a swarm. Just as each bee has its own motion relative to every other one, even though the swarm itself moves as one object, each of the Local Group galaxies has its own motion relative to every other galaxy. The relative motion of each individual galaxy is greater than the motion expected from the Hubble law. For example, M31, at a distance of 2.2 million ly, would be expected from Hubble's law to be moving away from us at about 30 km/sec. However, the galaxy's own motion within the Local Group, which is caused by the gravitational forces of its total mass, causes M31 to approach us at a speed of 275 km/sec. Galaxies outside our own local gravitational cluster *are* all receding from us, and it is an immediate consequence of this observation that we conclude that we live in an *expanding* universe. The only way to escape this conclusion would be to find an acceptable alternative interpretation of the observed redshift, and to date no one has succeeded in doing so.

YOU SHOULD DO DISCOVERY 23–2, THE EXPANSION OF THE UNIVERSE, AT THIS TIME.

CONSEQUENCES OF THE HUBBLE LAW

Because we see galaxies receding from us in every direction, it is tempting to conclude that we are in the center of the universe. This view is incorrect, and it is not difficult to show that any observer located anywhere in an expanding universe would see the same thing. A simple analogy illustrates this point. Consider galaxies to be like raisins in a rising loaf of bread (**Figure 23–2a**). Assume that as the bread rises, it expands uniformly; thus each raisin gets farther away from every other raisin in the loaf. After one hour, all dimensions have doubled (Figure 23–2b). The distances through which the raisins have moved during the hour, and their speeds, are shown in Figure 23–2c. You can readily see that the greater the distance between *any* two raisins, the more rapidly they will move away from each other. No matter which raisin you examine, the same velocity-distance relation will be observed. No true center can be defined from the observed expansion. The same holds true for galaxies in an expanding universe.

An additional consequence of the universe's expansion is that when we observe galaxies in space, we do not see a snapshot of the universe as it actually is now. The light we observe from distant galaxies left those objects millions or billions of years ago, so we are seeing them *where* and *as* they were then. To answer the question "Where is everything now?" is more complicated and requires a careful discussion of the geometry of space, which we will do later in this chapter.

Bear in mind that astronomers work under a fundamental limitation—we cannot see the whole universe!

There is a twofold reason for this. First, present technology allows us to detect objects only to a certain level of faintness (and the farther away, the fainter the objects are). Perhaps future space telescopes and refinements of technology will allow us to penetrate more deeply into space, but there will always be some limit. Second, and more fundamental, is the limitation imposed because light itself travels at a finite speed—we can only observe those parts of the universe from which light *has had time to reach us*. Clearly, we can never see the entire universe, although it is true that as the universe ages we are able to observe an ever-increasing fraction of the whole. But this takes place over billions of years and is of no help to us if we want to try to understand our universe within our lifetimes.

23.2 THE BIG BANG

If the distances between the galaxies increase with time, then in the past the galaxies must have been closer together. A consequence of this expansion is that the density of galaxies was higher, and the universe was more compressed in the past than it is now. Carrying this further back in time, we are forced to conclude that approximately 15 to 20 billion years ago all the matter of the universe was packed together into a state of astonishingly high temperature, pressure, and density.

Such extreme conditions could never be created in a laboratory, yet theory has enabled us to reconstruct a picture of what these conditions were like. The temperature once must have exceeded trillions of degrees and the density hundreds of trillions of grams per cubic centimeter. This hot, dense state is often referred to as the **primeval fireball**. It contained radiation in all forms, along with matter in the form of ordinary elementary particles such as photons and neutrinos, and a number of more exotic ones. As we will see, the particles we usually consider to make up ordinary matter—the familiar electrons, protons, and neutrons—formed directly from the energy in the fireball.

We hypothesize that the primeval fireball expanded outward rapidly in what we now call the **big bang**. The big bang was not an explosion in the usual sense of a loud noise. Furthermore, the universe is not expanding *into* something as usually happens with an explosion. The big bang did not happen somewhere; it happened everywhere. Referring back to the raisin bread analogy: just as the raisins are carried along with the overall expansion, galaxies are carried along with the general expansion.

The sequence of events that took place in the early moments of the big bang followed each other with ex-

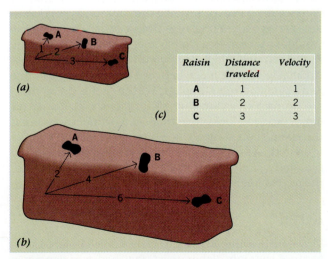

Raisin	Distance traveled	Velocity
A	1	1
B	2	2
C	3	3

FIGURE 23–2. An analogy to the expanding universe of galaxies: raisins in a loaf of bread. (*a*) Initial locations of raisins. (*b*) Positions after the loaf has risen for one hour. (*c*) Table of distances traveled and the velocities of raisins.

treme rapidity. Nevertheless, we are able to detect the effects of some of its stages in what we observe around us today.

The big bang is a *theory*, a hypothesis that has strong observational support. Our current state of knowledge is imperfect, but while questions and problems abound, they do not make the overall structure of the theory invalid. The history of science tells us that eventually the big-bang theory *as we know it today* will be replaced by something better. That does not mean that *all* our current ideas are wrong. Remember that while Newtonian physics was replaced by Einsteinian physics, Newton's ideas are still correct when applied to their appropriate realm of nature. The big-bang theory, too, may always be applicable in an appropriate realm. We examine the big-bang theory, and its verifying observations, next.

THE EARLY MOMENTS OF THE UNIVERSE

Discussion of the earliest moments of the universe stretches the mind as no other subject can. Unfortunately, there is nothing intuitive about it, and nothing in your previous experience can help you visualize it. The topic is extremely complex and mathematical. How, then, can we discuss these early times in a brief, introductory chapter? What we will do is present a coherent picture, along with some of the ideas and terminology used in modern discussions of the earliest times in the universe. While some details are presented for the sake of completeness, try to concentrate on the big picture so that you might gain the general flavor of the subject.

In its earliest moments, the universe was filled with high-energy radiation and material particles. The temperature was in excess of 10^{32} K. Two opposing processes, called **annihilation** and **pair production** (indicated schematically in **Figure 23–3**) were taking place. If a particle of matter collides with its antiparticle (which is, of course, antimatter), the particles annihilate each other, giving off a burst of radiation. The amount of energy produced is given by Einstein's equation $E = Mc^2$.

The equation is equally valid for the reverse process of changing energy into matter via pair production. During pair production, photons of sufficiently high energy (gamma rays) interact and create material particles and their corresponding antiparticles. The photons will have a sufficiently high energy if the temperature is high enough. For example, the temperature required to produce a proton is $10,889 \times 10^9$ K, while it is "only" 5.9×10^9 K for the less massive electron. The kind of particle formed depends critically on the temperature; the greater the temperature, the more massive the particle formed.

To understand what comes next, we must briefly discuss the **four fundamental forces** in the universe. We

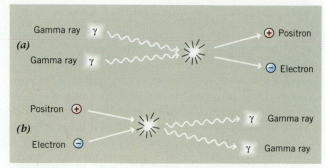

FIGURE 23–3. (*a*) Pair production. (*b*) Annihilation.

have already looked in some detail at two of them: the gravitational and the electromagnetic forces. While these forces act over long distances, they are not particularly strong. The **strong nuclear force** is a strongly attractive force that acts only over short distances and holds the nucleus of the atom together. At some 100 times the strength of the electromagnetic force, it is the force that brings particles together for fusion once they have overcome the repulsive electromagnetic force of the positively charged particles. The **weak nuclear force** comes into play only over distances less than the size of the nucleus and is responsible for radioactivity and the production of neutrinos. It is some 1000 times weaker than the electromagnetic force.

Theory suggests that prior to about 10^{-43} seconds after the big bang, the four fundamental forces behaved as a single unified force. All the particles present would have been massless particles, like the photon. These would have been created and annihilated in interactions with similar particles, just like the gamma rays and photons that produced positrons and electrons. The universe would have been in a simple and symmetrical state due to the extremely high temperature that prevailed then. An analogy can be made between this symmetrical state and a cloud of water droplets. Within such a cloud, there are no preferred directions. No matter which way we look, the cloud looks the same—it is symmetrical. But if the cloud cools somewhat, a snowflake may crystallize from it, thus breaking the cloud's symmetry. The idea of symmetry, and the breaking of symmetry, is crucial to the development of the universe.

GUTs

This state of symmetry can only be studied theoretically, because the only "laboratory" that has ever generated the high temperature required for it to exist was in the early moments of the big bang itself. Therefore it is difficult to test these ideas experimentally. At present, the ideas are highly speculative, although some scientists suspect that we may find that the mathematics of high-temperature physics allows little leeway for alternative possibilities.

At 10^{-43} second after the big bang, the highly symmetrical state of the universe was broken; just as the crystallization of a snowflake breaks the symmetry of a cloud of water vapor, the gravitational force "froze out" and separated itself from the other three forces. A new state of symmetry came into being, and from this moment on, gravity acted independently of the other forces. The strong, weak, and electromagnetic forces remained unified for a time as something we call a grand unified force. Physicists call this era the **GUT**, or **Grand Unified Theory**, era.

During this brief era, all material particles looked similar. Quarks, the fundamental building blocks that make up the protons and neutrons we are familiar with, were freely converted into particles such as electrons in processes that are so uncommon today they've never been observed. These processes (and their reverse processes) involved hypothetical particles that physicists often refer to as X particles (not to be confused with X-rays), which are cousins of some other particles that actually are observed. These hypothetical particles would require such enormous amounts of energy for their creation that even the most powerful atom-smashing machines fall many orders of magnitude short of being able to create them.

Nevertheless, it may be possible to study some of the processes of that era directly. One of the predictions of the grand unified theories is that protons are unstable particles and will eventually decay. Although the lifetime of a proton is enormously long—estimates are that a proton will live an average of perhaps 10^{34} years before decaying—it is not infinite. Given a sufficiently large mass of protons (the hydrogen nuclei in a large volume of water will do fine) it should be possible to detect occasional decays of a few protons within the mass. For example, a collection of 10^{34} protons should experience one decay per year, on the average. Efforts have been under way for many years to test the prediction of proton decay, but none has been seen. Perhaps we will one day have experimental evidence of the processes that are thought to have taken place during this remote time.

At 10^{-35} second, when the temperature had cooled to about 10^{28} K, another momentous event took place. The strong nuclear force "froze out" from the GUT force, while the weak and electromagnetic forces remained united as a single entity we call the **electroweak force**. From this point onward, low-mass particles such as electrons no longer felt the effects of the strong nuclear force, although quarks continued to be affected by elec-

tromagnetism. When the forces separated, a large amount of energy was released, just as energy is released when water changes its phase to become ice. The energy released by the breaking of the current state of symmetry caused the universe to enter a short "inflationary" era during which it expanded rapidly; the inflationary era is further discussed in Section 23.7.

We have considerable experimental evidence about the physics that dominated this era. The largest particle accelerators now operating are capable of energies high enough to produce the particles of the electroweak force. These particles have been detected, and the details of their physical interactions with matter are being thoroughly investigated.

At 10^{-10} second, the temperature had dropped to 10^{15} K, and the final "freezing-out" took place when the electromagnetic and weak nuclear forces separated. At this time, photons were still so energetic that they created quarks and antiquarks by pair production processes. A little later, after the universe had cooled a bit and the photons were less energetic, only electrons and their corresponding antimatter particles, positrons, could be created. The quarks then combined among themselves to form protons, neutrons, and mesons, which are particles having a mass between that of the massive proton and the low-mass electron.

During the entire process of cooling and expansion of the universe that we have discussed (which took 10^{-10} second), the creation of particles through pair production was balanced by the opposite process of annihilation, in which matter and antimatter particles combine to create radiant energy. The equilibrium between these two processes at each stage determined the balance between particles and radiation in the early moments of the universe. Before the universe was a tenth of a second old, the primeval mixture of matter and photons had cooled to a temperature of roughly 100,000 million degrees (10^{11} K) and a density over 100 million times that of water.

Because matter and radiation were in equilibrium, the spectrum of the radiation was that of a high-temperature blackbody.

MATTER FINALLY DOMINATES OVER RADIATION

By the time it was 1 second old, the expanding universe had cooled to 10^{10} K, and at the end of 3 minutes, the temperature was approximately 10^9 K. At this point, a modest amount of matter in the form of protons and electrons actually managed to survive for a time without conversion back to energy. Helium nuclei formed, and as the temperature cooled further, protons and helium nuclei combined with electrons to form hydrogen and

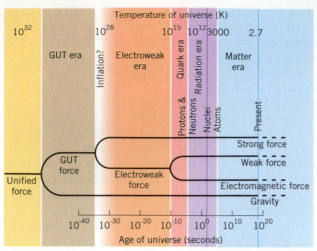

FIGURE 23–4. The development of the early universe. At first, all forces of nature were unified. At 10^{-43} second, the gravitational and GUT forces split. At 10^{-35} second, the strong and electroweak forces split, and an inflationary era began. At 10^{-10} second the electromagnetic and weak forces split, leaving the four forces as we know them today. At 10^{-3} second, quarks combined to make protons and neutrons, and at 3 minutes, nuclei of deuterium and helium formed. At 500,000 years the nuclei were able to capture electrons to form atoms.

helium atoms. Eventually there was a sufficient amount of atomic matter that we may begin to think of the universe as dominated by matter rather than radiation. This transition to a material universe occurred when the universe was approximately 500,000 years old, at a temperature of about 3000 K. **Figure 23–4** indicates the trend of the early universe with the progression of time and summarizes the discussion up to now.

It is worth noting that during the first few minutes in the life of the universe, the temperatures were high enough for hydrogen fusion to take place, creating deuterium and helium. The same reactions we studied earlier in connection with the centers of stars occurred then. But the universe was evolving rapidly at this time, and it quickly expanded and cooled to densities and temperatures below the necessary levels for fusion of any of the heavier elements to take place. There was an additional complication in that the fusion process comes to a halt at an atomic weight of 5, because all nuclei having a total of 5 protons and neutrons are unstable. Shortly after such a nucleus is assembled, it immediately decays. Special physical conditions are required for fusion to synthesize other elements of higher atomic weight; such conditions were not achieved in the universe until much later, with the formation of stars.

Inquiry 23–4 In stars, how is the "chasm" at atomic weight 5 overcome and how are heavier nuclei created? (Hint: Review the triple-alpha process discussed in Chapter 18.)

The heavier elements of the periodic table (and in particular the component atoms of the hydrocarbons so important for life) must have been created by synthesis of elements (nucleosynthesis) in the centers of stars. We do not merely *assume* that the early stages of the universe were devoid of heavy elements; the evolution of matter out of a photon-dominated era ensures that it was so.

Due to the chasm at atomic weight 5, and because the universe had only a short time interval in which the temperature and density were right to produce deuterium, the present abundance of deuterium in the universe is an important clue to the universe's early conditions. The abundance of light elements such as deuterium, helium, and lithium depends on the density of the universe at the time they were formed. Thus, as shown in **Figure 23–5**, the higher the density, the lower the abundance of deuterium, because higher density causes deuterium to be destroyed easily by nuclear reactions. The effects of density on the abundances of different elemental isotopes differ from element to element, as shown in Figure 23–5. Specific proportions of hydrogen, helium, and deuterium in the universe are predicted by the detailed calculations of the big bang. For example, the big-bang theory predicts that some 28% of the mass of the universe should be helium. Observations of the oldest stars and of the gases in nebulae show the actual helium abundance to be 25–30%, in excellent agreement with the predicted value.

The Hubble Space Telescope observed interstellar deuterium in the spectrum of the bright star Capella (**Figure 23–6**). From these data, astronomers found there to be merely 15 deuterium atoms for every million hydrogen atoms, in agreement with predictions of the big-bang theory. The agreement between the expected and observed elemental abundances provides some of the strongest evidence that the big-bang scenario sketched in the preceding paragraphs actually took place pretty much as described.

At the time when matter finally dominated over radiation (at about 500,000 years), the temperature had dropped to about 3000 K, low enough that ionized hydrogen (protons) and electrons were able to combine to form neutral hydrogen. Because neutral hydrogen is less effective in absorbing radiation than is the hydrogen ion, from this time onward radiation and matter stopped interacting and ceased to be in equilibrium with each

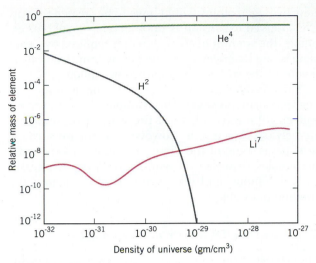

FIGURE 23–5. Dependence of the abundances of some light elements on the density of the universe as predicted by the big-bang theory.

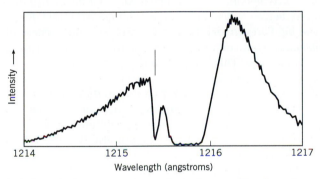

FIGURE 23–6. An absorption line of interstellar deuterium observed by the Hubble Space Telescope in the spectrum of the bright star Capella. The broad line is hydrogen.

other. We often speak of this event as the *decoupling* of radiation and matter. What this means is that radiation produced at this time would no longer be absorbed and scattered by the hydrogen, but could continue to travel throughout the entire expanding universe. In other words, radiation characteristic of the temperature at this time (3000 K) filled the universe. The universe was awash in radiation whose spectrum was that of a 3000-K blackbody.

The ideas we have just discussed allow us to make a number of important predictions based on the conditions at the time of the decoupling of radiation and matter. First, the universe as we see it today should be filled with the radiation that was present when the decoupling occurred. Second, this radiation should be coming

at us equally from all directions because it filled the entire universe. Third, observers on Earth should be able to detect this radiation; but, just as for distant galaxies, it should be considerably redshifted. Knowing that this radiation came from a time when the age of the universe was some 500,000 years, astronomers predict that the radiation should *appear* to come from a body 1000 times cooler; that is 3 K. We will check our predictions later, after we build up more of a background.

Inquiry 23–5 At what wavelength should 3-K radiation from decoupling be observable? What part of the electromagnetic spectrum is this?

ANTIMATTER IN THE UNIVERSE

Since its discovery in 1931, antimatter has been understood to stand in a symmetrical relationship with matter. There is no obvious reason why matter should be preferred over antimatter, yet in our universe this appears to be the case.

Inquiry 23–6 Swedish physicist Hannes Alfven has suggested that a simple way to explain the big bang is to think of the universe as starting out as an equal mixture of matter and antimatter. What observations should scientists make to test this theory?

Except for indirect evidence of some short-lived antimatter in the central regions of our own galaxy (see Section 20.3), no signs of significant amounts of antimatter have been found in the universe (although small amounts have been produced in particle accelerators on Earth). Even if large amounts do exist, in some configurations it would be difficult to detect. For example, an entire galaxy made up of antimatter would emit radiation indistinguishable from that emitted by stars in a matter galaxy. However, there are no good reasons at present for thinking that isolated systems of pure antimatter exist.

Recent developments in our understanding of the big bang may offer a solution to the question of why there is more matter than antimatter. Recall the X particles hypothesized to have existed during the GUT era of the universe. The grand unified theory predicts that the decay of these X particles into matter and antimatter is not exactly symmetrical. In particular, when the last X particles were decaying into the particles we see today, according to the calculations, for every billion and one quarks produced only a billion antiquarks would have been produced. After the antiquarks had been annihilated by quarks, there would remain an excess of quarks; that is, *matter*, which would persist to this day. If this idea is upheld, it would provide a neat solution to what has been for many years a vexing problem in cosmology.

23.3
AN ALTERNATIVE TO THE BIG-BANG THEORY

Some astronomers have objected to the big bang theory, partly because it seems to suggest that there was a special "creation" of the universe at some specific time in the past. Furthermore, our minds generally think that if there is a beginning, eventually there will be an end. The idea of an end to the universe is metaphysically unsatisfying to many people, because it leaves dangling such questions as "What was there before there was a universe? What will there be after the end of the universe?" Such questions may or may not be meaningful, but nevertheless a good deal of effort has been expended in constructing models for the universe that do not have a specific beginning or end.

THE STEADY-STATE HYPOTHESIS

One of the cleverest of these theories was put forth in the 1950s by astrophysicists Fred Hoyle, Hermann Bondi, and Thomas Gold, who suggested that the universe did not have a specific origin in time but has always existed and is infinitely old. This theory is known as the **steady-state hypothesis**. In a steady-state universe, the density will be constant. But the Hubble law tells us the universe is expanding. The only way an expanding universe can maintain a constant density is if new matter is being continuously created. Hoyle, Bondi, and Gold proposed that new matter was being created, not in large amounts, but at a rate just sufficient to keep the density of the expanding universe constant. New hydrogen atoms would spontaneously be created in space, out of "nothing," so to speak.

Without a big bang, what was making the universe expand? The spontaneous creation of matter would apply a pressure to the galaxies and cause an expansion.

We cannot reject this hypothesis out of hand, because the spontaneous creation of single atoms is surely no more absurd than the spontaneous creation of a whole universe, which is what the big-bang theory requires. Because each hypothesis will make different predictions of observable events, the hypotheses can be tested against each other. Because the time scale for the evolution of the universe is so long, the rate at which

new hydrogen atoms would have to be created to cause the observed expansion is unobservably small. For example, the amount of created matter required by steady state is one hydrogen atom every few hundred years in a volume the size of the Empire State Building.

The steady-state hypothesis does have observable consequences, however. For example, if the universe is indeed infinitely old, then there must be some extremely old objects in it. However, looking for such objects poses some of the same difficulties as looking for black holes, because they would have long since ceased to be luminous, and it is extremely difficult to detect nonluminous objects.

THE SPATIAL DENSITY OF GALAXIES

There is one easily tested difference between the two theories. In the steady-state theory, the present universe looks nearly identical to the universe of long ago. The expansion of the system is uniform and smooth, and new galaxies are continually being formed from the hydrogen that is created in space, in such a way that the *average* distance between galaxies does not change with time. But if the big-bang theory is correct, galaxies and clusters of galaxies were certainly closer together in the past than they are now. This suggests that we should examine the universe as it was in the distant past by observing the most distant galaxies. **Figure 23–7** compares these predictions.

The two competing theories predict not only that the density of galaxies in space would be different as we look to greater and greater distances, but also that the velocity-distance relationship would be different, as shown in **Figure 23–8**. In a steady-state universe, the velocity-distance relationship should be a straight line, as shown. This is so because the expansion of the universe would

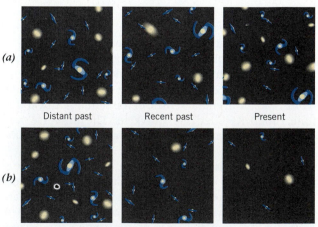

(a)

Distant past   Recent past   Present

(b)

FIGURE 23–7. Differences in galaxy counts between the *(a)* steady-state and *(b)* big-bang theories.

not slow down, because its speed would be constant through all time. Under a big-bang cosmology, the universe should have been expanding more rapidly in the past than now, so the slope of the Hubble relation would have been greater. Unfortunately, the two theories predict significant differences in the velocity-distance relation only at great distances, where objects are faint. In fact, the test suggested above is not really practical for optical telescopes, but because radio telescopes can penetrate considerably deeper into space, counts of radio galaxies have yielded much information about the variation in the number of galaxies with distance.

Sir Martin Ryle and other radio astronomers have made such counts, which indicate that the farther back one looks into the past, the greater is the density of galaxies. However, the radio observations were difficult and the interpretation of them was complicated and

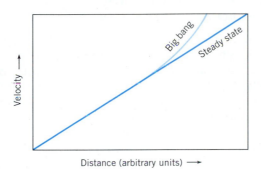

FIGURE 23–8. The velocity-distance relationship in various models of the universe.

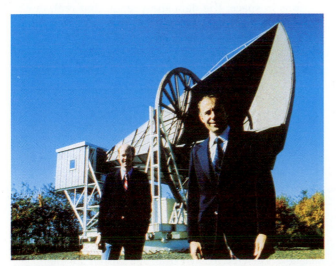

FIGURE 23–9. Penzias and Wilson, along with the radio receiver with which they discovered the 3-K background radiation.

subtle, and as a result many astronomers were not convinced that the counts of faint radio sources were capable of distinguishing between different cosmological theories. For this reason, in the middle 1960s the big-bang versus steady-state controversy was still very much alive.

Inquiry 23–7 The steady-state theory proposes that the universe should look the same in all places at all times. However, if the redshift of the quasars is entirely due to the Doppler shift of the expansion of the universe, then there are no nearby quasars, only distant ones. Is this consistent with the steady-state theory?

23.4
THE COSMIC BACKGROUND RADIATION

Sometimes scientists are fortunate enough to make a single discovery that provides critical evidence in deciding between competing hypotheses. We now examine such evidence in favor of a big bang cosmology.

THE FIRST DETECTION

One event that played a decisive role in distinguishing between the steady-state and big-bang models arrived by surprise in 1964, as a serendipitous discovery by two radio astronomers engaged in an unrelated area of research. Arno Penzias and Robert Wilson, working for Bell Telephone Laboratories, were using the 20-foot horn antenna shown in **Figure 23–9** to search for low-intensity radio signals from the sky. They wanted to push the sensitivity of the system as far as possible and look for extremely faint cosmic radio sources. To this end, they devoted a great deal of attention to reducing the receiver's internal noise (produced because of their equipment's high temperature), which obscured faint signals.

By cooling the equipment, among other techniques, Penzias and Wilson succeeded in reducing the static noise of their instrument to low levels. Tuned to a wavelength of 7.35 cm, they scanned the sky with it. They had not expected to find anything other than a few localized sources, but in fact they found a low-level signal coming from every direction in the sky. This signal had an amazing and entirely unexpected property—the intensity that they recorded was **isotropic**, that is, it was exactly the same no matter what direction in space they looked.

Penzias and Wilson made observations at more than one wavelength to construct the spectrum of this radiation, which, they found, looked like the spectrum of a blackbody. They found the wavelength of the spectrum's peak intensity and used Wien's law to compute the temperature that a body emitting such radiation would have, which turned out to be 3.5 K. It appeared that every point on the celestial sphere was emitting radiation as if it were a body at a temperature of 3.5 K. The radiation they detected is sometimes referred to as the "3-degree background radiation." **Figure 23–10** shows the spectrum of the radiation, along with the uncertainties of the measurements.

Astronomers quickly noted an important consequence of the isotropic nature of the radiation. Because precisely the same signal is received from all directions in space, the radiation cannot be associated with any specific source, either in our galaxy or outside it. It must be a property of the entire universe itself, and viewed in this light it has only one plausible interpretation: it is the remnant radiation expected from the decoupling of matter and radiation some 500,000 years after the big bang. Only in this manner can the isotropic nature of the radiation be explained. In the early epochs of the universe, all space was permeated by the primeval radiation, and that situation remains true today. What Penzias and Wilson accidently found was exactly what the big-bang theory predicted! For this discovery, they were awarded the Nobel Prize in physics in 1978.

Here at last we have the resolution of the debate between the two competing models of the universe. The big-bang theory leads us to expect an observable background of radiation left over from the primeval fireball. The fact that the prediction was made prior to its discovery adds to the weight of its importance. The steady-state theory has no natural way to explain such radiation and has therefore been falsified.

One additional interesting result comes from the observation of the background radiation. As Nobel Prize–winning physicist Steven Weinberg has pointed out in his excellent book *The First Three Minutes*, the fact that the presently observed temperature of the radiation is 1000 times less than the temperature that it had when matter and radiation ceased to be in equilibrium means that the universe has expanded by a factor of 1000 during that time interval. This means that the cosmic "static" that Penzias and Wilson observed was produced when the universe was 1000 times smaller than it is now. That makes it by far the oldest radiation we have ever received, having been emitted long before any of the galaxies or clusters of galaxies were formed.

> YOU SHOULD DO DISCOVERY 23–3,
> OBSERVING THE BACKGROUND
> RADIATION, AT THIS TIME.

THE UNIFORMITY OF THE BACKGROUND RADIATION

How uniform is the background radiation? Can any variations be detected? To test the isotropy of the background radiation, NASA launched the **Cosmic Background Explorer (COBE)** satellite in 1989. COBE was also to determine more accurately the temperature of the background radiation. **Figure 23–11** shows the resulting observed data as squares. From these data, we may compute the temperature of the blackbody that agrees best with the data, and thus derive the temperature of the cosmic background radiation. The resulting temperature is 2.74±0.06 K; the agreement at all frequencies is exceptional.

When the COBE data are examined in different directions in space, the data show small, but real, deviations from perfect isotropy. **Figure 23–12a** shows

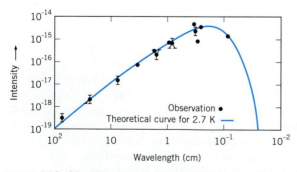

FIGURE 23–10. Spectrum of the 3-K background radiation.

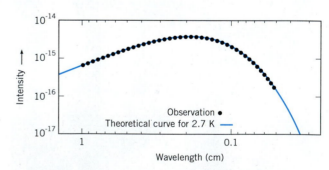

FIGURE 23–11. The spectrum of the 3-K background radiation as determined by COBE. The squares are the observed data points, while the curve is the best fitting blackbody curve.

COBE's direct measurements placed on a map of the Milky Way with the galactic equator along the center line. Color-coded pink regions have temperatures up to 0.003 K lower than average, while blue areas are as much as 0.003 K above average. These variations are not intrinsic to the background but are due to the Earth's motion of some 360 km/sec relative to the radiation. The motion is toward the blue region in the south galactic hemisphere in Figure 23–12*a*. Subtraction of this so-called **dipole anisotropy** produces Figure 23–12*b*, which shows the concentration of microwave emission *from the Milky Way* having temperature variations as small as 0.0003 K. Subtracting this local galactic emission results in Figure 23–12*c*, which shows temperature variations of 1 part in 100,000 for the universe as a whole. Although small, these observed variations are extremely important to our understanding of the formation of galaxies.

Some physicists suggested in 1992 that the fluctuations the COBE science team reported are not due to the structure of the early universe but are caused by gravity waves produced in the inflationary era. The existence of such waves is predicted by Einstein's general theory of relativity. Stay tuned for future developments on this topic!

HOW DID THE GALAXIES FORM?

We have seen that during its earliest phases the universe was largely radiation, but that after a few hundred thousand years matter became the most important component. The isotropy of the background radiation also implies that in its early years the universe was a homogeneous place, free of large variations in temperature and density. However, a smooth early universe poses problems for cosmologists, because virtually all parts of the universe that we examine today exhibit inhomogeneity rather than homogeneity—there are extreme variations in density, temperature, and composition.

Inhomogeneities include galaxies, clusters of galaxies, and large voids. To create them it was necessary for the matter in the universe to be clumpy and then to form condensations at some stage. It is for this reason that the COBE results are so significant; they show that there was some clumping present at the beginning from which galaxies might later form. Because we see galaxies today, a transition from smooth to clumpy clearly did occur (**Figure 23–13**). The required early clumpiness now seems to have been observed by COBE.

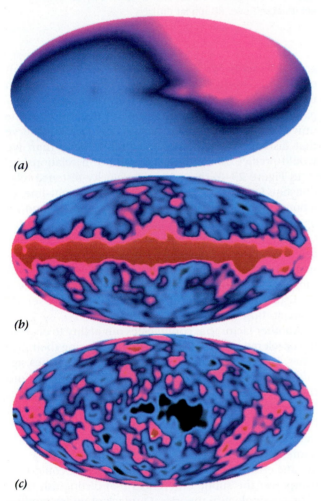

FIGURE 23–12. The variation in the strength of the 3-K radiation over the sky, shown in the coordinate system of the Milky Way. (*a*) The observed data in which blue represents the location to which the Sun is moving relative to the radiation; red is the direction opposite to the motion. (*b*) What remains after subtracting part *a*; the microwave emission from the Milky Way itself. (*c*) What remains after subtracting part *b*; the microwave background from the universe itself.

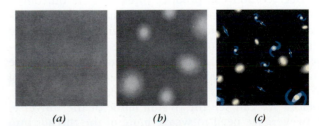

FIGURE 23–13. The formation of galaxies in the early universe. (*a*) Initial uniform distribution of gas. (*b*) Condensations form. (*c*) Condensations collapse and form into galaxies.

23.5

THE FUTURE OF THE UNIVERSE

Until now we have dealt with the past. It is natural to ask what the future holds. The logical first question is whether the expansion will continue forever. If it does, the future is fairly straightforward—the stars that we see now will burn out and die. Some newer generations of stars will form from the interstellar gas and dust between the stars in many galaxies, but it is only a matter of time until stellar evolution runs its course and most of the matter in galaxies is converted into white dwarfs, neutron stars, and black holes.

Strictly speaking, stars will take an infinite amount of time to reach zero temperature after their nuclear fires have died out. The cooling process slows down more and more as the temperature decreases, with each one-degree decrease in temperature taking longer than the previous one. However, it is clear what the eventual "end" state of the universe would be—matter would continue to expand until there was an almost infinite distance between the galaxies, and the galaxies themselves would become systems of hundreds of billions of dead stars, continuing forever in their gravitational orbits, with no potential ever again for luminous life. The universe would become rarefied, cold, and dead.

THE GRAVITATIONAL DECELERATION OF THE UNIVERSE

On the other hand, there is a fundamental force in the universe that fights this expansion—it is none other than gravity itself. Although the galaxies are moving away from each other, each of them attracts all the others with its gravity. The gravitational attraction of all parts of the universe for all the other parts is referred to as the "self-gravity" of the universe, and this self-gravity retards (decelerates) expansion. The expansion becomes slower and slower as the distance between galaxies increases and the force of gravity between galaxies weaker as time goes on (see **Figure 23–14**).

Deceleration can produce three possible outcomes. First, the deceleration could be so small that the universe will expand forever. Second, gravity could cause the expansion to slow and halt as the universe's age approaches infinity. Lastly, the deceleration could cause the expansion to halt and reverse direction. Discovering what will happen is a fundamental goal of cosmologists. Before we can see the overall fabric of the universe, however, we need to weave more of the threads into our story.

In theory, we should be able to observe the deceleration of the universe by looking backward in time (i.e., to

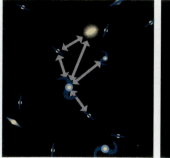

FIGURE 23–14. As the universe expands, the gravitational force between galaxies weakens (as shown by the arrows), yet it doesn't ever disappear entirely.

great distances). We have seen in the raisin bread analogy that the more distant galaxies will move away from us most rapidly, even if the entire universe were expanding uniformly. But if the rate of expansion of the universe is actually *slowing down*, we should observe that distant galaxies have a higher velocity than we would predict on the basis of a uniform relationship.

In Figure 23–8, the expected velocity-distance relationships for a steady state universe (the straight-line relationship) and universes in which a significant amount of deceleration has taken place were illustrated. Unfortunately, it is difficult to observe this deceleration, because we must observe distant galaxies that are extremely faint. The region where the curves diverge is at the working limit of present-day telescopes, and there is no firm consensus about the shape of the observed curve at those distances.

Another factor that complicates our ability to observe the deceleration is that galaxies evolve during their lifetimes. Evolution changes their observed properties significantly over time. At great distances, we look far into the past and see galaxies that are, on the average, *younger* than the nearby galaxies. Suppose, however, that young (distant) galaxies are systematically more luminous than nearby (older) galaxies (as the quasar phenomenon would suggest). This would introduce a systematic error into all distance estimates that could well overpower the subtle differences that distinguish competing cosmological theories. Our lack of understanding of galaxy evolution is perhaps the most important limitation to our understanding of cosmology.

THE AMOUNT OF MASS IN THE UNIVERSE

If we cannot observe deceleration directly, perhaps there is some other observable characteristic of the universe that will provide an indirect measurement of the deceleration and thus an answer to its future. For example, we can easily calculate the amount of mass that the universe would have to contain to bring the expansion to a halt. It works out to be approximately 10^{56} grams. If there is less mass than this, the self-gravity of the universe will not be able to stop its expansion, and it will continue to expand forever. If there is this much mass or more, the expansion will be brought to a halt. The amount of mass required to just halt the expansion is referred to as the **critical mass**.

Astronomers generally express this mass as a density. A critical mass of 10^{56} grams spread throughout the universe corresponds to a **critical density**[1] of about 5×10^{-30} gm/cm^3. (By comparison, the normal density of air in a room is about 10^{-3} gm/cm^3, and the density in the best vacuum that can be attained in an Earth laboratory is about 10^{-16} gm/cm^3.) The typical density of matter in the space between the stars in our own galaxy is about 10^{-26} gm/cm^3, about 2000 times the critical density. Because the density of matter between the galaxies is far less than this, the overall density of matter in the universe is much lower. Such numbers emphasize the emptiness of the universe. If the actual density of the entire universe is greater than the critical density, the expansion will halt.

Astronomers have been diligent in tallying all the mass that they can see. One method uses the abundances of deuterium discussed earlier. From the data presented in Figure 23–6, and the theoretical ideas in Figure 23–5, astronomers calculate that the density of the universe is only about 10% of that necessary to halt the expansion. However, we should not conclude from this that there is insufficient mass in the universe to bring the expansion to a halt, because our survey of the mass in the universe may be quite incomplete.

For example, as we look at distant galaxies, the visible light by which we see them is primarily the light from their brightest stars; yet most of the mass of the galaxy is composed of faint main-sequence stars. We can assume that these distant galaxies are like nearby ones in their stellar content and so estimate the amount of mass they have in the form of faint stars. However, we have said that galaxies may evolve significantly during their lifetimes. As we look to great distances, we see much younger galaxies, and the amount of energy each star emits may not be the same as similar stars in nearby galaxies.

Furthermore, how can we estimate the amount of mass that is not luminous at all? Black holes and other

dead stars are an obvious example of nonluminous material; depending on their age and the rate at which stellar evolution went on during past epochs, distant galaxies may contain a significant amount of mass in the form of black holes.

Inquiry 23–8 Suppose galaxies do indeed contain billion-solar-mass black holes in their centers. Would these black holes provide enough mass to bring the universe's overall density above the critical density? Why?

But it is not just dead stars that contribute to the total mass without contributing luminosity, for, as we have seen, hydrogen gas (the major constituent of the universe) can exist in states that are hard to detect. Cold neutral hydrogen atoms emit 21-cm radiation, but molecular hydrogen is hard to detect. Similarly, rarefied hydrogen gas at a temperature of about 1000 K is difficult to detect, as is hydrogen at 50,000 to 100,000 K. Much undetected gas could exist in external galaxies in these forms.

DARK MATTER

Astronomers do not look for additional mass in the galaxies merely because they hope that the expansion of the universe will come to an end. A persistent puzzle concerning clusters of galaxies suggests that there is indeed more mass to be found. Astronomers studying the motions of galaxies within clusters have found that the range of the individual galaxy velocities is larger than expected from the amount of *visible* mass. The observed velocities are so large that the *observed* amount of mass in the cluster could not prevent it from flying apart. But there are many reasons to think that clusters of galaxies should be stable. In fact, if they were as unstable as their velocities make them appear to be, there would not be any clusters of galaxies at all to be observed. Therefore, it is nearly certain that there is much more mass in these clusters than is observed directly. While often referred to as the "missing mass," it is more reasonably referred to as dark matter. Two questions concerning dark matter are "How much is there?" and "What kind of matter comprises it?"

Some astronomers claim to have detected a faint halo of emission around the large elliptical galaxy, M87, in the Virgo Cluster of galaxies (see Figures 21–2 and 22–8). They calculated that this faint halo, never before photographed, is due to matter at such a high temperature that it is difficult to detect. They also claim that the amount of mass responsible for the faint emission they

---

[1]In this calculation we took the size of the universe to be 2×10^{28} cm, the distance that a photon could have traveled over the age of the universe.

see is significantly more than the mass of the visible galaxy. Furthermore, they believe that if most galaxies prove to have such halos, there could be sufficient mass in the universe to stop its expansion. (Note that these galactic halos are the extended galactic coronas we discussed in Chapter 20, not the halos associated with population II stars.) As of this writing, it is not possible to offer a verdict on this claim, but recall the discussion of the mass of the Milky Way in Chapter 20. There we pointed out that analysis of the rotation curve of the Milky Way indicates that the mass of our system may actually be many times greater than astronomers had traditionally thought.

Other possibilities for additional matter include the idea that intergalactic matter exists in the form of a low-density distribution of gas *between* galaxies. While it is difficult to detect this gas directly because its density is low, cluster X-ray observations and the shapes of the lobes of radio galaxies moving through space (discussed in Chapter 22) show the presence of an intergalactic medium. If an intergalactic medium actually fills the enormous space between galaxies, it could represent a large amount of mass. The discovery of intergalactic clouds of hydrogen makes this a particularly attractive possibility. There may also be a large number of low-mass galaxies that are too small to detect.

Some scientists have hypothesized that neutrinos may not be massless particles but instead may have a small mass that might help resolve the dark matter problem. Although the mass of each neutrino must be extremely small, there should be a huge number of neutrinos in the universe left over from shortly after the big bang. In fact, it is estimated that there should be about 10^{10} neutrinos for each proton. If this is so, and if the neutrino has a small mass, the amount of mass in the form of neutrinos could be a substantial fraction of the universe's total mass. Such mass would tend to collect in the intergalactic space of large clusters of galaxies and could indeed be the "missing mass" for which astronomers have been searching so long.

Supernova SN 1987A in the Large Magellanic Cloud has provided us with data to check this hypothesis of neutrino mass. As we have seen, the collapsing core of the dying star emitted a burst of neutrinos that was caught by neutrino detectors here on Earth. The time of arrival of the pulse gave an *upper limit* to neutrino mass of approximately 20-millionths of the mass of the electron. If the mass of a neutrino must indeed be less than this value, then it would appear that neutrinos by themselves are insufficient to provide all the missing mass. Thus the hypothesis of neutrinos having mass is not verified.

At this point, we do not know whether the density of the universe exceeds the critical density. Although the

evidence is weak, some scientists conclude that we live in a universe that is close to the critical density. After adding more background information, we will return to a new aspect of this discussion in Section 23.7.

AFTER EXPANSION—WHAT?

There might be exactly enough mass in the universe to bring the expansion to a halt. In other words, although the universe would expand forever, as it did its expansion velocity would approach zero more and more closely as time approached infinity. However, it is unlikely that this *exact* amount of mass is present in the universe; it is more likely that either more or less mass than the critical mass exists. We have examined the consequences of less mass; now let us consider what might happen if there is more than the minimum amount needed to stop the expansion.

Let us suppose that, perhaps 100 billion (10^{11}) years from now, the universe stops expanding. Gravity, of course, still remains and acts on all matter. Mutual gravitational attraction would cause the universe to begin to contract. This contraction would proceed slowly at first; however, it would speed up as the decreasing distance between the galaxies increased the gravitational attraction, and hence the acceleration, between them. The speed with which the universe contracted would increase more and more, and the universe would collapse back in on itself. Our expanding universe could, therefore, be followed by an implosion (**Figure 23–15**), and its end would be marked by a "big crunch" that would be symmetrical to the big bang.

We can imagine certain aspects of the final stages of the implosion as the reverse of the explosion phase. Steven Weinberg points out that if the universe were to contract to about one-hundredth its present size, the radiation density in space would be so overwhelming that the night sky would be about as bright as the daytime sky is now. Less than 100 million years after that, the temperature of the universe would have risen so high

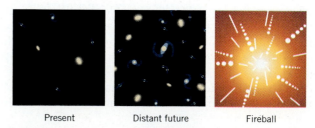

Present   Distant future   Fireball

FIGURE 23–15. The universe might not expand forever, but instead there might be enough self-gravity to cause it to contract again in the distant future. In this case, the universe would eventually collapse into a fiery "big crunch."

that molecules would begin to dissociate, and a few million years after that, atoms would be broken up into electrons and nuclei. Clearly, everything that had been built and created by all the civilizations of the universe would be destroyed.

Then things would begin to happen more rapidly. During the next 700,000 years, according to Weinberg, the cosmic temperature would rise to 10 million kelvins, and the objects of the universe—the galaxies, with their component stars, dust, and gas—would all be broken up into a continuous distribution of radiation and plasma. In only 22 more days, the temperature would increase another thousandfold, and nuclei would be broken up into protons and neutrons. Then, very quickly, the universe would return to a fantastic state of high temperature and density that can probably best be described as another primeval fireball.

A CYCLICAL UNIVERSE?

We would then arrive at a future state as extreme as the one in which the universe began, and it is almost impossible to say what would come after that. It has been suggested that from this "future fireball" the universe could simply rebound into another expanding phase with an entirely new set of galaxies, stars, planets, and, presumably, life-forms.

Such an **oscillating universe** does not single out any particular instant as a creation. In this picture, the universe does not need to be created, because it has always existed. It is infinitely old and will continue infinitely far into the future, periodically going through a fireball stage. For some people, such a situation is more logically satisfying. There could be long-term changes that might persist through the fireball stages, as opposed to the universe "starting from scratch" with each new explosion. In other words, it is conceivable that there is an enormously long-term evolution in progress as well, one that would be difficult for us to gain information about.

We have envisioned the cyclical universe as being in almost perpetual motion. Will there be losses of energy due to friction in the expansion and contraction process? Such frictional effects and other exotic considerations suggested by some theorists raise the possibility that there might be a small change in the ratio of photons to massive particles with each cycle, and this could change the nature of the universe's evolution. It would also suggest that there might not be an infinite number of cycles.

We are projecting our theories far into the future. We must admit that we are now at the stage where such conjectures are largely speculation. In our study of cosmology, we have made a number of important tacit assumptions. For example, we have assumed that the universe is **homogeneous** and **isotropic**. Homogeneity means that on large scales the universe appears the same everywhere. Isotropy in this context means that velocities are all away from us. Without these assumptions, our extrapolation to the future could become incorrect well before a future fireball develops. We have also assumed **universality**—that the fundamental laws of physics are valid throughout the universe, and that the constants of nature, such as those governing the strength of gravity and the speed of light, have not changed with time. If they have changed, the scenario we have sketched would be greatly altered.

How valid are these assumptions? Observations of the distribution of galaxies presented in Chapter 21 have shown that *over large scales* (hundreds of millions of light years), the universe is homogeneous and the motions are isotropic. The COBE results also show that to be the case. All our measurements of the fundamental constants of nature show them to be constant to a high level of accuracy. In fact, experiments have shown many physical constants to be constant to as much as 1 part in 10^{12}. Finally, we have no data indicating that the laws of physics as we understand them are not universally applicable.

Inquiry 23–9 What would be the effect on the expansion of the universe if the strength of the gravitational force decreased with time?

23.6
THE GEOMETRY OF SPACE AND THE SHAPE OF THE UNIVERSE

The title of this section may appear to make no sense at all, because our ordinary intuition leads us to believe that space is three-dimensional and infinite. Astronomers are often asked, "What is the universe expanding into?" After all, if the universe has an "end," there must be an "edge" and therefore something "outside" it. We will see that it is actually possible to have a universe that is finite in extent and yet still have nothing outside it. Certain pitfalls open up when we try to apply our three-dimensional powers of spatial visualization to the universe, and we will try to shed some light on how such concepts as curved or warped space can have meaning. In fact, we will discover that the concept of "outside" the universe is a meaningless one, no matter what model we believe in.

WHAT IS MEANT BY CURVED SPACE?

Most of us have studied the rules of geometry on a flat surface, which we can call *flat space*. Flat space has two important characteristics. The first is that in a simple triangle the sum of the angles is exactly 180° (**Figure 23–16a**). Second, if we have a line and a point not on the line, there is only one line that passes through the point and is parallel to the first line (Figure 23–16b), such that parallel lines never meet.

It is not difficult to give meaning to the concept of curved space. Let's consider the simple analogy of the appearance of the Earth to an observer standing on it. Although a line on the ground might appear straight, we know that it is actually a section of a great circle[2] on the Earth. **Figure 23–17** shows two great circles on the surface of the Earth. Observers at the equator examining each of these lines would have the impression that they were parallel, yet the curves eventually intersect at the North Pole. By examining only a local region of space you may not be able to see curvature over larger distances.

Let us pursue this analogy further. Suppose you were to start walking in any direction on the Earth. It would appear to you that you were traveling on a flat surface, and yet you would eventually return to the very point from which you departed, proving that the surface was curved. As you returned to your starting point, you would certainly observe that you had not reached any place that you could call the "end" of the Earth; yet the Earth's surface is indisputably finite in area. In an analogous way, it is quite possible for the universe to be finite in volume—never having anything that could be called either an "end" or an "outside."

The sphere discussed above is an example of a surface having a **positive curvature**. Although movement along this surface can continue forever, the surface is finite in extent. Because any two lines that start out parallel along great circles eventually intersect, the concept of parallel lines is meaningless. In addition, the angles of a triangle drawn on such a surface add up to more than 180°.

Other surfaces can have **negative curvature**, an example of which is a saddle (**Figure 23–18**). On such a surface, there are many lines parallel to the line *AB* (Figure 23–18a), and the sum of the angles in a triangle is *less* than 180° (Figure 23–18b).

The universe as a whole can be described in terms of these geometric concepts. A universe with a positive curvature is one that we speak of as a **closed universe**,

one that is finite. A closed universe is one whose density is greater than the critical density, so that expansion will stop and collapse will ensue. For this reason,

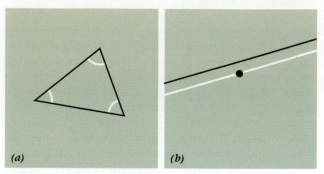

FIGURE 23–16. The geometric properties of flat space. (*a*) Properties of triangles. (*b*) Property of parallel lines.

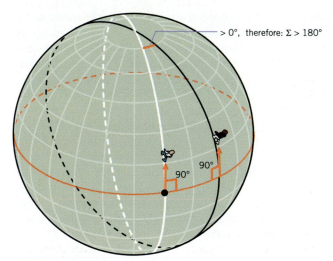

FIGURE 23–17. The geometric properties of a positively curved space. (*a*) Great circles intersect. (*b*) Properties of triangles.

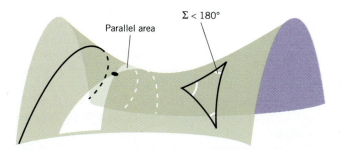

FIGURE 23–18. The geometric properties of negatively curved space. (*a*) Many lines are parallel to the line between points *A* and *B*. (*b*) Properties of triangles.

[2]A great circle on a sphere is a circle whose center coincides with the center of the sphere. An equator is an example. So, too, is a circle passing through both north and south poles.

we often speak of the critical density as that needed to "close" the universe. An **open universe** has a negative curvature. Its density is less than the critical density, so it is infinite in extent. For the case in which the density equals the critical density, the curvature is zero, and the expansion will stop when time equals infinity; this is a **flat universe**.

It is next to impossible to visualize the geometry of the universe satisfactorily. For one thing, we are talking about curved four-dimensional space-time, and humans are unable to picture the fourth dimension. Also, it is enormous. More significantly, we can see only those parts of it from which light has had time to reach us. We never see the whole. Furthermore, because light travels at a finite speed, we see different parts of the universe at different evolutionary epochs. Distant objects are seen as they were many years ago; their "present" appearance and location are certainly different.

Visualizing a curvature in space is the most difficult of all. One way to try it is to consider the trajectories of light rays as they move through space. It is in this sense that we can describe a black hole as "warping" the space in its vicinity, because it bends the trajectories of light rays passing near it and coming from it. In this way, the Sun warps the space around it. Concentrating on the trajectories of light rays is a useful way to get a feel for the geometry of space, because it is the light rays, after all, that bring us our information. This approach is consistent with science's goal of predicting what we will see.

HOW MIGHT WE MEASURE THE CURVATURE OF SPACE?

The various predictions that might help us distinguish between positively and negatively curved space all suffer from a common problem—they differ significantly only at great distances from the Earth, where objects are terribly faint and observations are difficult. The different geometries do predict different densities of objects at different distances from the Earth, and a different velocity-distance relationship, but it is not easy to distinguish between them with the observational data currently available.

Due to these difficulties, the problem of the geometry of the universe has not yet been solved, although we can hope that the Hubble Space Telescope will obtain observations that resolve the issue. However, it is possible that the issue could be resolved in another way, because the curvature of the universe is also related to its density as discussed above and summarized in Table 23–1.

A profound relationship exists between the curvature of space and the distribution of mass in the universe. The relationship was first formulated in a highly mathematical form by Albert Einstein in his general theory of relativity (1915). This theory, which has been experimentally verified by several different approaches, postulates an intimate relationship between space (or geometry) and mass (or gravity). It says that we may think of mass distribution as shaping space (or, as we discussed in the previous section, as determining the trajectories of light rays). Though this may be difficult for our inherently three-dimensional imaginations to picture, it makes sense in a fundamental way, because it is really not possible to satisfactorily define either matter or space alone without reference to the other.

Thus it is possible that in the near future the search for the missing mass will answer our questions concerning the curvature of space and the geometry of the universe. If this line of inquiry fails, however, there is always the "brute force" option of counting fainter and fainter objects until finally the answer becomes apparent. But to properly interpret such observations, we must have reasonable theoretical models of how the

TABLE 23-1

The Universe Under Various Models

Density	Nature of Space	Future of the Universe
Greater than the critical value (about 5×10^{-30} gm/cm³)	Positively curved, finite in volume, closed	The self-gravity of the universe will eventually stop the expansion; the universe will contact and end in a "future fireball."
Less than the critical value	Negatively curved, infinite in volume, open	The self-gravity of the universe is insufficient to halt the expansion of the universe, and it will expand forever.
Exactly the critical value (highly unlikely)	Flat, infinite in extent, open	The universe expands forever, but the velocity of the expansion gets closer and closer to zero the older the the universe becomes.

brightness of a galaxy changes throughout its lifetime. This in turn means that we will probably have to come to a firm understanding of what quasars are and how they fit into the overall scheme of galaxy evolution.

23.7

THE INFLATIONARY COSMOLOGICAL THEORY

The big-bang theory as we have described it (which, from here on out, we will refer to as the "simplest big-bang theory") is in complete agreement with events after the first second or so of the beginning. The theory predicts the observed expansion, the increase in the observed number of galaxies with distance, the observed abundances of light elements, and the strength and isotropy of the microwave background radiation. However, difficulties for the simplest big-bang theory arise for times earlier than one second.

For example, the *extreme* degree of isotropy expected for the background radiation is a source of some difficulty for the simplest big-bang models of the universe. We have seen that the temperature of the background radiation is the same even for parts of the visible universe that are now separated by large distances. The easiest way to explain this uniformity is to suppose that the different parts of the visible universe were once in physical contact with each other. Thus the hotter regions could warm up the cooler ones until all was at a constant temperature, much as heat flows from a hot object to a cool one when we put them in contact, until they are both at the same temperature.

The problem is that according to the simplest big-bang model, many regions of the visible universe were never in contact with each other. Rather, the regions of the visible universe in opposite directions from us have always been moving apart so fast that light beams from one region could not have reached the other in all the years the universe has existed, effectively isolating them. This makes the uniformity of the background radiation very hard to understand on the basis of the simplest big-bang model.

This model also predicts the formation of a large number of **magnetic monopoles**, exotic particles that correspond to an isolated north or south magnetic pole. As hard as physicists try, however, they have been unable to detect any. If the simplest big-bang theory were correct, magnetic monopoles should be abundant.

A final problem for the simplest big-bang theory is that because observational uncertainty prohibits observations from showing a definitive geometry, some scientists conclude the curvature of the universe must be extremely close to flat. Why should that be? To be flat requires the actual density to be *exactly* equal to the critical density, a situation that, if true, is surprising.

These problems with the simplest big-bang theory are solved by a modified model, originally proposed by MIT astrophysicist Alan Guth in 1980, of an **inflationary universe**. In the complex inflationary model, at the time of the GUT era (between about 10^{-43} and 10^{-35} second after the beginning) all parts of the universe were physically close enough to communicate their physical conditions to all other parts. It was at this time that the uniformity we currently observe came about. The phase of broken symmetry at 10^{-35} second, when the strong force separated from the electroweak force, occurred in such a way that a large number of bubbles of energy were produced. As each bubble cooled, a large pressure inflated the bubble to 10^{50} times its original size in 10^{-32} second. One of these bubbles became our universe.

The inflationary model proposes that the physical universe consists of a large number of separate entities, one of which comprises what we can detect and measure and therefore call our universe. After inflation, locations on opposite sides of the bubble are too far away from each other for communication to occur. Each bubble (universe) is, however, now uniform and geometrically flat, as observed.

Large numbers of magnetic monopoles are not produced in an inflationary universe, so the lack of observed monopoles is no longer a problem.

Inflation gives the universe the appearance of being exactly flat. As an analogy, suppose an ant on a small balloon is able to detect the balloon's curvature. As the balloon inflates, however, the space around the ant will become flatter until eventually the geometry around the ant will be indistinguishable from flat. A similar effect occurs for us when our bubble inflates.

An important part of the inflationary theory is that all possible initial conditions appear to produce the same state—the 3-K microwave background, the observed abundances of light elements, and a flat geometry. From all this, physicists appear to be finding that, contrary to rational thought and in agreement with the chapter opening quote, the universe can indeed have been formed from nothing!

23.8

THE HALLEY-OLBERS PARADOX

After delving into the details of how the geometry of space might be determined, it is refreshing to find that sometimes simple lines of reasoning lead to conclusions of great significance. For example, consider a question asked first by Edmund Halley and then later in 1826 by the German astronomer Heinrich Olbers: Why is the sky dark at night? If space is imagined to be without curvature and infinite in extent, this question is by no means trivial. **Figure 23–19** illustrates concentric circles around an observer in flat space such that the distance between each circle is the same. We will assume the stars are uniformly distributed. Because the volume of each region increases with the distance squared, the number of stars in each region also increases with the square of the distance. But the inverse square law of light tells us that the brightness of stars decreases with the distance squared. Because the increasing number of stars is exactly cancelled by their decreasing brightness, each zone, no matter how far away, adds the same amount as any other zone to the total light observed on Earth. The end result is that the night sky should look like wall-to-wall Sun, and we should be roasting in a 6000-K oven! The paradox, then, is that the night sky is dark when it should be bright.

How can we resolve the paradox that the night sky is dark when it might be expected to be bright? We cannot invoke the absorbing properties of interstellar dust to escape the apparent paradox, because within a finite amount of time the dust would come into equilibrium with the radiation and be heated to the temperature of the Sun's surface.

Might the expansion of the universe resolve the paradox? Photons from a receding galaxy will, of course, be redshifted out of the visual region of the spectrum. If you look farther, therefore, objects will appear fainter. Furthermore, the expansion reduces the density of photons in space the farther away you look. However, it can be calculated that the actual reduction of the energy flow is not large enough to relieve the paradox.

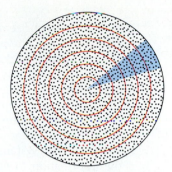

FIGURE 23–19. Concentric zones of equal thickness around the Earth have uniformly distributed stars whose number increases with the square of the distance.

One resolution could be that the universe is finite; if there are no radiating objects at great distances, the problem is immediately resolved.

The accepted solution to the paradox is that the universe is still young. Because light travels with finite speed, there is a physical limit to the *observable* universe—the distance light can travel since the universe formed. Therefore, the continuous zones in Figure 23–19 do *not* continue forever and the night sky is not bright.

We see that a simple question such as why is the sky dark at night can lead to complex considerations if its consequences are pursued to their logical end. We may not always be able to determine which explanation is correct, but considering many different possibilities remains one of the strongest tools of science. It is frequently impossible to predict from which direction the answer to a question will appear, and all avenues must be investigated.

DISCOVERY 23–1
THE AGE OF THE UNIVERSE

This Discovery will help you gain a further appreciation for the age of the universe, and the role played by humanity during that time. This activity may be done with at least one other person.

Obtain a standard-size roll of toilet paper. The length of this roll will represent the age of the universe. The package should specify the number of tissues on the roll. Use this information to determine the number of years represented by each sheet. Then, using a long hallway if available (if such a hall is not available, you can go back and forth within the longest area available), begin unrolling the paper, and mark each of the following events in the history of the universe: the formation of the 3-K background radiation, the formation of galaxies, the formation of the Milky Way's globular clusters, the formation of the Milky Way, the formation of stars, the formation of the solar system. As an added section, use the library if necessary to obtain the information to mark when life on Earth first appeared, when dinosaurs roamed the Earth, the appearance of humanoids, the time when the Neanderthals lived, the high point of the Greek civilization, and the time of your birth.

☐ **Discovery Inquiry 23–1a** On the scale of your model, how wide a line represents all of human history? The last 100 years?

DISCOVERY 23–2
THE EXPANSION OF THE UNIVERSE

When you have completed this Discovery you should be able to do the following:

■ Explain, using words and a graph, the reasoning astronomers use to conclude that the universe is expanding.

We can use Figure 21–9 to demonstrate the expansion of the universe. The five elliptical galaxies shown were selected from five different populous clusters of galaxies. Each is one of the largest and most luminous of the elliptical galaxies in its cluster. If we assume that these five objects are comparable in actual size and luminosity, then their *apparent* size is going to depend on their distance. The smaller the galaxy on the photograph, the farther away it is.

Measure the size of each galaxy with a millimeter scale, trying to read the size to a tenth of a millimeter. Calculate the quantity 1 divided by the size of the galaxy and make a table of your results. This quantity will get numerically larger as the galaxy's distance increases.

The spectrum of each galaxy is also shown in Figure 21–9, with the position of two absorption lines of ionized calcium (Ca II) in the spectrum indicated by an arrow. Measure the redshift shown by each galaxy by measuring the length of the arrow above each spectrum (in millimeters). Include these values in your table.

Plot your values of 1/size on the horizontal axis of a piece of graph paper and the corresponding values of the redshift (as measured by the length of the arrows in millimeters) on the vertical axis. Label the scales on your graph paper so that your plot extends over most of the page.

Clearly, the more distant objects show a larger redshift. If the redshift is interpreted as speed away from the Earth, this means that the more distant the galaxy, the faster it is receding. In other words, the clusters of galaxies are all moving away from each other; the universe is expanding. The only way to escape the logic of this argument is to find some other physical effect that will cause the redshift seen in the spectrum, and no one has come up with a successful alternate interpretation.

The photos that you are examining in this activity are comparable in quality to the data that were originally used to deduce the expansion of the universe. As a consequence, you can see that the actual experimental proof of this enormously significant idea is not difficult at all. The genius and the progress come from having the *idea* in the first place!

☐ **Discovery Inquiry 23–2a** Explain why your graph demonstrates the expansion of the universe.

DISCOVERY 23–3
OBSERVING
THE BACKGROUND RADIATION

We can detect the remnants of the big bang with an ordinary television receiver. Turn a TV set with a good antenna (not cable) to an unused channel. Set the contrast to its maximum, and then turn the brightness down to the point where the "snow" on the screen appears as white flecks on a black background. About one out of every hundred of the flecks of "snow" actually represents the detection of an individual photon from the remnants of the big bang. The other 99% represent primarily radio noise (static) generated in the receiver itself, and other environmentally produced interference. It was these interfering effects that Penzias and Wilson had to eliminate almost completely from their equipment before they could unambiguously detect the background radiation. Indeed, before recognizing the true source of the noise they detected with their equipment, they had to eliminate all other possibilities. At one time, they even thought that the droppings from a flock of pigeons that had nested in their antenna might have been responsible.

CHAPTER SUMMARY

In forming a comprehensive summary of cosmology, some applicable information is carried over from the two previous chapters.

OBSERVATIONS

- The **Hubble law** is a graph showing the relationship between the velocity with which a galaxy is moving away from the Milky Way versus its distance. The Hubble law is a linear relationship showing that the greater a galaxy's distance, the greater its velocity. The value of the Hubble constant is uncertain and generally thought to be between 15 and 30 km/sec/Mly.
- All quasars are generally considered to be at large distances with none nearby. Galaxy counts show there to be more galaxies at large distances than nearby.
- Helium accounts for some 25–30% of the mass of the universe, with hydrogen accounting for most of the rest.
- **Penzias and Wilson** observed a general background radiation to the universe at radio wavelengths. According to the **COBE** satellite, this radiation is almost perfectly isotropic and has a spectrum of a blackbody of temperature 2.74 ± 0.06 K. The COBE satellite has found small temperature fluctuations in the microwave background radiation.
- No detections have been made of either magnetic monopoles or the decay of the proton.
- Viewed on a large scale, the universe appears to be **homogeneous** and **isotropic**.

- The density of observable matter in the universe can be found by summing the mass of all observable matter and dividing by the observable volume, and from observations of deuterium in the universe.
- Galaxies in clusters have velocities larger than expected on the basis of the amount of observable matter within the clusters.

THEORY

- A summary of the history of the universe is in **Table 23–2.**
- The universe was once a dense, hot fireball that suddenly began expanding.
- The four forces of nature were unified until a short time after the big bang.

- The **steady-state theory**, in which the universe remains the same all over, requires the continuous creation of matter to hold the density of the expanding universe constant.
- The universe is probably decelerating because of the mutual gravitational attraction of each mass on every other mass.
- An open universe is one with a negative curvature and is infinite; a closed universe has a positive curvature and is finite. In a flat universe, the laws of Euclidean geometry hold. The critical density is the density of a flat universe. If the density is greater than the critical density, expansion will halt, and the universe may enter into a continuous state of oscillation between expansion and contraction.
- Observations of galaxies at large distances are capa-

TABLE 23–2

Summary of the History of the Universe[1]

Cosmic Time	Years Ago	Redshift $(\Delta\lambda/\lambda)$	Event
0	20×10^9	infinite	Big bang
10^{-43} sec	20×10^9	10^{32}	Particle creation
10^{-43}–10^{-35} sec	20×10^9		GUT era
10^{-35} sec	20×10^9		Inflation occurs
10^{-6} sec	20×10^9	10^{13}	Proton-antiproton pair annihilation
1 second	20×10^9	10^{10}	Electron-positron pair annihilation
1 minute	20×10^9	10^9	Helium and deuterium formed
10,000 years	20×10^9	10^4	Universe becomes matter-dominated
500,000 years	19.9997×10^9	10^3	Decoupling of matter and radiation; 3-K radiation comes from this time
1–2×10^9 years	18–19×10^9	10–30	Galaxy formation begins (bottom-up scenario)
3×10^9 years	17×10^9	5	Galaxy begins to cluster (bottom-up scenario), most distant quasar observed
4×10^9 years	16×10^9		Milky Way protogalaxy collapses
4.1×10^9 years	15.9×10^9		First stars form
5×10^9 years	15×10^9	3	Population II stars form
10×10^9 years	10^{10}	1	Population I stars form
15.3×10^9	4.7×10^9		Protosolar nebula collapses
15.4×10^9	4.6×10^9		Planets form
15.7×10^9	4.3×10^9		Intense cratering on Moon and planets
16.1×10^9	3.90×10^9		Oldest terrestrial rocks form
17×10^9	3×10^9		Microscopic life forms
18×10^9	2×10^9		Oxgyen-rich atmosphere develops
19×10^9	1×10^9		Macroscopic life forms
19.4×10^9	600×10^6		Oldest fossil record
19.8×10^9	200×10^6		First mammals
19.94×10^9	60×10^6		First primates
20×10^9	100,000		Homo sapiens

[1]After Barrow and Silk, *Scientific American*, April 1980.

ble of distinguishing between positive, negative, and flat universes.

- In studying the universe we make assumptions of homogeneity, isotropy, and universality.
- An upper limit to the age of the universe is given by $1/H$.

CONCLUSIONS

- While modifications to the big-bang theory are required to bring some of its details into agreement with observations, the general outline of the theory is in excellent agreement with three independent lines of evidence: (1) the abundances of the elements in the oldest stars; (2) the observed increase in the

number of galaxies with increasing distance; and (3) the agreement of the observed 3-K microwave background radiation with theoretical predictions.

- From COBE satellite observations we conclude that sufficient fluctuations were present within the big bang to allow galaxies to form.
- An extensive amount of dark matter in galaxy clusters is inferred from the motions of galaxies within them.
- Observations of distant galaxies are not accurate enough to distinguish between various models of the universe.
- The **inflationary model** of the universe accounts for suggested flatness, the lack of observed magnetic monopoles, and the high degree of isotropy.

SUMMARY QUESTIONS

1. What is the crucial role of measurement and observation in the definition of the universe?

2. What is the interpretation of Hubble's redshift observations?

3. What are the various observable differences to be expected between a steady-state universe, a big-bang universe that expands forever, and a big-bang universe that eventually collapses on itself? Your discussion should include observations of changes in density and expansion velocity, and the 3-K background radiation. What are the difficulties in making observations of each?

4. What are the main events during the expansion of the universe? What were the conditions prevailing when each event occurred? What evidence of the events can be observed today?

5. Why must elements heavier than helium have been created in the stars rather than during the big bang?

6. What is the future of the universe? How may we decide among the various possible outcomes?

7. How is the shape of the universe related to its history and the amount of mass it contains? What are some geometrical analogies to various types of universes?

8. What is the 3-K background radiation? Describe it both from the observational and theoretical viewpoints. What was the significance of the observations made by COBE?

9. What are the observations astronomers have made that lend credence to the big-bang theory?

10. What is the Halley-Olbers paradox? What are various methods by which it can be successfully resolved?

APPLYING YOUR KNOWLEDGE

1. What influence would the aging of galaxies have on observations of galaxy counts at greater and greater distances? Would such aging tend to make the universe seem to decelerate at a greater or lesser rate than it actually does?

2. Summarize the predictions the big-bang theory makes. What do observations have to say about each of these predictions?

3. The wavelength of a photon emitted in a transition is proportional to the velocity of light. Let's speculate that as light travels through the universe it becomes

"tired"—that is, it loses energy. Predict the consequences of tired light on our observations of both distant and nearby galaxies.

4. The wavelength of a photon emitted in a transition is inversely proportional to the mass of an electron. Let's speculate that the mass of an electron increases with time. Predict the consequences of increasing electron mass on our observations of the spectra of both distant and nearby galaxies.

5. Suppose that rather than beginning from a strongly condensed condition, the universe began as a ball of gas

having the size it has today. Suppose, further, that matter was spread out uniformly (except for a few local irregularities) and with zero motion (that is, it was static). Describe the subsequent history of this universe and the reasons for your description.

6.　Suppose the velocity-distance relation for the universe were as given in **Figure 23–20**. Explain the consequences of this velocity-distance relation.

7.　Creationists argue that because astronomers observe the mass within galaxy clusters to be insufficient to hold the clusters together, such clusters can exist only if the universe is considerably younger than astronomers say. How might you respond to such an argument?

8.　In Chapter 21 we saw that galaxy masses could be found from studies of rotation curves for spiral galaxies and velocity dispersion for elliptical galaxies. If galaxies containing massive black holes were studied with these techniques, would the computed masses include the masses of the black holes or would their masses be neglected? Explain. Would hypothesizing black holes at the centers of galaxies bring the universe's overall density above the critical density?

9.　Explain the Hubble law using points along a line, in a manner similar to that used in the raisin bread analogy.

10.　Compute how far a photon could travel over the age of the universe.

11.　Estimate the total number of atoms in the universe! (Hint: To solve this problem, you need to make a number of educated guesses.)

12.　Compute the maximum age of the universe (in years) for values of $H = 15$ km/sec/Mly and 25 km/sec/Mly.

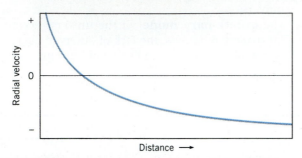

FIGURE 23–20.　A hypothetical velocity-distance relation.

ANSWERS TO INQUIRIES

23–1. Each million light-years corresponds to 15 km/sec; therefore, 300 million light-years corresponds to a velocity of $300 \times 15 = 4500$ km/sec.

23–2. The velocity is $0.16c = 0.16 \times 300,000 = 45,000$ km/sec. From the Hubble law, $d = V/H = 45,000/15 = 3,000$ Mly $= 3 \times 10^9$ ly.

23–3. Larger, because those galaxies are observed as they were when the universe was younger and expanding with a higher velocity.

23–4. The secret is that stars maintain a high density whereas during the big bang the density decreases rapidly. The problem is taken care of in the triple-alpha reaction (Chapter 18) when, for example, carbon is formed from short-lived beryllium. Over great periods of time, where the density remains high, the carbon abundances increase.

23–5. From $\lambda_{max} T = 3 \times 10^7$, we have $\lambda_{max} = 3 \times 10^7/3$ K $= 10^7$ Å $= 0.1$ cm, in the short end of the radio (microwave) region of the spectrum.

23–6. In Section 20.3, it was pointed out that matter-antimatter reactions produce strong gamma-ray photons. These have been searched for in the background radiation and have not been found. (See also Section 22.4 for a discussion of antimatter in quasars.)

23–7. No, because in a steady-state universe there would be as many nearby quasars as there are distant ones. In other words, they would be uniformly distributed.

23–8. No. Massive as they are, they would still be only about 1% of the total mass of a galaxy.

23–9. It would make it more difficult for gravity to halt the expansion of the universe and would increase the likelihood that it would expand forever.

APPENDIX A
USEFUL MATHEMATICAL
INFORMATION

1. THE POWERS OF TEN SHORTHAND FOR NUMBERS

No matter how carefully astronomers define appropriate units of measure, they will inevitably be involved with numbers that are both large and small, because they are concerned with sizes ranging from the smallest to the largest conceivable. A compact notation has been developed to represent this wide range of numbers conveniently and to facilitate arithmetic using them.

To start off with a familiar example, consider the number "10 squared." This of course means $10 \times 10 = 100$. In our compact notation, the number is expressed as 10^2, where the superscript after the ten (called the *exponent*) indicates how many times the number 10 is multiplied by itself. Thus, 10^2 can be read as "ten squared," or "ten raised to the second power," or "ten to the two."

Similarly, "10 cubed" is $10 \times 10 \times 10 = 1000 = 10^3$ (ten raised to the third power), and 10^6 (ten to the sixth power) is six 10's multiplied together, or one million.

Notice in these examples that the exponent gives the number of zeroes that follow the 1 when the number is written out in long form. Thus,

$$10^9 = \text{nine 10's multiplied together}$$
$$= 1,000,000,000$$
$$= 1 \text{ followed by } \textit{nine} \text{ zeroes.}$$

Powers of ten notation is a compact way to express large numbers. Thus we may write the number 1,723,000,000,000 as 1723×10^9, or by moving the decimal point around we could also express the same number in the following equivalent ways:

$$1723 \times 10^9$$
$$172.3 \times 10^{10}$$
$$17.23 \times 10^{11}$$
$$1.723 \times 10^{12}$$

The last form is actually the preferred form of expressing the number, although all the others are correct.

A further advantage of the powers of ten shorthand is that it makes arithmetic easy. Consider, for example, taking the product of 100 times 1000:

$$100 \times 1000 = 10^2 \times 10^3$$
$$= 10^{2+3}$$
$$= 10^5$$
$$= 100,000.$$

Thus, to multiply two numbers that are expressed in powers of ten notation, simply add the exponents together.

As a second example, consider $15,000 \times 3,000,000,000$. Expressing both numbers in powers of ten, we have

$$(1.5 \times 10^4) \times (3 \times 10^9)$$
$$= (1.5 \times 3) \times 10^{4+9}$$
$$= 4.5 \times 10^{13}.$$

If you follow this example, you cannot go wrong.

The powers of ten notation also handles small numbers by using negative exponents. By definition, when we write 10^{-2}, we mean the reciprocal of 10^2. That is,

$$10^{-2} = 1/10^2 = 0.01.$$

Similarly, $10^{-7} = 0.0000001$. Thus the exponent is the number of places the decimal point moves to the left of the 1.

The rule for adding exponents applies even if the exponent is negative. Thus we can multiply large and small numbers together with equal ease. Consider, for example, the product of 0.00000013 times $30,000,000,000,000,000,000 = 1.3 \times 10^{-7} \times 3 \times 10^{19}$. This becomes

$$(1.3 \times 3) \times 10^{-7+19} = 3.9 \times 10^{12}.$$

2. SOME USEFUL NUMBERS

Physical Constants

The velocity of light	$c = 3 \times 10^{10}$ cm/sec
	$= 3 \times 10^5$ km/sec
The gravitational constant	$G = 6.7 \times 10^{-8}$

when centimeters, grams, and seconds are used as the units of length, mass, and time.

The mass of the hydrogen atom	$m_H = 1.67 \times 10^{-24}$ gm
The mass of the electron	$M_e = 9.1 \times 10^{-28}$ gm
The angstrom	$1 \text{ Å} = 10^{-8}$ cm
Planck's constant	$h = 6.62 \times 10^{-27}$ erg·s

Boltzman's constant $k = 1.38 \times 10^{-16}$ erg deg^{-1}

Constant in Wien's law: $\lambda_{max}T = 2.9 \times 10^7$ Å·K

Astronomical Constants

The astronomical unit	1 AU = 1.5×10^8 km
The light-year	1 ly = 9.5×10^{12} km
The parsec	1 pc = 3.26 ly
Mass of Earth	$M_E = 5.974 \times 10^{27}$ gm
Mass of Sun	$M_S = 1.989 \times 10^{33}$ gm
Radius of Earth (equatorial)	$R_E = 6378$ km
Radius of Sun	$R_S = 6.96 \times 10^5$ km
Luminosity of Sun	$L_S = 3.827 \times 10^{33}$ ergs
	$= 3.827 \times 10^{26}$ watts

3. RELATIONS BETWEEN UNITS

Angles

A circle contains 360° (degrees)

1° = 60′ (minutes of arc)

1′ = 60″ (seconds of arc)

1 radian = 57.3°

Length

1 kilometer (km) = 1000 meters (m) = 0.62 mile (mi)

1 meter = 1.09 yards = 39.37 inches

1 centimeter (cm) = 0.01 meter ≈ 0.4 inch

1 millimeter (mm) = 0.1 cm = 0.001 m

1 micron (μm) = 10^{-6} m

1 mile = 1.6 km

1 inch = 2.54 cm

Mass/Weight

1 kilogram (kg) = 1000 grams (gm)

A mass of 1 kg on Earth has a weight of 2.2 pounds.

Temperature

$$°C = \frac{(°F - 32)}{1.8}$$

$$°F = (1.8 \times °C) + 32$$

$$K = °C + 273$$

(° is dropped when Kelvin scale is written)

The following graphs allow easy approximate conversion from K to °C or °F. **Figure A–1** is useful for Kelvin temperatures between 0 and 1000, while **Figure A–2** is for higher temperatures. For example, to convert 600 K to the other scales, find 600 on the horizontal scale; reading up to the Celsius line gives a temperature of about 340°C. Reading to the Fahrenheit line gives slightly over 600°F. To convert 30,000°F to K, find 30,000 on the vertical axis; moving horizontally to the Fahrenheit curve gives an equivalent temperature of about 17,000 K.

4. CONVERSION BETWEEN UNITS

Conversion between units is made easier if you write all units involved and separate each term with parentheses. As an example, to express 25 light-years in kilometers,

$$(25 \text{ ly}) \times (9.5 \times 10^{12} \text{ km/ly}) = 2.6 \times 10^{13} \text{ km.}$$

Note that the ly in the numerator cancels the ly in the denominator. To further convert to inches,

$$(2.6 \times 10^{13} \text{ km}) \times (10^3 \text{ m/km}) \times (10^2 \text{ cm/m})$$
$$\times (1 \text{ inch}/2.54 \text{ cm}) \approx 10^{18} \text{ inches.}$$

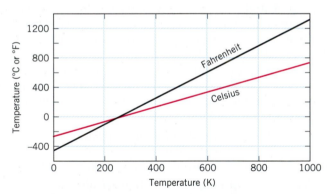

FIGURE A–1. Graph to convert between temperature scales for low temperatures.

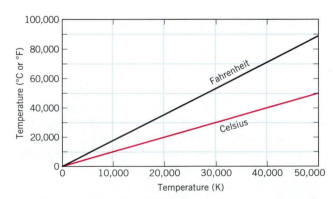

FIGURE A–2. Graph to convert between temperature scales for high temperatures.

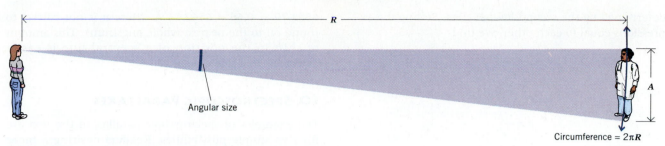

FIGURE A–3. Figure for the derivation of the angular size formula.

5. DERIVATION OF ANGULAR SIZE FORMULA

To find how true size, distance, and angular size are related, refer to **Figure A–3**, in which we draw about the observer a circle of radius R equal to the object's distance from the observer. From this drawing you can write the following simple proportion: The angular size of the person (in degrees) is to the true height of the person (A) as the complete angle around the circle (360°) is to the circumference of the circle ($2\pi R$). Symbolically,

$$\frac{\text{angular size}}{A} = \frac{360°}{2\pi R}.$$

Solving for the angular size, we find

$$\text{angular size} = \frac{360° \, A}{2\pi R}$$

$$= 57.3° \frac{A}{R}.$$

In other words,

$$\text{angular size} = 57.3° \times \frac{\text{true size}}{\text{distance}}.$$

6. COMPUTING RANDOM ERRORS

An average is not an exact number. How, then, can we express the uncertainty in a number? If the average of a set of measurements is, for example, 52.3, we express this as 52.3 plus or minus the random error. If such an error were 0.4, we would write it as

$$52.3 \pm 0.4 \, .$$

The meaning of this expression is seen in **Figure A–4**, which shows that there is a 68% probability that the true value is between $52.3 - 0.4 = 51.9$, and $52.3 + 0.4 = 52.7$. From probability theory, if we consider twice this range from the average, there is a 95% chance that the true value is between 51.5 and 53.1. While this is pretty good, it also means there is a 5% chance that the true value is *outside* these limits.

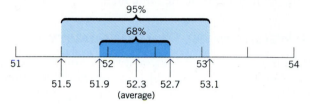

FIGURE A–4. The meaning of an error.

In a series of measurements, there will always be a largest and a smallest value, and the difference between these two numbers is called the *range* of the measurements. The range indicates the amount of variation about the average value that you are likely to encounter in a series of measurements, and it is therefore an indication of the amount of confidence that can be placed in any single measurement. Scientists characterize the variation about an average by a precisely defined quantity called the **standard deviation**, but for convenience we will adopt an approximate rule called **Snedecor's rough check**. This rule states that if you have five independent measurements of a quantity, the standard deviation of one measurement is approximately equal to the *range* of the measurements divided by two. If there are ten independent measurements, the standard deviation is approximately equal to the range divided by three. In general, dividing the range by the square root of $(N - 1)$, where N is the number of observations, will give you a number that is a good approximation to the standard deviation if N is small. The computed value is the number that follows the plus-and-minus symbol in presenting a result.

7. WHY BODIES OF DIFFERENT MASSES FALL AT THE SAME RATE

The reason bodies of different masses fall at the same rate can be seen by combining Newton's second law with his law of gravity. The force of gravity as given by Newton's universal law of gravity must equal the force determined by the second law, $F = ma$. Setting the expressions equal to each other, we find

$$ma = G\frac{mM}{d^2}.$$

Dividing each side by m, we find the acceleration:

$$a = \frac{GM}{d^2}.$$

The acceleration is thus independent of the mass of the falling body (m) and depends only on the size and mass of the gravitating body (M). In other words, bodies of different masses fall at the same rate!

8. WEIGHT AS THE FORCE OF GRAVITY

Because weight is the force gravity exerts on your body, you can express your weight using Newton's equation. If W is your weight, m_{you} is your mass, M_{Earth} is the mass of the Earth, and R_{Earth} is your distance from the center of the Earth, then

$$W = \frac{Gm_{you}M_{Earth}}{R^2_{Earth}}.$$

In a similar manner you could write your weight on the Moon:

$$W_{moon} = \frac{Gm_{you}M_{Moon}}{R^2_{Moon}}.$$

9. A FURTHER EXAMPLE OF THE DOPPLER EFFECT

Suppose a star is approaching Earth at a speed of 30 km/sec. Let's turn the problem around and determine by how much a spectral line will shift, and the wavelength at which the spectral line would be observed. Because the star is approaching Earth, we know that its observed wavelengths will be shorter than the rest wavelength. The ratio of speeds in the formula is

$$\frac{\text{relative speed between the star and Earth}}{\text{speed of light}}$$

$$= \frac{-30 \text{ km/sec}}{300{,}000 \text{ km/sec}}$$

$$= -10^{-4}$$

$$= \frac{\text{change in wavelength}}{\text{rest wavelength}}.$$

The change in the wavelength will be -10^{-4} of the rest wavelength. Suppose the rest wavelength is 6562.79 Å. The change in wavelength will be $6562.79 \times (-0.0001)$ $= -0.66$ Å, so the observed wavelength would be

6562.79 − 0.66 = 6562.13 Å. (Here it is not correct to round off to the nearest whole angstrom!) This amount of shift in the position of a spectral line is easily detectable in a star's spectrum.

10. SPECTROSCOPIC PARALLAXES

The examples of spectroscopic parallax in the text are for two simple possibilities. Readers desiring a more general and explicit mathematical treatment will find the following discussion useful. From the inverse square law we know that a star of luminosity L at a distance d will have an apparent (observed) brightness B given by

$$B \propto L/d^2.$$

If we have two stars, then we have

$$B_2 \propto L_2/d^2_2$$

$$B_1 \propto L_1/d^2_1.$$

Taking ratios of these two expressions we find

$$\frac{B_2}{B_1} = \frac{L_2/d^2_2}{L_1/d^2_1}.$$

(In taking the ratio, the proportionality becomes an equality because all constants cancel out.) Using the example in which α Centauri is star 2 and the Sun is star 1 we have

$$\frac{B_{\alpha\,Cen}}{B_{Sun}} = \frac{L_{\alpha\,Cen}/d^2_{\alpha\,Cen}}{L_{Sun}/d^2_{Sun}}.$$

$$\frac{1}{9 \times 10^{10}} = \frac{1/d^2_{\alpha\,Cen}}{1/d^2_{Sun}} = \frac{d^2_{Sun}}{d^2_{\alpha\,Cen}}.$$

Solving for $d_{\alpha\,Cen}/d_{Sun}$ we find $d_{\alpha\,Cen}/d_{Sun} = 3 \times 10^5$ as above.

Similarly, for the second example in which Canopus and α Cen appear equally bright we have:

$$\frac{B_{Canopus}}{B_{\alpha\,Cen}} = \frac{L_{Canopus}/d^2_{Canopus}}{L_{\alpha\,Cen}/d^2_{\alpha\,Cen}}.$$

or

$$\frac{B_{Canopus}}{B_{\alpha\,Cen}} = \frac{L_{Canopus} \times d^2_{\alpha\,Cen}}{L_{\alpha\,Cen} \times d^2_{Canopus}}$$

$$1 = 1200\frac{d^2_{\alpha\,Cen}}{d^2_{Canopus}}$$

$$d_{Canopus} = \sqrt{1200}\,d_{\alpha\,Cen}$$

$$= 35\,d_{\alpha\,Cen}.$$

APPENDIX B
CONSTELLATIONS

Constellation (Latin name)	Abbreviation	English name or description	Constellation (Latin name)	Abbreviation	English name or description
Andromeda	And	Princess of Ethiopia	Lacerta	Lac	Lizard
Antlia	Ant	Air Pump	Leo	Leo	Lion
Apus	Aps	Bird of Paradise	Leo Minor	LMi	Little Lion
Aquarius	Aqr	Water Bearer	Lepus	Lep	Hare
Aquila	Aql	Eagle	Libra	Lib	Scales
Ara	Ara	Altar	Lupus	Lup	Wolf
Aries	Ari	Ram	Lynx	Lyn	Lynx
Auriga	Aur	Charioteer	Lyra	Lyr	Lyre or Harp
Boötes	Boo	Herdsman	Mensa	Men	Table Mountain
Caelum	Cae	Graving Tool	Microscopium	Mic	Microscope
Camelopardalis	Cam	Giraffe	Monoceros	Mon	Unicorn
Cancer	Can	Crab	Musca	Mus	Fly
Canes Venatici	CVn	Hunting Dogs	Norma	Nor	Carpenter's Level
Canis Major	CMa	Big Dog	Octans	Oct	Octant
Canis Minor	CMi	Little Dog	Ophiuchus	Oph	Holder of Serpent
Capricornus	Cap	Sea Goat	Orion	Ori	Orion, the Hunter
Carina	Car	Keel of Argonauts' ship	Pavo	Pav	Peacock
Cassiopeia	Cas	Queen of Ethiopia	Pegasus	Peg	The Winged Horse
Centaurus	Cen	Centaur	Perseus	Per	He who saved Andromeda
Cepheus	Cep	King of Ethiopia	Phoenix	Phe	Phoenix
Cetus	Cet	Sea Monster (whale)	Pictor	Pic	Easel
Chamaeleon	Cha	Chameleon	Pisces	Psc	Fishes
Circinus	Cir	Compasses	Piscis Austrinus	PsA	Southern Fish
Columba	Col	Dove	Puppis	Pup	Stern of the Argonaut's ship
Coma Berenices	Com	Berenice's Hair	Pyxis	Pyx	Compass of the Argonaut's ship
Corona Australis	CrA	Southern Crown	Reticulum	Ret	Net
Corona Borealis	CrB	Northern Crown	Sagitta	Sge	Arrow
Corvus	Crv	Crow	Sagittarius	Sgr	Archer
Crater	Crt	Cup	Scorpius	Sco	Scorpion
Crux	Cru	Cross (Southern)	Sculptor	Scl	Sculptor's Tools
Cygnus	Cyg	Swan	Scutum	Sct	Shield
Delphinus	Del	Porpoise	Serpens	Ser	Serpent
Dorado	Dor	Swordfish	Sextans	Sex	Sextant
Draco	Dra	Dragon	Taurus	Tau	Bull
Equuleus	Equ	Little horse	Telescopium	Tel	Telescope
Eridanus	Eri	River	Triangulum	Tri	Triangle
Fornax	For	Furnace	Triangulum Australe	TrA	Southern Triangle
Gemini	Gem	Twins	Tucana	Tuc	Toucan
Grus	Gru	Crane	Ursa Major	UMa	Big bear
Hercules	Her	Hercules, Son of Zeus	Ursa Minor	UMi	Little bear
Horologium	Hor	Clock	Vela	Vel	Sail of the Argonaut's ship
Hydra	Hya	Water Snake (female)	Virgo	Vir	Virgin
Hydrus	Hyi	Water Snake (male)	Volans	Vol	Flying fish
Indus	Ind	Indian	Vulpecula	Vul	Fox

APPENDIX C
PROPERTIES OF
PLANETS AND SATELLITES

Orbital Data of the Planets

Planet	Semimajor axis (AU)	Semimajor axis (× 10⁶ km)	Sidereal orbital period (tropical years)	Sidereal orbital period (days)	Synodic orbital period (days)	Eccentricity	Inclination of orbital plane	Average orbital velocity (km/sec)	Inclination to orbital plane
Mercury	0.387	57.9	0.241	87.96	115.9	0.206	7°0'19"	47.89	0
Venus	0.723	108.2	0.615	224.7	583.9	0.007	3°23'41"	35.03	177° 18'
Earth	1	149.6	1	365.26	———	0.017	0°0'0"	29.79	23° 27'
Mars	1.52	228	1.881	686.98	779.9	0.093	1°51'1"	24.13	25° 12'
Jupiter	5.202	778.3	11.86	4333	398.9	0.048	1°18'17"	13.06	3° 07'
Saturn	9.54	1427	29.46	10759	378.1	0.052	2°29'9"	9.64	26° 44'
Uranus	19.28	2884	84.01	30685	369.7	0.045	0°46'23"	6.81	97° 52'
Neptune	30.22	4521	164.8	60188	367.5	0.007	1°46'15"	5.43	29° 34'
Pluto	39.83	5959	248.6	90700	366.7	0.255	17°7'30"	4.74	98°

Physical Data of the Planets

Planet	Radius (km)	Radius (Earth radii)	Mass (Earth masses)	Mass (kg)	Density (g/cm³)	Gravity (Earth = 1)	Albedo	Temperature (K)	Escape speed (km/s)
Mercury	2440	0.38	0.0562	3.30×10^{23}	5.43	0.38	0.06	100–700	4.2
Venus	6052	0.95	0.815	4.87×10^{24}	5.24	0.91	0.76	700	10.3
Earth	6378	1	1	5.974×10^{24}	5.52	1	0.4	250–300	11.2
Mars	3397	0.53	0.1074	6.419×10^{23}	3.94	0.39	0.16	210–300	5.1
Jupiter	71492	11.19	317.9	1.899×10^{27}	1.33	2.54	0.51	110–150	61
Saturn	60268	9.41	95.1	5.69×10^{26}	0.70	1.07	0.5	95	36
Uranus	25559	4.01	14.56	8.66×10^{25}	1.30	0.9	0.66	58	21
Neptune	24764	3.89	17.24	1.03×10^{26}	1.76	1.14	0.62	56	24
Pluto	1151	0.18	0.0024	1.5×10^{22}	2.10	0.06	0.5	40	2.1

Major Planetary Satellites

Planet	Satellite	Distance (10^3 km)	Distance (planetary radii)	Orbital period (days)	Radius (km)	Mass (Moon = 1)	Bulk density (g/cm^3)
Earth	Moon	384	60.2	27.32	1738	1.00	3.3
Mars	Phobos	9.38	2.76	0.319	$14 \times 11 \times 9$	1.3×10^{-7}	1.9
	Deimos	23.46	6.93	1.262	$8 \times 6 \times 6$	2.7×10^{-8}	2.1
Jupiter	Metis	128	1.79	0.29	20	1.3×10^{-6}	———
	Andrastea	129	1.8	0.3	$13 \times 10 \times 8$	2.6×10^{-7}	———
	Amalthea	181	2.55	0.5	$130 \times 83 \times 75$	5.2×10^{-5}	3.00
	Thebe	222	3.11	0.67	55×45	1.0×10^{-5}	———
	Io	422	5.95	1.77	1815	1.21	3.53
	Europa	671	9.47	3.55	1569	0.67	3.03
	Ganymede	1070	15.1	7.16	2631	2.02	1.93
	Callisto	1883	26.6	16.69	2400	1.47	1.79
	Leda	11,094	156	239	8	7.8×10^{-8}	———
	Himalia	11,480	161	251	93	1.3×10^{-4}	~1
	Lysithea	11,720	164	259	18	1.0×10^{-6}	———
	Elara	11,737	165	260	38	1.0×10^{-5}	———
	Ananke	21,200	291	631 (R)	15	5.2×10^{-7}	———
	Carme	22,600	314	692 (R)	20	1.3×10^{-6}	———
	Pasiphae	23,500	327	735 (R)	25	2.6×10^{-6}	———
	Sinope	23,700	333	758 (R)	18	1.0×10^{-6}	———
Saturn	Pan	133.58	2.23	0.58	10	———	———
	Atlas	137.67	2.28	0.602	20×10	———	———
	Prometheus	139.35	2.31	0.613	$70 \times 50 \times 40$	———	———
	Pandora	141.70	2.35	0.629	$55 \times 45 \times 35$	———	———
	Epimetheus	151.42	2.51	0.694	$70 \times 60 \times 50$	———	———
	Janus	151.47	2.51	0.695	$110 \times 100 \times 80$	———	———
	Mimas	185.52	3.08	0.942	196	6.2×10^{-4}	1.2
	Enceladus	238.02	3.95	1.37	250	1.0×10^{-3}	1.2
	Tethys	294.66	4.88	1.888	530	0.01	1.2
	Telesto	294.66	4.88	1.888	$17 \times 14 \times 13$	———	———
	Calypso	294.66	4.88	1.888	$17 \times 11 \times 11$	———	———
	Dione	377.40	6.26	2.737	560	0.01	1.4
	Helene	377.40	6.26	2.737	$18 \times 16 \times 15$	———	———
	Rhea	527.04	8.74	4.518	765	0.03	1.3
	Titan	1221.83	20.25	15.95	2575	1.82	1.88
	Hyperion	1481.1	24.55	21.28	$205 \times 130 \times 110$	2.3×10^{-4}	———
	Iapetus	3560.13	59.02	79.33	730	0.03	1.2
	Phoebe	12952	214.7	550.5	110	5.4×10^{-6}	———

Major Planetary Satellites (cont.)

Planet	Satellite	Distance (10^3 km)	Distance (planetary radii)	Orbital period (days)	Radius (km)	Mass (Moon=1)	Bulk density (g/cm³)
Uranus	Cordelia	48.75	1.91	0.34	13	——	——
	Ophelia	53.77	2.10	0.38	16	——	——
	Bianca	59.16	2.31	0.43	22	——	——
	Cressida	61.77	2.42	0.46	33	——	——
	Desdemona	62.65	2.45	0.47	29	——	——
	Juliet	64.63	2.53	0.49	42	——	——
	Portia	66.10	2.59	0.51	55	——	——
	Rosalind	69.93	2.74	0.56	27	——	——
	Belinda	75.25	2.94	0.62	34	——	——
	Puck	86.00	3.36	0.76	77	——	——
	Miranda	129.80	5.08	1.41	236	2.4×10^{-4}	——
	Ariel	191.20	7.48	2.52	579	0.02	——
	Umbriel	266.00	10.41	4.14	586	0.01	——
	Titania	435.80	17.05	8.71	790	0.08	~1.5
	Oberon	582.60	22.79	13.46	762	0.08	~1.5
Neptune	Naiad	48.23	1.95	0.29	29	——	——
	Thalassa	50.07	2.02	0.312	40	——	——
	Despoina	52.53	2.12	0.335	79	——	——
	Galatea	61.95	2.50	0.493	74	——	——
	Larissa	73.55	2.97	0.555	96	——	——
	Proteus	117.64	4.75	1.12	208	——	——
	Triton	354.8	14.33	5.88(R)	1350	1.82	——
	Nereid	5509	222.46	359.6	170	2.8×10^{-4}	——
Pluto	Charon	19.7	——	6.39	600	0.01	2.1

[1] (R) = Retrograde motion

APPENDIX D
THE 40 BRIGHTEST STARS

Star	Name	Apparent Magnitude	Spectral Type	Luminosity (Sun = 1)	Distance pc (ly)
α CMa A	Sirius A	−1.46	A1 V	21	2.7 (8.7)
α Car	Canopus	−0.72	F0 III	1300	30 (98)
α Boo	Arcturus	−0.06	K2 III	100	11 (36)
α Cen A	Rigil Kentaurus	0.01	G2 V	1.4	1.3 (4.2)
α Lyr	Vega	0.04	A0 V	49	8.1 (26.5)
α Aur	Capella	0.05	G8 III	134	13.8 (45)
β Ori A	Rigel	0.14	B8 Ia	53,000	276 (900)
α CMi A	Procyon	0.37	F5 IV-V	6.4	3.5 (11.3)
α Ori	Betelgeuse	0.41	M2 Iab	13,000	160 (520)
α Eri	Achernar	0.51	B3 V	640	36 (118)
β Cen AB	Hadar	0.63	B1III	9300	150 (490)
α Aql	Altair	0.77	A7 IV-V	10	5.1 (16.5)
α Tau A	Aldebaran	0.86	K5 III	150	21 (68)
α Vir	Spica	0.91	B1 V	1600	67 (220)
α Sco A	Antares	0.92	M1 Ib	8500	160 (520)
α PsA	Fomalhaut	1.15	A3 V	12	6.9 (22.6)
β Gem	Pollux	1.16	K0 III	31	10.7 (35)
α Cyg	Deneb	1.26	A2 Ia	53,000	490 (1600)
β Cru	Beta Crucis	1.28	B0.5 III	1.1	150 (490)
α Leo A	Regulus	1.36	B7 V	150	26 (85)
α Cru A	Acrux	1.39	B0.5 IV	3700	114 (372)
α Cen B		1.4	K4 V	0.4	1.3 (4.2)
ε CMa	Adhara	1.48	B2 II	6400	209 (681)
λ Sco	Shaula	1.6	B1 V	1900	95 (310)
γ Ori	Bellatrix	1.64	B2 III	2800	144 (469)
β Tau	Elnath	1.65	B7 III	300	92 (300)
β Car	Miaplacidus	1.67	A1 III	64	26 (85)
γ Cru	Gacrux	1.69	M4 III	230	67 (218)
ε Ori	Alnilam	1.7	B0 Ia	49,000	490 (1597)
α Gru	Al Na'ir	1.76	B7 IV	210	20 (65)
ζ Ori	Alnitak	1.79	O9.5 Ib	23,000	490 (1597)
ε UMa	Alioth	1.79	A0p	59	21 (69)
α Per	Mirfak	1.8	F5 Ib	8400	175 (571)
α UMa	Dubhe	1.81	K0 III	160	32 (104)
ε Sgr	Kaus Australis	1.81	B9.5 III	77	38 (124)
γ Vel	Suhail al Muhlif	1.83	WC8	37000	160 (522)
δ CMa		1.85	F8 Ia	120,000	644 (2100)
β Aur	Menkalinan	1.86	A2 V	41	27 (88)
α Cru B	Acrux	1.86	B1 V	1500	114 (372)
θ Sco	Sargas	1.86	F0 Ib	700	199 (649)

APPENDIX E
STARS NEARER THAN 4 PARSECS

Star	Parallax (arcseconds)	Distance pc (ly)	Spectral Type	Apparent Magnitude	Luminosity (Sun = 1)
Sun	——	——	G2 V	−26.7	1
α Cen A	0.750	1.3 (4.2)	G2 V	−0.01	1.6
α Cen B	0.750	1.3 (4.2)	K0 V	1.3	0.45
α Cen C	0.772	1.3 (4.2)	M5e	11.0	5.5×10^{-5}
Barnard's star	0.546	1.8 (5.9)	M5 V	9.5	4.5×10^{-4}
Wolf 359	0.419	2.4 (7.5)	M6e	13.5	1.9×10^{-5}
Lalande 21185	0.398	2.5 (8.2)	M2 V	7.5	5.5×10^{-3}
Luyten 726−8A	0.387	2.6 (8.8)	M6e	12.5	5.7×10^{-5}
Luyten 726−8B	0.387	2.6 (8.8)	M6e	13.0	4.2×10^{-5}
Sirius A	0.376	2.7 (8.8)	A1 V	−1.5	$2.4 \times 10^{+1}$
Sirius B	0.376	2.7 (8.8)	wd	8.7	2.9×10^{-3}
Ross 154	0.342	2.9 (9.5)	M4e	10.6	4.8×10^{-4}
Ross 248	0.314	3.2 (10.4)	M5e	12.3	1.1×10^{-4}
ε Eri	0.307	3.3 (10.8)	K2 V	3.7	3.0×10^{-1}
Ross 128	0.301	3.3 (10.8)	M5	11.1	3.6×10^{-4}
Luyten 789−6	0.294	3.4 (11.1)	M5e	12.2	1.3×10^{-4}
ε Ind	0.290	3.4 (11.1)	K5 V	4.7	1.4×10^{-1}
61 Cyg A	0.289	3.5 (11.4)	K5 V	5.2	8.2×10^{-2}
61 Cyg B	0.289	3.5 (11.4)	K7 V	6	3.9×10^{-2}
Procyon A	0.285	3.5 (11.4)	F5 IV	0.4	$7.7 \times 10^{+0}$
Procyon B	0.285	3.5 (11.4)	wd	10.7	5.5×10^{-4}
Σ 2398 A	0.284	3.5 (11.4)	M3.5 V	8.9	3.0×10^{-3}
Σ 2398 B	0.284	3.5 (11.4)	M4 V	9.7	1.5×10^{-3}
Groombridge 34A	0.282	3.5 (11.4)	M1 V	8.1	6.6×10^{-3}
Groombridge 34B	0.282	3.5 (11.4)	M6 V	11.0	4.2×10^{-4}
Lacaille 9352	0.279	3.6 (11.7)	M2 V	7.4	1.3×10^{-2}
τ Ceti	0.273	3.7 (12.1)	G8 V	3.5	4.5×10^{-1}
BD +5 1668	0.266	3.8 (12.4)	M4 V	9.8	1.4×10^{-3}
L725−32	0.265	3.8 (12.4)	M6e	11.6	3.2×10^{-4}
Lacaille 8760	0.260	3.8 (12.4)	M1	6.7	2.6×10^{-2}
Kapteyn's star	0.256	3.9 (12.7)	M0 V	8.8	3.9×10^{-3}
Kruger 60A	0.254	3.9 (12.7)	M4	9.8	1.6×10^{-3}
Kruger 60B	0.254	3.9 (12.7)	M5e	11.3	4.2×10^{-4}

APPENDIX F
THE LOCAL GROUP GALAXIES

	Type	Luminosity (Sun = 1)
M31 (Andromeda)	Sb	2.1×10^{10}
Milky Way	Sb/Sc	1.3×10^{10}
M33 = NGC 598	Sc	2.8×10^{9}
Large Magellanic Cloud	Ir	1.3×10^{9}
M32 = NGC 221	E2	2.8×10^{8}
NGC 6822	Ir	2.8×10^{8}
NGC 205	Spheroidal	2.6×10^{8}
Small Magellanic Cloud	Ir	2.3×10^{8}
NGC 185	Dwarf spheroidal	1.0×10^{8}
NGC 147	Dwarf spheroidal	8.5×10^{7}
IC 1613	Ir	7.0×10^{7}
WLM	Ir	3.4×10^{7}
Fornax	Dwarf spheroidal	2.3×10^{7}
And I	Dwarf spheroidal	4.1×10^{6}
And II	Dwarf spheroidal	4.1×10^{6}
Leo I	Dwarf spheroidal	3.7×10^{6}
DDO 210	Ir	3.1×10^{6}
Sculptor	Dwarf spheroidal	1.5×10^{6}
And III	Dwarf spheroidal	1.0×10^{6}
Pices	Ir	9.3×10^{5}
Sextans	Dwarf spheroidal	7.7×10^{5}
Phoenix	Dwarf Ir / Dwarf spheroidal	7.0×10^{5}
Tucana	Dwarf spheroidal	4.9×10^{5}
Leo II	Dwarf spheroidal	4.4×10^{5}
Ursa Minor	Dwarf spheroidal	2.8×10^{5}
Draco	Dwarf spheroidal	2.1×10^{5}
Carina	Dwarf spheroidal	8.5×10^{4}
EGB 0427+63	Dwarf Ir	———
Sagittarius	Dwarf spheroidal	———

APPENDIX G
STAR MAPS
FOR THE
NORTHERN HEMISPHERE

INDEX TO STAR MAPS

Choose the date and time; the appropriate map is then specified.

Date	8 P.M.	9 P.M.	10 P.M.	11 P.M.	12 P.M.	1 A.M.
Jan 05		1		2		3
Jan 20	1		2		3	
Feb 05		2		3		4
Feb 20	2		3		4	
Mar 05		3		4		5
Mar 20	3		4		5	
Apr 05		4		5		6
Apr 20	4		5		6	
May 05		5		6		7
May 20	5		6		7	
Jun 05		6		7		8
Jun 20	6		7		8	
Jul 05		7		8		9
Jul 20	7		8		9	
Aug 05		8		9		10
Aug 20	8		9		10	
Sep 05		9		10		11
Sep 20	9		10		11	
Oct 05		10		11		12
Oct 20	10		11		12	
Nov 05		11		12		1
Nov 20	11		12		1	
Dec 05		12		1		2
Dec 20	12		1		2	

Directions:
Hold map so that heading of observation is at the bottom.

January	20	8 p.m.
January	5	9 p.m.
December	20	10 p.m.
December	5	11 p.m.
November	20	12 p.m.
November	5	1 a.m.

Magnitudes

- 0 and brighter
- 1
- 2
- 3
- 4 and fainter

Map 1

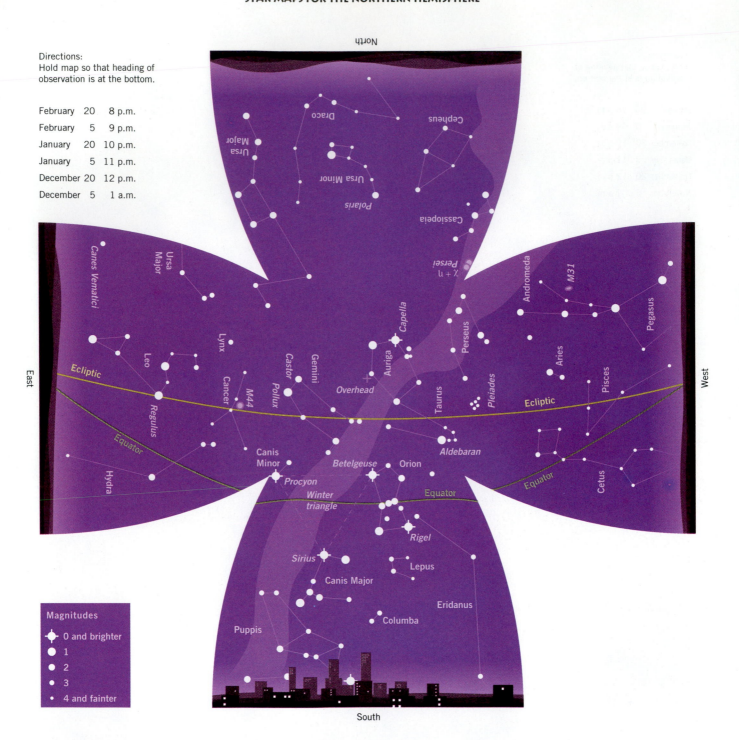

Directions:
Hold map so that heading of observation is at the bottom.

February	20	8 p.m.
February	5	9 p.m.
January	20	10 p.m.
January	5	11 p.m.
December	20	12 p.m.
December	5	1 a.m.

Magnitudes

- ✦ 0 and brighter
- ● 1
- ● 2
- • 3
- · 4 and fainter

Map 2

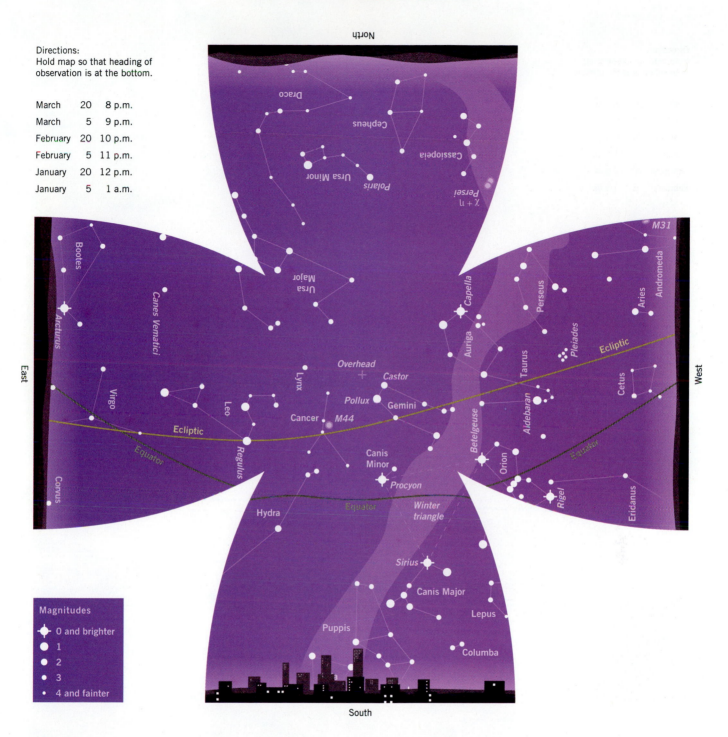

Directions:
Hold map so that heading of observation is at the bottom.

March	20	8 p.m.
March	5	9 p.m.
February	20	10 p.m.
February	5	11 p.m.
January	20	12 p.m.
January	5	1 a.m.

Magnitudes
- ✦ 0 and brighter
- ● 1
- ● 2
- • 3
- · 4 and fainter

Map 3

Directions:
Hold map so that heading of observation is at the bottom.

April	20	8 p.m.
April	5	9 p.m.
March	20	10 p.m.
March	5	11 p.m.
February	20	12 p.m.
February	5	1 a.m.

North

Perseus
χ + η Persei
Cassiopeia
Cepheus
Draco
Polaris
Ursa Minor

East

Hercules
M13
Corona Borealis
Bootes
Serpens
Arcturus
Canes Venatici
Virgo
Libra
Equator
Ecliptic
Hydra
Spica

Ursa Major
Lynx
Overhead
Leo
Regulus
Cancer
M44
Pollux
Castor
Gemini
Canis Minor
Procyon
Winter triangle
Capella
Auriga
Ecliptic
Perseus
Pleiades
Taurus
Betelgeuse
Aldebaran
Orion
Equator
Rigel
Eridanus
Lepus
Sirius

West

Equator
Hydra

Corvus

Canis Major
Puppis
Vela
Centaurus

Magnitudes
- ✦ 0 and brighter
- ● 1
- ● 2
- • 3
- · 4 and fainter

South

Map 4

Directions:
Hold map so that heading of
observation is at the bottom.

May	20	8 p.m.
May	5	9 p.m.
April	20	10 p.m.
April	5	11 p.m.
March	20	12 p.m.
March	5	1 a.m.

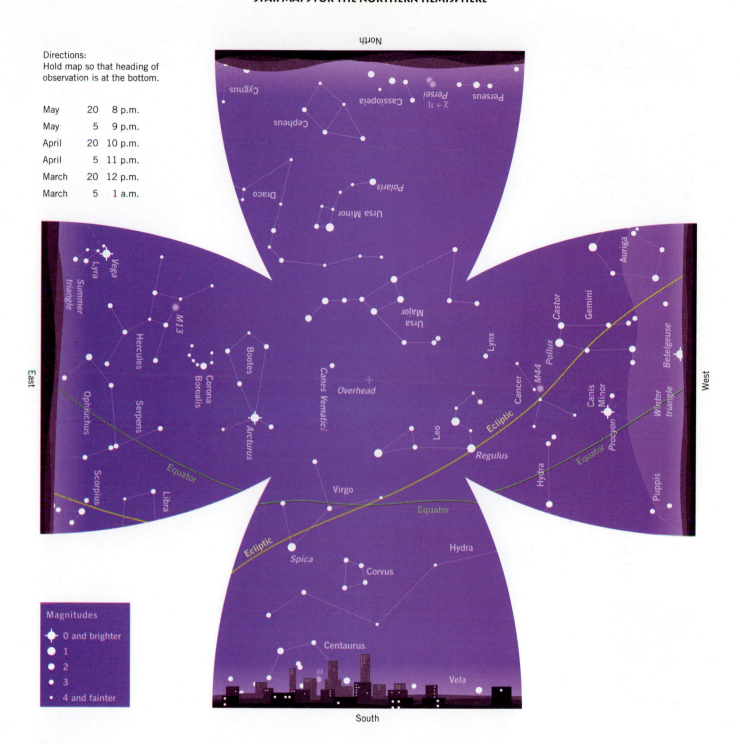

Magnitudes
- 0 and brighter
- 1
- 2
- 3
- 4 and fainter

Map 5

Directions:
Hold map so that heading of
observation is at the bottom.

June	20	8 p.m.
June	5	9 p.m.
May	20	10 p.m.
May	5	11 p.m.
April	20	12 p.m.
April	5	1 a.m.

Magnitudes
- 0 and brighter
- 1
- 2
- 3
- 4 and fainter

Map 6

Directions:
Hold map so that heading of
observation is at the bottom.

July	20	8 p.m.
July	5	9 p.m.
June	20	10 p.m.
June	5	11 p.m.
May	20	12 p.m.
May	5	1 a.m.

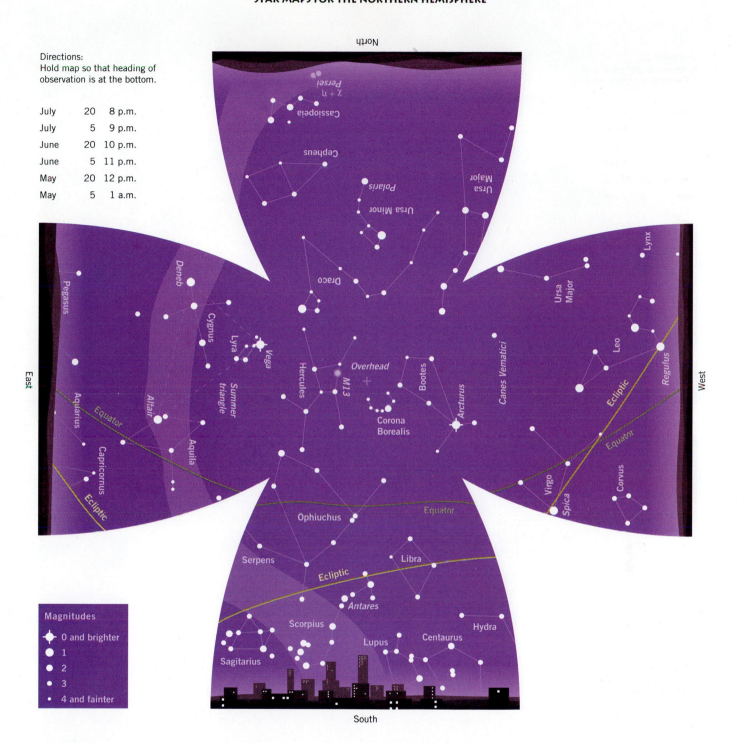

Magnitudes
- 0 and brighter
- 1
- 2
- 3
- 4 and fainter

Map 7

Directions:
Hold map so that heading of observation is at the bottom.

August	20	8 p.m.
August	5	9 p.m.
July	20	10 p.m.
July	5	11 p.m.
June	20	12 p.m.
June	5	1 a.m.

North

Perseï
χ + η
Cassiopeia
Polaris
Cepheus
Ursa Minor
Ursa Major
Draco
Canes Venatici
Leo

Andromeda
M31
Deneb
Cygnus
Lyra
Vega
Overhead
M13
Bootes
Arcturus

East

Pegasus
Pisces
Summer triangle
Hercules
Corona Borealis
Virgo

Aquarius
Altair
Aquila
Serpens
Equator
Equator
Ecliptic
Spica

Capricornus
Ecliptic
Equator
Ophiuchus
Hydra
Libra

West

Serpens
Libra
Ecliptic
Capricornus
Antares
Sagitarius
Scorpius
Lupus

Magnitudes
- 0 and brighter
- 1
- 2
- 3
- 4 and fainter

South

Map 8

Directions:
Hold map so that heading of observation is at the bottom.

September 20	8 p.m.
September 5	9 p.m.
August 20	10 p.m.
August 5	11 p.m.
July 20	12 p.m.
July 5	1 a.m.

Magnitudes
- 0 and brighter
- 1
- 2
- 3
- 4 and fainter

Map 9

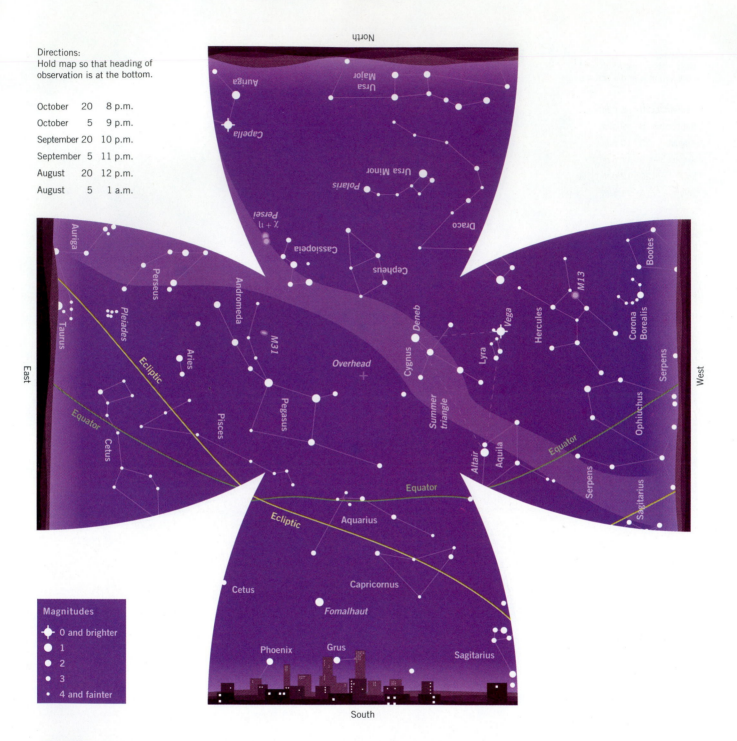

Directions:
Hold map so that heading of observation is at the bottom.

October	20	8 p.m.
October	5	9 p.m.
September	20	10 p.m.
September	5	11 p.m.
August	20	12 p.m.
August	5	1 a.m.

Magnitudes

- 0 and brighter
- 1
- 2
- 3
- 4 and fainter

North

East

West

South

Overhead

Summer triangle

Ecliptic

Equator

Map 10

North

Directions:
Hold map so that heading of
observation is at the bottom.

November 20   8 p.m.
November 5   9 p.m.
October   20   10 p.m.
October   5   11 p.m.
September 20   12 p.m.
September 5   1 a.m.

East

West

Ursa Major
Draco
Ursa Minor
Polaris
Cepheus
Gemini
Capella
Auriga
χ + η
Perseus
Cassiopeia
Vega
Hercules
M13
Lyra
Deneb
Betelgeuse
Orion
Perseus
Pleiades
Taurus
Aldebaran
Aries
Andromeda
M31
Overhead
Cygnus
Summer triangle
Rigel
Cetus
Pegasus
Aquila
Altair
Eridanus
Ecliptic
Equator
Pisces
Aquarius
Equator
Capricornus
Ecliptic
Cetus
Equator
Ecliptic
Aquarius
Capricornus
Fomalhaut
Phoenix
Grus

Magnitudes
⊕ 0 and brighter
● 1
● 2
• 3
· 4 and fainter

South

Map 11

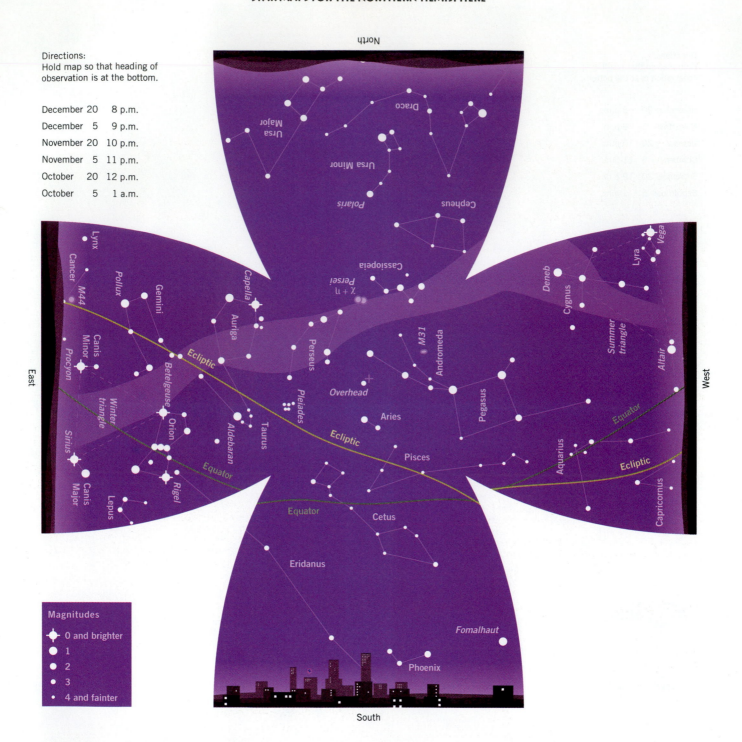

Directions:
Hold map so that heading of observation is at the bottom.

December 20	8 p.m.
December 5	9 p.m.
November 20	10 p.m.
November 5	11 p.m.
October 20	12 p.m.
October 5	1 a.m.

Magnitudes
✦ 0 and brighter
● 1
● 2
• 3
· 4 and fainter

Map 12

GLOSSARY

aberration of light: A small shift in the apparent direction to an object caused by the motion of the Earth and the finite speed of light. This effect was used by Bradley in 1727 to make the first historical proof that the Earth moved through space.

absorption-line spectrum: The spectrum formed by narrow regions in a spectrum when a continuous spectrum passes through a cooler gas.

acceleration: The rate of change of velocity with time (the change in velocity divided by the elapsed time).

accretion: An aggregation of material that is increasing its mass by gathering and incorporating smaller pieces of material—either by gravity or by collisions.

accretion disk: Material accreted by a larger mass and caused to spiral into a flattened "disk" shape by the conservation of angular momentum.

Active Galactic Nuclei (AGN): One of a class of galaxies with very energetic behavior observed in the nuclear region, including high levels of nonthermal continuous emission and unusual radio jets.

active optics: Telescopic systems that can adjust the shape and focus of their optical components rapidly, to compensate for distortion in the incoming signal due to turbulence and other motions in the Earth's atmosphere. Also called *adaptive optics*.

albedo: The fraction of incoming light that is reflected from a surface.

Algol: An Arabic name meaning "Demon Star." Algol is an eclipsing binary system whose variations in brightness can be seen by the naked eye.

Allende meteorite: A meteorite with unusual isotopic ratios suggesting that a supernova may have played a part in forming the early solar system.

Almagest: The Arabic title of Ptolemy's magnum opus.

almanac: A table giving the position at various times of a celestial body. Also called an *ephemeris*.

alpha–capture reactions: Nuclear reactions in which alpha particles are added to a heavier nucleus.

Alpha Centauri: A bright triple-star system close to the Sun.

alpha particle: The nucleus of a helium atom, composed of two protons and two neutrons.

altitude: The angular distance of an object above the horizon.

amino acids: Complex molecules that form the basic building blocks of protein molecules. Conditions inside the interstellar clouds are favorable for the formation of these molecules.

Anaximander: A member of the Pythagorean school, credited with early theories on the composition and motion of the universe.

Andromeda Galaxy: (also *Andromeda Nebula* and *M31*). This diffuse, naked-eye object is revealed by telescopic observations to be a giant, nearby spiral galaxy over 2 million light-years away.

angle of incidence: The angle between an incident ray of light and a perpendicular to the surface at the point of incidence.

angle of reflection: The angle between a reflected ray of light and the perpendicular to the surface at the point of reflection.

angstrom: (abbreviated Å) A unit for measuring the wavelengths of light. It is 10^{-8} cm. On this scale blue light is around 4500 Å.

angular momentum: A measure of the amount of spin or orbital motion possessed by a body due to its inertia. Physicists describe it as a "conserved" property. See *conservation*.

angular size: The angle at the eye subtended by rays drawn from the outside of an object. Same as *apparent size*.

annihilation: The total conversion of mass to energy brought about by the collision of matter with anti-matter.

annular eclipse: An eclipse of the Sun in which the angular size of the Moon's disk is not quite large enough to block the solar photosphere. In an annular eclipse, the Sun is not dimmed enough to allow the corona to be seen.

antielectron: The antiparticle of an electron.

antimatter: Material with charge and various other properties reversed when compared with normal matter. When matter and antimatter collide, they annihilate each other.

Ap (A-peculiar stars): A-type stars that have some anomalies in their chemical composition when compared with normal A-stars.

aphelion: The greatest distance from the Sun reached by an object orbiting the Sun. See also *perihelion*.

apogee: The most distant point in the orbit of an Earth-orbiting body. See also *perigee*.

Apollo asteroids: A group of asteroids with orbits that come close to the Earth.

Apollonius of Perga: Early Greek credited with the first suggestion of epicycles and deferents as a device to explain the motions of celestial bodies.

apparent magnitude: The brightnesses of celestial bodies measured on a scale first suggested by Hipparchus. In this system (which turns out to be logarithmic), fainter stars are characterized by larger numbers. Two stars that differ by one magnitude differ by a factor of 2.512 in their brightness ratio. A difference of 5 magnitudes corresponds to a factor of 100 in brightness.

apparent size: See *angular size*.

apparent solar time: Also called *sundial time*. The hours of the day as measured by the apparent progress of the Sun across the sky.

Ariel: One of the larger moons of Uranus.

Aristarchus: An early Greek astronomer who carried out estimates of the relative sizes and distances of the Earth, Moon, and Sun and espoused a sun-centered solar system model.

Aristotle: The most influential of Greek philosophers. He produced enduring studies in biology, geometry, and logic. His work in physical science and astronomy, however, was less successful.

asteroid: A body whose diameter ranges from that of a grain of sand up to that of Texas. Most asteroids orbit in the asteroid belt. Also called *minor planets*.

asteroid belt: A region of asteroids located between the orbits of Mars and Jupiter.

asthenosphere: A region of soft material in the Earth's upper mantle.

Astronomical Unit (AU): The average distance from the Earth to the Sun— about 150 million km (or 93 million mi).

asymptotic giant branch (AGB): A region in the H-R diagram of a globular cluster. It is a post-main-sequence evolutionary track followed by stars having both an H- and an He-burning shell.

atmosphere: The gaseous outer regions of a planet or a star.

atmospheric extinction: A reduction in the apparent brightness of the light from a celestial object due to the absorption and scattering of light out of the incoming beam of energy by the Earth's atmosphere.

atmospheric windows: Spectral regions of the atmosphere that are largely or partly transparent to radiation.

atom: The smallest subdivision of matter (the smallest piece) that still retains the chemical properties of the element.

atomic number: The number of protons in the nucleus of an element.

atomic weight: A scale of the relative weights of the atoms of different chemical elements. For an individual atom, the atomic weight would be the atomic mass number—the number of protons plus neutrons in the nucleus. For example, the most frequently occurring form of carbon has a nucleus with 6 protons and 6 neutrons and an atomic weight of 12. There is a less frequently occurring form of carbon with 6 protons and 7 neutrons and an atomic weight 13. The "average" atomic weight for carbon depends upon the frequency of occurrence in nature of the different types of carbon.

AU: Abbreviation for Astronomical Unit.

aurora: High-energy charged particles from space impact upon and excite atmospheric molecules causing them to radiate. This radiation is more commonly seen near the magnetic poles, since that is where the magnetic field lines tend to direct the incoming particles. The aurora borealis is seen around the north magnetic pole, and the aurora australis is seen in south magnetic regions.

auroral ring: Auroral radiation occurring in a ring-shaped region around the magnetic pole.

autumnal equinox: The moment when the Sun crosses the celestial equator from north to south, marking the beginning of fall. See also *equinox*.

average: From a series of measurements, the value most likely to be closest to the true value. It helps to negate the effects of high and low values.

axis: A line about which a body rotates.

azimuth: Angular distance along the horizon of a celestial object, measured clockwise (eastward) from north.

background 3 K radiation: Faint radio static with a highly isotropic distribution, thought to be remnant radiation from the early stages after the big bang.

Balmer lines: A set of spectral lines due to transitions of the hydrogen atom between level 2 and higher energy levels.

barium stars: Red giant stars showing abnormally large abundances of barium and other chemical elements not normally found in most stars.

Barnard's star: A star close to the Sun with "wiggles" in its motion, suggesting the possibility of other masses (in particular, planets) orbiting the star.

barred spirals: Galaxies with a "bar" of stars running through the nucleus. The spiral arms appear to start from the bar.

Barringer Crater: A one-mile diameter crater in Arizona created by meteoric impact.

basalt: Igneous rock (produced at high temperatures) that is common in the outer parts of the Earth and the Moon.

basin: Large, shallow depression seen on terrestrial planets, created by impact or plate tectonics.

Bayeux tapestry: Woven at the time of the Battle of Hastings, it shows a representation of the 1066 A.D. appearance of Halley's comet.

Becklin-Neugebauer source (B-N source): A bright point source of infrared emission seen in the Orion Nebula. It is probably either a new star or a new cluster of stars in the process of formation.

Beta Lyrae: A star whose rapid rotation changes the shape of its spectral line profiles.

Beta Persei: See *Algol*.

Beta Pictoris: A star circled by a disk of dusty material that may form planets.

big bang: A model that postulates the origin of the universe in a state of enormous temperature and density (see *primeval fireball*) that produced the expanding universe we observe today.

big crunch: What results when an expanding universe reverses direction, so that density and temperature again increase.

Bighorn medicine wheel: A construction of rocks high in the Wyoming Big Horn Mountains that may have been constructed by Plains Indians for astronomical, calendrical, and ritualistic purposes.

binary accretion model: Postulates that the formation of the Earth-Moon system occurred by accretion of both bodies from the same dust and gas cloud.

binary pulsar: A highly evolved binary pair of stars in which one of the two has become a pulsar.

binary star: Two stars in mutual orbit under the attraction of gravitational forces.

binary X-ray source: A highly evolved binary pair of stars in which one star has become an X-ray-emitting, collapsed object such as a white dwarf, neutron star, or black hole. The X-rays are thought to be emitted by a hot accretion disk in the system.

bipolar flow: Gas flowing outward in opposite directions seen in objects that may be young stellar objects (YSOs).

blackbody: A conceptualized "perfect radiator," which is found useful in theoretical studies of radiation. A blackbody is one that absorbs 100% of the radiation it encounters. It re-emits that energy in a spectral pattern that depends only upon temperature.

blackbody spectrum: A continuous distribution of energy given off by a blackbody. The amount of energy emitted at each wavelength by a blackbody is given by Planck's Law.

black holes: Collapsed objects whose gravitational field is so intense that not even light can escape from them. They cannot be observed directly, but their effect upon their environment can be studied.

blink microscope (blink comparator): An instrument that projects two different photos of the same area of the sky in rapid succession. Stars with brightness variations will be seen to "blink" through the viewer.

BL Lac objects: Active galaxies that show rapidly varying, highly polarized non-thermal emission from their nuclei.

Bohr model of the atom: An early "solar system" model for the atom. Negatively charged electrons were envisioned as orbiting the massive and positively charged nucleus of the atom. Only certain orbits were allowed.

Bok globules: Small dark regions seen in interstellar clouds, many of which are observed to surround a young star.

bolide: A meteor that explodes in the Earth's atmosphere.

bottom-up theory: A model of galaxy cluster formation in which galaxies form before clustering itself occurs.

bow shock: A shock wave of pressure and density created by material moving at supersonic speed through a medium. Sonic booms are a terrestrial example.

Brackett series: A set of spectral lines due to transitions of the hydrogen atom between level 3 and higher energy levels.

Brahe, Tycho: Sixteenth-century astronomer famous for his accurate naked-eye observations. Considered to be the father of modern observational astronomy. Collaborated with Johannes Kepler to complete the Copernican Revolution.

braided F-ring: One of Saturn's rings that shows time-variable shape and behavior.

breccia: A rock composed of different types of mineral fragments pressed together.

bright-line spectrum: Same as *emission-line spectrum*.

brown dwarf: A low-mass object that is larger than a typical planet but that cannot have hydrogen fusion in its core. As a consequence, it is an "almost-star" with low temperature and luminosity.

burster: Sporadic outbursts of high-intensity X-ray emission. The sources are still a matter for argument.

butterfly diagram: The pattern created when the latitude at which sunspots form are graphed as a function of time.

Callisto: The outermost Galilean satellite.

Caloris Basin: A large multi-ring basin on Mercury.

canali ("canals"): Long, dark features seen on Mars by some astronomers. Schiaparelli called them *canali*, which is Italian for channels. Other people made the transition to "canals" in their hypothesizing and went on to postulate life and modern civilizations on Mars.

capture model: Imagines the Moon to have been captured by the Earth's gravity.

carbonaceous chondrites: A group of meteorites with a relatively high carbon abundance and chondrules. See *chondrules*.

Cas A: A strong radio source that is the remnant of a supernova explosion that took place in 1667.

Cassegrain: A telescope arrangement that uses a secondary mirror and a small hole in the primary mirror to focus light behind the primary mirror.

Cassini division: A gap in Saturn's rings named after its discoverer—Giovanni Cassini.

cataclysmic variables: Stars that become unstable and violent in their final evolutionary stages.

CCD (charge coupled device): A two-dimensional semiconductor chip that is used as an extremely sensitive photon detector.

celestial equator: This great circle on the celestial sphere is the projection of the Earth's equator onto the plane of the sky. It is halfway between the celestial poles.

celestial pole: The projection onto the celestial sphere of the spin axis of the Earth. The North Celestial Pole (NCP) is currently located near the star Polaris, while the SCP is not near a bright star. The long, slow motion of the Earth called precession (*see below*) will cause these two celestial poles to slowly move with respect to the stars over a 26,000-year cycle.

celestial sphere: The "Bowl of the Sky." To the ancients, this is the apparent sphere of sky encircling the Earth on which the stars appear to be fixed.

Cen A: A strong radio source resembling an elliptical galaxy but with large regions of dust obscuration.

center of mass: The imaginary point on the sky about which two gravitationally bound bodies will orbit.

centripetal force or acceleration: A force or acceleration directed towards the center of a circular path.

Cen X-3: A powerful X-ray source in Centaurus composed of a giant star orbited by an X-ray-emitting neutron star.

Cepheid instability strip: The region of the H-R diagram where Cepheid variable stars and Cepheid-like variables are located.

Cepheid variables: Luminous giant stars whose energy output varies in a regular and recognizable manner. They are important for determining the distances to nearby galaxies.

Ceres: The first discovered and largest asteroid.

Chandrasekhar limit: The largest mass that theory will permit a white dwarf to acquire, roughly 1.4 solar masses.

Charon: Pluto's moon.

Chiron: An object discovered in 1977 in an orbit that lies mostly between Saturn and Uranus. Initially classified as an asteroid but later exhibited behavior more like the nucleus of a comet.

chondrites: See *carbonaceous chondrites*.

chondrules: Silicate nodules found in a particular class of meteorites that may be a good sample of primeval solar system composition.

chromatic aberration: A defect of telescopes made with lenses. Different colors of light are refracted (bent) by different amounts, so that they are brought to different focal points.

chromosphere: The layer of the Sun's atmosphere just above the photosphere where most of the absorption lines are formed.

circumpolar stars: Those stars that never go below the horizon in their 24-hour daily motion because of their proximity to the celestial pole.

Circus Maximus: A surface feature on Uranus's satellite, Miranda.

closed universe: A universe that possesses enough mass that the curvature of space would be positive, and the universe would be finite.

CNO cycle: A cycle of nuclear reactions in which H fusion produces He through reactions that use C nuclei as a catalyst.

COBE: **CO**smic **B**ackground **E**xplorer satellite.

collisional dissociation: The breakup of a molecule through a collision.

collisional ionization: An atom or an ion loses an electron due to an energetic collision.

coma: The bright diffuse halo of a comet. Also called the *head*.

comets: Small (a few miles in size) solar system bodies composed of dust and frozen gases that evaporate and glow as they near and go around the Sun. Some comets have regular and predictable reappearances, whereas others enter the solar system without warning and from unpredictable directions.

comparative planetology: An approach to the study of planetary bodies, having the goal of understanding their similarities and differences.

conduction: The transfer of heat by particle (often electron) collisions.

conjunction: Two bodies getting as close to each other in the sky as possible.

conservation: A conserved quantity is one whose numerical value stays constant in a system with no external forces present. Examples are linear and angular momentum, and energy. The momentum can be passed between bodies but it does not disappear.

constellations: Arbitrary groupings of stars that have passed down to us from ancient times. Different cultures on Earth historically evolved different constellations and myths related to them. For convenience, astronomers have adopted the ancient Greek constellations and given them well-defined boundaries.

constructive interference: When two similar waves encounter each other and interact in such a way as to combine their strength. See also *destructive interference*.

continental drift: A slow movement of the Earth's great land masses that has produced the present continents we now see from the breakup and drifting apart of one giant landmass.

continental plates: The large-scale, surface crust layers of the Earth.

continuous creation of matter: A theory that tries to explain the expansion of the universe as being due to the pressure caused by new matter that is hypothesized to be constantly and spontaneously appearing. See also *steady state hypothesis*.

continuous spectrum: A spectrum in which energy is emitted at all wavelengths. With the naked eye, we see a "rainbow" of color. It is distinct from the absorption- and emission-line spectrum.

convection: The transfer of heat energy by the movements of hot material.

co-orbital satellites: Satellites inhabiting the same orbits. Examples occur with Saturn's satellites Dione and Tethys.

Copernican hypothesis: The hypothesis put forth by Copernicus in 1543 that the planets all orbited around the Sun.

Copernicus satellite: An orbiting telescope for observing stars in the ultraviolet spectral region.

Cordillera Mountains: Lunar mountains forming the multi-ringed Orientale Basin.

coriolis effects: Observed effects due to living on a rotating frame of reference. Two examples would be (1) the deflections exhibited by artillery shells, and (2) the circulations of winds on the Earth.

corona: The Sun's tenuous outer atmosphere, which is visible only at the time of a total solar eclipse. It has a temperature near 2 million degrees.

coronagraph: A telescopic device to block most of the light from the bright surface of the Sun and reveal its fainter, outer layers without an eclipse.

coronal holes: Regions of the Sun's corona where strong loops of magnetic field produce a lower than normal density in the hot coronal plasma.

cosmic abundance: Most of the universe exhibits a chemical abundance that is (by number of atoms) roughly 90% hydrogen, almost 10% helium, and less than 1 or 2% of all the other elements. The planets form an exception to this generalization.

cosmic rays: Particles from space moving at close to the speed of light, impinging on the Earth with tremendous energies from all directions in space.

cosmology: The study of the structure and evolution of the universe.

coudé: A focal arrangement for a telescope that sends the light down the polar axis into an observing room. This is the longest possible focal length for a telescope and produces the largest possible image.

Crab Nebula: An energetic gas cloud with a pulsar in the center, which has been found to be the remains left over from the spectacular 1054 A.D. supernova explosion in Taurus.

crater: A depression in a planetary or satellite surface, created by the impact of a fast-moving body or by volcanism.

crater chains: A line of craters that would be unlikely to be produced by impact. In many cases, they provide evidence for some sort of "volcanic" activity. Some are caused by secondary impacts.

critical density: The particular value for the density of matter in the universe that would just bring the observed expansion of the universe to a halt.

critical mass: The mass of material required to reach the critical density.

curved space: According to Einstein's general theory of relativity, the distribution of mass in the universe determines the curvature of space. As bodies move, they must follow the local curvature of space, which will not necessarily be the familiar three dimensions of Euclidian geometry.

Cyg A: A powerful source of synchrotron radiation in Cygnus that may be two galaxies colliding with each other.

Cygnus X-1, X-3: X-ray sources in Cygnus that probably consist of a collapsed object emitting X-rays and orbiting a giant star.

dark cloud: An interstellar cloud of dust and gas that is dense enough to be opaque.

dark-line spectrum: See *absorption-line spectrum*.

dark matter: Not yet detected or explained, the existence of this material is revealed by the rapid accelerations of galaxies seen in clusters of galaxies. The amount of this material will be critical in determining the future fate of the universe.

Death Star: See *Nemesis*.

declination: The angular distance of a celestial object north or south of the celestial equator.

decoupling (of radiation and matter): The epoch early in the life of the universe when the matter becomes largely transparent to the radiation in the universe.

deductive reasoning (deduction): Inference from a general statement to a specific prediction.

deferent: In the ancient geocentric theory of the universe, this is a large circle approximately centered upon the Earth, along which a smaller circle, the epicycle, would move.

degenerate electron gas: Electrons compressed into a state of density so high that the ordinary gas laws break down. In the degenerate state, a change of temperature does not result in a change of pressure.

Deimos: One of the two small moons of Mars (the other is Phobos).

density: The amount of mass in a unit volume; mass divided by the volume (mass per unit volume).

density wave: A giant (galaxy-sized), spiral-shaped compression wave that may create spiral arm structure by compressing giant gas clouds and causing the onset of star formation.

destructive interference: If two similar interacting waves are out of phase with each other, the result will be a mutual cancellation.

deuterium: An isotope of hydrogen having one neutron. It is useful in age-dating celestial objects.

diffraction: The bending or spreading-out of waves when they pass through apertures or move close to edges of objects and obstacles.

diffraction grating: A surface that is marked with thousands of fine lines per inch. These lines diffract light as if they were many single slits working in unison, causing a spectrum to form.

diffuse nebula: A gas cloud excited to radiate by absorbing the ultraviolet emission from a hot star in or near the gas cloud. The gas is ionized, and as it recombines, it emits a bright-line spectrum.

dipole anisotropy: An apparent anisotropy in the 3K microwave background radiation caused by the motion of the Earth.

dirty snowball; dirty iceberg: A model for the nucleus of a comet that envisions its structure to be a frozen ball of gas, dust, and rocks that tends to break down when the comet nears the Sun. The evaporated material forms the tail of the comet.

discrete emission: Energy given off only at certain specific wavelengths.

dispersion: The separation of light into its component colors by the action of a prism or a grating.

dissociation: Breaking a molecule apart either by the absorption of a photon or a collision with something.

distance pyramid: A diagram showing how the methods for determining distances to distant objects depend on the distances to nearby ones.

Doppler effect: A shift in the observed wavelength of radiation caused by radial motion of the source of light with respect to the observer.

double-lined spectroscopic binary: A spectroscopic binary in which absorption lines from both stars are visible. Such systems provide vital information on the stellar masses.

double quasar: An optical illusion produced by a strong gravitational field in space. The light from a quasar is bent by the field, producing an extra, spurious image.

double shell source: When fusion occurs in two shells surrounding a star's nucleus.

DQ Herculus: The first stellar nova to be shown to be a binary star system.

dredge-up: Products of fusion reactions deep inside a star, sometimes carried up closer to the surface of the star by convective currents, changing the chemical composition of the material and affecting the reactions that are possible.

dwarf: A small, low-luminosity star. Sometimes used to describe main-sequence stars of luminosity class V.

dwarf elliptical: Small elliptical galaxies containing in some cases as few as a million stars.

dynamo model: A model that postulates the generation of magnetic fields in planets and other objects by the circulation of conducting fluids inside the object.

eccentricity: The deviation from circular shape (or "flattening") of an ellipse.

eclipse: When one celestial body passes in front of another, blocking its light as seen from the Earth.

eclipsing binary: A binary system where the plane of the orbit is nearly in our line of sight, causing each star to periodically pass in front of the other, blocking its light. The light curve of the system will exhibit dips at regular intervals.

ecliptic: The apparent path of the Sun on the celestial sphere with respect to the "fixed stars."

Einstein observatory: An X-ray telescope that orbited outside the atmosphere from 1978 to 1981.

ejecta blanket: Material excavated by an impact explosion and deposited over the surface.

electromagnetic radiation: Energy in the form of a wave, propogating through space at the speed of light.

electromagnetic waves: See *electromagnetic radiation.*

electron: A small, negatively charged particle. In the Bohr model of the atom, the electrons of an element orbit the positively charged, massive nucleus of the atom.

electroweak force: At the high energies permeating the universe in the initial moments after the big bang, the electromagnetic force and the weak nuclear force were initially combined into one force, the electroweak force.

ellipse: A geometric curve in the shape of a "squashed" circle. The shape is determined by the eccentricity. See *eccentricity.*

elliptical orbit: Bodies that are gravitationally bound to other bodies and moving in closed orbits will exhibit an orbital path that is an ellipse.

elongation: The angular distance between a planet and the Sun as viewed from the Earth.

emission-line spectrum: Energy emitted at specific, discrete wavelengths by transparent gases. Each chemical element has its own pattern of emission lines.

emission nebula: A rarified gas cloud excited to radiate by absorbing radiaton from a nearby hot star.

Enceladus: A satellite of Saturn that may be significantly affected by tidal friction.

encounter theory: Proposed explanations of the origin of the solar system in which a collision or near approach of another star to the Sun gravitationally pulls off some material from the Sun. This material is hypothesized to condense and form planets.

energy level diagram: A graphical display of the energy levels of an atom.

energy levels: A specific amount of energy associated with each possible electron orbit. Each chemical element has its own set of stable energy levels.

epicenter: A point on the Earth's surface from which seismic waves appear to radiate, located directly above the true center of the earthquake disturbance.

epicycle: In the ancient Ptolemaic (geocentric) theory of the universe, the epicycle is a small circle that moves around the deferent, while the planet moves around the epicycle. This arrangement can produce the retrograde motion shown by planets.

equatorial bulge: The equatorial diameter of the Earth is slightly larger than the polar diameter.

equatorial coordinate system: A coordinate grid on the sky analogous to latitude and longitude on the Earth. The north-south position of an object is called its declination, while the east-west location is called its right ascension.

equatorial mounting: A mount for a telescope that enables it to follow the motion of a celestial object across the sky. Most mounts will feature a polar axis that points toward the NCP.

equinox: Two points on the celestial sphere where the ecliptic plane and the celestial equator intersect. The spring (vernal) equinox occurs near March 21 when the Sun is crossing the celestial equator going from south to north. The autumnal equinox, at about September 21, occurs when the Sun crosses the celestial equator going from north to south. At these times, day and night are of equal length everywhere on the Earth (hence the name, meaning "equal night").

Eratosthenes: An ancient Greek astronomer who produced an accurate estimate of the size of the Earth by comparing shadows at two locations on Earth.

erg: A tiny unit of energy.

Eros: An Apollo asteroid that passes close to the Earth. Measurements of its parallax gave us our first accurate determinations of the scale of the orbits in the solar system.

escape velocity: If an orbiting body possesses a speed greater than this velocity, then it will escape from the body it is orbiting. Such orbiting bodies will move in parabolic or hyperbolic paths.

Europa: The smallest of the Galilean satellites of Jupiter.

event horizon (Schwarzschild radius): A critical distance characterizing a gravitating mass collapsing into a black hole. When the object shrinks beyond this critical distance, it is no longer visible to us, and light and other information is no longer able to escape from within the object itself.

evolutionary track: The path traced out with time on an H-R diagram by an evolving star, which is changing its temperature and luminosity.

excitation: The process of exciting an atom to a higher energy level, either by absorbing a photon (photo-excitation) or by collisions (collisional excitation).

excited state: An energy level (state) other than the ground state.

expansion of the universe: Everything in the universe is observed to be moving away from everything else.

extinction: The dimming of light observed when it passes through a medium. The interstellar medium causes dimming and reddening of starlight passing through it, as does the atmosphere of the Earth.

faint object camera; faint object spectrograph: Two instrument systems of the Hubble Space Telescope.

fireball: An unusually bright meteor.

fission: The nucleus of a heavy atom splitting apart and forming other, lighter nuclei (and giving off energy in the process).

fission model: A theory of the origin of the Moon, which proposes that the material for the Moon was thrown outward by a rapidly spinning Earth.

five-minute oscillation: An oscillation (expansion and contraction) of the outer layers of the Sun.

flare: Unpredictable energetic outbursts seen in some stars and on the surface of the Sun.

flash spectrum: The instant before a total solar eclipse when the dark-line spectrum is replaced by a spectrum of curved emission lines from the hot gases above the observable surface.

flat space: Ordinary three-dimensional Euclidean space.

fluorescence: A radiation process whereby an atom or molecule absorbs a photon and then re-emits the energy at longer wavelengths.

focal length: The distance from the primary mirror or lens at which the light from a distant source is brought to a focus.

focus: The point in space where the light gathered by telescope is formed into a clear image.

forbidden lines: Certain emission lines that are only seen in very low-density gases such as astronomical nebulae. These spectral features are not normally seen on Earth because of the higher densities (hence the misnomer "forbidden").

frequency: The number of waves passing the observer per second. The frequency of electromagnetic waves is measured in Hertz, which is the number of cycles per second.

full Moon: When the Moon is geometrically opposite to the Sun in the sky, Earth observers can see the disk of the Moon fully illuminated.

FU Orionis star: A small group of stars observed to increase brightness strongly in a short time. They are objects associated with star formation.

fusion: Light nuclei merging together to form a heavier nucleus (and give off energy).

galactic cannibalism: Smaller galaxies consumed by colliding with or being drawn into larger galaxies. This process may cause changes in the shape and form of the larger galaxy and change its evolutionary pattern.

galactic cluster: Small (a few hundred stars at most) groups of associated stars found in the galactic plane. They have formed from gas clouds in the plane and exhibit a variety of ages, from billions of year to recently formed.

galactic corona (galactic halo): State-of-the-art photography of many galaxies reveals the existence of extensive, spherical regions of emission surrounding the bright, obvious portions of galaxies. The Milky Way appears to have such a region. Depending upon physical conditions, these regions could contain large amounts of mass.

galactic equator: The great circle on the sky that traces out the fundamental plane of the Milky Way.

galactic plane (galactic disk): The flattened, pancake-shaped distribution of gas, dust, and young stars in our galaxy.

galactic rotation curve: A graph of the speed of the stars and other material versus their distance from the center of the galaxy.

galaxy: Giant aggregates of stars, gas, and dust in a variety of forms. They range in size from objects that contain perhaps a million solar masses worth of material up to perhaps a trillion stars.

galaxy cluster: Clusters of galaxies, ranging from small groups of only a few up to thousands of galaxies.

Galilean satellites: The four brightest, largest satellites of Jupiter.

gamma ray: Extremely high-energy photons with wavelengths on the order of the size of an atom (approximately 1 Å).

Ganymede: The largest of the Galilean satellites.

gas (ion) tail: A comet tail consisting of ions. Its structure and shape are determined by the material's interaction with the solar wind.

gegenschein: A small region of diffuse light, located 180 degrees away from the Sun in the sky, caused by sunlight reflected from dust particles in the ecliptic plane.

Geminga: A powerful source of high-energy gamma rays and pulsed X-rays.

general theory of relativity: Einstein's theory of gravity. General relativity is used to describe the motion of bodies in the presence of a strong mass, which warps the surrounding space and time, thus giving space a curvature.

geocentric theory: Early theories that imagined the Earth to be the unmoving center of the universe, and that the stars and planets revolved around the Earth.

geodesic: The shortest distance between two points. A generalization of the concept of a straight line.

geomagnetic axis: The line connecting the two magnetic poles on Earth.

giant elliptical galaxies: Huge extragalactic aggregates of stars that present an elliptical distribution of light. The largest of these may contain many trillions of stars.

giant impact theory: A theory for the formation of the Moon that postulates a collision between the young Earth and a Mars-sized object. Material from the Earth and the passing object combine to form a disk, which then condenses into the Moon.

giant molecular cloud: Aggregations of molecules into clouds of more than 100,000 solar masses of material, these objects are the largest discrete entities in the Galaxy. They are thought to be the site of star formation.

giant star: After leaving the main sequence, normal stars undergo changes in their physical state that cause them to swell in size, to sizes on the order of 10 solar radii. See also *supergiant stars*.

gibbous: The phase between the quarter phases and full in which the Moon's surface as seen from the Earth is more than half illuminated.

Global Oscillation Network Group (GONG): A worldwide network of solar telescopes for continuous monitoring of the Sun.

globular cluster: A star cluster, often containing more than 100,000 stars. Within the Milky Way, they orbit in elliptical orbits and have ages in excess of 10 billion years.

globules: See *Bok globules*.

gnomon: A stick or rod held perpendicular to the surface of the Earth for the purpose of casting shadows and determination of time.

Grand Unified Theory: A theory that attempts to unify the four forces of nature as different manifestations of a single force.

granulation: The cellular appearance of the solar photosphere, brought about by a boiling motion in which bright blobs of hot material are rising and cooler, dark material is sinking back into the Sun.

gravitation: The mutual attraction between any two bodies possessing mass. In nonrelativistic environments, the size of the attraction and the accelerations of the bodies involved can be calculated from Newton's Law of Gravity.

gravitational instability: A physical condition where material is on the verge of gravitational collapse.

gravitational lens: An object whose mass is sufficient to bend radiation from background objects and form images of the background object.

gravitational potential energy: The energy residing in gravitating masses that can be called upon to do work.

gravitational radiation (gravitational waves): Waves predicted by the general theory of relativity that should be emitted when large masses accelerate in strong gravitational fields.

gravitational redshift: A redshift caused by a photon losing energy when escaping from a large gravitational field.

Great circle: The largest circle one can trace on a sphere. Its center is at the center of the sphere.

Great Dark Spot: An atmospheric phenomenon observed in the atmosphere of Neptune.

Great Red Spot: A cyclonic storm in the upper atmosphere of Jupiter that has been observed for several hundred years.

Great Wall: In a figure of the distribution of galaxies with dis-

tance, an extensive alignment of galaxies appearing as a wall separating the nearby universe from more distant regions.

greatest (eastern or western) elongation: When the angular distance of Mercury or Venus from the Sun (as seen from Earth) achieves its maximum value.

greenhouse effect: The trapping of radiation near the surface of a planet due to the absorbing properties of the overlying atmosphere.

ground state: The lowest energy level of an atom.

GUT: See *Grand Unified Theory*.

half-life: The time it takes for one-half of a quantity of radioactive material to decay.

Halley's comet: A comet that approaches the Sun every 75–76 years. It has been observed repeatedly since ancient Chinese astronomers recorded it.

halo: A spherical region around the galaxy containing globular clusters and some stars. There may be other constituents of which we are unaware.

H alpha (Hα): The red line of the hydrogen Balmer series, located at 6563Å.

head-tail galaxy: A radio galaxy moving through intergalactic space will have its outer portions (where the radio emission comes from) swept backwards into the shape of a tail.

heat: The thermal energy of an object—the sum total energy of the motions of all of its particles.

Hebrew calendar: A lunar calendar that requires the insertion of additional months from time to time to keep pace with the solar calendar.

heliocentric theory: A model of the solar system with the Sun at the center and the planets (including the Earth) going around it. In ancient Greece, this hypothesis was put forward by Aristarchus.

helioseismology: The study of the interior of the Sun as revealed by vibrations (oscillations) seen at its surface.

helium burning: The fusion of helium into carbon via the triple-alpha process.

helium flash: An explosive episode of helium burning taking place in the degenerate core of a low-mass star.

Herbig-Haro objects: Emission regions observed to move away from young stellar objects in a bipolar flow.

hertz: A unit of frequency for electromagnetic waves, equal to one cycle per second.

Hertzsprung-Russell diagram: See *H-R diagram*.

Hidalgo: An asteroid with an elongated orbit that is inclined 40° to the ecliptic plane.

highlands: On the Moon, the regions of higher elevation that were not covered by the lava flows that cre-ated the maria.

high-velocity stars: Stars that show unusually large velocities relative to the Sun. They are part of the galactic halo and not moving in the plane of the galaxy as is the Sun.

Hipparchus: A Greek astronomer with many achievements: the first star catalogue, the magnitude scale, discovery of precession.

Hirayama asteroid families: Groups of asteroids with highly similar orbits.

H I region: Regions and gas clouds primarily composed of neutral atomic hydrogen.

horizon: The great circle that represents the intersection of the Earth and the sky.

horizontal branch: A region of the H-R diagram of globular clusters. It consists of evolved low-mass stars burning helium in the core and hydrogen in a shell.

H-R diagram: A graph of the luminosity of stars plotted against their surface temperatures (or spectral types, or color). The many forms of the diagram are often called collectively by the initials of its codiscoverers, Hertzsprung and Russell.

H₂: Molecular hydrogen—two H atoms bound together.

H II region: Regions and gas clouds primarily composed of ionized hydrogen, symbolized by H II. The clouds of ionized hydrogen are produced by nearby, high temperature O and B stars.

Hubble constant: The constant H in the mathematical expression of Hubble's law, $V = Hd$. See also *Hubble law*.

Hubble Space Telescope: The first complete observatory in space, equipped with all the instrumentation needed to observe celestial objects in many different ways.

Hubble law: Describes the expansion of the universe in which we see that the speed of recession of a galaxy is proportional to its distance from us.

Hyades: A nearby, open cluster of stars in the constellation Taurus.

hydrogen burning: Nuclear fusion in which 4 protons (hydrogen nuclei) are converted to helium with the emission of energy.

hydrostatic equilibrium: A stable condition where the inward force of gravity at every point in a star is balanced by the outward forces of gas and radiation pressure.

hyperbolic orbit: The mathematical curve that will be followed by an object that is orbiting at speeds greater than the velocity of escape.

Hyperion: A satellite of Saturn.

Iapetus: A satellite of Saturn, having its forward-facing hemisphere dark-colored and the other one light-colored.

ICE: International Comet Explorer satellite.

ideal gas: Under normal conditions, the pressure of a gas is directly dependent upon its temperature and density. Also known as a *perfect gas*.

igneous rock: Rock formed by the cooling of molten material.

inductive reasoning (induction): Reasoning from particular observations to general conclusions or principles.

inertia: The resistance to any change in an object's motion. Inertia is caused by the property of mass.

inferior conjunction: When Mercury and Venus come into conjunction with the Sun and are at their closest to Earth.

inferior planets: Planets closer to the Sun than the Earth—i.e., Mercury and Venus.

inflationary universe: The hypothesis that a period of extremely rapid expansion occurred early in the history of the universe (followed by the slower expansion we see today).

instability strip: The region of the H-R diagram where Cepheid variables and Cepheid-like variables are found.

interference: The interaction of two similar waves. See *constructive* and *destructive interference*.

interferometer: An instrument that uses the properties of interference to increase resolution. Radio interferometers consist of two or more telescopes examining the same source at the same time. The interference between the different dishes enables astronomers to examine the target with very high resolution.

intergalactic matter: Material in space between clusters of galaxies.

interstellar absorption lines: Sharp absorption lines seen in the spectra of some stars because of absorption of energy by interstellar gas.

interstellar extinction: Radiation removed from a beam of light due to its passage through the dust component of the interstellar material.

interstellar medium: The atoms, ions, and molecules in space between the stars.

inverse square law of light: The brightness of a radiating body varies inversely with the square of its distance from us.

Io: The innermost Galilean satellite of Jupiter, heated by gravitational and magnetic forces into an active volcanic state.

ion: An atom with a net charge, due to the loss or gain of one or more electrons.

ionization: The process in which an electron in an atom absorbs sufficient energy for the electron to escape.

ionization energy: The energy necessary to remove an electron from an atom or ion.

ionosphere: The layer of the Earth's atmosphere (at and beyond 100 km up) consisting of charged particles.

IRAS: The **I**nfra**R**ed **A**stronomical **S**atellite, which mapped most of the sky at long wavelengths. The data now serve as an archive for astronomical research.

iridium: A chemical element found in greater abundance in meteorites than on the Earth. It may allow us to locate ancient impacts more successfully and draw more conclusions about their consequences for life on Earth.

irregular galaxies: Galaxies with odd shapes that do not fit into the conventional schemes of classification by appearance.

Ishtar Terra: A large plateau in the northern hemisphere of Venus.

isotope: Two atoms having the same number of protons but different numbers of neutrons are said to be isotopes.

isotropy: Having equal value in all directions. Isotropic radiation would be radiation impinging on the Earth equally from all directions.

IUE: The International Ultraviolet Explorer satellite.

Jean's length: Condensations and perturbations in a gas that are larger than the Jean's length will be inclined toward gravitational collapse.

Jean's mass: The amount of material in a length equal to the Jean's length.

jet: A narrow, high-energy beam of particles and radiation.

jet stream: High-speed flows of material in the upper atmosphere.

Jovian planets: Solar system planets that are larger, rotate more rapidly, and are less dense than the terrestrial planets such as Earth. They have dense atmospheres and chemical compositions similar to that of the Sun.

Juno: The third asteroid discovered.

Kapteyn Universe: A model of the galaxy derived from star counts. The model, which suffered from systematic errors, showed the Sun to be at the center. Named after the 20th-century Dutch astronomer J. C. Kapteyn.

Keck telescope: As of 1994, the world's largest telescope with a 10-m diameter mirror.

kelvin: A temperature scale with no negative numbers. X°C is equal to X + 273 kelvins.

Kelvin–Helmholtz contraction time: The time required for a body to collapse to half its size under its own self-gravity.

Kepler: Johannes Kepler was the 17th-century astronomer who used Tycho Brahe's observations to discover the laws governing the motions of the planets in the solar system.

Keplerian (orbital) motion: Motion that follows Kepler's laws.

Kepler's laws: Three empirically determined laws, based on observations of Tycho Brahe, that describe the motions of planets around the Sun.

kinetic energy: Energy possessed by a body by virtue of its motion.

Kirchhoff's laws: Extensive laboratory work by Kirchhoff discovered several modes of emitting radiation (continuous, dark-line, and bright-line) and the conditions under which each would be produced.

Kirkwood gaps: Regions in the asteroid belt that contain no asteroids.

KL (Kleinman-Low) Nebula: A source of strong infrared radiation near the Orion Nebula.

KREEP: Lunar material with a composition high in potassium (K), rare earth elements (REE), and phosphorus (P).

Kuiper Airborne Observatory: A flying infrared telescope system.

Kuiper belt: A hypothesized disk-shaped region surrounding the solar system thought to contain short-period comets.

Lagrangian points: Five points in space where the combined gravitational attraction of two bodies will cancel each other out.

Large Magellanic Cloud (LMC): See *Magellanic Clouds*.

latitude: Angular distance north or south of the equator.

law of parsimony: Of two equally successful hypotheses, the simplest one is preferred.

law of superposition: When interpreting the surface of the Moon or other firm surfaces, the assumption that the topmost features are the most recently formed.

Leonids: An August meteor shower famous for its occasional, spectacular displays.

light-gathering power: A measure of the ability of a lens or telescope to gather light. It depends upon its area, which for a circular mirror is πR^2, where R is the radius of the lens or telescope.

light-year: The distance light travels in one year $(9.5 \times 10^{12}$ km).

limb: The edge of an object.

limb darkening: The edges of the Sun are less bright because the radiation we receive from there comes from outer, cooler layers. The center of the disk of the Sun is brighter because the radiation comes from deeper, hotter layers.

line profile: A detailed display with wavelength of the amount of energy absorbed at each and every point within an absorption line. Such details can reveal atmospheric density, rotation, magnetic fields, and other effects.

lithosphere: The solid, rocky layers of the Earth just above the aesthenosphere.

Local Group: Our galaxy is a member of a small group of about two dozen galaxies. This group also contains the Magellanic Clouds and the Andromeda Galaxy.

Local Supercluster: Our Local Group is also a member of a cluster of galaxy clusters dominated by the extensive Virgo Cluster of galaxies.

longitude: East-west location on the Earth along the equator, measuring from the circle that runs through Greenwich, England.

long-period variable star: A cool, giant star that may take more than a year to go through its cycle of brightness variation.

luminosity: The total energy emitted per second by a celestial object.

luminosity class: Stars of the same surface temperature show considerable variation in total energy output. The luminosity classification was invented to be able to label those differences. Class I contains superluminous supergiants while class V designates feeble stars from the bottom of the main sequence.

lunar eclipse: The Moon orbits through the Earth's shadow, and we see a circular line of shadow cross the Moon.

lunar occultation: The passing of the Moon in front of a star.

Lunar Orbiter: A series of satellites that provided an extensive survey of most parts of the lunar surface in the 1970s.

Lunar Ranger: The earliest explorations of the Moon featured rockets with TV cameras in the nose that would crash into the lunar surface, transmitting pictures as long as they were able.

Lyman series: Transitions of the H atom involving the

ground state. The resulting spectral lines are in the ultraviolet spectral region.

Maat Mons: A volcanic mountain on Venus.

Magellanic Clouds (Large and Small): Two small irregular galaxies that are probably bound gravitationally to the Milky Way. The SMC actually consists of two galaxies. They are naked-eye objects in the southern hemisphere.

Magellanic Stream: A ribbon of hydrogen gas running from the Milky Way to the Magellanic Clouds.

magma: Molten rock.

magnetic braking: A process in which particles ejected by a star interact with its magnetic field to slow the star's rotation.

magnetic field: A force field that can affect the motion of charged particles and magnetic materials.

magnetic lines of force: Lines of equal magnetic field intensity.

magnetosphere: A zone surrounding a planetary body in which solar wind particles are trapped by the planet's magnetic field.

magnification: An increase in the apparent size of an object. In a telescope system, the magnification is the focal length of the objective lens divided by the focal length of the eyepiece.

magnitude: A scale of apparent brightness used by astronomers since the time of Hipparchus. If two objects differ by one magnitude in apparent brightness, they differ in brightnesses by a factor of 2.512. (Note: this is not the same as "order of magnitude.")

main sequence: The region of the H-R diagram consisting of stable stars of different masses. A normal star spends most of its life on the main sequence.

major axis: The longest dimension of an ellipse, measured through the foci.

mantle: The region of the Earth between the crust and the core.

maria: *Singular:* mare. Latin for "seas," these are dark-colored lowlands on the Moon's surface that have been covered by lava flows.

Mariner: Spaceprobes to Venus and Mars in the late 1960s.

mascons: Concentrations of mass just under the Moon's surface discovered by the anomalous motions of Lunar Orbiters.

mass: A measure of the amount of material in a body. It is a measure of the concept of inertia.

mass-luminosity relationship: The luminosity of a star is approximately proportional to the 4 power of its mass; that is, $L \propto M^4$.

mass-radius relation: For white dwarfs. The more massive the white dwarf, the smaller its radius.

Maunder minimum: A decline in sunspot and other solar activity from 1645 to 1715, which had consequences for Earth weather.

meridian: The great circle of the sky through the two celestial poles and the zenith. It divides the sky into eastern and western halves.

metal-poor stars: Stars whose chemical composition is low in metal abundance relative to the Sun.

metal-rich stars: Stars with above average abundances of metals.

metamorphic rock: Rock that originally formed by igneous or sedimentary processes and changed in form due to the presence of high pressures or temperatures.

metastable states: Low-lying energy states where the atom will remain for an unusually long time before emitting radiation and undergoing transitions to the ground state.

meteor: A streak of light left by a meteoroid passing through the Earth's atmosphere.

meteorite: A meteoroid that survives the passage through the Earth's atmosphere and reaches the surface.

meteorite fall: A meteorite that is observed to hit the surface of the Earth and then is recovered.

meteorite find: A meteorite that attracts our attention by its unusual apearance and hence is discovered serendipitously.

meteoroid: A small chunk of material (rocky or metallic) that travels through interplanetary space.

meteor shower: An encounter between the Earth and a stream of particles that have evaporated from a comet but continued to orbit the Sun. This encounter produces a large number of meteors that appear to be radiating from one particular direction called the radiant.

micrometeorites: Minute particles of meteoritic material that are strongly decelerated by the atmosphere. They are able to radiate away their heat before striking the surface.

Milky Way: Originally, the prominent band of diffuse light that runs across the sky. More recently, the spiral system of perhaps 1 trillion stars in which our solar system is located.

millisecond pulsars: Pulsars whose signals vary in periods of milliseconds.

Mimas: A moon of Saturn showing a giant impact crater.

minor planet: An asteroid.

Mira variables: A long-period variable star, of which Mira is the prototype.

Miranda: A small Uranian moon with a highly "distressed" surface.

molecular bands: Groups of closely spaced absorption lines due to molecules that are seen in some spectra.

molecular clouds: Giant interstellar clouds with a large fraction of their matter in the form of molecules.

molecules: Two or more atoms bound together chemically.

momentum: For an object moving in a straight line, the *linear* momentum is mass times velocity. For an object moving in a circular orbit about the Sun, the *angular* momentum is mass times speed times distance from the Sun.

multiple mirror telescope: A telescope that acquires a large light-gathering power by collecting photons in many small mirrors that are oriented to bring all of the different beams of light to a common focal point.

nadir: The opposite direction from the zenith (straight down). The direction that is indicated by a weight hanging freely in the Earth's gravitational field.

nebula: A diffuse, nonstellar object.

nebular hypothesis: The theory suggesting that the Sun and planets were formed as condensations out of a cloud of dust and gas that was flattened into a disk of material by the rotation of the cloud.

negative curvature: A geometry in which two parallel beams of light would diverge from each other with distance and in which the sum of the angles in a triangle are less than 180 degrees.

Nemesis: A star hypothesized to orbit the Sun and to perturb the orbits of comets to produce a suggested 65-million-year period of mass extinctions on Earth.

Nereid: A satellite of Neptune.

neutrino: An elusive elementary particle with no electric charge and little if any mass. Neutrinos are produced in certain nuclear reactions and carry energy away from the star.

neutron: A nuclear particle with no electric charge and approximately the same mass as a proton.

neutron star: A star that derives most of its pressure support from degenerate neutrons. It is a collapsed object whose material approaches the density of the nucleus of an atom.

new moon: The phase where the Moon and the Sun are in conjunction but it is the back side of the Moon that is illuminated.

Newtonian reflector: A telescope that employs a small diagonal mirror to produce a focus outside the side of the telescope tube.

Newton's first law: A body will maintain its current state of motion until it is disturbed by an external force. Also called *law of inertia.*

Newton's second law: The famous $F = ma$ describes how forces cause accelerations. It shows that the acceleration a body experiences, $a = F/m$, is directly proportional to the force and inversely proportional to its mass.

Newton's third law: If object 1 exerts a force on object 2, then object 2 exerts an equal and opposite force on object 1—*action = reaction.*

Nix Olympica: A giant but now inactive volcano on Mars. Also known as Olympus Mons.

nonthermal radiation: Radiation produced by interactions between charged particles moving near the speed of light in a magnetic field. The intensity of the radiation does not follow the blackbody curves exhibited by thermal radiation and is polarized.

noon: Apparent noon is when the Sun reaches its maximum altitude and shows the shortest shadow for the day. Noon on the civil clock incorporates effects such as location in a time zone and Daylight Savings Time.

North celestial pole: See *celestial pole.*

nova: A star that increases its brightness dramatically in a short time, and then slowly declines in light output. Many novae are recurrent. Current models of novae postulate that they are binary systems where one star dumps mass onto a nearby collapsed object.

nuclear bulge: The mass concentration at the center of a galaxy.

nuclear reaction: An interaction between a nucleus and another particle that results in energy emission or absorption and a change in the type of particles present. See *fusion* and *fission.*

nucleosynthesis: The manufacture (synthesis) of elements through a variety of nuclear reactions inside stars.

nucleus of a comet: The small core of frozen gases and dust that evaporates when the comet is near the Sun to produce the bright halo and tail features.

nucleus of a galaxy: The central regions where one finds a concentration of mass and often violent activity.

nucleus of an atom: The central concentration of mass in an atom, composed of protons and neutrons and having a net positive charge. It is surrounded by a distribution of electrons that are bound to it by means of a negative electrical charge.

OB associations: Recently formed loose associations of O and B stars. The associations do not have enough mass to prevent them from drifting apart in a relatively short time.

Oberon: A satellite of Uranus.

objective: The large lens or mirror that gathers light in a telescope system.

objective prism spectrograph: A telescope-prism combination in which a large prism is placed in front of the telescope objective to produce a photograph of the spectra of numerous stars rather than one of the stars themselves.

Occam's Razor: See *law of parsimony.*

occultation: When one celestial body passes in front of another. See, for example, *lunar occultation.*

Olbers' paradox: In a uniform distribution of stars throughout an infinite universe, all lines of sight from the Earth would terminate on the surface of a star, thus causing the sky to be everywhere as bright as the surface of the Sun. This simple conjecture leads quickly to sophisticated cosmological conjectures.

Olympus Mons: A large extinct volcano on Mars.

Oort Cloud: A hypothesized cloud of primeval material surrounding the solar system at a considerable distance from the Sun and thought to be a source of comet nuclei, which are perturbed by stars to pass near the Sun.

open cluster: Clusters of stars that form in the galactic plane. Also called *galactic clusters.*

open universe: A universe that possesses insufficient mass to halt the observed expansion. The curvature of space would be negative, and the universe would be infinite.

opposition: When two objects are separated in the sky by 180 degrees as seen from Earth.

optical doubles: Two stars close together in the sky, which give the impression of being affiliated but are really at very different distances from the observer.

orbital angular momentum: See *momentum*.

orbital inclination: The angle between an orbital plane and some reference plane. For example, in the solar system the reference plane would be the ecliptic plane.

Orbiting Astronomical Observatory: One of the earlier telescope systems sent into high Earth orbit to escape the detrimental effects of the Earth's atmosphere.

order of magnitude: A concept that allows making approximate comparisons of numbers.

Orientale Basin (Mare Orientale): A giant, multi-ringed basin on the Moon.

Orion Nebula: A giant cloud of gas and dust illuminated by a cluster of hot stars in the cloud. The hot stars are young, and there is evidence that new stars are continuing to form inside the nebula. The Orion Nebula is visible to the naked eye as the middle star in the Sword of Orion.

oscillating universe: A theory of the evolution of the universe in which it expands and contracts endlessly, undergoing repeated fireballs.

outgasing: An expulsion of material from the inner, hotter regions of a planet.

ozone layer: A layer of the Earth's atmosphere in which sunlight ionizes O_2 molecules and creates O_3 (ozone).

pair production: The interaction of two high-energy gamma rays to produce a pair of particles—an atomic particle and its antiparticle.

Pallas: The second largest asteroid, discovered in 1802.

Pangaea: The name given to the hypothetical supercontinent of crustal material on the Earth that eventually separated and moved about to form the continents we see today.

parabola: Mathematically, a curve on which every point is equidistant from some point and some line. Dynamically, it is the orbit that would be followed by a body that just exactly achieved escape velocity.

paradigm: The state of a science at any particular time, comprising its current theories and observations and the conclusions drawn from them. A scientific revolution occurs when a field of science goes through a major change of paradigm, as for example when the Ptolemaic model was displaced by the Copernican model.

parallax: A shift in the apparent direction of an object due to a change in the position of the observer. Most commonly, the movement of nearby stars on the background of unmoving distant stars due to the motion of the Earth around the Sun. The term is often loosely used synonymously with distance.

parent molecule: A molecule inferred to exist from observations of the "daughter" molecules that we are able to see. The "daughter" is produced by a modification of the original "parent" molecules.

parsec: 3.26 ly, the distance at which the parallax of a star will equal one second of arc.

partial eclipse: When the eclipsed body is not totally covered by the eclipsing body.

Paschen series: The transitions of the hydrogen atom involving the third energy level and appearing in the near infrared.

penumbra: The region of partial shadow in an eclipse. Also, the lighter dark region surrounding a sunspot.

perfect gas: See *ideal gas*.

perigee: The nearest approach to the Earth by an Earth-orbiting body. See also *apogee*.

perihelion: The nearest approach to the Sun by a solar system object. See also *aphelion*.

period: The time interval required for a certain cyclic behavior to repeat itself. For example, the synodic period of Mars (the time interval between two successive oppositions) is 780 days.

period-luminosity relation: The correlation between the period of light varation and the luminosity of a Cepheid variable star. This property enables astronomers to judge the distances of nearby galaxies that have Cepheids in them.

Perseids: A meteor shower at its strongest in early to mid-August.

Pfund series: The series of transitions of the hydrogen atom that involve the fourth energy level. The lines occur in the infrared.

phases: The variation in the appearance of a solar system object seen by an observer as the object takes up different positions with respect to the observer and the Sun.

Phobos: One of the two small moons of Mars.

Phoebe: A moon of Saturn, consisting of a dark surface.

photodissociation: The separation of a molecule into two or more atoms or molecules caused by the absorption of a photon.

photoelectric effect: The process in which certain photosensitive materials will eject electrons when illuminated by blue light but not when illuminated by red light (no matter how strong the irradiation is). This behavior was explained by Einstein as implying that radiation comes in discrete particles, or chunks, known as photons.

photoionization: The absorption of a photon leads to the removal of an electron from an atom or ion.

photons: Discrete particles (bundles, chunks) of electromagnetic energy.

photosphere: The visible surface of the Sun. The region of the Sun's atmosphere that contributes most of the visible light that we see when we look at the Sun.

Pioneer 10, 11: 1970s spacecraft probes to Jupiter.

pixel: A picture element of a detector. It is the smallest detecting element on a photographic film or a CCD chip.

Planck's constant: The energy of a photon is $E = hf$, where f is the frequency of the radiation and h is Planck's constant.

Planck curve: A mathematical expression describing the emission given off by radiating bodies as a function of temperature. Describes the spectrum of a blackbody.

planet: The name means "wanderer" in Greek, since the ancients observed these bodies to move around with respect to the sphere of "fixed stars."

planetary nebula: An expanding shell of gas ejected from the outer atmosphere of red giant stars in a late stage of stellar evolution.

planetesimals: Aggregates of mass (about asteroid-size) that are thought to combine to form protoplanets in the early history of the solar system.

Plaskett's star: A star of high mass (approximately 70 times the mass of the Sun). It is perhaps the most massive star known.

plasma: A gas in which all or many of the components are charged particles (electrons and ions).

plate tectonics: Many changes in the surface of the Earth can be explained in terms of the slow movements of the Earth's crustal plates.

Pleiades: A small, open cluster of stars in Taurus that are easily visible to the naked eye and are known as the Seven Sisters. This star cluster has been singled out as worthy of observation by virtually every skywatching culture over the ages. Its appearances just before dawn in the east and its disappearances just after dusk in the west just happen to coincide with important phases of the agricultural cycles in the northern temperate zone.

polar axis: A line pointing toward the NCP about which telescopes can rotate to follow the daily motion of a star.

polar caps: Regions of ice and snow at the poles of a planet.

Polaris: The end star in the handle of the Little Dipper, Polaris is a second-magnitude star that coincidentally happens to be located within 1° of the NCP at the present time.

polarization: The alignment of the planes of vibration of electromagnetic waves.

polarization of starlight: Starlight that has been polarized by scattering from dust particles in space.

pole star: See *Polaris.*

Population I stars: Stars associated with the galactic plane that are (on the average) younger in age and have a higher percentage of heavy elements in their chemical compositions.

Population II stars: Older stars with a lower percentage of heavy elements in their chemical compositions, preferentially located in the galactic halo.

positive curvature: Spacetime that closes on itself, forming a finite configuration. If two parallel beams of light set out in positively curved space, the two beams will slowly approach each other, like lines drawn on the surface of the Earth.

positron: An antielectron, a unit of antimatter with the same mass as the electron but with opposite charge.

potential energy: Energy that is stored in some way and available to do work—e.g., a ball held in the air has potential energy; when it is released, that potential energy is converted to kinetic energy of motion. See *gravitational potential energy.*

precession of the equinoxes: The 26,000-year revolution of the Earth's rotation axis causes the equinox points to move slowly around the celestial sphere with the same period. Precession of the equinoxes produces a change in the coordinates of stars with time.

pressure: The force per unit area exerted by a gas or a liquid.

pressure wave (P-wave): Waves that are compressional in nature—in particular, a seismic wave in the Earth or a sound wave.

prime focus: The first focus that is produced by a lens or a mirror, unmodified by other extra reflecting surfaces. The prime focus allows short photographic exposure times and results in small images.

primeval fireball: The state of incredibly high temperature and pressure out of which our expanding universe began.

prominences: Solar material elevated above the surface by the magnetic forces found in active regions on the solar surface. Most easily seen at the limb.

proton: A positively charged nuclear particle of approximately the same mass as a neutron.

proton-proton (p-p) chain: A series of nuclear fusion reactions in low-mass stars. The chain produces energy in the process of creating a helium nucleus from four protons.

protoplanets: An early stage in the process of making a planet where the body is still accreting mass.

protostar: A cloud of gas and dust that is collapsing under its own gravity and destined to become a star when its central temperature becomes high enough to start nuclear reactions in its central regions.

protosun: The Sun when it was a protostar.

Proxima Centauri: The nearest star to the Earth.

Ptolemaic system: The model of the cosmos, passed on from ancient Greece, in which the Earth is the unmoving center of the universe and the planets and stars revolve around it.

pulsars: Objects that emit energetic and rapid pulses of radio radiation. They are thought to be rapidly rotating neutron stars surrounded by plasma and a strong magnetic field.

Pythagoras: An early Greek philosopher-scientist who formed a school arguing that numbers are the basis of all understanding.

Q stars: Hypothetical objects to explain certain observations in a way other than calling on the concept of the black hole. They consist of protons, electrons, and neutrons held together by the strong nuclear force rather than gravity.

quadrature: The configuration of a superior planet in which the Sun and the planet are 90 degrees apart in the sky, when viewed from the Earth.

quantum mechanics (quantum theory): A theory of physics, which says that, on the subatomic level, the properties of the universe are not continuous but are instead found in discrete units called *quanta*. A photon is an example of quantized energy.

quarks: A fundamental particle in subnuclear physics. It is proposed that other particles are composed of the union of different types of quarks.

quasar: A star-like appearing object observed to have a large redshift in a spectrum containing emission and absorption lines. The word is a contraction for quasi-stellar radio source but is applied even to objects that do not emit radio radiation. They are now known to be the nuclei of young galaxies.

radar observations: By bouncing radar waves off the surfaces of nearby celestial objects and determining the time it takes to return, details about the structure and elevation of the surface can be learned.

radial velocity: The relative line-of-sight velocity between a celestial object and the observer.

radian: A unit of angular measurement equal to 57.3 degrees.

radiant: A point in the sky from which objects in a meteor shower appear to diverge. It is not a physical point but an effect of perspective.

radiation: Historically, this term was applied to both particles (for example, beta radiation, which turned out to be emitted electrons) and energy propagating through space in the form of electromagnetic waves.

radiation belts: Equatorial regions of energetic charged particles trapped in the magnetic field of a planet. The strongest radiation belts in the solar system belong to Jupiter. For the Earth's belts, see *Van Allen Belts.*

radiation, cosmic 3 K background: See *background 3 K radiation.*

radiation, hydrogen 21-cm: Low-energy photons emitted at a radio wavelength of 21 centimeter by hydrogen atoms undergoing a "spin-flip" transition within the lowest energy level of the H atom.

radiation pressure: The "push" exerted by the impact of photons.

radiative transport: The movement of energy from one place to another by the emission and absorption of radiation.

radio galaxies: Galaxies that give off anomalously large quantities of radio emission.

radio interferometer: Two radio telescopes linked together electronically. By comparing the interference patterns between the two signals, fine details in the pattern of the object's radio image can be discerned.

radio jets: Material that appears to be ejected at high speeds from active regions in radio galaxies.

radio telescope: A telescope constructed to detect the emission of electromagnetic energy of wavelengths ranging from millimeters to meters.

radioactive dating: A method of dating rocks by examining the quantity of radioactive decay by-products in relation to the decaying material.

radioactive decay: A fission process in which a heavy element spontaneously splits into simpler components. The speed of the decay is governed by the "half-life" of the element. Every half-life, half of the remaining radioactive atoms will break down.

random error: An error in a set of measurements caused by random variations in the way a set of measurements is obtained. All measurements have random errors, which are measured by the *standard deviation.*

ray: A thin pencil of light. Also, a plume of material ejected radially from the point of a meteoric impact.

real image: An image formed at a definite point in space by the convergence of light rays. Energy is concentrated at this point and can be examined by eye or captured on a photographic plate or other detector.

recombination: An atomic process in which an electron and an ion reunite, giving off energy.

recurrent novae: Stars that undergo the nova phenomenon repeatedly.

reddening: In the scattering process that removes photons from a beam of radiation passing through a medium, the short wavelength radiation undergoes more scattering, leaving the color of the object to appear redder than it really is.

red giant: Low surface temperature, highly evolved stars of great size and luminosity.

redshift: The Doppler shifting of spectral features to longer wavelengths due to the relative motion of the source of light away from the Earth.

reflecting telescope: A telescope that uses a mirror to gather radiation.

reflection nebula: A dust cloud that reflects some starlight in our direction.

refracting telescope (refractor): A telescope that uses a lens to gather radiation.

refraction: The bending of light that passes from one medium into a different medium.

regolith: The crumbling and shattered top layer of lunar dirt created by constant bombardment of particles from space.

relativistic jets: Jets moving outward with speeds near that of light from active regions in galaxies.

resolving power: (resolution): The ability of an optical system to observe fine detail, such as two stars close together.

retrograde motion: The occasional westward movement in the motion of planets that generally move eastward relative to the background stars. The westward motions are caused by the Earth's movement in its orbit, when it laps or is lapped by another planet.

retroreflector: A reflector that will return a beam of light in exactly the same direction from which it came.

revolution: Orbital motion around a body or a center of mass.

Rhea: A cratered, icy moon of Saturn.

right ascension: The east-west coordinate in the equatorial coordinate system. It is the distance along the celestial equator (in hours of time running from 0 to 24) eastward from the vernal equinox.

rille: See *sinuous rille.*

ring galaxy: A galaxy containing a nucleus surrounded by a ring of material. Ring galaxies appear to form from galaxy collisions.

Ring Nebula: A well-known planetary nebula in the constellation Lyra.

ring system: A system of icy rocks and particles orbiting each of the giant planets.

Roche limit: A critical distance from a planet, inside of which tidal forces prevent material from accreting to form larger bodies. Inside the Roche limit, tidal forces are larger than the mutual forces between nearby gravitating bodies; outside the limit, mutual gravity allows bodies to accumulate more material and grow.

Roche lobe: An imaginary surface formed by the overlapping gravitational fields of the two stars in a binary system. The lobes look like the parts of a figure eight. Material inside each lobe is considered to belong to the nearest star; material can readily flow from one star to the other through the point of intersection.

ROSAT: An X-ray satellite currently in orbit. It is named for Wilhelm Röntgen, who discovered X-rays.

rotation: The movement of a body about an axis.

rotation curve: A graph of the orbital speeds of objects in a spiral galaxy as a function of distance from the center of the galaxy.

r-process (rapid process): A nuclear reaction that builds up heavy elements by the rapid accumulation of neutrons.

RR Lyrae stars: Variable stars whose luminosity varies in regular cycles of less than 24 hours.

Russell-Vogt theorem: The structure of a normal star in equilibrium depends only upon the mass, chemical composition, and age of the star.

SAGE: Soviet American Gallium Experiment, an experiment searching for solar neutrinos.

Sagittarius A (Sgr A): A group of several extremely energetic radio and infrared sources found at or near the center of our Milky Way galaxy.

scattering: The change in direction suffered by photons passing through gas atoms or dust particles.

Schmidt telescope: A telescope employing a spherical mirror and an aspherical correcting lens. Its advantages are short photographic exposure times and a wide field of view.

Schroeter's Valley: A lunar valley running from a crater to a mare. It is thought to be a channel for the flow of lava.

Schwarzschild radius: See *event horizon*.

Sculptor: A small spheroidal galaxy in the Local Group.

secondary craters: Craters formed by the impact of material ejected from the impact that formed the primary crater.

sedimentary rock: A rock formed when rock debris, transported to a new site by water, wind, ice, or gravity, settles and becomes cemented.

seeing: The distortion, blurring, and artificial enlargement of stellar images caused by the turbulent motions of gases in the Earth's atmosphere.

seismic waves: Waves caused by earthquakes that propagate through and around the Earth. See also *pressure wave* and *shear wave*.

seismometer: An instrument to detect seismic events.

selection effect: A bias in data collection that preferentially exhibits certain evidence and withholds other evidence from observers. A simple example: the assumption that most meteoroids are composed largely of iron, because iron meteorites are easier to recognize and discover when they are lying on the ground.

self-gravity: The gravitational attraction each part of an object has on all the other parts.

semi-major axis: Half the longest axis of an ellipse, which is also equal to the average distance of the orbiting object from the center of mass.

sexagesimal system: A number system devised in ancient Mesopotamia using a base of 60 rather than the base of 10 that is used in the decimal system. It is particularly applied to situations involving angles and time.

Sextans: A small spheroidal galaxy in the Local Group.

Seyfert galaxies: Spiral galaxies with brilliant, star-like nuclei exhibiting violent activity.

shear wave (s-wave): A type of seismic wave where the motion of the material is perpendicular to the direction of the movement of the wave.

shepherd satellites: Small moons that may be instrumental in shaping and maintaining the rings around giant planets.

shock wave: A compressional wave traveling at greater than the speed of sound creates a discontinuity propagating ahead of it. Sonic booms are a terrestrial example of a shock wave.

short-period comets: Comets having periods less than 200 years. They may have come from the hypothesized Kuiper belt.

sidereal day: The time interval between two successive meridian passages of a star (23 hours and 56 minutes).

sidereal period: The period of an object's movement with respect to the stellar background.

silicates: Rocks and compounds formed with silicon (e.g., sand).

single-line spectroscopic binary: A binary star in which the spectral lines of only one star are observed to shift back and forth due to the Doppler effect.

singularity: A point in space where extreme conditions are found and normal physical laws break down.

sinuous rille: A winding channel on a planetary or satellite surface, generally thought to be caused by past lava flows.

Sirius (Sirius A): The brightest star in the sky.

Sirius B: A white dwarf star in orbit with Sirius A.

Small Magellanic Cloud: See *Magellanic Clouds*.

Snedecor rough check: An approximation to the standard deviation. Given by the range in the observations divided by the square root of the number of observations minus 1.

S0 galaxy: A galaxy with a large nucleus and a flattened disk but no indication of spiral arms in the disk.

solar day: The time interval between two successive meridian passages of the Sun (= 24 hours).

solar neutrino problem: Theory predicts a larger number of neutrinos from the Sun than is measured by current detectors. This has spawned a variety of possible explanations, including the possibility that neutrinos actually have a small mass. As of 1994, the mystery was still not solved.

solar wind: An outward flow of high-speed particles streaming from the Sun.

solstice, summer and winter: The moment of the year when the Sun is as far north of the celestial equator (summer in the northern hemisphere) as it can be and as far south (winter) as it can be. The northern hemisphere summer solstice occurs about June 21 (the longest day of the year), and the winter solstice occurs around December 21.

space-time: In Einstein's general theory of relativity, time is no longer an absolute quantity. It becomes another coordinate like the spatial coordinates. Instead of discussing space and time, we must now speak of space-time.

Spacewatch Camera: A telescope system, at the University of Arizona, designed to search for asteroids that come near the Earth.

spectral class (spectral type): A classification scheme for spectra that uses letters and numbers to express the surface temperatures of stars.

spectral line: A feature seen at a discrete wavelength in a spectrum.

spectral sequence: OBAFGKM is a sequence of letters that order the stars by decreasing temperature in the astronomical classification scheme of spectral types.

spectrograph: An instrument that disperses the radiation from celestial objects into a spectrum, which is then recorded on film or electronically.

spectroheliograph: An instrument that produces an image of the Sun in one single wavelength, using a sophisticated filtering technique.

spectroscopic binary: An apparently single star that is discovered to be a multiple system by the detection of a variable Doppler shift in the spectrum.

spectroscopic parallax: A method of estimating the distances of stars from spectroscopic information, which provides a star's temperature and the luminosity class.

spectroscopy: The specialized study of spectra.

spectrum: *Plural:* spectra. The energy of different wavelengths radiated by celestial objects. See *bright line spectrum, continuous spectrum,* and *dark line spectrum.*

spherical aberration: Distortions in images formed by spherical mirrors and lenses.

spicules: Jets of bright gas dancing in the chromosphere of the Sun, looking like a waving field of wheat.

spin angular momentum: Angular momentum of a body due to its rotational motion about a spin axis.

spiral arm: A spiral-shaped region in a spiral galaxy, where bright gas clouds and recently formed, young stars are found.

spiral galaxies: Galaxies that show spiral patterns of brightness.

spiral tracers: Young objects that trace out the locations of the arms of a spiral galaxy.

s-process: A nuclear reaction involving the slow accumulation of neutrons.

SS 433: A complex star system exhibiting both redshifts and blueshifts in its spectrum. It may be a normal star losing matter to an accretion disk around a collapsed object, with rapidly moving jets of material in the system.

standard candle: An object whose luminosity is well-known, so that it can be used as a stable reference to estimate the luminosities, and thus distances, of distant objects.

standard deviation: A statistical measurement of the precision of a set of measurements. A measure of the random errors of a measurement.

starburst galaxy: A class of galaxy characterized by periodic outbursts of star formation over widespread regions in the galaxy.

starspots: Dark, magnetically active regions seen on stars that may be similar to sunspots in their origin.

steady state hypothesis: A hypothesis that proposes that the expansion of the universe is caused by the pressure of new matter that is continually being spontaneously produced. In this model, the universe would look the same at all times past, present, and future. It would be infinitely old, without a beginning or an end.

Stefan-Boltzmann law: The total energy per unit area emitted each second by a radiating object varies as the fourth power of its temperature ($E \propto T^4$).

stellar model: A theoretical model of the interior properties of a star. The model is expressed as a table or graph of important stellar properties (temperature, pressure, density, and so on) as a function of distance from the center.

stellar wind: An outflow of charged particles from a star; similar to the solar wind.

stochastic star formation: See *supernova-induced star formation.*

Stonehenge: An ancient and huge stone structure in England. Apparent astronomical alignments have been found in the structure, and many hypotheses have been put forward about the astronomical sophistication of the builders. Many of the hypotheses remain controversial.

stony-iron meteorite: A meteorite with some iron and nickel mixed into its rocky composition.

stony meteorite: A rocky meteorite that on casual examination looks like an Earth rock.

Straight Wall: A long cliff on the Moon.

stratosphere: A region of the Earth's atmosphere, between about 20 to 50 km, in which temperature increases with height.

strong nuclear force: The force responsible for binding particles together in the nucleus.

subduction: The process by which cold lithospheric material sinks into the asthenosphere.

sunspot cycle: A regular variation in the number of sunspots, which repeats itself every 11 years or so. If variations in magnetic orientation are included, the sunspot cycle is 22 years.

sunspots: Dark regions on the photosphere of the Sun that are cooler than their surroundings and appear dark by contrast. Sunspots also possess strong magnetic fields.

supercluster: A cluster of galaxy clusters.

supergiant star: Highly evolved star with low surface temperature, high luminosity, and very large size.

superior conjunction: When a planet is in conjunction with the Sun, but on the other side of the Sun, as viewed from the Earth.

superior planets: Planets with orbits larger than the Earth's.

supermassive black hole: A black hole containing millions, perhaps billions of solar masses. The Hubble Space Telescope has provided overwhelming evidence in 1994 of the existence of a supermassive black hole in the center of M87, the giant elliptical galaxy in the Virgo cluster of galaxies.

supernova: A star that undergoes a spectacular increase in its energy output due to a catastrophic explosion of the core.

supernova-induced collapse: The compressional wave from a supernova explosion can squeeze gas clouds that it encounters and induce them to begin a gravitational collapse.

supernova-induced star formation: A statistical mode of star formation in which the supernova explosion of a massive star triggers the formation of other massive stars, which evolve rapidly to become supernovae in turn, causing the process to continue.

supernova 1987A: A relatively nearby supernova in the Large Magellanic Cloud.

supernova remnant: The energetic and rapidly expanding gases produced by a supernova explosion. The appearance and the amount of radio emission as compared to visible wavelength emission depend upon the time since the explosion.

synchronous rotation: A satellite whose rotation period is exactly equal to its period of orbital revolution.

synchrotron radiation: Radiation emitted by charged particles moving in a strong magnetic field near the speed of light.

synodic period: For the Moon, the time required for the Moon to go from one phase back to the same phase (29.5 days). For a planet, the time required for it to go from a particular configuration with respect to the Sun back to that same configuration—e.g., from superior conjunction to superior conjunction.

systematic error: An error that causes a measurement to be always too large or too small.

T Tauri star: A young star that is still collapsing onto the main sequence.

technetium: A radioactive element with a cosmically short half-life that is seen in stellar atmospheres. Its very presence indicates that nuclear reactions are taking place in a star.

temperature: A measure of the speeds of the particles in a gas.

terminator: The line dividing light and dark on an illuminated body.

terrestrial planets: Small, high-density planets that are rocky in composition and have relatively thin atmospheres (Mercury, Venus, Earth, Mars).

Tethys: A satellite of Saturn with a crater 40% its size.

thermal equilibrium: An equilibrium in which the energy radiated by a star equals the energy produced in the interior.

thermal radiation: Energy emitted by a body as a consequence of its temperature; the same as blackbody radiation.

thermal runaway: An unstable situation where an uncontrolled energy buildup occurs.

3 K background radiation: See *background 3 K radiation*.

tidal force: A force resulting from unequal gravitational forces (called differential gravitational forces) on opposite sides of a body.

tidal friction: Water slowed by the tidal forces on it from the Moon will cause a drag force on the Earth by rubbing across the continental shelves.

Titan: Saturn's largest moon—large enough to retain an atmosphere of its own.

Titania: One of the five large moons of Uranus.

top-down theory: A model of galaxy cluster formation in which clustering occurs prior to the formation of the galaxies making up the cluster.

total eclipse: The complete blocking of one body by another when spatial alignments are just right as seen from the Earth.

transition region: The region between the chromosphere and the corona in which temperature rises dramatically.

Trapezium: A cluster of four young, luminous stars, ionizing and illuminating the central regions of the Orion Nebula.

trigonometric parallax: A method of determining the distance to an object by noting the angle of displacement it shows with respect to distant background objects when it is viewed from two different locations of the observer.

triple-alpha process: A fusion reaction that converts three helium nuclei into a carbon nucleus.

Triton: Neptune's largest moon.

Trojan group: A group of asteroids that orbit the Sun in Jupiter's orbit such that they maintain a 60-degree angle ahead of and behind Jupiter. The asteroids are located at two Lagrangian points.

troposphere: The lowest level of the Earth's atmosphere, where weather occurs.

Tully-Fisher relation: The larger the luminosity of a galaxy, the larger the width of its 21-cm emission feature.

Tunguska: A region in Siberia that suffered a massive impact of some sort in 1908.

tuning fork diagram: A spatial display of the various shapes and forms exhibited by galaxies.

turbulence: A chaotic, convective motion of material.

turnoff point: When the stars in a cluster are plotted in an

H-R diagram, they will generally exhibit a lower portion of a main sequence, terminated at the top end in a point where the evolution of more massive stars has caused them to move away from the main sequence towards the red giant region. This turnoff point can be used to determine the age of the cluster.

two-armed spiral shock model: Applying the density wave model to explain the simplest form of spiral structure that we observe—a plain, two-armed spiral system.

Tycho Brahe: The 16th-century astronomer who came to be called the "father of modern observational astronomy" by diligently obtaining observations that were as good as the naked eye can resolve. Brahe provided the observations that allowed Kepler to deduce for the first time the correct laws of motion for the planets.

Type I supernova: A supernova in which H lines are not seen in the spectrum. They show a sharp peak in brightness followed by a slow decline.

Type II supernova: A supernova in which strong H lines are seen in the spectrum. The brightness shows a broader peak up until a sharp decline in emission sets in.

ultraviolet radiation: Electromagnetic waves not visible to the eye, with wavelengths that are shorter than blue visible light.

Ulysses: The first spacecraft to orbit the Sun in a polar, rather than equatorial, orbit.

umbra: The region of total shadow in an eclipse. Also, the darkest part of a sunspot.

universe: The totality of everything that can conceivably be observed.

Valhalla: A multi-ringed basin on Jupiter's satellite, Callisto.

Valles Marineris: The extensive canyon system on Mars.

Van Allen belts: Regions of charged particles circling the Earth above the equator. The particles are captured and held by the Earth's magnetic field.

variable star: A star that varies in brightness.

velocity: Speed is distance traveled per unit time, but velocity is a vector quantity that includes both speed and direction of motion.

velocity dispersion: The range of velocities exhibited by some group of objects being studied—e.g., the stars in a cluster or the galaxies in a galaxy cluster.

velocity-distance relation: See *Hubble law*.

Venera: A series of Russian spacecrafts sent to study Venus in the 1970s.

vernal equinox: See *equinox*.

Very Large Array (VLA): A giant radio interferometer, consisting of 27 antennas mounted on railroad tracks upon which they may move.

Very Long Baseline Interferometry (VLBI): The process of combining the signals from radio telescopes on different continents (and possibly even telescopes in Earth orbit) to achieve high resolution radio views of celestial objects.

Viking: American spacecraft that landed on the surface of Mars in 1976.

Virgo Cluster: A very large cluster of galaxies relatively near to the Local Group.

virtual image: An image with no definite location in space. No energy passes through the apparent image.

viscosity: The property of a fluid that has a high resistance to flowing.

visual binaries: Double stars that can be seen as separate stars in a telescope.

voids: Large empty regions of intergalactic space.

volatiles: Material that condenses at relatively low temperatures.

volcanic domes: Raised regions on the Moon and planets that resembles the domes left on the Earth by volcanic activity.

Voyager: A pair of sophisticated interstellar probes that visited the Jovian planets in the 1980s.

Wavelength: The distance between two crests of a wave (or any two similar points on the waveform). It is usually represented in astronomy by the Greek letter lambda (λ).

wave-particle duality: A conclusion of quantum mechanics—that all objects have an admixture of both wave and particle properties. This duality is most easily seen in light, where it shows a wave nature (from diffraction and interference) and a particle nature (from the photoelectric effect).

weak nuclear force: The force responsibile for radioactivity.

weight: The force exerted on an object by gravity.

weightlessness: Not the lack of gravity, but the lack of acceleration relative to a local reference frame. For example, astronauts whose acceleration around the Earth is the same as that of their spacecraft will be weightless.

white dwarfs: A class of hot stars that are near the end of their evolution. They have lost their sources of nuclear energy and as a consequence have collapsed into extremely small configurations—the size of the Earth or smaller.

Widmanstätten patterns: Crystalline patterns seen in iron meteorites that indicate a past history of heating and slow cooling.

Wien's law: The wavelength at which the maximum amount of radiation is emitted is inversely proportional to the temperature of the radiating object.

winter solstice: See *solstice*.

Wolf-Rayet stars: Hot, large, and luminous stars with broad emission lines seen in their spectra caused by high-velocity stellar winds.

W Virginus stars: Cepheid variables that belong to Population II.

X-ray burster: Brief episodes of powerful X-ray emission from a celestial source.

X-rays: Highly energetic photons and electromagnetic waves of short wavelength (1 to a 100 angstroms).

Young Stellar Object (YSO): Point sources of infrared radiation associated with dark clouds and other phenomena connected with star formation.

Zeeman effect: The splitting of atomic energy levels caused by a magnetic field. The result is that numerous spectral lines are formed rather than a single line; when seen in low resolution, a line is broadened.

zenith: The point in the sky directly overhead, 90 degrees from the horizon.

Zero Age Main Sequence (ZAMS): The location of stars of different mass in the HR diagram as they leave the contracting phase and first settle into an equilibrium state powered by nuclear fusion.

Zodiac: Twelve constellations along the ecliptic that have been passed down from ancient cultures. These constellations divide the Sun's motion approximately into 12 months.

zodiacal light: A region of diffuse illumination stretching along the ecliptic and centered on the Sun—also known as the "False Dawn." It is caused by scattering of sunlight off of dust particles in the ecliptic plane.

Zone of Avoidance: A region near the galactic plane of the Milky Way where few galaxies can be observed, due to extinction caused by the dense dust clouds in the plane of the galaxy.

CREDITS

PHOTO CREDITS

Introduction

Courtesy NASA and The Space Telescope Science Institute.

Chapter 1

Opener: Photograph taken by W. Liller at the Eastern Island Station of NASA's International Halley Watch program. **Page 4:** Courtesy NASA. **Page 6:** (top) John Bova/Photo Researchers; (bottom) USGS, Flagstaff/Starlight. **Page 7:** (left and center right) Courtesy NASA; (bottom right) Courtesy Astronomical Society of the Pacific. **Page 8:** Courtesy National Optical Astronomy Observatory. **Page 9:** (top) Courtesy NASA; (center) Courtesy William Liller, NASA; (right) Courtesy Jet Propulsion Laboratory, California Institute of Technology. **Page 10:** (left) Courtesy California Institute of Technology and Carnegie Institution of Washington; (right) Kim Gordon/Photo Researchers. **Page 11:** (left) Courtesy Lick Observatory; (bottom) Courtesy NASA. **Page 12:** (top left) Courtesy Hale Observatories; (bottom) ©1984 Royal Observatory, Edinburgh, Anglo-Australian Telescope Board. **Page 13:** (left) Courtesy National Optical Astronomy Observatory; (right) Courtesy Alan Dressler.

Chapter 2

Opener: ©1988 Universal City Studios, Inc./Photofest. **Page 20:** Paul Mozell/Stock, Boston. **Page 21:** Courtesy National Optical Astronomy Observatory. **Page 24:** Courtesy Stephen Shawl, University of Kansas.

Chapter 3

Opener: The Granger Collection. **Page 36:** Courtesy New York Public Library. **Page 38:** Roger Ressmeyer/Starlight. **Page 39:** Earth Satellite Corporation/Photo Researchers. **Page 42:** Courtesy Hale Observatories.

Chapter 4

Opener: David Nunuk/Photo Researchers. **Page 46:** George Holton/Photo Researchers. **Page 47:** (top) Robin Scagell/Photo Researchers; (center) National Geographic Society; (bottom) Courtesy of the Collection Frederick R. Weisman Art Museum at the University of Minnesota, Minneapolis. **Page 48:** Courtesy William Jefferys. **Page 50:** Pekka Parviainen/Photo Researchers. **Page 52:** Jerry Schad/Photo Researchers. **Page 58:** Courtesy Whitin Observatory, Wellesley College, 1918. **Page 61:** (top) Courtesy Dennis di Cicco, Sky Publishing Corporation; (bottom) Rev. Ronald Royer/Photo Researchers.

Chapter 5

Opener: The Granger Collection. **Page 71:** Courtesy of The British Museum. **Page 76:** Courtesy M. Mahboubian. **Page 77:** Bettmann Archive. **Page 79:** The Granger Collection. **Pages 80 and 84:** Bettmann Archive.

Chapter 6

Opener and Page 100: Courtesy NASA. **Page 109:** Rick Hull/AstroStock.

Chapter 7

Opener: Photographs taken by W. Liller at the Eastern Island Station of NASA's International Halley Watch Program. **Page 121:** Bettmann Archive. **Page 122:** Courtesy Institute for Astronomy, University of Hawaii. **Page 126:** Courtesy Max-Planck Institute for Astronomy. **Page 127:** Courtesy Jet Propulsion Laboratory. **Page 130:** Michael Collier/Stock, Boston. **Page 131:** Courtesy NASA and the Space Telescope Science Institute. **Page 133:** Photo by James Baker, courtesy of Dennis Milon. **Page 135:** (top left) Courtesy NASA; (top right and bottom) Courtesy Science Graphics.

Chapter 8

Opener: Courtesy Jet Propulsion Laboratory. **Page 151:** Photo by J.D. Griggs, courtesy USGS. **Page 155:** Runk/Schoenberger/Grant Heilman Photography. **Page 156:** Courtesy Dr. L.A. Frank, University of Iowa. **Page 158:** Courtesy Lick Observatory. **Page 159:** (top) Courtesy Lick Observatory; (bottom) Courtesy Lunar and Planetary Institute. **Page 160:** Courtesy Lunar and

Planetary Institute. **Page 161:** (top) Courtesy Lunar and Planetary Institute; (bottom) Courtesy Astronomical Society of the Pacific. **Page 162:** Courtesy Lunar and Planetary Institute. **Page 163:** (top) Courtesy Lunar and Planetary Institute; (bottom) Courtesy USGS. **Page 164:** Courtesy NASA. **Page 165:** (left) Courtesy Brian J. Skinner; (right) Courtesy Lunar and Planetary Institute. **Page 169:** Courtesy William Benz.

Chapter 9

Opener: Courtesy Jet Propulsion Laboratory. **Page 177:** (left) Courtesy University of New Mexico, Department of Astronomy; (center) Courtesy Hale Observatories; (right) Dr. Robert Leighton/AstroStock. **Page 178:** Courtesy Lowell Observatory. **Page 180:** (top left) Courtesy Jet Propulsion Laboratory; (top right and bottom left) Courtesy NASA; (bottom right) USGS/Starlight. **Page 187:** (top) NASA/JPL/Starlight; (bottom left and right) Jay L. Inge/National Geographic Society. **Page 188:** Courtesy Astronomical Society of the Pacific. **Page 189:** (top left) Courtesy Jet Propulsion Laboratory; (top right and center) Courtesy NASA. **Page 190:** (top) Courtesy Astronomical Society of the Pacific; (bottom) Courtesy Jet Propulsion Laboratory. **Page 191:** Courtesy NASA. **Page 192:** (top, center, bottom) Courtesy Jet Propulsion Laboratory; (left) Courtesy NASA. **Page 193:** (left) Courtesy NASA; (right) Astronomical Society of the Pacific, photo courtesy of Stephen Shawl. **Page 194:** Courtesy Astronomical Society of the Pacific. **Page 195:** (top left and bottom) Astronomical Society of the Pacific, photo courtesy of Stephen Shawl; (top right) Courtesy Lunar and Planetary Institute. **Page 196:** (left) Courtesy Jet Propulsion Laboratory, California Institute of Technology; (right) Courtesy NASA. **Page 197:** Courtesy USGS. **Page 199:** Courtesy Astronomical Society of the Pacific.

Chapter 10

Opener: Courtesy NASA. **Page 207:** (top left) Hale Observatories/Photo Researchers; (top right) Roger Ressmeyer/Starlight; (center left) Courtesy National Optical Astronomy Observatory; (center right) Courtesy Lick Observatory. **Page 210:** Courtesy Astronomical Society of the Pacific. **Page 211:** (top and center) Courtesy Jet Propulsion Laboratory; (right) Courtesy NASA/Jet Propulsion Laboratory. **Page 213:** Courtesy NASA and the Space Telescope Science Institute. **Page 215:** (top) Courtesy USGS; (bottom) Courtesy Astronomical Society of the Pacific. **Page 216:** (left) Courtesy Jet Propulsion Laboratory; (right) Courtesy NASA. **Page 217:** (top left) Courtesy Jet Propulsion Laboratory; (center and right) Courtesy NASA. **Page 218:** (top, center and cen-

ter/right) Courtesy NASA and Finley Holiday Films; (bottom/left) Courtesy Jet Propulsion Laboratory; (center/left and bottom/right) Courtesy Astronomical Society of the Pacific. **Page 219:** Courtesy NASA and Finley Holiday Films. **Page 220:** (top, center) Courtesy Jet Propulsion Laboratory; (top left and right, bottom left and right): Courtesy Astronomical Society of the Pacific. **Page 221:** Courtesy Astronomical Society of the Pacific. **Page 222:** (top, center and bottom) Courtesy Jet Propulsion Laboratory; (center/left) Courtesy Astronomical Society of the Pacific; (center/right) Courtesy NASA and Finley Holiday Films. **Page 224:** (top and bottom left) Courtesy NASA and Finley Holiday Films; (bottom right) Courtesy Jet Propulsion Laboratory; (right) Courtesy Lowell Observatory. **Page 226:** Courtesy Stephen Shawl, University of Kansas. **Page 227:** Courtesy NASA and the Space Telescope Science Institute.

Chapter 11

Opener: Tony Stone Worldwide. **Page 239:** Runk/Schoenberger/Grant Heilman Photography. **Page 240:** Martin Dohrn/SPL/Photo Researchers. **Page 244:** From M. Cagnet, M. Francon, J. Thrier, *Atlas of Optical Phenomena*, Springer-Verlag, reproduced with permission. **Page 245:** Courtesy National Optical Astronomy Observatory.

Chapter 12

Opener: Courtesy European Southern Observatory. **Page 257:** Roger Ressmeyer/Starlight. **Page 260:** Courtesy McDonald Observatory, University of Texas. **Page 261:** Courtesy Astronomical Society of the Pacific. **Page 265:** (top left and center) Courtesy Stephen Shawl; (right) Courtesy Stephen Shawl and Ray White **Page 266:** Courtesy Astronomical Society of the Pacific. **Page 267:** Reproduced with permission, from the *Annual Review of Astronomy and Astrophysics*, Volume 5, Copyright ©1967 by Annual Reviews, Inc. Courtesy of Allen T. Moffet. **Page 268:** (top) Courtesy Astronomical Society of the Pacific; (bottom) Courtesy National Radio Astronomy Observatory and Astronomical Society of the Pacific. **Page 269:** (top) Courtesy National Optical Astronomy Observatories; (bottom) Courtesy USGS and EROS Data Center. **Page 271:** Courtesy NASA.

Chapter 13

Opener: National Geographic Society. **Page 275:** Courtesy Bausch & Lomb. **Page 276:** (top) Courtesy Helmut Abt, Kitt Peak National Observatory; (right) Courtesy Yerkes Observatory, University of Chicago. **Page 282:** Courtesy National Optical Astronomy Observatory. **Page 289:** Courtesy Association of Universities for Research in Astronomy, Inc., Sacramento Peak Observatory. **Page 290:** ©1961 California Institute of Technology and Carnegie Institution of Washington.

Chapter 14

Opener: Courtesy National Solar Observatory. **Page 298:** Courtesy Astronomical Society of the Pacific. **Page 299:** Courtesy G. de Vaucoulers. **Page 304:** Courtesy Helmut Abt. **Page 306:** Courtesy Astronomical Society of the Pacific. **Page 309:** Courtesy Lick Observatory. **Page 311:** (top) Photo by E. Olson, Kitt Peak National Observatory, courtesy of James Kaler; (bottom) Courtesy Stephen Shawl, University of Kansas. **Page 314:** Courtesy G. de Vaucouleurs.

Chapter 15

Opener: Courtesy Space Telescope Science Institute. **Page 329:** Courtesy Yerkes Observatory, University of Chicago. **Page 330:** Courtesy Hale Observatories. **Page 333:** Courtesy David Buscher, Chris Haniff, John Baldwin and Peter Warner, Mullard Radio Astronomy Observatory, Cambridge, UK.

Chapter 16

Opener: Courtesy NASA. **Page 343:** The New York Times. **Page 346:** Courtesy Astronomical Society of the Pacific. **Page 351:** Art by Peter Lloyd, National Geographic Society. **Page 352:** Courtesy Astronomical Society of the Pacific. **Page 353:** (top and bottom left) Courtesy Project Stratoscope, Princeton University, supported by NSF, NASA and ONR; (bottom right) Courtesy Association of Universities for Research in Astronomy, Inc., Sacramento Peak Observatory. **Page 354:** (top) Courtesy Astronomical Society of the Pacific; **Page 354:** (bottom) NASA/MSFC. Courtesy of David H. Hathaway. **Page 355:** NASA/MSFC. Courtesy of David H. Hathaway. **Page 357:** (left) Courtesy Peax Observatory, Air Force Cambridge Research Laboratories; (center) Courtesy Rhodes College and High Altitude Observatory; (bottom) Courtesy J.D. Bahng and K. L. Hallam. **Page 358:** Courtesy L. Golub, IBM Research and SAO. **Page 359:** (top) Hale Observatories/Photo Researchers; (center) Courtesy National Optical Astronomy Observatory. **Page 360:** (left) National Geographic Society; (right) Courtesy National Optical Astronomy Observatory.

Chapter 17

Opener: Courtesy NASA. **Page 366:** Courtesy David Malin, Anglo-Australian Telescope Board. **Page 373:**

Data obtained by Ronald Snell and Peter Schloerb, who used the 14 millimeter wavelength telescope of the Five College Radio Astronomy Observatory. **Page 376:** Courtesy Anglo-Australian Telescope Board. **Page 378:** (left) Courtesy Gopal Naravanan, University of Arizona; (right) Courtesy T. Buhrke and R. Mundt, Max-Planck Institute for Astronomy. **Page 379:** ©1961 California Institute of Technology and Carnegie Institution of Washington. **Page 380:** Courtesy David Malin, Anglo-Australian Telescope Board. **Page 381:** Courtesy Mark McCaughrean, Steward Observatory, University of Arizona. **Page 382:** Courtesy Infrared Processing and Analysis Center of Caltech's Jet Propulsion Laboratory.

Chapter 18

Opener: Courtesy David Malin, Anglo-Australian Telescope Board. **Page 394:** (left) Courtesy Dana Berry, Space Telescope Science Institute; (right) Bill Iburg/Milon/Photo Researchers. **Page 397:** Courtesy Lick Observatory. **Page 399:** Courtesy AIP Neils Bohr Library. **Page 401:** Courtesy Anglo-Australian Telescope Board. **Page 402:** Photo by Irene Little-Marenin.

Chapter 19

Opener: Courtesy NASA. **Page 409:** Courtesy Mount Wilson and Palomar Observatories. **Page 412:** Courtesy Lick Observatory. **Page 413:** (top, bottom left) Courtesy Anglo-Australian Telescope Board; (bottom right) Courtesy California Institute of Technology. **Page 414:** Courtesy National Radio Astronomy Observatory and Astronomical Society of the Pacific. **Page 418:** Courtesy Anglo-Australian Telescope Board and Astronomical Society of the Pacific. **Page 419:** Courtesy Royal Observatory, Edinburgh. **Page 421:** Courtesy Kitt Peak National Observatory. **Page 425:** Bettmann Archive.

Chapter 20

Opener: Courtesy Anglo-Australian Telescope Board. **Page 437:** (top left) Courtesy David Malin, Anglo-Australian Telescope Board; (top right) Roger Ressmeyer/Starlight; (center) Courtesy Anglo-Australian Telescope Board. **Page 442:** (left) Courtesy Anglo-Australian Telescope Board; (right) Courtesy David Malin, Anglo-Australian Telescope Board. **Page 444:** Photo by IPAC/ Caltech/ JPL, courtesy of Jay M. Pasachoff. **Page 449:** Courtesy NASA. **Page 450:** Courtesy G. Westerhout; data from the Netherlands Foundation for Radio Astronomy and the Division of Radiophysics, SCIRO, Australia. **Page 453:** (top) Courtesy NASA; (bottom) Courtesy Farhad Yusef-Zadeh, Northwestern University.

Chapter 21

Opener: Courtesy NASA. **Page 467:** (center) Courtesy Palomar Observatory; (bottom left) Courtesy NASA; (bottom right) Courtesy Anglo-Australian Telescope Board. **Page 469:** Courtesy Palomar Observatory. **Page 470:** (top left) Courtesy Anglo-Australian Telescope Board; (top right) Courtesy Hale Observatories and the AIP Neils Bohr Library. **Page 474:** Courtesy Palomar Observatory. **Page 478:** Courtesy V. Rubin, Carnegie Institute of Washington. **Pages 479 and 481:** Courtesy Anglo-Australian Telescope Board. **Page 484:** (top) Courtesy L. Hernquist, Princeton University; (bottom left) Courtesy John Tonry, Massachusetts Institute of Technology; (bottom right) Courtesy NASA. **Page 485:** (left) Courtesy Mt. Wilson and Palomar Observatories; (right) From P.E. Seiden and H. Gerola, Fund. Cosmic Physics 7, 241 (1982). **Page 486:** Courtesy Smithsonian Institution. **Page 487:** Courtesy T.A. Tyson, Bell Laboratories. **Page 488:** Courtesy P.J.E. Peebles, based on the Lick Observatory catalog by C. Shane and C. Wirtanen. **Page 490:** The Astrophysical Journal. Produced by Adrian Melott, University of Kansas. Courtesy of Adrian Melott.

Chapter 22

Opener: Courtesy NASA. **Page 497:** Courtesy Mt. Wilson and Palomar Observatories. **Page 498:** (top) Courtesy Dr. François Schweizer, Carnegie Institution of Washington; (bottom left) Courtesy Alar and Juri Toomre, Massachusetts Institute of Technology; (bottom right) Courtesy National Optical Astronomy Observatory. **Page 499:** Reproduced with permission, from the *Annual Review of Astronomy and Astrophysics*, Volume 13, Copyright ©1975 by Annual Reviews, Inc. **Page 500:** Courtesy David Malin, Anglo-Australian Telescope Board. **Page 501:** (top) Courtesy Palomar Observatory; (bottom) Courtesy NASA. **Page 502:** Courtesy Astronomical Society of the Pacific and the National Radio Astronomy Observatory. **Page 503:** (top) Courtesy NASA; (bottom) Courtesy Rudy Schild, Smithsonian Center for Astrophysics. **Page 504:** (top) Courtesy NASA; (bottom) Courtesy National Optical Astronomy Observatory. **Page 505:** Courtesy Mount Wilson and Palomar Observatories. **Page 506:** Courtesy M. Schmidt, California Institute of Technology. **Page 507:** Courtesy E. Ellingson and H.K.C. Yee, CCD image taken at Canada-France-Hawaii Telescope. **Pages 508 and 509:** Courtesy NASA. **Page 511:** Courtesy Rene Racine, University of Montreal at the Canada-France-Hawaii Telescope.

Chapter 23

Opener and Page 527: Courtesy NASA. **Page 525:** Courtesy Astronomical Society of the Pacific.

CHAPTER OPENING PHOTOS

Chapter 1. The universe from nearby to far away is shown in this photograph taken from Easter Island by astronomer William Liller. Halley's comet, the bright object on the left, is a nearby part of our solar system. The stars and the globular cluster Omega Centauri on the right side, are parts of our galaxy, the Milky Way. Below Halley's comet is the distant galaxy NGC 5128.

Chapter 2. Scene from *E.T.: The Extraterrestrial*.

Chapter 3. A depiction, in an English woodcut, of Halley's comet as observed in 1531 with the use of a cross-staff.

Chapter 4. Star trails around Delicate Arch, Arches National Monument. Polaris, our North Star, is the bright trail inside the arch.

Chapter 5. Galileo showing the heavens to the people of Venice through one of his telescopes.

Chapter 6. Protoplanetary discs, dubbed *proplyds*, around stars in the Orion Nebula were discovered with the Hubble telescope by C.R. O'Dell of Rice University.

Chapter 7. Halley's comet, as observed from Easter Island, on March 11, 1986.

Chapter 8. The Earth and Moon, photographed together by the Galileo spacecraft from a distance of 3.9 million miles. Antarctica is under the clouds at the bottom. The Moon's farside is visible.

Chapter 9. The surface of Venus, from a computer reconstruction of radar data from the Magellan spacecraft. The circular dome-like hills are about 15 miles across.

Chapter 10. The visible atmosphere of Jupiter contains colorful bands and storm systems. The Great Red Spot, shown here, is such a storm system that is larger than Earth.

Chapter 11. Rainbows, caused by the bending of light through raindrops, demonstrate the wave nature of light.

Chapter 12. An artist's view of the Very Large Telescope at the European Southern Observatory in Chile.

Chapter 13. The passage of white light through a prism, and its decomposition into component colors, was first understood by Isaac Newton.

Chapter 14. The dark lines of the spectrum of the sun allow astronomers to determine its chemical composition.

Chapter 15. Eta Carinae, taken by the refurbished Hubble Space Telescope. The red outer glow comes from material ejected in 1843. The bright inner material contains large quantities of dust, which reflects starlight.

Chapter 16. The Sun, as viewed in the ultraviolet by the Skylab spacecraft, shows one of the most energetic solar flares ever observed.

Chapter 17. A shockwave of gas, traveling at 148,000 miles per hour, was formed by material coming out of newly formed stars in the Orion Nebula. The plume is only 1,500 years old.

Chapter 18. The Helix Nebula, NGC 7293, is a planetary nebula—an object that results when a star like the sun evolves past its red giant stage and ejects its outer layers. Left behind in the center is a hot star that eventually cools to become a white dwarf.

Chapter 19. Rings of gas surrounding Supernova 1987A. The rings, which are undoubtedly in different planes, are surprising because astronomers expected to see an hourglass-shaped bubble being blown into space.

Chapter 20. An interstellar bubble formed by a high velocity stellar wind. This nebula is NGC 2359.

Chapter 21. The center of the peculiar galaxy NGC 7252, when observed with the Hubble telescope, reveals a "mini-spiral" structure of gas and stars only 10,000 ly across, and about 40 exceptionally bright, young globular clusters, which appear to have been formed in a collision of a pair of galaxies some 2 billion years age.

Chapter 22. An HST image of the optical jet coming from the giant elliptical galaxy M87. This is now known to have a massive black hole in the center.

Chapter 23. A cluster of galaxies that existed when the universe was two-thirds its present age. Taken with the Hubble Space Telescope, this cluster, known as CL 0930+4713, shows galaxies in collision with some tearing material away while others merging to form new galaxies.

TEXT CREDITS

Page 26. Figure 2–10. Reproduced with permission. © 1979 Pachart Publishing House. **Page 27.** Figure 2–11. Adapted with permission from Henry H. Bauer, *Scientific Literacy and the Myth of the Scientific Method*. University of Illinois Press. © 1992, University of Illinois Press. Courtesy of Henry H. Bauer. **Page 57.** Figure 4–19. Adapted from *Voyager* from Carina Software. **Page 61.** Figure 4–27. Adapted from *Voyager* from Carina Software. **Page 62.** Figure 4–28. Adapted from *Voyager* from Carina Software. **Page 99.** Chapter 6 opening quote. With permission, University of Michigan Press. **Page 114.** Table 6–3. Adapted with permission from Table 3.2, p. 45 of T. Encrenaz, I.-P. Bibring, and M. Blanc, *The Solar System*, Springer-Verlag. © 1990, Springer-Verlag. **Page 128.** Figure 7–12. After Dirk

Brouwer in *The Astronomical Journal*. Reproduced by permission. **Page 136.** Figure 7–22. Adapted from *The Nature of Asteroids* by Clark Chapman. Copyright © 1975 by Scientific American, Inc. All rights reserved. **Page 143.** Chapter 8 opening quote. Reproduced with permission of Graeme Edge of the Moody Blues. Copyright © 1967. Reprinted by permission of Threshold Records Limited. **Page 147.** Figure 8–2. Adapted from *The Dynamic Earth: An Introduction to Physical Geology* by B.J. Skinner and S.C. Porter. Copyright © 1989, 1992, by John Wiley & Sons, Inc. Adapted with permission of John Wiley & Sons, Inc. **Page 148.** Figure 8–4. Adapted from *The Dynamic Earth: An Introduction to Physical Geology* by B.J. Skinner and S.C. Porter. Copyright © 1989, 1992, John Wiley & Sons, Inc. Adapted with permission of John Wiley & Sons, Inc. **Page 149.** Figure 8–5. Adapted from *The Dynamic Earth: An Introduction to Physical Geology* by B.J. Skinner and S.C. Porter. Copyright © 1989, 1992, John Wiley & Sons, Inc. Adapted with permission of John Wiley & Sons, Inc. **Page 149.** Figure 8–7. Adapted from *The Dynamic Earth: An Introduction to Physical Geology* by B.J. Skinner and S.C. Porter. Copyright © 1989, 1992, John Wiley & Sons, Inc. Adapted with permission of John Wiley & Sons, Inc. **Page 154.** Figure 8–13. Adapted from *The Dynamic Earth: An Introduction to Physical Geology* by B.J. Skinner and S.C. Porter. Copyright © 1989, 1992, John Wiley & Sons, Inc. Adapted with permission of John Wiley & Sons, Inc. **Page 191.** Figure 9–18. Adapted from *The Dynamic Earth: An Introduction to Physical Geology* by B.J. Skinner and S.C. Porter. Copyright © 1989, 1992, John Wiley & Sons, Inc. Adapted with permission of John Wiley & Sons, Inc. **Page 200.** Figure 9–31. Adapted with permission from Cyril Ponnamperuma, *Comparative Planetology*, Academic Press. Copyright © 1978, Academic Press. **Page 212.** Figure 10–7. Adapted from figures in Joseph F. Baugher, *The Space-Age Solar System*. Copyright © 1988, John Wiley & Sons, Inc. Adapted by permission of John Wiley & Sons, Inc. **Page 299.** Figure 14–2. Courtesy G. deVaucouleurs. **Page 314.** Discovery Figure 14–1–1. Courtesy of G. deVaucouleurs. **Page 333.** Figure 15–19. After Nather, R.E., in *The Astrophysical Journal*. Copyright © 1970 by University of Chicago Press. Courtesy of R.E. Nather. **Page 368.** Figure 17–2. By special authorization of the European Space Agency (ESA). **Page 369.** Figure 17–3. Reproduced with permission from *Annual Review of Astronomy and Astrophysics*. Volume 13. © 1975 by Annual Reviews, Inc. Courtesy of L. Spitzer and E. Jenkins. **Page 375.** Figure 17–10. Adapted with permission of Kluwer Academic Publisher. Courtesy of Charles J. Lada, Smithsonian Astrophysical Observatory. **Page 377.** Figure 17–12. After Steven Stahler in *The Astrophysical Journal*. Copyright © 1970 by University of Chicago Press. Courtesy of Steven Stahler. **Page 377.** Figure 17–13. Adapted with permission from *The Publications of the Astronomical Society of the Pacific* and the *Annual Review of Astronomy and Astrophysics.* © 1987 by Annual Reviews, Inc. Courtesy of Bruce Wilking and Frank Shu. **Page 380.** Figure 17–19. After M. Walker in *The Astrophysical Journal.* © 1956, University of Chicago Press. **Page 381.** Figure 17–21. Courtesy of the Publications of the Astronomical Society of the Pacific. **Page 385.** Chapter 18 opener quote. Exerpt from "The Hollow Men" in COLLECTED POEMS 1909-1962 by T.S. Eliot, copyright 1963 by Harcourt Brace & Company. Copyright © 1964, 1963 by T.S. Eliot, reprinted by permission of the publisher. **Page 396.** Figure 18–20. Courtesy of Icko Iben, Jr. **Page 401.** Figure 18–27. After Allan Sandage, in *The Astrophysical Journal.* Original figure © 1957, University of Chicago Press. **Page 402.** Figure 18–30. Adapted with permission from *The Publications of the Astronomical Society of the Pacific.* Courtesy James E. Hesser. **Page 408.** Figure 19–1. After C. Payne-Gaposchkin and S. Gaposchkin. Reproduced by permission of Harvard College Observatory. **Page 420.** Figure 19–21. Courtesy of Professor A. Hewish. Reprinted by permission from *Nature,* vol. 217, p. 709. Copyright © 1968 Macmillian Journals Limited. **Page 428.** Figure 19–31. Adapted from *Sky and Telescope* magazine. Courtesy of Virginia Trimble. **Page 429.** Figure 19–32. Modified by permission from page 582 of *Understanding the Universe* by Phillip Flower. Copyright © 1990 by West Publishing Company. All rights reserved. **Page 440.** Figure 20–6. Adapted from a figure supplied by Harvey S. Liszt, and published data from Leo Blitz. **Page 448.** Figure 20–16. Courtesy of Alan Dressler and the Carnegie Observatories. **Page 448.** Figure 20–17. Courtesy of Bart Bok; Reprinted from *The Observatory.* **Page 449.** Figure 20–18. Adapted from *Astronomy and Astrophysics.* Courtesy of D.S. Mathewson. **Page 450.** Figure 20–21. Photograph by G. Westerhout; data from Netherlands Foundation for Radio Astronomy and Division of Radiophysics, CSIRO, Australia. **Page 461.** Discovery Figure 20–2–1. *Voyager* from Carina Software. **Page 485.** Figure 21–22. From P.E. Seiden and H. Gerola, *Fund. Cosmic Physics,* 7, 241 (1982). Reproduced with permission of the authors. **Page 489.** Figure 21–29. Courtesy of Luiz daCosta, Margaret Geller, and John Huchra. **Page 489.** Figure 21–30. Adapted from *Sky and Telescope* magazine. Courtesy of Alan Dressler and the Carnegie Observatories. **Page 498.** Figure 22–3. Courtesy of A. Toomre and J. Toomre. **Page 503.** Figure 22–14. Courtesy of Kitt Peak National Observatory. **Page 516.** Chapter 23 opener quote. From *The Collected Works of Ralph Waldo Emerson,* Volume III, Essays: Second Series, The Belknap Press of Harvard University Press, 1983. **Page 526.** Figure 23–10a. After P.J.E.

Peebles, in *Physical Cosmology.* © 1971 by Princeton University Press. Redrawn in *Frontiers of Astrophysics,* © 1976 by Smithsonian Astrophysical Observatory. Reproduced by permission of Princeton University Press and Smithsonian Astrophysical Observatory. **Page 538.** Table 23–2. Adapted from *The Structure of the Early Universe* by John D. Barrow and Joseph Silk. Copyright © 1980 by Scientific American, Inc. All rights reserved. **Pages A10, A11.** Appendices E and F. Data adapted from work of C.W. Allen.

INDEX